Marine Biology

Marine Biology

Function, Biodiversity, Ecology

SECOND EDITION

Jeffrey S. Levinton
State University of New York at Stony Brook

New York   Oxford
OXFORD UNIVERSITY PRESS
2001

Oxford University Press

Oxford   New York
Athens   Auckland   Bangkok   Bogotá   Buenos Aires   Calcutta
Cape Town   Chennai   Dar es Salaam   Delhi   Florence   Hong Kong   Istanbul
Karachi   Kuala Lumpur   Madrid   Melbourne   Mexico City   Mumbai
Nairobi   Paris   São Paulo   Shanghai   Singapore   Taipei   Tokyo   Toronto   Warsaw

and associated companies in
Berlin   Ibadan

Copyright © 1995, 2001 by Oxford University Press, Inc.

Published by Oxford University Press, Inc
198 Madison Avenue, New York, New York, 10016

Oxford is a registered trademark of Oxford University Press

Library of Congress Cataloging-in-Publication Data
Levinton, Jeffrey S.
   Marine biology : function, biodiversity, ecology / by Jeffrey S. Levinton.—2nd ed.
      p.   cm.
   Includes bibliographical refereces (p.).
   ISBN 0-19-514172-5 (alk. paper)
      1. Marine biology.   I. Title.
   QH91.L427 2001
   578.77—dc21   00-040636

Printing number: 9 8 7 6 5 4 3 2

Printed in the United States of America
on acid free paper

Harvest moon—
the tide rises
almost to my door

—Matsuo Basho, 1644–1694

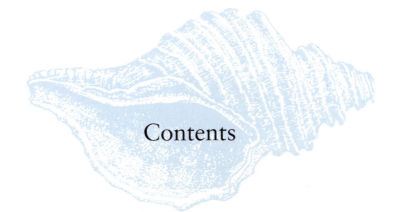

Contents

Color plate sections follow pages 84, 180, 276, and 372.

Preface

This new edition of *Marine Biology* reflects my excitement and philosophy of teaching as a gateway to marine biology, which is an active science with broad scope and lots of exciting recent achievements. Students must learn concepts and facts while appreciating the organismal diversity of the marine realm. Most importantly, they must feel the pulse of current happenings, and that is why I have included over twenty essays called **Hot Topics in Marine Biology** which are scattered throughout the text. These essays put students in contact with the current world of research, much as the enclosed **Marine Biology Explorations CD** and links to my **marine biology web page** link students to the world wide web of marine biology.

This edition retains the basics of the first edition, enhanced by current happenings in marine biology and a gateway to the myriad of marine biology connections on the internet. This text is designed for a one-semester course at the sophomore to senior level in four-year colleges. It would help greatly if the students have already taken a college-level biology course with coverage of organismal diversity. That said, I have successfully taught the sort of course for which this text is designed for many years, and many of the students taking it had no other background in biology. This book could also be used in a more advanced undergraduate course in marine ecology, if it were supplemented with journal articles. The new addition contains many updated references to the primary literature and the enclosed CD links the student to my MBREF web site, which includes a large variety of reference lists on special topics in marine biology. The **marine biology web page** also brings the student to a variety of world wide web links in many topics in marine biology, including research pages.

Marine biology is a subject in which the principles of cell biology, biomechanics, ecology, and so on are applied to marine biological problems. The interaction of these fields often leads to combinations that are uniquely marine biological. The text addresses three major themes: function, biodiversity, and ecology. **Function** refers to the way organisms solve problems and how physical and chemical factors constrain and select the solutions. What shape should a maneuvering fish have?

How does a cuttlefish stay at a specific depth? Of course, there are also ecological dimensions to these questions. **Biodiversity** is an essential part of marine biology, and I introduce the topic in this book both through chapters on marine organisms and through materials on creatures in various habitats. A separate chapter discusses diversity and the processes that regulate it, both ecological and evolutionary. This edition adds a great deal of coverage on recent advances, including our increased understanding of molecular methods of species and genetic identification, invasive species, and sibling species in the sea. **Ecology** is the interaction of organisms with their environment, studied usually by trying to understand the distribution and abundance of organisms. This involves a series of processes, which I introduce as a hierarchy, from individuals to ecosystems. It also involves a discussion of the processes along with accounts of major marine habitats and communities. This edition pays special attention to modern concepts of populations and species interactions including metapopulations, large-scale control of dispersal, and alternative stable states of communities.

Marine biology is such a diverse subject that some approach must be adopted to organize the subject. I firmly believe that principles must guide our understanding, rather than an accumulation of facts. The first part of the book introduces basic principles of how the ocean works and how marine organisms function, as individuals and at higher levels of the ecological hierarchy. Then we cover the organisms and the processes important in the water column. This is essential in order to understand the overall economy of the main part of the marine realm. Processes in the water column are also crucial for the benthos, which depend strongly upon the world above, both directly and indirectly. Next, we cover benthic creatures and the principles necessary to understand the biology of marine bottom organisms, which is followed by coverage of major marine habitats. The accompanying CD includes the application *Marine Biology Explorations*, which contains about 450 images (annotated with text). It is designed to illustrate specific ecological and functional relationships in a variety of major marine habitats including coral reefs, kelp forests, salt marshes, and mangroves. I have been selective and have emphasized those habitats that are not only important and interesting, but those where important principles can be illustrated to their best advantage. I look at the important gradient from the continental shelf to the deep sea, paying special attention to some of the fascinating newer discoveries about biological function and fascinating habitats, such as hot vents, in the deep sea. A chapter on gradients in biodiversity sums up larger scale variation in the sea and includes sections on invasive species, conservation of biodiversity, and conservation genetics. Finally, I tackle human interactions with the sea, as both a source of food, and unfortunately, a waste receptacle. I cover human effects on the ocean and include the important new field of global change.

This text has a series of features designed to help the student absorb a great variety of materials. Nearly every section has **summary heading sentences** that convey the essence of the material to follow. I find this of help to the student in anticipating what's ahead and in studying for exams in a subject that deals with so much biological and terminological diversity. **Text boxes** are included to explain a few equations and concepts and to keep the main body of the text free of excessive details. There are also more than twenty **Hot Topics in Marine Biology** essays scattered through the text. These essays are designed to point toward some recent advances in the understanding of marine biology or to discuss current issues, especially relating to new ecological debates, pollution, and fisheries. Each chapter is followed by a set of **review questions** and at the end of the book is a **glossary**. I have tried to write with very few embedded references to the scientific literature, referenced where necessary with numbered footnotes. At the end of each chapter, I have provided a number of references for further reading to allow the interested student to pursue a subject further, or even to get started on a term paper. This edition has far more references than the last, with **more references and leads on the CD**.

I have taught marine biology and ecology for over 25 years and have always been amazed at the diversity of students who take the course. Biology majors, geology majors, and humanities majors sit side by side. All learn a great deal and all seem to have that love for the ocean. You don't have to convince them to be there: They *want* to learn about marine biology. I do my best to keep that interest alive, and I find that field trips and use of many color slides help a great deal. I hope the enclosed CD will help instructors get students interested. I have taken many of my students with me to marine labs, and they have launched careers in science. I hope this text will help a wider audience to get excited about marine life.

Many people have helped me—too many to mention individually—with the preparation of this manuscript, and I absorbed a few, but probably not enough, of their helpful suggestions. I am especially grateful to the many individuals who shared their photographs and research experiences with me. The first edition was reviewed by Susan Bell, Paul Dayton, Alan Kohn, Larry McEdward, and Alan Stiven. I am very grateful for their constructive criticisms.

Most of this manuscript was prepared at Stony Brook, but I am very grateful as well to the staff of the Friday Harbor Laboratories and the Zoology Department, University of Washington, where I was able to complete the first edition. The second edition was completed during a stay at the Centre for Ecological Impacts of Coastal Cities, University of Sydney, and I am grateful for the inspiring intellectual environment that surrounded me during my stay. I am grateful to my wife Joan, who made many useful suggestions and tolerated from me, as usual, more than was appropriate. My sons Nathan and Andy have helped me continually by plunging, quite literally, into the subject of marine biology.

J. L.
Stony Brook, N. Y.

I

PRINCIPLES OF OCEANOGRAPHY AND MARINE BIOLOGY

1

Sounding the Deep

On every coast of the world, scientists work in field locations and in marine stations ranging from multi-million-dollar structures to small shacks with fanciful paintings of lobsters and crabs above the door. Some put out to sea in large ships, whereas others scarcely wet their knees (Figure 1.1). Some are content only when they are out sighting whales, whereas others peer through a microscope, patiently counting thousands of protozoa. It is the purpose of this textbook to give you an organized way of turning a fascination for the sea into an appreciation of the principles of marine biology that reflect the function and ecology of marine life.

Why study marine biology? I suppose mainly for the love of the sea and marine life. From a kayak, I once saw three killer whales jump up all at once and stand on their tails. At first I simply was overcome by what I saw, but I later wondered why they would do such a thing. You would be amazed at how much there still is to learn about why these whales behave as they do. Snorkel on a coral reef and you will see a wondrous beauty filled with never-ending change and almost unbelievable variety. But why? How did such variety come to be, and what is its meaning? How do all these creatures interact to form the seascape? Why do turtles, salmon, and even tiny larvae of snails and corals move thousands of miles? Such questions require an organized approach to a complex and somewhat foreign world. By the time you have finished your course and this textbook, you will be more familiar with that world.

✳ Marine biology is a subject mixing functional biology and ecology.

Marine biology is a diverse subject, but its main elements are functional biology and ecology. **Functional biology** is the study of how an organism carries out the basic functions such as reproduction, locomotion, feeding, and the cellular and biochemical processes relating to digestion, respiration, and other aspects of metabolism. Problems relating to function are quite varied. They might deal with questions such as: What skeletal and muscular arrangements do organisms as diverse as parrot fish and sea urchins use to eat seaweeds? When a whale dives for food to very great depths, how does it conserve oxygen? Why do some fish heat their heads only, whereas others heat the entire body? **Ecology**, on the other hand, is the study of the interaction of organisms with their physical and biological environments, and how these interactions determine the distribution and abundance of the organisms. For example, how does a snail living on a smooth exposed rock avoid being eaten by predatory birds? Why does one find patches on a rocky shore that are dominated by a completely different group of species? A major objective of the science of ecology is to understand the entire set of

Fig. 1.1   A fascination with marine creatures led Howard Sanders first to make major contributions to our understanding of the ecology of intertidal and shallow marine bottom communities. Later he pioneered American research in the deep sea, discovered marine animals previously unknown to science, and unlocked the secret of the deep-sea bottom's great biodiversity. (Photograph courtesy of the Woods Hole Oceanographic Institution.)

processes that underlie the distribution and abundance of organisms.

Although one can define a difference between functional biology and ecology, in specific cases it is usually difficult to make a clear distinction between the two. Almost all functional problems have ecological dimensions. The apparatus required to feed on a seaweed may depend on the mechanical toughness of the seaweed or on the seaweed's synthesis of toxic compounds for defense, among other factors. An organism's degree of protection usually corresponds to the number of effective predators that are present. Function thus has an ecological context. It is also pointless to study ecology without an understanding of an organism's functional biology. How can we determine how much a diving sperm whale is going to eat until we can determine its attack speed, its diving ability, and its jaw construction?

Because ecology is an environmental subject, the field of marine biology must cover the basic aspects of marine habitats. Without an understanding of the ocean in general, and of its specific habitat types in particular, it is not possible for you to understand just how and why marine organisms live the way they do, and how they survive in the particular habitats they occupy. We shall therefore spend considerable space explaining the various seascapes that are important to marine life.

Historical Background of Marine Biology

✳ **Marine biology began with simple observations of the distribution and variety of marine life.**

Of course there has always been a native lore of the biology of the sea, accumulated over thousands of years by those living near the shore and by fishing peoples. The earliest formal studies in marine biology date back to a time when there was little distinction among scientific specialties. Early biologists were "natural philosophers" who made general observations about anatomy and life habits. We owe the beginning of this tradition of natural philosophy to Aristotle (384–327 B.C.) and his Greek contemporaries, who recorded their observations on the distribution and habits of shore life. The next major steps forward took place in the eighteenth century, when a number of Europeans began to observe and classify living creatures. Most prominent among these was **Linnaeus** (1707–1778), who developed the modern means of naming species (see later). He described hundreds of marine animal and plant species, and developed larger-scale classifications. In the eighteenth century, the great French biologist **Georges Cuvier** (1769–1832) developed a major scheme under which all animals could be classified. He classified all the animals into four major classes of body plans: Articulata, Radiata, Vertebrata, and Mollusca.

The eighteenth century was also an important era of oceanic exploration and discovery. A number of expeditions circumnavigated the globe, bringing glory to explorers and new Pacific territory to European nations. Many of these expeditions had scientific components as well, and scientific staff were charged with collecting terrestrial and marine plants and animals and even charting. The voyage of French captain Nicolas Thomas Baudin explored the tropical Pacific, and numerous marine specimens, particularly mollusks and butterflies, were returned. Captain James Cook supervised the mapping of eastern Australia (1770), and his scientific staff also described and collected biological specimens.

Until the nineteenth century, most marine biology consisted of the description of anatomy and the naming and classification of species. Little was known about function and ecology. The only knowledge of open-ocean life was confined to experience with those animals that were fished or observed (or, in the case of mermaids and sea monsters, imagined) in the open sea. By the early 1800s, however, the study of natural philosophy had become popular, and a number of brilliant individuals devoted their lives to studying the ocean and its denizens.

✳ **In the nineteenth century, marine biology developed into a science, involving ecology and hypothesis testing.**

Edward Forbes (1815–1854) of the Isle of Man was the first of the great English-speaking marine biologists. After failing at art and failing his medical school studies, he set out to sea and participated in a number of expeditions in which a bottom sampler, known as a dredge, was used to dig into the seabed and collect organisms. He was hired as the naturalist on the *Beacon*, a ship that sailed on the Mediterranean Sea. He found that the number of creatures decreased with increasing depth, and he proposed what was probably the first marine biological **hypothesis**, or testable statement about the world of the sea: the **azoic theory**, which stated that no life existed on sea-beds deeper than 300 fathoms (1800 feet).

Forbes also discovered that different species live at different depths, and he proposed that the broader the depth zone of a species, the wider its geographic extent. Forbes opened up the ocean to scientific research and was appointed to the most prestigious post in natural philosophy of those times at the University of Edinburgh, Scotland. He published maps of geographic distributions of organisms, along with a natural history of European seas. During this time, Forbes was joined by many great pioneers from a number of European countries. In 1850, the Norwegian marine biologist **Michael Sars** disproved the azoic theory by describing 19 species that live deeper than 300 fathoms. The first plankton net was used during this period, and crude submersibles were developed. Marine biology was on its way.

Although he is usually remembered for his theory of evolution by means of natural selection, **Charles Darwin** (1809–1881) is the other great English father of marine biology (Figure 1.2). As a young man, he worked as naturalist on the H.M.S. *Beagle*, which sailed around the world in the years 1831–1836. He later wrote *The Voyage of the Beagle*, which was one of the best-selling travel books of the nineteenth century. Darwin made extensive collections of many types of marine animal and concentrated his own later efforts on the classification of the barnacles.

Fig. 1.2 Charles Darwin is best remembered for his theory of natural selection, but he made many important contributions to marine biology, including a book on coral reefs and a classification of barnacles that remains essentially unchanged to the present day.

Darwin formed a theory of the development of coral reefs. In this theory, he described their overall growth as a balance between the growth of corals upward and the sinking of the sea floor. If Forbes's azoic theory was the first important marine biology hypothesis, then Darwin's coral reef theory was the second. This subsidence theory was published in Darwin's first serious scientific book, and its brilliance was immediately recognized. Previously, most had believed that coral reefs in the open Pacific developed from the colonization and growth of corals on submerged extinct oceanic volcanoes. In contrast, Darwin argued that coral reefs developed around emergent rock that was slowly sinking, and this downward motion was balanced by upward growth of the corals. In this subsidence theory, as applied to the development of atolls (horseshoe-shaped rings of coral islands), Darwin was proven to be correct. About 100 years after the theory was developed, scientists drilled a hole in Enewetak Atoll in the Marshall Islands of the Pacific and bored through hundreds of meters of coral rock before hitting the volcanic rock basement below. Since corals grow only in very shallow water, this finding proved that the reef had been growing upward, for millions of years, as the island was sinking. Darwin was not completely right about coral reefs, however, insofar as he theorized that all reefs in the world are stages of subsidence leading to atolls. This has proven to be wrong:

many reefs are not subsiding, and atolls are special cases of reefs on volcanoes in oceanic crust (see Chapter 15).

Fisheries research began in earnest in the nineteenth century and became central in marine biological research. England was first at this activity, in 1863. Many nations began research efforts later in the century (see Chapter 18). In the United States, the Fish Commission sought to relate characteristics of the oceanic environment to the life history of fishes. Marine ecology became synonymous with fisheries research, and Canada used its fisheries effort to develop distinguished laboratories on both the Atlantic and Pacific coasts.

The last great advance in nineteenth-century oceanic exploration was initiated by the great biologists **W. B. Carpenter** and **C. Wyville Thomson**. Both had a passion for marine biology and convinced the British government to outfit the *Lightning*, a steam- and sail-powered ship that dredged the northern waters of the British Isles. Like the Norwegian biologist Michael Sars, they found marine life deeper than 300 fathoms and thus also helped to disprove Edward Forbes's azoic theory.

✴ The voyage around the world of the H.M.S. *Challenger* gave us the first global scale view of marine biology.

These expeditions set the stage for the great *Challenger* expedition (1872–1876) that would circumnavigate the globe and provide the first global perspective on the ocean's biotic diversity (Figure 1.3). The voyage was led by C. Wyville Thomson and by the great naturalist **John Murray**. The *Challenger* sampled the waters and bottoms of all seas but the Arctic, and 50 volumes were needed to describe the tremendous numbers of organisms that were recovered. On this expedition, the chemist J. Buchanan was able to disprove the existence of a so-called primordial slime, called *Bathybius*, that was supposed to be ubiquitous on the seafloor and capable of giving rise to higher forms of life. Buchanan discovered that the slime, which had been observed in collected samples of seawater, was merely an artifact of preserving seawater with alcohol.

During those same years, **Prince Albert I** of Monaco outfitted several yachts that sampled the ocean, and he eventually founded an oceanography institute and museum in Monaco. This facility came to be directed by the famous inventor–oceanographer Jacques-Yves Cousteau, who died in 1998. In America, the zoologist Alexander Agassiz led oceanographic expeditions, was the first to use piano wire instead of rope to lower samplers, and studied the embryology of starfish and their relatives. The now-famous Marine Biological Laboratory was founded on Cape Cod in 1886, and a number of marine stations were founded in Europe toward the end of the century. By the turn of the twentieth century, marine stations existed in many European countries. Marine laboratories such as the Marine Biological Laboratory in Woods Hole, Massachusetts, and Friday Harbor Laboratories in Washington made their appearance in the United States soon thereafter (Figure 1.4). Marine biology was now a full-fledged science, with a proud history of exploration and theorization.

Fig. 1.3 The H.M.S. *Challenger* at St. Paul's Rocks, a remote equatorial mid-Atlantic Island.

Fig. 1.4 Friday Harbor Laboratories, located in the San Juan Islands of Washington State, are a major site for marine biological research in rocky-shore ecology, biomechanics, larval biology, neurobiology, and many other areas of study.

✳ **Advances in modern marine biology included the development of major research institutions, faster ships, better navigation, and greatly improved diving technology.**

The early part of the twentieth century witnessed the founding of great oceangoing institutes, and a new technological ability to explore the ocean to its greatest depths. In America, the founding of the Scripps Institute of Oceanography in southern California (1903) and the Woods Hole Oceanographic Institution on Cape Cod (1930) gave the United States a unique ability to study the open sea. A large number of open-sea expeditions expanded our knowledge of marine life. The voyage of the Danish *Galathea* (1950–1952) was the last great deep-sea expedition of this era. As had happened in Europe toward the end of the nineteenth century, marine stations were opened in America, in every coastal state. Marine biology also flourished in many universities. Our knowledge of the ocean expanded during World War II, owing to the need for more navigational information. Advances in navigation, deep-sea bottom drilling, remote sensing, and other techniques led to a great expansion of our knowledge of the sea. A rich diversity of open-ocean and shore biology has since flourished, to the point that scores of journals now record the activities of a community of thousands of scientists. The numbers of such scientists in 1850 could have fit comfortably within a rather small room.

Technology in both the laboratory and the open sea has played an important role in the development of marine biology. Before the nineteenth century, poor navigation, inadequate sailing vessels, and generally crude bottom dredges and plankton nets prevented researchers from sampling the ocean systematically or completely. By the late 1800s, however, steam vessels allowed for the rapid lowering and raising of samplers, navigation was better, and vessels depended less upon the vagaries of the wind. In the twentieth century, modern diesel-driven ships such as the R.V. *Knorr*, ported in Woods Hole, Massachusetts, could navigate accurately by means of radio triangulation, and eventually by a very accurate satellite navigation system (Figure 1.5).

Before the mid-twentieth century, the deep-sea bottom could not be seen unless a piece of it was dredged and brought to the surface. This has changed dramatically, owing to the development of manned submarines, remotely operated vehicles, and SCUBA diving. In 1960, the spherical steel bathyscaph *Trieste* made a spectacular descent into the deepest oceanic trench, off the Marianas Islands. By the 1970s, a number of submarines routinely dived to depths of 2,000 m and more, and scientists were able to film and collect marine life (Figure 1.6). Mechanical arms made it possible to perform experiments, and accurate navigation systems permitted returns to remote sites in the ocean.

A number of smaller submarines allowed longer-term observation of depths of 300 m and less. One

Fig. 1.5 The R.V. *Knorr*, one of the U.S. oceanographic research fleet, has its home base at the Woods Hole Oceanographic Institution on Cape Cod, Massachusetts. (Photograph by Richard J. Bowen, 1984, courtesy of Woods Hole Oceanographic Institution.)

Fig. 1.6 The *Alvin*, a submarine capable of diving to several thousand meters, is equipped with accurate navigation and photography equipment, and underwater manipulators. (Photograph by Rod Catanech, Woods Hole Oceanographic Institution.)

of the more whimsical submersibles was used in a marine station near Nice, France. The steel hull was connected to the surface by an air hose, and the investigator sat inside on a bicycle seat, in a very cramped space. The first recorded observations in the Bay of Villefranche included one of a soup can on the murky bottom. Recently, researchers have used more modern submarines in the same area to observe spectacular bioluminescent plankton. To expand greatly the efficiency of deep-sea observation, **remote operating vehicles** (ROVs) have been developed. These vehicles are unmanned but can make precise surveys and even take samples (Figure 1.7). Nothing, however, has matched the importance of **SCUBA diving**, developed in the 1940s. This form of underwater exploration was not used often or effectively until the late 1950s, when the great coral reef biologist Thomas Goreau did some important studies. Now, direct observations and experiments could be done on the rich shallow-water marine biota.

While many advances have been made in diving and other technologies, the coming decades will see enormous strides toward remote sensing of the sea by satellite imaging. In the 1970s and 1980s an American satellite known as the *Coastal Zone Color Scanner* provided images and also sophisticated light-based estimates of water temperature, chlorophyll, and other parameters. Now a new satellite is up, with far more resolution. In conjunction with the new detectors, marine biologists are trying to use "ground-truthing" to produce equations that relate color information received by satellites to measurements

Fig. 1.7 Robert Ballard (left) of the Woods Hole Oceanographic Institution helped develop a remote underwater vehicle (in cradle) that surveyed the long-sunken ship *Titanic*. (Courtesy of Woods Hole Oceanographic Institution.)

taken at sea. In the long run, this will allow us to process worldwide data sets, a capability that is crucial in our current studies of global climate change.

Observation, Experimentation, and Hypotheses

＊ Marine biologists, like all scientists, use the scientific method, which is a systematic means of reasoning and observation.

Marine biology, like all science, depends upon a generalized scheme of observation and inference of the natural world, known as the **scientific method**. This sounds unduly stiff and distant. The scientific method is merely a systematic way to reason about and observe the world and universe in which we live. It depends upon observations, deduction, and prediction. We are constantly making observations about the natural world, and many of these are repeatable. For example, we might find that all fish we observe live in water (most do!). This would lead to a conclusion about the biology of fishes: fishes live in water.

The accumulation of specific observations to make a generalization is called **induction**. By contrast, we

might take some observation and generalize it to make a prediction that is logical. Such an inference, predicated on logical associations of conclusions with facts and premises, is a **deduction**. If you counted all the spectators in a football stadium drinking a beer, you might come to the conclusion that "At 2 P.M. during the game, 10 percent of the spectators drink beer." That is an induction. Instead, you might reason that most spectators like beer, but all could not be drinking all the time because drunkenness would set in, which spectators want to avoid. Therefore, you might deduce that only a fraction of the crowd will be drinking, but not constantly. If you knew the length of the game, how long it took to drink a beer, how many it took to get drunk, and how fast alcohol is metabolized to noninebriating products, you might be able to deduce how many spectators were holding a beer at any one time. This line of reasoning is far more valuable because it has led you to develop a prediction that could be applied to other stadiums. Deduction has the beautiful property of prediction.

Here's a more biological example. We might find that there is genetic polymorphism in a population and, knowing that the environment may change, we might predict that some variants will perform better and become more frequent in the population as the environment changes. This is Darwin's theory of natural selection, which uses the method of deduction. This form of inference always depends upon general premises and a logical pattern of reasoning to draw some specific conclusion. Most scientists strive to develop generalizations and theories, from which predictions follow by deduction. It may be well and good to count all the days of the year that are cloudy and then conclude that most of the year is cloudy; but it would be so much better if we had a set of premises, a predictive relationship, and a theory to understand why it is cloudy most of the year. Induction, however, is a necessary part of science and can even be an inspiration for deduction.

＊ Most marine biological research requires extensive observations and correlations, but experimentation is usually the most efficient way to answer a question.

Marine biological research involves a great deal of observation. In some cases the observation is general and not directed toward any specific research problem. It is essential, for example, to know the distribution of temperature, salt content, water depth, and other properties of seawater, because such information is required to solve a diverse array of specific problems. In other cases, observations are targeted toward more specific questions. To understand the migration of a species of fish, it may be necessary to sample the ocean with a sampling net at various times of the year and at various geographic locations and water depths.

In many instances, observations by themselves cannot solve a marine biological problem. One of the most common types of observation is a **correlation**, which is an observed relationship between one factor and another. You might discover an increase of fish abundance with increasing water depth. This would be a positive correlation between abundance and water depth, because both variables change in the same direction: as one increases, so does the other. On the other hand, you might discover a negative correlation, which in this case would mean that fish abundance decreases with increasing depth: as one variable decreases, the other increases. In either case, however, finding such a correlation does not prove that depth specifically is the cause of changes in fish abundance. The relationship might be coincidental. It might, for example, be due to a change of predator abundance with depth. The next year, the number of predators might be in a different relationship (correlation) with depth. This underscores a familiar saying among scientists: "Correlation does not prove causality."

Experimentation is a much sharper and more powerful way of establishing cause. Suppose that after finding a negative correlation between fish abundance and depth, you could perform an experiment and remove all the predators that are living in deeper water. If the prey fish then spread equally to all depths, you could reasonably conclude that the predators, and not water depth itself, were the cause of the negative correlation. Experimentation is an important tool for both laboratory and field studies. Unfortunately, many marine problems cannot be approached by experimentation; often, organisms and environments cannot be studied except by observation. This is especially true when the spatial scale is so great that it is impractical to perform experiments. Imagine changing the circulation of an ocean experimentally to study nutrient transfer and you get the idea. It is possible, however, to formulate hypotheses that employ tests using distributional data.

✳ Marine biological research involves the testing of hypotheses and may involve experimentation or sampling.

In solving problems in marine biology, additional observations beyond a certain point are not necessarily helpful. You could count all the fish in the ocean and still not know why they are abundant in some places, but absent in others. To solve a scientific problem in a satisfying way, a **hypothesis** must first be stated. A hypothesis is a statement that can be **tested**.

The following are examples of hypotheses:

- Predatory snails reduce the population size of mussels on the intertidal rocks on the coast of Nahant, Massachusetts.
- Increasing temperature increases the rate of oxygen consumption of crabs.

The following is not a hypothesis:

- Mermaids can never be observed, but they exist.

I hope you can see the difference easily. One can *test* a hypothesis. To test a hypothesis, one assumes that it is possible to produce an outcome that shows the hypothesis to be false. One makes a prediction, which must follow from the hypothesis. We, therefore, formulate an experiment, whose outcome will be consistent or not consistent with the hypothesis. If one has hypothesized that a population is controlled by predators, it is appropriate to remove the predator population and observe whether the prey population increases, as would be expected from the hypothesis.

It is also possible that the premises of the hypothesis are inappropriate. Take the following hypothesis:

- Predators cannot eat prey, and they therefore have no effect on marine populations.

Because the premise of the first clause is incorrect, the hypothesis is inappropriate. All hypotheses should be internally consistent and testable, and they should be based on correct premises.

Although some hypotheses are best tested by experiments, many cannot be. Sometimes the predictions will then be stated in terms of relationships, even if sometimes the relationships could conceivably have more than one explanation. For example, we might state the following hypothesis:

- When circulation of a very large water body deeper than 100 m has a current speed of less than 2 cm s^{-1}, then the oxygen there will decrease faster than it is replenished by circulation, and the water will lack oxygen.

We obviously cannot perform an experiment on such a large and deep water body. We might then look at current speeds in all water bodies and classify on the basis of current speed those that lack oxygen. If the results of the classification are consistent with the hypothesis, then we might look for any alternative hypotheses that could explain the same information. If none are obvious, then we might lean toward the correlation study as a weak test of the hypothesis.

✳ Hypothesis testing is most powerful when specific predictions for an experimental treatment can be contrasted with difference from a control.

The most difficult aspect of hypothesis testing is to formulate a hypothesis that lends itself to a specific program of experimentation or collection of data. Figure 1.8 shows a scheme that captures the best known way to think deductively and formulate hypotheses. All science usually derives from initial ob-

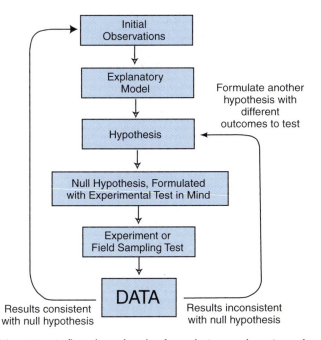

Fig. 1.8 A flowchart for the formulation and testing of hypotheses. (After Underwood and Chapman, 1995.)

servations that arouse curiosity. Finding a barnacle species only on the high shore would be an example. Then an **explanatory model** is formulated, which uses general principles to attempt to explain, for example, why a barnacle would be associated with a high shore location. One might argue that predation, which occurs only on the lower parts of the shore, prevents barnacles from surviving.

Now the crucial point arrives. One must formulate a hypothesis that is testable. The specific **explanatory hypothesis** here is that predation is more intense on the lower shore. But how does one know that there is an effect of shore height on predation? One way to find out would be to place a cage over the rocks that keeps predators out but allows the rocky-shore animals kept in the cage to function normally. We have now devised an experiment that manipulates predation intensity, but we must compare its results against one or several **controls**, which give us an idea of normal variation without manipulation of the specific experimental factor. We develop a **null hypothesis**, stating that there is no effect of caging, and we will reject the null hypothesis only if abundances are greater in low shore cages than outside. In our example we might do the following (Figure 1.9):

1. Place a steel mesh cage on the lower shore, to keep predators such as starfish and gastropods from penetrating the cage.
2. Place a steel mesh cage on the upper shore, to make sure that predation is absent at that level. (Control 1)
3. Follow open areas at both levels of the same area that is caged, as a control for comparison with the experimental cage treatment. (Control 2)
4. Construct a structure that has a roof but no sides. This would be a control to account for shading, since barnacles might do more poorly in cages with less light. Predators would have access through the open walls of the cage. (Control 3)

The null hypothesis in this case would state that after the cages have been in place for a set period of time, there would be no differences in barnacle population density between open and caged rocky shore, regardless of shore level. We have also attempted to study the effects of some aspects of cages, such as

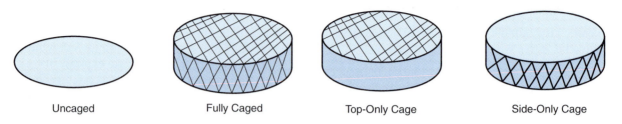

| Uncaged   Fully Caged   Top-Only Cage   Side-Only Cage |

Fig. 1.9 A caging scheme to test whether predation affects the abundance of a rocky-shore community at different tidal levels. Results from complete cages, which exclude predators at high and low levels, are compared with results from open areas. Controls that consider effects of cages include a topless cage, allowing predators in from the top, and a sideless cage, allowing predators in from the side.

their presence or absence, and shading effect. We thus have several controls. If barnacle densities between complete cages and open areas cannot be distinguished statistically, then the null hypothesis is not falsified. This may force us to make further observations or perhaps to find a new explanatory model to test. We might find, however that barnacles inside the cages on the lower shore had not changed in abundance, but those in the open treatments and in the roofed cages with no sides did decline. We might also find that there is no difference between caged and open areas in the upper shore, suggesting that predation not only is present in the lower shore but has no apparent effect on the upper shore. We would thus have falsified the null hypothesis, using three different types of control.

You can see that finding appropriate controls and the formulation of null hypotheses can become a complex procedure. For example, we did not discuss other caging controls, such as topless cages with normal mesh sides. If a null hypothesis is refuted, then one seeks to find more testable hypotheses that provide different sorts of prediction. One must always remember that even an experimental result is a correlation of outcome with experimental treatment. The treatment effect, however, may have nothing to do with the hypothetical effect being studied. For example, the open low-shore area might show a decline because of full exposure to the sun, while higher-shore animals have adjusted to this exposure. This may seem contrived, but it is consistent with the caging comparisons. The roof–caged control, however, gives us an opportunity to exclude this possibility.

In any test of a hypothesis, one must be aware of variation. **Statistics** is the field that deals with the calculation of trends and differences from repeated collections of information (e.g., measuring the height of all barnacles individually in the caging experiment and calculating the mean height per treatment) and assessments of variation. The difference in barnacle abundance between treatment and control may differ, but is the difference important? Two issues must be settled. First, a **test of statistical significance** must be established to determine whether the average barnacle density is greater within cages than outside cages. We need an estimate of variation and therefore need **replicates** of each treatment. If the variation among replicates is relatively low and the magnitude of mean difference high, then the difference may be significant (See Text Box 1.1).

The null hypothesis testing framework can be extended to nonexperimental studies. Consider the assessment of the environmental impact of the installation of a sewage treatment plant on the number of species of fish. Testing of an appropriate null hypothesis would require comparisons with control sites, preferably numerous, to estimate normal environmental variation. Ideally, the site of future impact would be studied before construction of the sewage plant, along with several control sites in similar environments. The null hypothesis would state that, following the construction of the sewage treatment plant, the number of fish species would be the same at the hypothetical impact site and the control sites. Owing to natural spatial and temporal variation, this is usually not a trivial problem, because variation among the control sites might be very great, making it hard to convincingly find a significant effect of the impact site, even if it were there.

TEXT BOX 1.1 A Glimpse into the Variation Problem: The Simplest of Statistical Tests

As we have discussed in the text, hypothesis testing involves the framing and testing of hypotheses. The simplest and most testable of hypotheses is the null hypothesis, which states that there is no difference between treatments you have chosen to study experimentally or by comparative observation. But how do we know that our outcome does or does not refute the null hypothesis? This requires a consideration of variation of observations. For any experiment, we must perform a series of replicates for each experimental treatment. To judge whether the results differ significantly we must compare the variation between treatments, as contrasted to the variability found within treatments.

Imagine the following two cases (Box Figure 1.1): a caging experiment is set out on a rocky shore, with 10 replicates each for caged and uncaged areas. After a time, barnacles are counted in each replicate. Box Figure 1.1a provides convincing support for the idea that barnacles were more abundant in the caged treatment. The mean (average) numbers are quite different, but the variation for replicates within a treatment is rather small. We could intuitively conclude that the difference is significant. In other words, the variation between treatments is much greater than the variation observed within treatments.

On the other hand, the outcome depicted in Box Figure 1.1b is not so clear. The mean differs, but there is a great deal of variation among replicates within a treatment. The difference in mean barnacle density may not be significant, but how would we know?

Analyzing variation to detect differences is the natural objective of statistics. We might at first calculate the mean, which is the sum of the numbers in all replicates for a given treatment (cages or uncaged), divided by the number of replicates.

Mean abundances of replicates ($N = 10$) for each treatment; the experiment was performed twice.

TREATMENT	RESULTS	
	BOX FIGURE 1.1A	BOX FIGURE 1.1B
Caged	11.7	10.1
Uncaged	4.7	8.2

We use a statistical test known as a t test to test the null hypothesis that the mean number of barnacles does not differ between the caged and uncaged treatments. Given the variation seen in the experiments, a value, known as t, can be calculated. If you have had a course in statistics, the following will make sense and will be familiar:

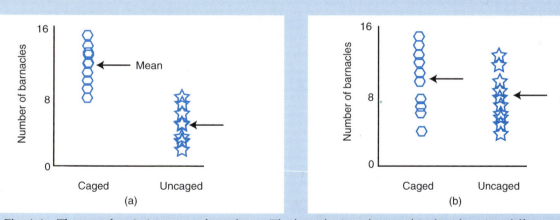

Box Fig. 1.1   The use of variation to test hypotheses. The hypothesis to be tested is that there is a difference in barnacle density in caged versus uncaged experimental treatments. (a) In this case it is clear that the differences of means (indicated by arrows) and range of numbers from a group of replicates for each experimental condition support the hypothesis of difference. (b) This set of results suggests that, while the mean number of barnacles does differ between treatments, the variation is too great to justify a firm conclusion that the different treatments cause different barnacle densities.

$$t = \frac{\text{sample mean for caged} - \text{sample mean for uncaged}}{\sqrt{(\text{variance of caged} + \text{variance of uncaged})/10}}$$

The variance is a measure of dispersion of points about the mean. This is saying that the higher the variance of data from the two treatments, the lower the value of t. Also, the lower the difference between sample means, the lower is the value of t. As t increases, there is a greater possibility that the null hypothesis of no difference in means is unlikely to be true.

Skipping over the details, it is possible to calculate critical values of t. If the sample t is greater than the critical value, then we can conclude that the null hypothesis is refuted. The threshold value of t is calculated at a **probability level**. If the probability level is 0.05 (probability is only 5 percent chance that the treatments are equal in effect),

the common value used by ecologists and statisticians, and if the sample t is greater than the threshold value, then we may conclude that there is less than 1 chance in 20 that the means are not different from each other. I am sparing you a number of assumptions behind this.

Now let's turn to our caging data. For the left-hand graph (Box Figure 1.1a), it turns out that the probability that the means are the same is 0.0002. This fits with our intuition. For the right-hand graph (Box Figure 1.1b), however, the probability is 0.10 that the means are different. Since we have chosen a threshold probability of 0.05 for our measure of significant differences, we conclude that the differences between caged and uncaged in Box Figure 1.1b are not significant. In turn, that means a further conclusion based on Box Figure 1.1b—namely, that the mean density in the caged is greater than that of the uncaged treatment—is pretty weak. The calculated mean is greater, but the variation is too great to sustain a refutation of the null hypothesis.

Habitats and Life Habits: Some Definitions

✳ **Some terms are necessary to describe life habits of marine organisms: neuston, plankton, nekton, benthos.**

Although all classifications are limited to some degree, it is useful to classify marine organisms by their general habitat (Figure 1.10). **Plankton** are those organisms that live suspended in the water. They may have some locomotory power, but not enough to counteract major ocean currents or turbulence. They include protists, animals, plants, and bacteria that are at most a few centimeters long. **Neuston,** organisms associated with the sea surface, include microorganisms that are bound to the surface slick of the sea. **Nekton** are animals that move in the water column, but they are capable of more powerful swimming and can move against a current or through turbulent water. They range from small shrimp, crabs, and fish, to the largest of whales. **Benthos** are those animals and plants associated with the seafloor. Some animals can burrow, or are infaunal, whereas others live on

the surface, or are **epifaunal**. Most clams are **infaunal,** whereas oysters and barnacles are epifaunal.

Figure 1.11 gives a general classification for marine habitats based upon water depth. The **intertidal zone** is the range of depths between the highest and lowest extent of the tides. In some parts of the world there is little or no tide, and wind mainly determines the vertical range of this fringing environment (see Chapter 2). The **subtidal zone** is the entire remainder of the sea, from the low-water tide mark to the greatest depth of the ocean. **Continental shelf** (or neritic) habitats include all seafloor and open-water habitats between the high-water mark and the edge of the continental shelf. Seaward of the shelf are a series of oceanic or **pelagic** habitats: the **epipelagic zone** includes the upper 150 m of water, the **mesopelagic zone** ranges from 150 to 2000 m depth, the **bathypelagic zone** ranges from 2000 to 4000 m depth, and the **abyssopelagic zone** ranges from 4000 to 6000 m depth; **bathyal benthic bottoms** range from 2000 to 4000 m depth, and **abyssobenthic bottoms** range from 4000 to 6000 m depth. **Hadal environments** include those in the seabed and the waters at the bottoms of the trenches.

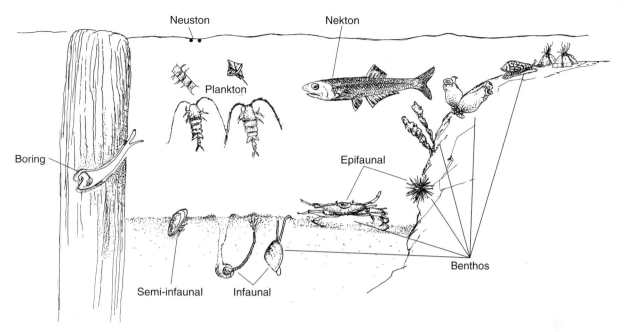

Fig. 1.10   General habits of marine organisms.

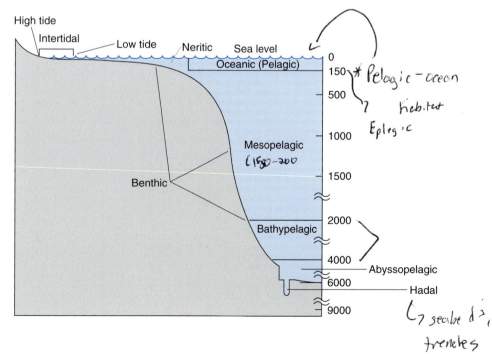

Fig. 1.11 A cross section of the ocean, from the shoreline to the deep sea, showing the location of major marine habitats.

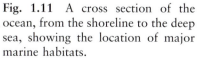

Further Reading

Carson, R. 1989. *The Sea Around Us*. New York: Oxford University Press. (Reissue of the classic 1950 book.)

Darwin, C. 1842. *On the Structure and Distribution of Coral Reefs*. London.

Deacon, M. 1971. *Scientists and the Sea: 1650–1900. A Study of Marine Science*. London: Academic Press.

Merriman, D. 1965. Edward Forbes—Manxman. *Progress in Oceanography*, v. 3, pp. 191–206.

Thomson, C. W., and J. Murray. 1884. *Report on the Scientific Results of the Voyage of the H.M.S. Challenger During the Years 1873–1876*. New York: Macmillan.

Underwood, A. J. 1981. Techniques of analysis of variance in experimental marine biology and ecology. *Oceanography and Marine Biology Annual Review*, v. 19, pp. 513–605.

Underwood, A. J. 1991. The logic of ecological experiments: A case history from studies of the distribution of macro-algae on rocky intertidal shores. *Journal of the Marine Biological Association of the United Kingdom*, v. 71, pp. 841–866.

Underwood, A. J., and M. G. Chapman. 1995. Introduction to coastal habitats. In A. J. Underwood and M. G. Chapman, eds., *Coastal Marine Ecology of Temperate Australia*. Sydney: University of New South Wales Press, pp. 1–15.

van Andel, T. 1981. *Science at Sea: Tales of an Old Ocean*. San Francisco: W. H. Freeman.

Review Questions

1. What was the azoic theory, and why could it be considered a testable hypothesis?

2. What might be the difference in potential contributions to marine biology by research done on the great oceanographic expeditions as opposed to research done at zoological stations on the coastline?

3. What was *Bathybius*, and why was its supposed existence of importance to the basic understanding of biology?

4. Why was the use of submarines so important in the development of marine science? Why was the use of SCUBA important in this development?

5. Distinguish between correlation and experimentation in the understanding of scientific relationships.

6. Devise a testable hypothesis about something in the room in which you are located now. How would you test this hypothesis?

7. Why is the following a poor hypothesis: Because whales are very small, they must be vulnerable to predation by snails.

2

The Oceanic Environment

The earth is a water planet. The ocean blankets about 71 percent of the earth's surface and strongly modifies climate. The ocean especially dominates the southern hemisphere, with about 80 percent coverage, whereas the northern hemisphere is only about 61 percent covered. On average, the ocean is 4,000 m deep, and over 84 percent of the ocean bottom is deeper than 2,000 m. The ocean reaches its greatest depth, about 11,000 m (nearly 7 miles) in the Marianas Trench of the western Pacific Ocean. Because, for all practical purposes, sunlight does not penetrate deeper than 1,000 m, the ocean below that depth is lightless and cold, usually less than 4°C. The upper, sunlit, part of the ocean is generally warm and is rich in living organisms.

The Ocean and Marginal Seas

✱ **The world's oceans can be divided into oceans and marginal seas.**

The Pacific is the largest ocean and is relatively little affected by differences in climate from region to region or by river input from the surrounding landmasses. Island chains are most numerous in the Pacific, and volcanic activity is pronounced around its margins. Think of Mount St. Helens in Washington, Mount Fuji in Japan, and Krakatoa in the south-

west Pacific and you'll soon understand what is meant by the idea that the rim of the Pacific is a ring of fire. The Atlantic is relatively narrow and is bordered by large **marginal seas** (e.g., Gulf of Mexico, Mediterranean Sea, Baltic Sea, North Sea). It drains many of the world's largest rivers (e.g., the Mississippi, the Amazon, the Nile, and the Congo). Its average depth is somewhat less than that of the Pacific. The Atlantic Ocean is affected to a larger degree by terrestrial climate and river-borne inputs of dissolved and particulate substances. The Antarctic Ocean is unique in its completely landless border with other water bodies. Its northern boundary is the Subtropical Convergence, where colder, more saline water descends northward. A general pattern of west winds in the range of 40–60 degrees south latitude generates a surface current circling eastward around Antarctica. This and other important surface currents are illustrated in Figure 2.1.

Most marginal seas have unique oceanographic characteristics, owing to their restricted connections with the open ocean, and their usually shallow water depths. For example, a shallow-water barrier, or **sill**, restricts exchange between the Mediterranean and the Atlantic Ocean, and a local excess of evaporation relative to precipitation increases the salinity within the Mediterranean. In the fairly recent geological past (5–6 million years ago), a global lowering of sea level severed the connection with the

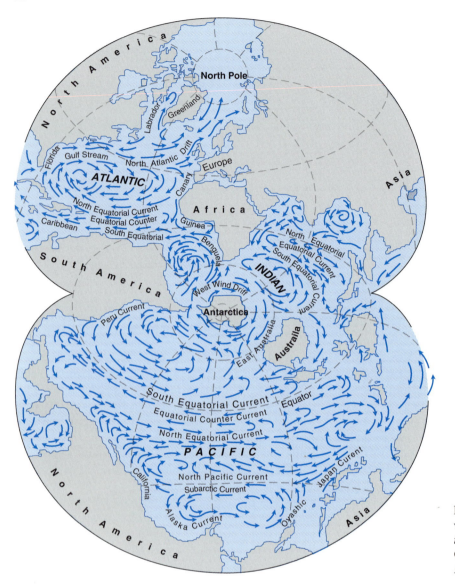

Fig. 2.1 Surface currents of the world's oceans. (From *The Circulation of the Oceans* by Walter Munk. Copyright © 1955 by Scientific American, Inc. All rights reserved.)

Atlantic, and extensive evaporation led to the formation of large-scale salt deposits. Most other marginal seas have had similar histories of strongly changing conditions.

Topography and Structure of the Ocean Floor

❋ **The oceans share three main topographic features: the continental shelf and slope, the deep-sea floor, and oceanic ridge systems.**

The **continental shelf**, a low-sloping (1:500, about 1 degree) platform, extends from roughly 10 km to over 300 km from the strand line (see Figure 2.2). Seaward of the **shelf-slope break** (usually a depth of 100–200 m), the continental slope increases in grade to about 1:20 (about 2.9°). The **continental slope** is usually dissected by **submarine canyons** that act as channels for down-slope transport of sediment. The foot of the slope merges with the more gently sloping **continental rise**, which descends 2–4 km to the **abyssal plain**, which averages 4,000 m depth. In some parts of the ocean (e.g., the west coast of South Amer-

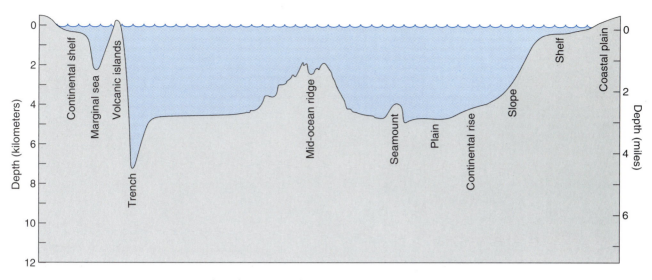

Fig. 2.2 Two examples of continental margins, showing various topographic features.

ica), trenches occur just seaward of the base of the continental slope and may be over 10,000 m deep. The trenches are long and narrow and may run parallel to the shoreline. Isolated **midoceanic islands** may rise from the deep-sea floor to the surface (e.g., the Hawaiian Islands). In contrast, **continental islands** are located on the continental shelf.

The **oceanic ridge systems** rise 2,000–4,000 meters from the ocean floor, and parts sometimes reach the sea surface, forming emergent islands (e.g., Iceland, on the Mid-Atlantic Ridge). In map view, ridges are linear features. The longest of the ridge systems runs the length of the middle of the north and south Atlantic Oceans, snakes around the southern tip of Africa, and runs northeastward into the Indian Ocean (Figure 2.3). Ridges are volcanic in origin and are cut by transverse faults, or breaks in the earth's crust. Rift valleys, at a ridge system's center, are parallel to the line of the ridge. The deep-ocean basin floor consists of volcanic rock, blanketed by soft sediment. The sediment contains varying combinations of mineralized plankton skeletons, clay, and other minerals deposited from continental sources, volcanic rocks, and precipitates, such as manganese nodules, which are scattered on the seabeds in certain areas. The skeletons of many organisms are composed of calcium carbonate, the material that makes up chalk. Open-ocean sediments are formed from a drizzle of calcium carbonate skeletons and make up, for ex-

ample, foraminiferan ooze (made of protistans known as Foraminifera). Calcium carbonate dissolves at great depth, however, and deep-ocean sediments become progressively dominated by clays and, at certain sites, a radiolarian ooze of silica skeletons. Beneath the soft-sediment covering lies the crust, made of sedimentary rocks and an underlying layer of volcanic rock.

✳ The earth's crust is in continual motion, and the main topographic features of the ocean reflect this motion.

As more and more of the ocean floor was mapped, it became obvious that the midoceanic ridges are large features, extending many thousands of kilometers. The ocean floor seemed to be divided into large sections, called **plates**, whose boundaries were ridges or large breaks in the crust, known as **faults**. Many geologists believed that some large-scale process caused the overall features of the oceanic crust.

In the 1960s, scientists discovered that several features of rocks and sediments occurred in bands that were parallel to the midoceanic ridges. To understand the significance of these bands, you need to know that the polarity of the earth's magnetic field has reversed repeatedly. Magnetic mineral grains are crystallized under the influence of the earth's magnetic field and reflect the polarity of the earth at the time

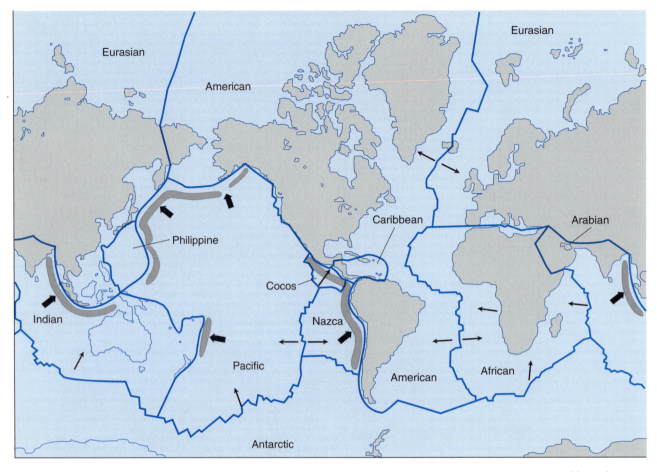

Fig. 2.3   The ocean floor, showing plate boundaries, midoceanic ridges, where new oceanic crust is created by volcanism (green lines with thin arrow), and trench zones (thick dark gray bands).

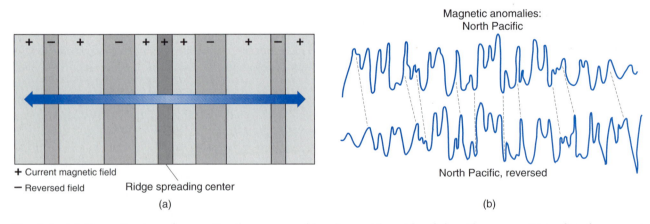

Fig. 2.4   (a) Generalization of anomalies into a map of bands on either side of the ridge center. Note that the current magnetic field of the earth (recorded at spreading center) is positive. (b) Trace of magnetic anomalies across a spreading ridge center, with transect reversed to show match of anomalies on either side of spreading center.

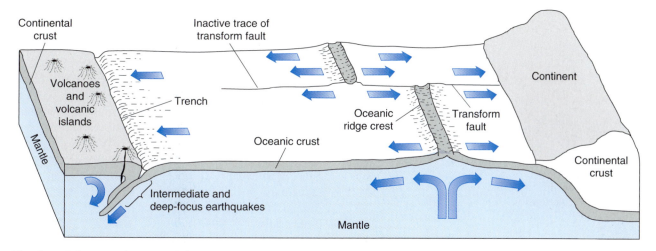

Fig. 2.5 Schematic diagram of the oceanic crust, showing the formation of crust at ridges and the downward transport and destruction of crust below trenches.

of formation. The series of bands discovered are symmetrical, and alternating bands of magnetically polarized seafloor rock occur in mirror image on either side of the ridges (Figure 2.4).

Nearest a given ridge, the polarity indicated by the oceanic crust matches that of earth's present magnetic field. In the next band on either side of the ridge however, the magnetic polarity is reversed. The next bands are reversed again, and so on (Figure 2.4). Why would such a curious pattern exist? Marine geologists could explain this by postulating that the oceanic crust, formed by volcanism near the ridge, was subsequently displaced from the ridge. This could be happening only if new crust was being formed through volcanic activity at the oceanic ridges and carried horizontally as if on a conveyor belt, away from the ridges in both directions. Because the bands could be dated by means of radioactive minerals and the width of the bands was known, it was possible to calculate a likely range of speed of seafloor spreading: 2–25 cm y^{-1}. It was later discovered that crustal material is dragged downward at trenches and melted into the upper mantle, the layer of the earth beneath the crust (Figure 2.5). The great depth of the trenches reflects the downward dragging of the crust. The mechanism of seafloor spreading is probably a convective process occurring within the mantle. Heat is generated within the mantle, probably mainly by radioactive decay. The heat melts rock in the upper mantle, and volcanism in the midoceanic ridges is the result.

Seafloor spreading explains many of the earth's topographic features. The continents consist of relatively low density crustal material, lying above a denser oceanic crustal layer. Continents move along with the spreading crust. If you consider time spans of millions of years, the rate of seafloor spreading is significant. For example, 200 million years before the present day, the Atlantic Ocean was probably less than 1,000 miles wide and South America and Africa were fairly close together. If the continental shelves are included, those two continents are found to fit quite well together, as pieces of a giant jigsaw puzzle. These two landmasses were probably once part of a supercontinent that divided, whereupon the resulting continents drifted apart. Seafloor spreading explains such continental drift. Throughout geological time, continents have had radically different arrangements, and even the major surface current systems have differed strongly from today's arrangement.

The Ocean Above the Seabed

Seawater

GENERAL PROPERTIES OF SEAWATER

✳ Many of the ocean's major features are due to the unique set of chemical and physical properties of water, including its dissolving power, high specific heat, transparency, and heat of vaporization.

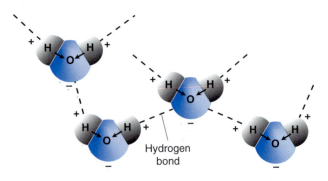

Fig. 2.6 The water molecule. A net negative charge on one side and a net positive side on the other is the basis of water's ability to form bonds with ions, making water an excellent solvent.

The formula for water, H$_2$O, indicates the component atoms of hydrogen and oxygen. Water molecules are asymmetric in charge, with a net positive charge on the hydrogen sides of the molecule and a net negative charge on the oxygen side (Figure 2.6). This overall polarity causes fairly strong attractions, called hydrogen bonds, between molecules. These attractions allow water to be liquid at atmospheric pressure and typical temperatures. Charge asymmetry also enhances the ability of water molecules to combine with ions, or particles. As a result, water is an effective and versatile solvent, and the ocean thus contains a diversity of dissolved substances.

Water, including typical open-ocean seawater, also has a very high specific heat; it takes 1 calorie to raise the temperature of a gram of pure water by 1 degree centigrade (at 15°C). Relative to most other liquids (and solids, such as soils), it takes a great deal of heat to change water temperature. Sea-water can store large amounts of heat, and moving currents can transport large amounts of heat, along with the water itself. Water has a high heat of vaporization, and large amounts of heat are absorbed when water vapor is formed by evaporation. As a result, the ocean has a very strong effect, through the atmosphere, on both oceanic and adjacent continental climates. It takes a great deal of heat to change the temperature of the ocean. The ocean thus tends to smooth out temperature variation, an effect that is especially noticeable on coastlines that receive oceanic breezes.

Temperature and Salinity

✴ Seawater temperature is regulated mainly by solar energy input and mixing of other water.

At low latitudes, there is a net capture by the earth of heat from solar energy, but at higher latitudes the earth tends to loses heat. As a result, there is a latitudinal temperature gradient of surface seawater (Figure 2.7). Surface transport and deep transport of heat energy are extensive, however, and the high heat capacity of water permits large transfers of heat among latitudes (see the following). Nevertheless, polar water temperatures hover about 0°C year round, whereas tropical water temperatures are almost constantly above 25°C. Local restriction of circulation causes further increases of temperature, sometimes approaching 35°C and higher in restricted lagoons and pools.

Seasonal temperature changes are maximal in the midlatitudes. Although temperature in high latitudes is low, the seasonal variation is minimal. Warm and seasonally constant temperatures are typical of the tropics. Seasonal change in the ocean is far more pronounced in the North Atlantic than in other oceans, owing to the strong influence of continentally derived weather systems. In coastal North Carolina, temperature ranges from roughly 3 to 30°C. By contrast, the ocean-moderated Pacific climate only an annual range of less than 5° along the coast of California, and summer maximum temperatures are far lower than at comparable latitudes in the Atlantic. Bodies of water with restricted circulation and shallow depth, such as the Mediterranean, have greater annual temperature ranges than does the adjacent open ocean.

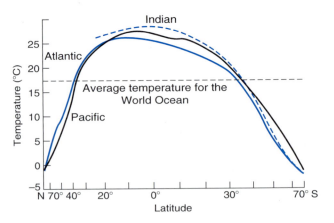

Fig. 2.7 Latitudinal variation in sea-surface temperatures in the Atlantic, Pacific, and Indian Oceans. (After Anikouchine and Sternberg, 1973.)

✳ **Salinity is a measure of the dissolved inorganic solids in seawater.**

The **salinity** of seawater is the number of grams of dissolved inorganic solids per thousand grams of seawater; it is expressed as parts per thousand (ppt, or ‰). Many of the elements in seawater (e.g., Na, Cl, Sr) are found to be in nearly constant ratios, even if the total salinity varies from place to place. The constancy becomes more pronounced with the increasing **residence time** of most elements in seawater. Residence time is the average time that a unit weight of a substance spends in the ocean, before it is lost to sediments or to the continents. In the ocean, the elements found to be in constant proportions have a residence time on the order of a million years, as opposed to the mixing time of seawater, which is on the order of thousands of years. The relatively short mixing time homogenizes the various elements relative to each other. Other elements have very short residence times. Good examples are nitrogen and sulfur, which are rapidly taken up and released by biological processes and vary greatly in proportion from place to place.

Seawater is a complex solution. The solutes dissolved in it consist of both dissolved inorganic and organic matter, including dissolved gases. Also, particles are suspended in seawater. The particles may be inorganic mineral grains, aggregates of organic particles, or living plankton. Dissolved inorganic matter enters the ocean mainly through river flow, but also through atmospheric precipitation. The **major elements** are defined as those that are present in concentrations that are greater than 100 parts per million (ppm). These include chlorine, sodium, magnesium, sulfur, calcium, and potassium. By far the dominant

elements are chlorine and sodium, and indeed, when seawater is evaporated, the salt sodium chloride is a very prominent residue. **Minor elements** are those present at concentrations between 1 and 100 ppm. These include bromine, carbon, strontium, boron, silicon, and fluorine. **Trace elements**, those present at less than 1 ppm, include nitrogen, phosphorus, and iron. A large number of elements occur in minute quantities, in concentrations of parts per billion (ppb).

The ratio of chlorine to other elements in seawater is essentially constant, despite overall changes of salinity. This fact enables you to estimate total salinity by measuring the chlorine content of seawater. **Chlorinity** is the total concentration of chloride per thousand milliliters of seawater (i.e., expressed in parts per thousand). Salinity is approximately 1.81 times the chlorinity. A simple **chemical titration** of chloride ions from seawater can therefore be used to estimate total salinity. Because a saline solution conducts electricity, **conductivity** (the ability to carry an electric current) can also be used to estimate salinity, though corrections must be made for temperature. Although it is accurate only to about 2 ‰, a simple **optical refractometer** can also be used to measure salinity, because increased salt content increases light refraction, the degree to which water bends light.

In the open ocean, salinity ranges from roughly 33 to 37 ‰. Locally, salinity is regulated by a balance of dilution (river input, rainfall, underground springs) and concentration (evaporation, sea-ice formation) processes. Latitudinal variation in the balance of precipitation and evaporation causes slight salinity maxima at 30° north and south latitude, with a minimum at the equator and declining salinities at higher latitudes (Figure 2.8). Most marginal seas dif-

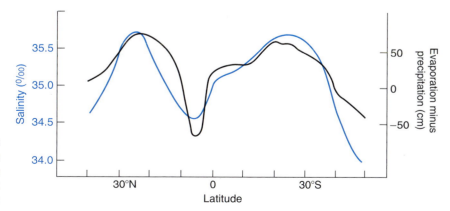

Fig. 2.8 Latitudinal variation in surface salinity of the open oceans (green curve). Balance of evaporation and precipitation also shown (black curve). (After Sverdrup et al., 1942.)

fer from the adjacent open ocean, owing to their restricted circulation and local factors such as excesses in rainfall, river input, or evaporation. The Baltic Sea has an excess of river runoff and restricted circulation with the Atlantic (via the North Sea). As a result, the salinity of much of the inner Baltic is 5 ‰ or less. By contrast, the Mediterranean has an excess of evaporation relative to precipitation and runoff, and its salinity is usually greater than 36 ‰, which influences the adjacent North Atlantic Ocean.

Oxygen in the Sea

✳ **Oxygen is added to seawater by mixing with the atmosphere and by photosynthesis; it is lost by respiration and by chemical oxidation of various compounds.**

Oxygen from the atmosphere dissolves in seawater at the sea surface. The amount that can be dissolved decreases with increasing temperature and decreases slightly with increasing salinity. When the wind is strong, larger amounts of atmospheric oxygen are mixed into the surface waters, which may then be mixed with deeper waters. As we shall show later in this chapter, the deep ocean is oxygenated because its waters begin as cold, oxygen-rich water at the sea surface of very high latitudes, and sink, owing to their relatively high density. In the tropics, there is an **oxygen minimum layer** at a depth ranging from 300 to 1,000 m. At this depth, sinking organic particles accumulate in a zone of strong water density change, and the stability of the water column reduces water exchange. Microbial activity there reduces the oxygen concentration.

The amount of dissolved oxygen in seawater is also affected by the balance of organism respiration and photosynthesis. Photosynthesis is the light-driven process that plants use to manufacture carbohydrates; oxygen is a by-product of the process. During respiration, the opposite process occurs: carbohydrates are oxidized to provide energy for the organism, and oxygen is consumed. An idealized equation for respiration is:

$$C_6H_{12}O_6(\text{glucose}) + 6O_2 \rightarrow 6CO_2 + 6H_2O + \text{energy}$$

In photosynthesis, this reaction goes to the left, although the biochemical details are very different.

In any body of water, the dissolved-oxygen content is a balance between two addition processes—**mixing with the atmosphere** and other oxygen-rich water bodies, and **photosynthesis**—and a subtraction process, **respiration** (chemical oxidation can sometimes be important). Respiration losses can usually be accounted for in terms of bacterial decomposition of organic matter and by animal and plant respiration. In some major water bodies, the supply of organic matter and its role in oxygen depletion outstrip the role played by the mixing of bottom waters of the basin with other oxygenated waters. For example, the bottom waters of many Norwegian fjords have restricted circulation with the open sea, owing to the presence of a sill at the entrance, which permits only surface water to leave the fjord (Figure 2.9). Decomposition of organic matter tends to consume the oxygen in the stagnant bottom waters. The deep waters of the Black Sea and the deeper parts of the Baltic (in summer) are also anoxic, or very low in oxygen.

Light

✳ **Environmental light originates mainly from the sun and is therefore strongest in surface waters.**

Light energy that reaches the earth emanates mainly from the sun, and, therefore, most light enters the

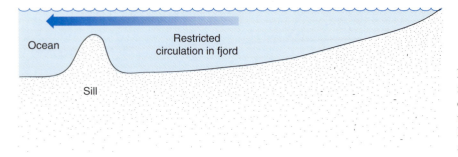

Fig. 2.9 The circulation between a fjord and the open sea. The lack of exchange of the fjord's bottom waters, combined with respiration, keeps the water low in oxygen, or anoxic.

ocean from above, at the surface. Light energy ranges from shorter-wavelength **ultraviolet light**, to light within the **visible spectrum**, to light in the **infrared** and longer-wavelength spectrum. At high angles near midday, very little sunlight is reflected from the sea surface. Much of the light is either **absorbed** by seawater and particles or **scattered** by particles in the water. Light intensity declines with increasing depth, as shown in Figure 2.10. The curvilinear decline is **exponential**. In general, the ultraviolet and infrared parts of the spectrum are very strongly attenuated with depth, so only the visible part of the spectrum reaches any appreciable depth. Blue light penetrates better than red light. That is why underwater photographs are often poor in reds and rich in blues.

Because it is the energy source for photosynthesis, light is crucial for life. Most marine animals also depend upon light to move about, detect prey, and spot predators. Light intensity near the surface is sufficient to depress biological activity, through the deactivation of protein and DNA. Ultraviolet light is damaging, especially in the clear waters of the tropics,

where significant amounts will penetrate to depths of 10 m. Some intertidal plants use calcium carbonate to reduce damage by absorbing the ultraviolet light. Some corals have pigments that absorb ultraviolet light. In more turbid waters, ultraviolet light is no real problem, and well over 90 percent of it is absorbed in the top meter of water.

The effect of sunlight diminishes greatly with depth. In turbid shelf waters, light is strongly limiting to plant growth even at depths of 10 m. At depths of 30–50 m, visibility is greatly reduced. This is especially true for color discrimination, and animals living in deep water do not have especially good color vision. In the open sea, at depths much greater than 1,000 m, there is very little sunlight. The animals in the deep oceanic world see only a faint glimmer from above, if anything at all.

Circulation in the Open Sea: Patterns and Causes

Surface Circulation

✳ **Overview: Surface oceanic currents are controlled by the interaction of the planetary wind system and the earth's rotation.**

Two factors interact to produce oceanic surface currents: the **planetary wind system** and the **earth's rotation**. More sunlight is captured near the equator than near the poles, and the heated air tends to rise and move to higher latitudes, and then sink. This overall effect drives the planetary winds, which drag over water and generate motion, or currents. The earth rotates from west to east, which deflects the air as it rises and sinks. This modulates the direction of the winds, which in turn affects surface currents. Now let's consider some of these circulation patterns.

The Coriolis Effect and Circulation Patterns

✳ **The earth's rotation causes a deflection in surface current direction called the Coriolis effect, which affects water flow on many geographic scales.**

The earth rotates once a day on its axis. A particle attached to the earth's surface at the equator must travel from west to east more rapidly than one attached near the pole, as it must traverse a longer dis-

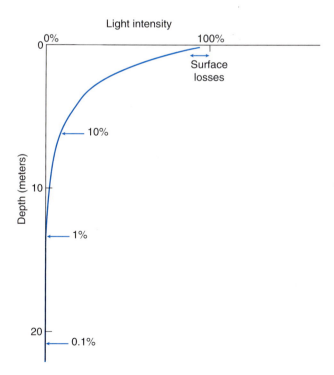

Fig. 2.10 An example of the pattern of exponential decline in light intensity with increasing depth in a coastal marine water column.

tance in each day's rotation. If a parcel of water not attached to the earth moves toward the north in the Northern Hemisphere, it is moving to a location of lower eastward velocity. Because the northward-moving parcel of water has greater eastward momentum than the water into which it is moving, it will have a relative deflection toward the east—that is, toward the right. If the parcel is moving south, then it moves from an area of lower eastward velocity into a region of increased eastward velocity. Relative to the local regime, the water will lag behind and will deflect toward the west—once again, that is, toward the right. A rightward deflection would occur even if the water were initially moving due east or west (Figure 2.11). This rotational effect is known as the **Coriolis effect**. It causes a deflection to the *right* for water traveling in the Northern Hemisphere. It causes a deflection to the *left* for water traveling in the Southern Hemisphere.

✳ The major oceanic surface currents are determined by the planetary wind systems, modified by the Coriolis effect.

The most conspicuous feature of the large-scale ocean surface is the presence of surface currents, which are driven by the planetary wind system (Figure 2.12).

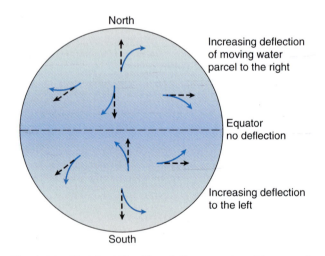

Fig. 2.11 The Coriolis effect deflects moving objects, such as the water parcels in this diagram. The dashed arrows show the paths the parcels would take if they moved undeflected in their original direction of motion. The solid arrows show their actual deflected paths. No deflection occurs at the equator, and deflection increases as the distance of the moving water parcel from the equator increases.

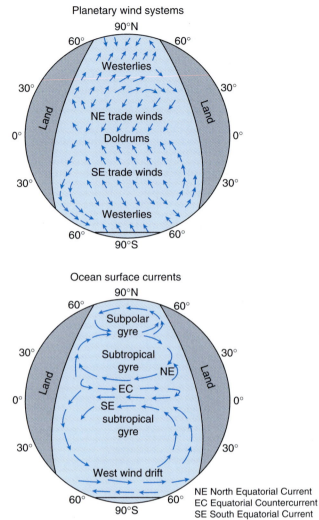

Fig. 2.12 The relationship of the ocean surface currents to the planetary wind system. (After Fleming, 1957, courtesy of the Geological Society of America.)

The winds are caused by the rise of air heated by the sun and the sinking of the air as it cools. These movements are influenced by the earth's rotation. In the latitudinal band of 30–60 degrees north and south, the **westerlies** move air toward the northeast (Northern Hemisphere) or the southeast (Southern Hemisphere). Between the equator and 30° north and south latitude, the **trade winds** blow toward the southwest (Northern Hemisphere) or the northwest (Southern Hemisphere). These winds, together with the direct action of the Coriolis effect, move tremendous volumes of surface water in large circular patterns known as **gyres**, centered around 30° latitude north or south. These gyres move clockwise in the Northern Hemisphere and counterclockwise in the Southern Hemisphere.

The Coriolis effect deflects water moving under the dragging force of the wind. The wind causes the water to move in surface sheets. Because of friction, these sheets drag on layers beneath. In the Northern Hemisphere, the Coriolis effect causes each successive sheet to deflect to the right, relative to the sheet above. Theoretically, this should produce a spiral circulation with increasing depth. In practice, the net direction of surface water is approximately 90° to the right of the wind in the Northern Hemisphere and 90° to the left of the wind in the Southern Hemisphere.

The eastward rotation of the earth piles and forms distinct water currents on the west sides of oceans. These currents are well defined, with strong thermal boundaries between the current and the surrounding oceanic water. The **Gulf Stream**, for example, originates from within the Gulf of Mexico, flows northward along the eastern coast of the United States, and moves eastward in a more diffuse current across the Atlantic toward Ireland and Great Britain. Large amounts of heat are stored in the current at its source, and this heat helps ameliorate the climate of the British Isles, which would otherwise be colder. As a result, subtropical vegetation grows well in Ireland and southern Scotland.

✳ Winds and the Coriolis effect combine to cause upwelling, which brings nutrient-rich deeper water to the surface.

The Coriolis effect strongly influences water movement on the continental shelves of the east sides of oceans. In the Northern Hemisphere (e.g., the coast of California), a wind blowing directly from the north will deflect surface waters to the right, or to the west. In the Southern Hemisphere, a wind from the south (e.g., off the coast of Peru), causes a leftward deflection of surface water, also to the west. These offshore surface water movements are compensated by the upward movement of deeper waters. This phenomenon, known as **upwelling**, brings nutrients from deeper waters and fuels large-scale phytoplankton blooms, which in turn ordinarily support major fisheries such as the anchovy fishery off Peru (Figure 2.13).

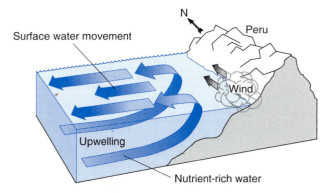

Fig. 2.13 Upwelling of coastal water caused by winds combined with the Coriolis effect.

✳ Every few years, persistent reversals of local current patterns occur in the eastern Pacific, causing a major shutdown of upwelling, known as El Niño. The effect reflects a global-scale cycle.

During most years, the surface layer of warm water is very shallow in the eastern Pacific, and cold nutrient-rich water rises to replace the surface waters that are driven offshore by winds and the Coriolis effect. Around December each year there is a local reversal, and warm water flows southward along the coast of Peru. In Latin America this effect is known as El Niño (named for the Christ Child, because it commences around Christmastime). Normally El Niño creates no particularly adverse effect, but at points in an irregular cycle of about 2–10 years the reversal is very strong. Then, nutrient-poor, warm surface water is driven eastward toward the coastline of the Americas. This change in the marine environment precipitates catastrophic declines in fish stocks: not only are nutrients reduced, but there are mass mortalities from the increased temperature, as well as coastal damage from the eastward movement of storm fronts. During El Niño of 1982–1983, about 600 people died from intense storms and Peru suffered fishing losses and storm damage amounting to about $2 billion. The effects were also felt along the California coastline, and many homes slid into the sea as a result of the severe storms. In a disastrous flood in Peru in 1728, the Piura River became raging rapids as a result of storms triggered by El Niño. Many people were swept away, and the survivors were mainly women who were wearing hoop skirts, which served as flotation devices by trapping air be-

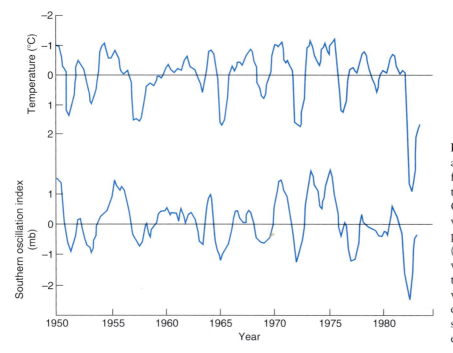

Fig. 2.14 The correspondence, over about 30 years, between departures from average sea-surface temperatures (top curve) and the Southern Oscillation Index (bottom curve), which is the difference in atmospheric pressure anomalies between Tahiti (relatively eastern Pacific) and Darwin, Australia (western Pacific). Note that since the temperature scale is inverted, positive pressure in the eastern Pacific corresponds to low sea-surface temperature. (After Shen et al., 1987.)

low the women's waists. In 1997–1998, El Niño was extremely damaging. During the height of this period, surface production in the equatorial Pacific was very low, but upwelling commenced at the end of the cycle and production of the plankton increased rapidly.

El Niño is driven by climatic oscillations that seem to be worldwide in extent but certainly involve an oscillation of the ocean–atmosphere system within the Pacific and Indian Oceans.[1] Every 2–10 years (Figure 2.14), there is a reversal of atmospheric pressure systems across the Pacific and Indian Oceans. In one alternative condition, there is a regional low-pressure system in the Indian Ocean and a high-pressure system over the eastern Pacific. At this time the waters in the eastern Pacific are relatively cool, and upwelling occurs there. This pattern reverses cyclically, however, resulting in what has been called the El Niño–Southern Oscillation, or ENSO. ENSOs affect the terrestrial and oceanic climate of the entire subtropical and tropical Pacific Ocean region. Since the trade winds are very weak during this time, the westward-flowing equatorial surface current is reduced in intensity, allowing a deeper eastward-flowing countercurrent to dominate. In effect, a wave of warm water is generated that propagates eastward across the Pacific. This movement brings eastward a lens of warm surface water that increases the sea-surface temperatures and shuts down the upwelling system of nutrient-rich cool water from below.

The movement of offshore low-nutrient water onto the normally nutrient-rich coast collapses rich fisheries. Warm waters often cause severe mortality of normally cold-water-adapted shellfish in the intertidal region of Peru and Chile. Peter Glynn[2] has documented extensive coral mortality in Panama during the 1982–1983 event, owing to the increase of sea surface temperatures. In general, during an ENSO, rain increases greatly on the western coasts of the Americas, but strong droughts dominate the climate of western Pacific areas such as Australia. The effects, actually, are broader in scope than just the Pacific–Indian Ocean regions—they are clearly worldwide. For example, it appears that when the ENSO part of the cycle occurs in the Pacific, the reverse condition occurs in the Atlantic.

❋ **Extraordinarily strong surface water motion is generated by storms and earthquakes.**

[1] For an excellent discussion, see Enfield, 1988, in Further Readings.
[2] See Glynn, 1988, in Further Reading.

Many other quite variable events can be as dramatic as the Pacific-wide El Niño. Cyclonic storms are quite common in both the Atlantic and the Pacific and may last for several weeks. The overall storm systems may move at speeds of a few kilometers per hour and can cover tens of thousands of square kilometers. Wind velocities of 160 km or more per hour are common, and wave surges of 10 m over coastal areas are possible. Hurricanes can completely destroy a coral reef and cause other terrible effects. In 1989, for example, Hurricane Hugo was powerful enough to toss large boulders several city blocks in the area of Charleston, South Carolina. Earthquakes or slumping of submarine sediment can also produce serious effects. They can generate waves known as **tsunamis** (often called tidal waves—a misnomer, given that they have nothing to do with the tide) that travel at speeds over 600 km h^{-1}. Their wave height may be very small in the open sea, but if a shallowing coastline causes the water to pile up, tremendous destruction can be caused by the resulting wave surge, which can exceed 5 m. In the late nineteenth century, tens of thousands of people on the coast of Japan were killed by one especially destructive tsunami.

Seawater Density and Deep Circulation

✳ Seawater density, an important property affecting the vertical movement of water, is controlled mainly by salinity and temperature.

The density of seawater (symbolized ρ, the Greek letter rho) is expressed as the number of grams per cubic centimeter. Density increases with decreasing temperature and increasing salinity (because of its salt content, seawater is denser than freshwater). Seawater density is only a few percent greater than that of distilled water, usually ranging between 1.02 and 1.07 g cm^{-3}.

Strong depth gradients in water density occur at many scales in the ocean. Vertical gradients in seawater density are known as **pycnoclines**. Because water density is affected significantly by variation in temperature, solar warming should play an important role in the vertical thermal structure of the ocean. This can be seen even at the scale of centimeters, if you dip your hands into a tidal pool on a hot day. The top few centimeters will be warm, but the wa-

ter beneath will be considerably colder. This principle works on the larger scale of a coastal water body with a depth of 30–50 m, such as Long Island Sound, which lies between Long Island, New York, and Connecticut. In the summer, solar warming from above produces a warm layer approximately 3–10 m thick, which is mixed by the wind and is isothermal (identical in temperature) throughout its depth (Figure 2.15). Beneath this mixed layer, a zone of rapid decrease in temperature, the thermocline, ends in a colder, deeper layer of and relatively constant temperature. Because of the effect of temperature on density, this arrangement is very stable, because the low-density water rides above the high-density water. Wind therefore does not mix the layers very effectively. As autumn sets in, this stratification of density begins to break down, and winds more easily mix the layers until the winter decrease of temperature results in a well-mixed and cold isothermal water column from the surface to the sea-floor.

On the largest spatial scale, the tropical ocean contains a surface warm layer, a deeper thermocline, and a still deeper cold layer. At the surface, the sun provides heat year round, and the deeper waters are kept cold by deep circulation. The precipitation–

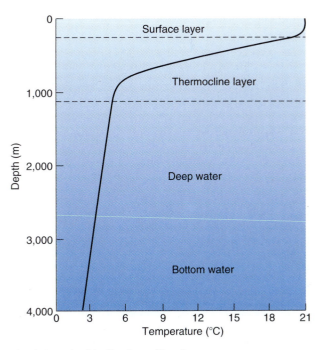

Fig. 2.15 An idealized profile of seawater temperature as a function of depth.

evaporation balance in the tropics tips toward precipitation, which further reduces the surface water density. Figure 2.15 illustrates a typical relationship between ocean temperature and depth.

＊ **Vertical circulation and deep-water circulation are regulated mainly by wind-driven mixing and differences in water density.**

In contrast to the major surface current features, deep oceanic circulation is characterized by movement of large **water masses** whose unique temperature and salinity characteristics are acquired at the sea surface at high latitudes. The high-latitude water masses sink, owing to their high density, and move to lower latitudes. Beneath the surface of the ocean, the water masses reach different levels, which are determined by the densities of the different water masses. These water masses are stacked like a layer cake, in descending order of increasing water density. Figure 2.16 shows an idealized cross section of the Atlantic Ocean. Note the continuity of deep-water masses from the surface at high latitude to lower latitudes and greater depths.

The vertical structure of the Atlantic (Figure 2.16) is explained as follows. In the Weddell Sea of Antarctica, the higher density of colder water is enhanced by slightly increased salinity generated by sea-ice formation. Thus the very dense Antarctic bottom water (AABW) is formed, sinks, and moves along the bottom toward the Northern Hemisphere. The North Atlantic deep water (NADW) forms in the Norwegian Sea, but it is not as dense as the AABW. It sinks and flows southward, moving above the AABW. The Antarctic intermediate water (AAIW) is formed near the Antarctic Circle but is not quite as cold and saline as the NADW. It sinks, but comes to rest just below a tropical surface layer of warm low-density water, and above the NADW. These movements form a series of layers of progressively lower-density water, as one moves toward the sea surface. Of course, there is considerable mixing at water mass boundaries, but the masses nevertheless can be traced by characteristic temperature–salinity combinations. The time of movement is on the order of hundreds of years.

The origin of these water masses is crucial for the maintenance of deep water life because the water originates at the surface, where it is saturated with dissolved oxygen. Consumption of oxygen during transport is insufficient to make the water masses anoxic, and the deep sea is therefore largely oxygenated. This allows deep-sea organisms to have an

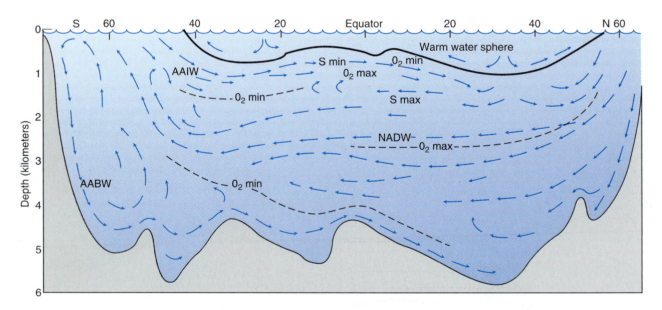

Fig. 2.16   Thermohaline deep circulation of the Atlantic Ocean. Water masses are as follows: AABW, Antarctic bottom water; AAIW, Antarctic intermediate water; NADW, North Atlantic deep water. (From Gerhard Neumann and Willard J. Pierson Jr., *Principles of Physical Oceanography*, copyright © 1966. Reprinted by permission of Prentice-Hall, Inc.)

ample supply of oxygen, usually. Water exchange is limited in some deep-sea basins, and these are anoxic.

Deep circulation is modified to a degree by the Coriolis effect, and some large-scale sedimentary features show deposition that follows the expected deflection. One might intuitively expect deep ocean currents to be sluggish, and this can be true. There are, however, some notable exceptions. Earthquakes whose foci are beneath the continental slope often set off rapid down-slope movements of a water–sediment mixture known as a **turbidity current**. These currents, which are of sufficient force to transport coarse sediment, were discovered by virtue of their ability to snap submarine communication cables. Even on the abyssal seabed, currents can be fast, on the order of 50–60 cm s^{-1}. The deep sea is thus not necessarily a monotonous and static environment. In some areas, large volumes of water are funneled through narrow openings in submarine ridge systems, and through narrow openings between marginal seas and the open ocean. There, current velocities are much higher than those on the rest of the seafloor.

The Edge of the Sea

Waves and the Shoreline

✳ Wind moving over the water surface sets up a series of wave patterns, which cause oscillatory water motion beneath.

Wind that moves over the water surface produces a set of **waves**, which appear on the surface as a series of **crests** and **troughs**, moving in the direction of the wind. This occurs even though no large current is in motion. During a storm, one first notices the appearance of a series of waves, and then an increase in the height of the waves. Wave height is proportional to the wind velocity, the duration of the wind, and the fetch, or distance over which the wind acts.

✳ Waves are defined by their period, wavelength, wave height, and velocity, as well as by the water depth.

Although waves are not always symmetrical and evenly spaced, a few dimensions adequately define their geometry (Figure 2.17). **Wave height** H is the vertical distance of crests from troughs. **Wavelength** L is the distance between successive crests, troughs,

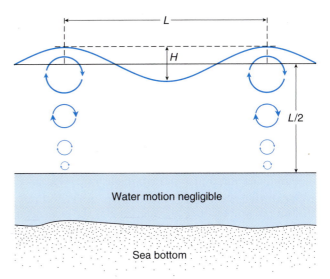

Fig. 2.17 Dimensions of ocean waves.

or other specified points. **Period** T is the time between passage of successive crests past a reference point. Finally, **velocity** V is the speed at which a crest, or other specified point, travels. The variables V, L, and T are related by the equation:

$$V = \frac{L}{T}$$

When waves are symmetrical, water particles move in orbits. The diameters of these orbits decrease with increasing water depth and become insignificant at depths greater than $L/2$.

✳ As a wave approaches shore, it becomes affected by the presence of the bottom when the depth is less than $L/2$; then the wave eventually breaks.

Surface waves set the water into orbital motion at depths of less than $L/2$. When the water becomes shallow enough, the orbital motion occurs at the bottom, and the waves at the surface begin to "feel the bottom," or be affected by its presence. The geometry of the wave begins to change and the wave height increases. These effects are due to the insufficient space available for complete orbits to occur; the water is effectively pushed upward as the wave approaches the shore. The orbital motion becomes more and more elliptical, and water and mobile bottom sediment are moved shoreward, sometimes with great force. When H/L exceeds 1/7, the wave becomes unable to sustain its own weight and collapses, or breaks.

✳ Shorelines greatly alter the pattern of waves and currents.

Waves often arrive with their crestline at an angle to the shore. This produces currents parallel to the shore, known as longshore currents (Figure 2.18). Occasionally **longshore currents** encounter an irregularity in the shoreline and suddenly run offshore in concentrated and quite dangerous **rip currents**. Longshore currents are responsible for extensive erosion and transport of beach sands along outer coast beaches. The effect often makes the dredging of channels into outer beach bars a hopeless task. Often, misguided beach engineers place groins on beaches to prevent erosion. This has the effect of choking off the sand supply further along the beach. As a result of such practices, many resort areas, most notably Miami Beach, Florida, have lost their bathing sand beaches in the past.

Irregularly shaped coastlines can strongly affect the direction and speed of incoming waves. As a wave front approaches the shore, horizontal water velocity diminishes as the water becomes shallower. Because the parts of a wave arriving on either side of a headland are still in somewhat deeper water, they will travel faster than the part of the wave striking the peninsula. This will result in the wave refracting about the peninsula (Figure 2.19). The degree of refraction may focus wave energy on the headland, and this will accelerate the erosion of the headland relative to the straighter coastline adjacent to the peninsula.

✳ The coast may be soft sediment or outcrops of rock under active erosion.

In quiet water, soft sediment accumulates as **sand or mud flats**, where sediment movement is minimal and the slope of the flats changes little throughout the year. Such flats can be hundreds to thousands of meters wide, as on the northern coast of France, or on the flats of the Bay of Fundy. On more exposed beaches, strong wave action causes extensive sediment transport and a seasonal change of the beach profile, owing to winter storms.

Figure 2.20 shows some features of an exposed beach. A series of windblown **dunes**, sometimes stabilized by vegetation, lies behind the beach. A relatively horizontal platform, the **berm**, extends to a break in slope. Owing to storms, the winter beach profile then increases strongly in slope to the low-tide

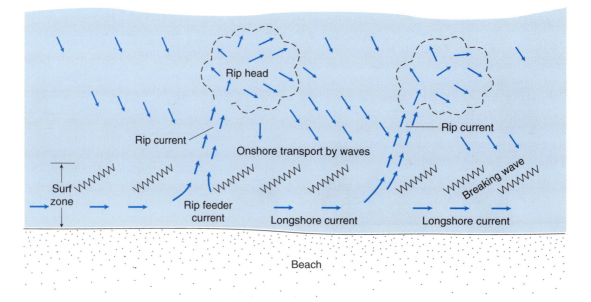

Fig. 2.18 Water transport adjacent to an exposed beach. (After Shepard, 1963. From *Submarine Geology*, 2nd edition. by Francis P. Shepard. Copyright © 1949, 1963 by Francis P. Shepard. Reprinted by permission of HarperCollins, Inc.)

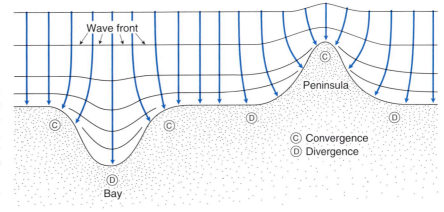

Fig. 2.19 Wave refraction. Note that wave action is concentrated at the headland, where the waves converge. (After Shepard, 1963. From *Submarine Geology*, 2nd edition. by Francis P. Shepard. Copyright © 1949, 1963 by Francis P. Shepard. Reprinted by permission of Harper-Collins, Inc.)

mark. Seaward, a complex of troughs and offshore sandbars develops. Often, emergent barrier islands develop that protect large **lagoonal complexes**.

Rocky coasts develop where outcrops of rock occur in geologically youthful terrains. The topography of such coasts depends on the local rock type and wave action. Poorly lithified sandstones, for example, are weathered and eroded into sand particles, which may produce a sandy beach at the base of a rocky outcrop (as at Santa Barbara, California). By contrast, highly cemented sedimentary deposits and crystalline rocks maintain their hard surface, and hard rock shores develop, as seen often in coastal areas of Maine and northern California.

Tides

* Periodic tides are caused by gravitational effects of the moon and sun, modulated by the earth's rotation and basin shape.

The force of gravity acting between any two bodies is proportional to the product of the masses of the two bodies and inversely proportional to the square of the distance between them. Both the sun and the moon exert significant gravitational attraction on the ocean, but the moon dominates the tides because it is closer to the earth. Tidal motion can be measured throughout the ocean, but it is especially noticeable at the shoreline in the form of tidal currents and vertical motion.

The extent of the tide is largely determined by the difference in gravitational attraction on either side of the earth. On the side closer to the moon the gravitational attraction pulls water toward the moon. On the opposite side of the earth, a minimum of gravitational attraction combines with the earth's spin to produce a net excess of centrifugal force, creating a tidal bulge away from the earth. Corresponding depressions (low tide) will exist on parts of the earth between the bulges, where there is no net excess of gravitational pull relative to centrifugal force. Because the moon "passes over" any point on the earth's surface every 24 hours, 50 minutes, or once each tidal day, ideally there should be two low and two high tides per day. Because the moon's position relative to the earth's equator shifts from 28.5° N to 28.5° S, the relative heights of high and low water differ geographically owing to changing vectors of gravitational attraction.

Tidal periodicity has both a semidiurnal component and a diurnal component. In semidiurnal tides, one finds two approximately equal high tides and two approximately equal low tides each day. Such cycles are characteristic of the east coast of the United States. At the other end of the spectrum, some coasts under some conditions (e.g., the Gulf of Mexico at

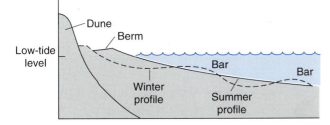

Fig. 2.20 Seasonal differences in the profile of a beach.

the time of the equinoxes when the sun's declination is zero and the moon's declination is nearly zero) are dominated by the **diurnal tide** component with a single high tide and a single low tide each day. Finally, **mixed tides** are also common, where the semidiurnal and diurnal components combine to give two high and two low tides each day, but with unequal ranges. On the west coast of North America, it is common to have one strong low tide and one very weak low tide each day. Figure 2.21 shows some tidal patterns over several days under different tidal regimes.

During times when the sun, earth, and moon are in line (Figure 2.22) the gravitational force exerted by the sun amplifies that of the moon, and maximal tidal range, at **spring tide**, is achieved. When sun, earth, and moon form a right angle, the gravitational effects tend to cancel each other out, and **neap tide** occurs, with the minimum vertical range. Two spring tides and two neap tides occur each lunar month (approximately 29.5 days).

Given the daily rotation of the earth, one might expect two equal high tides each day, separated by 12 hours. However, the presence of irregularly shaped basins, the tilt of the earth's axis, and the Coriolis effect tend to cause significant deviations from this idealized expectation. Each basin develops standing oscillations that resonate with tidal forces. If the travel times of these oscillations are short relative to the period of the tide-generating force, then the timing of the tide will be like that expected for an un-interrupted sphere covered with water. If travel time is long relative to the tidal period, the timing of the tide will be out of step with that expected for such a surface.

The Coriolis effect has a great influence on tidal heights and currents. When affected by the earth's rotation, tidal currents produce in the center of seas and oceans, rotating systems around **amphidromic points,** or points where no change in tidal height occurs over time.

In the open sea, both vertical tidal range and the strength of tidal currents tend to be modest. In coastal bays, however, volumes of water tend to be forced into ever smaller blind bays, and as a result currents and tidal heights are greater than on the adjacent open coasts. In areas with salt marshes, creeks tend to have higher tidal ranges than the adjacent open coasts. The most dramatic examples, however, are funnel-shaped basins such as the Gulf of California and the Bay of Fundy, where vertical tidal range exceeds 10 m. Tidal currents are also strong there and a pronounced incoming tidal wave, or bore, can be seen, for example, on the tidal flats of the Bay of Fundy or in certain rivers such as the Yangtse in China. Usually, there is a great difference of timing of the tides of bays, relative to the adjacent open coast. Tides can also be extremely slight, given the proper conditions. Because of the various interference effects of land-masses, areas such as eastern Denmark and the Caribbean have almost no vertical tidal movement.

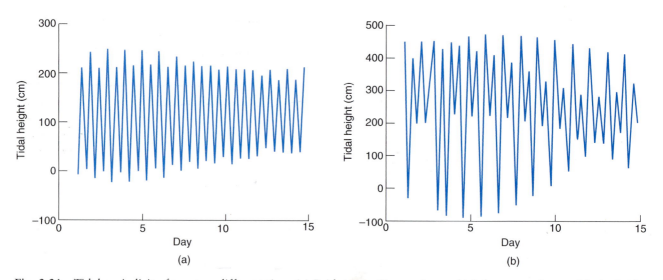

Fig. 2.21 Tidal periodicity from two different sites. (a) Bridgeport, Connecticut, which has a regular semidurinal tide. (b) Port Townsend, Washington, which has a strong diurnal component and strongly uneven low tides.

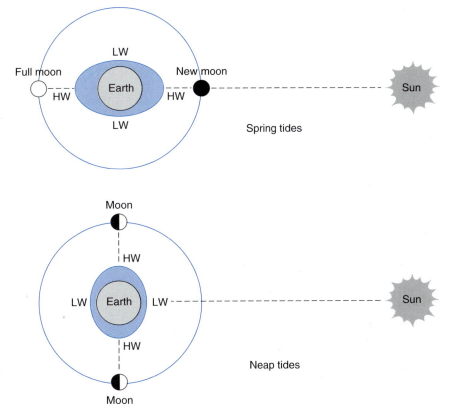

Fig. 2.22 Action of tidal forces at different alignments of the sun and moon: HW, high water; LW, low water.

Estuaries

✳ **Estuaries are coastal bodies of water where water of the open sea mixes with fresh water from a river.**

An **estuary** is a partially enclosed coastal body of water that has a free connection with the open sea but whose water is diluted by fresh water derived from a source of land drainage, such as a river. On the large end of the spectrum, Chesapeake Bay (Figure 2.23) drains five major rivers and includes a complex set of creeks, bays, and main channels. On the small end of the spectrum are individual rivers whose salinity changes with every tide. Nearly all large estuaries have been strongly influenced by the rise in sea level since the last glacial maximum. Chesapeake Bay's present extent stems from the drowning of the Susquehanna and adjacent rivers. Many estuaries, especially those on the east and Gulf coasts of the United States, owe their general form to the rise of sea level and the formation of **barrier bars**, which enclose lagoons of relatively lower and more variable salinity than the open ocean. The Outer Banks of North Carolina are barrier bars that protect an inland water system of great biological richness.

Water movement in estuaries depends on the amount of river discharge, tidal action, and basin shape. Some estuaries have remarkably predictable salinity gradients and circulation. Others are far less predictable, and salinity and circulation depend upon strong variations in freshwater flow from the watershed and variable connections with the open sea. Overall, there is a basic **estuarine flow**: relatively low-density river water flows downstream and eventually comes into contact with saline water mixing in from the coastal ocean. Owing to its lower density, the river water tends to rise above the denser saline water mass coming from the adjacent ocean. In a **highly stratified estuary**, the layers are quite separate. With moderate wind action and tidal motion, mixing occurs at all depths and vertical exchange occurs, causing the salinity of both the upper and lower layers to increase seaward. In such a **moderately stratified estuary**, a cross section would show that the **isohalines**,

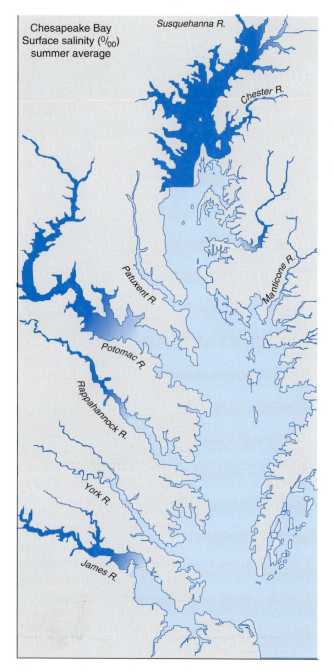

Fig. 2.23 The Chesapeake Bay estuarine system. Waters less than 10 ‰ in summer are darkened. (Modified from Carter and Pritchard, 1988.)

Vigorous tidal mixing tends to homogenize the vertical salinity gradient and results in a **vertically homogeneous estuary**. Because of strong tidal control, the salinity at any point in the estuary changes radically, depending on the state of the tide. At low tide, the salinity is dominated by downstream river flow, whereas at high tide the inrush of seawater increases the salinity. Such mixing can occur only in very shallow estuaries, such as the small freshwater rivers entering salt marshes.

Most estuaries experience strong seasonal effects and receive water, dissolved substances, and particulate matter from tributaries and runoff in the larger watershed drained by the estuaries and freshwater rivers. In the late winter and early spring, storms and snow melt can cause an increase in water flow, known as the **freshet**. This increase often brings into the estuary nutrients that sustain the productivity of the estuarine food web. If spring flow is relatively small, relative to the total volume of water in the estuary, as is true in Chesapeake Bay, then lines of equal salinity (isohalines) move down with the spring freshet, but only slightly.

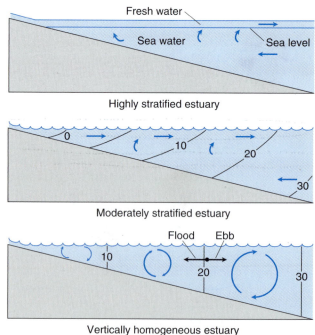

Fig. 2.24 Salinity variation in different types of estuary. Lines of equal salinity are indicated for moderately stratified and vertically mixed estuaries.

or lines of equal salinity, are inclined toward the sea (Figure 2.24). In such estuaries, water tends to flow up-estuary at depth and down-estuary at the surface. Such a flow pattern is found in the lower part of the Hudson River and in Chesapeake Bay.

Further Reading

Anikouchine, W. A., and R. W. Sternberg. 1973. *The World Ocean: An Introduction to Oceanography.* Englewood Cliffs, N.J.: Prentice-Hall.

Carter, H. H., and D. W. Pritchard. 1988. Oceanography of Chesapeake Bay. In B. Kjerfve, ed., *Hydrodynamics of Estuaries*, Boca Raton, FL: CRC Press, pp. 2–11.

Enfield, D. B. 1988. Is El Niño becoming more common? *Oceanography*, November, pp. 23–59.

Fleming, R. H. 1957. General features of the ocean. In J. W. Hedgpeth ed., *Treatise on Marine Ecology and Paleoecology, v. 1: Marine Ecology.* New York: Geological Society of America, Memoir 67, pp. 87–107.

Glantz, M. H. 1996. *El Niño's Impact on Climate and Society.* New York: Cambridge University Press.

Glynn, P. W. 1988. El Niño Southern Oscillation, 1982–1983: Nearshore population, community and ecosystem resonses. *Annual Review of Ecology and Systematics*, v. 19, pp. 309–345.

Hsü, K. J. 1983. *The Mediterranean Was a Desert.* Princeton NJ: Princeton University Press.

Kennett, J. 1982. *Marine Geology.* Englewood Cliffs, NJ: Prentice-Hall.

Kjerfve, B., ed. 1988a. *Hydrodynamics of Estuaries, v. 1: Estuarine Physics.* Boca Raton, FL: CRC Press.

Kjerfve, B., ed. 1988b. *Hydrodynamics of Estuaries, v. 2: Estuarine Case Studies.* Boca Raton, FL: CRC Press.

Knauss, J. A. 1978. *Introduction to Physical Oceanography.* Englewood Cliffs, NJ: Prentice-Hall.

Komar, P.D. 1998. *Beach Processes and Sedimentation*, 2nd ed. Englewood Cliffs, NJ, Prentice-Hall.

Maxwell, A. E. 1971. *The Sea, v. 4, parts I and II: New Concepts of Sea Floor Evolution.* New York: Wiley-Interscience.

McDonald, K. C., and P. J. Fox. 1990. The mid-ocean ridge. *Scientific American*, v. 262, pp. 72–79.

Officer, C. B. 1976. *Physical Oceanography of Estuaries (and Associated Coastal Waters).* New York: Wiley.

Pickard, G. L., and W. J. Emery. 1982. *Descriptive Physical Oceanography.* 4th ed. Oxford: Pergamon Press.

Ross, D. A. 1988. *Introduction to Oceanography*, 4th ed. Englewood Cliffs, NJ: Prentice-Hall.

Shen, G. T., E. A. Boyle, and D. W. Lea. 1987. Cadmium in corals as a tracer of historical upwelling and industrial fallout. *Nature*, v. 328, pp. 794–796.

Shepard, F. P. 1963. *Submarine Geology*, 2nd ed. New York: HarperCollins.

Stommel, H. 1966. *The Gulf Stream.* Berkeley: University of California Press.

Sverdrup, H. U., M. W. Johnson, and R. H. Fleming. 1942. *The Oceans: Their Physics, Chemistry, and General Biology.* Englewood Cliffs, NJ: Prentice-Hall. A classic text of oceanography.

van Andel, T. 1981. *Science at Sea: Tales of and Old Ocean.* San Francisco: W. H. Freeman and Co.

Von Arx, W. S. 1962. *An Introduction to Physical Oceanography.* Reading MA: Addison-Wesley.

Review Questions

1. Why is the ocean crust about the same age on opposite sides of a mid-oceanic ridge?

2. What process causes the formation of trenches?

3. Why is the ocean salty?

4. Why are the relative concentrations of some elements constant throughout the open ocean?

5. Why are ocean surface currents stronger and narrower on the west sides of oceans relative to the east sides?

6. Why does wind cause erosion of bottom sediments?

7. What is the main driving force of the ocean's surface currents? Of the deep circulation?

8. What processes cause the addition of oxygen to the ocean? Subtraction?

9. In what type of oceanic environment is ultraviolet light a problem for organisms?

10. What is the cause of upwelling along the western coasts of continents in the Atlantic and the Pacific Oceans?

11. What two factors play the largest role in determining seawater density?

12. Why is wave erosive action usually concentrated on headlands on the shore?

13. Why is the tidal range maximal two times each lunar month?

14. What is the principal cause of the gravitational pull that causes tides?

15. What is the major cause of vertical water stratification in an estuary?

16. Suppose surface circulation of the ocean ceased to take place. How would climate and near-shore oceanic life be affected?

17. How might recent ocean history, especially the rise and fall of sea level caused by glaciation, have affected the current biogeography of the ocean?

18. It has been argued that estuaries are not very old geologically and that the resident species therefore immigrated and adapted to them recently. Does this make estuaries unworthy of protection? Why or why not?

3

Ecological and Evolutionary Principles

To the average person, ecology means either the delicate balance between organisms and their environment or environmental protection. However, ecology is actually a field of science involving the study of all the interactions between organisms and their environments and how these interactions determine their distribution and abundance. These interactions can be divided into **biological interactions** and **abiotic interactions**. Biological interactions are those between organisms and include predator–prey interactions, biological dependencies between individuals of differing species, and parasite–host interactions. Abiotic interactions are effects of nonbiological factors, such as seawater chemistry, on the functioning of organisms. In reality, the two kinds of interaction cannot be easily separated. For example, low temperature might prevent a cold-blooded creature from moving very rapidly (an abiotic interaction), and this limitation might in turn reduce its chances of escaping a predator (a biotic interaction).

Ecological Interactions

✳ **Ecology is the study of interactions between organisms and their environment and the effects of these interactions on the distribution and abundance of organisms. Ecologists study the distribution and abundance of organisms and the mechanisms that determine the distributional patterns.**

Most ecological data collection starts with measuring the distribution and abundance of organisms. Observations lead to the answering of certain questions, such as: How many barnacles are packed onto a square meter of rock surface on a rocky shore? In what parts of the ocean can mussels be found? How deep can a sperm whale dive? Such observations reveal the patterns of nature that we seek to explain. From the knowledge of these patterns we seek to understand the mechanisms that explain them. As was discussed in Chapter 1, we translate our curiosity into testable hypotheses, the basis of good science. If we are lucky, we will be able to test the hypotheses with experiments that permit us to distinguish among alternative explanations.

✳ **Resources may be renewable or nonrenewable.**

The role of resources is important in the study of ecology. Both animals and plants require materials that can be in short supply. A **resource** is any material whose abundance in the natural environment can limit survival, growth, or reproduction. Food, space, and dissolved inorganic nutrients are all potentially limiting resources. Resources that can be depleted and are no longer available are **nonrenewable**. Those resources that will continue to become available are **renewable**. The issue of renewability can be resolved by scaling against the life span of the organism that is exploiting the resource. Over the lifetime of some attached organisms, space is a non-renewable resource, but it will of course be renewed once the individual dies. Protozoa as a food resource are renewable because they can grow back, even when grazed by larger and much longer-lived organisms.

The Ecological Hierarchy

✳ **Ecology is studied at many interacting hierarchical levels, including individual, population, species, community, ecosystem.**

Ecological processes should be studied at many levels of a hierarchy, or a nested series of sets. The important levels, from the most inclusive to most specific, are:

- Biosphere
- Ecosystem
- Community
- Population
- Individual

The Levels Defined

INDIVIDUAL LEVEL

An individual is an organism that is physiologically independent from other individuals. Examples include a single snail and an interconnected colony of coral polyps. At the individual level, we seek to understand how organisms survive under varying physicochemical conditions, and how individuals manage to find shelter and mates, avoid predators, and locate food.

POPULATION LEVEL

A **population** is a group of individuals of the same species that are responding to the same environmental factors and freely mix with one another—for example, in mating. An example is a group of crabs living and interbreeding freely within a small estuary. At the **population level**, we want to understand issues such as how large a population must be to produce enough young so that the population will persist. We are interested also in whether the population has sufficient genetic variability to allow evolutionary adaptation to a changing environment. These questions are different from those we ask at the individual level of the hierarchy.

A **species** is a single population, or a group of populations that is genetically isolated from other species—or in other words, is reproductively isolated from other species. Because species are genetically distinct, we are especially interested in this level of the hierarchy, especially as it relates to the origin of new species and the extinction of living species. An example of an appropriate question at this level of the hierarchy is: Will a change in sea temperature cause a species to become extinct? Although we do not include species per se in the ecological hierarchy, they are crucial in understanding long-term evolutionary directions of ecosystems.

Biogeography is the study of the distribution and abundance of species throughout the ocean. As species have arisen, the changing geography and climate of the ocean have had strong effects on subsequent spread and distribution of the species. Biogeographers first examine the pattern of distribution and then seek mechanisms to explain this.

COMMUNITY LEVEL

A **community** is a group of populations, each belonging to different species and all living in the same place—for example, all the barnacles, snails, seaweeds, starfish, and other species that live together on a rocky shore. At this level, we are interested in **interspecies interactions** that might cause changes in population size of the interacting species. For example, a predator species might overexploit a prey species and then be starved and decline sharply in population size; an immigrating species that was very efficient at exploiting a limited resource might reduce opportunities for a resident species, whose population might decline rapidly as a result.

ECOSYSTEM LEVEL

An **ecosystem** is an entire habitat, including all the abiotic features of the landscape or seascape and all the living species within it—for example, an estuary and its inhabitants. The definition is somewhat arbitrary and we can choose the boundaries. For example, we can define a coral reef ecosystem, but we sometimes might want to define a coral reef–open-ocean ecosystem, if we want to understand the processes affecting the many species that broadcast larvae into the open sea. At this level, we would be interested in water currents, reproductive timing of species, and anything else that explains the overall structure of an ecosystem.

BIOSPHERE LEVEL

The biosphere is the entire set of living things on Earth and the environment with which they interact.

The biosphere level may seem too abstract to warrant our concern, but we have learned in recent years that interactions at the biosphere level may be crucial to human welfare. For example, the carbon budget of the earth depends upon the amount of forest, the burning of fossil fuel, and the amount of photosynthesis and circulation in the ocean. The more carbon dioxide we add to the atmosphere, the more the atmosphere may trap heat and cause a global change in climate.

Interactions Among the Levels

The various levels of the hierarchy cannot always be studied separately; hierarchical levels do interact. For example, changes in climate at the biosphere level may affect an individual snail's ability to escape a predator. This would be an example of **downward causation** (or a **top-down process**), in which changes at a higher level of a hierarchy affect elements at lower levels. As another example, the efficiency of photosynthesis of individual phytoplankton may sum up to a major change in the nutrient cycling of a marine ecosystem. Such an effect is an example of **upward causation** (or a **bottom-up process**), in which changes at a lower level of the hierarchy affect upper hierarchical levels.

Interactions on the Scale of Individuals

At the scale of individuals, both abiotic and biotic interactions are quite important. We can define **ecological niche** as the range of environments over which a species is found. The range of environments has both biological and physicochemical dimensions, such as species interactions, water depth range, and salinity range. In Chapter 4, we shall focus on the abiotic interactions, but now we discuss interactions between species on the individual level.

✻ Many ecological interactions occur between individuals and may be classified on a plus–minus–zero system, depending on whether an individual benefits, suffers because of, or is not particularly affected by the interaction.

A plus–minus–zero system may be used to characterize ecological interactions. Plus (+) interactions benefit a species. Minus (−) interactions harm it. Zero (0) interactions do not affect it in important ways. Table 3.1 summarizes the basic interactions and the plus–minus–zero classification for the organisms involved in each. Note that an interaction is generally classified using two symbols (e.g., + −) to represent the effect on both kinds of organism involved in it.

Territoriality

✻ Territoriality is the maintenance of a home range, which is defended.

Territoriality is the maintenance of a home range and its defense against intruders. An individual may maintain a territory to protect (1) a feeding area, (2) a breeding site, or (3) a specific nest site. In most cases, territoriality is intraspecific, although some species defend their turf against many potentially intruding species. An example is the maintenance of breeding territories by many species of damsel fishes, who actively defend areas of bottom against other individuals of the same and sometimes different species.

Predation

✻ Mobile and stationary predators search for prey, using chemical, mechanical, and visual stimuli; some lure prey by using various "deceptions."

Table 3.1 Types of ecological interaction

TYPE	NATURE OF INTERACTION	PLUS–MINUS–ZERO CLASSIFICATION
Predation	Beneficial to one and detrimental to another	+ −
Parasitism	Beneficial to one and detrimental to another	+ −
Competition	Beneficial to one and detrimental to another, or detrimental to both individuals	+ − or − −
Territoriality	Beneficial to one and detrimental to another, or detrimental to both individuals	+ − or − −
Commensalism	Beneficial to one, but no effect on the other	+ 0
Mutualism	Beneficial to both individuals	+ +

A successful predator must locate prey organisms, entrap them, and have the means to ingest and assimilate them. In the marine environment, predators may be either **stationary** or **mobile**. Stationary predators include anemones (Figure 3.1) and other coelenterates, while mobile predators include fishes, starfish (Figure 3.2), gastropods, birds, and crabs. All predators share the ability to locate prey, although individual species use very different methods. Despite the diversity, there are some organizing principles relating to prey handling and capture and to interactions between predators and prey.

✻ Mobile predators may adjust their hunting behavior to optimize the rate of ingestion of prey.

Efficient predators will consume more prey, which in turn will increase growth and reproductive output. Therefore, one might expect natural selection to optimize the organism's efficiency, either in maximizing the amount of energy gained per unit time or in minimizing the time spent feeding, so that there is more time to carry out other vital functions such as reproduction. **Optimal foraging theory** was developed to predict the decision rules used by predators to optimize their food intake.

Many predators are able to consume a variety of prey items. Many drilling snails, for example, can consume a variety of barnacle and mussel species. Sea otters dive and retrieve urchins, abalones, and other large benthic invertebrates. Some species are of greater nutritional value than others, and the ques-

Fig. 3.2 A (barely) mobile predator, the starfish *Pisaster ochraceus*, consuming a cockle. Note the extended tube feet that are attached to the bivalve shell of the prey, via suction. (Photograph by Paulette Brunner, with permission from Friday-Harbor Laboratories.)

tion is whether to specialize on the good items or resort to feeding on the poorer morsels. When a predator encounters a prey item that is not very rewarding, should the predator feed on the item or pass it up to find something better? Optimal foraging theory predicts that, when overall food density is high, it pays to specialize on the good items and to ignore the choices of lower food quality. As overall food density decreases, it pays to become less choosy and broaden the range of prey. This conclusion can be altered if there is some cost in learning to switch from one prey item to another. For example, a snail might develop olfactory imprinting on a given prey type. It might cost more to change this imprinting than to continue to hunt for the original prey item. Satiation, or the limits of digestive activities, may also be important. A predatory animal might pass up a prey item if the predator's gut is full and it can digest no more for the moment.

The **time spent in a food patch** is also an important area of decision that affects the predator's total intake of prey. Consider the eastern Pacific starfish *Hippasteria spinosa*, which specializes on the subtidal sea pen *Ptilosarcus gurneyi*. The sea pen is found in dense patches of different size. If the starfish spends a long time in a sea-pen patch that it has already partially exhausted, it may invest a great deal of time harvesting an area with little reward per unit time. Alternatively, the starfish might abandon the first patch and find another, richer patch. Optimal forag-

Fig. 3.1 A stationary predator, the anemone *Anthopleura xanthogrammica*, consuming a mussel. (Photograph by the author.)

ing theory predicts that the time spent in a patch of prey should increase with an increase of travel time between patches. This makes intuitive sense, because an increase of travel time reduces the overall opportunity to gain food. It is not worth finding a new patch unless the food in it justifies the travel time. Imagine a green turtle feeding on a bed of sea grass. If there are many other unexploited beds of sea grass within easy reach, then it would pay to move to one of these beds as soon as the blades in the current patch become harder to find. If the distance between grass beds is great, however, it would pay to stay longer, because travel to another patch is relatively costly in terms of time lost to feeding.

The **choice of a best-sized prey** is a good example of the optimal foraging approach. Prey organisms that are too large might take an inordinate amount of time to consume. Imagine a starfish spending two days to open and consume a large mussel. That time might be more profitably spent on somewhat smaller mussels, whose relative ease of opening would compensate for the reduced reward. It might also be relatively unprofitable to select very small mussels, because too much time would be invested in handling and opening prey with little reward. Figure 3.3, which illustrates this argument graphically, shows the results of a study of crabs feeding on mussels. A large mussel provides a big meal for the crab, but the length of time required to crack such a mussel open makes it more profitable to select smaller mussels. Mussels that are too small are not worth the handling time. As a result, the crab selects intermediate-sized mussels.

Foraging behavior is also strongly affected by the presence of predators. When bivalve mollusks are siphon-feeding, they must expose their siphons. This makes them prey to benthic carnivorous fishes. Exposure of the siphon may thus lead to its loss, and the inability to feed until the organ has regenerated. Foragers such as these bivalve mollusks must therefore modulate their feeding behavior to accommodate the potential damage that could be done to them by predators. Partially, this can be accomplished by modulating the frequency of feeding, but predation has also brought about natural selection for a variety of predator avoidance traits.

Predator Avoidance

✳ **Resistance to predators increases individual fitness and is therefore enhanced by natural selection.**

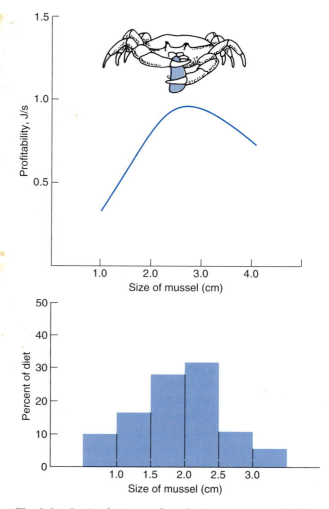

Fig. 3.3 Optimal strategy for selecting prey mussels. Top: a theoretical cost–benefit analysis for the reward of a mussel prey as a function of prey size (in terms of energetic return in joules obtained per second). As expected, the shore crab *Carcinus maenas* selects intermediate-sized mussels. Bottom: the actual sizes selected by the crab.

The popular image of the lion and the baby antelope gives one a false sense of helplessness of the prey. Many marine creatures have a large variety of traits that have evolved expressly to deter predators. For example, the large majority of tropical sponges are highly poisonous. This might be expected of a sessile group, with no other obvious defenses against predators.

Like other adaptations, antipredator defenses originate as variations in natural populations of prey. Predators increase the frequency of the resistant variants by preferentially selecting more vulnerable prey.

Fig. 3.4 Flounders can match a variegated background by means of chromatophores, which can rapidly alter their color. (Photograph by the author.)

The presence of any deterrent morphology or poison would enhance survival, and the individuals that possess such traits would contribute their genes to later generations. As a result, intense predation selects for species with extensive defenses.

✳ **Marine organisms avoid predators by means of crypsis, deceit, and escape responses.**

A most obvious strategy to avoid predation is **crypsis,** or blending with the background. A variety of animals, including many fishes, crustaceans, and cephalopods, employ **chromatophores,** which are cells that can rapidly alter their color. Flounders, for example, can change in a few seconds the color pattern of their dorsal surface to match sand or a mottled bottom (Figure 3.4).

Most of these cryptic species have fixed colors that are drab and blend in well with the background. For example, periwinkles usually blend with the background of their rocky substrata; for example, I have even seen a population of orange snails in an area of orange granite in Scotland. Many species blend with the background by means of camouflage coverings. To camouflage its dorsal surface, a decorator crab (spider crab) picks up fragments of seaweed, sponges, bryozoa, and compound ascidians, and whole anemones (Figure 3.5). The rostrum and sides of its carapace are festooned with hooked setae, which anchor the attached materials. The crab usually spends the day motionless and pressed against the bottom, with its claws folded under the ventral surface. At night, when it is safe to do so, it moves openly and feeds.

A number of species exhibit deceptive coloration and behavior. Many smaller reef fishes have large posterior spots. Predators are fooled into attacking

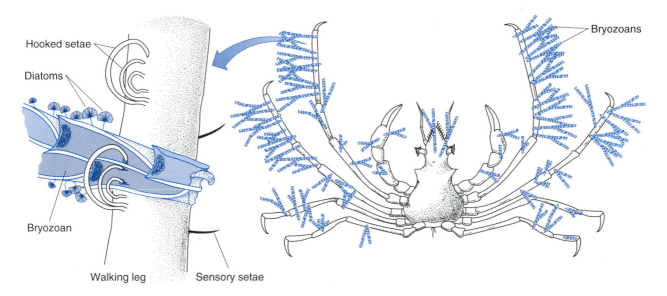

Fig. 3.5 Spider crabs have hooked setae on the sides and rostrum (anterior pointed section) and on the sides of the carapace. The California crab *Podochela hemphilli* may carry fragments of bryozoan colonies, which are ensnared in the hooked setae and camouflage the crabs from the view of predators. (After Wicksten, 1982.)

the posterior of the fish as it is swimming to escape in the opposite direction. Squid, cuttlefish, and the sea hare *Aplysia* squirt a dark ink, which conceals their escape.

Many species respond to predators by means of very specific escape responses, and some are specialized to detect and avoid certain predatory species. Fishes and crabs detect predators visually and can move away rapidly. Many sluggish benthic invertebrates have stereotyped escape responses. For example, when in contact with starfish, scallops es-cape by clapping their valves rapidly and expelling water through jet holes on either side of the hinge. Some anemones react to starfish by lifting off from the substratum and swimming into the water column. To escape from the large starfish *Pycnopodia*, the large Pacific sea cucumber *Parastichopus californicus* violently contracts its body wall muscles and springs up from the bottom, into the water. The burrowing clam *Spisula* violently flips its foot and can literally hop over an approaching starfish (Figure 3.6).

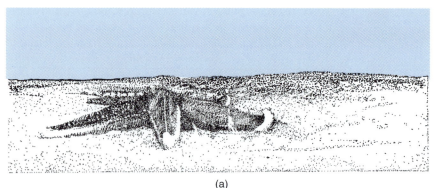

(a)

(b)

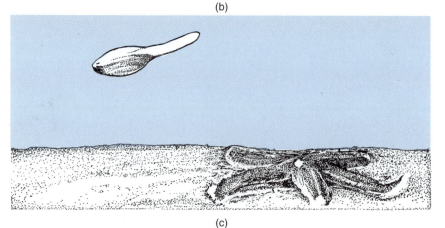

(c)

Fig. 3.6 The hopping response of the bivalve *Spisula* as it confronts an attacking starfish. (a) The starfish arm touches the clam's siphon. (b) The clam extends its foot and (c) jumps to escape. (Drawn after Feder, 1982.)

✳ **Many marine organisms can produce various morphologic features to discourage predator attacks (e.g., spines, strengthened skeletons, and other devices). In many cases, the defenses are induced by the presence of predators.**

Mechanical defense is one of the most common adaptations in defending against predators. In some cases, simple toughening of the body wall or stiffening by means of internal structures proves very effective. In many tropical seaweeds (e.g., *Halimeda, Pennicillus*), the thallus is strengthened with calcium carbonate. Combined with chemical defenses in some cases, this strengthening effectively deters predators. Many gastropods have a thickened shell, which deters predatory fishes. A large number of fish species have spines, some of which are poisonous. For example, members of the family Scorpaenidae, including scorpion fishes and stone fishes, are armed with poisonous spines. In the case of the Pacific coral reef stone fish, the poison is quite virulent and can kill an adult human. The Caribbean urchin *Diadema antillarum* has long sharp spines with reversed barbs. These spines deter many predators, but some fishes bite at the urchin and apparently survive piercing by its spines.

In some species, predators induce the growth of mechanical defenses. Such **inducible defenses** provide an advantage to the prey, which might otherwise waste resources when predators are absent. These resources might be used for growth or reproduction instead. Inducible defenses are especially useful for sessile species, which can deploy them in the direction of predator attack. For example, the sessile bryozoan *Membranipora membranacea*, which lives as a sheet of individuals on hard surfaces, is often attacked by a specialized sea slug, *Doridella steinbergae*. An attack induces the production of a peripheral zone of colony members whose skeletons are armed with spines (Figure 3.7). The spines reduce predation by about 40 percent, but the whole colony grows more slowly than colonies not exposed to predators. Inducible defenses can also be used to deter intraspecific and interspecific competitors.

Some of the kinds of conflict an organism faces show up well in the acorn barnacle, *Chthamalus anisopoma*, which lives in the Gulf of California. This barnacle occurs in two forms, conical and bent (Figure 3.8). The conical form is typical of most acorn barnacles. In the case of the bent form, the barnacle grows with the rim of the aperture oriented perpendicular (rather than parallel) to its base. This protective growth form is induced by the presence of the carnivorous snail *Acanthina angelica*, which is spatially variable in abundance. One might ask why the

(a)   (b)

Fig. 3.7 Inducible defenses. (a) Spines induced by the predatory sea slug *Doridella steinbergae* on a colony of the bryozoan *Membranipora membranacea*. Scale is 1 mm. (Courtesy of Drew Harvell.); (b) Stolons armed with nematocysts (light band), induced when unrelated colonies of the hydroid *Hydractinia* come into contact. (Photograph by Richard Grosberg.)

Fig. 3.8 The conical (right) and bent (left) forms of the acorn barnacle *Chthamalus anisopoma*. The animal develops the bent form if predatory snails are present. (Courtesy of Curtis Lively.)

barnacle does not always produce the bent form. Apparently, the bent form is produced at the expense of somatic growth and fecundity. Thus, there is an advantage to being conical if predation is low. This situation stabilizes the coexistence of the two-form strategy.

Many species lack inducible defenses, but there may be genetic variation in the degree of mechanical strengthening. This is true in the three-spined stickleback *Gasterosteus aculeatus*, a common fish in near-shore, estuarine, and lake environments. Three common forms are known, ranging from a strongly plated form to one with virtually no bony lateral plates (Figure 3.9). The plated form occurs in waters where carnivores are more common. This fish also has a set of dorsal spines that can be stiffened, and these help to deter predatory fishes and snakes.

In many cases it is not clear whether the thickened shell or the presence of spines is an evolved genetic variant favored by natural selection over unprotected forms, or an inducible defense. We will come back to this later in our discussion of morphological variation in marine populations.

* Many marine organisms are defended chemically by toxic organic secondary compounds, acid secretions, and toxic metals.

The active substances in common spices and stimulants used by people are present as a result of the evolutionary response of many plants to their predators. Nicotine, caffeine, and mustard are just a few of these many substances. Because these compounds are often not employed by the plants in typical metabolic processes, at least at the high concentrations that are used for chemical defense, they are often called **secondary compounds**.

Production of toxic compounds includes the secretion of acid by seaweeds and tunicates, and the manufacture of toxic organic compounds by many species of marine higher plants, seaweeds, and animals. These substances are usually synthesized by the organism, although some animals can eat toxic plants and store the toxic substance within themselves, a practice that deters the animals' predators. For example, the sea hare *Aplysia* can graze on the alga *Laurencia* and sequester this organism's halogen-bearing terpenes, which are highly toxic. The sea hare is thus also toxic.

* The presence of toxic defense substances is often associated with conspicuous coloration.

When the lion arrives, it might seem to make sense to hide, but species that contain or produce poisons might profit from advertising the presence of those substances. If a conspicuous color can be associated behaviorally with an unpleasant dining experience, a predator might avoid the prey upon other encounters. Natural selection would thus increase variants that have conspicuous coloration, but only if predation attempts largely failed and allowed the prey to escape. Otherwise, the conspicuous prey could not live to reproduce and spread the conspicuous color trait in the population. Thus conspicuous color is as-

Fig. 3.9 Three common forms of the stickleback *Gasterosteus aculeatus*. From left to right: most fully armed and most resistant to predators, to least resistant. (Courtesy of Michael Bell.)

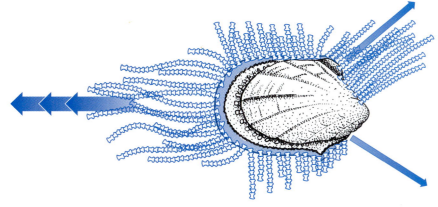

Fig. 3.10 Escape response of the swimming bivalve mollusk *Lima hians*.

sociated with the presence of toxic or bad-tasting materials that deter predators.

Many of the most poisonous marine organisms are conspicuous rather than cryptic. The poisonous black tunicate *Phallusia nigra* is conspicuous against its usual background: white coral reef or sand. Its tunic can contain as much as one percent vanadium, a highly toxic metal, and the tunicate also can produce vacuoles of sulfuric acid. Many toxic tropical species are bright red or yellow. The acidic Panamanian tunicate *Rhopalea birklandii* is a bright electric blue. Some bivalves have tentacular structures that are conspicuous and almost appear to be temptations for fishes. Species of the bivalve *Lima* may have mantle tentacles that can be autotomized, or shed, and are distasteful (Figure 3.10). Fishes bite such tentacles and spit them out, leaving the animal unharmed. It then swims away.

✳ Mechanical and chemical defenses against predation change in frequency with latitude, habitat, and oceanic basin.

There are a number of regional trends in the presence of traits that help prey organisms resist predators. The proportion of sponges and sea cucumbers that are toxic increases toward the tropics, and can reach 100 percent on tropical coral reefs. Mechanical adaptations of snails to resist crushing by crabs also increase toward the tropics. These trends reflect greater predation pressure in the tropics, which enhances natural selection for increased defense. Although predation is also often intense at some high latitudes, the high diversity of predators in low latitudes may impose the greater selective force.

The correspondence between the morphology of

predators and prey probably evolved through a history of natural arms escalation. Two evolutionary routes are possible. First it is possible that natural selection favors genetic variants that are more resistant to predation. Thus any variant that would produce a thicker shell would be favored when a crushing predator is present. It is also possible that all members of a population have the ability to respond to the presence of predators by thickening the shell. We mentioned earlier this response in a barnacle that grows in a bent-over position, which deters entry into the plates by predatory gastropods. An interesting change has occurred fairly recently in some Maine populations of the rocky-shore periwinkle *Littorina obtusata*. In locales where the snail has been exposed to the predatory crab *Carcinus maenas*, shells are lower spired and thicker, but shells are thinner where the crabs are rare or absent. Natural selection was hypothesized[1] to increase the frequency of thicker-shelled, low-spired forms only where crabs were present. Why wouldn't all snails have the same deterrence against predators? The increased cost in shell deposition was otherwise selected against. These observations, however, are also consistent with another hypothesis, which has been tested successfully.[2] In the presence of crab predation, snails might grow thicker shells with lower spires as a nongenetic response to predation.

✳ Microhabitat can strongly affect a creature's vulnerability to predators.

[1] See Seeley, 1986, in Further Reading, Predation and Optimal Foraging.
[2] See Trussell, 1996, in Further Reading, Predation and Optimal Foraging.

Marine animals may be able to avoid predators by simply retreating to inaccessible habitats. In some cases, the organism lives in a **refuge** that is permanently inaccessible to most predators. Marine animals may also alternate between a microhabitat that provides a refuge from predators and one used for feeding or reproduction. Rocky intertidal shores have strong gradients of desiccation and temperature. This is a special problem for mobile predators such as asteroid starfish and drilling gastropods, which require long periods of time to subdue and consume their prey. As a result, predation intensity is far less intense in the highest part of the shore. Mark Hay[3] has noted that many small herbivorous invertebrates, such as amphipods, feed on seaweeds that are otherwise toxic to larger herbivores such as mobile fishes. The smaller herbivores may have evolved a preference for toxic plants, which escape removal by larger herbivores and thus provide a refuge and food source for the smaller animal species.

Commensalism

✳ Commensal relationships benefit one species only. The benefit usually relates to food, substratum, or burrow space.

Commensal species acquire a benefit from another species, but return no benefit. Commensal relationships may be facultative or obligatory. A facultative commensal species is not completely dependent upon a certain single species, but may live on a variety of species. Barnacles, for example, may settle and live on a variety of species of mussel, or on other barnacles, seaweeds, or even rock. On the other hand, **obligatory commensals** can live only with certain other species. The western North Atlantic parchment worm *Chaetopterus* often contains a commensal crab *Pinnixa chaetopterans*, which settles and invades the worm tube as a larva. The crab eventually grows too large to leave the tube, and eats material swept in by currents generated by the worm's parapodia. Burrows of the eastern Pacific echiurid worm *Urechis caupo* often contain a gobiid fish, a polynoid polychaete, and a pinnotherid crab (Figure 3.11). The polychaete feeds on some of the mucus bag constructed by *Urechis* for suspension feeding. The fish and polychaete probably derive protection from predators and also probably feed on detritus and prey in the burrow.

    Some commensals use other species as attachment sites. A wide variety of animals and plants are attached

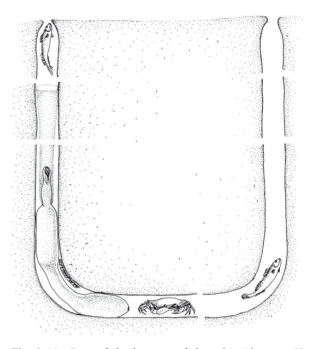

Fig. 3.11 Part of the burrow of the echiurid worm *Urechis caupo*, showing the following commensals: the goby *Clevelandia*, the polynoid polychaete *Harmothoe*, and the pinnotherid crab *Scleroplax*. (After Fisher and MacGinitie, 1928.)

to fish, marine mammals, and sea turtles. Whale barnacles of the family Coronulidae attach to whales and feed on phytoplankton as the whales swim. Remoras are fish with dorsal fins that are strongly modified into suckers, which are attached to sharks or sea turtles. Remora attach to pelagic sharks, but can rapidly detach themselves to feed on the shark's food. By contrast, pilot fish lack a specialized sucker, but congregate around sharks and eat their leavings.

Mutualism

✳ A mutualism is an evolved association among two or more species that benefits all participants.

Many pairs of marine species are often found where the mutual benefit is obvious. Such relationships probably began as facultative interactions, but genetic variation allowed complete dependence. Genetic variants that are interdependent might have had more offspring than those that did not participate in

[3] See Hay, 1991, in Further Reading, Predation and Optimal Foraging.

a relationship with another species, whereupon dependent variants would have taken over the population, and the relationship would have become obligatory. The obvious disadvantage of an obligatory relationship is the danger that one of the species will become extinct. Obligatory mutualisms thus depend to a large degree on the simultaneous stability of populations of two species.

✱ **Mutualism often reduces the risk of predation or disease, or provides food for one member of the species pair.**

Many mutualisms are a trade-off between protection against predation on the part of one species and some other benefit on the part of the other participant in the mutualism. An association between the coelenterate *Hydractinia* and the hermit crab *Pagurus* is a good example. The coelenterate lives as a colony on hermit crab shells and the relationship is species specific. The hermit crab is protected against predators and fouling by the *Hydractinia*. The hermit crab occupies a relatively fresh shell that serves as a substratum for the coelenterate. Mutualisms are found widely among crabs. A number of crab species carry anemones on their claws, and some species have clearly defined rows of teeth on the claws for holding the anemone. The Chilean actiniarian *Actiniloba reticulata* will move actively toward the legs of the crab *Hepatus chilensis*; upon reaching its destination, it proceeds to creep along the crab's body, eventually coming to rest on the claws. In other cases, the crab collects the anemones. When disturbed, such anemone crabs wave their claws and threaten intruders with the stinging tentacles of the attached anemone. Some hermit crabs have their shells enveloped with an anemone, which protects the crab against predators.

One of the most remarkable mutualisms in coral reefs is that between cleaner shrimp or cleaner fishes and a large number of fish species. Cleaner shrimp and fishes feed by picking ectoparasites off fishes, which approach them regularly. The cleaning fish *Labroides dimidiatus* maintains cleaning stations that are visited by about 50 species of fishes each day. "Customers" are attracted to the undulating movements of the cleaning fish. The fish *Aspidonotus taeniatus* mimics this undulation, but, instead of picking parasites, it attacks the approaching fish and takes a bite out of its fins (Figure 3.12).

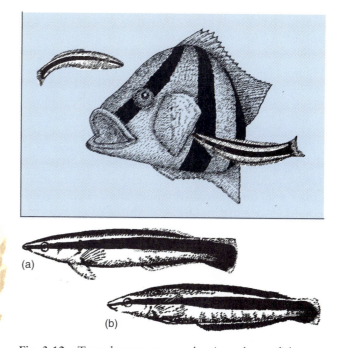

(a)

(b)

Fig. 3.12 Top: cleaner wrasses cleaning a larger fish on a coral reef. Bbottom: the cleaner wrasse *Labroides dimidiatus* (a) and its mimic *Aspidonotus taeniatus* (b).

Parasitism

✱ **Parasitism occurs when members of one species live at the expense of individuals of another, ideally without consuming the hosts totally as food and thereby killing them.**

Parasites live at the expense of other species and may get nutrients or shelter by damaging their hosts. **Ectoparasites** live attached to or embedded within the external body surface. They include animals attached to gills, body walls, and other surfaces. By contrast, **endoparasites** live within the body and may occupy circulatory vessels or ramify within certain organs or tissues. Parasites live on an exquisite edge of doom. If they are ineffective in utilizing their host, other parasites may enter and displace them by competition. If they are too effective, they may kill their host or even drive the host population to extinction. Because of this, parasitic species probably evolve through cycles of varying virulence.

It is often difficult to draw an exact distinction between commensals and ectoparasites. Barnacles, for example, are often attached to a wide variety of species of fishes and marine mammals. They are probably harmless when sparse in density. In great numbers, however, they create sufficient projections to increase drag and thus impede the host's swimming. Other species that have more specialized dependencies on their hosts may nevertheless border on being commensals. The pea crabs, members of the family Pinnotheridae, live in the mantle cavities of oysters and mussels. The crab has a highly modified morphology and a carapace too soft to allow it to survive independently of the host bivalve. There is some controversy over the detrimental effects of the crab on its host. Several studies have shown that the crab damages the host's gills, reduces its growth and increases its mortality, but the effect is unpredictable.

Endoparasites live within host tissues and have highly modified morphologies that adapted them to life within body cavities and to food uptake and absorption of fluids. Organs needed for free life, such as sensory structures and locomotory appendages, are usually lost. The life stage of the parasite that resides in the host can seem barely related to its actual relatives, which may be typical free-living forms. In contrast to the overall degeneration, the reproductive organs of such parasites are usually hypertrophied and acquire a central importance.

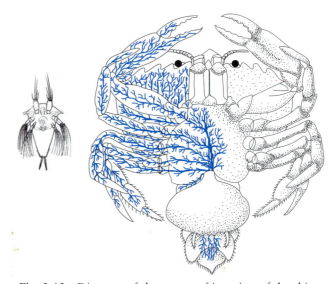

Fig. 3.13 Diagram of the extent of invasion of the rhizocephalan barnacle *Sacculina* into the body of a crab. Swimming larvae (left) invade a crab host and inject cells that reproduce and propagate a nutrient-absorbing tissue within the crab's body (right). (After Nicol, 1967.)

✳ Parasites of invertebrates often affect the reproduction of the host.

Some parasites seem to affect the reproductive organs of their hosts more than they affect any other organ. As a result, the hosts often survive, but are sterile. The parasitic barnacle *Sacculina*, for example, has a typical crustacean planktonic larva that invades the fatty tissues of the reproductive organs of its crab hosts and can spread (Figure 3.13). The parasite uses the fat reserves for its own reproduction, at the expense of the host, which may not have functional gonads as a result. Many animals are often in a race to grow and mature before the parasite load becomes too high for reproduction or even survival. This is a special problem for the eastern American mud snail *Ilyanassa obsoleta*, which reproduces in its third year. By this time, females in many populations are densely parasitized by several species of trematodes and may not be able to reproduce.

✳ Parasites often have complex life cycles that depend upon more than one host species.

Because the host dies eventually, or because its death may be accelerated by the presence of parasites, the parasites must have a means of dispersing to other hosts. As a result, parasites often have complex life cycles, with very different morphologies adapted to function in widely differing microenvironments. There is a danger in the dependence upon multiple hosts, because one of the hosts might be absent or difficult to locate.

Many parasites have life stages suitable for specialized parasitic existence, for dispersal, and for location of hosts (Figure 3.14). The crustacean isopod group Epicaridea may have two hosts. For example, the parasite *Portunion maenadis* has a larval stage that attaches to the copepod *Acartia*, a free-swimming stage, and a second parasitic stage, which lives in the visceral cavity of any of a number of crabs. The isopod parasite of the shore crab *Carcinus maenas* becomes a saclike sheath and bears no resemblance to a typical free-living isopod. In the phylum Platyhelminthes, or flatworms, a number of trematodes also have complex life cycles. Many species have a stage that inhabits mollusks, a free-swimming stage, and a terminal stage that invades fishes or birds. The fishes and birds often pick up the parasites while preying on the bodies of mollusks, or even the siphons of clams.

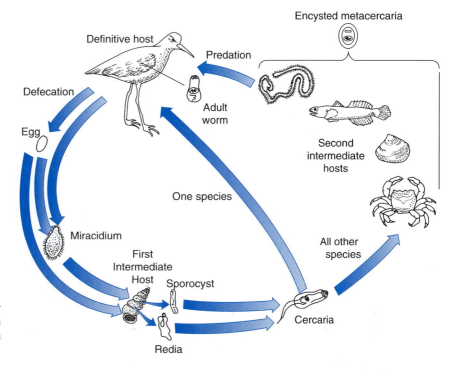

Fig. 3.14 Marine parasitic trematodes have complex life cycles, with several intermediate hosts. (From Sousa, 1993.)

The Population Level

Most ecological problems require a knowledge of a variety of numerical aspects of natural populations. **Population size** refers to the number of organisms in a defined area. Usually, we seek to define populations whose individuals interact and increase and decrease independently of other populations of the same species. A practical example of this is the definition of fishery stocks, which are a series of populations, each with its separate spawning and feeding areas (see Chapter 18 for further discussion). **Population density** refers to the number of individuals per unit area (e.g., per square meter-animals m^{-2}), or per unit volume (e.g., per cubic meter-animals m^{-3}) and gives an idea of the degree of crowding, or of the degree of access of individuals to scarce resources such as food or space. To calculate the amount of material in a population, we often measure density in terms of **biomass**, which is the mass of individuals per unit area (e.g., in units such as grams per square meter: g m^{-2}).

✳ A population is a group of individuals that are affected by the same overall environment and are relatively unconnected with other populations of the same species.

A species can be divided into a series of **populations**. The individuals in a population share the same general influence of the physical and biological environment. Within the population, it is much more likely that individuals will breed with each other, as opposed to members of other populations of the same species. This implies that the members of the same population can move freely throughout the geographic range of the population but are isolated from other populations. Geographic barriers such as peninsulas or sudden breaks in the environment might divide the species into a series of populations. For example, Cape Cod, Massachusetts, is a major barrier along the coast of the eastern United States, and many species do not have extensive dispersal across this barrier. Water temperature increases greatly from the north to the south of Cape Cod, and the same applies to Point Conception in southern California. The geographic range of many species ends at such barriers.

✳ Population change stems from survival, birth, death, immigration, and emigration.

Most marine populations are dynamic, and extensive change is the rule. **Survival of adults** is a major factor in population change. If survival is high, then the

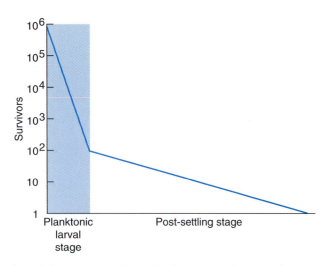

Fig. 3.15 Expected survivorship curve for a marine invertebrate species with planktonic larvae.

current population size plays a major role in explaining the population size in future time periods. **Birth** and successful **survival** of young are also of obvious importance. **Generation time** is the mean time between birth and the age of first reproduction. The existence of more generations per unit time will produce more offspring and a greater potential rate of population increase. Many marine species are capable of producing hundreds of thousands of eggs per female. This is testimony to the low survival rate of adults and the extremely low survival typical of juveniles. Juveniles are often planktonic larvae, and the variability of ocean currents often dooms them to failure in that they never find the proper habitat in which to settle. Food limitation may also limit reproduction. **Immigration** and **emigration** of adults can also affect the change of population abundance.

We can chart the probability of survival of different-age classes by using a graph known as a **survivorship curve**. Figure 3.15 shows an expected survivorship curve for a species (e.g., a crab) with a planktonic larval stage and a postsettling adult stage. We begin with a starting cohort and follow the mortality of these animals with increasing age. The survivors are plotted on a logarithmic abundance scale, and the slope of the line is therefore a rate of mortality. As can be seen, the rate of mortality for the planktonic stage is far greater than for the postsettling stage. Survival can be estimated by sampling a population repeatedly, as long as immigration is slight and one can distinguish newly born individuals.

Reproduction is usually seasonal and corresponds to increases of food for reproducing adults and to environmental factors such as temperature and salinity. Because of this, birth is also seasonal. Different **year classes**, or sets of individuals born in the same year, can usually be identified by distinct sizes, because animals of one year class have an entire year's head start on growth, relative to the next year class. It is possible sometimes to determine the age of marine organisms whose date of birth is unknown. **Growth rings** can be found in the otoliths ("earbones") of fishes, in the skeletons of corals, and in the shells of clams and snails.

✳ **Limiting resources may affect population growth.**

If resources were limitless and if there were no natural catastrophes such as storms, then a population could continue to increase indefinitely. In the real world, food or space will eventually run out. As the resource becomes scarce, **resource limitation** of survival, growth, and reproduction will occur. Figure 3.16 shows several cases of population change. In case a, the population increases by the same proportion with the passing of a given amount of time: this is **exponential growth**, which might continue indefinitely if resources were limitless. By contrast, case b shows a limit to the maximum population, or **carrying capacity**, that the environment's limited resources can sustain. As the population size approaches the carrying capacity, the rate of population growth decreases. When above carrying capacity, the population is too great for the available resources and it declines. These situations involve **intraspecific** (within-species) **competition** for resources. Many natural

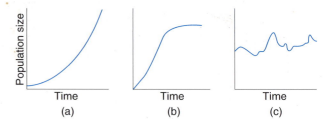

Fig. 3.16 Some examples of population change. (a) Exponential growth, a continuing proportional increase. (b) Resource-limited growth, where a population's increase decelerates as carrying capacity K is approached. (c) Random growth, where population-controlling factors are too complex to form any pattern.

populations resemble case c, which looks random. In this case, the factors regulating population size are too complex to show any simple pattern.

Many species occur together and require the same resources or at least overlap strongly in their resource use. This leads **to interspecific competition** for resources, and the carrying capacity of any one species is reduced, owing to the similar resource requirements of other species. Interactions of populations are discussed later, at the hierarchical level of the community.

✳ **Populations are often metapopulations, which are a series of interconnected subpopulations, some of which may contribute disproportionately large numbers of individuals to the metapopulation as a whole.**

It makes sense that population dynamics and genetic differences will be strongly controlled by the interrelationships among populations. It stands to reason that a complete mixing of individuals each generation will depress genetic differences between populations, simply because any changes occurring within a generation will be swamped out during the dispersal event. A widely dispersing planktonic larval stage might be an example of such a situation. The only differences between subpopulations therefore must develop within the time span of a generation. For example, one might imagine predators differing from one population to the next. In one population, large predators might kill all but the smallest prey items; but in another population, the presence of smaller predators might confer immunity on the largest individuals. Therefore you might find only very small prey animals in the first population and only large animals in the second. The important point is that in every generation, a strong dispersal stage will mix these subpopulations. Local adaptation to the different types of predator is, therefore, very difficult. It might be possible for the snails to evolve the phenotypic plasticity that permits any individual to react and defend differently against different types of predators.

Populations that are living in discrete habitats but are nevertheless connected by dispersal are known as **metapopulations**. Two different types of metapopulation illustrate the different consequences of connected local populations (Figure 3.17a). The isolated population model is represented by a series of local populations, each in one of a set of selective regimes. You can imagine, for example, an isolated popula-

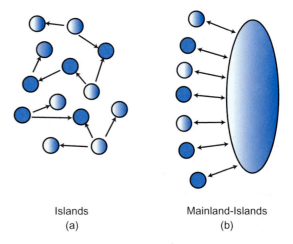

Islands
(a)

Mainland-Islands
(b)

Fig. 3.17 Two common ways of structuring a metapopulation into subpopulations: (a) a series of subequally sized subpopulations with connections by means of dispersal, and (b) an island–mainland scheme, where a large area is occupied by a large subpopulation and is close to a series of small areas occupied by "island" subpopulations.

tion of snails that is exposed to a visual predator. In such a place, there might be strong selection for shell color that matches the background environment. In other sites predators might be absent and there would be no selection. If the two habitat types were equal in frequency, and dispersal random and homogeneous throughout the populations, then the product of selection in the subpopulations with predators would be exported randomly to those in which no natural selection has occurred. In other words, high dispersal rates among subpopulations of the metapopulation might work against local adaptation. In the opposite extreme, dispersal might be very restricted between subpopulations. Here the local subpopulations subjected to strong selection to match substrate color would in fact evolve and differentiate successfully.

An alternative metapopulation structure might involve a large continuous population located near to a series of much smaller separated subpopulations; the subpopulations would appear to be islands near a "mainland" of the main population. A number of marine examples would come to mind. For example, a series of small patch reefs might lie slightly separated from the main reef. There might very well be a strong overlap of species in the two habitat types. If there are more individuals in the "mainland" reef subpopulation and if dispersal is strong and contin-

Fig. 3.18 (a) *Littoraria filosa*, shell height approximately 2 cm. (b) An isolated mangrove island in Queensland, where the snail was collected.

uous, then all processes will probably be dominated by dispersal from the "mainland" subpopulation to the patch reefs (Figure 3.17b). The island subpopulations may produce individuals, but they will not contribute nearly as many progeny to the metapopulation, especially to the main reef.

If you consider any particular subpopulation, it may be a net source or sink in the metapopulation. A **source** is a subpopulation that contributes disproportionately more individuals to the metapopulation. This may occur when reproduction and dispersal are unusually high in the local subpopulation. A **sink** occurs when a subpopulation receives immigrants from other populations but does not contribute individuals to the metapopulation. An obvious example is provided by the death of all immigrants into the sink subpopulation and failure of local individuals to reproduce.

Metapopulation thinking is very appropriate for marine systems, where dispersal of larvae between relatively isolated subpopulations is common. A fascinating example was recently discovered in a study of gastropods on Mangroves of central Queensland, Australia. The intertidal arboreal snail *Littoraria filosa* (Figure 3.18a) occurs commonly on mangrove leaves and is obviously very resistant to desiccation. The snail has separate sexes, which copulate, and the planktotrophic dispersing larvae swim in the water for about a month. Stephen and Ruth McKillup[4] have followed populations for a number of years and found that the snail appeared to be an annual: adults

died during the period of larval dispersal and settling, and there was little temporal overlap between successive generations. They were surprised when they began to investigate individual mangrove trees (see Chapter 14 for a description of mangrove forests) and found that snails on isolated trees (e.g., snails on isolated trees only 10–15 m from the forest) were not annuals at all, and continued to live after reproduction. It was apparent that the snails in the main mangrove forest were not "programmed to die"; but, then, why were they dying at all? As it turned out, an as-yet undescribed species of flesh-eating fly was the main cause. After female flies oviposited a crawling larva near a snail shell, the larva would crawl into the shell and consume the snail's body. In continuous mangrove forests, this process was so effective that no snails survived. Worse than that, upon further inspection it was clear that none of these snails ever reproduced successfully. The settlement of larvae had to come from some other source.

The isolated mangrove trees apparently provide the answer. Snails there are rarely parasitized by the fly, perhaps because wind prevents flies from locating the isolated trees. The snails reproduce well. Now if we return to our metapopulation idea, we can liken this situation to a mainland–islands scheme in reverse (Figure 3.19). The main population distributed within large patches of mangrove forest is actually a

[4] See McKillup and McKillup 2000, in Further Reading, The Population Level.

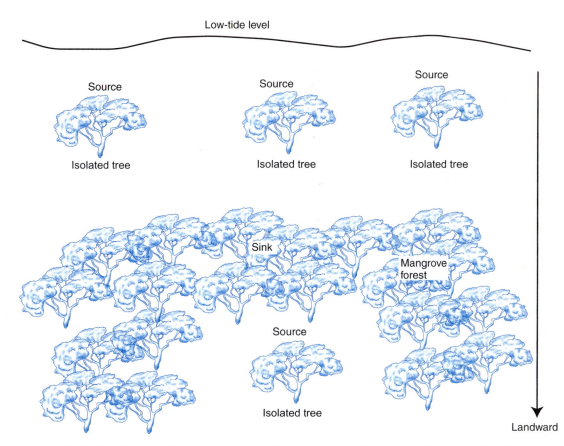

Low-tide level

Source

Isolated tree

Source

Isolated tree

Source

Isolated tree

Sink

Mangrove forest

Source

Isolated tree

Landward

Fig. 3.19 Metapopulation structure in the mangrove leaf gastropod *Littoraria luteola*.

sink. Larvae that arrive there will not reproduce. Incongruously, it is the series of "island" isolated trees that constitute sources and may be responsible for supplying the entire population. Snails found on mudflats may also contribute to the larger metapopulation.

Spatial Variation

✳ Spatial distribution is a measure of the spacing among individuals in a given area.

Spatial pattern is a useful feature of natural populations. The spatial distribution is the measure of the type of spacing among individuals. Consider a square meter of rock on a shoreline that has a population of barnacles. If a barnacle has the same chance of being located in one spot as in any other spot, then the population has a **random distribution**. Figure 3.20a shows such a distribution, which has the appearance of randomly sprinkled grains of salt. If more barna-

cles occur in a given subarea than is expected by chance, then other areas will be depleted of animals, giving an **aggregated** or **patchy** distribution of clusters and empty space (Figure 3.20c). If every equal subarea contains a constant number of individuals, or at least a more nearly constant number than is expected by chance, then there is a **uniform spatial dis-**

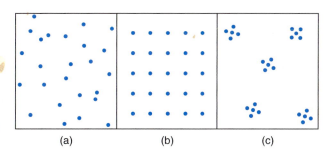

(a)   (b)   (c)

Fig. 3.20 Patterns of spatial distribution: (a) random, (b) uniform, and (c) aggregated.

tribution, which ideally looks like a grid (Figure 3.20b).

Spatial distributions are useful because they suggest hypotheses about the mechanisms affecting natural populations. It is rare for populations to have a random distribution. This usually occurs when larval settlement from the water column is random or animals are moving about randomly, as when mud snails move on the surface of a mudflat. Uniform distributions occur when animals are maximizing the distance between neighbors. This implies the establishment of territories. For example, when larvae of the tube worm *Spirorbis borealis* settle on seaweed from the water column, they usually crawl away from other settled larvae before metamorphosing into adult worms. As a result, one eventually may see a uniform array of tiny spiral tubes on the seaweed frond. Aggregations usually imply some sort of patchiness about the environment, but organisms might be socially attracted to each other for mating, or to form fish schools or other aggregations to protect against predators.

✳ A population may show a regular change in density along a sampling line.

If sampling is done along a transect line, many populations change in a definable pattern. The population density might increase or decrease, or might show a pattern of increases and decreases. A population of plant-eating snails, for example, might decrease regularly with increasing water depth, because the food source also declines with depth. As another example, a beach might be covered with ripple marks, and certain species might be abundant in the troughs but rare on the ripple crests. If the density of a population at one site can be predicted from the density at neighboring sites, we speak of the environment as having **spatial autocorrelation**. A spatial autocorrelation might occur if (1) there is a change in the environment that affects survival or causes differential subhabitat selection; (2) the population is moving in a defined direction (the change in density might reflect, for example, the tail end of a migrating population of fish); or (3) a random process occurs, which occasionally can cause a nonrandom spatial pattern.

Species and Classification

✳ Taxonomic classification involves successively nested grouping of species.

It is impossible to consider all the marine species without using some sort of classification system. Biologists universally accept the system devised by Linnaeus, which gives a species a binomial (two-name) description. Every species is described by its **genus** (plural, genera) plus **species** names. For example, the killer whale is named *Orcinus orca*. Note that species names are published in italics or written and then underlined. Some species are divided into subspecies and three names are then used. One can abbreviate the genus portion of a species name (e.g., *O. orca*).

Organisms are classified into groups larger than the species level. Each high-ranking group is made up of a cluster of groups of the next lower level. The major taxa, or classification ranks, are as follows (from the lowest to the highest): **species, genus, family, order, class, phylum, kingdom**. For example, the sea mussel *Mytilus edulis* belongs to the genus *Mytilus*, has the species name *Mytilus edulis*, is a member of the family Mytilidae, the order Fillibranchia, the class Bivalvia, the phylum Mollusca, and the kingdom Animalia. Note that only the genus and species names are italicized.

Species are grouped by their overall evolutionary relationships. Species in the same genus are hypothesized to be more closely related by descent to each other than to species belonging to other genera. Genera within one family are usually believed to be more closely related to each other than to those in another family.

✳ Evolutionary relationships can be used to construct trees of relationship. Taxa are grouped by means of shared evolutionary derived characters.

All members of a given group have shared evolutionary characters, which unite them by descent and distinguish them from other groups. Thus, mollusks have an external calcium carbonate shell, differing in that respect from members of other phyla. Arthropods, such as insects, horseshoe crabs, and shrimps, all have an external cuticle, a distinct segmentation, and jointed appendages. We argue that the more unique characters members of a group may share, the more likely it is that the group evolved from a common ancestor. This notion allows us to construct evolutionary trees of relationship, or **cladograms**, like Figure 3.21, which shows the relationships of some purely hypothetical critters. Note that we cannot be sure about the exact history, such as who the ancestors might be. We can only say who is more closely

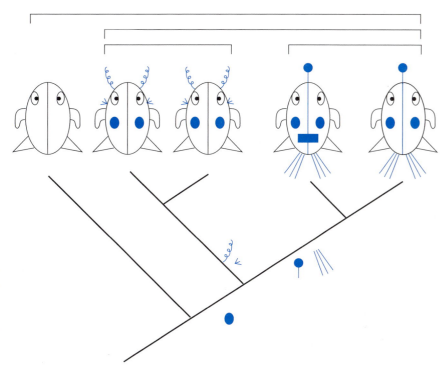

Fig. 3.21 Organisms that share more and more unique characters are likely to derive from common ancestors. Brackets on top indicate hierarchy of groups; characters used to unite groups are shown along branches of tree.

related to whom. More direct evidence such as a fossil record might help to determine ancestry.

An analysis of real critters may lead to a good deal of complexity (Figure 3.22) but also to an impartial analysis of evolutionary relationships. Groups are united by sets of uniquely shared evolved characters. Thus as you go "upstream" to the base of the tree, you are encountering organisms that have traits that unite all the downstream groups.

Genetic Basis of Organismal Traits

✳ **Organismal features can be explained on the basis of a combination of genetic and nongenetic components.**

Marine organisms are universally variable in form, color, and biochemistry. A **polymorphism** is any variation that can be identified in terms of a series of discretely different forms. In other cases, variation can be measured only as **continuous variation**, such as differences in body size, or the proportional size of a fin. We must distinguish between genotype and phenotype. The **genotype** refers to the genes that characterize an individual, or to those that control a particular trait, such as eye color. The **phenotype**, by contrast, is the form the organism takes. For exam-

ple, it is possible that all the brown-haired individuals in a human population will not have the same genotype. By contrast, people with gray or blond hair may be of identical genotype. The gray-hair phenotype is mainly associated with age and not genetic heritage, although all too many people seem to think that they are genetically disposed to being prematurely gray.

Morphological variation in a population can be explained with a simple equation:

Morphological variation = variation explained by genetic factors + variation explained by environmental factors + an interaction between genetic and environmental factors

If we see variation (e.g., a range of lengths) in a population, we need to be aware that much of it has nothing at all to do with genetic variation. Environmental effects such as nutritional status and microclimate alter the course of growth and development of animals with identical genotype. Much of the variation we see in natural populations, however, exists because of the inheritance of different genes. Shell color is a conspicuous example of this in many mollusks. Usually, however, both genetic and nongenetic components contribute to determine a trait. Body size

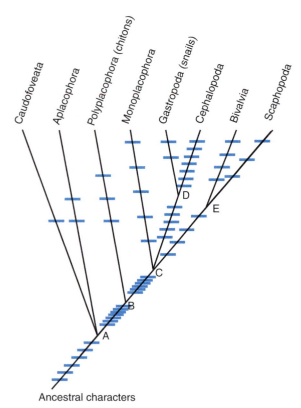

Fig. 3.22   A tree of evolutionary relationships for the phylum Mollusca. Traits (horizontal lines) between nodes unite all taxa downstream of the node higher on the page. On the stem of the tree a number of characters unite the entire phylum, including reduction of the coelom, presence of an open hemocoelic circulatory system, and production of spicules or shell by a mantle shell gland. "Downstream" of every node (locations A–E) are traits that uniquely identify individual groups. Thus the Bivalvia–Scaphopoda line is downstream of node E and the two groups are united by head reduction and decentralization of the nervous system, among other traits. (Modified from Brusca and Brusca, 1990).

is a useful example. It is almost always controlled partially by genes, but the environment also exerts a large effect.

It is extremely important to realize that having a given gene does not guarantee that the form of an organism will be constant. The same genotype may have a different phenotype when raised in different environments. This is known as a **genotype-by-environment** interaction.

In a few cases, variation is due to a **genetic locus**, or single location on the genetic material, or DNA.

In such a case, an individual has two genes for the trait, one inherited from the father and one from the mother.[5] The genes might be identical, or they may be different variants, or **alleles**. If there are two alleles, *a* and *b*, then there are three possible genotypes: *aa*, *ab*, and *bb*. All three variants may look different, or one allele may be **dominant**. For example, if the *a* allele is dominant, then *aa* and *ab* genotypes may have identical phenotypes.

The mussel *Mytilus edulis* can be blue-black or brown, owing to the control of a single genetic locus, with two alleles. The *brown* allele is dominant over the *blue-black*, and the heterozygote, which inherits one *brown* allele and one *blue-black* allele, is therefore colored brown. In most traits, several or many genetic loci are in control. Body size, for example, is usually controlled by many loci. In such cases, the genetic component can be found by studying the degree of resemblance among relatives. The correlation of a trait between parents and offspring, for example, can give evidence of a genetic component. Figure 3.23 shows the correspondence between number of vertebrae in mothers and in offspring of the eelpout *Zoarces viviparus*. The correlation is high, and we conclude that the variation in the trait is therefore controlled largely by genetic variation.

There are several types of common variation observed in marine populations. **Chromosome number can be variable in natural populations.** The Atlantic drilling snail *Nucella lapillus* is variable in chromosome number when found in different degrees of wave exposure. Many species have **color polymorphisms**. These polymorphisms may be explained mainly by genetic variation. Many **morphological characters** (e.g., body size, number of fin rays on a fish) are variable and are controlled, at least in part, by genetic variation. Variation in **biochemical and physiological traits** (e.g., presence of specific proteins, different levels of oxygen consumption) is common, and **polymorphisms of enzymes** occur widely. **DNA sequences** are now being used actively to determine whether populations have their own unique sequences, which would be an indication that they are isolated somewhat from other populations. This would be the beginning of an isolating process, resulting eventually in speciation.

What maintains genetic variability in populations? **Natural selection** is the process whereby individuals

---

[5] With the interesting exception of sex-determining chromosomes.

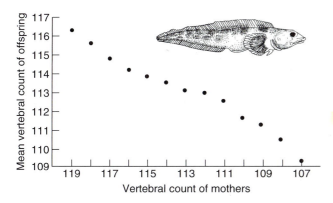

Fig. 3.23 The mean vertebral count of mothers and their offspring in the eelpout *Zoarces viviparus*. (Data from Schmidt, 1920.)

with certain genes survive and reproduce more efficiently than others; this leads to dominance in the population by certain genetic variants. The relative survival and reproduction of a given genotype is its **fitness**. **Adaptation** occurs when natural selection causes evolutionary change in a population, which results in an increase in the ability of a typical member of the population to perform in that environment. We usually judge performance with respect to a given function, such as resisting heat shock.

Variation can be maintained by shifting of environments, which may favor one genetic variant, then another. Alternatively, a complex environment can favor several genetic variants, but in different microhabitats. In some cases, a gene is favored simply because it is rare, which would cause a dynamic shifting back and forth of gene frequencies. This process, known as **frequency-dependent selection**, should work commonly when rare morphs are favored in mating. Finally, gene-level variation can be maintained if heterozygotes are favored in natural selection. This tends to keep alleles in the population, because selection for a homozygote, or organism with identical alleles for a trait at a genetic locus, would favor one allele at the expense of others. Finally, immigration of different genetic variants from adjacent populations can increase variability.

Although natural selection is ubiquitous in natural populations, **random events** can also influence the genetics of marine populations. If the population is very small, chance events may cause the loss of certain variants from the population. This outcome, which is more likely when there is little difference in fitness

among genotypes, has been claimed especially for some biochemical variation in proteins.

✱ Single genotypes may have the capacity to develop into distinctly different morphologies.

A given genotype make take different forms under different circumstances, controlled strictly by environmental circumstances. This phenomenon is known as **phenotypic plasticity**, which is the ability of a single genotype to develop into different forms, usually as a response to environmental circumstances. One can rightly say that a population has evolved individuals that are plastic and capable of responding to individual circumstances. Every individual has, at the ready, the capacity to respond to local circumstances. We encountered such phenotypic plasticity in our discussion of inducible defenses. Many marine organisms can grow spines, increase shell thickness, or change morphology completely in response to predators. Work by Dianna Padilla[6] and colleagues has demonstrated that herbivorous snails can completely alter the morphology of their feeding teeth simply by switching from one algal food to another (Figure 11.18).

It is of great interest to ask why some species show plasticity and can adapt to all circumstances, whereas in other cases genetically distinct morphs coexist in natural populations, with each morph being better suited to function under different circumstances. This is quite a profound question, and the evolution of phenotypic plasticity is an active area of research.[7] In both cases you can imagine that there might be a great cost. If you were phenotypically plastic, you might be able to generate a range of morphologies, with none of them quite right. In other words, you would be a jack-of-all-trades but master of none. If the environment is very unpredictable and it is not clear that you will or will not encounter a given situation (e.g., predators), then phenotypic plasticity might be selected for. In a stable set of microhabitats, on the other hand, a genetic polymorphism for specialized individuals might be selected, assuming that the specialized morphs had greater efficiency than could be achieved by the phenotypically plastic form.

[6] See Padilla and others, 1996, in Further Reading, Species, Classification, and Genetic Control of Traits.
[7] See Via and others, 1995, in Further Reading, Species, Classification, and Genetic Control of Traits.

✳ The geographic change in the frequency of genetic variants is called a cline.

It is common for members of a species to differ from place to place in morphology, color, or size. When such differences are clear and discrete, species can be divided into **subspecies**. More often, variation is more continuous. For example, the color polymorphism in mussels, described earlier, shows an increase in the brown form as one moves southward. Such geographic variation is also common in biochemical polymorphisms. Any change in frequency with geography is known as a **cline**. Figure 3.24 shows the change in frequency of an allele at an enzyme locus as one samples the crested blenny *Anoplarchus purpurescens* at different latitudes in the Puget Sound region.

Clinal variation can be found at many geographic scales. Differences in color and biochemical protein genetic variants can be found at different levels of an intertidal rocky shore. This variation may be due to short-term natural selection on larvae that have settled all through the shore, but survived differentially, depending upon genotype. On a larger scale, clines are found commonly along scales of 100–1,000 km along coasts. Such clines may also be due to differential natural selection along the coast, but isolation between adjacent coastal populations may cause them to be exaggerated. Because most coastlines in the ocean have a north–south orientation, genetic differences in populations at different latitudes may often relate to differential performance at different oceanic temperatures.

✳ New species usually originate after a species is divided by a geographic barrier.

For new species to originate, it is usually necessary for a barrier to isolate a species into two or more populations. If the barrier is short-lived, the populations will reconnect. If the barrier is longer-lived, and especially if the populations diverge genetically, then they may be relatively incompatible when reconnected. Offspring of population crosses between populations will be less fertile than crosses within populations. This would cause selection for mating with one's own kind, leading to further genetic differences between the populations, whereupon separate species would evolve. In some cases we can see the recent effect of such barriers. Many pairs of closely allied species are found on either side of the Isthmus of Panama, which is only about 3 million years old. A number of pairs of coastal invertebrate species are found on the north and south sides of Cape Cod. Isolation by the coastal barrier may have been magnified by evolution of local populations to adapt to the very different temperature regimes north (cold) and south (warm) of the Cape. In many cases, newly evolved species are so similar that they are identical, or nearly indistinguishable, morphologically. Such species, known as **sibling species**, are very common among marine species.[8] Although sibling species may have separate geographic ranges, many cases of co-occurrence have been discovered. For example, the mud-dwelling polychaete annelid worm "species" *Capitella capitata* is now known to consist of several closely related species, which cannot be distinguished easily as adults but are quite different in the larval stage, and also in chromosome number (Figure 3.25).

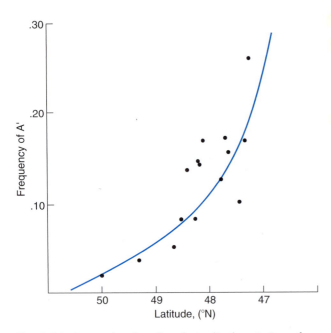

Fig. 3.24 Example of a cline: latitudinal variation of an allele, *A′*, at a genetic locus coding for the enzyme lactate dehydrogenase in the crested blenny, *Anoplarchus purpurescens*, in Puget Sound, Washington. (After Johnson, 1971.)

[8] See Knowlton, 1993, in Further Reading, Species, Classification, and Genetic Control of Traits.

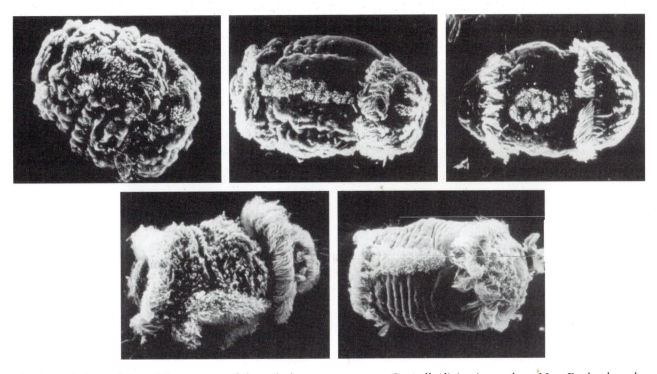

Fig. 3.25 Larvae of five sibling species of the polychaete worm genus *Capitella*, living in southern New England muds. (Courtesy of K. J. Eckelbarger and J. P. Grassle.)

Marine Biogeography

* **The geographic range of species is limited by geographic barriers and habitat limitations.**

No marine species occurs worldwide. There are two main factors that limit the geographic range of species. First, most species can survive and reproduce only in a limited range of habitats. For example, most clam species require soft sediment and survive poorly on rocks or in the water column. Because the distribution of soft sediment is restricted, most clam species cannot live everywhere. Even if soft sediment were ubiquitous, many **barriers to dispersal** would limit movement. On coastlines, large estuaries and peninsulas might restrict dispersal. Many species require shallow water bottoms as adults but have limited dispersal stages. Such species cannot usually traverse deep oceans. The combination of coastal barriers and deep-ocean barriers breaks up the distribution of coastal species. Many deep-sea species have far greater geographic ranges because the seabed consists of great expanses of relatively homogeneous mud, with fewer important barriers to dispersal. In the deep ocean, midoceanic ridges are often biogeographic barriers and sometimes divide oceanic realms of differing water bottom temperature.

Marine species do occasionally breach dispersal barriers, and they may then spread rapidly. The periwinkle *Littorina littorea* probably arrived in North American several times, and shells have been found in a nearly thousand-year-old Viking boat on the north shore of Newfoundland. But 200 years ago or so, the species arrived again on the shore of Nova Scotia and began to spread south. Now, it is the dominant shore snail from Nova Scotia to New Jersey. It has a planktonic larval stage, and shore currents have carried larvae farther south with every passing year. Thus the species has spread about 1,500 km in 100 years.

* **The marine biota can be divided into geographic provinces.**

The combination of barriers to dispersal of marine organisms breaks up the marine world into a series of relatively distinct assemblages of species. The geographic regions containing these assemblages are known as **provinces**. Although provinces may have their own unique species, many marine species may occur in adjacent provinces. Many provinces can be defined statistically.

Temperature changes with latitude tend to sharpen differences of species assemblages along coastlines. As was explained earlier, many capes and other coastal irregularities are the locations of significant temperature change. Cape Hatteras, North Carolina, for example, is the location of a large change in ocean temperature, and many northern species cannot survive the warm temperatures south of this cape. On the west coast of North America, Point Conception, California, has a similar effect. Very different assemblages of coastal species are found north and south of this point. The general north–south orientation of our coastlines, combined with the latitudinal change of climate, breaks up coastlines into high-latitude, boreal, temperate, subtropical, and tropical regimes, each with its own assemblage of species (although there are species that bridge the provincial borders). Currents often also serve to isolate species on either side of a barrier. The southeastern tip of Florida is an especially important isolating barrier in North America, separating Gulf coastal populations and species from those of the Atlantic coast.

✳ The present distribution of species is the end result of speciation, dispersal, and extinction.

Although marine species today can be organized as members of a group of biogeographic provinces, the current distribution of species can be quite misleading with respect to their evolutionary origins. Some of the patterns make obvious sense, however. For example, consider the many pairs of species of various invertebrate groups on either side of the Isthmus of Panama. As it turns out, the isthmus arose 1–3 million years ago, and this event isolated populations of formerly continuous marine species on either side of a land barrier. The isolated populations eventually evolved into distinct species, so there is a large number of such pairs today.

On the other hand, a species that is a conspicuous, even dominant, member of a marine community may have only recently dispersed to its present site. Good examples on the east coast of North America are the periwinkle *Littorina littorea* and the shore crab *Carcinus maenas*. Both dispersed from Europe about 100 years ago, and both are major species in eastern North American shore communities. On a much larger scale, species or taxonomic groups that are restricted today to a small region may have been far more widespread in the past. An obvious example is found in the articulate brachiopods, or lampshells, which today are observed in deep water, in some high-latitude regions, and spottily in coral reefs. Their distribution today is a paltry remainder of their former dominance of shallow seas in the Paleozoic Era. At the end of this era, the group suffered a massive extinction, and we find only small remnants of them today.

The Community Level: Interspecies Interactions

✳ **Distribution and abundance of species populations in a community are determined by the combined effects of the following processes: (1) dispersal of larvae, spores, and adults, (2) competition, (3) predation and herbivory, (4) parasitism, (5) disturbance, and (6) facilitation.**

Physical features of the environment, such as temperature, salinity, dissolved oxygen, and nature of substratum, may determine the maximum environmental range of a species. However, a series of dynamic community-level processes strongly affects distribution and abundance and usually reduces the maximum physical range possible. These processes are as follows:

1. **Dispersal to appropriate habitats** via movement of larvae, spores, or adult dispersal
2. **Competition** with other species for limiting resources
3. **Predation and herbivory**
4. **Disease and parasitism**
5. **Disturbance,** or effects of storms, sudden temperature change, and other factors that cause rapid mortality
6. **Facilitation,** or the conditioning of the environment by one species, allowing another species to invade

We cannot isolate the effects of biological and physical factors because the interactions are so significant and complex. For example, an animal enfeebled by temperature stress may fall victim to a predator sooner than a healthy individual might. In this section we shall examine how biological interactions affect the composition of marine communities.

Larval Access

* **Larval recruitment patterns strongly reflect the species composition of marine communities.**

In the ocean, many species of fishes and invertebrates have planktonic larvae, which can disperse great distances, sometimes across an ocean. Larval settlement can be the limiting factor in the composition of marine communities. In most marine habitats, there are "good" and "bad" recruitment years. Some of this variation may result from the effects of local ocean currents, which may sweep larvae out to sea, or keep them near the shoreline. Feeding larvae may be short of food in some years, with a consequent reduction in successful settlement. The phenomenon of good and bad years for larvae leads to wide swings in adult population size.

Interspecific Competition

* **Competition within and between species centers around the limiting resources of space and food.**

Competition occurs when two individuals of the same (intraspecific) or different (interspecific) species exploit a limiting resource (Figure 3.26). **The two prime limiting resources are space and food.** As we discussed previously, resources may be renewable or nonrenewable. Nonrenewable resources are monopolized at least for the lifetime of the exploiting organism. The space occupied by the basal plate of an acorn barnacle is a good example. By contrast, renewable resources that are exploited may later become available again to the exploiter. In most cases, renewable resources are themselves living. Planktonic copepods, for example, graze the renewable resource of microalgae, but the latter can recover through rapid population growth. Within the same habitat, copepod excretion of ammonia provides a renewable resource for phytoplankton nutrient uptake. In sediment-dwelling bacteria, the recovery rate of the bacteria outstrips the ability of grazer populations to reduce bacterial standing stock. Because the resource is renewable, organisms are not exactly competing for the same item at any given time. Rather, competitive success usually depends upon which organism is more efficient. Competition for fixed resources, such as territories, often involves various forms of aggression.

The study of competition must be focused on limiting resources. A **guild** is a group of species that exploit the same resource. Guilds need not include closely related species. Consider the occupation of space on a rocky shore. In a typical site on the Pacific American coast, several hundred sessile species, including stalked barnacles, acorn barnacles, mussels, brown seaweeds, green seaweeds, and crustose coralline algae, share the same space resource. In a study of competition, it would make no sense to study mussels without also considering seaweeds. Indeed,

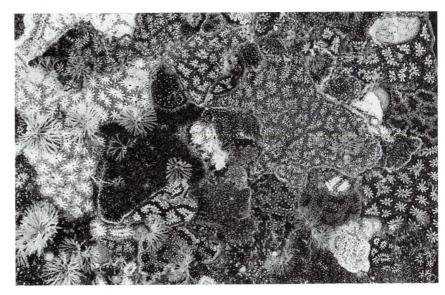

Fig. 3.26 Competition affects abundance when resources are limiting. In this example of intraspecific competition, a number of colonies of the colonial tunicate *Botryllus schlosseri* dominate a surface. (Courtesy of R. Grosberg.)

mussels may compete with seaweeds for space on rocky shores. In pelagic environments, many seabirds compete with carnivorous fish for smaller fish prey.

* Competition between species may involve direct displacement, preemption, or differential efficiency in the use of resources.

Competition between species can proceed in several different ways. Often when space is the limiting resource, one species may succeed through direct **displacement** of another. In such a case, we must assume that all encounters between species A and species B have the same outcome (e.g., A displacing B). By contrast, a species that holds space by colonizing a bare spot may then **preempt** invasion by another species. Its priority in arrival to the site gives it precedence there. Finally, if two species compete by virtue of requiring and exploiting the same resources, there might not be any direct behavioral interaction. In such **scramble competition** (or **resource competition**), the more efficient species might gain more food and gradually increase in population size at the expense of the other.

* Competition has been demonstrated in marine communities by experimental removals of abundant species, followed by expansions of competitors.

One can be overwhelmed by the variety and complexity of nature. We often cannot explain patterns in marine communities very easily. If a species is absent, are predators the cause of its absence? Has the species lost out in interspecific competition? A similar problem arises when we examine zonation, the most common feature of rocky shores, where dominant species may occur in a series of horizontal bands.

One commonly observes, especially in quiet waters, a series of horizontal bands. Classically, they consist of, in order from high to low intertidal, lichen, barnacle (sometimes equal to), limpets, and mussels. Why such dominance by single species? You might be tempted to explain this by competition, solely on the basis of thinking that one species must be especially superior in its own band. You would not be entirely right; it turns out that efficient predation at the low end of the rocky shore is often an important factor in zonation.

What should be done to resolve such cases? Marine ecologists, inspired principally by the pioneering works of Joseph Connell and Robert Paine, have approached the problem through systematic **manipulative field experiments**. The experiments involve removals of hypothesized predators or competitors or caging of areas against predators, with careful observation of the consequences. For example, for many years, Robert Paine removed the voracious starfish *Pisaster ochraceus* from a rocky shore off Cape Flattery, Washington. It was more than 10 years before a significant change took place in the distribution of bed of the mussel *Mytilus californianus*, which extended downward and overgrew several species of seaweeds (see Chapter 14).

The experimental approach allows direct observations of **process**. This is important because a simple phenomenon such as zonation may be explained by one of several processes or by their combination. With this tool, we can now explore the consequences of biological interactions.

Field experiments may be prohibitively difficult because the organisms are microscopic or because the manipulation is hard to interpret. One couldn't carry out a field removal of a species of bacterium. Many small, abundant organisms cannot be manipulated efficiently. Some field caging experiments, furthermore, change the experimental microenvironment in unacceptable ways. Cages built to protect soft bottoms from mobile predators also alter the depositional environment, and fine sediment settles within the cage. The experimenter is then altering two factors at once. This kind of situation may preclude field experiments; but laboratory experiments may be quite informative as long as some element of realism permits one to relate the laboratory results to field conditions. For example, laboratory work with soft-sediment deposit feeders would require simulation of field conditions of sediment organic content, microbial abundance, and sediment stability.

As preferable as experiments may be, in many cases biological oceanographers are limited strictly to one form or another of remote sensing. The deep sea is a case in point. Although it is possible to explore the deep sea with submersibles and even to perform a few experiments there, the possibility of long-term continuous manipulations and observations is very limited. In the plankton, it usually is impossible to manipulate natural populations.

Occasionally, so-called **natural experiments** are encountered. We may discover, for example, that in most sites two species are found together. However, we might find that, in some locations, one species is naturally absent, and the other species has expanded in abundance. We might provisionally conclude that the first species normally affects the other's abundance. Although that is a fairly safe conclusion, we should remember that this is not a controlled experiment. The factor that removed one species may also have enhanced the other's abundance. For example, many fish species decline in estuaries, but mullets are often very abundant. One might be tempted to explain this set of circumstances on the basis of relaxed competition between mullets and other fish species. However, mullets are detritivores, and estuaries often have increased supplies of detritus. Reduced salinity may have independently eliminated the other fish, while detrital supply enhanced the mullet.

In other cases, natural experiments allow us to come to fairly safe conclusions. For example, the depth range of algal-grazing fish tends to expand when the depth range of the algal-grazing Caribbean sea urchin *Diadema antillarum* contracts. *D. antillarum* is a very effective herbivore, and its removal results in dramatic growth of benthic algae. The shift in relative range can therefore be explained convincingly in terms of shifts in algal abundance.

✳ Competition for space involves mechanisms of overgrowth, standoffs, and various means of aggression.

On hard substrata, space is usually the prime limiting resource. For suspension feeders, access to the water column makes competition even more intense, whereas light access is crucial to photosynthetic animals and plants. Food, especially suspended phytoplankton and zooplankton, exerts an important feedback by permitting strong growth and displacement of competitors. Mechanisms of competition for space include the following:

1. Simple overgrowth and undercutting of neighbors
2. Secretion of poisons
3. Presence of special aggressive structures
4. Shading of competitors (in photosynthetic organisms)

Hard-substrate organisms often outcompete others by virtue of their growth rate. In many cases, species with the most rapid growth rate are the most successful competitors. Rapid growth should be advantageous to photosynthetic organisms, such as seaweeds and reef-building corals. This would result in

the shading of neighbors that grow more slowly. Many marine invertebrates have one of a series of structures that are specifically adapted to resisting overgrowth. Bryozoans and coralline algae deploy spines and upward projections to stem the tide of competitors. Bryozoans are polymorphic colonies, and individual zooids in some species are beaklike avicularia, which can pinch competitors and predators. Some anemones and scleractinian corals have special sweeper tentacles that inflict wounds on potential competing neighbors. Many seaweeds are erect, and when wave action causes them to whip from side to side, the whiplashing tends to damage neighboring seaweeds and invertebrates. This in turn reduces space competition.

✳ Interspecific competition may lead to dominance by single species in a given microhabitat.

We can use our experimental tools to demonstrate that interspecific competition often leads to dominance, or **competitive exclusion**, by a single species that monopolizes the habitat indefinitely, short of a major perturbation. In a classic experiment, Robert Paine removed the predatory starfish *Pisaster ochraceus* from a rocky shore where a number of species of bivalves, barnacles, and a brachiopod coexisted within the same horizontal tidal level. After the removal, the mussel *Mytilus californianus* recruited to the rocks and displaced most of the coexisting sessile species.

✳ Competition combined with differential success in different microhabitats results in niche structure.

We define **niche structure** to be any predictable partitioning of a habitat into subhabitats. This must also involve the differential utilization of limiting resources. Ecologists have long believed that no two species can coexist on the same limiting resource. Although this turns out not to be always true theoretically, the presence of coexisting species on different resources has been used as evidence of the action of interspecific competition. Many of these studies are observational only. For example, species of the carnivorous snail genus *Conus* live associated with coral reefs throughout the Pacific. A. J. Kohn[9] found that species of *Conus* in subtidal coral reef habitats with high species numbers were highly specialized and tended to eat different foods. Overall, coral species

---

[9] See Kohn, 1967, in Further Reading, Competition.

numbers increase toward an Indo-Pacific center near Malaysia and the Philippines, and species of *Conus* that depend on coral refuges tend to increase in number toward the coral center of diversity.

By contrast, the single species of *Conus* found on the coast of California has much more general food preferences. In the much more homogeneous intertidal smooth platform habitat, species diversity does not increase toward the Indo-Pacific region (Figure 3.27). Although the evidence is circumstantial, it suggests that some niche structure exists and results from competition among *Conus* species or prey.

* Some assemblages of natural species show extensive coexistence of presumed competitors despite apparent resource limitation.

Unless there is an opportunity for niche displacement, one would expect a competitive dominant species to displace all other species. However, this "law of competitive exclusion" often does not seem to apply to all natural communities. For example, many species of phytoplankton coexist, despite apparent resource limitation. In the open-ocean tropics, scores of species of phytoplankton coexist even though dissolved nutrients such as nitrogen are undetectable in the water column. Why have all inferior species not been outcompeted by species superior at taking up nutrients from the water? The great ecologist G. Evelyn Hutchinson termed this coexistence "the paradox of the plankton."

A number of processes can explain such a lack of competitive exclusion. These include the following.

1. There may be complex competitive interactions, combining multiple means of competitive superiority with no clear competitive dominant. It is possible that species A is competitively superior to species B but inferior to species C, leading to different outcomes of dominance, depending upon which species come into contact (Figure 3.28).
2. Adult sites may be limited, but colonization is from a random larval pool in the water column. If an animal dies, a larva might settle from the water column and establish a territory, but the

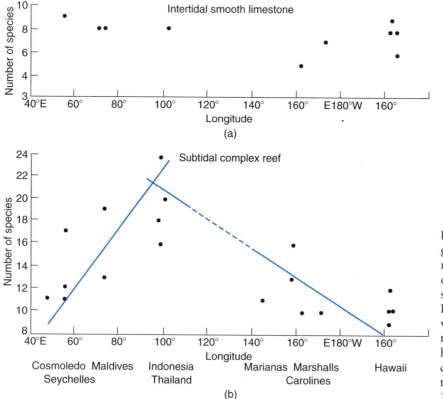

(a)

(b)

Cosmoledo Maldives · Indonesia · Marianas Marshalls · Hawaii
Seychelles · Thailand · Carolines

Fig. 3.27 Environmental heterogeneity promotes coexistence of many species by providing several distinct environments within which species may exploit unique resources. Diversity increases toward the southwest Pacific for the carnivorous gastropod *Conus* in complex subtidal hard substrata (b) but does not change in topographically simpler intertidal platforms (a). (After Kohn, 1967.)

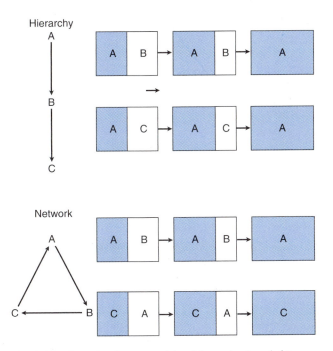

Fig. 3.28 In a simple competition hierarchy (top left), one species is superior to all others, and any given species is superior to another at a lower hierarchical level. In a network (lower left), however, species A may outcompete B and B may outcompete C, but C may outcompete A. This can happen only if a distinctly different mechanism of competitive superiority for the C–A interaction exists to delay the eventual dominance of any particular species. The lower figure shows that networks create a variety of outcomes when the species are combined, perhaps by larval settlement of pairs of species.

particular species of the colonist would be determined by currents and chance. The species composition of the community would be determined more by colonization than by interactions of the resident adults. Peter Sale[10] explained the coexistence of several territorial fish species by means of the **lottery hypothesis,** which emphasizes random mortality and colonization by planktonic larvae. The fishes occupying any given site result from a random drawing from among the pool of planktonic fish larvae that happen to be in the water column when a benthic territory becomes open. Since gains made by a competitor are often lost because of random mortality, such random processes of adult extinction and recolonization delay or completely prevent the competitive displacement of one species by another.

3. Complex patterns of disturbance may preclude the rise of any one species to dominance. Disturbance may be so common (as discussed shortly), that competitive dominance is prevented.
4. There may be habitat complexity, which permits coexistence of many species. As discussed earlier, habitat differences may allow species to coexist by specializing on slightly different microhabitats.

Predation and Herbivory

✱ **Predation may prevent the domination by a superior competitor and may strongly affect species composition.**

The experimental removal of *Pisaster ochraceus* that resulted in the dominance of the mussel *Mytilus californianus* suggests a common effect—namely, that predation delays the competitive displacement of competitively inferior species by the competitive dominant. Herbivory often has the same effect on competing species of seaweeds or sea grasses. Experimental removal of sea urchins usually results in dominance by one rapidly growing seaweed species over the others. In many natural communities, predation is so intense that the competitive dominant is only slightly more numerous than competitively inferior species.

✱ **Predators are not omnipotent: prey can move into various refuges, escaping in space or time, or they can outgrow the predator.**

In Chapter 13, we will describe some of the means by which benthic predators locate, immobilize, and consume their prey. Although certain species are remarkably potent predators, many prey species have **refuges,** which are means of escape from predators. Some refuges exist in space or time; others relate to prey size. A successful prey organism might hide in a crevice, be active at night when a predator is inactive, or grow rapidly to be too large to be overpowered by the predator.

✱ **Seasonal influxes of predators in shallow water and in the intertidal zone may devastate local communities.**

[10] See Sale, 1977, in Further Reading, Competition.

While many habitats have a local permanent population of predators, the spring and summer often bring on invasions of large populations of migratory predators, with devastating consequences for prey populations. In the intertidal zone, the most prominent example of such predators consists of shore birds, whose migrations may extend for thousands of kilometers. These birds often have favored feeding grounds on muddy or sandy beaches, which they visit successively during their migration. Fish often come inshore in summer and devastate local invertebrate populations.

Disturbance

✳ **Disturbance opens up space in the community. Its frequency may regulate long-term aspects of species composition in a habitat.**

Marine populations suffer extensively from storms, continuous wave action, and unstable sediments. Intertidal populations often crash, owing to large swings in temperature and humidity. Ice crushing (in high latitudes) and the bashing of floating logs are also major problems. Even in subtidal habitats, large swings in temperature may occur, as in the great increases in temperature during El Niño events. Any of these general physical effects is known as **disturbance**. Mobile animals may also cause mortality, unrelated to predation. Such effects are known as **biological disturbance**. For example, while moving along rocky surfaces, limpets often bulldoze newly settled barnacle larvae from the rocks.

The effects of disturbance resemble those of predation because competitors are reduced to low population levels. However, predation is usually a one-at-a-time removal process, although some predators come in devastating waves. Disturbance, by contrast, usually operates on larger spatial scales, removing patches of the community. Disturbance often initiates an orderly sequence of dominance of species, known as succession (to be discussed later).

✳ **Species diversity may be maximized at intermediate levels of disturbance and predation.**

Let's consider first a gradient from very low to very high predation rates; we can apply the same set of causes and effects to disturbance intensity.

If there are no predators, we might expect a competitive dominant, if present, to displace all compet-

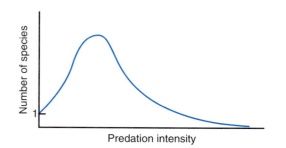

Fig. 3.29 If there are competing species and no predation, one superior species may take over. With predation that is random or targeted toward the competitively superior species, more species can coexist. With very high predation pressure, all individuals and species may decline. Thus, we might expect an intermediate graphical hump in the number of coexisting species, as a function of increasing predation pressure. A similar effect is found for levels of disturbance.

itively inferior species. But, as predation increases, resource space will be opened up and more competing species may be allowed to coexist. As predation intensity increases further, however, nearly all individuals of all species will be removed, and the species diversity will decline relative to the intermediate disturbance levels. Thus, species diversity tends to be maximized at intermediate levels of predation.

Disturbance is likely to work in a similar way, since it will most likely pare down the abundance of the competitive dominant. But if disturbance is very low, then the competitive dominant will win and species diversity will no longer exist. If disturbance is extremely strong, all species, including the competitive dominant, will be eliminated. These combined effects are known as **the intermediate disturbance–predation effect** (Figure 3.29).

It is important to realize that this relationship is fueled by a large recruitment rate of new individuals into the area we are considering. With recruitment of all species low, coexistence might occur even at very low levels of disturbance because there would not be a sufficient influx of larvae of a competitive dominant to recruit to displace other species.[11]

While the intermediate disturbance–predation hypothesis seems logical, it is not clear how often it works in nature. In some habitats, as we shall discuss, it is rare for a single competitive dominant to

[11] See Dial and Roughgarden, 1998, in Further Reading, Community, Level Interactions: Disturbance.

exist. Furthermore, inevitably, many species will not be eliminated by a competitively superior species. Often, as we have mentioned, species whose arrival is first will preempt the takeover by other species, even in the face of heavy recruitment.

Parasites and Disease

* Parasites are common and can affect their hosts by reduction in growth and reproduction or by enfeeblement.

Parasites are ubiquitous and reduce the growth and reproduction of their hosts by robbing them of nutrients. In some cases, like the common tapeworm, dissolved nutrients are absorbed directly from the host's digestive tract, but the parasite does not destroy host tissues. Many other organisms, like the flatworms that invade invertebrates, destroy host tissues. Such attacks may reduce growth or reproduction, especially because some parasites attack the reproductive organs, as noted earlier. Many snails cannot reproduce after a few years because their reproductive organs are destroyed by parasites. They are alive, but can contribute no offspring to the population.

As mentioned previously, parasites have complex life cycles, often with several hosts. Unless all hosts are present, the parasite is bound to become extinct, and parasite populations therefore have complicated dynamics. Parasites also may compete within hosts: the redia stages of trematode parasites of invertebrates are free living, have a mouth, and may prey on other parasite species. Many marine parasites have several possible hosts, and the host composition is a function of recruitment of mobile parasite life stages and competition among parasitic species within a given host.

* Diseases in marine organisms are poorly understood, but they can cause swift population declines.

Marine biologists have documented many swift population declines, and it is now obvious that disease is a major cause of massive and widespread mortality. Unfortunately, our understanding of marine pathogens is very limited. For example, a marine protozoan is known to be the main cause of a disease that devastates populations of the eastern oyster *Crassostrea virginica*. This parasite caused the collapse of the very profitable oyster industry in Dela-

ware Bay and has recently been a major cause of decline in Chesapeake Bay. Despite its obvious commercial importance, however, we still have no complete idea of the parasite's life cycle. In the late 1980s, many dolphins were found dead on the shores of the northeastern United States, and it seemed that a bacterial infection was the cause. Several rapid declines of common Caribbean species have also been related to disease.

Facilitation

Many species cannot invade an environment unless another species appears first. Such facilitation is of necessity a form of cooperation among species, but the species that benefits may eventually displace the facilitating species. For example, many small colonial animals cannot live on a bare mud bottom; they require a growth of seaweed on which they can colonize. Eventually, the organisms may spread and smother the seaweed. In some cases, species facilitate each other. In soft sediments, several species might burrow and oxygenate the sediment, thus making it more hospitable for all burrowing species to live within the sediment (see Chapter 13 for more on the effects of burrowing species on sediment properties).

Succession

* Succession is a predictable ordering of arrival and dominance of species, usually following a disturbance.

Many people are familiar with the fate of small ponds in forests. The ponds fill in with sediment and are colonized by vegetation. Eventually, the soil and biota come to resemble those of the surrounding area. Succession comprises all the processes that are involved in such progression. **Succession** is a predictable ordering of appearance and dominance of species, usually following an initial **disturbance**. A predictable final state, or **climax community**, may eventually develop. Succession is explained as either (1) trend toward a more stable assemblage of species or (2) the simple sum of the colonization and persistence potentials of the species. Succession is not necessarily inevitable, nor is the rate of change predetermined. Much research on succession suggests that it is often more like a net trend than a closely integrated sequence of biological events.

Several factors are at work in varying degrees to determine the pattern of succession, even if the sequence is more or less predictable. They are as follows.

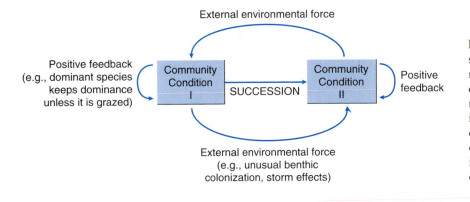

Fig. 3.30 Traditional models of succession would suggest that environments move through a series of community states. However, external changes or conditions, such as incursions of grazers, seasonal changes, or storms, may cause sudden shifts in community states. Positive feedbacks may keep the condition from changing.

1. Differential rates of colonization might result in the early arrival of certain disturbance-dependent species. Such species have high reproductive rates and short generation times. They are adapted to locate in newly disturbed environments, but such "weedy" species often are poor at holding on for very long.

2. Conditioning of the environment by resident species might facilitate the appearance of other species or prevent others from colonizing.

3. There may be monopolization of the habitat until some event (e.g., grazing) eliminates the dominant species and permits further colonization.

4. There may be irregularity in the time course of succession, depending upon events such as the arrival of predators or variation in recruitment to the site. On rocky shores, for example, filmy green algae often arrive first. They are often replaced by species with tougher holdfasts and compounds that deter herbivores. These species, in turn, often cannot colonize unless herbivores such as snails and urchins have eaten the green algae.

5. There may be an eventual dominance of species that are relatively resistant to predation and competitively superior to early succession species, at least under the conditions found late in succession.

✳ **Succession may bring a community from one condition to another; but other forces may also change community composition in a profound way, and local feedbacks may preserve the change.**

As we discussed earlier, succession is not an entirely predictable process. For example, it is highly likely that forest ponds will be filled in with sediments and eventually will succeed to a terrestrial forested habitat. We usually think of the final state of succession as having a series of properties that deter change to earlier stages of succession, although we have already discussed situations in which even early stages of succession may prevent later stages from commencing. Often in marine communities, however, major external disturbances or even differences in the time of year may cause major shifts from one community condition to another (Figure 3.30). For example, a major recruitment of larvae of grazing urchins might devastate an assemblage of seaweeds, which may have been the final state of a successional sequence. Grazing of these seaweeds might lead to the colonization of other seaweeds dominant in early succession. By then, however, the urchins may have starved because they were unable to survive on this new seaweed assemblage. In that case, grazing would be eliminated and this early-stage seaweed would be preserved (assuming that only grazing had been endangering the population).

John Sutherland[12] studied benthic colonization of ceramic plates and showed that the community composition of sessile animals depends strongly on the time of year the plates were placed in the water. Some species, such as the colonial hydroid *Hydractinia*, would colonize and could resist further incursions of species that might settle late in the year. On the other hand the tunicate *Styela* would colonize predictably, usurping space from colonial bryozoans. Most communities therefore do not fit the neat mold of succession as a predictable process of community condition *a* going to community condition *b*, and so on. Sutherland termed the locally persistent assem-

---

[12] See Sutherland, 1974, in Further Reading, Community-Level Interactions: Succession, Food Webs, and Ecosystems.

blages of organisms multiple stable points, but we shall call them **alternative stable states**. We shall see that this concept applies well to marine communities, and we discuss this in Chapter 14 in our treatment of rocky shores and salt marshes and in Chapter 15 in connection with kelp succession and coral reef structure.

The Ecosystem Level

✳ **An ecosystem is a group of interdependent biological communities in a single geographic area, capable of living nearly independently of other ecosystems.**

An ecosystem consists of a group of communities that interact with the physical–chemical environment in a specific geographic ecosystem. Within the ecosystem, nutrients recycle between organisms and the environment, some of the species manufacture organic molecules using only solar energy and inorganic chemical sources (e.g., plants), and the system can continue independently of other systems. Under this definition, a large lake and its immediate drainage comprise an ecosystem, because the organisms in the lake can survive indefinitely. A coral reef and its immediate surrounding water also qualify as an ecosystem, because no import is necessary to sustain the system. In reality, all ecosystems exchange nutrients with other ecosystems. It is crucial, therefore, to determine the boundaries of an ecosystem and the places where losses and gains may occur.

✳ **Nearly all ecosystems have primary producers (mainly plants), which are consumed by herbivores, which in turn are eaten by carnivores. Material escaping this cycle passes through the saprophyte cycle.**

The manufacture of organic molecules is accomplished **by primary producers**. Plants are the most familiar of these organisms, but many bacteria can also manufacture organic substances with the aid of light or chemical energy. Plants are consumed by herbivores, and these are, in turn, consumed by carnivores. In many marine ecosystems, most of the plant material produced is never consumed by herbivores; rather, much of it falls to the seafloor and is decomposed by bacteria and fungi, producing dissolved nutrients. The dissolved nutrients are then available for primary producers. This pathway is known as the **saprophyte cycle**.

Biomass, Productivity, Primary Productivity, and Secondary Productivity

Recall that biomass is the mass of organisms present in a defined area or volume (expressed in units such as grams per square meter: $g \ m^{-2}$). Biomass is to be distinguished from productivity, the amount of living material produced per unit area per unit time (e.g., $g \ m^{-2} \ y^{-1}$). Productivity may be expressed in units of body mass or in terms of the carbon content of the organisms.

In a natural environment, all organisms depend upon plants, which use light energy in the process of photosynthesis to convert carbon dioxide and water into sugars and other essential compounds. **Primary productivity** is the amount of living material produced in photosynthesis, per unit area per unit time. In contrast, **secondary productivity** refers to the production of plant consumers, or herbivores, per unit area per unit time. The productivity of carnivores, or consumers of herbivores, is **tertiary productivity**. A **food chain** is a set of connected feeding levels of primary, secondary, and tertiary (and so on) sources of productivity. An example of a simple food chain is:

$$seaweed \rightarrow snail \rightarrow shorebird$$

Each organism (primary producer seaweed, secondary producer snail, tertiary producer shorebird) occupies a **trophic** (or food) **level**. In more complicated systems, a simple chain cannot be constructed, and a more complex **food web** is a better description. We discuss transfer through food webs in Chapter 10.

In general, primary production is greater than secondary production, which in turn is greater than tertiary production. Secondary production depends upon consumption of primary producers, but this process is not perfectly efficient. Some material is never eaten, and even the eaten fraction may not be digested completely. Finally, not all the food that is digested is used for growth (i.e., production). In the case of carbohydrates, for example, a large fraction of the carbon content is respired in the form of carbon dioxide. As a result of such processes, material is lost through food chains.

✳ **Some predatory species at the apex of food webs exert strong effects on the overall ecosystem.**

Predators at the top of food webs may exert strong effects not only on competitive interactions but also on entire ecosystems if there are strong interactions between the trophic levels. In Chapter 15 we will discuss the strong effects of sea otters on urchins, which, in turn, strongly affect the kelps and other seaweeds upon which they graze. A predator at the top of a food web exerting such strong effects is known as a **keystone species**, a distinction first recognized by rocky-shore ecologist Robert T. Paine. When linkages among trophic levels are strong, changes in abundance of the top predator causes a **trophic cascade** through the trophic levels. To return to our sea otter example, what would you guess would happen if we removed all the sea otters from a nearshore kelp forest?

＊ **Ecosystem studies usually account for the processes that affect movement of materials and energy through food webs and through the nonliving part of the ecosystem.**

Studies at the ecosystem level attempt to account for the processes that control the **throughputs of materials or energy** through the system. For example, an ecosystem study might focus on the control of movement of nitrogen through a marine planktonic ecosystem. Clearly, this is a biologically complex problem, involving anything from microbial control of conversion among different forms of nitrogen to movement of nitrogenous materials from the water to plants, herbivores, and carnivores. Ecosystem studies involve a search for general features of material flow, and species are usually treated only as **functional groups** (e.g., herbivores). In some cases individual species have extraordinary effects on ecosystems. Beavers, for example, can completely change the structure of a stream by building a dam and creating a large pond. In Chapters 13 and 16 we shall discuss the strong effects of marine burrowing organisms on the structure of bottom sediments and the benthic ecosystem.

Further Reading

ECOLOGY AND EVOLUTION, INTRODUCTORY AND GENERAL

Alcock, J. 1993. *Animal Behavior; An Evolutionary Approach*. Sunderland, MA: Sinauer Associates.

Begon, M., J. L. Harper, and C. Townsend. 1990. *Ecology*, 2nd ed. Oxford: Blackwell Scientific Publications.

Futuyma, D. J. 1998. *Evolutionary Biology*, 3rd ed. Sunderland, MA: Sinauer Associates.

Jumars, P. A. 1992. *Concepts in Biological Oceanography: An Interdisciplinary Primer*. New York: Oxford University Press.

Krebs, J. R., and N. B. Davies. 1993. *Introduction to Behavioural Ecology*. Sunderland, MA: Sinauer Associates.

Maynard Smith, J. 1989. *Evolutionary Genetics*. Oxford: Oxford University Press.

Ricklefs, R. E. 1993. *The Economy of Nature*. Chicago: University of Chicago Press.

PREDATION AND OPTIMAL FORAGING

Elner, R. W., and R. N. Hughes. 1978. Energy maximization in the shore crab *Carcinus maenas* (L.). *Journal of Animal Ecology*, v. 47, pp. 103–116.

Feder, H. M. 1982. Escape responses in marine invertebrates. In *Life in the Sea*. San Francisco: W. H. Freeman, pp. 163–170.

Hughes, R. N. 1980. Optimal foraging theory in the marine context. *Oceanography and Marine Biology Annual Reviews*, v. 18, pp. 423–481.

Palmer, A. R. 1979. Fish predation and the evolution of gastropod shell sculpture: Experimental and geographic evidence. *Evolution*, v. 33, pp. 697–713.

Palmer, A. R. 1981. Predator errors, foraging in unpredictable enironments and risk: The consequences of prey variation in handling time versus net energy. *American Naturalist*, v. 118, pp. 908–915.

Seeley, R. H. 1986. Intense natural selection causes a rapid morphological transition in a living marine snail. *Proceedings of the National Academy of Sciences (USA)*, v. 83, pp. 6897–6901.

Trussell, G. C. 1996. Phenotypic plasticity in an intertidal snail: The role of a common crab predator. *Evolution*, v. 50, pp. 448–454.

Weissburg, M. J. 1993. Sex and the single forager: Gender-specific energy maximization strategies in fiddler crabs. *Ecology*, v. 74, pp. 279–281.

Wicksten, M. K. 1982. Decorator crabs. In *Life in the Sea*. San Francisco: W. H. Freeman, pp. 171–177.

PREDATOR DEFENSE

Bakus, G. J., N. M. Targett, and B. Schulte. 1986. Chemical ecology of marine organisms: An overview. *Journal of Chemical Ecology*, v. 12, pp. 951–957.

Duffy, J. E., and M. E. Hay. 1990. Seaweed adaptations to herbivory. *BioScience*, v. 40, pp. 368–375.

Harvell, C. D. 1990. The ecology and evolution of inducible defenses. *Quarterly Review of Biology*, v. 65, pp. 323–340.

Hay, M. E., W. Fenical, and K. Gustafson. 1987. Chemical defense against diverse coral-reef herbivores. *Ecology*, v. 68, pp. 1581–1591.

Hughes, R. N., and R. W. Elner. 1979. Tactics of a predator, *Carcinus maenas*, and morphological responses of the prey, *Nucella lapillus*. *Journal of Animal Ecology*, v. 48, pp. 65–78.

Paine, R. T. 1976. Size limited predation: An observational and experimental approach with the *Mytilus–Pisaster* interaction. *Ecology*, v. 57, pp. 858–873.

Paul, V. J., and M. E. Hay. 1986. Seaweed susceptibility to herbivory: Chemical and morphological correlates. *Marine Ecology Progress—Series*, v. 33, pp. 255–264.

Steinberg, P. D. 1985. Feeding preferences of *Tegula funebralis* and chemical defenses of marine brown algae. *Ecological Monographs*, v. 55, pp. 333–349.

Stoecker, D. 1978. Resistance of a tunicate to fouling. *Biological Bulletin*, v. 155, pp. 615–626.

MUTUALISM

Boucher, D. H., ed. 1985. *The Biology of Mutualism: Ecology and Evolution*. New York: Oxford University Press.

Karplus, I. R. 1974. The burrows of alpheid shrimp with gobiid fish in the northern Red Sea. *Marine Biology*, v. 24, pp. 259–268.

Kawanabe, H., J. E. Cohen, and K. Iwasaki, eds. 1993. *Mutualism and Community Organization: Behavioural, Theoretical, and Food Web Approaches*. New York: Oxford University Press.

Mariscal, R. N. 1970. The nature of the symbiosis between Indo-Pacific anemone fishes and sea anemones. *Marine Biology*, v. 6, pp. 58–65.

COMMENSALISM

Duffy, J. E. 1993. Host use patterns and demography in a guild of tropical sponge-dwelling shrimps. *Marine Ecology—Progress Series*, v. 90, pp. 127–138.

Fisher, W. K., and G. E. MacGinitie. 1928. The natural history of an echiuroid worm. *Annual Magazine of Natural History*, Series 10(1), p. 204.

Sakashita, H. 1992. Sexual dimorphism and food habits of the clingfish *Diademichthys lineatus*, and its dependence on host sea urchin. *Environmental Biology of Fishes*, v. 34, pp. 95–101.

PARASITISM

Esch, G. W., A. O. Bush, J.M. Aho, eds. 1990. *Parasite Communities: Patterns and Processes*. London: Chapman & Hall.

Kuris, A. M. 1974. Trophic interactions: Similarity of parasitic castrators to parasitoids. *Quarterly Review of Biology*, v. 49, 129–148.

Lively, C. M., C. Craddock, and R. C. Vrijenhoek. 1990. Red Queen hypothesis supported by parasitism in sexual and clonal fish. *Nature*, v. 344, pp. 864–866.

Price, P. W. 1980. *Evolutionary Biology of Parasites*. Princeton, NJ: Princeton University Press.

Sousa, W. P. 1993. Interspecific antagonism and species coexistence in a diverse guild of larval trematode parasites. *Ecological Monographs*, v. 63, pp. 103–128.

THE POPULATION LEVEL

Hanski, I. 1991. Single-species metapopulation dynamics: Concepts, models and observations. *Biological Journal of the Linnaean Society*, v. 42, pp. 17–38.

Hanski, I. 1998. Metapopulation dynamics. *Nature*, v. 396, pp. 41–49.

Krebs, C. J. 1994. *Ecology: The Experimental Analysis of Distribution and Abundance*. New York: HarperCollins.

McKillup, S. C., and R. V. McKillup. 2000. The effects of two parasitoids on the life history and metapopulation structure of the intertidal snail *Littoraria filosa* in different-sized patches of mangrove forest. *Oecologia*, v. 123, pp. 525–534.

Pulliam, H. R. 1988. Sources, sinks, and population regulation. *American Naturalist*, v. 132, pp. 652–661.

SPECIES, CLASSIFICATION, AND
GENETIC CONTROL OF TRAITS

Anderson, C. M., and M. Treshow. 1980. A review of environmental and genetic factors that affect height in *Spartina alterniflora* Loisei (salt marsh cord grass). *Estuaries*, v. 3, pp. 168–176.

Bowen, B. W., A. B. Meylan, J. P. Ross, C. J. Limpus, G. H. Balaza, and J. C. Avise. 1992. Global population structure and natural history of the green turtle (*Chelonia mydas*) in terms of matriarchal phylogeny. *Evolution*, v. 46, pp. 865–881.

Brusca, R. C., and G. J. Brusca. 1990. *Invertebrates*. Sunderland, MA: Sinauer Associates.

Eckelbarger, K. J., and J. P. Grassle. 1987. Interspecific variation in genetical spine, sperm, and larval morphology in six sibling species of *Capitella*. *Bulletin of the Biological Society of Washington*, no. 7, pp. 62–76.

Endler, J. M. 1980. Natural selection on color patterns in *Poecilia reticulata*. *Evolution*, v. 34, pp. 76–91.

Grassle, J. P., and J. F. Grassle. 1976. Sibling species in the marine pollution indicator *Capitella* (Polychaeta). *Science*, v. 192, pp. 567–569.

Johnson, M. S. 1971. Adaptive lactate dehydrogenase variation in the crested blenny *Anoplarchus*. *Heredity*, v. 27, pp. 205–226.

Knowlton, N. 1993. Sibling species in the sea. *Annual Review of Ecology and Systematics*, v. 24, pp. 189–216.

Padilla, D. K., D. E. Dittman, J. Franz, and R. Sladek. 1996. Radular production rates in two species of *Lacuna tur-*

ton (Gastropoda: Littorinidae). *Journal of Molluscan Studies*, v. 62, pp. 275–280.

Palumbi, S. B. 1992. Marine speciation on a small planet. *Trends in Ecology and Evolution*, v. 4, pp. 114–118.

Schmidt, J. 1920. Experiments with *Zoarces viviparus* L. *Comptes Rendus Des Travaux du Laboratoire Carlsberg*, v. 14, pp. 1–14.

Via, S., R. Gomulkiewicz, G. DeJong, S. M. Scheiner, C. D. Schlichting, and P. H. van Teinderen. 1995. Adaptive phenotypic plasticity: Consensus and controversy. *Trends in Ecology and Evolution*, v. 10, pp. 212–217.

COMMUNITY-LEVEL INTERACTIONS
Competition

Buss, L. W. Competition within and between encrusting clonal invertebrates. *Trends in Ecology and Evolution*, v. 5, pp. 352–356.

Keddy, P. A. 1989. *Competition*. London and New York: Chapman & Hall.

Kohn, A. J. 1967. Environmental complexity and species diversity in the gastropod genus *Conus* on Indo–West Pacific platforms. *American Naturalist*, v. 101, pp. 251–259.

Sale, P. F. 1977. Maintenance of high diversity in coral reef fish communities. *American Naturalist*, v. 3, pp. 337–359.

Tilman, D. 1982. *Resource Competition and Community Structure*. Princeton, NJ: Princeton University Press.

Predation

Abrams, P. A., C. Hill, and R. Elmgren. 1990. The functional response of the predatory polychaete *Harmothoe sars*, to the amphipod, *Pontoporeia affinis*. *Oikos*, v. 59, pp. 261–269.

Hay, M. E. 1991. Marine–terrestrial contrasts in the ecology of plant chemical defenses against herbivores. *Trends in Ecology and Evolution*, v. 6, pp. 362–365.

Hay, M. E., and W. Fenical. 1988. Marine plant–herbivore interactions: The ecology of chemical defense. *Annual Review of Ecology and Systematics*, v. 19, pp. 111–145.

Levinton, J. S. 1971. Control of tellinacean (Mollusca: Bivalvia) feeding behavior by predation. *Limnology and Oceanography*, v. 16, pp. 660–662.

Schneider, D. C. 1978. Equalisation of prey numbers by migratory shorebirds. *Nature*, v. 271, pp. 353–354.

Vermeij, G. J. 1976. Interoceanic differences in vulnerability of shelled prey to crab predation. *Nature*, v. 260, pp. 135–136.

Disturbance

Dayton, P. K. 1971. Competition, disturbance and community organization: The provision and subsequent utilization of space in a rocky intertidal community. *Ecological Monographs*, v. 41, pp. 351–389.

Dial, R., and J. Roughgarden. 1998. Theory of marine communities: The intermediate disturbance hypothesis. *Ecology*, v. 79, pp. 1412–1424.

Petraitis, P. S., R. E. Latham, and R. A. Niesenbaum. 1989. The maintenance of species diversity by disturbance. *Quarterly Review of Biology*, v. 64, pp. 393–418.

Sousa, W. P. 1979. Disturbance in marine intertidal boulder fields: The nonequilibrium maintenance of species diversity. *Ecology*, v. 60, pp. 1225–1239.

Sousa, W. P. 1984. The role of disturbance in natural communities. *Annual Review of Ecology and Systematics*, v. 15, pp. 353–391.

Disease and Parasitism

Esch, G. W., A. O. Bush, and J. M. Aho, eds. 1990. *Parasite Communities: Patterns and Processes*. London: Chapman & Hall.

Nicol, J. A. C. 1967. *The Biology of Marine Animals*. London: Sir Isaac Pitman and Sons.

Smith, F. G. W. 1941. Sponge disease in British Honduras, and its transmission by water currents. *Ecology*, v. 22, pp. 415–421.

Sousa, W. P. 1993. Interspecific antagonism and species coexistence in a diverse guild of larval trematode parasites. *Ecological Monographs*, v. 63. pp. 103–128.

Facilitation

Bertness, M. D. 1985. Fiddler crab regulation of *Spartina alterniflora* production on a New England salt marsh. *Ecology*, v. 66, pp. 1042–1055.

Bertness, M. D. 1989. Intraspecific competition and facilitation in a northern acorn barnacle population. *Ecology*, v. 70, pp. 257–268.

Stephens, E. G., and M. D. Bertness. 1991. Mussel facilitation of barnacle survival in a sheltered bay habitat. *Journal of Experimental Marine Biology and Ecology*, v. 145, pp. 33–48.

Succession, Food Webs, and Ecosystems

Connell, J. H. and R. O. Slatyer. 1977. Mechanisms of succession in natural communities and their role in community stability and organizations. *American Naturalist*, v. 111, pp. 1119–1144.

Golley, F. B. 1977. *Ecological Succession*. Stroudsburg, PA: Dowden, Hutchinson, and Ross.

Pickett, S. T. A., and P. S. White. 1985. Patch dynamics: A synthesis. In S. T. A. Pickett and P. S. White, eds., *The Ecology of Natural Disturbance and Patch Dynamics*. New York: Academic Press, pp. 371–385.

Power, M. E., D. Tilman, J. A. Estes, B. A. Menge, W. J. Bond, L. S. Mills, G. Daily, J. C. Castilla, J. Lubchenco, and R. T. Paine. 1996. Challenges in the quest for keystones: Identifying keystone species is difficult—but essential to understanding how loss of species will affect ecosystems. *BioScience*, v. 46, pp. 609–620.

Sommer, U., ed. 1989. *Plankton Ecology: Succession in Plankton Communities*. Berlin and New York: Springer-Verlag.

Sutherland, J. 1974. Multiple stable points in natural communities. *American Naturalist*, v. 108, pp. 859–873.

Review Questions

1. Describe the ecological hierarchy.

2. Distinguish between a population and a community.

3. What is the difference between renewable and nonrenewable resources?

4. If the distance between exploitable patches increases, should the time spent by a forager in a patch increase or decrease? Explain your answer.

5. Under what conditions might a marine creature have color that matches the background? When might it have strongly visible coloration?

6. What is the advantage of an inducible defense, as opposed to a fixed trait that is always available for defense?

7. Distinguish between commensalism and mutualism.

8. Why are parasites likely to have complex life cycles?

9. How might a resource limit population growth?

10. How might several genetic variants be maintained over time in a single population?

11. What is the main determinant of biogeographic provinces in coastal marine environments?

12. What is the major effect of predation in communities of competing prey species?

13. What are the major processes that contribute to determining the relative abundance of species in a community?

14. Define ecological succession.

15. Distinguish between biomass and productivity.

16. Some species consist of genetically identical individuals, all of which are very flexible in their ability to live in dfferent subhabitats, whereas other species consist of individuals each of which is distinctly different and specialized for a given subhabitat, but inflexible. Under what conditions might each species be favored?

II

MARINE ORGANISMS:
FUNCTION AND ENVIRONMENT

4

The Chemical and Physical Environment

In some parts of the oceans, particularly the deep sea, environmental characteristics such as temperature and salinity change little throughout the year. Most coastal habitats change rapidly, however, which presents a challenge to marine organisms. With a class, I once found that a strong rainstorm brought the salinity of a Florida tide pool from normal marine salinity to a freshwater state in about an hour. In a small tidal estuary, salinity can change greatly with every rise and fall of the tide, as the water is dominated alternately by incoming seawater and outgoing fresh water. The purpose of this chapter is to show how marine organisms respond to changes in the chemical and physical aspects of their environment.

Measures of Physiological Performance

✴ Measures of organismal response include whole-organismal, behavioral, physiological, and biochemical factors.

During its lifetime, any organism is subjected to a range of environmental variation. An organism's ability to survive environmental change is ultimately determined by its genes and is modified by environmental effects on physiological condition. When an environmental change, such as a rise in temperature,

occurs, the individual must first have **receptors** to sense the change. This information must then be conveyed and translated into an **adaptive response,**[1] such as shivering in the cold, or crawling into a cool, wet burrow when conditions are hot and dry. There is usually a hierarchy of possible responses, depending on the extremity of the environmental change. For many invertebrates, reproductive behavior is carried out within a narrower range of temperature than is the case for feeding behavior. Feeding behavior, in turn, occurs within a narrower range of temperature than is the case for a state of immobility that is designed to permit survival under certain conditions: think of a snail withdrawing into a shell and remaining there to avoid heat and desiccation shock.

An adaptive response need not be confined to behavior. Many crucial adjustments are **biochemical** and **physiological.** If an animal is exposed to toxic metals, chances are that the concentration of a metal-binding protein called metallothionein will increase within hours. Various enzymes, hormones, and other vital molecules are maintained at concentrations that maximize performance. These are examples of **biochemical changes.** Responses involving **physiological change** include such processes as ciliary beating and transport across membranes. All these responses re-

[1] An adaptive response is a plastic response that has evolved to increase fitness.

79

quire energy, and the **metabolic rate** is the rate of energy use by the organism. Metabolic rate is often used to demonstrate an overall physiological response to an environmental change.

* Organisms respond to environmental change by reaching a new equilibrium through a process known as acclimation.

Organisms rarely have fixed responses to a given environmental condition. For instance, if mussels are collected from Massachusetts in winter in near-freezing seawater, they will die quickly if placed in seawater at 20°C. No mortality occurs, however, if mussels collected in summer are transferred into water at this same temperature. This is because the animal has adjusted physiologically between winter and summer conditions and has developed a capacity to function in a warmer environment. A change of function and tolerance that results in a change of response to new physicochemical conditions is known as **acclimation**. We assume that an organism improves its functioning in a new environment by shifting an array of physiological and biochemical processes. Acclimation can be studied by changing the environment of a laboratory animal. After an environmental change, an **immediate response** is followed by an adjustment period, which is then followed by a new **steady state** (Figure 4.1a).

If the external environment changes, the individual may have a battery of adaptations to maintain constancy in body temperature, for example. Such organisms are said to be **regulators**. Organisms whose body temperature or cellular salt concentration changes in direct conformance with environmental change are **conformers**. Some species will regulate their response to variation in some environmental factors (e.g., salt content of cellular fluids of fishes), but conform to others (e.g., temperature). Figure 4.1b shows the responses of a complete regulator and a complete conformer to changes in salinity.

* Scope for growth is a measure of the food intake that can be used for growth and reproduction, beyond the cost of metabolism.

As an animal consumes and digests food, energy is acquired that can be used to fuel either cellular energy needs of biochemical reactions or growth and the production of gametes. The greater the energetic cost of overall cellular reactions, or **cost of metabolism**, the less energy is available for growth and reproduction. **Scope for growth** is the difference between the amount of energy assimilated from the animal's food and the cost of metabolism. This implies there is a zero point of **maintenance metabolism** for any given animal. At this point an animal will be gaining enough energy to just balance its energetic needs. If scope for growth is positive, then there is energy available for growth and reproduction. If scope for growth is negative, however, the animal has a cost of metabolism greater than can be matched by energetic intake of food. Such an animal will burn off reserve carbohydrates, lose weight, and, if the process is prolonged, eventually die.

To estimate scope for growth, we need to measure the feeding rate, the percentage of food assimilated, and the respiration rate, which is an estimate of the cost of metabolism. When surplus energy is available, energy may be partitioned between somatic growth and production of gametes. If food is abundant, scope for growth will increase. Owing to increased chemical reaction rates, increasing temperature often increases the energy need of cells. Increased temperature may therefore decrease scope for growth (Figure 4.2).

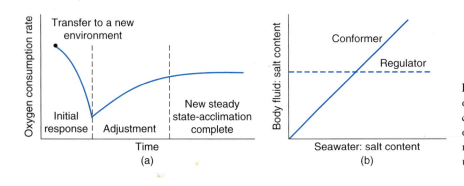

Fig. 4.1 (a) Acclimation response of oxygen consumption, following a change of temperature. (b) Response of body fluid salt content to environmental salinity variation of a regulator and of a conformer.

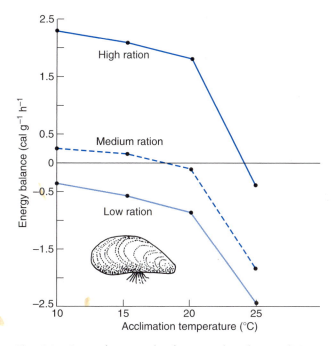

Fig. 4.2   Scope for growth of a mussel under conditions of high, medium, and low food ration, and varying temperature. Note that as temperature increases or food decreases beyond a certain point, scope for growth, in terms of energy balance, goes below zero.

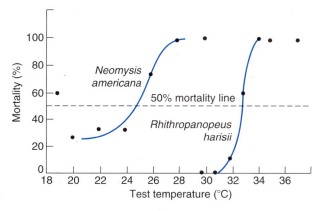

Fig. 4.3   One can measure differences in temperature tolerance by exposing groups of animals to different temperatures and measuring mortality after 24 hours. The LD_{50} is the temperature at which 50% cumulative mortality occurs. In this example comparing eastern U.S. arthropods, a mysid shrimp, *Neomysis americana*, was taken from near the southern end of its geographic range. It proves to be less tolerant of high temperature than the mud crab *Rhithropanopeus harisii*. (Modified from Mihursky and Kennedy, 1967.)

* **Mortality rate can also be used as a measure of the effect of environmental change.**

The effect of extreme environmental conditions on mortality is more commonly assessed than such measures as scope for growth. Experimental populations are usually kept at a standard laboratory condition for a period permitting acclimation. Then, lethal temperature, for example, can be determined (a) by inducing a slow decline or rise of water temperature or (b) rapidly transferring the laboratory-acclimated individuals to a constant, extreme temperature. The lethal dose required to kill 50% of the experimental population after a specified time (24 hours is a common period) is the LD_{50}. To find the LD_{50}, one experimentally varies a factor, say, temperature, and observes a series of mortality rates. This produces a series of points relating the factor to percentage mortality. The LD_{50} can then be interpolated (e.g., Figure 4.3).

Temperature

* **Temperature affects the latitudinal distribution of most marine species.**

Continental coastlines are mostly oriented in an approximately north–south direction, and temperature tends to increase toward the tropics. This latitudinal temperature gradient is especially steep on the east coast of North America, because of the continental influenced climate. Seasonality is also especially pronounced on this coast, relative to the American Pacific Northwest, where there is little seasonal change in temperature. Because marine species have thermal limits, geographic limits are fixed latitudinally, and the ranges of many marine species coincide with regions bounded by major latitudinal jumps in temperature (e.g., Cape Hatteras, North Carolina; Point Conception, California).

Homeotherms and Poikilotherms

* **Homeotherms regulate their body temperature, whereas the body temperature of poikilotherms conforms to the external environmental temperature.**

There are two extremes of temperature regulation. **Homeotherms** are organisms that regulate body temperature to a constant level, usually above that of the ambient environment. A constant and relatively high body temperature enables biochemical reactions to occur in a relatively constant internal environment and at a relatively high rate. Most birds have a body temperature of about 40°C, whereas the temperature of most marine mammals is about 38°C. Because such temperatures are much higher than that of most seawater, marine homeotherms lose heat rapidly to the surrounding environment.

There is another completely different style of living. **Poikilotherms** are organisms whose body temperature conforms to that of the ambient environment. All subtidal marine invertebrates and most fishes fit into this category. There is an interesting intermediate status, in which body temperature is usually somewhat higher than ambient temperature. Strong-swimming fishes, such as skipjack and yellowfin tuna, have this intermediate status. Their temperature rise above ambient conditions stems from metabolic heat generated by muscular activity (swimming), combined with a heat retention mechanism. The temperature rise is probably necessary to generate the increased biochemical reaction rates that are necessary for sustained activity. In contrast, some intertidal animals are not true poikilotherms in that they maintain themselves at lower than ambient body temperature, using both evaporation and circulation of body fluids to avoid being heated at low tide by the sun. Their body temperatures therefore differ from that of an inanimate object that might be placed on the shore. Intertidal organisms absorb and lose heat directly, to the air. Darker-colored forms can absorb more heat than can light-colored forms, and variation in color can therefore reflect differences in adaptation to the capture of solar energy at different latitudes.

✳ Homeotherms use restriction of circulation, insulating materials, and a countercurrent mechanism to reduce heat loss to the environment.

As we discussed in Chapter 2, ocean temperatures are usually less than 27°C, and may be less than 0°C in some locations and during some seasons. Most homeothermic mammals and birds must therefore lose heat continuously to the environment. The skin is the main surface of heat loss, especially by direct conductance of heat from the skin to the contacting colder water. Because animals have a circulatory system, **convective heat loss** also occurs as warm interior blood is transferred and moves into contact with the periphery of the body. The body also radiates

heat, usually in the infrared part of the spectrum. Finally, as animals exhale, the resulting **evaporation** of water involves a considerable loss of heat.

The first line of defense against heat loss is a well-insulated body surface. Marine birds deal with this problem by means of specially adapted feathers. A series of interlocking **contour feathers** enclose a thick layer of **down feathers** that traps stationary air, which in turn acts as an insulating layer. Whales, porpoises, and seals are insulated against the lower sea temperatures by a thick layer of subcutaneous fat. Sea otters lack such a layer, but they constantly preen and fluff up a relatively thick layer of fur. Such mechanisms are only partly successful, however, and to generate more body heat to maintain a constant temperature, marine mammals usually must have a higher metabolic rate than similarly sized terrestrial animals.

In marine mammals that have limbs, the limbs are the principal sources of heat loss, because they expose a relatively greater amount of body surface area per unit volume to cold water. However, warm arterial blood must be supplied to limbs, such as the flipper of a porpoise. Heat loss in porpoises is minimized by a **countercurrent heat exchanger**. The arteries are surrounded by veins, within which blood is returning to the core of the animal (Figure 4.4). At any contact point, the artery is warmer than a surrounding vein, so heat is lost to the returning venous blood flow (Figure 4.5). Heat is thus reabsorbed and returned to the porpoise's body core. This spatial relationship of circulatory vessels minimizes heat loss to the flipper and thence to the water. Although the anatomical details are quite different, fishes such as skipjack tuna have a circulatory anatomy based on the same overall design. Arteries and veins in the near-surface musculature are in contact, and in arteries and veins, respectively, blood flows in opposite directions (see Chapter 8).

Temperature and Metabolic Rate

✳ In poikilotherms, metabolic rate increases with increasing temperature.

In poikilotherms, an increase in temperature usually increases metabolic and behavioral activity. Oxygen consumption is a convenient expression for overall metabolic activity, because most poikilotherms consume oxygen as they burn carbohydrates for energy. With an increase of 10°C, the corresponding change

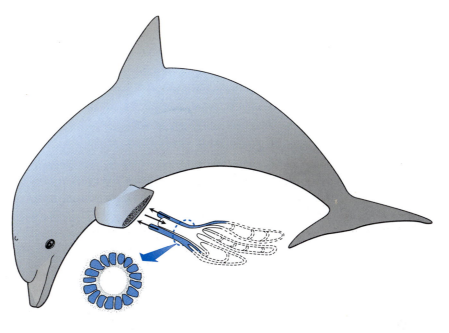

Fig. 4.4 The pattern of veins (green) wrapped around arteries in the flipper of a dolphin. This arrangement permits the venous blood to be warmed by heat transfer from arterial blood before it reenters the body core. (From Schmidt-Nielsen, 1975.)

in metabolic rate as measured by oxygen consumption is called the Q_{10}. For most poikilotherms, Q_{10} ranges from 2 to 3. Thus there is a doubling to tripling of oxygen consumption with a 10°C rise in temperature.

For a poikilotherm, an increase in temperature causes an increase in metabolic rate. This cost will reduce the animal's energy reserves, and will therefore reduce the scope for growth. Refer again to Figure 4.2, which shows the scope for growth in the mussel *Mytilus edulis* at different temperatures and at low, medium, and high rations. At low ration,

scope for growth is negative at all temperatures, whereas it is also negative at 25°C for all food levels. At such high temperatures, the animal cannot compensate for the high metabolic cost of living at high temperature. Over a wide intermediate temperature range, however, physiological acclimation results in a stable scope for growth (see the following).

✳ Poikilotherms can compensate for changes of temperature by means of an acclimation process.

If temperature is changed relatively rapidly (e.g., on the scale of minutes), most poikilotherms are condemned to increase or decrease their metabolic rate accordingly. If you place a tropical crab in a kitchen refrigerator, it will move very slowly, if at all. Such a relationship of temperature to metabolic rate would pose problems for poikilotherms living in seasonal environments. To avoid this restriction, many poikilotherms are able to acclimate to seasonal changes in temperature. Consider a graph of metabolic rate as a function of temperature for an animal living in high summer temperatures (Figure 4.6). If temperature is suddenly decreased, then metabolic rate will be very low and the animal will not be able to generate the energy needed for activity. To get around this problem, the animal acclimates by shifting its metabolism–temperature curve to the winter acclimation form shown in Figure 4-6. If an animal acclimated to winter temperatures experiences a sudden

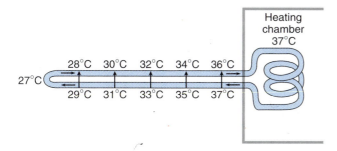

Fig. 4.5 A model of a countercurrent heat exchanger. Warm fluid leaves the hot water bath and loses heat to the external environment, thus reducing the temperature toward the left. Because, however, the return flow gains heat by exchange with the adjacent outflow tube, some of the heat is recovered and returned to the heating chamber.

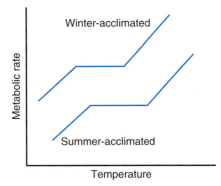

Fig. 4.6 Expected response curve of metabolic rate versus the temperature for an animal acclimated to winter low temperature and the same animal acclimated to summer high temperature.

temperature increase, metabolic rate may rise to such a point that not enough energy would be available. Given enough time, however, the animal will respond by adjusting its metabolism to the "summer acclimation curve."

Acclimation to changing seasonal temperature can be shown by collecting animals that have been acclimating to the temperature in one season, then shifting them rapidly to a new temperature. For example, exposure of winter-acclimated animals to high temperature should result in metabolic rates greater than those expected for summer-acclimated animals. At 3°C, winter-collected burrowing mole crabs, *Emerita talpoida*, consume oxygen at a rate four times greater than do summer-collected animals tested at the same temperature. A seasonal change in tolerance usually accompanies acclimation of metabolic rate; the upper lethal temperature is greater in summer-collected animals than in members of the same species collected in winter.

Acclimation to seasonal temperature change therefore serves to maintain activity and to strike a favorable energy balance. With seasonal—that is, relatively slow—changes in temperature, many marine invertebrates acclimate and adjust the metabolism–temperature relationship to new conditions. This adjustment can result in the metabolic rates being constant for acclimated individuals over a wide range of temperatures. Two favorable results are achieved. The organism depresses the metabolic cost of living in warm water and yet can also adjust me-

tabolism upward, so that the organism can be active in winter.

Similar types of compensatory mechanism dampen the range of metabolic activity expected within a species living in a latitudinal thermal gradient. At low temperature, oxygen consumption tends to be greater in animals living at high latitudes than in members of the same species living at low latitudes. Similarly, geographically separate populations of oysters (*Crassostrea virginica*) and sea squirts (*Ciona intestinalis*) have varying temperature optima for breeding, depending upon latitude. Populations with compensatory responses that enable them to function in different parts of the latitudinal temperature gradient have been termed **physiological races**. These races may be genetically different, or the individuals in them may merely become acclimated at different temperatures.

Tolerance to temperature is an important factor regulating the distribution of marine organisms. Because intertidal environments tend to have much greater daily and seasonal temperature ranges, intertidal organisms tend to tolerate a broader temperature range than do subtidal marine species. Furthermore, acclimation in seasonal habitats results in greater tolerance to high temperature in summer and lower temperature in winter. The geographic ranges of marine species indicate that natural selection has shifted the optimum response of species to that of their native temperature regime. The Antarctic fish genus *Trematomus*, for instance, has representatives that live in water temperatures close to −1.9°C throughout the year. These fish will die at an upper limit of only 6°C. Many Arctic species cannot tolerate the "high" temperature of 10°C! In the mitochondria of the Antarctic *Trematomus bernacchii*, synthesis of adenosine triphosphate, **ATP**, the general source of energy for living cells, fails at the lowest temperature of any marine animal species thus far measured.[2] At any one geographic location, the marine biota consists of an assemblage of species whose optimum temperature ranges are different. Thus after the severe winter of 1962–1963 in Great Britain, more tropically adapted species showed great mortality, whereas Arctic-adapted elements suffered no ill effects.

---

[2] See Weinstein and Somero, 1998, in Further Reading.

Plate I. Marine Invertebrate Planktonic Larvae

I.1. *Left,* Planula larva of the coral *Pocillopora damicornis.* Note the rows of zooxanthellae. (Photograph by Robert Richmond)

I.2. *Middle,* Sempers (Zoanthina) larva (Order Zoanthidea, Phylum Cnidaria). (Photograph by Rudolph Scheltema)

I.3. *Bottom left,* Planktotrophic larva of the hairy triton, *Cymatium parthenopetum,* a teleplanic larva found in the tropical Atlantic. (Photograph by Rudolph Scheltema)

I.4. *Bottom right,* Veliger larva of the gastropod *Xylophaga atlantica.* (Photograph by Rudolph Scheltema)

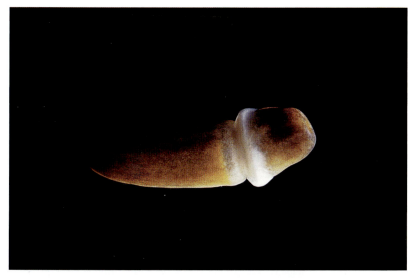

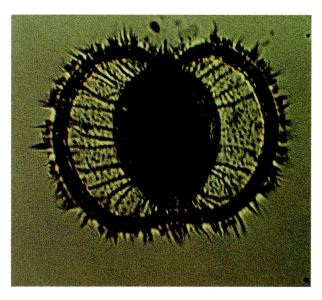

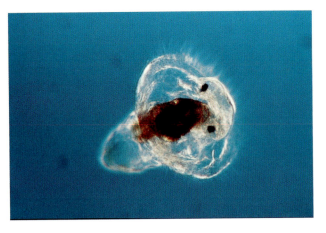

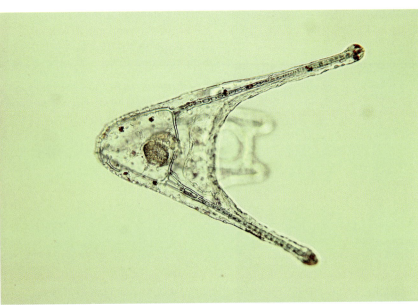

Plate II. Marine Invertebrate Planktonic Larvae

II.1. *Top left,* Mesotroch larvae (left) of the polychaete *Spirobranchus giganteus;* the adult lives in a calcareous tube on reef-building corals. (Photograph by Rudolph Scheltema)

II.2. *Top right,* Megalopa larva of the shame-faced crab, *Calappa* sp., taken from the Gulf Stream, but adults are normally found in the Caribbean Sea. (Photograph by Rudolph Scheltema)

II.3. *Middle,* Pluteus larva of the sea urchin *Lytechinus* sp., Florida. (Photograph by Will Jaeckle)

II.4. *Bottom,* A brachiolaria larva of an unidentified asteroid sea star. (Scanning electron micrograph by Will Jaeckle)

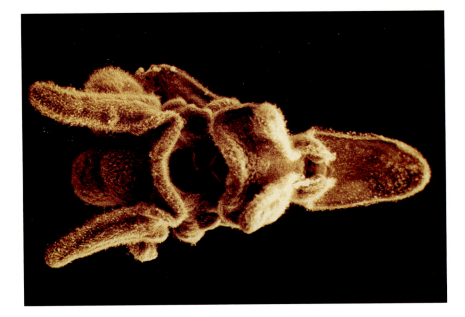

Plate III. Plankton

III.1. *Top,* Paired "Bongo" plankton nets are cast off the side of the R.V. *New Horizon,* principally to collect zooplankton. Note the sampler at the lower left, who is holding a protractor that enables her to estimate the angle of entry of the wire holding the nets. (Courtesy of Scripps Institute of Oceanography)

III.2. *Middle, Thalassiosira* sp. A common diatom in the spring phytoplankton bloom of temperate–boreal waters, usually occurring in chains of a few cells. (Photograph by George Rowland)

III.3. *Bottom left, Chaetoceros* sp. This diatom occurs in chains of cells armed with spines and is not preferred by suspension-feeding bivalve mollusks; often irritates the gills of fishes. (Photograph by George Rowland)

III.4. *Bottom right,* The diatom *Asterionella japonica,* often a dominant form of the phytoplankton. (Photograph by George Rowland)

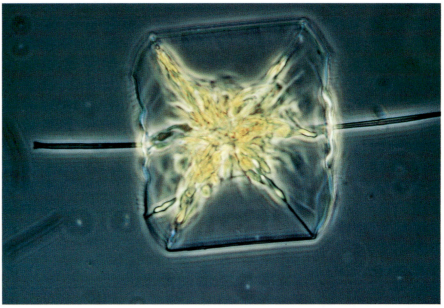

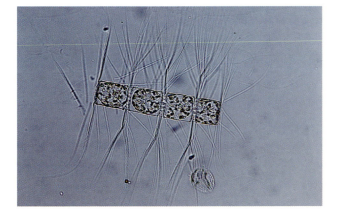

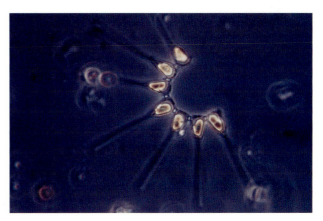

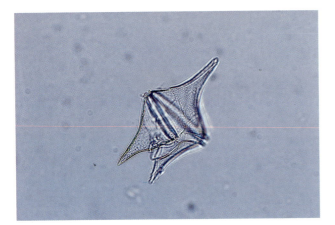

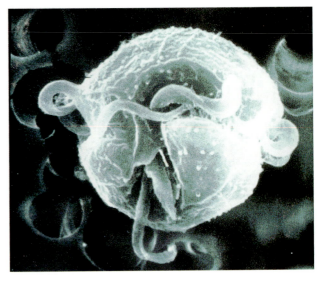

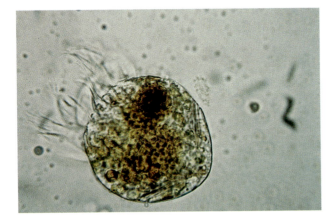

Plate IV. Plankton

IV.1. *Top left, Peridinium* sp. This is a dinoflagellate common in estuarine and shelf waters. It is about 50 μm across. (Photograph by George Rowland)

IV.2. *Top right, Pfiesteria piscicida.* This dinoflagellate has become famous for its many life history stages (greater than 20) and its implication in many fish kills in east coast United States estuaries. It is highly toxic. (Photograph courtesy of Joann Burkholder)

IV.3. *Bottom right,* Ciliate *Strombidium conicum.* This common oligotrich ciliate feeds on algae and retains chloroplasts and also feeds on smaller protists by means of the ciliary tufts on top. (Scanning electron micrograph by Diane Stoecker)

IV.4. *Bottom left,* Ciliate *Strombidium capitatum.* This ciliate, about 30 μm in size, is common in shelf and estuarine environments and is one of the protistans important in the microbial loop. (Photograph by Diane Stoecker)

Plate V. Some Gelatinous Zooplankton

V.1. *Top left,* The jellyfish *Aequorea victoria*, Puget Sound region. (Photograph by Claudia Mills)

V.2. *Below,* The trachymedusan jellyfish *Benthocodon pedunculata*, collected at a depth of ca. 900 m offshore of the Bahamas. (Photograph by Claudia Mills)

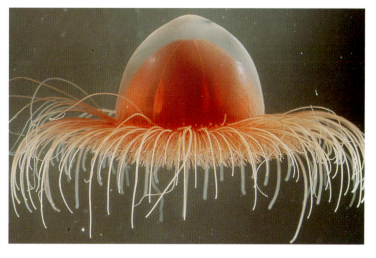

V.3. *Bottom,* The anthomedusan *Pandea conica*, common in continental slope waters of the western Atlantic. A predator of other gelatinous zooplankton, it can be found up to 30 mm high. (Photograph by Laurence P. Madin)

V.4. *Right,* The siphonophore *Physophora hydrostatica* can be found up to about 50 mm high. (Photograph by Laurence P. Madin)

Plate VI. Zooplankton

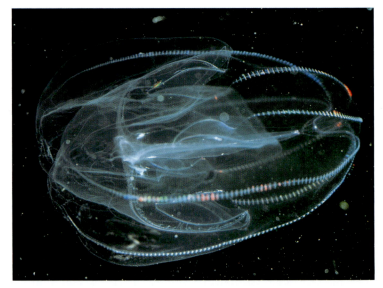

VI.1. *Left,* An undescribed oceanic ctenophore. (Photograph by Marsh Youngbluth)

VI.2. *Below, Bolinopsis vitrea,* a lobate ctenophore, common in the Caribbean and other subtropical regions can be found up to 60 mm high; preys on copepods and other small crustaceans. (Photograph by Laurence P. Madin)

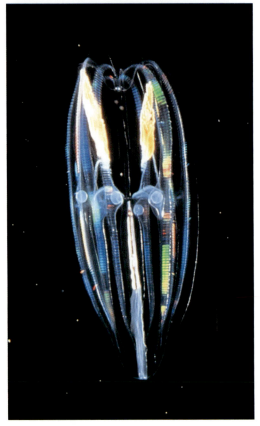

VI.3. *Left,* An undescribed ctenophore. (Photograph by Marsh Youngbluth)

VI.4. *Above, Gleba cordata,* a sea butterfly. (Photograph by Marsh Youngbluth)

Plate VII. Crustacean Zooplankton

VII.1. *Right,* An egg-bearing female of the copepod *Euchaeta elongata,* taken from Dabob Bay, Washington. (Photograph by Steve Bollens)

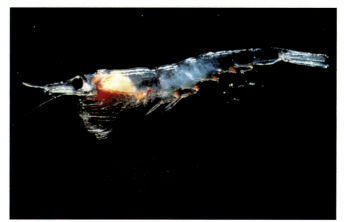

VII.2. *Above,* The copepod *Euchaeta norvegica.* Note the long mechanosensory hairs on the first antennae, which allow detection of approaching prey. (Photograph by Jeannette Yen)

VII.3. *Middle,* The planktonic amphipod *Themisto compressa,* about 25 mm long. (Photograph by Marsh Youngbluth)

VII.4. *Right,* The krill *Meganyctiphanes norvegica,* about 35 mm long. (Photograph by Marsh Youngbluth)

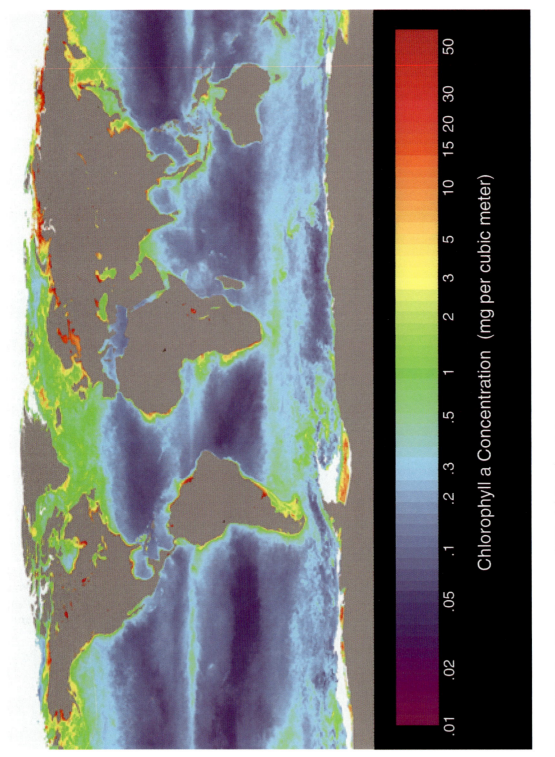

Chlorophyll a Concentration (mg per cubic meter)

| .01 | .02 | .05 | .1 | .2 | .3 | .5 | 1 | 2 | 3 | 5 | 10 | 15 20 30 | 50 |

Plate VIII. Satellite Image of World Productivity

The worldwide, year-round integrated estimate of primary production, derived from data collected by the SeaWiFS program (sea-viewing wide-field-of-view sensor), run by the National Aeronautics and Space Agency. "False colors" are used to contour the data with the accompanying scale.

Temperature and Physiological Performance

* Large temperature changes can reduce physiological performance by affecting physiological integration.

At the biochemical level, heat death must be related to protein function, particularly the catalytic efficiency of enzymes. To function, enzymes must bind to substrates efficiently, catalyze reactions, and then release substrates and products that will allow further reactions to occur. The binding sites of enzymes can be likened to furrows in a complex three-dimensional molecule, within which the substrate must fit for a reaction to be catalyzed properly. At temperatures that are too high, the binding sites are, in effect, too wide open and enzyme function is inefficient.

High-temperature shock not only causes dysfunctional increases in binding sites, it can cause total unfolding of the protein, producing irreversible damage. Animals respond by producing heat shock proteins, which forestall the unfolding of a protein's three-dimensional structure and in some cases reestablish folding to make a protein functional after temperature damage. If damage is extreme, the low molecular weight protein ubiquitin binds to degraded proteins, which are then degraded by cellular proteolytic enzymes. Individuals can acclimate to varying temperature by changing the production of heat shock proteins and ubiquitin. In the short term, animals such as mussels produce sharply increased levels of heat shock proteins in response to the temperature increases observed typically in the rocky intertidal zone. The same trend should be expected between species living in different thermal regimes: low-latitude species should be adapted to higher temperature and should have protein structures that do not fail as easily at higher temperatures. Hoffmann and Somero[3] found that a high-latitude Pacific U.S. coast mussel species, Mytilus trossulus, responded to increased temperature by producing higher levels of ubiquitin and heat shock proteins. The lower-latitude M. galloprovincialis responded less, which suggests that its soluble proteins function better at high temperatures.

Most likely, high temperature leads to failure of a series of interdependent metabolic reactions. This is known as a loss of physiological integration, or a de-

crease in the degree of coordination among interdependent biochemical reactions. If various metabolic processes are influenced differently near the upper temperature limit, then coordination among them will be unbalanced.

The disruption of physiological integration can sometimes be observed at the cellular level by the investigation of coordinated ciliary activity of epithelial cells that can be examined over a range of temperature. A failure of ciliary activity is a special case of a more general problem of maintenance of cell membrane function. If a poikilotherm is placed in a lower temperature, the packing of the structural phospholipids in its membranes will increase, causing more ordering of the membrane structure. Conversely, an increase in temperature will reduce packing and therefore membrane order. If phospholipids are packed too poorly or too well, functions such as transport across the cell membrane will be harmed. Membrane order can be measured by creating an artificial small vesicle, into which phospholipids are introduced. Membrane order is proportional to the phospholipids' polarization of light.

Intertidal invertebrates often experience both seasonal changes and tidal changes in temperature, sometimes exceeding a range of 20 degrees. As a consequence, we expect seasonal and tidal temperature changes to result in acclimation of membrane order. In the California sea mussel Mytilus californianus, seasonal temperature change is met with acclimation by means of changing membrane order. As temperature increases, mussels acclimate by maintaining membrane order.[4] In summer, high intertidal mussels can alter membrane order rapidly, which allows them to respond to the rapid changes in temperature experienced on the upper shore.

The effects of low temperature, particularly in tropical organisms, probably also involve enzyme function and physiological integration mechanisms similar to those discussed for the upper lethal limit. In tropical fishes, cold depression of respiratory systems can lead to anoxia and death. However, in high-latitude marine environments, and in winter in some midlatitudes, freezing presents a severe environmental problem. Larvae of many fishes, and Foraminifera, can be found encased in pack ice in Antarctic

[3] See Hofmann and Somero, 1996, in Further Reading.
[4] See Williams and Somero, 1996, in Further Reading.

waters. Many fleshy algae and barnacles survive the winter under freezing conditions in the intertidal zone. However, the formation of ice can shear and distort delicate structures and may increase the cellular salt content of the remaining fluids. It is possible that the salt content allows cellular fluids to become supercooled—that is, to remain in the liquid state below the freezing point of water. In most cases investigated, however, tidal animals show varying degrees of freezing under subzero temperatures. At progressively lower temperatures, increasing percentages of the body fluids are frozen. Nevertheless, intertidal fleshy algae can survive extended periods of freezing.

Dissolved salts lower the freezing point of seawater and of cellular fluids. In winter, temperatures in polar waters reach the freezing points of seawater and of the cellular fluids of many invertebrates and fishes. The problem is especially acute for bony fishes, whose cellular fluids have a lower salt content of seawater. Because salt lowers the freezing point of seawater, the surrounding seawater may actually be in a liquid state at a temperature at which water would normally freeze within the fish. Deep-water fishes out of contact with ice can remain supercooled. The presence of ice crystals in shallow water, however, would initiate cellular freezing, especially because fishes drink seawater laden with tiny ice crystals. Shallow-water Antarctic fish of the suborder Notothenioidei (Figure 4.7) can synthesize special **glycoproteins,** which behave much like automobile antifreeze and depress the freezing point of cellular fluids.

Temperature also affects growth and reproduction in marine organisms. Most marine species grow and reproduce over a narrower range of temperature than the range that permits individual survival. Within an intermediate range, growth is usually faster at higher temperatures. In bivalves, members of the same species have been found to grow more slowly, but they survive to older age and reach larger size in high latitudes than in low latitudes. Growth in seasonal habitats is greater in warmer times of the year, but the increased growth may also reflect the greater availability of food.

Planktonic larval stages may drift to thermally unfavorable habitats. In spring, larvae of the mussel *Mytilus edulis* are often carried from the north to rocks and dock pilings around Cape Hatteras, North Carolina. They settle, metamorphose, and then grow quite rapidly. In summer, however, the temperature exceeds the thermal limit for the species and the mussels die. The high-latitude geographic limit of a species is limited by the maximum summer temperature, whereas the low-latitude extent is limited by minimum winter temperature.

Temperature often sets the timing and can determine the style of reproduction. Many invertebrate species will spawn only when a given temperature is reached. Some species may switch from asexual to sexual reproduction as temperature increases. Even sex may be determined by environmental temperature in some marine species. For example, in the Atlantic silverside, *Menidia menidia*, embryos develop into females at low temperature, but into males at higher temperature. In general, seasonal changes in the timing and amount of egg and sperm production and release are highly correlated with temperature (as well as food and photoperiod). Figure 4.8 shows seasonal gonad changes in the sea star *Pisaster ochraceus*.

✳ The seasonal extremes of temperature have different effects, depending upon the location of an individual within the latitudinal range of the species.

The summer is the most stressful time of the year at a species's lowest latitudinal limit; high temperatures in some years probably go beyond many individuals' physiological capacity (Figure 4.9). At the lower latitudinal limit (high-temperature limit), winter may be

Fig. 4.7 The Antarctic "ice fish" *Pagothenia borchgrevinski* lives in Antarctic waters, just beneath the ice. It has antifreeze glycoproteins, which help prevent freezing of tissue fluids. (Photograph by Ian McDonald.)

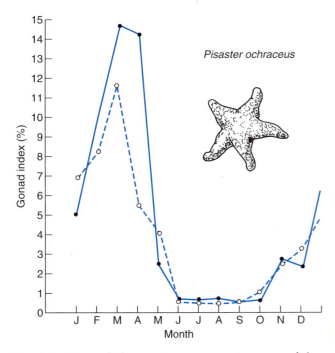

Fig. 4.8 Seasonal changes in two successive years of the gonad development of the intertidal sea star *Pisaster ochraceus*, on the coast of California. (After Boolootian, 1966.)

the only time that reproduction is possible. For example, the tomcod is a northern Atlantic estuarine fish that reaches its southerly latitudinal extreme in the Hudson River; there, it reproduces only in the winter, because only then is the temperature appropriate for such a cold-water species to spawn. The tomcod can survive, however, for the rest of the year

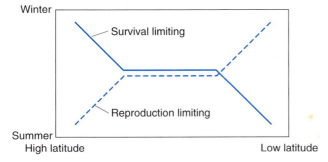

Fig. 4.9 Marine species often live along coasts that extend over a large latitudinal gradient. The effects of the highest and lowest latitudinal limits of a species range act to limit reproduction (dashed line) and survival (solid line), depending upon season.

at higher temperatures. Figure 4.10 shows the case of a bryozoan near the southern end of its range. Colonies are most abundant in winter and are rare in summer, because high temperature is stressful. At the highest latitudinal limit, the winter would be stressful, whereas reproduction may be possible only in summer.

Salinity

✳ **Salinity can change rapidly, which may have a detrimental effect on marine organisms.**

In near-shore habitats, salinity may change rapidly over very short spatial scales. This poses a challenge for marine organisms, which are generally adapted to a narrow salinity range. To operate efficiently, using a specific set of biochemical reactions, marine organisms must maintain rather constant chemical conditions within cells. Any process that causes significant changes in cellular chemistry will harm marine organisms. Significant changes in dissolved cellular inorganic constituents such as potassium and sodium will strongly affect the function of important proteins. The quantity of salts affects catalytic rate and the interaction of substrates with enzymes. Membrane transport depends on the precise regulation of inorganic constituents such as potassium. Salinity variation may cause such change, unless the organism can respond to dampen the change.

✳ **Changes in salinity affect marine organisms through the processes of osmosis and diffusion.**

When salinity changes, marine organisms may face difficulties owing to **osmosis**, which is the movement of pure water across a membrane that is permeable to water but not to solute (material dissolved in the water). If the salt content differs on the two sides of the membrane, pure water will move across the membrane in the direction of higher salt content. In an enclosed container, this creates an **osmotic pressure**, which can be counteracted by pushing a plunger, as shown in Figure 4.11a. A solution is said to be hyperosmotic if water will flow into it across a semipermeable membrane, and hypoosmotic if water leaves that solution. The osmotic effect generated by dissolved substances in body fluids can be estimated by the amount of depression of the freezing point below 0°C. Greater solute concentration, which pro-

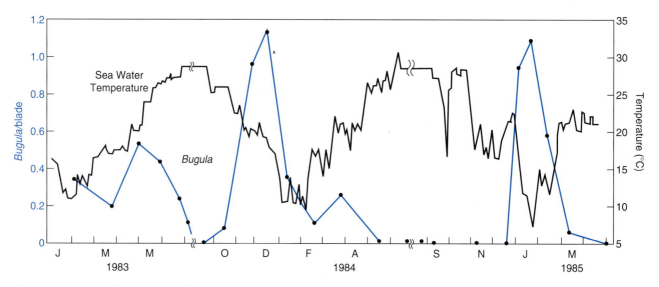

Fig. 4.10 Seasonal variation in the abundance of colonies of the bryozoan *Bugula neritinea* in the northern Gulf of Mexico, near the lowest latitudinal limit of its range. Because the organism is near the warm limit of its distribution, its abundance is greater in the colder part of the year. (Courtesy of M. Keough.)

duces greater osmotic strength, causes a greater freezing point depression.

An osmotic pressure develops when an animal is exposed to a change in salinity. Let's consider a worm that does not differ osmotically from the external seawater environment and whose body is permeable to pure water. If we expose the worm to a lower salinity and no regulation occurs, then pure water will move from the external environment, across the body

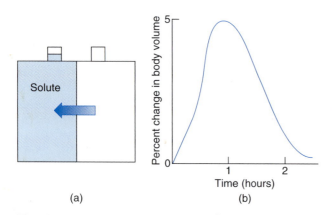

(a)   (b)

Fig. 4.11 (a) Osmosis: movement of pure water occurs across a membrane in the direction of higher solute concentration. (b) Change in the original body volume of the sipunculid *Golfingia gouldii* when transferred into diluted seawater at time = 0 hours.

wall, into the worm. Inevitably, the worm's body volume will increase. If the external salinity is low enough, one might suppose that the body volume will increase rapidly, causing the animal to explode. (After all, the osmotic pressure effect will continue until the salt content is equal on both sides of the body wall!) Figure 4.11b shows what really occurs in such an experiment, performed on a sipunculid worm. At first, the body volume does increase, but gradually it then decreases to approximately the same volume the worm had before it was plunged into the lower-salinity water. At first, osmosis results in pure water entering the body across the permeable body wall. However, the animal regulates its body volume by excretion of salts through its nephridiopores. As salts are lost, water moves osmotically across the body wall into the seawater medium, and the body volume returns to normal.

Marine organisms must also counteract the process of **diffusion**, which is the random movement of dissolved substances across a permeable membrane. The diffusion occurs until the concentration equalizes on either side of the membrane. If the salinity decreases suddenly, then salts leave the body for the external seawater. This is bad for the organism, because for cells and biochemical reactions to proceed efficiently, the overall cellular concentration of salts, or the concentration of particular constituents, must

Table 4.1 Ionic composition of seawater and of fluids of marine animals (mmol kg^{-1} of water)

	ION					
SOURCE	Na	Mg	Ca	K	Cl	SO$_4$
Seawater	478.3	54.5	10.5	10.1	558.4	28.8
Jellyfish (*Aurelia*)	474	53	10	10.7	580	15.8
Polychaete (*Aphrodite*)	476	54.6	10.5	10.5	557	26.5
Sea urchin (*Echinus*)	474	53.5	10.6	10.1	557	28.7
Mussel (*Mytilus*)	474	52.6	11.9	12	553	28.9
Squid (*Loligo*)	456	55.4	10.6	22.2	578	8.1
Isopod (*Ligia*)	566	20.2	34.9	13.3	629	4
Crab (*Maia*)	488	44.1	13.6	12.4	554	14.5
Shore crab (*Carcinus*)	531	19.5	13.3	12.3	557	16.5
Norwegian lobster (*Nephrops*)	541	9.3	11.9	7.8	552	19.8
Hagfish (*Myxine*)	537	18	5.9	9.1	542	6.3

Source: Modified after Potts and Parry, 1964, with permission from Pergamon Press, Ltd.

be regulated. Table 4.1 shows the concentrations of common ions in seawater and in some marine animals. For the most part, there is a strong similarity between cellular concentrations and those of the open marine environment. However, the hagfish and the Norwegian lobster maintain magnesium concentrations far different from that of seawater. Jellyfish actively eliminate sulfate. Diffusion would tend to equalize these concentrations, to the detriment of the organisms, which must therefore regulate their cellular concentrations to counteract this effect.

Most marine organisms are not completely permeable to the external seawater medium. For example, the outer skin of fishes and the thicker part of the chitinous body wall of arthropods are relatively impermeable to dissolved substances and to pure water. These organisms are able to localize exchange of dissolved salts at special sites, such as gills and excretory organs.

* Marine organisms regulate both inorganic and organic cellular constituents to adjust to changing salinity.

As mentioned earlier, most organisms must precisely regulate the cellular concentrations of inorganic constituents such as potassium and sodium. As a result, most organisms do not vary inorganic concentrations to adjust to osmotic stress. Instead, they regulate the concentrations of a variety of **organic osmolytes**, which are small carbon-based molecules generated by cellular reactions. Many unicellular algae, seaweeds,

and salt-tolerant plants employ glycerol, mannitol, and sucrose as osmolytes. These substances have little effect on protein function, despite wide changes in cellular concentration. **Free amino acids** are also used by several phyla of marine invertebrates (including mollusks and polychaetes), bacteria, and hagfishes. These organisms employ only those amino acids (e.g., glycine, alanine, taurine, proline, β-alanine) that have little effect on protein function. Other amino acids, such as arginine, have strong effects and are not regulated to adapt to osmotic stress. Urea is used by cartilaginous marine fishes (e.g., sharks) and by coelacanths to counteract osmotic stress. Urea affects protein function, but this effect is counteracted by the presence of various methylamines, which usually occur in a 1:2 ratio with urea. Adjustments to new salinity can be slow, and as much as several weeks may be needed for complete acclimation.

* The body fluids of fish have low osmotic strength, and therefore fish must regulate salt content in full-strength seawater.

Bony fishes maintain their body fluids at concentrations of only one-third to one-fourth that of normal seawater. The reason is unclear, but it may be related either to optimal functioning of fish enzymes or to the evolution of fishes in seawater that is less than full concentration (e.g., in an estuary). In any event, in the absence of the organisms' ability to regulate the composition of their body fluids, the process of osmosis would cause a continual loss of fresh water

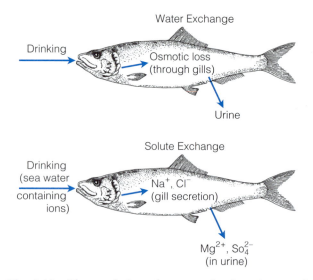

Fig. 4.12 The regulation of water and solutes by a typical marine teleost fish. Because the body fluids of the teleost are osmotically more dilute than the surrounding seawater, the fish must drink substantial amounts of seawater and excrete the excess salts across the gills. (After Schmidt Nielsen, 1975.)

to the external environment (Figure 4.12). Salts are usually taken up, as teleosts drink to maintain water balance; they then are actively eliminated to maintain a lower (hypoosmotic) salt content. The gills help to maintain the salt balance by actively excreting salts. In elasmobranch fishes, such as sharks and rays, a high concentration of urea is used to maintain osmotic balance, as in the case of free amino acids for bivalve mollusks. Sharks and rays also actively eliminate inorganic ions, such as sodium.

Many fishes migrate between water bodies of widely differing salinity. The Atlantic eels *Anguilla rostrata* and *A. anguilla* reproduce in the Sargasso Sea, and juveniles return to salt marshes and other inshore water habitats. They mature and can then live in fresh water. Salmon hatch in freshwater rivers, migrate to the ocean for 1–3 years, and then return to fresh water to spawn. The eastern American killifish *Fundulus heteroclitus* can live in fresh water and in seawater. These species have a great capacity for regulation of cellular ionic content; this capacity is coordinated hormonally.

Oxygen

✳ **Oxygen is required by most organisms to oxidize energy-yielding compounds.**

All organisms require energy to fuel chemical reactions within the cell. Although some organisms (e.g., some protozoa) can live in the complete absence of oxygen, most require oxygen for the manufacture of necessary reserves of **ATP**. As a result, the availability of oxygen and mechanisms for its uptake are of great importance in the understanding of the function of most marine organisms.

Consumption Rate of Oxygen

✳ **Although total oxygen consumption increases with increasing overall body mass, the weight-specific consumption of oxygen decreases with increasing overall weight.**

Oxygen consumption is usually expressed as milliliters consumed per unit body mass per unit time (mL O_2 h^{-1} mg^{-1}). This is the **mass-specific oxygen consumption rate**. Although large individuals consume more oxygen than do small members of the same species, the mass-specific oxygen consumption rate is inversely related to body size. Thus on a per-unit-mass basis, small snails consume more oxygen than do large snails. We can put this another way: if oxygen consumption of the whole animal is graphed as a function of body size, the curve will rise, but the rate of increase will decelerate (Figure 4.13). This change can be quantified (see the Text Box 4.1).

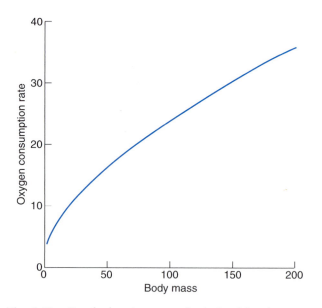

Fig. 4.13 Graph showing general relationship of oxygen consumption rate to body mass in a typical animal. Line is concave down, indicating that the rate of oxygen consumption decelerates with increasing body mass.

TEXT BOX 4.1 Quantifying the Relationship Between Body Size and Oxygen Consumption Rate

The relationship of oxygen consumption rate to body mass can be expressed as follows:

$$\text{mL O}_2 \text{ consumed} = kW^b$$

where b is a fitted exponent, W is body mass, k is a constant, and oxygen consumption rate is expressed in milliliters of oxygen per hour per gram: mL O_2 h^{-1} g^{-1}). Most marine poikilotherms have b values less than 1.0 (typically, 0.66–1.0), indicating that their metabolic rate does not increase linearly with increasing body mass. The rise of the curve, therefore, decelerates.

There are several possible reasons why metabolic rate does not increase proportionately with body mass. In protozoa, the decrease in surface area/volume with increasing size is a probable limiting mechanism. In organisms equipped with respiratory and circulation systems, other mechanisms may be important. An increase in the proportion of non-respiring mass (nonliving parts of skeletons, fat, and connective tissue), changing activity and growth patterns, and limitations of respiratory organs such as gills, may be important.

✳ Oxygen consumption rate increases in animals with increased activity.

Active species require more oxygen for energy and therefore consume more oxygen, assuming that other factors, such as body mass, are kept constant. Figure 4.14 depicts such a difference among benthic crustacea of differing swimming ability. Sponges, sea squirts, and most bivalve mollusks consume much less oxygen than do decapod crustaceans, cephalopods, and fishes. Activity may change within the life span of a single individual. Thus, species moving and feeding during the day require more oxygen during that time.

✳ Some organisms are obligate anaerobes, but most organisms require oxygen. Aerobic animals, however, may rely on a varying mix of metabolic pathways with or without the need for oxygen.

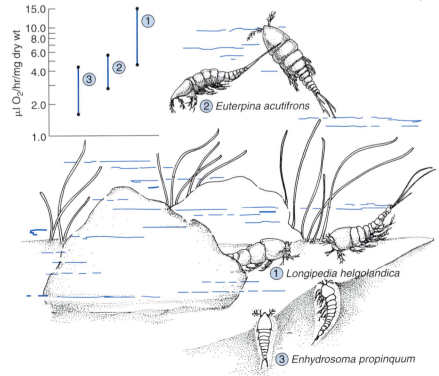

Fig. 4.14 Variation in respiratory rates of harpacticoid copepods living in different habitats. *Euterpina* is a swimming species; *Longipedia* is an active benthic form; *Enhydrosoma* is a sluggish benthic form. (After Coull and Vernberg, 1970.)

Aerobic environments are those that contain oxygen, whereas **anaerobic** (or **anoxic**) environments lack oxygen. Organisms that require oxygen are known as **aerobes**, whereas those that live in the absence of oxygen are **anaerobes**.

Nearly all eukaryotic organisms (those having a nucleus, double membrane, cell organelles, and, often, true cilia or flagella) require oxygen for life. There are a few exceptions to this general rule. In 1970, Fenchel and Riedl described a biota of microscopic benthic animals, living in the anoxic interstitial water of sediments (see Chapter 13 for a description of the sediment habitat). Their claim is controversial, but Tom Fenchel[5] has found ciliates that can live only in the absence of oxygen. Typically, these ciliates have endosymbiotic bacteria that metabolize substrates anaerobically and provide nutrition for the ciliate. These forms are quite small—certainly less than 1 mm in length. Energy acquisition is very inefficient in the absence of oxygen, so such a lifestyle is restricted to small and relatively slow-moving forms.

Although all larger eukaryotic organisms require oxygen, many use a mix of sources to obtain energy to manufacture ATP, the main source of energy in the cell. If oxygen is available, it is used to oxidize carbohydrate, and the energy is stored in the energy-rich ATP, which can be later used as a source of energy for muscle contraction, ciliary movement, and so on. During intense muscular action in vertebrates, energy demands can exceed the available oxygen. During these periods, glucose or glycogen is broken down anaerobically. This process, known as glycolysis, has less than 10 percent of the efficiency of aerobic breakdown. The end product of glycolysis is lactic acid, which accumulates in muscle tissues. Whales and dolphins that are in the process of diving provide an extreme case of oxygen in short supply. See Chapter 8 for a description of diving.

In the intertidal zone at the time of low tide, animals requiring submersion in water for oxygen uptake are subjected to a protracted period of oxygen depletion. In high latitudes during the winter, respiration is also depressed and low temperature causes lowered transport rates of oxygen to cells. During these times, many marine invertebrates sidestep the need for oxygen by using metabolic pathways involving the nonaerobic breakdown of organic materials. As in the vertebrates, anaerobic breakdown is less efficient than breakdown in the presence of oxygen. At the time of low tide, end products of anaer-

obic metabolism (alanine and succinic acid) build up in tissues. In mollusks, a portion of the succinic acid is neutralized by dissolution of part of the calcium carbonate shell. In winter, the inner layer of the shell of the marsh mussel *Geukensia demissa* becomes pitted because of this sort of dissolution process.

✳ **Animals only a few millimeters thick rely upon diffusion for oxygen uptake; larger animals have specialized organs, such as gills, for this purpose.**

Mechanisms of oxygen uptake vary with body size, phyletic origin, habitat, age, and activity. Many small organisms, such as protozoans, nematodes, and embryos and larvae in general, rely on diffusion of oxygen across the body wall. This mechanism of uptake cannot be employed for bodies thicker than a few millimeters. Larger polychaetes, mollusks, and most crustacea use gills and circulatory systems for respiratory exchange. All gills have a large respiratory surface and may be moved rapidly or have water currents passed over them for oxygen exchange. Burrowing species, such as polychaetes, always create a water current to irrigate burrows with oxygenated water. Fishes have respiratory gills on gill arches. In many cases, gills play a role in feeding (e.g., in mollusks) and in ion exchange (e.g., in crustaceans, fishes), as well as in respiration. Mammals have lungs with an enormous surface area, to enable them to acquire oxygen rapidly. A whale has a lung surface area on the order of 1,000 m^2.

✳ **Animals respond to lowered oxygen by regulating oxygen consumption, but their eventual response to very low oxygen levels is to leave the habitat if possible, or to reduce activity levels.**

In many marine habitats, marine organisms are exposed to quite varied oxygen conditions, sometimes on very short time scales. For example, a fish that swims in oxygenated surface waters may then hunt prey in low-oxygen bottom environments, where exchange with oxygenated waters is very low. When confronting low-oxygen conditions, the simplest response for mobile organisms is to "vote with their feet" and leave the area. In many cases, such as in estuaries in summer, this option is not available be-

[5] See Fenchel, 1993, and Fenchel and Bernard, 1993, in Further Reading.

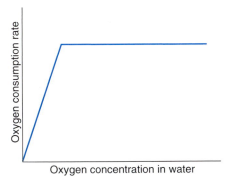

Fig. 4.15 A typical relationship between the rate of oxygen consumption of an animal and the concentration of oxygen in the water. Over a range of relatively high oxygen concentrations, the animal regulates respiration rate, but the rate declines past a lower threshold and declines steadily with decreasing environmental oxygen.

cause whole basins may be low in oxygen. The response of a typical aerobic animal is shown in Figure 4.15. There is usually a broad zone of environmental oxygen concentration within which oxygen consumption is relatively constant; within that zone, the animal can apparently regulate its oxygen consumption. However, at some minimum environmental oxygen concentration, the rate of respiration will decline with decreasing oxygen. Beyond that point, the animal cannot regulate its oxygen consumption efficiently and may respond by reducing its activity levels, which in turn reduces its requirements for oxygen to fuel aerobic muscular metabolism.

Regulation of oxygen consumption at low oxygen levels can occur only if more oxygen can be transported into an uptake structure such as a gill, or if more oxygen, once taken up, can be retained and delivered to the needy tissues. At low oxygen concentrations, crabs may increase ventilation rates and bivalves may increase heartbeat rate, both good enhancers of transport rates of water across the gills. An increase of efficiency of retention is accomplished by changes of blood chemistry and certain blood pigments, which are discussed in the following section.

Oxygen-Binding Pigments

✳ **Animals requiring great amounts of oxygen or living in environments where oxygen is difficult to acquire may have blood pigments that greatly increase the blood's capacity for oxygen transport.**

Many animals contain oxygen-binding pigment compounds that greatly increase the capacity for oxygen in blood and other tissue fluids. They are often colored—hence the name **blood pigment**. One of the most common of these compounds is hemoglobin, which is a combination of a protein unit (globin) and a unit bearing iron. Such pigments increase the carrying capacity for oxygen by a hundredfold. They are widespread; hemoglobin, for example, is found in many phyla. The list that follows (modified after Schmidt-Nielsen, 1975) provides information on these pigments.

- **Hemocyanin.** Copper-containing protein, carried in solution having a molecular weight (MW) of 300,000–9,000,000. Present in mollusks, arthropods.
- **Hemerythrin.** Iron-containing protein, always in cells, nonporphyrin structure, MW 108,000. Present in all sipunculids; some polychaetes, priapulids, and brachiopods.
- **Chlorocruorin.** Iron-containing porphyrin protein, carried in solution, MW 2,750,000. Present in some polychaetes.
- **Hemoglobin.** Iron–porphyrin protein, found in solution or in cells, MW 17,000–3,000,000. Present in echinoderms, arthropods, annelids, nematodes, flatworms, protozoa, and plants, and in all vertebrates except some Antarctic fishes.

The Caribbean clam *Codakia orbicularis* contains hemoglobin and is a resident of shallow soft sediments rich in decaying turtle grass and other organic debris. Hemoglobin is found in other invertebrates living in anoxic sediments, but is found universally in animals requiring large oxygen supplies (e.g., fishes, birds, mammals).

The binding and release of oxygen to and from a blood pigment like hemoglobin (Hb) can be described by the following equation:

$$Hb + O_2 \rightarrow HbO_2$$

As the blood oxygen concentration increases, more and more oxygen binds to hemoglobin until the hemoglobin is saturated. At a given concentration of oxygen, a definite proportion of the hemoglobin present is bound to O_2. We can draw an **oxygen dissociation curve** (Figure 4.16a) that portrays the percentage of the hemoglobin bound with oxygen as a function of dissolved oxygen concentration. If the

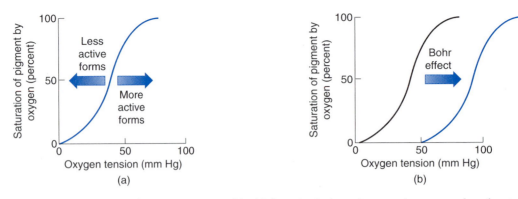

Fig. 4.16   (a) Oxygen association–dissociation curve (black) for a typical respiratory pigment, such as hemoglobin. Lowered pH (increased acidity) due to an increase of carbon dioxide released into capillaries by respiring cells causes the Bohr effect (green curve), which reduces the hemoglobin's capacity to hold oxygen. Then oxygen is released and is absorbed by needy cells. (b) Relationship of oxygen dissociation curves for animals of varying activity.

oxygen concentration is lowered, the hemoglobin releases oxygen.

The pH of the blood and coelomic fluids also affects hemoglobin-binding characteristics. Lowering the pH can shift the oxygen dissociation curve to the right in what is called the **Bohr effect** (Figure 4.16b). Oxygenated hemoglobin tends to release oxygen into the blood when CO_2 is abundant (thus lowering the pH) and when O_2 is at lower concentration. This effect is adaptively significant: as oxygenated blood reaches tissues that are poor in oxygen and rich in carbon dioxide, the hemoglobin releases oxygen to the blood, thus making oxygen available for the cells.

We would expect active animals requiring large supplies of oxygen to have their dissociation curves to the right of the curves for less active forms (Figure 4.16a). This appears to be the case for invertebrates containing the copper pigment hemocyanin. Cephalopods, for example, have high levels of hemocyanin and rely greatly on the substance for oxygen transport. Inactive animals living in environments low in oxygen concentration should have dissociation curves shifted to the left. Blood pigments may serve as oxygen reservoirs for burrowing animals living in sediments that contain little oxygen. Bivalves living in tropical low-oxygen soft sediments, for instance, tend to have blood pigments, whereas pigments are not present in species living in environments that are more highly oxygenated.

✳ **Animals with blood pigments can respond to low oxygen levels by changes in the character of the oxygen-carrying molecule.**

Marine animals with blood pigments not only can carry more oxygen in the blood, but can also use changes in their oxygen-carrying molecules to respond to low-oxygen conditions. Under such conditions, bony fish hemoglobin and crustacean hemocyanin are responsive to changes in blood chemistry and become capable of binding more efficiently to oxygen. During shortages of oxygen, both calcium ion and lactate tend to build up in the blood, which, in crustacea, has a direct effect on the three-dimensional structure of hemocyanin, raising its affinity for oxygen. Hemocyanin can occur in hexamers (six joined repeats of variants of the protein unit) or in dodecamers (12 joined repeats). Under low-oxygen conditions, the hexamers have higher affinity for oxygen and are produced by crabs in greater proportions. In the horseshoe crab *Limulus polyphemus*, increased lactate concentration is usually correlated with exposure to low environmental oxygen, which causes a **reverse Bohr effect**—a higher binding of hemocyanin to oxygen.

✳ **The highest intertidal zone and oxygen minimum layers pose special problems for aerobic animals.**

High-intertidal animals may spend more time exposed to air than immersed in seawater. Several bivalve species open their valves when exposed to air. In the marsh mussel, such gaping allows direct access to the air for a relatively large surface area of water trapped in the mantle cavity. This consumption can occur to a meaningful extent. The eastern Pacific coast mussel *Mytilus californianus* consumes oxygen in air at rates comparable to its rates of respiration in water.

Organic matter may accumulate at the base of thermoclines in the open ocean; breakdown of this organic matter by oxygen-consuming bacteria causes a zone of reduced oxygen—**an oxygen minimum layer**. The oxygen minimum layers in the tropical oceans cause low-oxygen stress. Some zooplankton pass through this zone daily during vertical migrations (see Chapter 7). Some of the zooplankton spend much of their time at this depth. The deep-water mysid *Gnathophausia ingens* moves water rapidly across its gills and can extract 80 percent of the oxygen from the water in which it is located. The animals have larger heart and arteries than do similar-sized animals from more highly oxygen-enriched habitats.

Light

* Light in the ocean comes mainly from the sun: sunlight is the source of photosynthesis in the surface layers and adds heat at the surface. It also allows animals to use vision to help them function.

Without the sun's light, there would be no marine biology, because solar energy is the main source of the ocean's heat and the basis for its biological production. The sun emits 2×10^{45} photons each second, but the ocean captures 1/10,000 millionth of the sun's energy. Only a thin surface layer, some 50–100 m thick, receives sufficient light for photosynthesis, yet this drives the biological engine of the entire ocean. All environments depend upon this surface production.

Phytoplankton that live at the surface die and sink to the deepest part of the ocean, where their remains are eaten by deep-sea creatures. The sun is also the ocean's main source of heat. As discussed in Chapter 2, the great heat capacity of seawater also makes the ocean a storehouse for solar heat, which can be transferred thousands of kilometers horizontally when the water is driven by the winds.

Vision is impossible without light, which declines in intensity with increasing depth (see Chapter 2). In the surface layers, nearly all animals have well-developed eyes, or at least some ability to sense the presence of light, and they depend upon vision to detect prey, avoid predators, and find mates. In the deepest parts of the ocean where surface light is essentially extinguished, many organisms make their own light, and eyes are everywhere.

* Marine animals may detect images with the aid of a simple layer of sensory cells, lenses, concave mirrors, or even a structure functioning like a pinhole camera.

To see, an animal must have a means of gathering light and a degree of nervous integration to form and interpret an image. In the air, the human cornea and air combined bend light and help focus images on the retina, with the aid of a lens. Light bends when it moves between two substances of differing density. Light bends between air and the human cornea, which allows focusing of light, but it is pretty useless as a focusing mechanism in water because the density of seawater is close to that of the cornea. As a consequence, marine organisms must use a lens almost exclusively to bend and to focus light if they are to form a sharp image.

The simplest kind of eye forms no image at all. Many animals, including marine larvae and flatworms, simply have a small pit lined with light receptor cells. All the animal receives is a signal: light! They may also be able to monitor the light intensity and use the information to move away from or toward the stimulus, depending on how the animal processes the signal. On an even simpler level, light captured by plants stimulates hormone production, which, in turn, causes the plant to grow in the direction of the light.

Marine animals use any of three basic ways of forming an image. Fishes, squids, some annelids, and some arthropods (e.g., lobsters, crabs) use a lens or a series of lenses much like a telescope to focus light gathered by the eye (Figure 4.17). A fish lens differs from a glass lens and consists of material of continually increasing density in the direction of the center of the lens. This variation allows the lens to bend light and focus an image in a very short distance; it also reduces distortion. Fishes have muscles that move the lens forward and back to adjust for focus (as opposed to mammals, which alter the shape of the lens itself). Lobsters and their allies, a few clams, and even some annelids have a large number of adjacent lenses, which together form a compound eye. A brain or some other part of the nervous system must integrate all the individual images into a mosaic image.

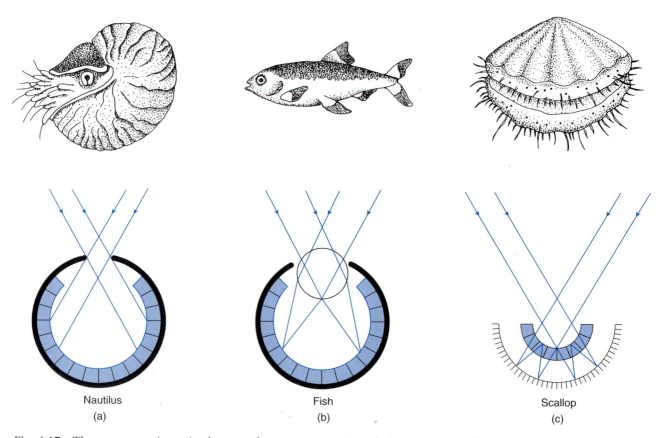

Fig. 4.17   Three ways marine animals use to focus an image: (a) pinhole camera, used by the nautilus; (b) lens, used by fish, squid, and lobsters; and (c) curved reflector, used by scallops.

The most cleverly designed marine eye belongs to the pearly nautilus, a relative of squids and cuttlefish that secretes a spiral external shell and uses its arms to capture animal prey. Light enters through a tiny hole and rays directly hit on the retina. The hole acts like a pinhole camera, permitting an inverted image to form on the back of the eye (Figure 4.17a). The nautilus uses muscles to adjust the pinhole opening, but if the opening is too large, image quality decreases. The major disadvantage of a pinhole camera (and of a nautilus eye) is that the design allows vision only in high light. A lens would allow more light to enter, but somehow the nautilus and its ancestors never evolved a lens, over some 400 million years.

A small number of marine animals form an image much like that formed by an astronomical reflecting telescope. If you look at a live scallop, you will probably notice hundreds of shiny bluish spots around the periphery of the shell. These are eyes, and the silvery blue is a concave surface. Light passes first through a crude lens, then through a nearly transparent retina (layer of light receptors), and reflects from the shiny layer, to be detected on return by the retina (Figure 4.17c). A deep-sea arthropod species has a similar eye structure, which, in the description of Sir Alistair Hardy, resembles the reflectors of automobile head lights.

✳ Light is an important cue in behavioral adaptation of marine organisms.

Many mobile intertidal animals use positive and negative responses to light to adjust their optimum position relative to tidal height (see Chapter 14). Diurnal vertical migrations in response to diurnal light changes are well known in zooplankton and are discussed in Chapter 7. Many intertidal fishes, such as blennies, use well-developed vision to navigate excursions and returns to preferred shelters. Some fishes and invertebrates are believed to use the sun as a compass to accomplish migrations on tidal flats or to and from feeding grounds. Parrot fishes living in Bermudian coral reefs migrate from offshore caves to feeding grounds near shore by monitoring the orientation between their direction of movement and the sun's location. This behavior is depressed at night and on cloudy days.

Many salmon species and sea turtles accomplish spectacular migrations over thousands of kilometers. The role of solar navigation in these migrations is questionable, however. Salmon accomplish their long journey, from freshwater spawning grounds to the open sea and back, under all conditions of cloud cover. Green turtles are extremely myopic in air and can see stars from under water only in perfectly calm seas. Chemoreception is a more likely mechanism for the navigation.

* Bioluminescence, a widely distributed property among marine organisms, is produced by specialized structures in animals, or by (symbiotic) bacteria.

The sun is not the only source of light in the ocean. Many organisms are **bioluminescent**; that is, they can produce their own light. In the deep ocean, this is the probably the only light encountered by marine organisms. Bioluminescence occurs in bacteria, algae, protozoans, sponges, comb jellies, coelenterates, annelids, crustaceans, cephalopods, echinoderms, and fishes. Many animals exhibit **intracellular luminescence**. Squids and cuttlefish, for example, have elaborate photophore cells with focusing lenses and reflectors. Comb jellies have luminescent structures along the eight radial canals and the tentacles. At night, one can see them by the thousands, emitting a pale greenish light. Some animals emit a luminous mucus secretion. Some squid and fish species have special sacs, within which lie populations of symbiotic luminescent bacteria.

The biochemical mechanisms of light emission usually involve the reaction of a protein with some other substance. Most widespread is the reaction of **luciferin**, which emits light when it reacts with the enzyme **luciferase**. The jellyfish *Aequorea* contains a protein, aequorin, which lights up when in contact with calcium ion.

Bioluminescence is very common in species living in midwaters of the open ocean. The function of the bioluminescence is not always clear, but it has been related to a mechanism of confusing predators and even finding mates. A group of phytoplankton known as dinoflagellates can startle and confuse potential predators by glowing. Jellyfish and comb jellies can glow on and off rhythmically, which may also confuse predators. Luminescence may sometimes serve the function of camouflage. Many deep-water animals have luminescent organs on the ventral surface, and this may confuse predators hunting from below, perhaps because they mistake the luminescence for pale sunlight from above.

Cycles: Physiological and Behavioral Response

Throughout this chapter we have mentioned incidentally the influence of various cycles on the physiology and behavior of marine organisms. Clearly annual–seasonal, tidal, and diurnal changes exert both direct and indirect effects on organisms; the organisms correspondingly adapt to these changes in a variety of ways. It is easy to appreciate the important influence of cycles on living creatures. Seasonal changes are more pronounced in middle to higher latitudes and in shallow water. As may be expected, tropical shallow water shows less temperature variability throughout the year than water in higher latitudes. Polar regions also show depressed seasonal temperature variation. Thus midlatitudes present the maximum amount of seasonal contrast in temperature. Light, however, obviously shows the greatest winter–summer variation in polar regions between winter and summer. Consequently, tropical regions are, on balance, seasonally the most constant environments for marine organisms.

The interaction of seasonal changes of temperature and light can have important consequences for the physiology of marine animals. In northern latitudes, winter is a time of low temperature and diminished feeding. Growth is therefore reduced. Reduced feeding, however, may also lead to altered metabolism; many invertebrates rely more on anaerobic pathways to metabolism because of the lack of food necessary to fuel aerobic metabolism. At the other end of the scale, summers in midlatitudes may be stressful because there is not enough food to fuel the metabolic requirements that accompany the high temperature. A switch to a predominance of anaerobic metabolism may therefore occur.

Seasonal variation in temperature and weather results in a rapid increase of temperature and available food in spring in midlatitudes. Therefore, somatic growth and reproduction are usually coupled to seasonal cycles. The transfer of food to deeper water may result in a seasonal component even at depths where temperature variation is damped. Thus, growth cycles can be coupled with seasonality in deeper water as well. In the deep sea, however, growth cyclicity is not closely related to season.

Tidal cycles similarly exert a pervasive effect on marine organisms. Obviously the intertidal zone is the habitat most affected by tidal variation. Exposure to air at each low tide presents a challenge and an opportunity to organisms, depending on their morphology, behavior, and physiology. At low tide, for instance, animals that must draw oxygen from the water or must have a moist gill to take up oxygen are subjected to a long period of oxygen depletion. When submerged at high tide, the bivalve *Mercenaria mercenaria* actively respires oxygen, but at low tide, anaerobic pathways predominate and end products of anaerobic metabolism (such as succinic acid and alanine) are found. The cyclicity in metabolism is recorded in the shell: during anaerobic periods some of the shell is dissolved, leaving a zone of organic matrix. In contrast, the marsh mussel *Geukensia demissa* is capable of air breathing at low tide. Presumably, exposure to air has not nearly as much physiological impact for *G. demissa* as for *M. mercenaria*.

Exposure to air similarly presents contrasting challenges to animals' burrowing and feeding behavior. For example, at low tide, many burrowing polychaetes and crustaceans retreat into moist burrows to avoid desiccation. At this same time, however, predators and grazers are less active as well. Thus, certain bivalves of the family Tellinacea actively protrude their inhalant siphons and feed at the sediment–water interface; bottom-feeding fishes would easily spot and consume the siphons if the bottom were covered by water. Many species of benthic diatoms migrate a distance of about 2 mm from within the sediment to the sediment surface at the time of low tide. In some cases, the presence of both air and daylight is required for this rhythmic behavior. In others, the rhythmic behavior is coupled strictly to the tides and is reinforced by an endogenous rhythm that decays if the diatoms are brought into the laboratory and maintained under constant submersion. This general behavior may be adaptive insofar as it brings the diatoms to the surface for photosynthesis when light is maximal and grazers are inactive. Apparently, perhaps as a response to avoid surface grazers, diatoms are capable of rapid withdrawal below the sediment–air interface on mechanical disturbance.

The general increase of feeding activity at low tide suggests a cycle of growth; this factor is well established for mollusks. Bivalve mollusks actively secrete calcium carbonate plus organic matrix when submerged. Yet when exposed to air, clams produce acidic products of anaerobiosis, and the most recently deposited calcium carbonate is dissolved. Because the organic matrix remains, we can see an alternating series of growth increments consisting of organic matrix and a calcium carbonate–organic matrix complex. The fortnightly component of the tidal pattern results in a cyclicity of thickness of growth increments.

Finally, diurnal changes are of great importance. Clearly, daily changes in light have a strong influence on photosynthetic organisms, on animals dependent on light for detection of prey, and on prey whose visibility is heightened during the day. Chapter 7 discusses one of the most fascinating and mysterious behavioral responses—the daily vertical migrations of planktonic animals.

Further Reading

Bayne, B. L., ed. 1976. *Marine Mussels: Their Ecology and Physiology*. Cambridge: Cambridge University Press.

Boolootian, R. A. 1966. *Physiology of Echinodermata*. New York: Interscience Publishers.

Childress, J. J. 1971. Respiratory adaptations to the oxygen minimum layer in the bathypelagic mysid *Gnathophausia ingens*. Biological Bulletin (Woods Hole), v. 141, pp. 109–121.

Coull, B. C., and W. B. Vernberg. 1970. Harpacticoid copepod respiration: *Enhydrosoma propinquum* (Brady) and *Longipedia helgolandica* (Klie). *Marine Biology*, v. 5, pp. 341–344.

Dudgeon, S. R., I. R. Davison, and R. L. Vadas. 1989. Effect of freezing on photosynthesis of intertidal macroalgae: Relative tolerance of *Chondrus crispus* and *Mastocarpus stellatus* (Rhodophyta*). Marine Biology*, v. 101, pp. 107–114.

Eastman, J. T. 1993. *Antarctic Fish Biology*. San Diego, CA: Academic Press.

Fenchel, T. 1993. Methanogenesis in marine shallow water sediments. The quantitative role of anaerobic protozoa with endosymbiotic methanogenic bacteria. *Ophelia*, v. 37, pp. 67–82.

Fenchel, T., and C. Bernard. 1993. Endosymbiotic purple non-sulfur bacteria in an anaerobic ciliated protozoan. *FEMS Microbiology Letters*, v. 110, pp. 21–25.

Fenchel, T., and R. Riedl. 1970. The sulfide system: A new biotic community underneath the oxidized layer of marine sand bottoms. *Marine Biology*, v. 7, pp. 255–268.

Herring, P. J., A. K. Campbell, M. Whitfield, and L. Maddock, eds. 1990. *Light and Life in the Sea*. Cambridge: Cambridge University Press.

Hoffman, G. E., and G. N. Somero. 1996. Interspecific variation in thermal denaturation of proteins in the congeneric mussels *Mytilus trossulus* and *M. galloprovincialis*: Evidence from the heat-shock response and protein ubiquitination. *Marine Biology*, v. 126, pp. 65–75.

Kinne, O. 1970. Non-genetic adaptation to temperature and salinity. *Helgoländ Wissenschaften Meeresuntersuchungen* v. 10, pp. 433–458.

Maitland, D. P. 1986. Crabs that breathe air with their legs—*Scopimera* and *Dotilla*. *Nature*, v. 319, pp. 493–495.

Mihursky, J. A., and V. S. Kennedy. 1967. Water temperature criteria to protect aquatic life. *American Fishery Society Special Publication*, v. 4, pp. 20–32.

Newell, R. C., ed. 1976. *Adaptation to the Environment: Essays on the Physiology of Marine Animals*. London: Butterworths.

Potts, W. T. W. and G. Parry. 1964. *Osmotic and Ionic Regulation in Animals*. Oxford, Pergamon Press.

Schmidt-Nielsen, K. 1975. *Animal Physiology*. Cambridge: Cambridge University Press.

Somero, G. N. 1995. Proteins and temperature. *Annual Review of Physiology*, v. 57, pp. 43–68.

Southward, A. J. 1964. The relationship between temperature and rhythmic cirral activity in some Cirripedia considered in connection with their geographical distribution. *Helgoländer Wissenschaften Meeresuntersuchungen*, v. 10, pp. 391–403.

Weinstein, R. B., and G. N. Somero. 1998. Effects of temperature on mitochondrial function in the Antarctic fish *Trematomus bernacchi*. *Journal of Comparative Physiology B*, v. 168, pp. 190–196.

Williams, E. E., and G. N. Somero 1996. Seasonal-, tidal-cycle-and microhabitat-related variation in membrane order of phospholipid vesicles from gills of the intertidal mussel *Mytilus californianus*. *Journal of Experimental Biology*, v. 199, pp. 1587–1596.

Yancey, P. H., M. E. Clark, S. C. Hand, R. D. Bowlus, and G. N. Somero. 1982. Living with water stress: Evolution of osmolyte systems. *Science*, v. 217, pp. 1214–1222.

Review Questions

1. Distinguish between conformance and regulation.

2. What is the advantage of homeothermy? What is the cost?

3. What in principle underlies the operation of a countercurrent heat exchanger? What is its purpose in dolphins?

4. Why is it that some animals that maintain activity in winter temperatures would cease to be active in summer if moved suddenly to the same low temperature?

5. What are the advantages and disadvantages of exposing a poikilotherm to higher temperatures in the spring?

6. At what time of year is a species liable to reproduce at the low-latitude extreme of the species range?

7. Distinguish between the processes of osmosis and diffusion.

8. What do most invertebrates do to acclimate to changing salinity?

9. What special osmotic problem do marine bony fishes have?

10. Why do some migrating fish have special salinity problems?

11. What are the major ways in which organisms take up oxygen from the environment?

12. What is the Bohr effect, and why is it important?

13. What factors principally explain variation in oxygen consumption rates among species?

14. How do marine animals deal with lowered oxygen in the environment?

15. What are the three main ways in which marine animals focus light to create an image?

16. Why may a marine organism in a pool of water on sea ice experience problems as more and more ice forms in the pool?

5

Reproduction, Dispersal, and Migration

Reproduction, dispersal, and migration are the fundamental processes that allow living populations to grow and exploit new habitats. **Reproduction** is the replication of individuals and is necessary for population growth. Nearly all species have some form of **sex,** which allows exchange of genetic materials among individuals. Sex and reproduction are intimately related, but many organisms can reproduce without sex, so the two processes must be discussed separately. **Dispersal** is the spread of progeny to locations that may differ from that of the parent. Dispersal is a one-way process and often is controlled by the vagaries of water currents. Because water is a supportive medium and because water circulates rapidly over great distances, many newly hatched young can disperse over large distances, often as much as hundreds of kilometers. As a result, marine species often have broad geographic distributions and the capability of rapid extension of their geographic ranges. Moreover, many species migrate. **Migration,** a directed movement between specific areas, allows an expansion of habitat use and can increase a species's efficiency at exploiting the best resources at optimal feeding and spawning sites. In this chapter, we will discuss the important factors in sex, reproduction, dispersal, and migration.

Ecological and Evolutionary Factors in Sex

Benefits of Sexual Reproduction

✳ Sex is a nearly universal characteristic, despite the considerable costs to organisms of maintaining it.

Sex is a species property whereby different individuals have the capacity to exchange or combine DNA, which causes offspring to differ genetically from their parents. Sex may involve simple transfer of DNA between bacterial cells or the union of gametes to produce zygotes. Nearly all organisms have some form of sexuality, ranging from the mating types of bacteria to the separate sexes and mating dance of humpback whales. The near universality of sex turns out to be a paradox, because there is a measurable cost to sex, at least in typical marine organisms characterized by Mendelian inheritance. In a typical diploid organism, each parent contributes the same number, n, of chromosomes to the offspring, giving the proper diploid number of $2n$. However, the mother usually invests much more in her offspring than does the father, even though half the genes are his (think of the energy put into egg production and the greater amount of parental care usually contributed by the female). Why not devote all her energy to nurturing offspring that carry only her own genes?

There are other costs to maintaining sex. Finding a mate can be costly in terms of time and energy. In fiddler crabs (genus *Uca*), the males have two claws, one much larger than the other, which comprises over 40% of the body weight (Figure 5.1). This larger claw is used only for sexual displays and for combat with other males. Because the claw is not used for feeding, males must compensate for this handicap by feeding longer and faster. Why should so much cost be attached to sex and sexual differences such as these?

The most obvious benefit of sex is the **generation of genetic diversity**. Genetic variation allows a population to live in a broader variety of habitats. There is also an increased **evolutionary potential** when the environment changes. Without sex, all offspring are genetically identical to the parent. No evolutionary change is possible except by means of mutations, which occur only rarely. Asexual populations are therefore clones, with little or no genetic diversity. In contrast, sex provides continually new combinations of genes and changes of gene arrangements due to chromosome crossover during meiosis. Crossing over increases intragenic changes, but also produces new combinations of alleles on different chromosomes.

This gives the flexibility required for the appearance of offspring with new gene combinations, which may be better able to exploit the environment and superior in ability to meet a changing environment. The process of natural selection results in the eventual dominance of these new and adaptively superior genetic variants. Genetic variation may also allow some of the population to survive the continual onslaught of diseases that confront a species. It would be better to possess a gene for resistance to a new disease, rather than only one that determines a phenotype susceptible to the disease. A study of a species the snail genus *Potomopyrgus* demonstrates this well.[1] This species, living in New Zealand freshwater, has populations that reproduce sexually and asexually. The asexual populations are far less able to resist invasions of parasites.

Once sex exists, the differences between the sexes can be enhanced by **sexual selection**, the selection for secondary sex features (e.g., the antlers of deer, the large claw of fiddler crabs) that increase mating success. Genetic variants that have the most successful mating abilities will increase in the population. Success might involve being the largest in overall body size, or having the largest antlers or the most conspicuous colors. For example, fiddler crab males with larger claws may be better at attracting mates or holding high-quality territories where females can incubate eggs. This superiority would set in motion a selective process that could lead to selection for larger and larger claw size.

Types of Sexuality and Their Value

✳ **Sexes may be separate (gonochoristic species), simultaneous, or sequential in the same body.**

Organisms of nearly all kinds (with the exception of bacteria and some protozoa) have two sexes. Why only two is a story in itself, but we shall focus on the types of sex. Species that have separate sexes, or **gonochoristic** species, are the most familiar. In animals, this characteristic entails the need for a mechanism of sperm transfer. The presence of seawater permits the simple shedding of sperm (and possibly eggs) in the water. Planktonic gamete production often involves simultaneous spawning of males and fe-

Fig. 5.1 Displaying male of the fiddler crab, *Uca pugilator*. Note the large major claw, which is used for displays and male–male combat.

[1] See Lively and others, 1990, in Further Reading, Sex, Fertilization, and Life Cycles.

males, which is often keyed into tidal or lunar rhythms. To avoid interspecific matings, the simultaneous presence in the water of eggs and sperm of many species requires extremely precise egg–sperm chemical recognition. Recent studies of abalone species on the west coast of the United States show that sperm–egg recognition proteins are extremely specialized to the degree that interspecific zygote formation is nearly precluded.[2] In sea urchins, sperm that penetrate an egg's jelly coat have a protein that binds specifically with a receptor of eggs of its own species only. It may be that speciation in externally fertilized species is strongly controlled by such recognition proteins, and the rapid evolution of the recognition proteins may be the major factor in species isolation.

A **hermaphrodite** can produce gametes of both sexes during an individual's lifetime. In **simultaneous hermaphrodites**, sex cells for both eggs and sperm are active at the same time. Despite this, self-fertilization is rare, probably because inbreeding often produces inviable offspring, which imposes selection against such behavior. Being a simultaneous hermaphrodite means never having to say you're sorry, because an encounter between any two individuals guarantees that mating can occur. For example, acorn barnacles need to copulate to reproduce, but they are stuck with a sessile lifestyle and can reach only nearby barnacles. Simultaneous hermaphroditism is a great advantage because the nearest individual will always be of the complementary sex. An extraordinarily long penis allows a barnacle to reach relatively distant fixed mates. One of the most fascinating aspects of simultaneous hermaphroditism is the common continual trading of mating roles. In many simultaneously hermaphroditic fishes, for example, members of a mating pair change sex roles every few minutes, first producing sperm, then spawning eggs (as the other individual, who also has changed roles, now deposits sperm).

Sequential hermaphrodites start their mature sex life as one sex, and then they transform into the opposite sex. If they are male first, they are said to be **protandrous**, while if they are female first, they are **protogynous**. Many invertebrates, including some oysters, polychaetes, prawns, and coelenterates, are protandrous, whereas some fishes, particularly coral reef fishes, are protogynous. In some cases, such as the diminutive protandrous polychaete species of the genus *Ophryotrocha*, sex change can involve a mere

Fig. 5.2 The sequentially hermaphroditic snail *Crepidula fornicata*, stacked on a gravel beach. Females are on the bottom (right). Males are the smallest and topmost individuals; transitional forms are in the middle.

switch from the manufacture of oocytes to the manufacture of spermatocytes. In protogynous reef fishes, however, the sex change also involves color, size, and morphological transformation. The snail *Crepidula fornicata* usually occurs in stacks, with larger and older females below and smaller and younger males on top (Figure 5.2). In their case, the stimulation to change sex involves contact with other individuals. Top-layer members of the stacks are oriented with right anterior margin in contact with the same margin of the lower member. This allows insertion of the penis of the upper snail into the female gonopore of a lower snail.

✻ **The relative contributions of different sexes to zygotes in the next generation determine the value of hermaphroditism, and the relative sizes of males and females.**

Why be a simultaneous hermaphrodite? In a sense, one might turn the question around and ask: Why aren't all organisms simultaneously hermaphroditic, given the obvious advantage of finding a potential mate in every individual? The answer may lie in the limitations of being a jack-of-all-trades and master of none, versus being a specialist. Simultaneous hermaphrodites must invest not only in an apparatus to form gametes, but in the secondary sexual attributes necessary to attract another mate. If one started with

[2] See Swanson and Vacquier, 1998, in Further Reading, Sex, Fertilization, and Life Cycle.

a population in which genetic variants for hermaphroditism and for separate sexes coexisted, one might find that sexual selection would favor the "pure" males and females, which they can devote all their respective resources to either male or female mating structures and activity. The hermaphrodite would lose out in the competition because it serves both functions, but each less efficiently. Hermaphroditism is favored only when the value of always being able to find a mate compensates for the disadvantage of not being a sexual specialist. The acorn barnacle may be just such a case. Simultaneous hermaphroditism is far more common in freshwater invertebrates, where it is generally harder to find a mate owing to the lower population densities and the uncertainties of freshwater existence.

Sequential hermaphrodites would seem to have all the advantages, given that they can be both sexes at different times. Unlike simultaneous hermaphrodites, however, they lack the advantage of always being able to find a mate in another individual. Then what is the advantage of changing sex? Here, the answer lies in the advantage of being a male or female at different ages or sizes. Consider protandry, or the quality of first being a male, and later a female. As a general rule, producing eggs costs more energy than producing sperm. It is also usually true that a larger and older animal has more energy at its disposal than does a smaller and younger one, simply because of the former's size. Then it would be best to be a male while still small, when a relatively small investment in sperm could produce many offspring. Above a certain threshold size, however, females can parent more offspring, because the available energy can be used to produce more eggs. The threshold size of switching (Figure 5.3) should be that size at which the number of offspring parented would be the same regardless of the sex of the animal. Below that threshold size, the animal would sire more offspring if it were male; above that threshold size, it would parent more offspring were it a female. The considerations are thus simply those of a cost–benefit game.

Commercial fishers inadvertently performed an experiment that tested certain aspects of the theory explaining the size at which sex switching occurs. Species of the prawn genus *Pandalus* are fished widely in both the North Atlantic and Pacific Oceans. The prawns mature first as males, but, after a variable period of time, switch to being females for the rest of their lives. A fishery in which such prawns

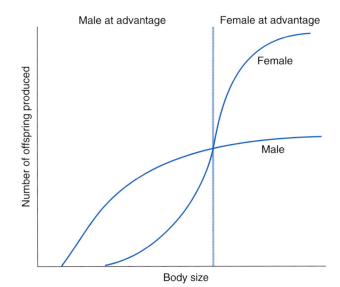

Fig. 5.3 Graph showing how to find the optimal size for a switch from male to female in a protandrous hermaphrodite. The curves shown offspring an individual would parent if it remained either male or female for its entire life. The curves cross once, and output as a female surpasses that for a male as the crossover is reached and passed. For sizes smaller than the crossover, output is greater as a male.

were caught commenced in Danish and Swedish waters about 1930. Following the establishment of this enterprise, the average body size of the catch decreased and the threshold size of switching to females also decreased. This development can be interpreted as follows: increased mortality (from fishing) will cause females to become relatively rare if males continue to switch sex only when they have reached a large body size. If a new genetic variant appears that switches at a smaller size, it will produce more offspring. Owing to natural selection for these variants, the size of sex switch will decline over time. If fishing pressure is very severe, it might "pay" for some individuals to mature first as females, rather than going through a male stage. Such populations exist.

Why in general should an organism be female first, then male? Here we must consider the common role of males as competitors for matings with females. Success as a male often entails agility at combat, bright colors, rapid swimming, and so on. Most of these traits are enhanced with increased size and experience. Many territorial coral reef fish have flashy males and are protogynous. For example, the cleaner wrasse *Labroides dimidiatus* can usually be found in a group of 10–15 fish. All but the largest fish in the group are relatively dull in color and are females. By contrast, the largest fish is brightly colored and male. The male constantly tries to prevent the females from mating with interlopers. If this male is removed from his harem, the largest female will transform herself into a male and begin to serve the male function. In other species of wrasses, individuals can either be terminal-phase males (Figure 5.4), which hold territories, initial-phase males, or females. Initial-phase males do not hold territories and either spawn in groups with females, or attempt to sneak into the territories of the flashier terminal-phase males. Both initial-phase males and females are capable of changing into terminal-phase males. The proportion of initial-phase males varies from place to place. This suggests that there are places where being a small male results in successful matings.

Although it is possible to classify types of sequential hermaphroditism, it is important to realize that many natural populations contain considerable geographic variation in sex change. This is especially true of coral reef fish. Certain species may be locally protogynous but gonochoristic in other sites. Hermaphroditism is not necessarily a fixed trait of a species and may be under active natural selection in different directions in different populations.

Fig. 5.5 A female brittle star, *Amphilycus androphorus*, carrying a dwarf male. The two are attached mouth to mouth. (From Hyman, 1955.)

* Dwarf parasitic males are found in some species that live in situations where it is difficult to find mates.

Where the sexes are separate and mate location is difficult, males may be small animals that either attach to larger females or lreside very close to them. Good examples are found among some barnacles, where tiny dwarf males attach within the mantle cavity of normal-sized females (e.g., the stalked general *Scalpellum* and *Ibia*, and the boring Acrothoracica). Many deep-water fish have dwarf males. Males show varying degrees of modification and may be parasitic upon females. In some cases, the males is a miniature version of the female, except for the reproductive organs. This is common in some brittle stars (Echinoderm class Ophiuroidia), whose dwarf males cling to the much larger females (Figure 5.5).

Fertilization Success

* Fertilization success is affected by the mode of sperm transfer, the volume of gamete production, the distance between males and females, water turbulence, timing, and behavior.

Many animal species use a variety of specialized means of sperm transfer involving direct male-to-female contact. In the simplest cases, found in some polychaetes and fishes, the male applies sperm to the body surface of the females or to an egg clutch as the

Fig. 5.4 A terminal-phase male of the bluehead wrasse *Thalassoma bifasciatum* courting a female. The two are about to mate. (Photograph courtesy of Robert R. Warner.)

eggs are released by the female. Males of many species (e.g., gastropods, crustacea, fishes) have copulatory structures. This mode of sexual contact increases the probability of fertilization and allows for mate selection. Many species, ranging from migratory fishes to intertidal drilling gastropods, form breeding swarms at the time of mating to further ensure the finding of a mate.

Although direct contact and copulation guarantee sperm transfer, many marine species take advantage of the presence of water and shed their gametes directly into it. At best, the gametes will mix and fertilization frequency will be high, because the sperm are so much more numerous than the eggs. However, turbulence or large distance between spawning individuals greatly reduces fertilization success. Studies by Don Levitan[3] and colleagues show that fertilization success in sea urchins can be greatly reduced by low volume of sperm production, combined with turbulence and great distance between males and females. In one respect, however, turbulence can be a positive factor, because it permits sperm to be mixed with eggs shed by females. However, turbulence also dilutes gamete concentration and may reduce the probability of fertilization. Some species, such as the Pacific sea star *Acanthaster planci*, produce such large volumes of gametes that high rates of fertiliza-

tion can occur between males and females separated by tens of meters (see Hot Topics 5.1).

Spawning by one or a few individuals may induce mass spawning of an entire local population. Mussels will spawn in response to phytoplankton in the water; the mussel larvae then can eat the phytoplankton. Timed spawning is usually keyed into lunar tidal cycles. In some reef sponges, all male individuals spawn simultaneously, spreading a fog of sperm over the reef. Many polychaete annelids can change morphology radically to transform themselves into epitokes, individuals that are essentially swimming sacs of gametes. Operating on a lunar cycle, they may swim to the surface and perform a nuptial dance, in which males and females swim rapidly about each other while releasing gametes into the water. South Pacific peoples take advantage of this behavior by harvesting swimming polychaete Pololo worms. Many coral reef fish shed their gametes, but the fishes often locate themselves strategically on the downstream end of patch reefs, to minimize turbulence and gamete dilution. As a result, fertilization success is surprisingly high, often exceeding 90 percent.

[3] See Levitan and others, 1992, in Further Reading, Reading, Fertilization, and Life Cycles.

Hot Topics in Marine Biology 5.1 When Gametes Collide: Fertilization and the New Molecular Wave

As we discuss in this chapter, marine animals have a bewildering diversity of sexualities. If there is anything truly unique about the ocean, it is the frequent occurrence of species that indulge in **free spawning**: the release of gametes into the water by both males and females. In most of these cases fertilization is followed by the development of swimming larvae that may remain in the water column for weeks and sometimes even months.

The phenomenon of the free spawning of gametes has some advantages, but also some strong challenges. At first glance one notices an obvious advantage of broadcasting gametes: one need not invest energy to find another mature individual of the appropriate sex with which to copulate. With free spawning the game changes, because both eggs and sperm must encounter each other, sometimes based upon hydrodynamics as opposed to sex appeal. Don

Levitan has pointed out[4] that in free spawners, the eggs suddenly are part of the gamete collision equation and there might be strong selection for egg traits that maximize sperm encounters and even attraction.

But recent research also defines significant costs of this form of gamete release.

1. Fertilization success is often low when distance is great between males and females and when turbulence is strong.[5] Near bottom, water turbulence mixes gametes in all directions and therefore effectively dilutes the con-

[4] See Levitan, 1996, in Further Reading, Sex, Fertilization, and Life Cycles.
[5] The interested student should consult Vogel and others, 1982, and Styan, 1998, in Further Reading, Sex, Fertilization, and Life Cycles.

centration of gametes encountered by an individual of appropriate sex. At low population densities and with few spawning individuals, one might expect to see the effect of sperm limitation, and this has been demonstrated in the eastern Pacific temperate sea urchin *Strongylocentrotus franciscanus* (see Box Figure 5.1). In the Pacific coral reef crown-of-thorns starfish, a large volume of sperm is released, making fertilization effective even over a distance of many meters from the source male.[6] This may solve the problem, but a heavier investment of resources into gametes is involved, as well.

2. When males and females spawn, there is a timing problem: gamete release must be simultaneous, meaning that an intraspecific signal to ensure simultaneous spawning must be evolved, such as a response to a lunar cycle. Alternatively, some species of mussels are known to respond to the spawning of nearby individuals. This may mean that certain local individuals will make a disproportionate contribution to the next generation, especially when a male invertebrate spawning millions of sperm successfully fertilizes a large number gametes of nearby females.

3. What if several species evolve a response to the same lunar cycle? Many species would be spawning at the same time. Gametes would likely encounter those of other species. If reproductive isolation has occurred, this would cause lowered fitness unless there was extreme specificity of gametes. For example, Peter Harrison and colleagues[7] described mass spawning by over 20 species of corals in the Great Barrier Reef of Australia in a single night.

Our discussion in this chapter concludes that sperm limitations may figure importantly in reproductive success, especially when gametes are diluted by turbulence. A number of experiments performed with sea urchin eggs and sperm by Levitan showed very strong dilution of sperm supply, especially when there was turbulence. The work of Phil Yund and Michael McCartney[8] demonstrated that, in her-

[6] See Babcock, Mundy, and Whitehead, 1994, in Further Reading, Sex, Fertilization, and Life Cycles.
[7] See Harrison, and others, 1984, in Further Reading, Sex, Fertilization, and Life Cycles.
[8] See Yund and McCartney, 1994, in Further Reading, Sex, Fertilization, and Life Cycles.

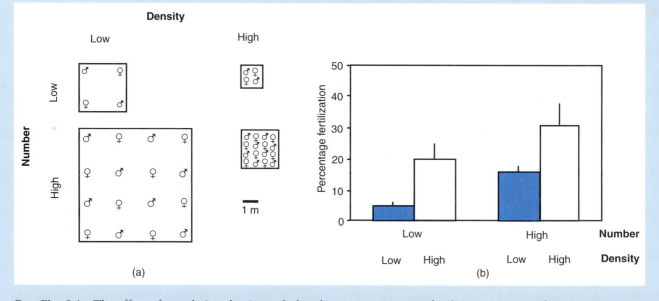

(a)

(b)

Box Fig. 5.1 The effect of population density and abundance on percentage fertilization success of eggs using experimental populations (actually cups of urchin eggs) of the Pacific sea urchin *Strongylocentrotus franciscanus*, which were exposed to sea urchin sperm supplied by a syringe. "Population size" was either high or low and the size of the patch with a given population was either small or large, making population density high or low for the two population sizes. (a) Experimental design. (b) Percentage fertilization. (After Levitan et al., 1992.)

maphroditic colonial ascidians and bryozoans, the degree of devotion of resources to male function may limit fertilization function. Distance, population density, and turbulence all severely limit the effectiveness of broadcast spawning. High population density of males might select for more sperm production, because male genotypes are competing with each other for matings, and increased sperm release might increase the production of successful fertilizations.

So more sperm is better. Or is it? If sperm concentration is very high, then polyspermy, which entails the beginning of zygote formation by two or more sperm and one egg, may develop. When this happens, successful fertilization is usually foiled. All known animals have elegant blocks to polyspermy, some of which act within seconds to prevent another sperm from attaching to the egg. But the phenomenon is still possible and perhaps likely in a concentrated cloud of sperm, as might be produced by some invertebrates. I once measured sperm volume in a mussel and recovered about 250 mL.

A simple fertilization experiment in laboratory containers with free-spawned gametes works a bit like a chemical titration and gives us an understanding of the evolutionary consequences of gamete encounter and compatibility. Steven Palumbi and students have taken a number of very closely related sea urchin species of the genus *Echinometra* (Box Figure 5.2), as well as geographically separated populations of the same species, and followed the percentage fertilization success as gamete concentration increases. If the response is a high percentage successful fertilizations, we can conclude that the gametes come from the same species. Mixes of gametes of different species have very low fertilization rates.

The fertilization rates suggest that some aspect of fertilization biology is the key to the mechanism that isolates species from each other. Indeed, in free-spawning marine animals, how else can isolation efficiently occur? The union of gametes is as close to mating as we can get here.

We can see several steps at fertilization that might be disrupted when gametes of different species come into contact:

1. Receptors in the egg coating induce the sperm acrosome reaction, which is an initiation of attachment.
2. Sperm and egg attach, owing to the presence of complementary binding (sperm) and receiving (egg) proteins.
3. Further reactions cause the uniting and fusing of gamete plasma membranes.

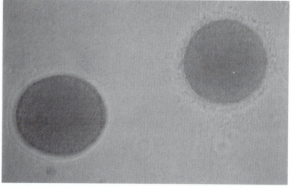

Box Fig. 5.2 The moment of sperm–egg contact in a Pacific sea urchin of the genus *Arbacia*. The egg at the right is surrounded by active sperm of the same species, ignoring the egg of a closely related species. (Courtesy of Steve Palumbi.)

We have information on only two major groups of organisms—abalones (Phylum Mollusca, Class Gastropoda, Order Archaeogastropoda, Family Haliotidae) and sea urchins (Phylum Echinodermata, Class Echinoidea, several genera)—but a surprising similarity has emerged. Both have a highly specific sperm–egg binding system. In urchins, sperm have the protein bindin, which unites with a bindin receptor in the egg membrane. In abalones, gamete recognition is controlled by the sperm protein lysin, which creates a hole in the egg envelope, but only for eggs of the same species as that of the sperm. In the abalone egg envelope, a receptor binds specifically to lysin. Hybrids are thus rare, even though seven species of California abalone free-spawn gametes and overlap in geographic range and breeding season. Different species of urchins rarely if ever have gametes that will fertilize each other.

The gametes and larvae of abalones, urchins, and many other marine groups spend time in the plankton. Larvae may spend so much time that there is little opportunity for populations to be isolated from each other. While isolation of populations has been found in the sea, it is very much true that species with planktonic feeding larvae tend to have much broader geographic ranges than species with lecithotrophic larvae, or direct release. Isolation in distant populations of many marine species has been found by the discovery of genetic markers such as fixed differences in DNA sequences.

The fascinating situation here is that some remarkably simple property of free-spawned gametes is the gatekeeper that admits compatible gametes and bans inappropriate mates. No, it is not feelings or love at first sight, but the amino acid sequences of the recognition proteins. Bindins from different species have different amino acid sequences,[9] and this makes the crucial difference in the sperm–egg recognition process. The same is true for lysins in abalones.

Proteins are typically folded in a complex manner. Some of the amino acids are involved in regulating the three-dimensional structure, while others are involved in binding with other molecules. Thus it is only a portion of the amino acid chain that is involved in gamete recognition.

Studies of species-specific recognition proteins yield very surprising results, and we have only scratched the surface of this fascinating field.

1. **High divergence between species.** Amino acid sequence divergence between lysins in different species is surprisingly high, suggesting that some sort of positive natural selection force may be at work, since random processes would not permit such a degree of divergence. This is especially a surprise because one might expect the evolution of amino acid sequences to be very conservative to assure gamete compatibility. The pressure for high rates of lysin evolution in abalones may be the constant and high rates of evolution of the receptor, a protein known as VERL. A poorly understood process known as concerted evolution tends to duplicate certain sequences in a protein, and this apparently is happening in VERL. Such change would require evolutionary change in the sperm protein lysin, to enable it to keep up its function.[10] In effect, the lysin is evolving much as prescribed by the Red Queen in *Alice in Wonderland*: one must keep on moving in order to stay in the same place.

2. **Polymorphism within species.** In some species of the Pacific sea urchin genus *Echinometra* there is intraspecific polymorphism in the gamete recognition portion of bindin sequences.[11] This is a big surprise. If bindin is so specific to its binding site, how can the molecule be variable and yet still exhibit very high successful fertilization rates? It is possible that the bindin egg binding site is flexible in gamete recognition, allowing a number of bindin genetic variants to coexist in a population. This is not found universally, since species of the urchin genus *Arbacia* have no intraspecific variation in bindin.[12] Nevertheless, the results of *Echinometra* studies show how variation in bindins might arise in a population. If two populations of an urchin species were isolated, then one can imagine one variant being fixed in one population and another being fixed in the other population. With further evolution, these two populations might become incompatible, resulting in speciation.

These latest results show that the evolution of gamete recognition proteins is identifiable in living populations, and we can therefore see traces of the dynamic process of speciation in the sea, even in species with free-spawned gametes. It may well be that geographic divergence of gamete recognition proteins is the key to understanding speciation. Much more needs to be learned about mechanisms of gamete compatibility, but we are being led in the right direction as a result of newly understood molecular mechanisms and methods.

[9] Remember that a protein consists of a string of amino acids of very specific sequence.
[10] See Swanson and Vacquier, 1998, in Further Reading, Sex, Fertilization, and Life Cycles.
[11] See Metz and Palumbi, 1996, in Further Reading, Sex, Fertilization, and Life Cycles.
[12] See Metz and others, 1998, in Further Reading, Sex, Fertilization, and Life Cycles.

Parental Care

✳ **Parental care is nonexistent in many marine animal species, but in some cases females or males care for the young.**

Marine species often exhibit absolutely no parental care. Many of these species spawn eggs and sperm, which unite to form zygotes, which, in turn, develop into free-swimming larvae that are completely at the mercy of the seas. However, in some groups, such as the fiddler crab *Uca*, the female incubates the eggs for a couple of weeks and then releases swimming larvae into the water. Many species care for their young to an even greater degree. In some species, the young are reared to a juvenile stage within the mother. This is of course true for all marine mammals, but it is also the case for certain species of many of the invertebrate groups and fishes. The Atlantic and Pacific intertidal clam *Gemma gemma*, which is only a few millimeters long, nurtures its young within the mantle cavity and releases them as shelled juveniles.

In some species, courtship is strongly related to male parental care. The male of the three-spined stickleback, *Gasterosteus aculeatus*, has a vivid red belly during courtship. This anatomical feature serves to lure females to lay eggs in its hidden nest, usually among rocks on the bottom. Many aquarium enthusiasts have observed the male dance in a zigzag motion, to lure the female to the nest. The great behaviorist Niko Tinbergen showed that the reproductive females have a stereotypical response: they are strongly attracted even to silver disks painted red on the bottom. Behaviorist Susan Foster[13] has observed that in some locations the female participates, perhaps to a dominant degree, in mate selection and often swims up rapidly to a male and jumps onto on his back! During the courtship process, a female enters the nest and lays her eggs, which are fertilized externally by the male. Subsequently, the male aerates the eggs until they hatch. It is in the male's interest to rear the young of all females that have laid eggs in his nest, because they are his progeny. When a female enters, however, the eggs that have already been laid there by other females are—well—just food. Commonly, an intruding female will gobble up the eggs already present and bolt.

Nonsexual Reproduction

* **Nonsexual reproduction permits the same genetic type to increase rapidly in an open environment.**

Although sexuality is nearly universal in marine organisms, many of them are capable of reproducing without the formation of a zygote. Asexual reproduction lacks the cost of sexual reproduction and permits the spread of a genotype that has successfully colonized a given habitat. A population of genetically identical individuals, all deriving from one founder, is known as a **clone**.

The exact style of asexual reproduction varies, depending upon the biology of the individual group.

Diatoms, a major group of marine phytoplankton, consist of cells or cell chains and can reproduce asexually by fission. The diatoms are usually unicellular and have a silica skeleton composed of two valves. At cell division, each daughter cell inherits one valve. Although the rate of cell division varies, most forms fall within the range of 0.6–6 doublings per day under optimal laboratory conditions. The genetic clone with the highest combination of survival and doubling rate will come to dominate the local population.

In multicellular organisms, fragmentation can serve the same function as fission. Some annelids fragment, while others divide by forming head segments midway down the body and then splitting into two new individuals. Many seaweeds (e.g., *Gracilaria*, *Polysiphonia*) and many corals (e.g., the Caribbean staghorn coral *Acropora cervicornis*) may reproduce mainly by fragmentation. Tropical storms may benefit a staghorn coral population by dispersing newly created fragmented individuals.

Colonial invertebrates may reproduce asexually by fission of whole individuals in the colony. As opposed to the production of new independent individuals, this process produces genetically identical and morphologically similar **modules**. This mode is utilized by such colonial animals as encrusting sponges, coelenterates, and bryozoans. Marine algae and angiosperms include many **vegetatively reproducing** species. The marsh grass *Spartina alterniflora* usually consists of a large number of plants connected together by a rhizome system. It can be demonstrated by biochemical marker techniques that genetically identical plants, all of which have derived from one progenitor, may cover tens of square meters. A marsh meadow is thus typically a series of clones.

The value of vegetative spread varies with the biology of the individual group. In some corals and sponges, the overall form of the colony is important in the collection of food. In certain cases, the strength of the colony is helpful in resisting strong currents. Large colony size may also be an advantage. Larger colonies will have a greater ratio of living surface to periphery exposed to moving sediment. They will

---

[13] See Foster, 1990, in Further Reading, Sex, Fertilization, and Life Cycles.

therefore be superior to small colonies in ability to survive the movement of sediment along the bottom. For the same sort of reason, colonial forms may survive the attacks of predators, at least those attacking from the side.

Group living is of greatest value when it benefits related individuals. In this case, all individuals increase the survival of their own genes by group living. In groups that consist of unrelated individuals, some of the disadvantageous aspects of group living actually decrease an individual's probability of passing on its own genes. Consider the poor members at the periphery of a colony, which are more likely to succumb to predators. Their residence in the colony would be counterproductive to them as a group if the individuals in the center were unrelated to them. Because they are related, however, the exposure of some helps protect the individuals farther inside, and helps ensure the continued existence of the genetically identical group. As the famous biologist J. B. S. Haldane once said, "I would give my life for my brother, or for eight of my cousins." (See Hot Topics 5.2.)

Colonies of the colonial ascidian *Botryllus schlosseri* are usually founded by a single sexually produced larva, which attaches to a hard surface, metamorphoses (transforms into an immature adult), and then reproduces asexually. However, sometimes the larvae settle in aggregations. Using a unique genetic marker at an enzyme locus, Richard Grosberg and J. F. Quinn demonstrated[14] that such aggregations consist of larvae that are genetically similar in tissue compatibility. These larvae can fuse and form larger initial colonies, which are more resistant to predation and, because the initiation of reproduction is size dependent, may reproduce sexually at an earlier age.

The phenomenon of immunological recognition of genetically related individuals seems to be widespread among groups that may benefit from group living. On eastern Pacific shores, the anemone *Anthopleura elegantissima* (Figure 5.6, Plate XVI.3) lives in large clones of several hundred to several thousand individuals. All derive from a single colonizing larva that repeatedly divides asexually. At any time, it is usually possible to catch some individuals in the middle of fission. Lisbeth Francis[15] found that contacts between individuals of different clones result in a stereotyped aggressive response. The affected anemones raise their tentacles and expose acrorhagi, which sting

Fig. 5.6 The rocky shore intertidal anemone *Anthopleura elegantissima* (top) often occurs as clones of a few hundred, but different clones have a bare zone between them (arrow, lower center of photograph). When an individual at the edge of one clone encounters members of another (bottom), they rear up and expose acrorhagi, which sting the individuals of the other clone. (Courtesy of Lisbeth Francis.)

individuals from the other clone. No such aggression occurs between individuals within the same aggregation. This behavior can be interpreted as the defense of a communal territory, but it is not known what benefit is obtained from the defense.

[14] See Grosberg and Quinn, 1986, in Further Reading, Larval Dispersal and Recruitment.
[15] See Francis, 1973, in Further Reading, Sex, Fertilization, and Life Cycles.

Hot Topics in Marine Biology 5.2 When Shrimp Socialize in the Extreme

George Williams's classic 1966 book *Adaptation and Natural Selection* emphasized the role of individuals and genes in the process of natural selection. Selection paid little attention to groups of individuals, but rather to the functioning of individuals that might leave different numbers of offspring into the next generation. A gene has no loyalty but to itself, so the individuals carrying a given gene may or may not increase in frequency, depending upon that gene's contribution to fitness. Owing to this crucial individuality, Williams argued that selection would not go on for the "good of the group," especially because the genes in individuals were not confined to specific groups.

This generalization applies to most populations in which individuals are at best distant cousins of each other. But what happens if all individuals in a population are closely related, as in the case of siblings? If so, they will share on average 50 percent of their genes. In social bees, workers are sterile and defend the colony, often dying in the process. But if individual fitness always prevails, why this apparently **altruistic behavior**? William Hamilton first explained this seeming anomaly in terms of the relatedness of members of a colony. The workers are extremely close relatives of the queen; their role in doing the work of the colony therefore increases not only the genes of the queen but also their own. Hamilton used the term **inclusive fitness** in discussing the larger groupings of closely related individuals that may form to cooperatively proliferate the same genes. One might expect inclusive fitness to result in the evolution of altruism, as long as the members of a group are closely related.

The evolutionary force leading to the evolution of cooperative behavior among close relatives to proliferate similar genes is known as **kin selection**. The extreme case of an entire colony of individuals behaving in cooperative fashion is known as **eusociality**. This behavior usually is found in a group of individuals that serve a queen's reproduction. It is extremely rare beyond social ants, bees, wasps, and a few other groups of insects. Naked mole rats (*Heterocephalus glaber*) are also eusocial: there is one "queen," several mating males, and numerous other closely related nonmating siblings or very close relatives that gather resources and take care of the tunnels in which the mole rats live.

The key to eusociality is the ability to construct a functioning integrated colony, which is often enclosed in a structure built or occupied by the eusocial organisms. Naked mole rat colonies, for example, maintain an elaborate system of burrows, which must be defended. The queen does all the reproduction for the colony, whereas the reproducing males and nonreproducing members all perform communal roles in tunnel maintenance and defense. If the colony lives for a long time in one place, then there is an obvious advantage for closely related individuals to maintain and defend the colony, occasionally producing queens to found new colonies.

Some marine environments lend themselves to discrete colonies of animals. If you cut a large sponge in the tropics, you will immediately encounter a large number of invertebrates that live in the interstices of the sponge colony. Indeed, marine biologists have been showing their students such assemblages for many years. Large numbers, often of a single species of crustacean, were found commonly, especially species of snapping shrimps of the family Synalpheidae. These sponge-associated populations are liable to be separated from those in other sponges, because movement between sponges would expose them to predators.

After cutting open many sponges in Belize, Emmett Duffy[16] came upon a discovery that had probably been overlooked by many others. Groups of individuals of the shrimp *Synalpheus regalis* were found in sponges of Belize, averaging 150 individuals per sponge (Box Figure 5.3). Usually a large female was found with many smaller males or immature animals. A simple experiment demonstrated altruistic behavior. When species of other snapping shrimp were introduced into colonies of *S. regalis*, groups of resident shrimp used their powerful major claws to attack the intruders, eventually killing them. If resident shrimp were removed and then reintroduced, however, defense behavior evaporated, suggesting the existence of a recognition system among colony members.

Genetic analyses further add evidence for eusociality in these synalpheid shrimp. Using enzyme genetic polymorphisms as markers, Duffy demonstrated that the average difference between shrimp in a colony was close to that expected for siblings. One large individual was a female, and she is probably analogous to the queen of mole rat colonies. Such a colony and its sponge cavity habitat fits well with the nesting areas of many other eusocial species,

[16] See Duffy, 1996a, in Further Reading, Sex, Fertilization, and Life Cycles.

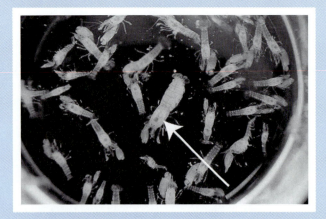

Box Fig. 5.3   A eusocial shrimp. Members of a colony of the sponge-dwelling shrimp *Synalpheus regalis* in Belize. Sole female ("queen") is at center, recognizable by clutch of eggs under abdomen. Surrounding individuals are juveniles and adult males. The colony may number up to 350 individuals within a sponge (Courtesy of J. Emmett Duffy.)

including many ants, termites, naked mole rats, and other species. Sponges are long-lived, and shrimp colonies could similarly persist over many years. Selection for eusocial behavior is strongly favored in such cases, and Duffy has

now found a number of species of eusocial synalpheid shrimp species.[17]

Because there are many species of apparently eusocial synalpheids, the question immediately arises: Did eusociality evolve only once in this group, or is selection and the response to selection strong enough that this behavior may have arisen in a number of species associated with sponges? Duffy[18] constructed an evolutionary tree, combining data from DNA sequences and morphological characters. The evidence was clear: three eusocial species were scattered among the tips of the tree. Eusociality had apparently evolved several times from noneusocial ancestors (Box Figure 5.4). Again, the association with the interstices of discrete and long-lived sponge colonies is the driving force for the several evolutionary appearances of eusociality. More generally, specialization for living in specific sponge hosts has been a general driving force in the evolution of the Synalpheidae.[19]

[17] See Duffy, 1998, in Further Reading, Sex, Fertilization, and Life Cycles.
[18] See Duffy et al. 2000, in Further Reading, Sex, Fertilization, and Life Cycles.
[19] See Duffy, 1996b, in Further Reading, Sex, Fertilization, and Life Cycles.

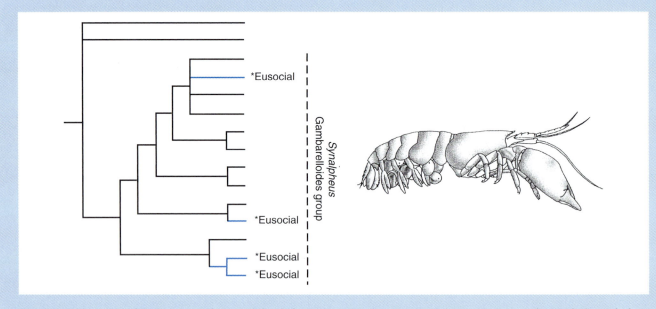

Box Fig. 5.4   An evolutionary tree for synalpheid shrimp species, based upon DNA sequence and morphological characters. Note the starred eusocial species, which apparently arose several times in independent lineages. (Courtesy of J. Emmett Duffy.)

Reproduction, Demography, and Life Cycles

Natural Selection of Reproduction

❋ **Age of first reproduction, reproductive effort (resources devoted to reproduction), and longevity may be partially determined by patterns of mortality and predictability of reproductive success in the population.**

Not all marine organisms spawn for the first time at the same age, nor do all devote the same proportion of available resources to reproduction. Consider the several Pacific species of salmon. All spend three years or so at sea while feeding as adults. They then make the long return trip to their spawning grounds: the tributary from which they originally came. During the trip upriver, they change morphology significantly, and, by the time they spawn, they have no ability to return to the sea. They are so weak after spawning that they lie on the spawning beds and soon die. By contrast, Atlantic salmon spawn, but then may go through at least one more migration–spawning cycle.

Why should there be such a difference among species? Additional differences among species can be found in the age of first reproduction, proportion of energy devoted to reproduction, or **reproductive effort**, and whether the animal spawns only once (**semelparity**) or more than once (**iteroparity**). The theory explaining these differences is based partially upon the premise that there is a cost of reproduction. We would expect evolution to maximize the total reproductive output over the lifetime of an organism. If there is a cost to reproduction, it then matters how reserves are allocated to reproduction versus somatic growth. If more reserves are devoted to reproduction, there will be fewer that can be devoted to growth. If mortality during reproduction is heavy, then any investment of resources into growth may be wasted, because the organism will probably die before getting a chance to reproduce again. High adult mortality therefore selects for earlier reproduction. If adult mortality is low, it may pay to invest in additional growth to ensure repeated reproduction, because larger adults can often produce more gametes and are often more experienced at winning mates than are smaller adults. Low adult mortality therefore selects for later age of first reproduction, and for repeated reproduction.

Commercial fishing pressure should strongly influence the distribution of life histories of the exploited fish species. After all, species with low reproductive effort and late age of first reproduction will be the first to be seriously affected by intensive fishing. By contrast, species with early reproduction and high reproductive effort would be best suited to withstand the onslaught. It also stands to reason that life histories may evolve as a response to fishing. Again, fishers have inadvertently performed an experiment for us. The spiny dogfish *Squalus acanthias* is fished heavily in Europe, but until recently was hardly fished at all off the American Atlantic coast. As a result, spiny dogfish populations were far more dense in American waters. The European populations have an earlier age of first reproduction and higher reproductive effort in the first spawning season. Assuming that all other things are equal, this would indicate that fishing pressure has selected for reproductive tactics that ensure more reproduction. Recent increased fishing pressure on North American populations will probably yield the same effect.

Environmental uncertainty is also an important factor and can be related to the potential success of a new year class. If environmental uncertainty is high, a new group of juveniles may not survive. This imposes a selective force for repeated reproduction, to ensure that some year class of juveniles will carry on the population. In effect, a female engaged in such repeated reproduction is hedging bets against a bad year. Table 5.1 compares age of first reproduction and reproductive span with variation in spawning success for five species. Fishes with relatively high spawning variability tend to have a longer reproductive span than those with low variability.

Migration

❋ **Fishes, crustaceans, turtles, and marine mammals often migrate between spawning and feeding grounds.**

Many marine species have a reproduction and migration cycle similar to the scheme in Figure 5.7. Juveniles drift from a spawning area to a nursery ground and then move to an adult feeding ground as they grow older. Adults then migrate back to the spawning area, The cycle of spawning, drift, and active migration back again reflects a maximization of reproductive success by feeding and spawning in the optimal sites. The movement from nursery grounds to different adult feeding grounds may reflect com-

Table 5.1   The relationship between variation in spawning success (recruitment of young in the best year divided by recruitment in the worst year) and age of first reproduction and total reproductive span

POPULATION	AGE AT FIRST MATURITY (years)	REPRODUCTIVE SPAN (years)	VARIATION IN SPAWNING SUCCESS[1]
Herring (Atlantic–Scandinavian)	5–6	18	25
Herring (North Sea)	3–5	10	9
Pacific sardine	2–3	10	10
Herring (Baltic)	2–3	4	3
Anchoveta (Peru)	1	2	2

1. Variation in spawning success seems to cause natural selection for increased age of first reproduction and prolonged total reproductive life span.

Source: Data from Murphy, 1968.

petition between juveniles and adults for limited food resources, or it may result from differing food requirements of the different age classes. In many cases, fishes spawn for more than one reproductive season, so there is repeated migration between the adult stock area and the spawning area.

Migratory patterns are classified on the basis of the location of spawning and adult feeding grounds. **Anadromous** fishes (e.g., salmon, shad, sea lamprey) are those that spend most of their time in the sea but breed in fresh water. **Catadromous** fishes (e.g., eels of the genus *Anguilla*) spend their adult lives in fresh water but move to the sea to breed. Many species (e.g., herring, cod, plaice) feed and breed in the open sea, though they migrate between different localities.

Migratory species vary in their degree of homing. Pacific salmon species are born in freshwater tributaries and migrate to the sea after a few months. They then return several years later, almost always to the same tributary. By contrast, the early larval stages of the herring *Clupea harengus* (Figure 5.8) drift shoreward from the spawning grounds. As they grow, the herring move to deeper water and feed upon larger zooplankton. They return to a spawning ground after a year, but homing is not exact. The herring (like cod and other species), often occur in distinct **stocks**,

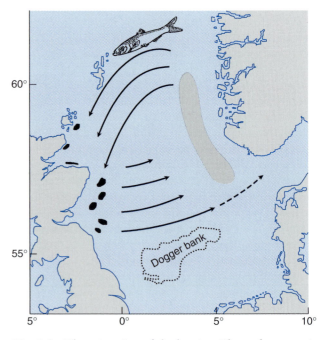

Fig. 5.8   The migration of the herring *Clupea harengus* in part of the North Sea, between spawning grounds (dark ovals) and feeding grounds (gray area). (After Harden Jones, 1968.)

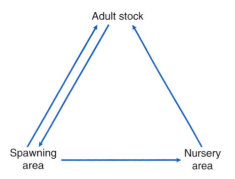

Fig. 5.7   A schematic pattern of migration. The two-way arrow between the spawning area and the adult stock area applies to cases of adults that migrates more than once to spawn. (After Harden Jones, 1968.)

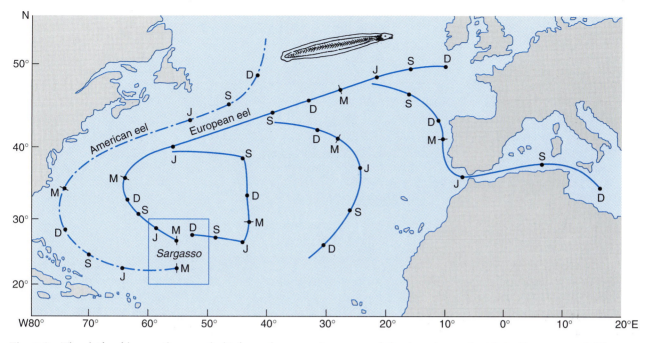

Fig. 5.9 The drift of larvae (leptocephali) from the spawning areas of the American eel and the European eel. The geographic positions of larvae hatched in March (M) in the two spawning area are plotted at quarterly intervals: June (J), September (S), December (D), and March (M). (From Harden Jones, 1968.)

or separated populations that maintain separate migratory routes.

The American eel *Anguilla rostrata* and the European eel *Anguilla anguilla* undergo one of the world's most spectacular and enigmatic migrations (Figure 5.9). Both species migrate from freshwater rivers, ponds, and estuaries in eastern American or European waters to spawn in partially overlapping areas in the Sargasso Sea. The trip to the spawning site is very poorly known, but adults are believed to swim in deep water. After spawning and zygote formation, the larvae drift north and eastward in the Gulf Stream. American eels then move from a range of latitudes, westward across the continental shelf, toward shallow-water eastern American rivers and estuaries, whereas European eels continue to drift across the Atlantic and are carried by currents into shallow waters there. The two species are distinctive in chromosome count, in various biochemical genetic markers, and even in the number of vertebrae (which is determined at birth). Other species of *Anguilla*, such as the Australian *A. reinhardti*, also undergo very long migrations that depend upon currents to carry larvae back to adult habitats.

It is not clear at all why these eel species migrate so far to their spawning grounds. Could these particular areas be optimal for spawning? If so, how were they "discovered" at such a great distance? One wonders whether the migration cycles of the North Atlantic species started millions of years ago, when the Atlantic was narrower. The movement of larvae must be passive, so the further passage of European eels must be determined genetically. A fascinating population has been discovered in Iceland, which appears to be intermediate in morphology and genetic markers. These individuals may be hybrids of *A. rostrata* and *A. anguilla*.

Green turtles (*Chelonia mydas*) are still something of an enigma to marine scientists. It is clear that they migrate hundreds to thousands of kilometers between feeding grounds and beaches where females lay eggs. Turtles that lay eggs at Tortuguero on the Atlantic coast of Costa Rica, for example, can be found at other times throughout the Caribbean, including Venezuela, Mexico, Cuba, and Puerto Rico. At least some tagged females return repeatedly to the same beach to lay eggs (Figure 5.10). It is unclear whether this behavior starts at birth, when hatchlings move

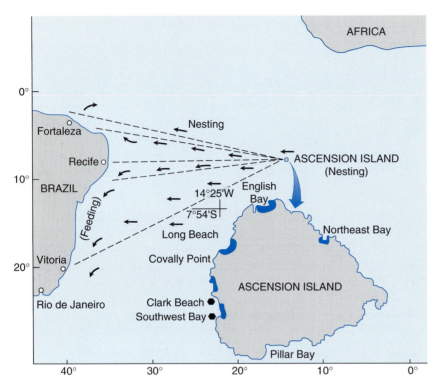

Fig. 5.10   The location of egg-laying and feeding grounds of the Ascension Island nesting populations of the migrating green turtle *Chelonia mydas*. Nesting sites are shown in green at lower right on the map of Ascension Island, disproportionately enlarged in relation to the mainland. (After Koch et al., 1969.)

to the sea, or whether females just follow others to good beaches to lay eggs. Right now, the evidence favors the hypothesis of imprinting at birth.[20] Female turtles from different beaches are genetically distinct, but turtles from the same beach are very similar. This can be shown by studies of the DNA in their mitochondria, which are inherited maternally. Populations from different beaches have probably been isolated from each other long enough to have diverged genetically.

* Anadromous fish species are more common in high latitudes, and catadromous species dominate low latitudes. This pattern is related to the availability of adult food.

Fish migration is one of the most perplexing phenomena in marine biology. Some species breed in fresh water but migrate, sometimes thousands of miles, to feed in salt water. Others do the reverse. Why should some species move in one direction, whereas others take the opposite tack?

Migration should make sense as a way to increase the growth rate of a fish population. After all, if migration had costliness as its only characteristic, ge-

netic variants that did not migrate would win out in the evolutionary race. And indeed, migration costs a lot. Such a journey increases the risks of starvation, predation, and simply becoming lost! It must be that migration produces increased benefits in terms of increased reproduction at a suitable spawning ground and increased growth in a good feeding area.

As it turns out, there is a systematic variation in the relative abundance of anadromous and catadromous fish species in different regions. Anadromous fishes are more common in high latitudes, whereas catadromous fishes seem to dominate tropical habitats. Despite some glaring exceptions—the Atlantic eels are catadromous but live as adults in the North Atlantic—overall, the pattern holds well.

Zoologist Mart R. Gross and colleagues[21] discovered an important key to success in migration. Food abundance is obviously important for fish survival and growth. As it turns out, high-latitude rivers and streams have lower overall productivity, and therefore less fish food, than the adjacent oceans. Thus anadromous fish spend their time feeding in the ocean

[20] See Meylan and others, 1990, in Further Reading, Migration.
[21] See Gross and others, 1988, in Further Reading, Migration.

but come into freshwater bodies briefly to spawn. By contrast, tropical oceans are very low in productivity, but tropical rivers and streams are very productive. This suggests that the difference in migration is not really paradoxical. Fishes are feeding where the food is. An interesting experiment confirms the hypothesis. The Arctic char is anadromous, but experimental additions of food to fresh waters decreased the migrations back to the sea.

Larval Dispersal: The Long and the Short Haul

Modes of Dispersal

* **Dispersal and range extension of marine populations are strongly controlled by currents that transport larvae and may have important ecological consequences.**

Planktonic dispersal does not guarantee the ability to survive a long trip across the ocean, or to colonize new habitats successfully. The overwhelming majority of planktonic larvae die in the water column. Some are swept to inappropriate habitats, others are eaten by predators, and still others starve. Most recruits, even of planktonic larvae, probably come from nearby. Figure 5.11 shows the general possibilities of larval transport. Depending upon the local current regime, larvae might be swept off shore, or they might even be brought onshore by wave trains (see following discussion on larval mortality). Longshore drift is common, and larvae usually are moved parallel to coastlines to some degree. This may permit rapid extension of a species's biogeographic range. Finally, in-

homogeneities in the coastline often create eddies, which may concentrate planktonic larvae (see further discussion in Chapter 7 on the movement of plankton by currents).

Although average dispersal distance per individual larva is probably short, the potential exists for long-distance dispersal in a significant fraction of the planktotrophic larval cohort, and this potential must increase with the planktonic larval life span and with the favorable nature of ecological opportunities at the distant site. Rudolph Scheltema discovered that many planktonic larvae of coastal invertebrate species live in the open sea (Figure 5.12), particularly in major transoceanic surface currents. He termed such larvae **teleplanic larvae** and found transoceanic similarities in species whose larvae were common in the open ocean. Another particularly interesting case is that of the larvae of the coral *Pocillopora damicornis*, which is widespread and dominant throughout Pacific coral reefs. Robert Richmond has shown that these larvae can live in the plankton and are competent to settle and metamorphose for periods greater than 100 days. This may explain the coral's broad geographic distribution. The larva's symbiotic zooxanthellae (see Chapter 15) photosynthesize and provide food, which fuels the larva's journey over long distances across the open sea.

If some ecological opening exists, arriving larvae may flourish, reproduce, and continue to extend the species's geographical range along a coastline. Such invasions have happened several times during the last century, and the sudden expansions may have been aided by commercial shipping. In the mid-nineteenth century, the shore periwinkle *Littorina littorea* arrived on the shores of Nova Scotia and spread southward. It had arrived at Cape Cod, Massachusetts, by the turn of the century and has now reached the Middle Atlantic states. It is unlikely that the species will spread much further southward, given its thermal and geographic ranges in Europe. Many invasive species also have spread rapidly along a new coastline, once having arrived.

* **Despite the potential for dispersal, planktonic larvae often come to settle quite near their origin, owing to cyclonic currents.**

Given the long lives of many planktotrophic larvae, one expects that dispersal should occur over great distances. A larva with a 30-day life span and a steady

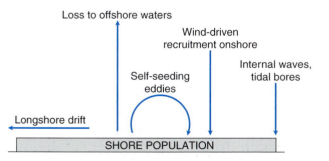

Fig. 5.11 General types of movement of larvae of benthic invertebrates with respect to the coastline.

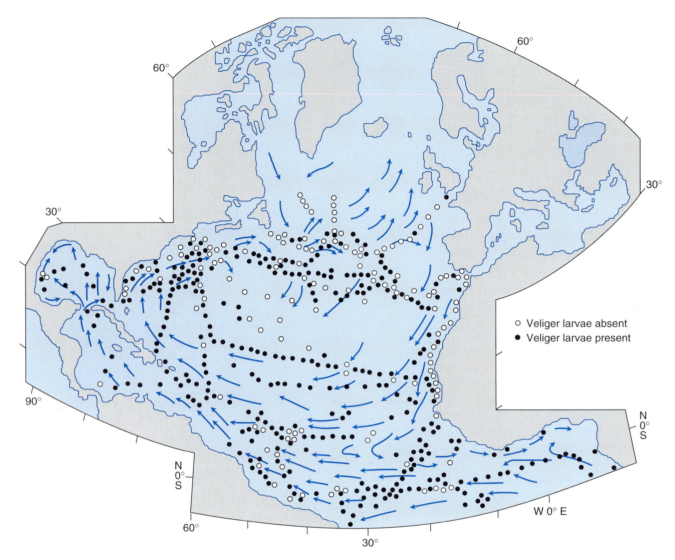

Fig. 5.12 Geographic distribution of teleplanic gastropod veliger planktonic larvae, based on samplings taken in the tropical and North Atlantic Ocean. (After Scheltema, 1971.)

and straight current of 2 cm s^{-2} should disperse over 50 km. This would suggest extensive mixing of marine species with planktotrophic larvae. The assumption of straight and steady currents is the problem with such a prediction. Not all currents move steadily or in one direction. Shapes of basins and large obstructions to flow, such as islands, often create eddies that cause larvae to travel in cyclonic flow patterns, sometimes to their parent populations.

Surprisingly reduced dispersal of planktonic coral larvae was demonstrated for corals by Paul Sammarco (1988), who monitored larval settling in the

vicinity of an isolated patch reef in the Great Barrier Reef of Australia. Settling plates were placed at various distances from the patch reef, and a detailed study of local water currents allowed a prediction of sites where larvae might be concentrated. Settling occurred mainly in the close vicinity of the patch reef (Figure 5.13a), suggesting that most larvae come from within the patch system itself. Settlement was greatest in a vortex of currents, where water was trapped and recirculated for a while before being swept to sea.

A similar story emerges from the dispersal and cur-

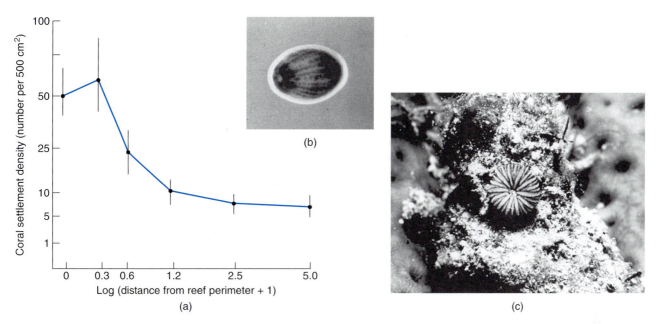

Fig. 5.13 (a) Diagram illustrating settlement onto settlement plates of planktonic coral larvae around a patch reef on the Great Barrier Reef of Australia, as estimated by counts of newly settled coral colonies (distance in meters). Note that settlement was highest near the patch reef, indicating that most settlers came from within the patch itself. (After Sammarco, 1991). (b) A coral planula larva. (c) A newly settled coral, only a few millimeters across. (Photographs by Robert Richmond.)

rent patterns associated with a population of the European lobster *Nephrops norvegicus* in the western Irish Sea. Alan Hill and colleagues[22] discovered a cyclonic flow in spring and summer. This flow lies above a large patch of mud, which has the main population of the lobster in this region. In spring, newly hatched larvae spend approximately 50 days in the water column, and successful recruitment depends upon finding the mud patch. Larval distributions show a strong correspondence to a stratified water body enclosed by the small gyre system. This ensures that the larvae will settle in the vicinity of the mud patch (Figure 5.14).

✳ **Marine invertebrate offspring may be (1) brooded or released as small adults, (2) dispersed only to a small degree by means of relatively short-lived, yolk-dependent lecithotrophic larvae, or (3) dispersed great distances by means of longer-lived plankton-feeding planktotrophic larvae.**

The hallmark of larval production and dispersal is a complex life cycle. An adult stage gives rise to a dispersing larva, which moves to a new site and com-

pletes the cycle by establishing itself and eventually growing to reproductive maturity. Figure 5.15 shows an example of such a cycle. Although there is a continuum of dispersal distances, several qualitatively different types of release and spread of larvae result in modes of dispersal distances. **Direct release** of individuals next to adults is the shortest type of dispersal. Many species are **viviparous**; that is, they bear live embryos and then release juveniles as crawling miniature adults. Such young may crawl directly on the mud or on the rocks. The Atlantic periwinkle *Littorina saxatilis* broods fertilized eggs in a modified oviduct, and fully shelled young snails are later released. Many Antarctic species of feather stars (comatulid crinoids) develop their eggs in brood chambers located near the gonads in the arms at the pinnule bases. Other species are **oviparous**, producers of young that hatch from egg cases. Several species of the drilling genus *Nucella* lay egg capsules that they attach to rocks. Embryos develop in the eggs and hatch as fully shelled juveniles. Some of the eggs

[22] See Hill and others, 1996, in Further Reading, Larval Dispersal and Recruitment.

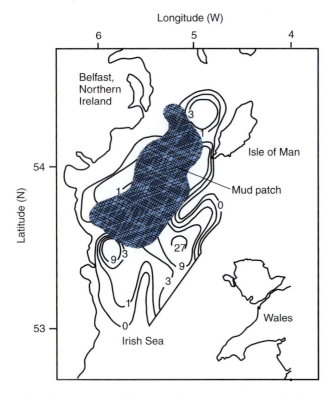

Fig. 5.14 The distribution of stage III larvae of the lobster *Nephrops norvegicus* in the western Irish Sea. Larvae are retained in this area by a cyclonic flow pattern and thus find their way back to a large patch of mud (hatched area) suitable for adult survival. (After Hill et al., 1996.)

may serve as nurse eggs, to be consumed by the hatchlings. It is extremely common for meiofauna to have direct release, since any sort of current-driven dispersal will likely take them to inappropriate habitats. The small size of meiofauna also restricts them to releasing very few young, which puts a premium on maximizing survival. Dispersal entails great risks (see following discussion), so direct release is favored in these forms.

Most invertebrates have larvae that swim for varying amounts of time before settlement and metamorphosis. **Lecithotrophic larvae** (Figures 5.16, 5.17) are swimming larvae that depend on nourishment from the yolk provided in a relatively large egg; the larvae have no feeding or digestive structures. Capable of limited swimming, they spend a few hours to a day or so moving either along the bottom or up in the water column. This mode of larval development cannot permit dispersal for much more than 10–100 m. Lecithotrophic larval development occurs in species of many phyla, including mollusks, annelids, and ectoprocts.

Planktotrophic larvae feed while they are in the plankton. They usually have specialized larval feeding structures and digestive systems. They feed on planktonic bacteria, algae, and smaller zooplankton and usually drift for one to several weeks in the water column (Figures 5.17, 5.18). Because of the time

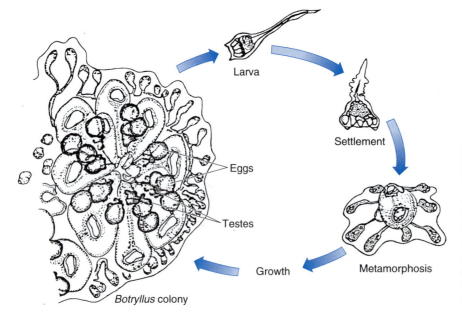

Fig. 5.15 Example of a complex life cycle of a marine invertebrate with a dispersing phase. The colonial sea squirt *Botryllus schlosseri* is a colonial hermaphroditic species. Sperm move through the water from one colony and fertilize a nearby neighbor. Larvae emerge and move only a couple of meters to settle, metamorphose, and establish a new colony.

Fig. 5.16 An example of a lecithotrophic larva: a "tadpole" larva of the colonial ascidian *Botryllus schlosseri*. This larva lives for only a few hours, is a poor swimmer, and moves only a few meters from the parent colony. (Courtesy of Richard Grosberg.)

they spend in the open waters, they have a great capacity for long-distance dispersal. In most groups, development proceeds through a number of planktonic stages. Until it reaches a terminal stage of development, the larva cannot find the adult habitat and is incapable of metamorphosing into the benthic adult stage. This requires a minimum time to be spent in the plankton. If a suitable adult substratum is not found, the larva may be capable of delaying metamorphosis.

Lecithotrophic larvae are often covered with ciliary bands, which are used for locomotion. However, planktotrophic larvae use ciliary bands or tufts to move and to feed upon the planktonic food (Figure 5.19). The small size of the cilia and the low velocities of ciliary movement cause planktonic larval feeding to operate at low Reynolds number (see Chapter 6), meaning that viscosity dominates inertial forces in fluid behavior. The cilia must beat to propel water and food particles. Particles do not move through water like a sinking stone; rather, the domination of the role of viscosity makes them stop when water movement stops. Although there is a great deal of variation in morphology, a surprising number of taxonomic groups of organisms have ciliated bands (Figure 5.19). The cilia beat and create a current across the band. To capture a particle, it appears that a cilium suddenly reverses its rowing motion toward downstream. The reverse beat captures the particle and traps it upstream of the ciliary band.

The blue mussel *Mytilus edulis*, which lives on intertidal and shallow subtidal rocks and cobbles, is a

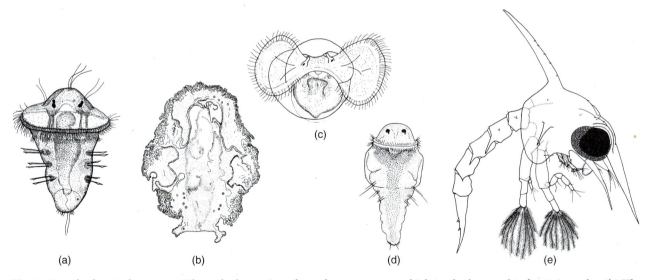

(a)   (b)   (c)   (d)   (e)

Fig. 5.17 Planktonic larvae: (a) The polychaete *Spirobranchus giganteus*, which is planktotrophic for 1–2 weeks. (b) The sea cucumber *Auricularia nudibranchiata*, which is planktotrophic and teleplanic, capable of many months in a plankton. (c) The gastropod *Ilyanassa obsoleta*, which is planktotrophic for 12 days to 6 weeks. (d) The polychaete *Spirorbis spirorbis*, which is lecithotrophic for less than one day. (e) A planktotrophic zoea-stage larva of the crab *Portunus sayi*. Adults live on pelagic *Sargassum* seaweed. (Drawings a–d by R. Scheltema. Drawing e by I. P. Williams.)

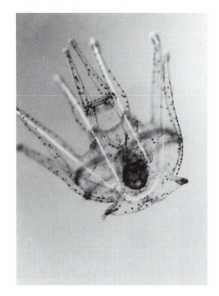

Fig. 5.18 Planktotrophic larva of the red sea urchin *Strongylocentrotus franciscanus*. Dark pigment spots are concentrated near the ciliary bands used for feeding and swimming. (Courtesy of Richard Strathman.)

✳ **Gamete production and larval life must often be timed precisely to allow settlement and promote dispersal, to avoid being swept from inappropriate habitats, and to counter predation.**

The commitment of either gametes or larvae to the water column entails great dangers of extensive and maybe total mortality. We discussed earlier the problems associated with ensuring that fertilization is maximized. In estuaries, larvae in the surface low-salinity layer may be lost to sea (see later). Predation on gametes and larvae also can strongly depress reproductive success.

These dangers require precise timing in many cases to avoid gamete and larval death. For example, animals living or breeding in the highest part of the intertidal zone must have precise larval release schedules because the animals may be inundated only once every 2 weeks during spring high tides. The Atlantic marsh snail *Melampus bidentatus* lives in the highest part of the tide zone and breathes air, but it lays eggs that develop into swimming larvae. These larvae are launched at a spring high tide and return to settle during the next spring high tide. The Pacific American grunion *Leuresthes tenuis* comes onto sand flats to spawn at spring high tide at night. Females dig a pit in the sand and lay eggs at the highest level of the tide, while males curl around and deposit sperm. For about 2 weeks, the high tides fail to reach this level, but the next spring high tide washes out the eggs, which hatch into swimming larvae.

Tidal environments are most challenging because gametes and larvae may be trapped in the surface layer during the day when visual predators are active. One might therefore expect strong selection for larvae to be moving in the surface waters at night.

good example of a typical planktotrophic larva. Sexes are separate, and the animals release eggs and sperm into the water. Within 9 hours after fertilization, the larvae are completely ciliated and are strong swimmers. Within 5–7 days, fully developed feeding veliger larvae develop. Normally, larval life may be 4–5 weeks. By contrast, the Atlantic mud snail *Ilyanassa obsoleta* produces fertilized eggs laid in cases that are attached to rocks. Larvae emerge from the cases and swim away. Planktotrophic larvae may also be released directly from the mother, as in the acorn barnacles.

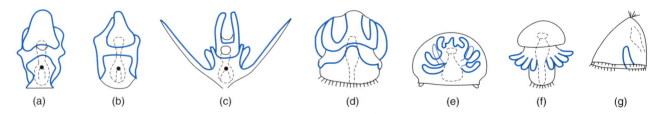

Fig. 5.19 Many different groups have evolved ciliary bands (green) for feeding on plankton: (a) sea star, (b) sea cucumber, (c) brittle star, (d) hemichordate, (e) inarticulate brachiopod, (f) phoronid worm and (g) bryozoan. (Courtesy of Richard Strathman.)

This has been found in the coral reef-flat damselfish *Pomacentrus flavicauda*, which lays eggs at dawn; its embryos hatch as larvae on the fourth sunset. Furthermore, egg production is timed so that hatching occurs at the time of the month when sunset coincides with ebb tide, which at once lowers predation and sweeps the larvae offshore.[23]

✳ Variation in egg size is considerable among marine species, and this may relate to different consequences in terms of mortality.

Although variation in age of reproduction is considerable, equally noticeable is the great variation in egg size in marine animals, sometimes between closely related species. One expects larger eggs to have more yolk nutrients, but this is only broadly true, and there are exceptions. In large measure the variation remains unexplained, but there are some general patterns. As mentioned earlier, lecithotrophic larvae develop from eggs that are much larger than planktotrophic eggs. Eggs of directly released juveniles are largest of all. As a general rule, planktotrophic eggs produced per female are also much more numerous than lecithotrophic eggs, which are in turn more numerous than the eggs of directly releasing species. This suggests a strategy, which in turn may determine egg size. Planktotrophic larvae swim in the water far longer and probably have much higher mortality rates than do lecithotrophic larvae, which are in the water for only a few hours. In turn, directly releasing species deposit their young in appropriate habitats (namely, those of the adults), thus further minimizing mortality. It seems likely then that the female, by producing a large number of planktotrophic eggs, "hedges her bet," because far fewer young will survive, relative to lecithotrophic eggs. Some have suggested that the intermediate strategy of producing an intermediate size egg with intermediate numbers may offer the worst of both worlds, producing high mortality with relatively low numbers of eggs and young. Thus, we expect to observe either the low-number/quality-care approach or the high-number/leave-them-alone approach.

Larval Settlement, Metamorphosis, and Early Juvenile Survival

✳ Larval settlement and metamorphosis involves active choice, aided by chemical and physical cues of the substratum.

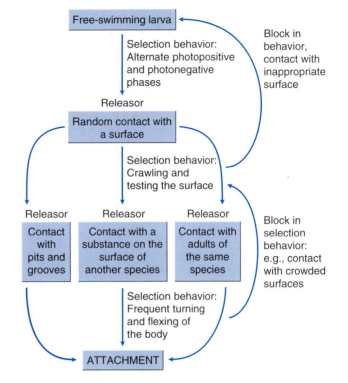

Fig. 5.20 Stages in the selection of a suitable substratum by planktonic larvae. (Modified after Newell, 1979.)

Planktonic larvae (Plates I and II) reach a stage at which they must locate the adult habitat, where they will settle and attach before metamorphosis. Although the vast majority of larvae die before reaching a potential adult habitat, the road ahead is still dangerous for the survivors. Natural selection and evolution favor any larval features that enhance location and settlement on the proper substratum. Figure 5.20 provides a general outline of the stages of larval selection.

In the plankton, larvae pass through one or more stages of photopositive and photonegative behavior. These permit the larvae to remain near the sea surface to feed, and then to drop to the bottom and settle on the proper habitat. Larvae of intertidal animals are photopositive during their entire larval life, so that they may feed in the surface waters and settle in the intertidal zone. Richard Grosberg,[24] who exam-

[23] See Doherty, 1983, in Further Reading, Sex, Fertilization, and Life Cycles.
[24] See Grosberg, 1987, in Further Reading, Larval Dispersal and Recruitment.

ined the water column distribution of intertidal barnacle species in California reported that the highest intertidal species was to be found at the surface during larval life; species living in the mid-intertidal, however, were found a meter or two below the surface. Intertidal larvae are apparently capable of subtle adaptations to maintain the proper tidal height at the time of settlement. Subtidal species, by contrast, are first positively phototactic, but then switch to negative phototaxis just before the time of settling. Larvae are also capable of responding to pressure differences that might enable them to select a specific depth during the period of feeding, dispersal, and settlement.

Planktonic larvae must maintain their proper depth in the water column during the dispersal phase, either to feed on phytoplankton that are abundant near the surface or simply to avoid sinking before development is completed. Most larvae have slightly negative buoyancy, but simple turbulence retards sinking. Larvae can be simple and ciliated, but planktotrophic larvae usually have a swimming organ (e.g., the velum of gastropod larvae, which is lost at the time of metamorphosis). Young postlarval mussels (e.g., *Mytilus edulis*) secrete monofilamental byssal threads more than a hundred times longer than the larval shell. The filaments exert viscous drag force sufficient to reduce greatly the sinking rate, thereby enhancing dispersal.

Amazingly enough, deep-sea hot-vent environments (see Chapter 16) have species with planktotrophic larvae. As judged from submarines, hot vents are scattered throughout the deep-sea floor and are ephemeral, because the hot nutrient-rich sources give out after a time. Dispersal might thus be expected. The protoconch (larval shell) size of hot-vent bivalve mollusks is of the typical small size for planktotrophic larvae, and isotope data from the larval shell (isotope ratios of oxygen in the shell can be used as a sort of thermometer) show that the larvae spend a time in water that is warm—like surface oceanic water and unlike the deep sea. Vent bivalve larvae must make the seemingly impossible trip to the surface waters, and then some manage to locate an incredibly rare appropriate deep-bottom environment. By contrast, archaeogastropods (a group of snails) near the vents have large yolky eggs and probably have larvae with quite limited dispersal. It may be that these eggs hatch as swimming but nonfeeding

Fig. 5.21 Preferential recruitment of barnacles into cracks, on a rocky shore near Nahant, Massachusetts. It is not clear whether this pattern results from preferential larval settlement, from differential postsettlement mortality, or from a combination of these processes.

larvae, with strong dispersal abilities, relative to their nonvent relatives.

After touching the bottom, larvae must be able to determine whether the substratum is suitable for adult existence. Larvae use chemical cues and mechanical cues to detect suitable settling sites. Almost all larvae prefer surfaces coated with bacteria. Sterilized sand or rock usually inhibits larval settling. Larvae of some sand-dwelling polychaetes can select sand grains of the appropriate size, whereas rock-dwelling barnacles have larvae that move preferentially to surfaces with pits and grooves, which provide a secure attachment against wave action, predators, and competing barnacles that might overgrow or undercut them. Although barnacles are often found in cracks, it is often not clear whether this is the result of preferential settlement or differential survival after settlement (Figure 5.21).

In **gregarious settling**, larvae settle on adults of their own species. This usually requires direct contact of larvae with adults, because the chemical cue is a relatively insoluble molecule, such as arthropodin in barnacles. Zimmer-Faust and Tamburri[25] discovered that the cue for settling of oyster larvae is a soluble peptide (which is a string of amino acids). Such a waterborne settling cue exists in effective concen-

[25] See Zimmer-Faust and Tamburri, 1994, in Further Reading, Larval Dispersal and Recruitment.

trations only within a few millimeters of the bottom, but larvae sinking within the bottom boundary layer can detect this cue and settle. The greatest concentration of such a settling cue would likely emanate from the water coming from adult oysters, which would concentrate settling of oyster larvae on adults of their own species. Larvae are often entrained in currents too strong to swim against, but they frequently drop down toward the substratum and make arclike swimming patterns. On encountering, the cue they can slow down, settle, and metamorphose.

Gregarious settling allows a larva to settle in a site where adults have already settled and survived. There is the disadvantage, however, of being eaten by adults of the same species. There also is the problem of how, in a gregarious species, a "pioneer" can ever manage to colonize a new site. Toonen and Pawlik[26] discovered that the colonial tube-dwelling polychaete worm *Hydroides dianthus* has two distinct types of planktonic larva: one seeks adults of its own species and the other is specifically a "pioneer" larva that seeks new, bare, hard substratum. The cue for gregarious settlement is a waterborne compound emanating from benthic adults.[27]

Planktonic larvae may also use the chemical characteristics of other benthic species as cues. For example, adults of the hydroid family Proboscidactylidae live on the tubes of members of the polychaete annelid family Sabellidae. At settling time, the larvae can detect and settle on the tentacles of the worm. After settling they move down the tentacles and live as adults on the tube. The mechanism probably involves **chemical attraction**. Some bryozoan larvae are attracted to seaweeds, and organic substances can be extracted from some seaweeds that will attract larvae onto another nonliving surface. In other species, **mechanical attraction** is involved. Planktotrophic larvae of the mussel *Mytilus edulis* are attracted to filamentous red algae, but the larvae will also settle on fibrous rope that has approximately the same texture. This attraction is used to great advantage in mussel mariculture.

After settling, larvae may move a short distance, no more than a few centimeters, to a better site. This is be very important for larvae of sessile species, whose movements at this point commit them to life at one location. Newly settled barnacle larvae can move to a small degree, locate optimal microsites, and space themselves away from other barnacles. The

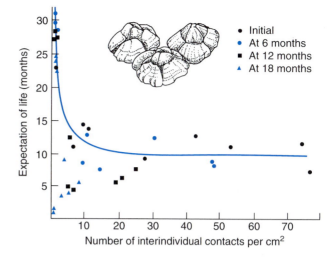

Fig. 5.22 Relation between crowding, as measured by number of interindividual contacts per square centimeter, and future expectation of life of the intertidal barnacle *Semibalanus balanoides*. Expectation of life is shown for barnacles that were first observed at first settling, 6, 12, or 18 months of age. (After Deevey, 1947.)

tubeworm *Spirorbis borealis* settles randomly on the seaweed *Fucus serratus*, but individuals then crawl and space themselves evenly. Avoidance of crowding reduces overgrowth and competition for food. In barnacles, among which settlement is often very dense, the reduction of crowding that follows a period of mortality after settlement greatly enhances the subsequent expectation of life (Figure 5.22).

When they reach the substratum, marine planktonic larvae encounter a complex physical and chemical environment. As we have discussed, many chemical cues can be used by larvae to select an appropriate site upon which to settle. However, there are probably a number of chemical cues that may lead to avoidance of a local site. Unfortunately, we know very little about these cues, even though they may be a major means of local site selection. Sara Ann Woodin and colleagues[28] demonstrated that noxious chemicals released by benthic animals may reside in the sediment pore waters and can inhibit

[26] See Toonan and Pawlik, 1994, in Further Reading, Larval Dispersal and Recruitment.
[27] See Toonen and Pawlik, 1996, in Further Reading, Larval Dispersal and Recruitment.
[28] See Woodin and others, 1993, in Further Reading, Larval Dispersal and Recruitment.

settlement by larvae. In soft-sediment intertidal flats of Washington, the infaunal terebellid polychaete *Thelepus crispus* is locally abundant and harbors toxic bromine-containing aromatic compounds. These compounds are found in the sediment pore waters adjacent to the worm. Woodin and colleagues demonstrated that such sediments are actively rejected by settling larvae of the nereid polychaete *Nereis vexillosa*, a common resident of sandy shores in the area. This result suggests that negative chemical interactions may also be important in determining the settling and metamorphosis of planktonic larvae.

Finally, a larva commits itself and metamorphoses into an adult stage. (Recall that metamorphosis is the process whereby a larva changes dramatically into the adult form.) In the case of barnacles, for example, the football-shaped cypris develops adult appendages, and a basal plate is laid down. The energetic cost of metamorphosis often is so severe that the animal must feed immediately afterward. Some sea star species die if they fail to find a prey item within a couple of days after metamorphosis. If larvae of the plaice *Pleuronectes platessa* are starved of zooplankton for more than 8 days, a point of no return is reached and they will not have the energy to feed subsequently. Older larvae can survive at least 25 days before the point of no return is reached. It may be that early larval development in these fish is also very costly.

* Planktonic larvae suffer extensive mortality from predation, transport away from appropriate bottom substrata, and food shortage.

The vast majority of planktonic larvae never make it to their destination. One gets an inkling of this by counting the number of eggs produced by most female invertebrates that spawn and produce planktotrophic larvae. In the small Atlantic American bivalve mollusk *Mulinia lateralis*, females can mature at lengths of less than 1 cm and produce hundreds of thousands of eggs. Although larvae of this species can settle in spectacular densities locally, one cannot account for the total larval output in terms of successful settlement. This is the general picture for most marine organisms. Where do the larvae go?

The problem is easiest to visualize for organisms that have restricted habitats. Larvae of rocky-shore invertebrates are released in currents and may easily

be swept to inappropriate habitats. Along the southern New England coast, a longshore current would tend to sweep larvae from the rocky coasts of Rhode Island to the sandy south shore of Long Island, New York, and most larvae of the rocky-shore species would be doomed. Species with nonplanktonic larvae dominate the Galapagos Islands, off Ecuador, where currents carry surface waters offshore for a large portion of the year. As a result, species with planktonic larvae are relatively rare in the Galapagos.

Currents may cause net movement toward the shore. General eastward currents and winds tend to keep eastern Atlantic larvae pressed against the shorelines of the Atlantic coast of Europe. Off most open-ocean coasts, one can observe surface slicks that contain jetsam that seems to move toward shore. Surface floats off the California coast are often carried 1–2 km shoreward in 2–3 hours. Planktonic invertebrate and fish larvae are 6–40 times more concentrated in these slicks than in the surface waters between the slicks. The slicks are formed by the interaction with the sea surface of tidally driven internal waves (waves within the water column). Planktonic larvae may therefore be "trapped" against the shoreline. In recent years, episodic recruitment has been discovered to be the rule for settling megalopa larvae of the blue crab *Callinectes sapidus*, both on the east and Gulf coasts of North America. In some cases, these recruitment events last only a day and may be simultaneous over several hundred kilometers of coastline. Recruitment in the mid-Atlantic states region appears to be correlated with onshore transport caused by wind stress and by spring flood tides, although other factors may also be important.

Many species have apparently taken advantage of such current effects. Many estuarine species spawn within the estuary, but larvae and juveniles may spend some period of time in coastal waters (Figure 5.23), perhaps to avoid predation within the estuaries (see the following). Because they are entirely dependent upon currents for transport, they must take advantage of the various wind- and tide-driven transport sources discussed earlier to return to the estuary from adjacent coastal waters.

As was discussed in Chapter 2, surface water flows seaward in moderately stratified estuaries, whereas the more saline bottom layer moves landward. If estuarine planktonic larvae held their position near the surface, they would be carried out to sea. Although we know far too little as yet, it appears that many

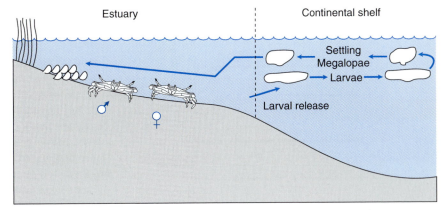

Fig. 5.23 Reproduction of many invertebrates and fishes, such as the blue crab *Callinectes sapidus*, occurs within the estuary. Larvae and juveniles of many species, however, move into coastal waters, and later move back into the estuary to spend their adult lives. (Courtesy of Steven Morgan.)

estuarine species move actively to avoid the seaward surface flow. Larvae of the eastern oyster (*Crassostrea virginica*) studied in the James River estuary, Virginia, differ in depth distribution from inert particles and are found in surface waters more frequently at the time of flood tide, which would tend to move them upstream. Larvae of the mud crab *Rhithropanopeus harrisii* gradually concentrate up the estuary during the larval dispersal season. This suggests that they are actively moving up and down the water column with the tide, to avoid being swept from the estuary to the coastal ocean. Mud crab larvae appear to spend more time at the bottom at the time of low tide, moving higher in the water column during the flood. Larvae (especially the later developmental stages) that are retained within estuaries tend to spend more time near the bottom, which places them in the landward moving layer of the estuarine water column. As mentioned before, other species have larvae that start within the estuary, but stay in the surface waters and are dispersed out to the coastal zone. They may return to the estuaries after a larval life in coastal waters, and larval recruitment depends very strongly on often quite irregular onshore winds.

The evidence suggests that there are two alternative strategies for planktonic larval habits of estuarine species (Figure 5.24). Some species have larvae with adaptations for **estuarine retention,** rising on the flood and moving to the bottom during ebb tide. A consequence may be increased predation by large populations of smaller planktivorous fishes in the estuary. Alternatively, larvae may rise on the ebb to move to coastal waters and then depend upon wind-driven currents to recruit back to the estuary. While such larvae would be living in waters of lower pre-

dation, the strategy is plagued by a higher probability that the larvae will never return to the estuary. The evolution of these alternative strategies is complicated by the presence of populations of many estuarine species along quieter coasts of the open sea. It is therefore not entirely advantageous to be restricted to recruiting within the estuary.

The water column is filled with predators that cause considerable planktonic larval mortality. In temperate waters, planktivorous fishes and ctenophores are important grazers in the spring and summer. Ctenophores are especially abundant in the inner shelf waters and are major predators of zooplankton. Predation is a major problem for larvae that are adapted to be retained in estuaries. Steven Morgan[29] has shown that three species of crustacea retained in a North Carolina estuary have evolved pronounced spines. The mud crab *Rhithropanopeus harrisii*, for example, has movable spines, which are erected when fishes come close (Figure 5.25). Other crustacean species whose larvae spend time in coastal waters experience less predation and have less completely developed spination.

Although primary production is often sufficient to feed larval populations, shortages or failures of phytoplankton can devastate a planktonic feeding larval population that might be produced only once a year for a brief period. Harold Barnes encountered a bad phytoplankton year in British waters in 1951 (Figure 5.26). Early larval stages of barnacles could be found at first, but these soon disappeared because of starvation. It would be best for larvae if their release

[29] See Morgan, 1990, in Further Reading, Larval Dispersal and Recruitment.

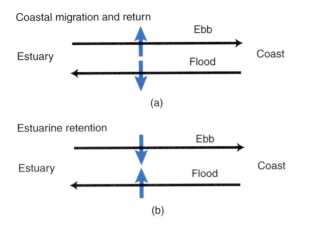

Coastal migration and return

Estuary   Coast

(a)

Estuarine retention

Estuary   Coast

(b)

Fig. 5.24 Two alternative strategies for larval recruitment in estuaries: (a) movement of larvae to offshore waters with a subsequent return and (b) retention of larvae within estuaries.

could be coupled to the spring phytoplankton bloom. Such coupling has been found in sea urchins and mussels living in the St. Lawrence estuary of eastern Canada, which spawn in response to high concentrations of phytoplankton. The response is similar when urchins and mussels are exposed to only the

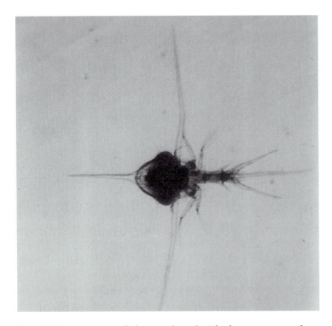

Fig. 5.25 Larvae of the mud crab *Rhithropanopeus harrisii* have erectible spines, which make them very difficult for fishes to attack. This is a necessary defense for larvae that are retained within estuaries, where predation by fishes is intense. (Courtesy of Steven Morgan.)

filtered water in which the phytoplankton lived. The response is also advantageous because zooplankton predators tend to be rare at the peak of the phytoplankton bloom, so larvae can eat and not be eaten.

Sometimes the presence of even copious amounts of phytoplankton does no good, because they are either indigestible or inedible. Such has been the case in recent years in coastal bays of New York, New Jersey, and Rhode Island. So-called **brown tides,** or dense blooms of small-celled phytoplankton, have developed, but these cannot be digested very efficiently by larvae or adults of the bay scallop *Argopecten irradians.* Some other as yet unknown factor also causes these phytoplankton to harm the scallops. As a result, the scallop fishery has nearly collapsed in the Peconic Bay (eastern Long Island, New York), an area formerly rich in these scallops.

There may be cases in which larvae do not suffer especially from food shortages. Echinoderm larvae may survive well despite an absence of phytoplankton, and many bivalve mollusk larvae are also resistant to the effects of food shortages. It is possible that these groups are able to survive on dissolved organic matter in seawater. (The interested student should consult Olson and Olson, 1989, for more on this subject.)

✳ **Larval recruitment is the combined result of larval habitat selection and early mortality.**

In March, the rocks in Long Island Sound waters begin to be covered by small red specks—the cypris stages of the barnacle *Semibalanus balanoides*—which soon metamorphose into tiny white-shelled barnacles. **Recruitment** represents the residue of those larvae that have (1) dispersed, (2) settled at the adult site, (3) made some final movements toward the adult habitat, (4) metamorphosed successfully, and (5) survived to be detected by the observer. Recruitment, therefore, could obscure much of the life cycle of the dispersing larva. When observers arrive on the scene, they usually know only how a combination of larval selection and mortality has resulted in the final distribution of recruits.

If we find a set of recruits living under the most favorable adult conditions, it is potentially fallacious to conclude that the larvae are good at finding the proper adult habitats. The final match between recruit and optimal environment may result from extensive mortality just after settlement in suboptimal

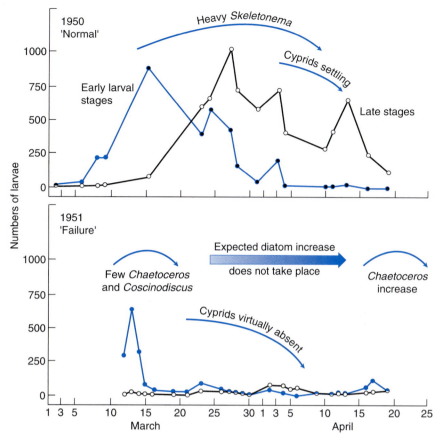

Fig. 5.26 Successful settlement and failure in the barnacle, *Semibalanus balanoides*, in good and bad phytoplankton seasons. (After Barnes, 1956.)

sites. Consider, for example, the lecithotrophic larvae of the bryozoan *Bugula neritinea* on sea grass blades in the Gulf of Mexico. Colonies were found mainly at the tips of the blades, where suspended food was more plentiful. By transplanting young colonies away from the blade tips, Michael Keough[30] showed that survival was reduced. Settlement occurred mostly on the blade tips, but some larvae settled in suboptimal sites and did poorly. Recruitment in this case was mainly the result of selective larval settlement, but mortality enhanced the concentration of recruits at the blade tips. There is a similar complexity in recruitment patterns of the Atlantic oyster *Crassostrea virginica*. Resident adults, such as sea squirts, tend to inhibit settling, but larvae often avoid the residents and settle on adjacent open space. Even though sea squirts eat oyster larvae, and other species preclude settling, there may be so much space that overall recruitment is not greatly affected.

Biogeography and Function of Larval Dispersal

❋ **The geographic range of a species with planktonic dispersal is greater than the range for species without planktonic larvae.**

Planktonic dispersal gives some species the opportunity to invade distant shores merely by being passengers on the ocean's transport system of currents. We rarely see this process in action, but several species have invaded new coasts in the past 100 years, and their spread has been documented. We mentioned earlier in this connection the shore periwinkle *Littorina littorea*, which invaded Nova Scotia waters

---

[30] See Keough, 1989, in Further Reading, Larval Dispersal and Recruitment.

in the late 1800s and has now spread as far south as New Jersey.

It stands to reason that current systems that bring water toward shore should also contribute toward increased larval recruitment. We have noted, for example, that the Galapagos Islands stand in a current system that is not conducive to successful larval transport. Most coasts are likely to vary much more, owing to variations in current transport. On the west coast of North America, for example, areas of upwelling are often localized, often on a scale of tens of kilometers. When an upwelling system is close to shore, larvae may be transported mostly out to sea in surface waters, which would strongly reduce recruitment. In El Niño years on these coasts, one might expect a great increase in recruitment, since the influx of warm surface water will trap larvae at the California coast. and even transport coastal water northward. During El Niño years, the usual isolation that is enforced by current systems at Point Conception is broken down and coastal water is transported northward, bringing larvae to the coast of northern California, often of subtropical forms. Connolly and Roughgarden[31] found that barnacle recruitment increased strongly over a broad region in central and northern California from 1996 to 1997, a probable result of El Niño. Thus infrequent changes in climate provide opportunities for breakthroughs in dispersal between previously isolated biogeographic zones.

While there are some exceptions, invertebrate species with planktonic larvae appear to have far greater geographic ranges than those species with direct release. Figure 5.27a shows a series of biogeographic zones on the Atlantic coast of North and South America, identified for the distinctness of their respective benthic species. Note that species with non-planktonic dispersal tend to occupy two to three zones, whereas those with planktonic larvae occupy four to five zones (Figure 5.27b). Although planktonic dispersal increases the geographic range, teleplanic larvae with dispersal times of many weeks or months do not seem to have any broader range than those whose larvae spend only 2–6 weeks in the plankton.

[31] See Connolly and Roughgarden, 1999, in Further Reading, Larval Dispersal and Recruitment.

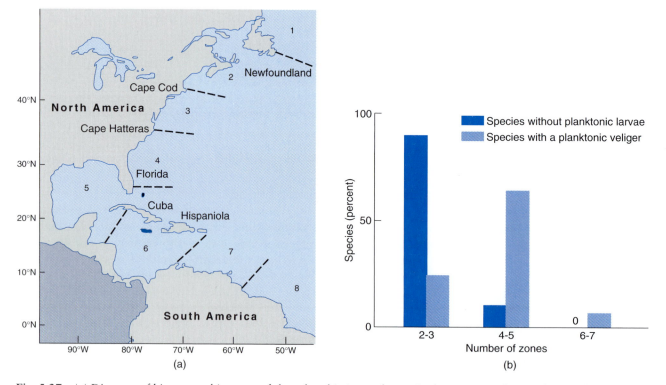

Fig. 5.27 (a) Diagram of biogeographic zones of shore benthic invertebrates in the western Atlantic. (b) Number of zones occupied by invertebrate benthic species with and without planktonic larvae. (After Scheltema, 1989.)

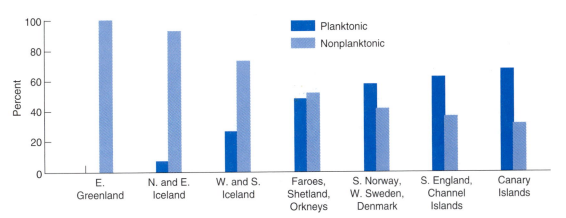

Fig. 5.28   Latitudinal variation in the abundance of prosobranch gastropod species with planktonic larvae. (After Thorson, 1950.)

✳ **Why disperse? Dispersal of planktonic larvae ensures that local habitat destruction will not lead to extinction.**

Given the dangers of planktonic dispersal, one might wonder why this mode evolved at all. Richard Strathman (1987) has pointed out that many evolutionary lines have lost the ability to produce feeding planktonic larvae simply because it is fairly easy, in the evolutionary sense, to lose a swimming or plankton-feeding structure, but hard to reacquire it. However, the great majority of marine invertebrates have planktonic feeding larvae. Planktotrophic dispersal is rare only in high latitudes (Figure 5.28), where the phytoplankton season is very short and the water temperature is low. This may stem from the danger of failing to synchronize reproduction with the short phytoplankton production season, combined with the very long development times due to low temperature, which would increase the danger of predation and of being swept to sea.

Avoidance of crowding is one significant benefit of dispersal. Many marine habitats are severely space-limited, and dispersal to an open habitat would ensure population increase. The rate of destruction of local marine habitats may also make dispersal beneficial. All marine habitats experience major alterations. Coral reefs, which are often represented to students as benign environments, are continually disturbed by major storms, and some are affected by temperature fluctuations. No rocky shore long escapes the force of waves, ice, hot and dry weather, or an influx of predators. If an organism's offspring

could not disperse, the chances are that its genes would become extinct. Dispersal therefore ensures that some of the progeny will escape a catastrophe. It is of interest in this regard that paleontologists have discovered that mollusk species characterized by long-distance dispersal have lower extinction rates. One should not forget, however, that many species with reduced dispersal manage to survive. They have the advantage of giving their young access to a habitat that has at least proved suitable to the parents.

A study on the southern Australian soft-sediment snail *Nassarius pauperatus* makes a convincing case for the dispersal value of planktotrophic larvae.[32] In some sand flats the mud snails are clearly in a poorer nutritive state than in others. This can be estimated by a measure of how responsive snails are to food, which is in effect an estimate of hunger. As food availability decreases, *N. pauperatus* produce more eggs per capsule, suggesting a shift to investment in dispersing progeny. This response can be explained in terms of life history theory, which predicts that adult mortality will select for one-time reproduction (such species are said to be **annuals**) and heavy investment in that singular reproduction. By contrast, snails living in flats with abundant food invest less in progeny, because selective pressure is not intense enough to warrant massive devotion to early reproduction. Some reserves are instead devoted to growth to more reproductive seasons (such populations are known as

---

[32] See McKillup and Butler, 1979, in Further Reading, Larval Dispersal and Recruitment.

perennial). Planktonic dispersal therefore serves the successful dispersal of progeny of food-starved populations to sites that may be better trophically.

Long-distance dispersal also has the advantage of spreading the young over a variety of habitat types. In carrying out such a dispersal, a parent is hedging its bets on the success of its offspring. Instead of investing heavily in one site, dispersal allows settling on many sites. In any year, one site may be very poor for settling, whereas another might be good. In the long run, survival is averaged over all habitats.

The minimum time required for a planktotrophic larva to develop and the distance the larva travels in the water column during this period is puzzling. Why should a larva travel for 2 weeks or more, when it could sample a diversity of new sites only a few hundred meters (and a couple of days) away? Strathman[33] suggests that planktotrophic larvae might not have evolved as an adaptation for long-distance dispersal, but that the larvae may have evolved planktotrophy to take advantage of the phytoplankton food source, much as a juvenile fish exploits a nursery ground. The exact length of dispersal may therefore be something of an evolutionary accident, even if wider dispersal does have the consequence of lowered species extinction and the potential for rapid colonization.

Although long-distance dispersers are common, many colonial animals, such as bryozoans and coelenterates, are known to have short-distance-dispersing lecithotrophic larvae. Indeed, this form of dispersal has been underemphasized in the past. The bryozoan *Bugula neritinea* lives in Florida on blades of turtle grass. Areas with dense populations are self-maintaining, but nearby areas without *Bugula* also exist for several years without any significant colonization. Thus, marine populations in certain areas may be limited by dispersal potential, especially in the case of species with no planktonic larvae or with lecithotrophic larvae.

[33] See Strathman, 1985, in Further Reading, Larval Dispersal and Recruitment.

Further Reading

SEX, FERTILIZATION, AND LIFE CYCLES

Andersson, M. 1994. *Sexual Selection.* Princeton, NJ: Princeton University Press.

Babcock, R. C. and C. N. Mundy. 1992. Reproductive biology, spawning and field fertilization rates of *Acanthaster planci. Australian Journal of Marine and Freshwater Research*, v. 43, pp. 523–534.

Babcock, R. C., C. N. Mundy, and D. Whitehead. 1994. Sperm diffusion and in situ confirmation of long-distance fertilization in the free-spawning asteroid *Acanthaster planci. Biological Bulletin*, v. 186, pp. 17–28.

Bacci, G. 1965. *Sex Determination.* New York: Pergamon Press.

Berglund, A. 1990. Sequential hermaphroditism and the size-advantage hypothesis: An experimental test. *Animal Behavior*, v. 39, pp. 426–433.

Christy, J. H. 1983. Female choice in the resource-defense mating system of the sand fiddler crab, *Uca pugilator. Behavioral Ecology and Sociobiology*, v. 12, pp. 169–180.

Denny, M. W., and M. F. Shibata. 1989. Consequences of surf-zone turbulence for settlement and external fertilization. *American Naturalist*, v. 134, pp. 859–889.

Doherty, P. J. 1983. Diel, lunar and seasonal rhythms in the reproduction of two tropical damselfishes: *Pomacentrus flavicauda* and *P. wardi. Marine Biology*, v. 75, pp. 115–124.

Duffy, J. E. 1996a. Eusociality in a coral reef shrimp. *Nature*, v. 381, pp. 512–514.

Duffy, J. E. 1996b. Species boundaries, specialization, and the radiation of sponge-dwelling alpheid shrimp. *Biological Journal of the Linnaean Society*, v. 58, pp. 307–324.

Duffy, J. E. 1998. On the frequency of eusociality in snapping shrimps (Decapoda: Alpheidae), with description of a second eusocial species. *Bulletin of Marine Science*, v. 62, pp. 387–400.

Duffy, J. E., C.L. Morrison, and R. Rios. 2000. Multiple origins of eusociality among sponge-dwelling shrimps. *Evolution*, v. 54, pp. 503–516.

Foster, S. 1990. Courting disaster in cannibal territory. *Natural History*, November, pp. 52–60.

Francis, L. 1973. Intraspecific aggression and its effect on the distribution of *Anthopleura elegantissima* and some related sea anemones. *Biological Bulletin*, v. 144, pp. 73–92.

Ghiselin, M. T. 1987. Evolutionary aspects of marine invertebrate reproduction. In A. C. Giese, J. C. Pearse, and V. B. Pearse, eds, *Reproduction of Marine Invertebrates, v. IX: General Aspects: Seeking Unity in Diversity.* Palo Alto, CA: Blackwell Scientific, pp. 609–665.

Harrison, P., R. Babcock, G. D. Bull, J. Oliver, C. Wallace, and B. Willis. 1984. Mass spawning in tropical reef corals. *Science*, v. 223, pp. 1186–1189.

Hughes, D. J., and R. N. Hughes. 1986. Life history variation *in Celloporella hyalina* (Bryozoa). *Proceedings of the Royal Society of London B*, v. 228, pp. 127–132.

Hughes, R. N. 1989. *A Functional Biology of Clonal Organisms*, London: Chapman & Hall.

Hyman, L. 1955. *The Invertebrates: Echinodermata*. New York: McGraw-Hill.

Kirkpatrick, M. 1982. Sexual selection and the evolution of female choice. *Evolution*, v. 36, pp. 1–12.

Levitan, D. R. 1996. Effects of gamete traits on fertilization in the sea and the evolution of sexual dimorphism. *Nature*, v. 382, pp. 153–155.

Levitan, D. R. 2000. Optimal egg size in marine invertebrates: Theory and phylogenetic analysis of the critical relationship between egg size and development time in echinoids. *American Naturalist*, v. 156, pp. 175–192.

Levitan, D. R., and C. Peterson. 1995. Sperm limitation in the sea. *Trends in Ecology and Evolution*, v. 10, pp. 228–231.

Levitan, D. R., M. A. Sewell, and F.-C. Chia. 1992. How distribution and abundance influence fertilization success in the sea urchin *Strongylocentrotus franciscanus*. *Ecology*, v. 73, pp. 248–254.

Lively, C. M., C. Craddock, and R. C. Vrijenhoek. 1990. Red Queen hypothesis supported by parasitism in sexual and clonal fish. *Nature*, v. 344, pp. 864–866.

Maynard Smith, J. 1971. What use is sex? *Journal of Theoretical Biology*, v. 30, pp. 319–335.

Metz, E. C., and S. R. Palumbi. 1996. Positive selection and sequence rearrangements generate extensive polymorphism in the gamete recognition protein bindin. *Molecular Biology and Evolution*, v. 13, pp. 397–406.

Metz, E. C., R. E. Kane, H. Yanagimachi, and S. R. Palumbi. 1994. Fertilization between closely related sea urchins is blocked by incompatibilities during sperm–egg attachment and early stages of fusion. *Biological Bulletin*, v. 187, pp. 23–34.

Metz, E. C., G. Gómez-Gutiérrez, and V. D. Vacquier. 1998. Mitochondrial DNA and bindin gene sequence evolution among allopatric species of the sea urchin genus *Arbacia*. *Molecular Biology and Evolution*, v. 15, pp. 185–195.

Murphy, G. I. 1968. Pattern in life history and the environment. *American Naturalist*, v. 102, pp. 391–403.

Peterson, C. W. 1991. Variation in fertilization rate in the tropical reef fish *Halochoeres bivattatus*; correlates and implications. *Biological Bulletin* (Woods Hole), v. 18, pp. 232–237.

Pianka, E. R., and W. S. Parker. 1975. Age-specific reproductive tactics. *American Naturalist*, v. 109, pp. 453–464.

Robertson, D. R. 1972. Social control of sex reversal in a coral-reef fish. *Science* v. 177, pp. 1007–1009.

Stearns, S. C. 1976. Life history tactics: A review of the ideas. *Quarterly Review of Biology* v. 51, pp. 3–47.

Styan, C. A. 1998. Polyspermy, egg size, and the fertilization kinetics of free-spawning marine invertebrates. *American Naturalist*, v. 152, pp. 290–297.

Swanson, W. J., and V. D. Vacquier. 1998. Concerted evolution in an egg receptor for a rapidly evolving abalone sperm protein. *Science*, v. 281, pp. 710–712.

Vahl, O. 1981. Age-specific residual reproductive value and reproductive efforts in the Iceland scallop, *Chlamys islandica* (O. F. Moller). *Oecologia*, v. 51, pp. 53–56.

Vogel, H., G. Czihak, P. Chang, and W. Wolf. 1982. Fertilization kinetics of sea urchin eggs. *Mathematical Biosciences*, v. 58, pp. 189–216.

Warner, R. R. 1975. The adaptive significance of sequential hermaphroditism in animals. *American Naturalist*, v. 109, pp. 61–82.

Yund, P. O., and M. M. McCartney. 1994. Male reproductive success in sessile invertebrates: Competition for fertilizations. *Ecology*, v. 7, pp. 2151–2167.

Williams, G. C. 1975. *Sex and Evolution*. Princeton, NJ: Princeton University Press.

MIGRATION

Gross, M. R., R. M. Coleman, and R. M. McDowall. 1988, Aquatic productivity and the evolution of diadromous fish migration. *Science*, v. 239, pp. 1291–1293.

Harden Jones, F. R. 1968. *Fish Migration*, 2nd ed. London, Edward Arnold.

Koch, A. L., A. Carr, and D. W. Ehrenfeld. 1969. The problem of open-sea navigation: The migration of the green turtle to Ascension Island. *Journal of Theoretical Biology*, v. 22, pp. 163–179.

Meylan, A. B., B. W. Bowen, and J. C. Avise. 1990. A genetic test of the natal homing versus social facilitation models for green turtle migration. *Science*, v. 248, pp. 724–727.

LARVAL DISPERSAL AND RECRUITMENT

Barnes, H. 1956. *Balanus balanoides* (L.) in the Firth of Clyde: The development and annual variation of the larval population, and the causative factors. *Journal of Animal Ecology*, v. 25, pp. 72–84.

Butman, C. A. 1987. Larval settlement of soft-sediment invertebrates: The spatial scales explained by active habitat selection and the emerging role of hydrodynamic processes. *Oceanography and Marine Biology Annual Reviews*, v. 25, pp. 113–165.

Butman, C. A., J. P. Grassle, and E. J. Buskey. 1988. Horizontal swimming and gravitational sinking of *Capitella* sp. 1 (Annelida: Polychaeta) larvae: Implications for settlement. *Ophelia*, v. 29, pp. 43–57.

Connolly, S. R., and J. Roughgarden. 1999. Increased recruitment of northeast Pacific barnacles during the 1997

El Niño. *Limnology and Oceanography*, v. 44, pp. 466–469.

Cronin, T. W. 1982. The estuarine retention of larvae of the crab *Rhithropanopeus harrisii*. *Estuarine and Coastal Shelf Science*, v. 15, pp. 207–220.

Deevey, E. S. 1947. Life tables for natural populations of animals. *Quarterly Review of Biology*, v. 22, pp. 283–314.

Doherty, P. J., and D. McB. Williams. 1988. The replenishment of coral reef fish populations. *Oceanography and Marine Biology Annual Review*, v. 26, pp. 487–551.

Eckman, J. E. 1983. Hydrodynamic processes affecting benthic recruitment. *Limnology and Oceanography*, v. 28, pp. 241–257.

Gaines, S. D., and J. Roughgarden. 1987. Fish in offshore kelp forests affect recruitment to intertidal barnacle populations. *Science*, v. 235, pp. 479–481.

Grosberg, R. K. 1987. Limited dispersal and proximity-dependent mating success in the colonial ascidian *Botryllus schlosseri*. *Evolution* 41, pp. 372–384.

Grosberg, R. K., and J. R. Quinn. 1986. The genetic control and consequences of kin recognition by the larvae of a colonial marine invertebrate. *Nature*, v. 322, pp. 456–459.

Hill, A. E., J. Brown, and L. Fernand. 1996. The western Irish Sea gyre: A retention system for Norway lobster (*Nephrops norvegicus*)? *Oceanologica Acta*, v. 19, pp. 357–368.

Hoek, C. van den. 1987. The possible significance of long-range dispersal for the biogeography of seaweeds. *Helgoländer Wissenschaften Meeresuntersungen*, v. 41, pp. 261–273.

Houde, E. D. 1987. Fish early life dynamics and recruitment variability. *American Fishery Society Symposium*, v. 2, pp. 17–29.

Hunt, J. H., W. G. Ambrose Jr., and C. H. Peterson. 1987. Effects of the gastropod *Ilyanassa obsoleta* (Say) and the bivalve *Mercenaria mercenaria* (L.) on larval settlement and juvenile recruitment of infauna. *Journal of Experimental Marine Biology and Ecology*, v. 108, pp. 229–240.

Innes, D. I. 1984. Genetic differentiation among populations of marine algae. *Helgoländer Wissenschaften Meeresuntersuchungen*, v. 38, pp. 401–417.

Judge, M. L., J. F. Quinn, and C. L. Wolin. 1988. Variability in recruitment of *Balanus glandula* (Darwin, 1854) along the central California coast. *Journal of Experimental Marine Biology and Ecology*, v. 119, pp. 235–251.

Keough, M. J. 1989. Dispersal of the bryozoan *Bugula neritinea* and effects of adults on newly metamorphosed juveniles. *Marine Ecology—Progress Series*, v. 57, pp. 163–172.

Levin, L. A. 1986. The influence of tides on larval availability in shallow waters overlying a mudflat. *Bulletin of Marine Science*, v. 39, pp. 224–233.

Lutz, R. A., D. Jablonski, and R. D. Turner. 1984. Larval development and dispersal at deep-sea hydrothermal vents. *Science*, v. 226, pp. 1451–1453.

McConaugha, J. R. 1988. Export and reinvasion of larvae as regulators of estuarine decapod populations. *American Fisheries Society Symposium*, v. 3, pp. 90–103.

McEdward, L., ed. 1995. *Ecology of Marine Invertebrate Larvae*. Boca Raton, FL: CRC Press.

McKillup, S. C., and A. J. Butler. 1979. Modification of egg production and packaging in response to food availability by *Nassarius pauperatus*. *Oecologia*, v. 43, pp. 221–231.

Mileikovsky, S. A. 1974. Types of larval development in marine bottom invertebrates: An ecological scheme. *Thallassia Jugoslavica*, v. 10, pp. 171–179.

Miller, S. E., and M. G. Hadfield. 1990. Developmental arrest during larval life and life-span extension in a marine mollusc. *Science*, v. 248, pp. 356–358.

Morgan, S. P. 1990. Impact of planktivorous fishes on the dispersal, hatching and morphology of estuarine crab larvae. *Ecology*, v. 71, pp. 1639–1652.

Mullineaux, L. S., and C. A. Butman. 1991. Initial contact, exploration, and attachment of barnacle cyprids (*Balanus amphitrite*) settling in flow. *Marine Biology*, v. 110, pp. 93–103.

Newell, R. C. 1979. *Biology of Intertidal Animals*, 3rd ed. Faversham, England: Marine Ecological Surveys, Ltd.

Olson, R. R. 1985. The consequences of short distance dispersal in a sessile marine invertebrate. *Ecology*, v. 66, pp. 30–39.

Olson, R. R., and M. H. Olson. 1989. Food limitation of planktotrophic marine invertebrate larvae: Does it control recruitment success? *Annual Review of Ecology and Systematics*, v. 20, pp. 225–247.

Osman, R. W., R. B. Whitlatch, and R. N. Zajac. 1989. Effects of resident species on recruitment into a community: Larval settlement versus post-settlement mortality in the oyster *Crassostrea virginica*. *Marine Ecology—Progress Series*, v. 54, pp. 61–73.

Pawlik, J. R. 1992. Chemical ecology of the settlement of benthic marine invertebrates. *Oceanography and Marine Biology Annual Reviews*, v. 30, pp. 273–335.

Pawlik, J. R., C. A. Butman, and V. Starczak. 1991. Hydrodynamic facilitation of gregarious settlement of a reef-building tube worm. *Science*, v. 251, p. 421–424.

Peterson, C. H., and H. C. Summerson. 1992. Basin-scale coherence of population dynamics of an exploited marine invertebrate, the bay scallop. Implications for recruitment limitation. *Marine Ecology—Progress Series*, v. 90, pp. 257–272.

Pineda, J. 1991. Predictable upwelling and the shoreward transport of planktonic larvae by internal tidal bores. *Science*, v. 253, pp. 548–551.

Pineda, J. 1994. Spatial and temporal patterns in barnacle settlement rate along a southern California rocky shore. *Marine Ecology—Progress Series*, v. 107. pp. 125–138.

Richmond, R. H. 1985. Reversible metamorphosis in coral planula larvae. *Marine Ecology—Progress Series*, v. 22, pp. 181–185.

Sammarco, P. W. 1988. Localized dispersal and recruitment in Great Barrier Reef corals: The Helix experiment. *Science*, v. 239, pp. 1422–1424.

Sammarco, P. W. 1991. Geographically specific recruitment and postsettlement mortality as influences on coral communities: The cross-continental shelf transplant experiment. *Limnology and Oceanography*, v. 36, pp. 496–514.

Sammarco, P. W., and M. L. Heron, eds. 1994. *The Bio-Physics of Marine Larval Dispersal*. Washington, DC: American Geophysical Union.

Scheltema, R. S. 1971. Larval dispersal as a means of genetic exchange between geographically separated populations of shallow-water benthic marine gastropods. *Biological Bulletin*, v. 140, pp. 284–322.

Scheltema, R. S. 1989. Planktonic and non-planktonic development among prosobranch gastropods and its relationship to the geographic range of species. In J. S. Ryland and P. A. Tyler, eds., *Reproduction, Genetics, and Distributions of Marine Organisms, 23rd European Marine Biology Symposium*. Fredensborg, Denmark: Olsen and Olsen, pp. 183–188.

Shanks, A. L. 1983. Surface slicks associated with tidally forced internal waves may transport pelagic larvae of benthic invertebrates and fishes shoreward. *Marine Ecology—Progress Series*, v. 13, pp. 311–315.

Starr, M., J. H. Himmelman, and J.-C. Therriault. 1990. Direct coupling of marine invertebrate spawning with phytoplankton blooms. *Science*, v. 247, pp. 1071–1074.

Strathman, R. R. 1974. The spread of sibling larvae of sedentary marine invertebrates. *American Naturalist*, v. 108, pp. 29–44.

Strathman, R. R. 1985. Feeding and nonfeeding larval development and life-history evolution in marine invertebrates. *Annual Review of Ecology and Systematics*, v. 16, pp. 339–361.

Strathman, R. R. 1987. Larval feeding. In A. C. Giese, J. C. Pearse, and V. B. Pearse, eds., *Reproduction of Marine Invertebrates, v. IX: General Aspects: Seeking Unity in Diversity*. Palo Alto, CA, Blackwell Scientific, pp. 465–550.

Thorson, G. 1950. Reproductive and larval ecology of marine bottom invertebrates. *Biological Reviews*, v. 25, pp. 1–45.

Toonan, R. J., and J. R. Pawlik. 1994. Foundations of gregariousness. *Nature*, v. 370, pp. 511–512.

Toonan, R. J. and J. R. Pawlik. 1996. Settlement of the tube worm *Hydroides dianthus* (Polychaeta: Serpulidae): Cues for gregarious settlement. *Marine Biology*, v. 126, pp. 725–733.

Woodin, S. A., R. L. Marinelli, and D. E. Lincoln. 1993. Allelochemical inhibition of recruitment in a sedimentary assemblage. *Journal of Chemical Ecology*, v. 19, pp. 517–530.

Young, C. M., and F.-S. Chia. 1987. Abundance and distribution of pelagic larvae as influenced by predation, behavior, and hydrographic factors. In A. C. Giese, J. C. Pearse, and V. B. Pearse, eds., *Reproduction of Marine Invertebrates, v. IX: General Aspects: Seeking Unity in Diversity*. Palo Alto, CA: Blackwell Scientific, pp. 385–463.

Zimmer-Faust, R. K., and M. N. Tamburri. 1994. Chemical identity and ecological implications of a waterborne, larval settlement cue. *Limnology and Oceanography*, v. 39, pp. 1075–1087.

Review Questions

1. Why are not all sexual species hermaphroditic?

2. What is the "cost of sex," biologically speaking?

3. What might be the benefit of sex to organisms?

4. Under what circumstances does it make sense for a hermaphrodite to be protandrous? Protogynous?

5. When might you expect the presence of dwarf males attached to females?

6. What is the advantage of clonal reproduction? The disadvantage?

7. Why do fishes such as Pacific salmon species reproduce only once?

8. What might be the value, if any, of long-distance dispersal across an ocean?

9. What are the advantages of having planktotrophic larvae capable of settling and metamorphosing upon any substratum? What are the disadvantages?

10. Why are species with planktotrophic larvae more common in the tropics than at polar latitudes?

11. How might planktonic larvae be able to find adults of their own species?

12. What is the value of planktonic-feeding larval development?

13. What are the potential sources of mortality for planktonic larvae?

14. What effect does planktotrophic larval dispersal tend to have on the geographic range of a coastal marine invertebrate species?

15. Do you think that the total evolutionary life spans of species with planktotrophic larvae are liable to be greater than those with lecithotrophic larvae? Explain your answer.

16. Anemones often occur in clones of large numbers, that have arisen by fission from a founder individual. What experiment might be performed to determine the benefit of large numbers of adjacent anemones, as opposed to smaller groups or solitary individuals?

6

Life in a Fluid Medium

Nothing is more fundamental about marine life than the medium of seawater itself. It is the purpose of this chapter to introduce some of the physical properties of seawater that affect the functioning of marine organisms. Many of the effects involve the movement of fluids or the movement of structures of marine organisms through the fluid. We should distinguish between primary and secondary effects of water movement. **Primary effects** are the direct results of the properties of the water and its motion, such as current speed. For example, water velocity and turbulence strongly influence the morphology and taxonomic composition of marine organisms and communities. Variation in water movement has **secondary effects** on food, nutrients, and oxygen availability. Stagnant waters will decrease in oxygen content and in the supply of planktonic organisms for suspension feeders. Sluggish currents will also be inefficient at dispersing pelagic larval stages of marine species (see Chapter 5).

✳ **Whatever is part of the flow will not cross streamlines in a flowing fluid.**

The movement of fluids can be readily diagrammed as a series of approximately parallel streamlines as in Figure 6.1—but keep in mind that all is not usually so simple. Whatever is part of the flow will not cross the streamlines. One effect of this simple rule is that fluids and entrained particles (e.g., protozoans, dye) move in the same direction. This helps greatly in tracing fluid motion because particles can be used to characterize flow; it's easier to watch dye, for example, than to try to look at the movement of transparent water! Thus, an important question is: Under what conditions will a particle be entrained in the flow? That is the subject of the next few sections.

Physical Properties: Density, Viscosity, and Velocity

✳ **Water is relatively dense and viscous.**

Water has physical properties far different from those of air. Seawater bathes marine organisms and protects soft moist tissues from drying. Oxygen can be obtained from solution. However, water is also a more supportive medium than air because it is denser. This eliminates the need for the strong supportive skeleton required by large terrestrial organisms. Water is also a viscous medium, and this poses some unique challenges. Imagine a frigate bird diving for fish through 10 m of water, instead of through the same distance in air. Viscosity makes movement through the water far more difficult. This chapter will serve as an introduction to some aspects of life in flu-

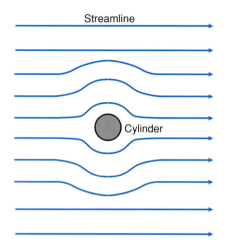

Fig. 6.1 Water flow can be visualized as streamlines, which indicate the path that individual particles would take. Particles move along streamlines, not across them. In this illustration, water is flowing around a fixed cylinder, which is viewed in cross section.

ids, and as a resource for material to be presented in Chapters 7, 8, and 13.

The important properties of density and viscosity must be defined more clearly if you are to understand how flow affects marine organisms. **Density** is the mass per unit volume and is expressed as grams per cubic centimeter (g cm^{-3}). The salt content makes seawater somewhat denser than fresh water. The density of seawater varies inversely with temperature. **Dynamic viscosity** is a measure of the molecular "stickiness" between layers of a fluid. Honey has greater dynamic viscosity than water, and water is more dynamically viscous than air. Dynamic viscosity decreases with increasing temperature.

✳ **The Reynolds number is an estimate of the relative importance of viscous and inertial forces in a fluid.**

In any fluid, there are two basic competing forces: **viscous forces** and **inertial forces**. Viscous forces are "sticky." They keep the fluid together and flowing in smooth streamlines. In a fluid, particles affected by viscosity move, or stay still, depending upon the movement of the fluid. As viscosity increases, molecular stickiness keeps different parts of a fluid from separating easily, and any object in the fluid will be less able to move unless the surrounding fluid is also moving. Inertial forces are those that relate to iner-

tia, the tendency of a moving object to continue moving, when no force is applied to it. If you throw a steel ball through air, for example, it continues to move; that is an example of inertia. Inertial forces make a fluid break up into uneven streamlines, or allow an object to "drop" through a fluid like a stone in water—in other words, to *not* go with the flow.

The **Reynolds number** (*Re*) is a measure of the relative importance of inertial and viscous effects of a fluid and on objects in a fluid. As the Reynolds number increases, the inertial forces come to dominate. Under high Reynolds number, objects in a fluid are dominated by inertia, that is, they tend to keep on moving when a force is applied to them. Under low Reynolds number, objects do not move unless a force is applied, because viscous forces dominate.

Reynolds number, *Re*, is simply the product of the velocity *V*, size *l*, and density *ρ*, divided by the dynamic viscosity *v*.[1] As long as we are dealing with seawater, we can take the density and the dynamic viscosity to be constants. We therefore must measure the remaining two variables, velocity *V* and size of the object *l*. Reynolds number increases with an increase of either velocity or size.

There are two different approaches that we can take to measure the velocity, which is needed to calculate the Reynolds number (Figure 6.2). First, if we

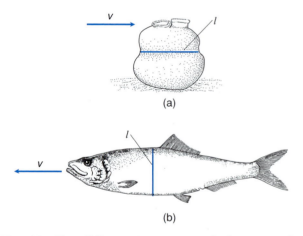

Fig. 6.2 Two different situations in which one can find the size *l* and velocity *V*, parameters necessary to determine the Reynolds number. (a) An object (a sea squirt) is stationary in moving water. (b) An object (e.g., a fish) is moving through the water.

[1] The units cancel out and *Re* is therefore a number with no dimensions.

Table 6.1 Reynolds number for a range of swimming organisms of different sizes and swimming velocities

ANIMAL AND VELOCITY	RE
Large whale swimming at 10 m s^{-1}	300,000,000
Tuna swimming at 10 m s^{-1}	30,000,000
Copepod swimming at 20 cm s^{-1}	30,000
Sea urchin sperm swimming at 0.2 mm s^{-1}	0.03

Source: Data from Vogel, 1981.

place an object in a moving fluid, and keep it fixed to the bottom, we can measure the fluid's velocity past the object (e.g., water flowing past a coral). As an alternative method, we can take a nonmoving fluid and measure the velocity of motion of an object through this stationary fluid (e.g., measuring the swimming speed of a fish).

Table 6.1 gives *Re* for a range of organism sizes and swimming velocities. Empirical research has shown that when *Re* is less than about 1,000, viscous forces predominate. If *Re* is much greater than 1,000, inertial forces predominate. The threshold (actually a broad band of transition) makes all the difference in terms of how organisms move in and react to their environment. Because the viscosity of seawater can be assumed to be constant, we need worry only about velocity and size. A small object traveling at a relatively low velocity is dominated by viscous forces; in other words, it is living in what amounts to a sticky medium. That is why a paramecium swimming through still water can stop seemingly instantaneously. The instant the protozoan ceases to move its cilia, it is entrained in the still water and stops. By contrast, a supertanker may require several miles to come to rest from full speed. The supertanker, which operates at high *Re*, is only modestly affected by viscosity, and considerable inertia must be overcome to bring the vessel to a stop. The protozoan, by virtue of its minuscule size and low velocity, lives in a world completely different from that of a ship or a fast and large fish, even though all move in the same fluid. If a protozoan ceases movement, it stops instantaneously, because, at low Reynolds number, inertia is unimportant relative to viscous forces. To swim, a protozoan must therefore continuously exert a force against the surrounding medium; it cannot depend upon inertia for move-

ment. Measured in terms of body lengths per unit time, paramecium should have a better reputation for being an excellent swimmer, at least in relation to its body size. A protozoan can sustain swimming speeds of 100 body lengths per second, whereas a tuna cannot swim much faster than 10 body lengths per second.

✳ Laminar flow is regular, whereas turbulent flow is irregular.

We can distinguish between two main types of flow. **Laminar flow** is regular, and lines describing movement of water molecules characterized by such flow are parallel (Figure 6.3). By contrast, **turbulent flow** is characterized by lines that are very irregular, and the overall direction of flow can be determined only as an aggregate of individual irregular motions. The famous hydrodynamicist O. Reynolds (for whom the Reynolds number is named) discovered that flow in a pipe becomes irregular (i.e., turbulent) if velocity, pipe diameter, or fluid density increases beyond a certain point. These factors contribute to increasing the Reynolds number. (In this case, *l* is equal to the pipe diameter.) The same principles can also apply in open water. When the Reynolds number is high, a fluid encountering an object may change velocity rapidly, and inertia may cause the fluid to break up into complex vortices and wakes behind the object.

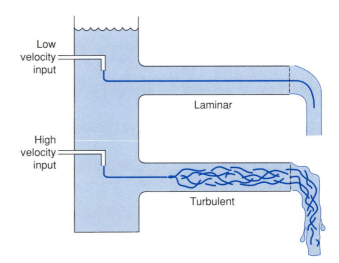

Fig. 6.3 Laminar and turbulent flow. Green lines represent paths of flow. (After Vogel, 1981.)

Water Moving Over Surfaces and Obstructions, Such as Organisms

❋ As a fluid moves over a solid surface, velocity steadily decreases with depth, the water reaching a standstill at the solid surface.

Consider water moving over a perfectly smooth-surfaced bottom. It is a hydrodynamic necessity that water velocity will decrease to zero at the bottom (Figure 6.4). This is called the **no-slip condition**. In effect, the grab of the bottom is perfect at the surface but rapidly loses hold as you go into a **mainstream**, where the current velocity is not affected by the solid surface. The exact decline of velocity is complex but is determined in large measure by the velocity of the fluid, and by the dynamic viscosity.

If the flow is regular and the bottom is level, very close to the bottom there is always a **boundary layer**, where velocity declines approximately linearly with decreasing distance to the bottom surface. Outside this boundary layer, velocity increases asymptotically toward the mainstream current. The border of the boundary layer is better described as a transition zone than as having a precise depth. Within the boundary layer, viscosity must be factored into studies of flow. The thickness of the boundary layer, relative to the size of an object, decreases with increasing Reynolds

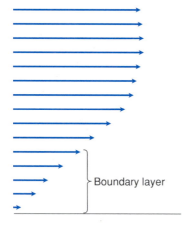

Fig. 6.4 Flow over the bottom. Water velocity maintains an average mainstream velocity well above the bottom, but velocity decreases to zero at the bottom surface. Near the bottom is a thin boundary layer, where velocity is relatively low but increases linearly from the bottom surface. (Not to scale: boundary layer is often less than a centimeter thick.)

number. When *Re* is small, the boundary layer is relatively thick, making it something of a barrier to the exchange of materials and energy. A microscopic organism, such as a protozoan, lives enmeshed in a relatively viscous environment. It used to be believed, for example, that cilia on the gills of bivalves or setae on the feeding appendages of copepods trapped food particles between the fibers, much like a sieve, as the particles drifted by on currents and impacted on the cilia or setae. The boundary layer prevents this, however. As the particles approach the feeding microstructures, they may even be deflected from the boundary layer. If they are trapped within the boundary layer, the cilia must reach out and touch or grab the particles, or the particles must collide directly with the cilia or other collecting fibers. The new understanding of this process has changed the way we look at suspension feeders, or animals that collect particles as food from seawater (see Chapter 13).

❋ The principle of continuity allows one to calculate flow velocity in a biological circulatory system.

Consider a unit volume of fluid flowing through a rigid pipe. Assume that the fluid is incompressible; therefore, if a liter volume enters at one end, an equal volume will leave at the other. The product of the velocity and the cross-sectional area always remains constant (neglecting friction). If the fluid is then forced to flow through a pipe of half the cross-sectional area, its velocity will be doubled (Figure 6.5). This **principle of continuity** applies equally to changes in cross section of a single pipe and to the case in which a pipe splits into several smaller pipes. If, for example, a pipe splits into several equal subsections,

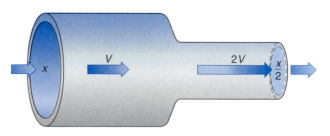

Fig. 6.5 Flow velocity through a pipe and the principle of continuity. The product of cross-sectional area and velocity is constant. Therefore, if cross-sectional area *x* decreases by half, the velocity doubles.

the product of the velocity and cross-sectional area of the main pipe will equal the sum of the products of the velocity and the sum of the cross-sectional areas of the smaller pipes.

The principle of continuity permits organisms to regulate water velocity. The principle can be applied, for example, to understand flow through a simple sponge (Figure 6.6). Sponges consist of networks of chambers, each of which is lined by flagellated cells known as choanocytes. The chambers are all connected to one or several main water expulsion channels, which guide wastewater from the sponge. If you dive, an application of food coloring to a sponge will quickly show you that the velocity of these excurrent openings is great, on the order of 10 cm s^{-1}. How can a collection of choanocytes, each able to produce a water velocity of only 50 μm s^{-1}, manage to produce such a rapid exit speed together? The total cross-sectional area of choanocyte chambers adds up to several thousand times the cross-sectional area of the excurrent canal. The velocity through the excurrent canal must increase proportionally. Suspension-feeding polychaetes and mollusks employ a similar principle to drive water at fairly high speeds through interfilamental openings or siphons. Tens of thousands of cilia may operate at low individual velocity, but the total cross-sectional area is great, relative to (for example) that of the exhalent siphon of a bivalve mollusk.

Bernoulli's Principle

✳ **Bernoulli's principle states that pressure varies inversely with fluid velocity.**

Bernoulli's principle applies the principle of conservation of energy to pressure changes in pipes and burrows, or along surfaces. If total energy must be constant, then **pressure will vary inversely with the velocity of the fluid**. The simplest case is represented by a pipe whose diameter changes somewhere along its length. If the diameter decreases, then velocity in that section of the pipe will increase, as just discussed. Pressure, however, will decrease in this section. If you punctured a pipe of this sort filled with flowing gas and then lit a match, the flame would be higher in the thick section of the pipe, owing to the higher pressure.

This principle has broad biological applications. The design of a cross section of a wing is based directly on an application of Bernoulli's principle. The lower surface is flat, whereas the upper surface is curved. As air encounters the wing, it moves more rapidly along the upper surface and the pressure is thus lower than along the lower surface. The pressure difference creates lift. The same principle applies to a flatfish (Figure 6.7a). As the fish pushes through the water, greater pressure is exerted on its flatter bottom than on the curved top, and lift develops.

Pressure differences on either side of a tube can

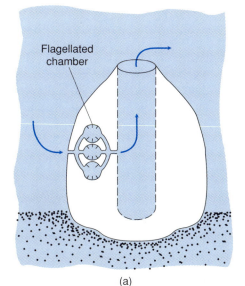
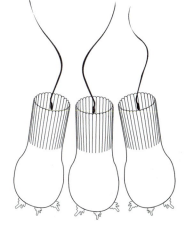

Fig. 6.6 How a sponge generates a relatively high exit velocity through its excurrent channels. (a) The low velocity of the water from flagellated cells in flagellated chambers is compensated by the far greater total cross-sectional area of the flagellated chambers relative to the excurrent opening of the sponge. (b) Diagram of flagellated sponge cells. (After Vogel, 1981.)

(a)   (b)

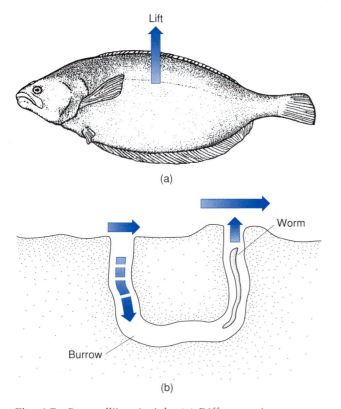

Fig. 6.7 Bernoulli's principle. (a) Differences in pressure above and below a moving flatfish creates lift. (After Vogel, 1981.) (b) A raised mound on one end of a buried U-shaped tube places it in a slightly higher current velocity relative to the other opening, which is flush with the sediment. Water moving past the two holes creates a pressure difference, with lower pressure on the raised area, and this drives water through the tube. Length of arrows is proportional to water velocity.

also be used by an organism to create a current. Consider a U-shaped tube in the mud (Figure 6.7b), with one entrance to the tube opening to a small rise, higher than the adjacent entrance. Because the entrance in the rise is probably exposed to a slightly higher current speed, the pressure will be lower than at the other entrance. As a result, water will flow passively through the tube. This principle also reduces the work needed to drive water through worm tubes or through sponges, because a moving current above the sponge reduces the pressure at the exit.

Drag

✳ **Water moving past an object creates drag, a force that operates differently at different Reynolds numbers.**

Consider a blade of eelgrass sticking up into the water from the bottom. The flow will be disrupted and will exert a force, or **drag**, upon the blade. Drag can be dissected into two components, either of which may dominate depending upon the Reynolds number. At low Reynolds numbers (more sticky, or viscous, situations), **skin friction** dominates. It is a force resulting from the interlayer stickiness of the fluid, and it acts parallel to the surface of the object and in the local direction of fluid flow. The more surface is exposed to the flow, the more skin friction there is. Although skin friction never disappears, it becomes far less important at higher Reynolds number.

Pressure drag is mainly the result of inertial forces in the fluid, and it dominates total drag at high Reynolds numbers. It occurs because pressure exerted on the upstream part of the object (e.g., a stationary coral in a current) is not exactly counterbalanced by an equal pressure on the downstream side. The object is effectively pushed along by the stream. Pressure drag increases proportionately to the cross-sectional area exposed to the current, and with the square of the current velocity. For example, a flat plate oriented perpendicular to a current exerts a maximum amount of pressure drag; the pressure drag is minimized when the plate is parallel to the current. Anyone who has driven an old van knows that its flat front creates sufficient drag to increase fuel consumption. Modern vans have much more streamlined shapes to reduce drag, but it's still more efficient (and fun) to drive a highly streamlined vehicle, such as a Ferrari.

The best way for an engineer to minimize pressure drag is to orient elongated objects parallel to the current and give them a long tapering tail on the downstream side. This allows the fluid, after passing over the front of the object, to decelerate gradually in the rear. The object is pushed forward by the closure of the fluid around the object toward the rear. This principle explains the streamlined teardrop shape of fast-swimming fish, such as skipjack tuna, and the shape of submarines. (See further discussion in Chapter 8.) Many organisms are fixed to the bottom, and pressure drag on them may be considerable. Seaweeds, corals, sea pens, and sea anemones all project into the flow from the bottom. The work of Miriam Koehl[1] has contributed much to our understanding of how flexible organisms can reduce drag, both by

[1] See Koehl, 1976, in Further Reading.

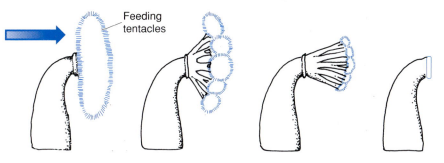

Fig. 6.8 Changes (left to right) of the sea anemone *Metridium senile* as the current increases.

the structure of their body wall and by alteration of their behavior. Drag can be reduced by flexibility, and by bending over in strong flow, much as palm fronds conform to a strong wind.

When currents and drag are too strong, some animals, such as anemones and feather stars, either contract the body or retreat to a crevice (see Chapter 13). The sea anemone *Metridium senile*, for example, normally protrudes its tentacles into the flow to feed, but this creates drag. When the current increases and pressure drag is too great, the animal withdraws its tentacles, which greatly reduces drag (Figure 6.8).

Let's consider a specific example of how pressure drag may affect an organism. If we visit an exposed rocky shore, we immediately see waves crashing against the organisms attached to the rocks. In many cases, the mobile invertebrates such as drilling snails and limpets found on exposed coasts are much smaller than members of the same species inhabiting protected rocky shores. It is possible that larger animals would simply be swept away by the waves. Larger animals project above the surface to a greater degree and are simply more exposed to the drag effect of passing waves. Michael Judge[2] tested the latter idea by gluing vertical copper plates to the west coast rocky shore limpet *Lottia gigantea*. The limpets with the plates spent less time moving around than those lacking plates, which gave the plated limpets

less time to feed. The loss of feeding time means fewer resources for growth, which may also explain the smaller size of the exposed shore limpets.

✳ Fish streamlining is a compromise between drag and skin friction.

In Chapter 8 the form of fish will be discussed. However it is worthwhile at this point to think about how the characteristics of a fluid contribute to influencing the swimming efficiency of a fish. Most fast-swimming fishes move by rhythmic contractions, which pass through the body as a wave. At any one time, part of the fish body is pushing against the water, propelling the fish forward.

As a fish moves forward, the forces on the fish surface include pressure and frictional effects. As water in streamlines passes over the fish, friction causes the water to lose some kinetic energy. This loss prevents the water from penetrating the steep pressure gradient behind the fish and the water that leaves the surface behind the fish forms a wake (Figure 6.9). If a fish is short and squat, there is a very steep pressure gradient from front to rear, and this leaves a large wake as the fish moves along. The difference in pressure creates drag. In effect, the fish is being pushed

[2] See Judge, 1988, in Further Reading.

Fig. 6.9 Drag on a fish is affected greatly by streamlining. (a) If a fish is well streamlined, the wake is reduced, streamlines are maintained behind the fish, and the drag is much reduced. (b) If a fish is poorly streamlined, a wake is created at the rear, producing a pressure gradient and drag.

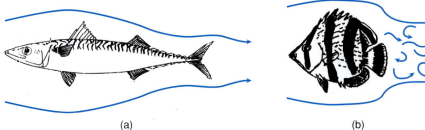

(a)   (b)

HOT TOPICS IN MARINE BIOLOGY 6.1 Flow Is a Drag, but It Sure Can Smell Good

The first thing that we usually learn about organisms is that they are rarely controlled by only one factor in nature. Not only are several different effects usually at work (e.g., temperature, salinity, water flow), but the organism has numerous needs (e.g., to find food, mates, a proper substratum). This complexity is readily illustrated by a study of the effects of flow on marine epifaunal invertebrates such as crabs and gastropods.

Consider the carnivorous blue crab *Callinectes sapidus* (see Figure 13.17d), which is a common coastal crab on the east and Gulf coasts of North America. Its favored prey is the infaunal clam *Mercenaria mercenaria*, which is found in soft intertidal and shallow subtidal bottoms. The crab walks actively about and can locate its prey by means of chemosensory detectors, located on hairs that recognize molecules carried to it in bottom currents. Where do these molecules come from? They are propelled through the *Mercenaria* clam's exhalant siphon and become entrained in the flow, creating an odor plume that eventually reaches the crab (Box Figure 6.1).

The crab's problem is as follows: How can it locate the origin of the detected molecules (even clams have a scent, which must be explained by various excretory products such as peptides and urea)? It can walk randomly until the scent (or, more properly rate of stimulus of the chemosensory hairs) becomes stronger, but this would be rather inefficient. It helps that the sensory hairs are on both the right and left antennae. As the crab moves, it is able to judge whether to move left or right, depending on the relative strength of the signal that reaches the sensory hairs. (The sensory hairs are quite small, and a viscous boundary layer of water develops around them. That's why crabs and lobsters often flick their antennae to reduce the boundary layer and to allow flowing water and the entrained molecules to impact upon the chemosensory hairs.)

The crab also derives information directly from the flow. The molecules are moving downstream, and so it makes sense to use water flow information to aid in locating higher concentrations of those molecules. That is why crabs also have mechanoreceptor hairs on the antennae.

Armed with both chemosensory and mechanosensory hairs, the crab's prey location problem appears to be solved, and it might seem that any nearby clams are doomed. Or are they? The odor plume emitted by a clam can provide useful information only if there is a regular decline of concentration with distance of the material in the plume responsible for the strength of the odor. That occurs if there is a well-developed boundary layer. Then water is propelled through the clam's exhalant siphon and moves regularly up into the mainstream. From an initial position downstream, the crab walks upstream toward the source of the odor and eventually finds its way to the exhalant siphon by successively encountering increasing odor intensity, much as a heat-seeking missile might find its way down a warm chimney.

If, on the other hand, the bottom is irregular, or if the near-bottom water current is very fast, the water flow near the bottom will be turbulent and there will be a poorly developed boundary layer. Marc Weissburg and Richard Zimmer-Faust[3] investigated whether a crab manages to follow the odor plume to its source under such circumstances.

[3] See Weissburg and Zimmer-Faust, 1993, in Further Reading.

(a)   (b)

Box Fig. 6.1   Dye traces of odor plumes. The blue crab *Callinectes sapidus* must locate the odor plume emanating from the exhalant siphon of the clam *Mercenaria mercenaria*. (a) If the current is relatively weak, the flow is laminar and the odor plume maintains its integrity. (b) If the current is strong, or if the bottom is rough, the boundary layer becomes turbulent, and the crab will have trouble following the odor to its source. (Courtesy of Marc Weissburg.)

They placed crabs in a flume with a sediment box containing the clam *Mercenaria mercenaria*. As flow was increased and as larger particles were placed on the sediment–water interface, the flow in the boundary layer became more and more turbulent and odor plumes more and more irregular. Under these conditions, crabs often failed to locate the odor plumes generated by the clams. In the natural estuarine habitats of the crabs, such as Chesapeake Bay, the current velocities are often in the range at which a boundary layer is well developed, and recent research shows that the plumes are regular enough to permit the crabs to follow them to the source.

When an animal detects an odor plume, we wonder if the creature responds to the concentration of a substance before moving toward its source, or whether it is rather the rate of arrival of the substance that determines the response. It is also possible that rate of arrival and concentration do not work independently. Instead, flux (a product of current speed and concentration) might determine responses to odors. Therefore a high concentration in a slowly arriving current might be no more effective than a low concentration in a rapidly arriving current. If flux is important, then it is the rate of arrival of odor molecules that matters. Richard Zimmer-Faust tested this hypothesis by varying current speed and concentration in dissolved nitrogenous compounds reaching the scavenging marine snail *Ilyanassa obsoleta*. The hypothesis was confirmed: neither current speed nor concentration predicted responses of the snails as well as flux.

Blue crabs are extremely efficient walkers and can rapidly propel themselves against considerable current speeds. However, many other benthic carnivores, such as starfish, nudibranch sea slugs, and shelled snails move far more slowly and project into the flow, which can generate considerable drag.[4] The nudibranch[5] slug *Tritonia diomedea*, which lives on shallow-water soft bottoms in the northern Pacific, is a carnivore and may locate prey much as the blue crab does, using a combination of chemoreceptors and mechanoreceptors (Box Figure 6.2a). However, the slug is propelled quite slowly by cilia at the bottom of its flattened molluscan foot. Also, its weighs very little in seawater. Once it is dislodged, even a modest current can cause it to tumble along the bottom, unable to reattach to the substrate.

To counter this rather disadvantageous event, it would benefit the sea slug to turn in the direction that minimizes the pressure drag created by the currents. Both in the field and in the laboratory, slugs point directly into the current. Dennis Willows investigated this problem by placing a series of slugs[6] on an underwater turntable and measuring drag as a function of the current velocity that passed by each experimental subject. A force transducer on a wire connected

---

[4] See Weissburg and Zimmer-Faust, 1993, in Further Reading.
[5] See Willows, 1978, in Further Reading.
[6] Nudibranchs are gastropods without shells. Most spend their time on the surface and have a muscular foot, but propel themselves along by means of cilia on the ventral surface. They are nearly all carnivores and may feed on relatively immobile invertebrates, such as sponges, soft corals, and sea pens.

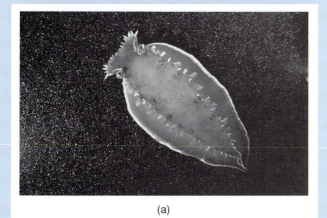

(a)

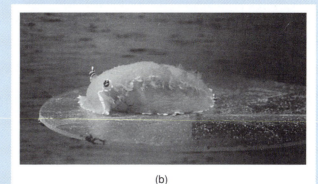

(b)

Box Fig. 6.2   (a) The sea slug *Tritonia diomedea*, fully expanded, has been placed in a flume in a strong current (ca. 15 cm sec^{-1}). (b) To avoid being swept away, it has lowered its veil and flattened the anterior part of its foot, to reduce the cross-sectional area presented to the flow. (Courtesy of James Murray.)

to the turntable indicated the amount of drag. The more the transducer was stretched, the greater the drag on the slug. As it turned out, drag was minimized when the slug was oriented with its anterior–posterior axis parallel to the flow, facing head first. In this orientation the slug has a teardrop shape (with the wide end of the teardrop pointing upstream), which directs flow about the animal most efficiently and minimizes drag. If flow increases dramatically, then the animal lowers its oral veil, a curtainlike structure that drapes over its anterior end. As the current strength increases, the slug also flattens the entire foot (Box Figure 6.2b). The flattening reduces and streamlines the cross-sectional area that faces the current, and thus reduces drag. Eventually the slug stops moving. Other species use the tactic of burrowing into the sediment to avoid being carried away in the flow.

Tritonia sea slugs use both mechanosensory and chemosensory aids to orient into the flow and to locate prey. A rhinophore projects upward from either side of the head, and chemosensory neurons can detect molecules emanating from prey, which include sea pens and sea whips (both of the phylum Cnidaria). The oral veil has mechanoreceptors that detect flow. Nerves extend into the left and right sides of the oral veil. If both nerves are surgically severed, the slug's ability to detect flow is impaired. When a slug finally encounters prey, mechanoreceptors on the oral veil help the slug orient the prey so that it can be grasped by the jaws and rasped by the radular apparatus, a behavior that is typical of nearly all gastropods.

The same mechanisms that *Tritonia* slugs use to orient themselves in a flow are used to locate odor plumes of their sea pen and sea whip prey. Dennis Willows[7] found that the *Tritonia* orient in a Y-shaped chamber orient toward the branch of the Y in which sea pen prey had been placed. As in the case of the blue crab, chemosensors detect the strength of odors, and orientation may be aided by differential lateral sensation. If the current emanating from prey is increased, the sea slug moves faster toward the prey. Moving upstream toward the odor serves the objectives of both locating food and orienting to minimize drag.

[7] See Willows, 1978, in Further Reading.

---

back to an extent as it swims. Through streamlining, the wake is diminished and the drag is reduced greatly. That is why fast and continuously swimming fish exhibit some variation of the classic shark or tuna shape.

The question arises: If length and slenderness reduce drag, why aren't fast-swimming fish much longer and more slender than they actually are? Why don't they have an eel shape? The answer has to do with the friction caused by the surface of the fish in contact with the water. The longer and narrower the fish, the more fish surface there is over which the water must move, and the greater the friction for a given body weight. The shape of a shark or a tuna is therefore a compromise between considerations of drag and friction minimization.

The Use of Flumes

✳ **Flumes are useful for studying the effects of moving fluids on organisms, although flumes must be scaled carefully.**

It is usually quite tricky to study the effects of flow in the marine environment. Some clever investigators have devised field current meters and have been able to characterize the flow field around an organism. Because of inaccessibility, this becomes impractical for studying the streamlining of a tuna or the flow about a deep-sea organism. Even in accessible habitats, it is very difficult to measure flow on the smallest of scales. Electronic devices, such as flow meters based upon thermistor sensors, are usually difficult to use accurately in the field.

In still waters, microcinematography allows us to study the behavior of very small creatures at low Reynolds numbers. In moving waters, flumes of various types are used to study the effects of flow. A flume is a device that includes a source of moving water, a working area where the organism and flow field are characterized, and a drain–return system (Figure 6.10).

Most flume designers seek two objectives: maintenance of laminar flow and maintenance of scaling by Reynolds number (and there are also other parameters beyond the scope of this text). A long flume is desirable because it takes a while for the flow over the bottom surface to stabilize and produce a predictable boundary layer and velocity profile above

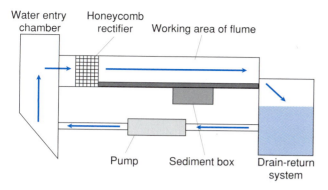

Fig. 6.10 A flume designed to study the effects of flow on small epibenthic animals. In this design, water is recirculated by means of a pump. Water enters a honeycomb-like material to rectify the flow and then flows into the working area of the flume, where organisms are placed. Water leaves the working area and drains into a sump, from which it is pumped to the water entry chamber.

the bottom. A wide flume, relative to water column height, prevents effects of the walls on flow. Scaling by Reynolds number is also essential to keep the proper ratio of inertial to viscous forces. This refinement has an advantage, however, in that you can study a very small object, such as a copepod, by making a larger model and placing it in a more viscous medium.

Using a flume, one can study the hydrodynamic forces at work on a biological object. For example, I have studied[8] the reaction of the siphon of a sediment-ingesting bivalve mollusk, *Macoma secta*, which feeds much like a vacuum cleaner. At low current velocity, the siphon is protruded into the water,

and swirls around, picking up sand grains. If the current increases to about 15 cm s^{-1}, the pressure drag on the siphon makes the organ difficult to control, and it is withdrawn. The animal feeds on sediment within the burrow. At velocities above 35 cm s^{-1}, the bottom sediment is stirred up and the bottom is very unstable. At that point the animal ceases to feed. Such qualitative observations can be carried out quite easily by using a flume with a video recording device.

Conflicting Hydrodynamic Constraints

✱ Hydrodynamic forces often present conflicting constraints.

Hydrodynamics may suggest simple rules for both behavior and morphology, but many marine organisms find themselves having to live with conflicting needs. Consider the blade of sessile eelgrass discussed earlier. It projects upward from the bottom and thereby is able to capture light and nutrients. However, this upward projection causes a significant pressure drag. There is a conflict of different functional requirements. The size and velocity scaling of hydrodynamic effects create additional conflicts. As an organism grows in length, the Reynolds number increases. Consider Figure 6.11, which shows patterns of flow downstream of a cylinder under conditions of varying Reynolds number. When *Re* is about 30, a pair of attached vortices reside just downstream of the cylinder. This might give a small coral the "opportunity" of feeding on particles that are relatively sta-

[8] See Levinton, 1991, in Further Reading.

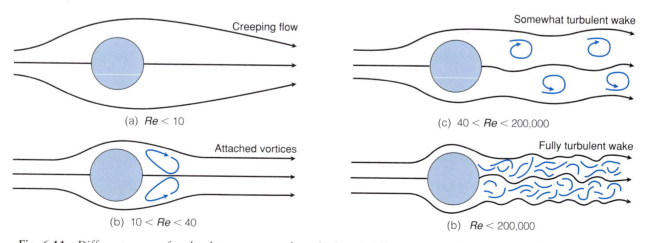

Fig. 6.11 Different types of wake down-current of a cylinder, at different Reynolds numbers. (After Vogel, 1981.)

tionary. As the coral grows larger (or as the current increases in velocity), these vortices become more erratic, however, and food may not be held in a pre-

dictable pattern. Size and velocity can increase together; as an organism grows larger, it often projects into a rapid "mainstream" current.

Further Reading

Denny, M. W. 1988. *Biology and the Mechanics of the Wave-Swept Environment*. Princeton, NJ: Princeton University Press.

Denny, M. W., T. L. Daniel, and M. A. R. Koehl. 1985. Mechanical limits to size in wave-swept organisms. *Ecological Monographs*, v. 55, pp. 69–102.

Judge, M. L. 1988. The effects of increased drag on *Lottia gigantea* (Sowerby 1834) foraging behavior. *Functional Ecology*, v. 2, pp. 363–369.

Koehl, M. A. R. 1976. Mechanical design in sea anemones on the flow forces that they encounter. *Journal of Experimental Biology*, v. 69, pp. 87–105.

Levinton, J. S. 1991. Variable feeding behavior in three species of *Macoma* (Bivalvia: Tellinacea) as a response to water flow and sediment transport. *Marine Biology*, v. 110, pp. 375–383.

Murray, J. A., R. S. Hewes, and A. O. Dennis Willows. 1992. Water-flow sensitive pedal neurons in *Tritonia*: Role in rheotaxis. *Journal of Comparative Physiology A*, v. 171, pp. 373–385.

Purcell, E. M. 1977. Life at low Reynolds number. *American Journal of Physics*, v. 45, pp. 3–11.

Vogel, S. 1981. *Life in Moving Fluids*. Boston: Willard Grant Press.

Vogel, S. 1992. *Vital Circuits*. New York: Oxford University Press.

Vogel, S. 1993. *Life in Moving Fluids*, 2nd ed. Princeton, NJ: Princeton University Press.

Vogel, S., and W. L. Bretz. 1972. Interfacial organisms: Passive ventilation in the velocity gradients near surfaces. *Science*, v. 175, pp. 210–211.

Wainwright, S. A., W. D. Biggs, J. D. Currey, and J. M. Gosline. 1976. *Mechanical Design in Organisms*. Princeton, NJ: Princeton University Press.

Weissburg, M. J. 2000. The fluid dynamical context of chemosensory behavior. *Biological Bulletin*, v. 198, pp. 188–202.

Weissburg, M. J., and R. K. Zimmer-Faust. 1993. Life and death in moving fluids: Hydrodynamic effects on chemosensory-mediated predation. *Ecology*, v. 74, pp. 1428–1443.

Willows, A. O. D. 1978. Physiology of feeding in *Tritonia*. I. Behavior and mechanics. *Marine Behavior and Physiology*, v. 5, pp. 115–135.

Yates, G. T. 1986. How microorganisms move through water. *American Scientist*, v. 74, p. 358–375.

Review Questions

1. What is dynamic viscosity? Give an example of a fluid that is more dynamically viscous than water.

2. What information does the Reynolds number provide?

3. In seawater, what are the principal factors that vary to determine the Reynolds number?

4. Why can a ciliate stop nearly instantaneously, whereas a large swimming fish takes much more time and a higher number of body lengths to stop?

5. What is the no-slip condition?

6. What conditions are different within the boundary layer, relative to mainstream flow conditions?

7. How can a sponge have an exit velocity through its exhalant siphon that is orders of magnitude greater than the intake velocity?

8. How might an attached organism reduce drag, even when it protrudes into the mainstream flow?

9. How might a burrowing worm take advantage of the Bernoulli principle to enhance the flow of water through its burrow?

10. What are the advantages and disadvantages encountered by a sessile marine organism that projects its body into the mainstream flow?

III

ORGANISMS OF THE OPEN SEA

7

The Water Column: Plankton

Introduction and Definitions

The **plankton** are those organisms that live in the water column and are too small to be able to swim counter to typical ocean currents. **Phytoplankton** are photosynthetic planktonic protists and plants, and usually consist of single-celled organisms or of chains of cells. Although some have locomotory organs such as flagella, phytoplankton movements in the water column are nearly completely controlled by water turbulence and currents and by the bulk density of the organisms. **Zooplankton** comprise nonphotosynthetic planktonic protists and animals, ranging from single-celled forms to smaller vertebrates, such as larval fishes. Although they may be able to swim, major water currents, turbulence, and bulk density determine their major movements. Some plankton do not fit either the zooplankton or the phytoplankton mold. Some protists for example, are photosynthetic, but also can ingest other plankton. Such plankton are said to be **mixoplankton**. **Meroplankton** are zooplankton that spend only part of their lifetime in the plankton, but are benthic for the remainder. They include the planktonic larval stages of many benthic invertebrate groups. **Holoplankton**, by contrast, are those planktonic organisms that spend all their time in the water column. **Neuston** are the plankton associated with the water surface, such as bacterial films, whereas **pleuston** are plankton that live at the surface but protrude into the air. The latter include animals that have floats, such as the Portuguese man-of-war.

Planktonic organisms are also classified by their size. These classes are **ultraplankton** (<2 μm), **nannoplankton** (2–20 μm), **microplankton** (20–200 μm), **macroplankton** (200–$2,000$ μm), and **megaplankton** ($>2,000$ μm). Most plankton nets (Figure 7.1) catch plankton of microplankton size and larger.

Life in the Open Sea

✳ **The vertical position of plankton is determined by the bulk density of the organism, structures that create drag, water motion, and swimming ability.**

Planktonic organisms usually depend upon the surface waters for survival. Phytoplankton will die unless they are near a source of sunlight for photosynthesis. They will be inviable if they sink below a depth of 50–100 m in the open sea but are unable to photosynthesize even in much shallower depths in estuaries and inshore waters. Zooplankton usually depend upon phytoplankton, or at least upon animals that consume phytoplankton. Zooplankton, therefore, also must remain in the surface waters. How do plankton keep near the surface?

Observation of a pot of soup can reveal much

Fig. 7.1 Paired plankton nets are placed overboard. The length of wire and angle of entry are measured to find the depth of sampling. A calibrated propeller gives the volume of water sampled. (Courtesy of Scripps Institution of Oceanography.)

Table 7.1 Sinking rate of various particles and living organisms in the plankton

GROUP	SINKING RATE (m day^{-1})
Phytoplankton	
Living	0–30
Dead, intact cells	<1–510
Fragments	1,500–26,000
Protozoans	
Foraminifera	30–4,800
Radiolaria	350
Other Zooplankton	
Chactognaths	435
Copepods	36–720
Pteropods	760–2270
Salps	165–253
Fecal pellets	36–376
Fish eggs	215–400

Source: Modified from Parsons and Takahashi 1973, with permission of Pergamon Press, Ltd.

about the distribution of plankton in a natural water column. If you stir up the soup, many pieces will be lifted up from the bottom into the fluid, but will eventually settle again as the broth calms down. Some pieces are less dense than the broth, and they will float. Hopefully, nothing is swimming in your soup, but that is the only other way for a particle to remain above the bottom of the pot in still water. Application of these points to seawater reveals that a particle will remain suspended if (1) it is less dense than seawater; (2) it has a shape that increases drag and reduces settling velocity; (3) it can swim; or (4) water turbulence keeps the particle suspended in the water column. In the sea, turbulence is usually generated by wind, which mixes the water column and forestalls the sinking of plankton that are denser than seawater. All four mechanisms combine to keep plankton from sinking rapidly to the bottom or from sinking to depths greater than the depth at which they can photosynthesize or survive for other reasons.

Many planktonic organisms are somewhat denser than seawater and will therefore sink in a quiet water column (see Table 7.1). The silica skeleton of diatoms, for example, makes their bulk density greater than that of seawater. Most crustaceans are denser than seawater and will surely sink unless they swim upward. Not all plankton have negative buoyancy, however. Flotation structures are commonly employed to keep the animal suspended. In the extreme

case, for example, siphonophores like the Portuguese man-of-war *Physalia* have a large gas-filled sac that acts as a float from which the rest of the colony is suspended. In some species, the gas can be withdrawn to permit the colony to sink below the surface during storms, thereby avoiding the surface turbulence. A second overall mechanism is the **regulation of bulk density** through variations in chemical composition. The dinoflagellate *Noctiluca*, for example, accumulates ions of low specific gravity, which reduces their density and some blue-green bacteria are believed to have vacuole-like structures that contain low-density gaseous nitrogen. Many zooplankton become more neutrally buoyant by replacing dense magnesium, calcium, and sulfate ions with lower-density ammonium, sodium, and chloride.

Gas secretion is used by many animals as a means of maintaining neutral buoyancy. The nektonic cephalopods are able to regulate the gas content of their bodies and thereby achieve neutral buoyancy under different conditions of depth and feeding. In the pearly nautilus, the gas content of the inner chambers of the shell can be filled or emptied, to suit individual conditions. In the cuttlefish *Sepia*, the inner spaces of the cuttlebone, an internal skeleton, can be regulated similarly. In fish, the swim bladder is used to adjust bulk density (see Chapter 8).

Many planktonic creatures have shapes that retard sinking. The bell-shaped jellyfish is an obvious example. Rhythmic contractions expel water from the bell and propel the jellyfish upward, but the jellyfish's flat bottom and tentacles combine to create a drag that slows sinking relative to a sphere of the same shape and bulk composition. Some diatoms, such as

Hot Topics in Marine Biology 7.1
The Water Revolution:
Water Flows, or Does It?

A newfound appreciation for the physics and chemistry of water flow has been revolutionizing the thinking of marine biologists about water and turning many of their notions upside down. Much of the new thinking relates to the issues of size and speed. Small objects moving slowly through water simply do not behave like large ones moving rapidly. A protozoan that is smaller than a pinhead and that moves very slowly is stuck in a honeylike world. If it stops making an active effort to move, it will not continue to move, but will stop. There is nearly no inertial effect for the protozoan.

An understanding of life at low Reynolds number has changed how we think about the process of filter feeding, which is probably the single most common form of animal feeding in the sea. Planktonic animals such as copepods use feeding appendages covered with hairs to feed on particles. Scientists once thought that these hairs filtered out particles that were larger than the distance between the hairs, much as a kitchen sieve might catch peas. On the small scale of these organisms, however, water doesn't behave like that, and it has become clear that the particles are "plucked" out of the water, much as one would pick a peppercorn out of a salami. If feeding is more of grabbing action than a sieving process, it follows that copepods would be rewarded if they could sense particles individually. When this was realized, scientists searched for, and found, smell receptors on the feeding appendages. Suddenly, copepods seemed to be more like sense-equipped, decision-making human beings at the dinner table than like mechanical sieves automatically trapping particles.

Chaetoceros, form rather twisted chains, which spiral as they sink slowly. Many tropical zooplankton have elaborate projections, which also retard sinking but may in fact serve primarily as a defense mechanism against predators.

Swimming is the final major means of avoiding sinking. Although flagellated phytoplankton can swim to a degree, this mechanism is best developed in zooplankton. In the pteropods (Plate VI.4), for example, the foot has two lateral winglike projections, and the snails flap through the water. Jellyfish move in pulses, through rapid compressions of circular muscles, which compress the bell and force water backward. Crustaceans such as copepods use assorted appendages to push against the water to move forward, while arrow worms swim by undulation, as do many fish. All these swimming methods are effective means of countering sinking.

✴ **Small zooplankton and phytoplankton live in a world dominated by low Reynolds number.**

If a body is denser than seawater, and the water is perfectly still, then the body will sink to the bottom. In Chapter 4, we discussed the conditions of a body that is very small and moving at low velocity. Under such conditions of **low Reynolds number** ($Re < 0.5$), there is a boundary layer around the body, which causes it to be dragged by the surrounding water. Under these conditions, if the particle is small and the velocity slow, the sinking velocity will increase with increasing weight of the particle. These organisms will reach a constant terminal velocity. This phenomenon is known as Stokes's law. Heavier organisms will sink faster than lighter organisms. Remember, though, that this applies only to nannoplankton and smaller, relatively slow-swimming microplankton.

Phytoplankton

Diatoms

Diatoms (Class Bacillariophyceae, Phylum Chrysophyta) dominate the phytoplankton in waters from the temperate to the polar zones. They either occur as single cells or form chains (Figure 7.2) in the size range of nannoplankton and microplankton. Each cell is encased in a silica shell (actually a cell wall impregnated with silica) of two valves that fit together much like a pillbox. The shell may be covered with

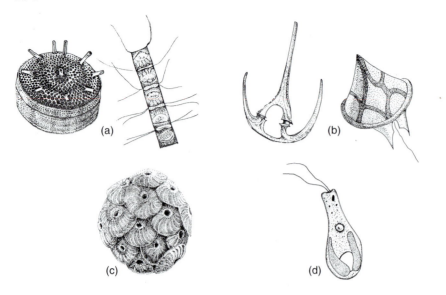

Fig. 7.2 Some members of the phytoplankton: (a) diatoms, (b) dinoflagellates, (c) coccolithophore, and (d) the microflagellate *Isochrysis*.

spines, or it may be ornamented with a complex series of pores and ridges. The pores are the only connection between the cell and the external environment. Planktonic diatoms are usually radially symmetrical (shaped like a pinwheel) and are known as centric diatoms (Plate III), in contrast to pennate diatoms, which are bilaterally symmetrical and usually live in sediments.

Diatoms reproduce by binary cell division, with one valve going to each daughter cell and serving as the larger valve. A smaller valve is then formed. After several successive generations of cell division, cell size usually decreases to a lower threshold, when gametes are often formed. The smaller diatoms may also form an auxospore, which enlarges and casts off the valves, forms new large valves, and then undergoes asexual divisions. Many diatom species can form asexual resting spores, which settle to the seabed and may be regenerated later as a planktonic form. Because of their asexual mode of reproduction, diatoms can increase in population size rapidly, with doubling rates on the order of 0.5–6 per day.

Dinoflagellates

The dinoflagellates (Class Dinophyceae, Phylum Pyrrophyta) are unicellular forms that often dominate the subtropical and tropical phytoplankton and also may dominate the summer and autumn phytoplankton of the temperate and boreal zones. They have two flagella and are in the size range of nannoplankton and microplankton (Plates IV.1–2). The transverse flagellum is located in a groove (the girdle) that divides the cell into two subequal parts (as shown in Figure 7.2). The other flagellum is oriented perpendicularly to the transverse flagellum and extends toward the posterior. The organism is generally covered with a series of contiguous cellulose plates, the theca, whose pattern is usually diagnostic of the species.

Dinoflagellates reproduce asexually by binary fission. Like diatoms, they have a capacity to reproduce up to several times per day. During cell division, the theca may be shed, or some of the theca may be inherited by each daughter cell. Sexual reproduction also occurs, and dinoflagellates can form resting stages, or cysts, which sink to the bottom and can be regenerated later as planktonic forms. Some dinoflagellates are bioluminescent, and the gentle flickering of shallow waters is often due to luminescence by *Noctiluca*.

Some dinoflagellates, such as the genera *Protogonyaulax*, *Gonyaulax*, and *Gymnodinium* are known to be the organisms involved in **red tides,** which are caused by rapid increases of dinoflagellate populations of sufficient magnitude to color the seawater a dull red-brown. Owing to their production of the deadly neurotoxic substance saxitoxin, these increases are dangerous to people. Commercially exploited shellfish consume the dinoflagellates and concentrate the toxin. Saxitoxin depresses sodium transport and inhibits nervous transmission. You can usually tell immediately that you have eaten a saxitoxin-contaminated shellfish because your lips feel numb. Within 12 hours of ingestion of a toxic bivalve, human respiration is inhibited and cardiac arrest may follow. Saxitoxin also affects the shellfish, potentially causing reduced activity and death.

The origin of red tides, like many other sudden blooms of phytoplankton, is an incompletely explained phenomenon. Red tides are often associated with sudden influxes of nutrients or washout of nutrients from land sources into the sea. Storms may remobilize populations of cysts buried in the bottom sediment, setting the stage for red tides.

Coccolithophores

Coccolithophores (Phylum Chrysophyta) are unicellular and usually nannoplanktonic. They are important in the tropical pelagic phytoplankton. They are nearly spherical and are covered with a series of calcium carbonate plates, or coccoliths, which often blanket the deep pelagic seabed (refer again to Figure 7. 2). Also in this group are a large number of naked forms that are poorly preserved. They can be quite an important component of some phytoplankton assemblages.

Silicoflagellates

Silicoflagellates (Phylum Chrysophyta) are unicellular and biflagellate; they have numerous chloroplasts and an internal skeleton of silica scales. They are usually less abundant than diatoms but are notable in Antarctic phytoplankton and in the plankton of many other open-ocean locales.

Cyanobacteria

The cyanobacteria, which are members of the Cyanophyceae (often called blue-green algae, although they are not true algae), occur in and may dominate nearshore waters of restricted circulation, as well as brackish water. The filamentous *Trichodesmium* (Figure 7.3) is found in the nutrient-poor waters of the warm oceanic gyres. Tiny (roughly 1 μm) unicellular forms have been found ubiquitously throughout the ocean and may be the food of smaller zooplankton. Cyanobacteria can undergo **nitrogen fixation,** in which gaseous nitrogen is converted to NH_4^+, or ammonium ion, and is then available for incorporation into amino acids and proteins.

Green Algae

The true green algae (Class Chlorophyceae, Phylum Chlorophyta) are rare in marine waters but can dom-

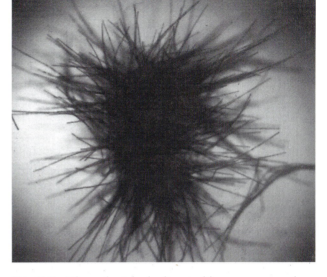

Fig. 7.3 The colonial planktonic blue-green cyanobacterium *Trichodesmium thiebautii*. (Photograph by Edward Carpenter.)

inate the phytoplankton of enclosed estuaries or enclosed lagoons, especially in late summer and fall. They can be flagellated or nonmotile. Several species cause nuisance phytoplankton blooms associated with coastal pollution. Green algae are much more important in the benthos and dominate intertidal soft-bottom seaweed assemblages.

Cryptomonad Flagellates

Members of the class Cryptophyceae (Phylum Cryptophyta) are widespread and locally abundant in estuaries. They are unique in having chlorophyll types *a* and *c*, as well as the photosynthetic light-absorbing pigment phycobilin.

The Zooplankton

Crustacean Zooplankton

Crustaceans comprise a class of the phylum Arthropoda, whose distinctive features include (1) an external skeleton of chitin, a flexible but stiff material, that is relatively impermeable to the external environment, and (2) some degree of segmentation, with paired, jointed appendages (e.g., legs, antennae). The crustaceans possess antennae, mandibles, and maxillae as head appendages, and usually have compound eyes. They include the crabs, shrimp, lobsters, cray-fish, and sow bugs (which are land organisms). They are notable for their mobility, armored exoskeleton, and generally good vision.

COPEPODS

There are more copepods than any other kind of zoo-plankton, and these organisms dominate nearly all the oceans and marginal seas. The crustacean order Copepoda is by far the largest group of crustaceans in the zooplankton. They range in length from less than 1 mm to a few millimeters. The calanoid cope-pods dominate and are free living. By weight, there are probably more calanoid copepods than any other planktonic animal group. The harpacticoids usually have benthic adults, but the larvae of some species may dominate the estuarine zooplankton. Members of the Lernacopodoida parasitize marine mammals and fish. They are wormlike and can project from their host over 30 cm into the water.

Calanoid copepods (Figure 7.4, Plates VII.1 and VII.2) are usually barrel shaped, and the body is com-posed of head, thorax, and abdomen. They swim by

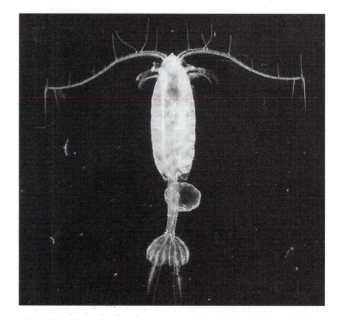

Fig. 7.5 The planktonic copepod *Euchaeta norvegica* de-tects prey food particles ahead of means of sensory hairs, which are concentrated on the large first antennae. (Cour-tesy of Jeanette Yen.)

means of rhythmic strokes of the first pair of anten-nae and the five posterior pairs of thoracic ap-pendages. Calanoids have a median naupliar eye but lack the compound eyes found in many other crus-taceans. In the genus *Calanus*, the female lays eggs in clutches of about 50, every 10–14 days. Larvae go through a series of naupliar and copepodite molts be-fore the adult stage.

Calanoids feed mainly on phytoplankton, some or-ganic particles, and smaller zooplankton. Detection of food occurs first on the first antennae, which are festooned with sensory hairs (Figure 7.5). They trap particles on hairlike maxillary setules, located on se-tae, which, in turn, are located on the maxillipeds. Recall the conditions for low Reynolds number (see Chapter 6), where water velocity is low and the size of the objects in the fluid are small. In these circum-stances, water does not flow freely past a copepod's feeding appendage. Rather, viscous flow dominates around the appendage. The animal flaps four pairs of appendages (Figure 7.6) to propel water past it-self. As a diatom comes near the copepod, the max-illiped reaches out and grabs it, rather than sieving it through the setules (as was originally thought). Ac-tually, because of the viscosity of its environment, the animal grabs the diatom with a surrounding enve-

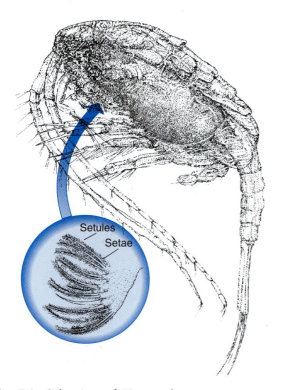

Setules
Setae

Fig. 7.4 Side view of *Temora longicornis*, a common calanoid copepod of inshore waters of New York. The sec-ond maxilliped is shown in detail. (From Ninivaggi, 1979.)

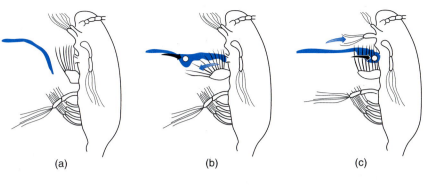

Fig. 7.6 Left-side view of a calanoid copepod, showing (a) the feeding current created by flapping of the appendages, (b) the outward stroke to reach and the particle (diatom), and (c) the inward stroke to capture the particle. (After Koehl and Strickler, 1981.)

lope of water, which may have to be strained off the particle before it is passed to the mouth parts. In the larger copepods, Reynolds number is about 1 during feeding and there is some contribution of inertial forces. Under these circumstances, sieving by the setules may occur. The animal must detect the presence of a diatom by means of chemosensors on the feeding appendages before reaching out for the food.

KRILL

Krill (Figure 7.7a, Plate VII.4), which are members of the crustacean family Euphausiidae, are shrimplike

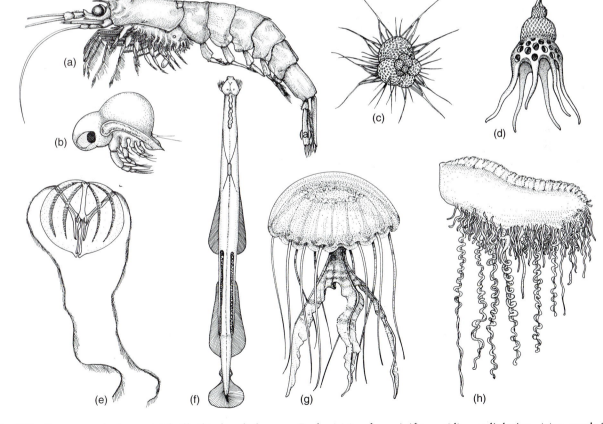

Fig. 7.7 Some zooplankton: (a) krill, (b) the cladoceran *Podon*, (c) a foraminiferan, (d) a radiolarian, (e) a comb jelly, (f) an arrow worm, (g) a scyphozoan jellyfish, and (h) a pleustonic siphonophore, the Portuguese man-of-war *Physalia physalis*.

creatures ranging up to 5 cm long. They dominate the zooplankton of much of the Antarctic Ocean but are also common in pelagic waters of high productivity throughout the world, such as in the Benguela Current off Africa. In the Antarctic, they are the main food of baleen whales.

Krill feed on phytoplankton and smaller zooplankton, by means of six long limbs attached to the cephalothorax and serving as a basket. In the Antarctic *Euphausia superba*, the basket is closed on all sides by means of interlimb contact. Like the calanoid copepods, krill have appendages adorned with setae and setules, which capture the food particles. Sperm are transferred in spermatophores and eggs are carried in the thoracic basket or in ovisacs attached to the ventral thorax.

All but a single genus of euphausiids are luminescent. The luminescent material is intracellular and is located in light-producing photophores, usually on the upper end of the ocular peduncle.

CLADOCERA

Cladocera are more important in fresh water than in seawater, but they are sometimes abundant in estuaries. The genus *Podon* (Figure 7.7b) preys on other zooplankton.

OTHER CRUSTACEANS

A few mysids, ostracods, and cumaceans are truly planktonic but rarely dominate the zooplankton. Some mysids rise into the plankton at night but live on the bottom during the day. A few amphipods (Plate VII.3) are holoplanktonic (such as the genera *Euthemisto* and *Hyperia*) and are at times important members of the zooplankton in many parts of the world.

Gelatinous Zooplankton

COELENTERATES (PHYLUM COELENTERATA OR CNIDARIA)

The class Scyphozoa—the true jellyfish—is divided into several taxonomic orders. As mentioned earlier, jellyfish swim by rhythmically contracting the bell (Plate V.1), which propels them forward (Figure 7.7g). They capture zooplankton on their tentacles, by means of stinging and with sticky structures called nematocysts. Some nematocysts bear poisons, and some of the sea wasps (Cubomedusae) and the Portuguese man-of-war *Physalia physalis* (Figure 7.7h) can sicken and even kill human beings.

The siphonophores are a specialized colonial and polymorphic group of cnidarians (Plate V.4). Individuals of differing morphology serve the functions of feeding, reproduction, and floating. In the Portuguese man-of-war *Physalia physalis*, one individual is a pneumatophore, or float, and may be as long as 10–30 cm. The tentacles dangle below. A small fish, *Nomeus gronovii*, lives among the tentacles of *Physalia* and is well protected against predators. It is probably a commensal, but some have suggested that the fish attracts predators into the tentacles of its stinging host. The by-the-wind sailor *Velella velella* is smaller but has a remarkable sail-like structure, which catches the wind (Figure 7.8). These organisms can be seen in the millions off the Pacific coast and are often washed ashore from the outer coasts of California to Washington. The animal has a prominent skirt around the gas-filled float, which stabilizes the animal as the wind hits the sail. The sail is a low-profile flat sheet that takes advantage of drag for downwind sailing. Unlike modern tall triangular yacht sails, *Velella's* sail does not use the Bernoulli principle to create lift. The closely related *Porpita porpita* seems also to have a stabilizer, in the form of radiating tentacles (Figure 7.9), but it lacks a sail.

THE COMB JELLIES

Members of the phylum Ctenophora are gelatinous, nearly transparent, and egg shaped (Plate VI.1–3). Comb jellies are distinguished by eight external rows of meridional plates, but some have long tentacles (Figure 7.7e). They are carnivorous and are especially effective at consuming copepods. *Pleurobrachia* captures copepods on its long looping tentacles and draws the tentacles toward its mouth (Figure 7.10).

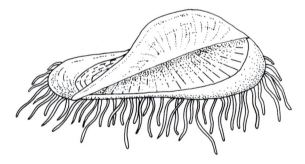

Fig. 7.8 The by-the-wind sailor *Velella* has a sail, which catches the wind. The skirt stabilizes the animal as the wind strikes the sail. Millions of these creatures may be found stranded on North American coasts from California to Washington.

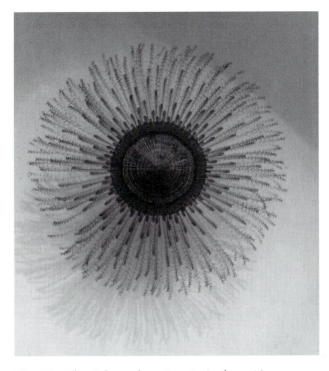

Fig. 7.9 The siphonophore *Porpita* is about 10 cm across and is stabilized against wind and turbulence by tentacles on the water surface.

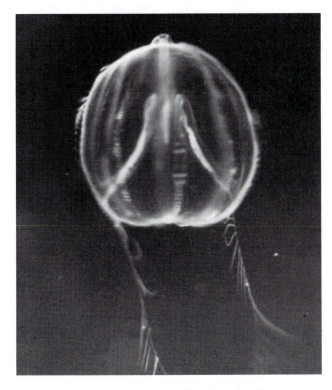

Fig. 7.10 The comb jelly *Pleurobrachia*.

Mnemiopsis lacks the long tentacles, and cilia move the prey into grooves, where they are entangled by a row of short tentacles, which pass the food to the labial trough, a structure that leads directly to the mouth. The feeding mechanism is very efficient, and comb jellies are believed to cause strong declines in shellfish larval populations. A recent invasion of *Mnemiopsis leidyi* into the Black Sea resulted in the decline of a number of species of fish, owing to predation on eggs and larvae. Gametes are usually shed into the water, and the embryo develops into a larva that resembles the adult. Comb jellies are often strongly bioluminescent and light up like flashing bulbs when disturbed.

SALPS

The Thaliacea (Phylum Urochordata; see Chapter 11) are specialized for a free-swimming planktonic existence. Unlike their relatives the benthic sea squirts, they have intake and exit siphons at opposite ends of the body. *Salpa* is barrel shaped and solitary. *Pyrosoma*, by contrast, is colonial and shaped like a cylinder closed at one end. Colonies reach over 2 m in length, and each individual has its intake siphons oriented outside, whereas its exit siphon empties into the central cavity. Salps strain phytoplankton and fine particulate matter on a ciliary mucus net. Salps feed on particles in the range of 1 μm to 1 mm. Some salps are important predators on fish larvae.

LARVACEA

The larvacea (Phylum Urochordata; see Chapter 11) are another group of specialized planktonic gelatinous zooplankton. Usually only a few millimeters in length, these organisms have some features, such as a tail, that are typical of tunicate swimming larvae. The animal constructs a "house," to which it is attached or by which it is enclosed. The beat of the tail generates a current through the house. The current is strained by a grid of fine fibers, stretched across the anterior house opening, that trap food. Particles are passed to the mouth and are sorted in the pharynx.

Other Zooplankton

ARROW WORMS

Arrow worms (Phylum Chaetognatha), are torpedo shaped (length is 4–10 cm), with one or two pairs of lateral fins (Figure 7.7f). They swim rapidly by means of rapid contractions of longitudinal trunk muscles. Armed with grasping spines, the head is well adapted for grabbing prey, and arrow worms feed voraciously on other zooplankton. Individual species are often confined to specific water masses, and certain of the species can be used to distinguish pelagic from neritic waters. They are hermaphroditic, and eggs may be shed into the water or attached to floating objects. The larva develops directly into a free-living juvenile.

PTEROPODS

Pteropods are holoplanktonic snails that swim by means of lateral projections from the foot, which is otherwise reduced relative to the bottom-living snails (Plate VI.4). Pteropods are sometimes quite abundant, and the shells of one group—the thecosomes—sink to the bottom in great abundance and form sediments known as pteropod ooze.

PLANKTONIC POLYCHAETES

A few families of polychaetes are holoplanktonic and have well-developed locomotory appendages (parapodia) and sense organs (e.g., the genus *Tomopteris*).

Protistan Plankton

PROTISTS

Single-celled animals are common in the plankton. Protists are of major importance in the plankton, especially because they consume very small creatures, such as bacteria, that are largely unavailable to most other zooplankton. Protists are consumed by larger zooplankton and are therefore a major link between microbial forms and the rest of the planktonic food chain (see Chapter 9).

Ciliates (Phylum Ciliophora), are ubiquitous in the plankton and often very abundant. They are elongate usually and are covered with rows of cilia, whose coordinated beating propels water, thus setting the protistans in motion. Ciliates have an oral opening (Figure 7.11), and cilia around this opening move food particles into the body, where they are engulfed by food vacuoles and digested. Ciliates usually feed on bacteria and phytoplankton. Some species may take up phytoplanktonic cells but retain the chloroplasts, which remain functional within the ciliate. Photosynthetic products are transferred into the body of the ciliate.

Members of the phylum **Foraminifera** (Figure 7.7c) are common throughout the ocean. They range from less than 1 mm to a few millimeters in size. They usu-

Fig. 7.11 Scanning electron micrograph of a marine planktonic oligotrich ciliate *Strombidium conicum*. Oligotrichs are the dominant ciliates in the plankton, but some, which retain chloroplasts as they feed on phytoplankton, are photosynthetic. (Courtesy of Diane Stoecker.)

ally secrete an external calcium carbonate skeleton, which is divided into chambers. High-latitude forms usually have simple spiral groupings of spherical chambers, whereas low-latitude species include forms with elaborately spined sculpture. Cytoplasm occupies the chambers and streams out through perforations in the shell to form contractile pseudopodia, which capture phytoplankton and bacteria.

Reproduction of foraminiferans usually involves several cycles of asexual cell division, alternating with a cycle of gamete formation. After the gametes have fused, the zygote produces a microspheric shell, which is smaller than the megalospheric form, which reproduces asexually by fission. Some tropical Foraminifera have intracellular algal symbionts, which contribute dissolved food to the animal in the partnership.

Foraminifera are abundant in the open sea, and certain species are good indicators of water masses. Forams sink to the bottoms in great numbers. In depths less than 2,000 m, where calcium carbonate does not dissolve, they form deep-sea sediments known as globigerina ooze, named after a common foraminiferan genus.

The **Radiolaria** (Figure 7.7d) range from less than 50 μm to a few millimeters in size, and colonial forms can attain several centimeters. They are common, especially in tropical pelagic waters. A membrane of pseudochitin separates the body into a central capsule and an extracellular cytoplasm, the calymma. Straight, threadlike pseudopodia (axopods) radiate from the central capsule. The silica skeleton is usually a combination of radiating spines and spheres, producing a complex lattice of great beauty.

Radiolarians feed much like foraminiferans, and some species have symbiotic algae, known as zooxanthellae, within the calymma. The radiolarians derive some nutrition from these algae. Radiolarians reproduce asexually by binary fission, but gametes are also produced. Radiolarians are sufficiently abundant in some parts of the ocean to sink to the bottom and dominate the sediment, forming radiolarian ooze.

Patchiness and Vertical Migration

Patchiness of the Plankton

✳ **Although the vertical distribution of phytoplankton is often determined by turbulence and sinking of organisms, zooplankton can swim sufficiently well to aggregate, sometimes in deep scattering layers.**

Both phytoplankton and zooplankton are aggregated in space, on many spatial scales. In some cases, discrete patches are encountered. A plankton net will capture large numbers of a species in one location, but none may be caught in an adjacent location. More frequently, patchiness can be defined only statistically: individuals are found more frequently together than one could predict from an expectation of randomness (see Chapter 3). The following are the major causes of patchiness in the plankton:

1. Spatial changes in physical conditions, such as light, temperature, and salinity
2. Water turbulence and current transport
3. Grazing in some areas and reduced grazing in others
4. Localized reproduction
5. Aggregating social behavior

Wind moving over the water surface generates spatial structure over a wide range of spatial scales. **Langmuir circulation** results from the creation of vortices by wind-driven water movement. The vortices are small in scale and result in small divergences and convergences of water (Figure 7.12). Phytoplankton may be concentrated in the convergences. Zooplankton may be trapped in an upward current while attempting to swim downward to avoid light originating from the surface. Langmuir circulation forms linear convergences at the surfaces (Figure 7.12a). Wind-generated turbulence can generate medium-scale patchiness in coastal waters (ca. 10^2–10^3 m). Apparent differences in abundance between sampling stations may reflect temporal changes that are likely to occur in any particular place. In the shallow waters of Nova Scotia, time-dependent variation of chlorophyll (which is used to estimate phytoplankton abundance) is consistent with a turbulence model using patch scales of 1–1,000 m. Water currents can also, however, cause persistent spatial patterns in circulation. Obstructions such as islands and narrow passes at the mouths of estuaries can strongly alter flow patterns.

A concentrated patch of phytoplankton must inevitably diffuse outward because of the transfer of wind and current energy into kinetic energy that moves a phytoplankton cell away from a specific point. Dye that is placed in even relatively quiet water will gradually disperse in all directions (Figure 7.13a). If the dye is placed in a current, the dispersion of dye will be mainly in the direction of the current (Figure 7.13b). However, a phytoplankton population is obviously not a jar of dye; phytoplankton reproduce, often with a few generations per day. As the rate of water dispersal decreases, and the rate of phytoplankton reproduction increases, there is an increased likelihood of appearance of a concentrated patch of phytoplankton. Thus, the introduction of a rapidly growing phytoplankton population into a sluggish body of water is the probable condition necessary for a rapid growth of phytoplankton, called a **bloom**.

These conditions have application to the problem

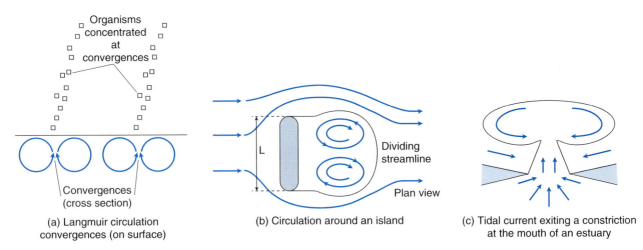

Fig. 7.12   Inhomogeneities generated by wind and moving water. In all these cases, plankton would be entrained in these currents and eddies. Streamlines are in green. (a) Langmuir circulation cells, which are circular in cross section. Converging currents, however, produce a series of parallel linear aggregations of plankton at the surface. (b) Vortices formed downstream of a current moving past an island. (c) Vortices formed as a tidal current moves through a narrow pass, such as the mouth of an estuary.

of **nuisance blooms,** or blooms of phytoplankton that are potentially toxic. The phenomenon of red tides (the water is actually colored red or some other color), which are potentially toxic blooms of certain dinoflagellates, was discussed earlier. Other blooms consist of algae that may be indigestible to shellfish or may be harmful because they shade out benthic plants such as eelgrass. Recent evidence suggests that many of these blooms are initiated by periods dur-

ing which a body of water is contained in a restricted area, whether because of low winds or because of differences in density from adjacent water bodies. In the case of dinoflagellates, benthic resting stages may be swept into such sluggish water bodies, and dense blooms may be initiated afterward.

In the zooplankton, swimming behavior and other forms of adjustments of water depth are the major causes of patchiness. When sonar was first employed, **deep scattering layers** were discovered at depths of 5–1,500 meters. These consisted of dense populations of vertically migrating fish and cephalopods that reflected sonar. Because of regular daily migrations (to be discussed in the following section), many zooplankton groups are aggregated as they move up and down.

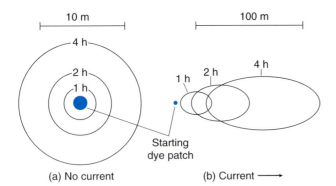

Fig. 7.13   The spread of a concentrated patch of dye, when added to the water: size and shape of patch are shown at the start and at several subsequent times. (a) Case of no current and random turbulence; dye is diffusing equally in all directions. (b) In a unidirectional current, shape of dye patches becomes larger and elongated in the direction of the current.

Diurnal Vertical Migration of the Zooplankton

✳ **Many planktonic organisms undergo diurnal vertical migrations: they move toward the surface during the night and descend during the day.**

In the 1950s the Norwegian explorer Thor Heyerdahl set off across the Pacific in an outrigger canoe, much like those used by ancient Polynesian explorers. At night in the open ocean, he noticed a fantastic abundance of jellyfish and other zooplankton near the sea surface, but far fewer at midday. Zooplank-

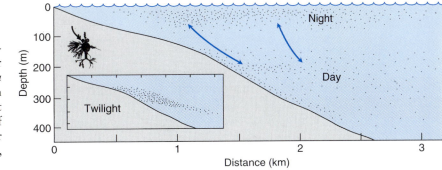

Fig. 7.14 Diurnal vertical migration. A typical migratory pattern for a planktonic shrimp species, *Sergia lucens*, shows a strong concentration of animals at the surface at night (green), but a more diffuse spread of animals over a wide zone of deeper water during the day. (After Omori, 1974.)

ton were migrating to the surface waters at night but spent the daytime at some depth beneath the surface.

The phenomenon of **diurnal vertical migrations**, which is widespread throughout the ocean, has been noticed and quantified by many marine biologists (Figure 7.14). The first complete scientific observations were made during the famous *Challenger* expedition of 1872–1875. Copepods were then found only in surface samples taken at night. In 1912, Esterly found that copepods off the Pacific coast of California could migrate as deep as 400 m during the day and return to the surface at night. Since the early twentieth century, diurnal migrations have been observed in a wide variety of zooplankton, including copepods, euphausiids, jellyfish, ctenophores, and arrow worms.

Diurnal vertical migrations are common in coastal shallow waters, but they also have been observed to depths of greater than 1,000 m, where light is not detectable from above. Some animals, such as copepods and jellyfish, can migrate 400–800 m in a single day. During the day, passive sinking can be as important as active downward swimming. At night, however, the animals must swim upward continuously and actively.

To measure swimming speeds, Hardy and Bainbridge developed the **plankton wheel** (Figure 7.15), a circular glass tube that is rotated continuously at constant speed. If the speed against the direction in which the animal swims is adjusted, a zooplankter will appear to be stationary, and this speed is the animal's swimming speed. Copepods of the genus *Calanus* move upward at a velocity of 15 m h^{-1} and downward at 100 m h^{-1}, while the holoplanktonic polychaete annelid *Tomopteris* can swim 200 m h^{-1}. These speeds are well within the range predicted by field observations. There is, however, much variabil-

ity in the results of field studies. On the northeast coast of North America, for example, diurnal vertical migrations of copepods can be 100 m in one place and virtually absent in others. Vertical migrations are often strong in spring and early summer, but copepods frequently move to greater depths in the autumn, and vertical migration is dampened somewhat.

Diurnal variation in light seems to be the major pacer of the migrations. Onset of dawn and dusk would seem to be the only stimuli that can set the clock. There are zooplankton populations, however, for which downward migration begins before dawn and whose surface movements commence before dusk. Direct response to variation in light intensity is not sufficient to explain the diurnal pattern of movement. There may be an internal biological clock,

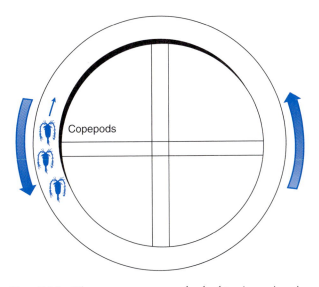

Fig. 7.15 The measurement of planktonic swimming speed in a Hardy–Bainbridge plankton wheel.

which is probably reset continually by the day–night cycle.

EXPLANATIONS FOR DIURNAL VERTICAL MIGRATIONS

The magnificent universality of diurnal migrations is matched only by the great mystery of why such migrations occur at all! Why should zooplankton move several hundred meters down, only to rise to their original depths again a few hours later? Even if it is known that diurnal light change is required to continually reset a biological clock, this information does not reveal what forces caused the evolution of the behavior. Given its universal occurrence, such migrations may be presumed to have an adaptive value. Many adaptive explanations have been proposed. They fall into the following overall categories.

1. **Strong light hypothesis.** Zooplankton are adversely affected by strong light and therefore leave surface waters during the day.
2. **Phytoplankton recovery hypothesis.** Zooplankton exploit the phytoplankton but dive to allow the phytoplankton to photosynthesize and to recover during the day, so that they can be exploited again the following night.
3. **Predation hypothesis.** Predators (e.g., fish, diving birds) use vision to capture prey, so zooplankton leave the surface waters during the day to avoid being seen. They return at night to the surface waters to exploit the phytoplankton.
4. **Energy conservation hypothesis.** It is energetically advantageous to spend the day in colder deeper water, where metabolic rate and energy needs are lower. The animals come up at night to feed on phytoplankton or on other zooplankton that come to the surface.
5. **Surface mixing hypothesis.** The zooplankton move downward during the day, in the hope of returning to surface waters that have been driven from another locale by the wind. The new surface waters are likely to contain a new supply of phytoplanktonic food. The mixing of waters might also promote mixing of differently evolved populations and provide opportunities for the use of new genetically based variants to evolve and adapt to changing environments.

Let us now consider the merits of each of these hypotheses in turn. Light intensity probably has nothing to do with the evolution of diurnal vertical migrations, because the animals migrate to depths far greater than those at which surface intensity can be damaging. The phytoplankton recovery hypothesis is also unlikely, because cooperation among zooplankton organisms would be required for a complete recovery to occur. However, natural selection favors individuals, and those that would "cheat" by remaining in the surface waters could take advantage of the remaining phytoplankton. The surface mixing hypothesis has the same logical problem.

The predation hypothesis is consistent with the idea of avoidance of visual predators, which would be unable to detect prey at night. In Gatun Lake, Panama, the calanoid copepod *Diaptomus gatunensis.* avoids fish predation through a diurnal vertical migration. The lake has no vertical gradient in temperature, a fact that eliminates the energy conservation hypothesis from consideration, at least in the lake. Copepods in Dabob Bay, Washington, actually perform a reverse vertical migration, spending daylight hours in the surface waters and nighttime at depth. This seems to be a response to arrow worm predators, which themselves carry out a typical vertical migration (upward at night, downward during the day).

There are some shortcomings to the predation hypothesis, however. Many zooplankton migrate far deeper than is needed to avoid predators. Also, many vertically migrating species bioluminesce at night, which would seem only to attract predators. A third objection is that members of species that are relatively invisible (e.g., transparent gelatinous zooplankton) are among the strongest in vertical migration.

The energy conservation hypothesis depends on the increase of metabolic rate with increasing temperature in poikilotherms. By spending some time in the cooler deep waters, an animal can save energy. This gain would have to be balanced against the energy required to migrate and against the time lost from feeding at the surface. Apparently, the energy expended in downward and upward movements is quite modest and not a consideration. Dawidowicz and Loose[1] measured reproduction and growth rate in moving and in stationary water fleas and found no differences between the two groups. McLaren[2] cal-

---

[1] See Dawidowicz and Loose, 1992, in Further Reading, Vertical Position and Patchiness.
[2] See McLaren, 1963, in Further Reading, Vertical Position and Patchiness.

culated that copepods should have an energetic advantage in spending time in deeper waters where temperatures are cooler and metabolic costs lower. This energetic advantage would be realized in the production of additional eggs and greater potential population growth. Copepods often retire to deeper waters after the spring peak of phytoplankton production. It may be that this is an adaptation to lower energy expenditure that comes into play once the surface food source is no longer abundant.

Defense Against Predation

❋ Predation is intense in the plankton. Both phytoplankton and zooplankton have evolved armature, chemical defenses, and transparency to avoid predation.

The composition of planktonic communities must be affected by differential susceptibility of planktonic species to predation. However, few studies have demonstrated this process to be important in marine systems. In freshwater plankton, differential susceptibility is probably a common underlying explanation for the relative abundance of species. The development of body spines and armature in many phytoplankton and zooplankton is probably an adaptation to increase the difficulty of capture and ingestion. Common diatoms, such as those of the genus *Chaetoceros*, have large projecting spines that increase the effective body size (and may also aid in flotation). Many planktonic crustaceans are similarly armed with sometimes elaborate spines. The presence of spines may place the prey out of the size range of predators or may make the prey difficult to handle and seize.

Many planktonic organisms are nearly transparent. Jellyfish, comb jellies, arrow worms, and many other groups are difficult to spot in turbulent water. The animals become inconspicuous under these conditions and tend to be overlooked by predators. It is likely that this feature has evolved in many groups independently.

Bioluminescence is a common feature of many planktonic phytoplankton and zooplankton species. Dinoflagellates commonly luminesce under turbulent conditions. Calanoid copepods avoid luminescent dinoflagellates and favor grazing on nonluminescent forms. As copepods feed or as they swim among swarms of luminescent dinoflagellates, the resultant microturbulence induces flashes of light. These flashes may endanger the copepods by making them visible to fish predators. Many deep-sea fishes have opaque digestive tracts that may protect against the obvious signal that a luminescent prey in a transparent fish could create for the fish's potential predators.

Many phytoplankton species, notably the cyanobacteria and dinoflagellates, are toxic to grazers. This toxicity undoubtedly influences the distribution of species of phytoplankton in grazed systems. Species of the chrysophyte alga *Phaeocystis* produce large amounts of acrylic acid, an effective antibiotic capable of sterilizing the guts of consumers. It may inhibit microbially mediated digestion and could induce future avoidance of the alga. Bioluminescence in some dinoflagellates may serve as a "warning" system that has coevolved with the production of toxins and the presence of grazers.

Further Reading

PHYTOPLANKTON

Anderson, D. M., and D. Wall. 1978. Potential importance of benthic cysts of *Gonyaulax tamarensis* and *G. excavata* in initiating toxic dinoflagellate blooms. *Journal of Phycology*, v. 14, pp. 224–234.

Raymont, J. E. G. 1980. *Plankton and Productivity in the Oceans*, 2nd ed., v. 1: *Phytoplankton*. Oxford: Pergamon Press.

Round, F. E., R. M. Crawford, and D. G. Mann. 1990. *The Diatoms. Biology and Morphology of the Genera*. Cambridge: Cambridge University Press.

Sarjeant, W. A. S. 1974. *Fossil and Living Dinoflagellates*. London: Academic Press.

ZOOPLANKTON

Alldredge, A. L., and L. P. Madin. 1982. Pelagic tunicates: Unique herbivores in the marine plankton. *BioScience*, v. 32, pp. 655–663.

Emlet, R. B., and R. R. Strathman. 1985. Gravity, drag, and feeding currents of small zooplankton. *Science*, v. 228, pp. 1016–1017.

Francis, L. 1991. Sailing downwind: Aerodynamic performance of the *Velella* sail. *Journal of Experimental Biology*, v. 158, pp. 117–132.

Harbison, G. R., L. P. Madin, and N. R. Swanberg. 1978. On the natural history and distribution of oceanic ctenophores. *Deep-Sea Research*, v. 25, p. 233–256.

Hardy, A. C. 1954. *The Open Sea, Its Natural History: The World of Plankton*. London: Collins.

Koehl, M. A. R., and J. R. Strickler. 1981. Copepod feeding currents: Food capture at low Reynolds number. *Limnology and Oceanography*, v. 26, pp. 1062–1073.

Marshall, S. M., and A. P. Orr. 1955. *The Biology of a Marine Copepod Calanus finnmarchicus* (Gunnerus). Edinburgh: Oliver & Boyd.

Ninivaggi, D. V. 1979. Particle retention efficiencies in *Temora longicornia* (Muller). M.S. thesis, State University of New York, Stony Brook.

Paffenhöffer, G.-A., J. R. Strickler, and M. Alcaraz. 1982. Suspension-feeding by herbivorous calanoid copepods: A cinematographic study. *Marine Biology*, v. 67, pp. 193–199.

Raymont, J. E. G. 1983. *Plankton and Productivity in the Oceans*, 2nd ed., v. 2: *Zooplankton*. Oxford: Pergamon Press.

Yen, J., B. Sanderson, J. R. Strickler, and A. Okubo. 1991. Feeding currents and energy dissipation by *Euchaeta rimana*, a subtropical pelagic copepod. *Limnology and Oceanography*, v. 36, pp. 362–369.

VERTICAL POSITION AND PATCHINESS

Cowles, D. L., and J. J. Childress. 1988. Swimming speed and oxygen consumption in the bathypelagic mysid *Gnathophausia ingens*. *Biological Bulletin*, v. 175, pp. 111–121.

Dawidowicz, P., and C. J. Loose. 1992. Cost of swimming by *Daphnia* during diel vertical migration. *Limnology and Oceanography*, v. 37, pp. 665–669.

Enright, J. T. 1977. Diurnal vertical migration: Adaptive significance and timing. I: Selective advantage: A metabolic model. *Limnology and Oceanography*, v. 22, pp. 873–886.

Gibbons, M. J. 1993. Vertical migration and feeding of *Euphausia lucens* at two 72h stations in the southern Benguela upwelling region. *Marine Biology*, v. 116, pp. 257–268.

Hardy, A. C., and R. Bainbridge. 1954. Experimental observations on vertical migrations of plankton animals. *Journal of the Marine Biological Association, United Kingdom*, v. 33, pp. 409–448.

McLaren, I. A. 1963. Effects of temperature on growth of zooplankton and the adaptive value of vertical migrations. *Journal Fisheries Research Board of Canada*, v. 20, pp. 685–727.

Ohman, M. D. 1990. The demographic benefits of diel vertical migration by zooplankton. *Ecological Monographs*, v. 60, pp. 257–281.

Okubo, A. 1984. Critical patch size for plankton and patchiness. In S. A. Levin and T. G. Hallam, eds., *Lecture Notes in Biomathematics*, v. 55. Berlin: Springer-Verlag, pp. 456–477.

Omori, M. 1974. The biology of pelagic shrimps in the ocean. Advances in *Marine Biology*, v. 12, pp. 233–324.

Parsons, T. R., and M. Takahashi. 1973. *Biological Oceanographic Processes*. Oxford: Pergamon Press.

Vuorinen, I., M. Rajasilta, and J. Salo. 1983. Selective predation and habitat shift in a copepod species—Support for the predation hypothesis. *Oecologia*, v. 59, pp. 62–64.

Yen, J. 2000. Life in transition: Balancing intertial and viscous forces by planktonic copepods. *Biological Bulletin*, v. 198, pp. 213–224.

Review Questions

1. Distinguish between holoplankton and meroplankton.

2. Why is it important for planktonic organisms to remain in the surface waters?

3. What major process in the ocean retards sinking of planktonic organisms?

4. What adaptations of planktonic organisms may retard sinking from the surface waters?

5. What is the relative value of the Reynolds number (high or low) for most plankton? Why does this matter for their biology?

6. What is the difference between the hard parts of diatoms and dinoflagellates?

7. Why do some dinoflagellates represent potential difficulty for other organisms?

8. How does low Reynolds number influence the feeding of marine planktonic copepods?

9. Where are krill especially dominant in the zooplankton?

10. How do siphonophores stabilize themselves and keep from tipping over in a heavy wind?

11. Upon what element do foraminiferans depend in the making of their skeletons?

12. Describe a mechanism by which plankton may be concentrated in the surface waters by wind action.

13. Distinguish between the factors that may cause planktonic animals to migrate through the water column, and describe the possible adaptive advantages for such migrations.

14. In some marine habitats copepods undergo strong diurnal migrations, whereas in others migrations are very weak. Why may this be so?

15. Describe two means by which plankton defend against predation.

16. Many tropical zooplankton have elaborate projections and spines, which have been interpreted in two distinct ways: as a defense against predators and as a means of increasing drag to slow sinking from surface waters. How would you test between these two general hypotheses for a copepod that has elaborate spines on its first antennae. Suggest another hypothesis to explain this. (*Hint:* Think in terms of feeding.)

8

The Water Column: Nekton

✳ Nekton are those animals capable of swimming to the degree that they can overcome many ocean currents. Nekton usually live in a world dominated by high Reynolds number conditions.

Nekton include fishes, cephalopods, marine mammals, birds, and reptiles. Unlike the plankton, they can swim, often against strong currents. This allows them to move great distances within a day. Some nekton migrate over thousands of kilometers.

Objects that are large and of high velocity, such as most nekton, live according to the laws of high Reynolds number (see Chapter 4). Viscosity matters less for these forms than it does for nannoplankton. Nekton have momentum while moving through the water. In other words, if a fish thrusts through the water, it will not stop immediately when it ceases to undulate. There is no relatively thick boundary layer around it to interact viscously with the surrounding fluid. This fact allows nekton to dart about, coasting along on the strength of periodic thrusts. Contrast this with the situation for nannoplankton, which are dominated by viscosity. When a nannoplankter moves through the water, it effectively drags a small envelope of water along with it.

Cephalopods

✳ Cephalopods belong to the phylum Mollusca, are nearly always carnivorous, and are characterized by complex behavior, a well-organized nervous system, a circle of grasping arms, and a powerful beak.

The cephalopods (Figure 8.1) belong to the phylum Mollusca (see Chapter 11 for a more complete introduction to this phylum) and include squids, cuttlefish, pearly nautilus (Figure 8.2), and octopus. They all grasp prey by means of a circle of arms (plus tentacles in squids), which are often covered with suckers. The mouth is armed with a powerful beak. The **mantle** encloses a volume of water, which can be expelled through a **siphon**, which can be pointed in any of many directions and rapidly propels the animal through the water. The nervous system is well organized, and most cephalopods have extremely well-developed eyes, resembling the eyes of vertebrates in their overall organization. Cephalopods are the largest of the invertebrates. The open-ocean squid *Architeuthis* exceeds lengths of 15 m, including the two long tentacles. Squids and octopods have an **ink gland**, which produces an ink that is expelled and af-

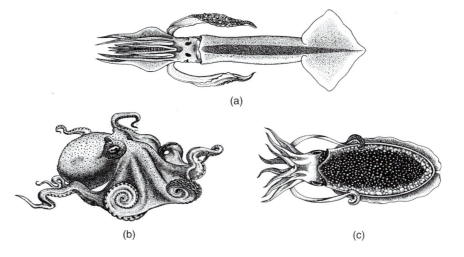

Fig. 8.1 Some cephalopods: (a) the squid *Rossia pacifica*, (b) the Pacific octopus *Octopus dofleini*, and (c) the western Pacific cuttlefish *Sepiella japonica*.

(a)

(b)

(c)

fects olfaction in predators. The ink of *Sepia* is used commercially.

Cephalopods are the most mobile of the mollusks, and their greater nervous organization facilitates their advanced locomotion. Cephalopods are carnivores. Squids can attack prey by contracting the mantle muscles and expelling water through the siphon. The siphon must be directed so that the animal attacks with the arms first. It then bites into the prey, using its beak. The prey may be seized with the arms, which, in the case of squids and octopods, are lined with suckers. Squid also use two long tentacles to draw prey to the beak. The nautilus lacks suckers but has over 90 arms, as opposed to 8 for octopods and 10 for squids (eight shorter arms and two very long tentacles). The cuttlefish *Sepia* usually feeds on bottom crustacea by shooting a jet of water at the sand and uncovering its prey. As its prey move, they are attacked.

Cephalopods have great powers of color change. Using combinations of three cell color types, squids and cuttlefish are capable of nearly instantaneous and spectacular color changes, which often are related to camouflage against visual predators. The most remarkable change is a moving pattern of ripples, which pass rapidly along the body surface. The device is believed to be a contrivance to resemble the ripple effect caused by light dappling on the seabed.

Cephalopods reproduce following transfer of a spermatophore from male to female and internal fertilization. In most forms, the eggs are attached to a substratum or are free floating. The female of the pa-

per argonaut, *Argonauta argo*, spreads two arms and secretes a beautiful spiral shell, within which the eggs are deposited.

✳ **Some cephalopods can regulate buoyancy by altering the gas content within a rigid structure.**

Some cephalopods are able to use gas production and absorption to regulate their depth. The chambers of the shell of the pearly nautilus are filled with gas and water in varying proportions, which are controlled

Fig. 8.2 The pearly nautilus in its natural midwater habitat. (Courtesy of Peter Ward.)

by the animal to set the depth. In squid and cuttle-fish, an internal chambered and rigid structure also holds varying proportions of gas and water.

The cuttlefish use a rigid structure, the **cuttlebone**, made of calcium carbonate. The cuttlebone lies on the dorsal side of the cuttlefish, within the body. It consists of a series of lamellae, separated by pillars that help to form a series of chambers (Figure 8.3). The cuttlebone is surrounded by a yellow secretory membrane. Because the cuttlefish can dive between the surface and a depth of about 200 m, it must have a mechanism for gas and water exchange between the rigid chambers and the external medium. The animal uses an **osmotic pump** for this purpose. As we discussed in Chapter 3, distilled water will diffuse across a membrane until the overall salt content is equal on either side. To remove water from the cuttlebone, salt is actively pumped into small ampullae adjacent to the cuttlebone chambers, forcing the water to diffuse out from the cuttlebone. The reverse process forces water to move across a membrane into the cuttlebone chambers. The gas content of the chambers is essentially a by-product of diffusion of gas into the chambers as water is removed, or displacement, as the water is added to the cuttlebone chambers. As a result of these processes, cuttlefish can regulate the density of the cuttlebone to vary between about 0.5 and 0.7 of seawater.

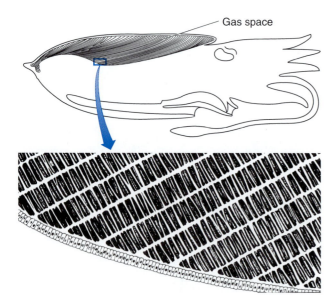

Gas space

Fig. 8.3 Top: the location of the cuttlebone of a cuttlefish. Bottom: cross section showing chambers, which may be filled with water or gas.

The overall mechanism is similar in the chambered nautilus, although the details are quite different. As the nautilus grows, it lays down new calcareous septa within the shell, creating new chambers. At first, the chambers are filled with fluid, which is similar in salt content to seawater. Gradually, however, some of the fluid is removed as the animal grows and the body and shell become heavier. A calcified tube, the **siphuncle**, connects all the chambers. The tube is coated with a soft tissue, which contains large numbers of small canals. These canals are filled with fluid of high salt content, creating an osmotic gradient between themselves and the shell chambers. Fluid flows from the shell chambers into the soft tissue chambers and then empties further into the nautilus' body until neutral buoyancy is achieved.

Fish

* **The marine fishes, which occupy nearly all marine habitats, include mainly the cartilaginous fish (sharks, skates, and rays) and the bony fish.**

Fishes are ubiquitous members of marine communities, and it is impossible to do justice to them in a short section. There are about 20,000 species, which can be found in coral reefs, estuaries, submarine canyons, and midwater bathyal environments, and on the deep-sea bed. It is difficult to distinguish between water column and bottom fish owing to the great mobility of the many bottom-living fish and to occupation of multiple habitats.

All fishes are members of the phylum Chordata, subphylum Vertebrata, and are characterized by a spinal cord, an internal skeleton, a complete digestive system, and various arrangements of fins to control movement. The cartilaginous fishes (**Chondrichthyes**) all have a skeleton of cartilage, whereas the bony fishes (**Osteichthyes**) secrete a true bony skeleton. Although they are a much smaller group, the cartilaginous fishes (Figure 8.4) include the predatory sharks, the bottom-feeding rays, and the whale shark, a huge animal that cruises through the water and strains zooplankton. Cartilaginous fishes usually have tooth rows that are replaceable, whereas bony fishes have fixed teeth, fused with the jaw. Far more diverse in form and habitat, bony fishes (Figure 8.5, Plate IX) include salmon, herrings, and others with greater maneuverability and speed and a greater

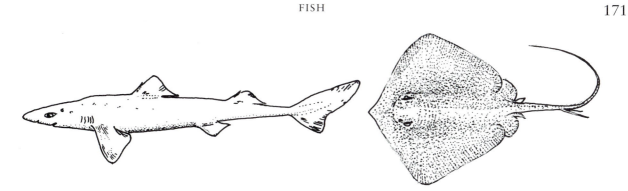

Fig. 8.4 Cartilaginous fishes: the shark *Squalus acanthias* and the ray *Dasyiatis akajei*.

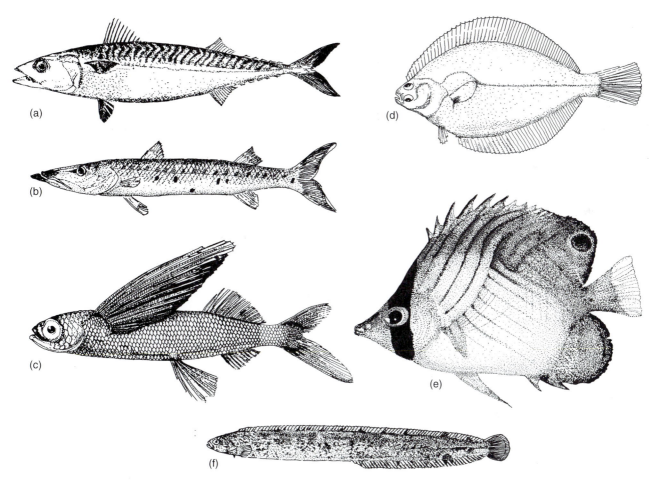

Fig. 8.5 Variation in the form of bony fishes: (a) rover predator, (b) lie-in-wait predator, (c) surface-oriented fish, (d) bottom-feeding flatfish, (e) deep-bodied fish, and (f) eel-like fish.

diversity of feeding adaptations. The mouth usually has a specialized jaw, ranging from tooth rows to highly fused crushing plates. Whereas the jaw of a cartilaginous fish is usually ventral, a bony fish's jaw coincides with the midline.

Classification in Terms of Form and Ecology

✳ **The body form of fishes is often a good indicator of their feeding and habitat requirements.**

Fishes can be classified by their overall form and ecology (Figure 8.5).

Rover predators (e.g., tuna, marlin) are usually long and torpedo shaped, with the fins spaced along the body for maneuverability. Lie-in-wait fish predators are also torpedo shaped, but the fins are often concentrated toward the rear, to help provide the sudden thrust necessary to capture prey. **Surface-oriented fishes** (e.g. flying fishes) often have the mouth oriented upward, to capture surface-bound prey. **Bottom fishes** are quite variable in shape: the flounders, plaice, and soles are notable for their flattened shape, and some bottom-roving sharks have strongly flattened heads. Sculpins have modified pelvic fins that allow them to adhere to the bottom. **Deep-bodied fishes** are flattened laterally and are excellent at maneuvering, although they are relatively slow swimmers. They are often small and able to maneuver among grass blades or among crevices in coral reefs. The fins of many are armed with spines, which deter predators. **Eel-like fishes** are well adapted for moving in crevices, as the moray eel does in coral reefs.

Swimming

✳ **Swimming is often accomplished by means of undulatory body movements, with the aid of individual appendages.**

Most fishes swim by throwing the body into undulatory waves, producing a component of thrust tangential to the body surface and a normal component (Figure 8.6) that exerts the forward push. Gars and barracuda use their strong flattened caudal region for rapid acceleration. Rays use flapping winglike projections for swimming, and a few fishes, such as the ocean sunfish *Mola mola*, rely upon the flapping of fins for propulsion through the water.

There are three main functional components to swimming (Figure 8.7). **Acceleration** is maximized

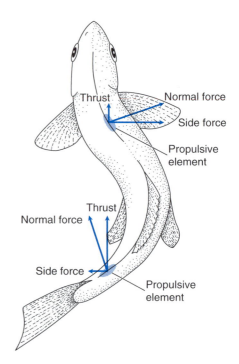

Fig. 8.6 Swimming in fishes. Components of force generated by undulation are shown.

by the propulsion of a strong caudal fin. This allows rapid movements, such as escapes or strikes at prey. A deep body allows maximum contact with the water and maximum thrust. By contrast, **cruising** involves continued undulation of the body. The skip-

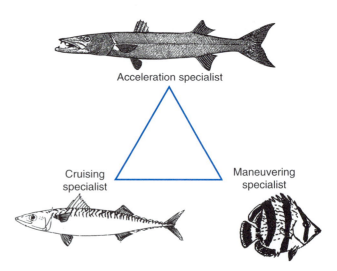

Fig. 8.7 The overall form of fishes can represent a compromise among the forms that would be ideal for the separate activities of accelerating, steady cruising, and maneuvering

jack tuna is a specialist at cruising. It has a stiff body and a streamlined shape, with the greatest body depth about midway down the body length. **Maneuvering** is best accomplished with a disk shape, which permits body flexure and sudden changes of direction. The presence of oscillatory fins permits further refinements in maneuverability. The butterfly fish is a good example of this general type. Although some species of fishes are specialized for one of the functions, most fishes have compromise forms, to permit use of all three component functions to some degree.

To undulate efficiently, swimmers must have some sort of rigid skeleton. Most fishes have a vertebral column, but sharks have a helical external meshwork of collagen against which muscular action can propel the fish. Fast swimmers are streamlined, to reduce drag. The fastest forms, such as swordfishes and tunas, have quarter-moon-shaped tails, which "shed" water easily as the fish moves, thus minimizing disruptive turbulence at the posterior of the fish. Although most fishes move by undulation, some (e.g., triggerfish) use the pectoral and other fins extensively in locomotion.

Fins also play an important role in locomotion. In fish requiring short-range maneuverability, the **pectoral fins** are usually located up on the sides, but, for fast-moving fishes, fins tend to be found more toward the ventral surface. In sculpins, which are dependent upon the bottom, the pectorals are broad. The shape of the **caudal fin**, or tail fin, is related to swimming speed. As mentioned earlier, fast swimmers such as tuna have a quarter-moon-shaped caudal fin, which is attached to the body by a narrow peduncle. The peduncle tends to reduce drag as the body is flexed during swimming. Forked tails are associated with the most active forms. Basking sharks, blue sharks, and threshers—all capable of rapid bursts of speed—have this characteristic tail. Although basking sharks swim slowly while they are feeding on zooplankton, their tails aid in rapid swimming and even leaping into the air from the water.

Oxygen Use and Buoyancy

✳ Fish gain oxygen by passing water over gills, within which blood flows counter to the external water current.

Fish are active creatures and nearly all have very high oxygen requirements to fuel their activity and nerv-ous system. Fish muscle is composed of **red muscle fibers**, which are responsible for rapid contraction and **white muscle fibers**, which are responsible for slow contraction. Active fishes, such as tuna and sailfish, have much higher red muscle fiber contents than do inactive fishes, such as deep-sea fishes, as discussed later. As fishes swim or actively ventilate by flapping the gill covers, water passes over structures called **gills**. The **gill arch** branches off gill filaments, which are in turn covered with thin secondary lamellae (Figure 8.8). As the water flows over the lamellae, oxygen diffuses into the gill. Blood flows in a direction opposite to that of the external water. Oxygen canot diffuse into the gill, however, unless its concentration in the water is greater than its concentration in the blood. By moving the blood against the direction of the external water flow, a new supply of relatively deoxygenated blood will be exposed to more highly oxygenated water. This is the classic countercurrent exchange mechanism that is used so often in biological systems to maximize exchange (see Chapter 4). Hemoglobin in the blood allows the efficient uptake and transport of oxygen by the circulatory system to needy tissues.

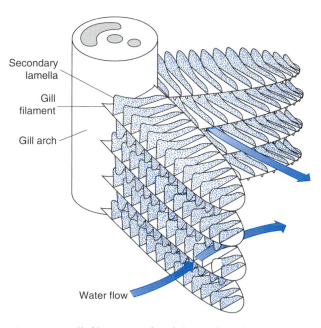

Fig. 8.8 Gill filaments of a fish. Within the secondary lamellae, blood flows counter to the direction of the external oxygenated water, to ensure that the external water will be higher in oxygen and that oxygen will therefore diffuse into the blood. (After Hughes and Grimstone, 1965.)

Secondary lamella

Gill filament

Gill arch

Water flow

* Fishes differ in buoyancy owing to bulk composition, but bony fishes use a swim bladder to alter buoyancy by changing the gas content.

The maintenance of neutral buoyancy can save considerable energy. A fish can remain suspended in the water column at no energy cost to itself. Moreover, the energy expended to avoid sinking when swimming can be reduced. When a fish is swimming at one body length per second, 60 percent of the total power of movement is expended in avoiding sinking. This percentage is considerably reduced at higher swimming speeds, but a normal cruising speed of 3–4 body lengths per second still would involve a 20 percent cost to avoid sinking. Neutral buoyancy therefore confers a considerable reduction of energetic cost.

Alterations of **bulk chemical composition** may contribute to buoyancy in bony fishes. The low salt content of fish cellular fluids reduces bulk density. Sharks have **high lipid content** in the liver, an organ that may account for up to 25 percent of the body volume. The use of the lipid squalene in some elasmobranch fishes can reduce bulk density considerably. Bony fishes mainly use a **swim bladder** to adjust bulk density. The organ is located middorsally, below the vertebral column, and occupies about 5 percent of the animal's volume. When gas-filled, this volume makes up for a fish's excess density of 5 percent relative to seawater. The swim bladder absorbs or secretes gas to adjust the depth at which the fish is neutrally buoyant. Fishes have no muscular control over the swim bladder. Rather, they secrete variable amounts of gas to keep the bladder at a constant volume with changing depth. Some fishes have a connection between the esophagus and the swim bladder and can swallow air at the surface. The pressure increases by approximately 1 atmosphere with each increase of 10 m depth. Thus, when a fish descends, the increased pressure will compress the swim bladder unless the fish secretes more gas. Similarly, when a fish ascends, the lowered pressure will result in the expansion of the bladder's volume unless the gas is removed actively. The fish's secretory apparatus therefore adds or removes gas to keep the swim bladder volume constant. If a deep-water fish is brought rapidly to the surface, the swim bladder expands rapidly and may rupture.

A swim bladder is illustrated in Figure 8.9a. The **rete mirabile** (Figure 8.9b) consists of two intertwined networks of capillaries, in which fluid is flowing in opposite directions. For simplicity, the diagram shows only one contact between arterial and venous capillaries. The swim bladder usually contains nitrogen, oxygen, and carbon dioxide in varying proportions, depending upon the species. Gas uptake is effected by a **gas gland**, often against a large pressure gradient. The gas gland secretes lactic acid, which causes the hemoglobin to release its oxygen (this process is discussed shortly). As this oxygen leaves in the venous blood, its pressure is greater than in the cap-

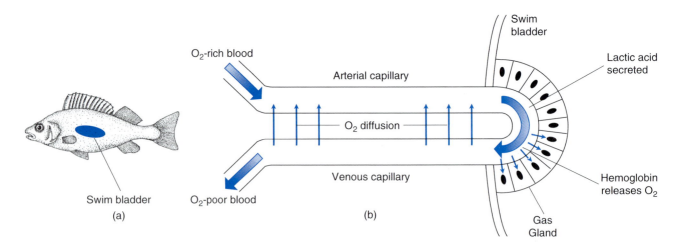

Fig. 8.9   (a) Location of the swim bladder of a teleost fish. (b) Function the rete mirabile, showing the fate of incoming arterial blood with 10 mL of O_2 per 100 mL of blood. Countercurrent flow retards loss through venous capillary return. (After Schmidt-Nielsen, 1975.)

illaries carrying the arterial blood, and oxygen passes across into the arterial capillaries. The oxygen is then carried back and deposited into the swim bladder. This countercurrent exchange mechanism works because the oxygen tension is always greater in the venous than in the arterial blood, thus causing the diffusion.

The chemistry of the capillary blood also facilitates the exchange of oxygen. A reduction of pH (due to an increase in glycolysis) in the capillaries reduces the maximum capacity of hemoglobin to bind oxygen. Oxygen is thus released into the blood. However, the solubility of oxygen in the capillary blood is reduced, owing to an increase in the concentration of lactate. By an osmotic process, the presence of the small lactate molecules effectively forces dissolved oxygen out of solution. It is the sum of these two effects that forces more oxygen into the swim bladder.

Gas can also be retained in the swim bladder by the same overall mechanism. As gas in the arterial blood flows toward the swim bladder, the arterial capillaries are in intimate contact with the venous capillaries. Gas tending to diffuse out from the venous capillaries diffuses into the arterial capillaries and is available for uptake into the swim bladder. This tends to maintain the gas content of the swim bladder.

Not all fishes have swim bladders. As might be expected, many bottom-living fish lack them. Sharks, rays, and mackerel also lack them. This may be advantageous to them, because rapid vertical locomotion would be difficult for an animal obliged to contend continually with the changing gas pressures of a swim bladder. The nearly horizontal pectoral fins of sharks provide some lift by means of the Bernoulli effect. The asymmetrical tail (the upper lobe is much larger) also causes sharks to move upward while swimming.

Feeding

* Fishes feed by means of ram feeding and suction.

Most fish feed by **suction**. The rapid opening of the mouth cavity sucks both water and prey into the mouth. This versatile feeding mechanism can be used to capture reasonably large prey and also can be used to draw much smaller prey, which are captured on extensions of the gill known as gill rakers. Fish that feed on smaller plankton often have finely arrayed and long rakers. Larger carnivorous fish may be **ram**

feeders. They move forward and their open jaws encounter the prey directly, which can hurt! Sharks, for example, lift the upper jaw and depress the lower jaw as they attack. Protrusion of the jaw and enlargement of the mouth cavity coincide with the widening of the jaw. As the encounter with a prey animal occurs, the jaws are closed. The enlargement of the mouth cavity causes suction. Prey capture by sharks is therefore a combination of ram and suction feeding.

* Fishes use teeth to scrape algae from surfaces, tear prey, and crush shells. Many fishes use gill rakers to suspension-feed.

Feeding in fishes includes scraping of algae from surfaces (an action carried out, for example, by parrot fishes and surgeon fishes), crushing of shelled prey (by puffers), filtering of organic detrital particles (by mullets), filtering of zooplankton (by basking sharks), and active seizure of prey and carrion (by bass, tuna, mackerel, and sharks). Mouth morphology varies extensively. Active carnivores have rows of sharp teeth. The jaw teeth of parrot fishes (Family Scaridae) are fused into plates that can nip off chunks of coral skeletons. A spiny puffer (*Diodon*) can take a shell into its mouth and by a series of apparent inhaling and exhaling motions, position it on crushing plates embedded in the jaws. The shell is crushed repeatedly until the soft parts can be separated and swallowed. Some sculpins have teeth that puncture snail shells, allowing digestive juices to penetrate the shell (Figure 8.10). In filterers such as the basking shark, modified gill rakers strain out zooplankton, and the jaw and dentition are relatively weak.

Suspension-feeding fishes are still understood incompletely, though the feeding mode is widespread. Some are **continuous ram feeders**, and water is continuously strained as the fishes move through the water. Basking sharks (Figure 8.11), for example, have a series of gill arches, on which gill rakers are located, which are in turn festooned with tiny projections that catch particles. Herrings, manta rays, and some mackerels also feed in this way. By contrast, some fishes apply **suction** within the mouth and strain food particles, such as zooplankton, across the gill rakers. Laurie Sanderson and colleagues,[1] who used a surgical telescope to study blackfish, found that particles are guided by the gill rakers to a degree, and

---

[1] See Sanderson and others, 1991, in Further Reading, Fishes.

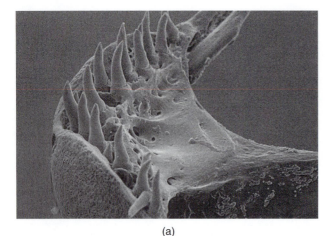

(a)

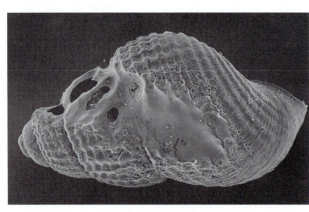

(b)

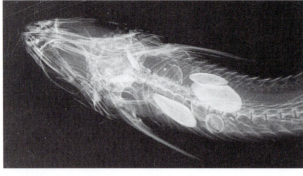

(c)

Fig. 8.10 (a) Close-up of the operative mouthpart of the shell-crushing sculpin *Asemichthys taylori*, found in the U.S. Pacific northwest. A snail shell with holes punched by sculpin teeth. (c) X-radiograph showing bivalve mollusks in a predator's gut. (Courtesy of Stephen F. Norton.)

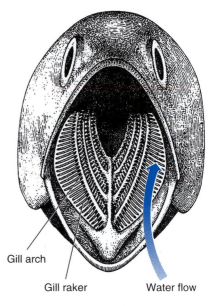

Gill arch

Gill raker   Water flow

Fig. 8.11 Suspension feeding of a basking shark. Water is directed across the gill rakers as the fish is swimming, and plankton is caught on barbs projecting from the gill rakers. (Modified after Sanderson and Wassersug, 1990.)

also by a dorsal fleshy organ toward the rear of the mouth.

Sensory Perception

✳ **Fishes perceive the environment by means of eyes and a lateral line system. Olfaction and hearing are also important.**

Fish usually have excellent sensory perception This is accomplished by means of eyes, a **lateral line system** that detects water movement and, in many species, well-developed olfaction and hearing. The inner ear is used both for hearing and balance. **Otoliths** are suspended and in contact with hairlike fibers. The pressure of the otolith against the fibers provides the fish with information about its spatial orientation. The lateral line system consists partially of a series of mechanoreceptors forming a line along each of the fish's sides, with separate nerves leading to the brain. As the fish moves, slight disturbances created by the presence of rocks or prey aid the fish in navigation. Blind freshwater cave fish can maneuver around stones on the bottom by means of the lateral line system alone. In some fishes, there also is a system of electroreceptors that can detect extremely small voltages. Because moving objects (e.g., predators) generate electrical currents, this ability is very useful. Cartilaginous fish can detect prey electrically at distances of greater than a meter. Vision is developed in fishes to varying degrees. In well-lit shallow-water habitats, many species have excellent color vision, and mating signals often involve displays of color.

Schooling

✳ Schooling is aggregation by mutual attraction.

Schooling (Figure 8.12) is a common phenomenon in fish and squid. Schooling is aggregation by mutual attraction, as opposed to aggregation by currents, or by a common attraction of individuals to an attractive microhabitat. In nearly all cases of active movement, schooling is confined to a single species. But feeding schools may consist of one or many species, especially in coral reef environments.

Schooling may benefit fish by providing protection against predation. Schools can confuse predators, and fishes may either form a ring around the predator or swim in an undulating line. Most tightly schooling fishes have silvery sides, which serve to reflect sunlight and confuse predators. Fishes wandering from the school are usually more prone to predation. Schools approaching predators may confuse the predator and buy enough time to reduce the predator's efficiency in making a decision to attack a specific fish before the entire group has escaped. In some cases fish aggregate into approximately spherical schools known as **fish balls**, which minimize the peripheral area exposed to predators.

Schooling also provides the benefit of reduction of drag. When an entire school swims in a tight pack, the fishes within the pack or at the rear experience less drag. A member of the school usually drops behind and slightly to the side of the fish before it, which locates it in a vortex at which drag is minimal. This may permit the aggregation to move with

less energy expended per individual than would be the case if all individuals swam alone and dispersed.

Body Temperature

✳ Tunas, some billfishes, and the butterfly mackerel are warm-blooded, which aids in sustained activity and in maintaining temperature while moving into environments at which temperatures vary.

Nearly all bony fishes are ectotherms. Their body temperatures hover about 1–2°C around the ambient water temperature. Elevated muscle temperature allows production of more power and is an advantage to fishes that must swim for extended periods at high velocity. However, being warm blooded requires expenditure of a tremendous amount of energy, and only a very few species groups have evolved this character. Endothermy occurs in the group Scombroidei, an assemblage of large oceanic fishes that includes the tunas. It also occurs in the shark families Lamnidae and Alopidae. It seems likely that the sharks and the scombroid teleosts evolved endothermy independently. Otherwise, all fishes would have had at some point to evolve the trait, and many lineages would have then had to lose it to produce the pattern of endothermy observed today in such distantly related groups. It is not clear, however, whether endothermy evolved more than once within the scombroids.

For endothermy to evolve from an ectothermic ancestor, a mechanism must be developed to elevate metabolic activity (to generate heat). Such activity is aided by the delivery of oxygen by hemoglobin to muscle cells. There also must be a mechanism to reduce heat loss. In most fishes, the blood is delivered through main vessels that lie just beneath the vertebrae, with branches that deliver blood to the outer part of the body. Tunas, however, have their blood vessels just beneath the skin, with smaller vessels delivering blood to the interior. This is the reverse of the typical condition, and it allows heat generated in the body to be conserved. In tunas, heat is generated by red muscle, which is located near the axial region, instead of closer to the periphery of the body (an arrangement that characterizes most other fishes). The central placement of red muscle may be related to the thrusting of tunas, which involves maintaining a relatively stiff body while flexing the tail and

Fig. 8.12 Schooling in fishes: (a) traveling school and (b) feeding school.

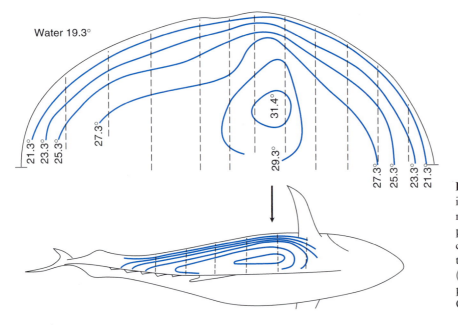

Water 19.3°

Fig. 8.13 Temperature distribution in a bluefin tuna. Long thermistor needles were used to measure temperature shortly after the tuna was caught. Isotherms, or lines of equal temperature, show the differential (>10°C) between the core body temperature and the environment. (After Carey et al., 1971.)

peduncle. Heat loss in tunas is reduced through the whole body with the aid of a countercurrent heat exchanger (see the Chapter 4 discussion of countercurrent flow in dolphins). The heat exchange involves a series of arteries whose flow is in the direction opposite to that of the adjacent veins. The exchanger allows tunas and sharks to maintain body temperatures of greater than 10°C over ambient temperature (Figure 8.13).

Billfishes, such as the swordfish, heat only the brain in a process that is accomplished by passing blood through a specially adapted eye muscle. The butterfly mackerel uses a different eye muscle to accomplish heat production. In both cases, however, a countercurrent mechanism of heat exchange is used to retain heat.

Physiologist Barbara Block[2] has used molecular sequences to analyze the evolutionary relationships among the tunas, billfishes, butterfly mackerel, and other scombroid groups. Apparently, tunas, billfishes, and the butterfly mackerel have evolved endothermy independently. In all three cases, this evolution has accompanied an expansion of habitat into cool temperate oceanic habitats. This suggests that the evolution of endothermy was driven by expansion of habitat exploitation, as opposed to merely an increase of potential for activity. This is most apparent in the billfishes and the butterfly mackerel, which heat only

the brain. These "hot-headed" fishes control and sustain metabolic activity in the brain, abilities that afford them maximum sensory control while hunting over a wide range of seawater temperatures. For example, warming the brain allows swordfish to hunt squids that migrate vertically across a great range of depth and temperature.

Mesopelagic Fishes

✳ Mesopelagic fishes are adapted for low light, rarity of mates, avoidance of predation by counterillumination, and consumption of rare prey.

Mesopelagic fishes live in depths between 150 and 2,000 m. Trawls from the mesopelagic depths usually bring up an astounding assortment of fantastic fish, some with mouths that are enormous in comparison to the overall fish size, and an array of protruding structures. Fishes of this zone have well-developed musculature and feeding apparatus, and are commonly excellent hunters. Many species have tubular eyes that point upward to spot zooplankton prey. Fishes of these species have relatively short and small mouths, which they use to ingest small prey rapidly. The spectacular angler fishes, on the other hand, seem to be all mouth, and they have a lure pro-

[2] See Block and others, 1993, in Further Reading, Fishes.

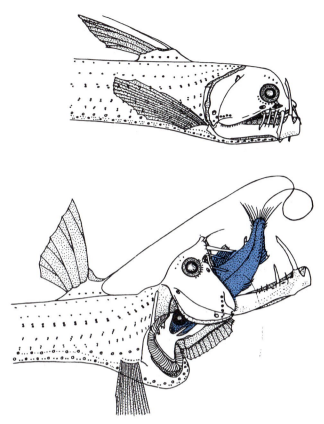

tach to females. Many other species are simultaneous hermaphrodites.

Many midwater fishes possess **photophores,** probably for several functions, including attracting mates and luring and illuminating prey. The photophores of many bioluminescent mesopelagic fish are concentrated on the ventral surface (Figure 8.15). This array, known as **counterillumination,** confuses predators looking upward into the dim light. Many other mobile mesopelagic fish have a silvery color that blends with the downstreaming light. Lie-in-wait predators, such as many angler fishes, often have opaque, black digestive tracts, which may serve to help conceal them from predators. The black digestive tract may also block off the light produced by bioluminescent prey as they are swallowed by the angler fish. This would prevent an obvious cue for yet another predator, who might otherwise obtain a double meal!

✳ **Bathypelagic and abyssopelagic fishes exhibit strong reduction of activity.**

Bathypelagic fishes (depths of 2,000–4,000 m) and abyssopelagic fishes (depths of 4,000–6,000 m) inhabit a world of great food impoverishment (see Chapter 16). There is therefore little payoff for the development of the strong skeleton and musculature observed in fishes in shallower water. Indeed, extensive musculature consumes oxygen and is therefore a disadvantage when food is so scarce. Fishes in this zone therefore have poor musculature and incompletely ossified skeletons. Swim bladders are usually absent. These fishes are usually inactive, feeding only occasionally. Abyssal benthic fishes are somewhat more active than those in the abyssal water column; this may reflect a greater food supply. The fish fauna of deep-sea bottoms is dominated by eel-like forms. The eel shape may be related to the development of a good lateral line system, which is important in a dark environment.

Fig. 8.14 The deep-water fish *Chauliodus* uses its specialized backbone to enable the opening of its enormous mouth in order to consume large fish.

truding upward from the dorsal surface, between the eyes. This lure is a fin ray modified by evolution into a structure that attracts other fish. *Chauliodus* has a specialized backbone, which accommodates the enormous mouth as it is opened (Figure 8.14). The relative rarity of prey in this zone has probably brought about this selection for an evolutionary increase in mouth size. Because of the difficulty of finding a mate, the angler fishes have dwarf males, which at-

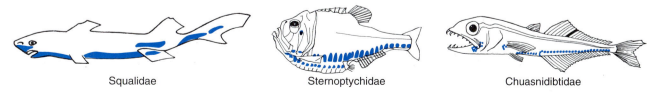

Squalidae   Sternoptychidae   Chuasnidibtidae

Fig. 8.15 Ventral illumination areas in members of three families of mesopelagic fishes. (After McAllister, 1967.)

Mammals

Whales and Porpoises

* **The mammalian order Cetacea includes the toothed group Odontoceti and the baleen-bearing group Mysticeti. All are streamlined and highly adapted to a fully marine existence.**

The mammalian order **Cetacea** (Phylum Chordata, Class Mammalia) includes the whales and porpoises. The group includes the suborder **Odontoceti,** the toothed whales (e.g., the sperm whale) and porpoises, and the **Mysticeti,** the baleen whales (e.g., blue whale). In both groups, the body is nearly hairless, elongate, and streamlined, properties that reduce drag as the animal swims. Unlike their terrestrial relatives, these animals have forelimbs that have been reduced to stabilizing paddles. The hind limbs have been completely reduced and can be identified only by vestigial bones, which do not protrude from the main trunk. The posterior is strongly muscularized and ends in a pair of **flukes,** which resemble the horizontal tail of an airplane. A strong twisting motion causes the flukes to push against the water and propel the animal forward. This body design enables very efficient swimming. Some large whales can travel continuously at speeds of over 10 miles per hour, and baleen whales can sustain burst speeds of over 20 miles per hour. (Estimates come from investigators on whaling ships, who observed whales attempting to escape.)

Cetaceans are homeothermic, and all have a thick subdermal layer of fat that retards heat loss. In extreme cold, blood circulation can be restricted to the trunk and brain. Heat is also conserved in the limbs by a countercurrent exchange system (see Chapter 4).

Cetaceans reproduce much the same as terrestrial mammals. Some species have elaborate courtship rituals, and in a few species a third whale appears to aid in copulation by leaning against the copulating couple. After a gestation period of usually several months, as much as a year for killer whales and blue whales, the young are born live and underwater, which is unique for marine mammals. The young suckle, and the "milk" has a far higher fat content than that of terrestrial mammals. It has a fishy flavor and is cheesy in consistency. Birth size is also quite large; a blue whale, for example, is about 12 m long at birth. By the time of weaning, after 7 months or so, it has already doubled in length. Cetacean mothers care for their young.

Cetaceans are air breathers and must return to the surface for oxygen. The nasal opening is the **blowhole,** located on the back of the head. When a whale reaches the surface, there is often a characteristic loud sneezing sound as carbon dioxide is expelled through special flaps that seal off the nasal opening when the animal dives.

TOOTHED WHALES AND PORPOISES

The suborder **Odontoceti** includes species such as the sperm whale, the killer whale (Figure 8.16), the beluga whale (Figure 8.17), and porpoises. All actively hunt large prey, such as fishes and smaller marine mammals. They have typical mammalian teeth, which are modified to various degrees. There is a single blowhole, and most species are excellent divers, a characteristic that is probably related to their hunting behavior for mobile prey. Sperm whales can dive routinely to depths of 1,000 m, and one has been spotted by sonar methods at a depth of 2,250 m! Sperm whales are squid eaters, and they probably dive to great depths, wait there, and then attack unwary squid. Toothed whales and dolphins range in size from just a few meters in length (the dolphins), to lengths of 15–20 m (the sperm whale).

Most odontocetes are capable of sophisticated oral communication and can generate a series of sonic and ultrasonic clicking signals. In killer whales, pods (groups of whales) often use sounds different from other pods. The clicks are also used as a means of

Fig. 8.16 A killer whale breaching the surface, San Juan Islands, Washington State. (With permission of Friday Harbor Laboratories.)

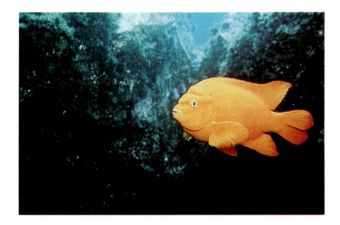

Plate IX. Some Marine Fishes

IX.1. *Top left,* The garibaldi, *Hypsypops rubicundus,* is a temperate damselfish common in kelp areas in the shallow subtidal regions of California. Males are very territorial and maintain the nests where females lay eggs. (Photograph by Carol Eunmi Lee)

IX.2. *Top right,* The French angel fish, *Pomicanthus paru,* which feeds mainly upon sponges, is commonly found in shallow Caribbean coral reefs. (Photograph by Carol Eunmi Lee)

IX.3. *Middle,* The ocean sunfish *Mola mola* is found in all tropical and temperate seas (seen here in waters off Long Island, New York), feeds upon jellyfish, and is enormous, weighing up to 2 tons. It is related to puffer fish. (Photograph by Sam Sadove)

IX.4. *Bottom,* Embryo of the swell shark, *Cephaloscyllium ventriosum.* Note the large yolk supply, which is protected, along with the embryo, by a rigid outer shell, known as a "widow's purse." The embryo is about 10 cm long. (Photograph by Jeff Levinton)

Plate X. Marine Mammals: Cetacea

X.1. *Left,* A humpback whale breaching, near Montauk Point, Long Island, New York. (Photograph by Sam Sadove)

X.2. *Middle,* Fluke of a humpback whale that is "lobtailing." (Photograph by Sam Sadove)

X.3. *Below,* The bottle-nose dolphin, *Tursiops truncatus.* (Photograph by Sam Sadove)

X.4. *Above,* Hauling and butchering of a fin whale, *Balaenoptera physalus,* at a whaling station in the 1980s. (Photograph by Sam Sadove)

Plate XV. Rocky Shores

XV.1. *Top left,* A tidal pool on an exposed shore of the Pacific Northwest. Note the presence outside the pool of desiccation-resistant barnacles and mussels, and more sensitive anemones, hydrocorals, and starfish within. (Photograph by Jeff Levinton)

XV.2. *Top right,* Coexistence of the mussel *Mytilus californianus* and the stalked barnacle *Policipes polymerus* on a rocky shore at Tatoosh Island. (Photograph by Jeff Levinton)

XV.3. *Middle,* The intertidal Pacific coast anemone *Anthopleura elegantissima* occurs in groups that arose from the successive fission of a single colonizing individual. (Photograph by Jeff Levinton)

XV.4. *Bottom,* The intertidal brown seaweed *Fucus gardneri* often completely dominates protected rocky shores of the Pacific coast of the United States and Canada. (Photograph by Jeff Levinton)

Plate XVI. Rocky Shores

XVI.1. *Top left,* The "mussel glacier," Tatoosh Island, outer Pacific coast of Washington. The lower border of the *Mytilus californianus* mussel bed is maintained by predation by the starfish *Pisaster ochraceus* (on rock wall below mussel bed). (Photograph by Jeff Levinton)

XVI.2. *Top right,* Aggregation of the starfish *Pisaster ochraceus* at the periphery of a *Mytilus californianus* mussel bed, Shi Shi Beach, outer coast of Washington. (Photograph by Thomas H. Suchanek)

XVI.3. *Middle,* The chiton *Catherina tunicata* (ca. 12 cm long) at Tatoosh Island. Chiton grazing strongly affects the distribution and abundance of seaweeds. (Photograph by Jeff Levinton)

XVI.4. *Bottom,* Aggression between nonclonemates of the sea anemone *Anthopleura elegantissima* by means of nematocyst-laden acrorhagi (white structures). If not removed, the right animal would eventually kill the one on the left. (Photograph by Jeff Levinton)

Fig. 8.17 Beluga whales reach several meters in length and prey upon fishes and smaller marine mammals. (Photograph by the author.)

echo-location; the animals can accurately estimate distance on the basis of the travel and return time of the clicks. Many species have a bulbous structure, the **melon**, on the anterior part of the upper skull. The melon is filled with a fine oil that was once prized by whalers for use in oil lamps. Some suggest that the melon serves as a device to focus returning sound waves, though others argue that the melon is a storage site for gaseous nitrogen, so that the animal will not develop the "bends" as it surfaces (see the section on diving later in this chapter). It is true that the auditory canal is reduced in toothed whales, so sound is likely transmitted through another route, such as the bones of the lower jaw, which are thin and seem to be good transmitters.

Communication and social interaction are of major importance in these species. In most of the species, traveling in small pods is the rule. Killer whales (*Orcinus orca*) travel in pods of usually less than 10, with a dominant male, several females, and a few subordinate males. The dominant male usually mates with the females, but the subordinate males are sometimes successful maters. Pods usually consist of a mother, her offspring, and sometimes members of the third generation. When the group reaches a large size, it tends to split into two pods, so a local group of pods can therefore be placed on a tree of descent. Bottlenose dolphins travel in herds of about 15, and sometimes actively spread out and then corner schooling anchovies in a confined area, to facilitate capture. Some evidence exists for specialization

within different populations of the same species. In the coastal waters of Washington and British Columbia, resident pods of killer whales consist of fish specialists, whereas more mobile pods are specialized for feeding on pinnipeds. It would be interesting to know whether this difference has led to language differentiation and mating isolation among the two feeding groups.

BALEEN WHALES

The **Mysticeti**, or **baleen whales**, are distinguished by the replacement of teeth in the adult by horny baleen plates (Figure 8.18), derived from dermal tissue. Such plates are attached to the upper jaw, from which project sheets composed of individual long straining bristles. The sheets are maintained normal to the axis of swimming, and the bristles strain out larger zooplankton, such as krill. The right whales, such as bowhead and right whales proper, are continuous ram suspension feeders. Their baleen plates are longer and the whales swim relatively slowly through the water, continuously taking in water and straining finer zooplankton, such as copepods. By contrast, **rorqual whales**, which include blue whales, are **intermittent ram suspension feeders**. Periodically, the blue whale closes its mouth and the tongue is raised, thus forcing water out through the baleen plates, which are suspended from the roof of the mouth. The zooplankton trapped on the baleen are then swallowed. In the blue and fin whales, the lower mouth is a furrowed ventral pouch. When closed, the lower mouth appears externally as a series of parallel ridges and furrows below the mouth opening. As the mouth opens, the pouch expands, and may enclose as much as 70 tons of water (Figure 8.18).

Baleen whales do not feed continuously. Whales feeding on small fish known as capelin feed only on dense schools. The apparent threshold of feeding activity[3] above a certain fish density must be a behavioral adjustment to increase foraging gains relative to the considerable costs of increased drag while a whale is swimming with its mouth open and baleen exposed to the water.

Baleen whales range in size from rorquals and gray whales, which are 10–15 m long, to the blue whale, which can exceed 30 m. Baleen whales often migrate great distances. Gray whales spend the summer in the Bering Sea and in the Arctic Ocean. They winter in breeding grounds in bays on the Pacific coast off Baja

[3] See Piatt and Methuen, 1993, in Further Reading, Mammals.

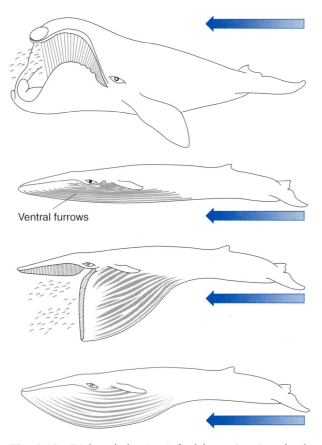

Ventral furrows

Fig. 8.18 Right whales (top) feed by swimming slowly through the water with the mouth open and baleen plates suspended. Rorqual whales, such as the blue whale, periodically open the lower jaw (bottom sequence) and extend an enormous ventral pouch. When the pouch closes, water is forced through the baleen plates, which are shorter than those of right whales. Zooplankton are trapped on the plates. (From Pivorunas, 1979.)

California and off Korea and Japan. There are several independent migrating populations of humpback whales. One, for example, winters in the Hawaiian Islands and spends the summers in Alaskan waters. Another population divides its time between waters off of California and Mexico.

SEALS, WALRUSES, AND SEA LIONS

✳ **Pinnipeds, which include seals, sea lions, and walruses, have hair but lack the fat layers of cetaceans. Sea otters are marine mammals belonging to the family Mustelidae that are coastal carnivores.**

Members of the suborder **Pinnipedia**, mammalian order Carnivora, include the seals, sea lions, walruses, and elephant seals. Pinnipeds have hair and lack the large amount of fat so characteristic of whales, but they are streamlined and expert swimmers. The rear legs are modified as flippers and can propel the animal through the water. The group ranges in size from the relatively small (1 m long) sea lions (Figure 8.19) to the enormous southern elephant seal, which can reach 4000 kg (males) and 6 m in length. Seals spend a great deal of time out at sea diving for prey. Most spectacular are the elephant seals. At sea, males spend a great deal of their time under water and one has been traced to a depth of over 1,500 m.

Seals are divided (Figure 8.20) between the true seals (Family Phocidae) and the eared seals (Family Otariidae). True seals include the Weddell seal and the elephant seal; they have a small ear hole, and short backward-pointing hind flippers that are used for propulsion in water. These seals do not get about on land very well, in comparison to their elegant propulsion in the water. By contrast, eared seals, which include the Australian fur seal, have an external ear and relatively longer necks; they use the anterior flippers for swimming. They can fold the rear flippers forward, sit on them, and use them for propulsion on land. In the walrus, the upper canine teeth have become modified into large tusks. In most pinniped species, the teeth are simpler and less

Fig. 8.19 Stellar's sea lion, a resident of North Pacific waters.

HOT TOPICS IN MARINE BIOLOGY 8.1 The Evolution of Whales and Their Relatives: Debate

The order Cetacea[4] includes a fascinating diversity of mammals, all permanent residents of the open marine environment. The whales and dolphins are remarkable for the highly specialized morphology that serves them well for movement and feeding in the marine environment. They are extraordinary divers, sometimes descending to depths exceeding 1,000 m. In addition, they are streamlined for efficient movement through the waters, having flukes, flippers, and even fins to help propel and guide them through the water.

Yet, the cetaceans are mammals, and it is clear that they must have arisen from terrestrial ancestors. The evidence of ancestry can be seen even within many of the living whales; baleen whales lack typical mammalian teeth, but such features appear in the embryo briefly, only to disappear later in development. It is as if the ancestral features have been retained as evolutionary ancestral whisperings within the animal's current form and lifestyle. Cetaceans share many of the characters of placental mammals, including placental development, live birth, homeothermy, mammalian skin, bone structure, and skeletal features.

Despite the similarities between cetaceans and terrestrial mammals, it has been difficult to establish a link of morphologies between the whales and dolphins and any other group of mammals. It is clear that the whales arose from terrestrial ancestors, which required a complete restructuring of morphology (imagine a cow swimming deep below the surface and catching fish, and you get the idea that a lot of evolutionary change was required). To get a whale from a terrestrial quadruped, one must have a skeleton that allows axial swimming, which must derive from a four-legged land animal. Most mammalogists have linked the general skeletal features of cetaceans to certain groups of terrestrial mammals, particularly a fossil group known as the mesonychids, which were probably running and carnivorous forms. The challenge has always been to find intermediates. After all, it is not easy to imagine what intermediate forms could bridge the gap between fully terrestrial running carnivores and fully swimming forms, with no external trace of hind limbs and even horizontal flukes that propel the animal through the water.

In the last few years the gap has been filled with a number of intermediate-looking fossils (Box Figure 8.1), although the exact path from ancestor to descendant is not yet known completely. Paleontologist Philip Gingerich described the skull (that is all that has been recovered of this fossil) of *Pakicetus*, a fossil from the Eocene epoch of Pakistan (about 52 million years old).[5] Even though this seems like an incomplete find, many aspects of morphology of the skull can be related to ecology and even function of the postcranial skeleton. Its teeth resemble those of the presumed mesonychid ancestors, but many other features of the skull resemble those of the Cetacea. *Pakicetus* lacked, however, some of the features associated with deep diving and hearing in whales. Fossils of another genus, named inappropriately with the reptilian-sounding name *Basilosaurus*, were more complete, but the hind limbs were extremely small: too small, in fact, to serve a locomotory function.

More important was the 1994 report of a fossil in Pakistani sediments younger than those that yielded *Pakicetus*. The new fossil, *Ambulocetus natans*, clearly had rear limbs that could function in walking, yet the whole animal was elongated and definitely aquatic (Box Figure 8.1). Its hind limbs were sufficiently robust to suggest that it probably ambled about on land (its estimated weight is 300 kg), while being primarily an aquatic creature. It had a fairly typical mammalian tail, which is absent in the modern cetaceans. Thus one of the largest gaps in evolutionary history has now been filled with some intermediate forms, although there still are many questions about evolutionary relationships among the cetaceans, whose relationships (Box Figure 8.2) to the mesonychids are at least well understood.[6] The connections between the primitive whales, the archaeocetes, and the modern toothed and baleen whales is still poorly understood.[7]

More surprises and an important controversy have emerged from the study of evolutionary relationships of whales. Characters in the various cetacean species can be grouped to form trees of evolutionary relationship, as we discussed in Chapter 3. Groups are identified by the presence of unique characters that they have in common. Thus humans have DNA sequences very similar to those of chimpanzees, gorillas, and orangutans and therefore are closely related.

Two controversies have arisen from studies using DNA sequences. One relates to the determination of the closest relatives of the entire Cetacea. As we mentioned, morpho-

[4] See Berta, 1994, in Further Reading, Mammals.
[5] See Gingerich and others, 1983, in Further Reading, Mammals.
[6] See O'Leary and Uhen, 1999, in Further Reading, Mammals.
[7] See Thewissen and others, 1994, in Further Reading, Mammals.

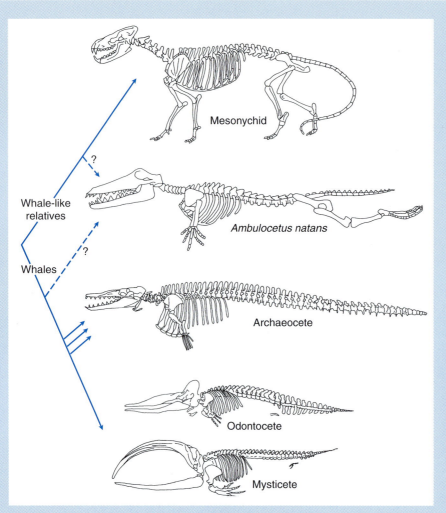

Box Fig. 8.1 Some of the ancestors and relatives of the modern toothed whales (Odontoceti) and baleen whales (Mysticeti). The hypothesized ancestral mammalian mesonychids might have appeared as in the top sketch. Also shown are the intermediate form *Ambulocetus natans* and an example of the archaeocetes, a primitive whale group preserved as fossils. (After Berta, 1994.)

logical evidence based upon fossils allies the early Cetacea with the fossil mesonychids. These two groups are thus believed to be more closely related to each other than either one is related to the mammalian group Artiodactyla, which includes the hippos, pigs, cows, and camels. The morphological fossil evidence, therefore argues that the ancestors of whales, the mesonychids, have some distant characteristics in common with artiodactyls, but that is about all. Molecular evidence can only produce family trees of living groups, but some recent evidence has produced some surprising results.[8] Molecular sequences ally the whales closely with hippos, which conflicts with the fossil evidence that suggests that the Cetacea are more distant relatives of the entire Artiodactyla. Instead, the molecular evidence suggests that the hippos are the ancestral group of whales and dolphins. At present, this controversy has not been resolved.

The second controversy involves our understanding of relationships within the cetaceans. In this chapter, we mention the existence of two cetacean groups, toothed whales (Odontoceti) and baleen whales (Mysticeti). These groups have always been regarded as separate evolutionary groups. The presence of carnivorous teeth and associated jaw structure, among a large number of other morphological structures, unites the toothed whales and dolphins as a group distinct in evolutionary history from the baleen-bearing whales. Milinkovitch, Orti, and Meyer[9] used DNA sequencing to acquire an independent set of characters to assess evolutionary relationships, and the results were

[8] See Nikaido and others, 1999, in Further Reading, Mammals.
[9] See Milinkovitch and others, 1993, in Further Reading, Mammals.

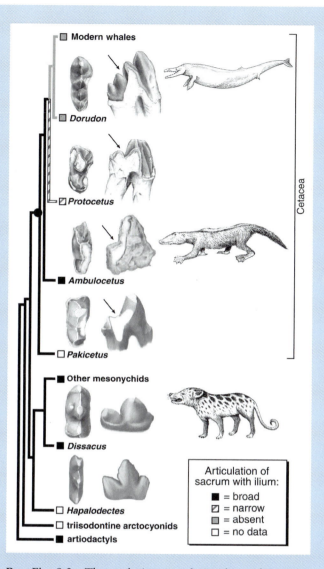

Box Fig. 8.2 The evolutionary relationships of Cetacea, especially with their sister group, the mesonychids. (Courtesy of Maureen O'Leary, after O'Leary and Uhen, 1999).

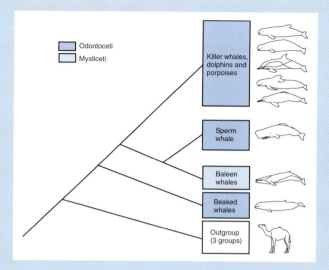

Box Fig. 8.3 The evolutionary relationships of dolphins, sperm whales, baleen whales, and bowhead whales, as inferred from differences in DNA sequences of a number of different genes. These sequences are analyzed with reference to sequences of an "outgroup" of terrestrial mammals related to the ancestral group. Note that the closest relatives of the dolphins are baleen whales, and not sperm whales, as suggested by the old myticete–odontocete taxonomic classification. (Drawn from data courtesy of Axel Meyer.)

The relationships revealed by DNA sequencing are completely unexpected. Although the suggestion that toothed forms are ancestral is not surprising, the close relationship of sperm whales and baleen whales suggests that the highly specialized baleen lifestyle of feeding on plankton has arisen relatively recently (25 million years ago, instead of the previous estimate of 40 million years). That is, the evolutionary split is not nearly as ancient as was formerly supposed.

New controversial hypotheses are always useful challenges to old ideas and sometimes succeed in changing the course of scientific investigation. But this one is proving to fail the test of consistency. Heyning[10] has done an extensive analysis of the evolutionary relationships of whales, using a large number of morphological characters to establish associations. The characters are diverse, numerous, and have clear-cut relationships. The more unique characters in common, the more a grouping can be justified. The morphological analysis is unambiguous. As the traditional

surprising and controversial (Box Figure 8.3). According to the molecular analysis, baleen whales appear to be closely related to the toothed sperm whales, and the dolphins stand as a relatively distantly related group whereas the morphological analysis strongly allies the toothed groups together. Recent evidence employing sequencing of more genes is consistent with this surprising result. Given the currently available data, the bowhead whales remain difficult to relate to the others.

[10] See Heyning, 1997, in Further Reading, Mammals.

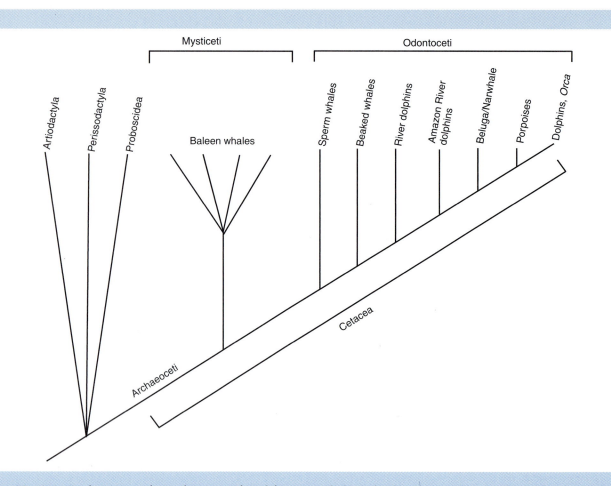

Box Fig. 8.4 Evolutionary relationships are inferred from morphological information. These results conform to the traditional, and expected, unity of the toothed whales. (After Heyning, 1997.)

hypothesis suggests, toothed whales (sperm whales, porpoises, and others) form a natural evolutionary group, which, in turn is the closest relative of the baleen whales (Box Figure 8.4). This makes sense, since we find a number of fossils of clear mysticetes that had teeth as adults, and we can construct a rather complete morphological series from the ancient archaeocetes to modern mysticetes, such as the blue whale.

If the molecular data conflict with the morphological data, how can we choose between them? One might suggest using all the data combined, but given the degree of conflict in the present case, this might only muddle things, especially if one analysis is possibly especially weak. Heyning criticizes the utility of the molecular data, which, he argues, do not have a strong enough signal to define the groupings with statistical confidence. A special problem is finding a group to make a comparative analysis within the

Cetacea. One needs the closest related group, or sister group. Divergence within the Cetacea, in molecular or morphological characters, is made with reference to this sister group. It is well known that the choice of a sister group can radically change an analysis, because the sister group's characters determine the polarity of change from the group to groups within the whales. It also is often difficult to do analyses of DNA characters of groups that diverged many millions of years ago. The combination of problems makes us suspicious of the molecular analysis, and the morphological analysis is so strong that it cannot be ignored. Thus we should tentatively accept the traditional view of cetacean relationships and discard the conclusions generated by the DNA analysis. The morphological evidence convincingly allies the toothed groups and calls into question some aspect of the DNA analysis. Likely the sister group DNA sequence is the problem.

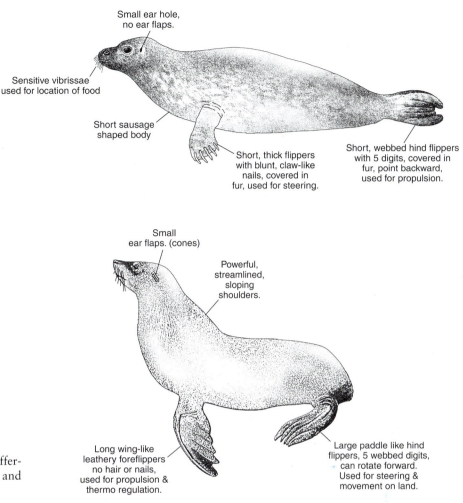

Small ear hole, no ear flaps.

Sensitive vibrissae used for location of food

Short sausage shaped body

Short, thick flippers with blunt, claw-like nails, covered in fur, used for steering.

Short, webbed hind flippers with 5 digits, covered in fur, point backward, used for propulsion.

Small ear flaps. (cones)

Powerful, streamlined, sloping shoulders.

Long wing-like leathery foreflippers no hair or nails, used for propulsion & thermo regulation.

Large paddle like hind flippers, 5 webbed digits, can rotate forward. Used for steering & movement on land.

Fig. 8.20 General external differences between true seals (top) and eared seals (bottom).

specialized than those of terrestrial mammals. Pinnipeds are carnivores and are usually excellent divers.

Pinnipeds are well known for their mating and reproductive habits. Although most species spend the majority of time at sea, they come to large beach and rocky-shore areas to mate and rear young. Usually, large males arrive first and establish territories. This is true for species such as the northern fur seal and the elephant seal. The females and subordinate males then arrive, and the dominant males maintain harems of one to several females. Subordinate males risk severe injury if they attempt to mate with a harem female. Dominant males maintain their territories by means of fierce displays and pitched battle. In the elephant seal, male combat often results in severe wounds with profuse bleeding. Subordinate males usually congregate on the fringe of the main territories of the dominant males.

Sea otters (Figure 8.21) are distantly related from seals and belong to the mammalian family Mustelidae, which includes otters and badgers. Sea otters have evolved a number of traits for life at sea, including streamlined body, modified appendages, and hair that can be preened into a thermal barrier to retard heat loss to seawater. They are marine carnivores, preying mainly on benthic species such as sea urchins and mollusks, and on fishes. They often dive tens of meters to pull abalones and urchins from hard bottoms. They bring these prey animals to the surface and often crush them with the aid of rocks.

Fig. 8.21   The sea otter, *Enhydra lutris,* is an important predator in kelp forests.

SIRENIA

* Sirenians include manatees, dugongs, and sea cows. They are hairless and usually herbivorous.

Members of the mammalian order Sirenia are generally hairless and streamlined, and they superficially resemble whales. They have blunt and broad muzzles. They are generally sluggish and live in shallow water, where they eat seaweeds or other aquatic vegetation. Manatees (Figure 8.22) live in the shallows of the Caribbean and in Florida; they feed upon wa-

Fig. 8.22   A manatee in the Crystal River in Florida. (Photograph by Patrick M. Rose, courtesy Florida Marine Research Institute, Department of Environmental Protection.)

ter hyacinths in fresh water and green seaweeds in saline water. Because of their slow movement and continuous grazing on vegetation in shallow water, they are often damaged by the propellers of boats. Dugongs live in shallow waters from the Red Sea to Australia, but their oil and meat are highly prized, and hunting has severely reduced their numbers. Steller's sea cow was discovered by explorers in the North Pacific, but hunters exterminated them by the 1700s.

Diving by Marine Mammals

Oxygen Debt and Circulation Problems

* Diving causes problems of oxygen shortage, which require a number of conservation responses.

As mentioned earlier, marine mammals can be excellent divers. Toothed whales dive regularly to depths greater than 1,000 m, but deep diving is also common in elephant seals, otters, and many other species. The Weddell seal, a resident of Antarctica, dives deeper than 500 m, and can stay below for nearly an hour, and, we have already noted that elephant seals spend much of their time under water. Diving is clearly valuable because it gives access to prey living at great depths, but the animals must deal with several physiological challenges. Over such long periods under water, oxygen is consumed but carbon dioxide and lactic acid build up in the blood.

Marine mammals do not have gills, hence must get their air supply at the surface. Moreover, these animals have relatively high oxygen consumption rates. One might therefore suppose that they risk running out of oxygen during extended dives. Special mechanisms, however, have evolved to accommodate these circumstances. Seals, for example, can carry much more oxygen than humans by making use of the following mechanisms: (1) increased volume of arteries and veins, (2) storage of oxygen attached to hemoglobin in muscles, (3) ability to carry more oxygen per unit volume of blood, owing to increased red blood cell concentration, (4) decreased heartbeat and oxygen consumption rate, and (5) restriction of peripheral circulation and circulation to abdominal organs (to reserve the oxygen supply for essential functions, such as nervous transmission and operation of main circulatory arteries).

It may be surprising that seals and toothed whales

do not have extraordinary lung capacity. Uptake of large amounts of air at the surface would be dangerous to them because then toxic amounts of oxygen might be released at depth. Also, nitrogen taken up at low pressure might then be released, giving the animal the bends (see the following). The animals are, however, extremely efficient at extracting oxygen during breathing. Oxygen combines with hemoglobin in a process that, in the case of cetaceans, is extremely efficient. During the dive, oxygen in the blood decreases, and a "debt" builds up. After surfacing, diving mammals usually increase their metabolic rate and provide needed oxygen to tissues.

GAS BUBBLE PROBLEMS

When a diving animal ascends, the pressure decreases about 1 atmosphere for every 10 vertical meters. If the bloodstream is saturated with gas, such a decrease of ambient pressure will cause the gas to bubble out. In humans, bubbles accumulate in the circulatory system at the joints, causing the painful and potentially fatal syndrome known as the bends. The only possible treatment is to place the victim in a recompression chamber, where pressure is first increased, then lowered very slowly, so that the gas leaves the bloodstream slowly, without the formation of large bubbles. This is important because bubbles that accumulate in capillaries of the nervous system can suddenly cut off blood supply to part of the brain and cause death. Seals and whales avoid this problem to a large extent. Not only is their lung capacity relatively small, but they are able to constrict blood flow between the lungs and the rest of the circulatory system, and to allow their lungs and associated structures to collapse, to prevent storage of any extra gas that might be released during drops in pressure. These physical characters reduce the amount of nitrogen that enters the circulatory system. They also do not breathe compressed air at depths the way scuba divers do, so their bubble problems are fewer.

Birds and Reptiles

Seabirds

Seabirds are birds that travel some great distance across the sea and typically breed on offshore islands or coastal areas. Seabirds live throughout the oceans and consist of a surprisingly diverse array of adaptive types. They have a salt gland that efficiently excretes salts gained from seawater and food. They range from the flightless cormorant to the frigate bird, which is completely dependent upon long-term flight. They range from species feeding upon small zooplankton to those, such as pelicans, that feed on large muscular fish. Some are faithful to a relatively small feeding and breeding area, whereas others migrate for thousands of kilometers. Seabirds are long lived, and albatrosses probably reach the age of 50 years rather often. In most seabirds, apart from gulls, evolution has involved a strong degree of loss in the ability to walk efficiently on land. Many species breed in colonies, often comprising several thousand birds, and some tern colonies probably number in the millions of pairs. Breeding often occurs only in isolation from predators, and large breeding groups are often found on remote islands, which has made many species vulnerable to human disturbance.

✳ **Seabirds include the penguins, the petrels and their allies, the pelicans, and the gulls and their allies.**

Seabirds can be conveniently divided into four major groups (Table 8.1, Figure 8.23). **Penguins** are flightless, and their flippers are modified forewings that derive from flying ancestors. They live in cold Antarctic and subantarctic waters in colonies that vary from a few pairs to thousands. They are well insulated from cold air and water by means of a layer of blubber and a thick layer of feathers. They dive from the surface for their food, which usually consists of small fish. They range in size from the little blue, about a kilogram or two and standing about 40 cm, to the majestic emperor, which stands over a meter tall and weighs about 30 kg. Protection against the cold is a major factor in penguin biology. Penguins such as the emperor huddle in aggregations to keep warm, the outer birds often moving into the middle of the pack. Father penguins incubate the mother's single egg throughout the entire Antarctic winter; the baby bird hatches in spring, to take advantage of the increase of plankton and fish. King penguin chicks do not leave the care of their parents until about 14 months after hatching from the egg. To keep warm in winter, thousands of chicks huddle together to retain heat.

Petrels and their allies—the albatrosses, petrels, shearwaters, and diving petrels—have large external nostrils, which may be useful to smell prey, and a hooked bill. The albatrosses can have wingspans of

Table 8.1   The main groups of seabirds

ORDER	FAMILIES	NUMBER OF SPECIES	COMMON NAMES
Sphenisciformes	Spheniscidae	16	Penguins
Procellariformes	Diomedeidae	13	Albatrosses
	Procellaridae	55	Fulmars, prions, petrels, shearwaters
	Hydrobatidae	20	Storm petrels
	Pelecanoididae	4	Diving petrels
Pelecaniformes	Phaethontidae	3	Tropic birds
	Pelecanidae	7	Pelicans
	Phalacrocoracidae	27	Cormorants, shags
	Fregataidae	5	Frigate birds
Charadriiformes	Stercorariidae	6	Skuas
	Laridae	87	Gulls, terns, noddins
	Rynchopidae	3	Skimmers
	Alcidae	22	Auks

over 3 m and are superb gliders, taking advantage of the steady winds in the southern oceans. They are nearly all colonial, breeding on open and windswept ground. They range from the giant petrel, which preys on other birds to the small-fish-eating puffins to the zooplankton-straining prions, which have comblike plates on each side of the mouth. They may nest in colonies from several thousand to just a few, and some species participate in long-ranging migrations.

Pelicans and their relatives—the boobies, gannets and cormorants—are often heavy and include many brightly colored and ornamented species. They are mainly tropical, but some species nest in the Arctic and Antarctic. While some, such as frigate birds, fly far out to sea, most of this group stays closer to land.

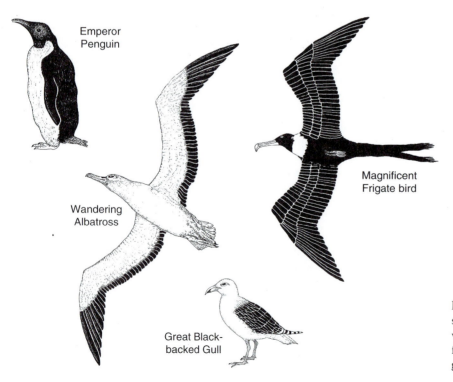

Emperor Penguin

Wandering Albatross

Magnificent Frigate bird

Great Black-backed Gull

Fig. 8.23 Some of the diversity of seabirds: (a) emperor penguin, (b) wandering albatross, (c) magnificent frigate bird, and (d) black-backed gull.

They are diverse in hunting methods, from the plunging dives of gannets to the underwater pursuit of cormorants. As opposed to the broad gliding habit of albatrosses, frigate birds are capable of tight maneuvering in the air. Feeding in this group is restricted to fishes. Most species of this group tend to breed much earlier than the likes of petrels.

Gulls, terns, and auks comprise by far the most diverse group of seabirds. They can be found in the millions, although their breeding colonies are in the thousands and not particularly larger than others, such as gannets. The herring gull, a complex of closely related species, extends over vast areas of the Northern Hemisphere and can be found breeding in a wide variety of shoreline and island habitats. Terns are smaller and more marine than gulls and are most abundant in diversity in the tropics, although the Arctic and Antarctic terns nest in extemely high latitudes. The auks and puffins are to be found in high latitudes. Puffins nest in holes or dig their own. Auks can be found in very large breeding colonies, but some species nest as single breeding pairs. They feed mainly on small fish and zooplankton.

BREEDING BEHAVIOR, NESTING GROUNDS, AND MIGRATION

✳ **Seabirds are nearly always monogamous and nearly always breed communally.**

Care of young is almost always shared by the mother and father, and this may explain the ubiquity of monogamy in seabirds. Monogamy appears to assure earlier breeding and increased survival of young. Even egg incubation may be shared by the male; the male emperor penguin incubates a single egg for the entire Antarctic winter. Nearly all species nest in colonies, some numbering in the thousands. **Colonial breeding** (Plate XI.2) may allow sharing of information about feeding grounds among foraging parents, but the practice more likely arose from the shortage of suitable sites that are remote from predators of nesting birds and their eggs. Communal defense, such as the "dive-bombing" of herring gulls, also is an effective deterrent against predators. Colonial living, however, has the disadvantage of putting birds in intense competition for limited resources near the breeding colony.

✳ **Territoriality and combat are common in seabirds because high-quality nesting sites are in short supply.**

Dense colonies of seabirds are found on remote islands, and space on these islands is usually in short supply. It should be no surprise, therefore, that there is intense competition for limited breeding sites. Seabird pairs often spend much time selecting and defending sites, even well before nesting occurs. When a bird moves through a colony—for example, to take off from a high promontory—it inevitably moves through the protected areas of many nesting pairs. This usually involves gripping of beaks, pushing, and flailing around. Still, a male without a nesting site will continually search, and attempt to find an unoccupied location.

When it first lands, a bird is often confronted aggressively by its mate until recognition occurs. Fighting is usually more intense in species that nest on cliffs or in holes, where space is more limiting than, say, on open-ground sites. In many cases, fights are avoided by stylized displays that demonstrate the mood or size of the potential combatants (Figure 8.24). Gulls, for example, can adopt an upright posture, which signifies a readiness to peck rapidly and downward. Gannets bow continuously during nesting to signify their ownership of a site. When one

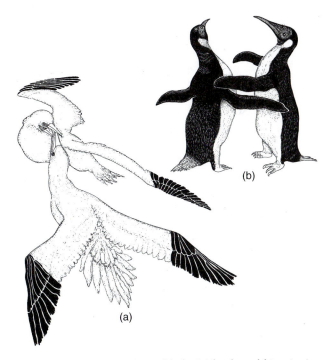

Fig. 8.24 Aggression in seabirds: (a) beak grabbing in Atlantic gannets and (b) flipper fighting between male penguins. (After Nelson, 1979.)

member of a mated pair lands, the two greet by crossing bills and calling. To see this occur by the thousands on a colony is among the most exciting sights one can ever witness.

* Mating pairs may be formed before or after a nesting site is chosen. Courtship often includes elaborate displays, which may involved groups of males.

Mate selection can be intimately involved with choice of a nest site. In strongly territorial species such as gannets, the male first establishes a nesting site and displays to attract females. But female choice clearly involves both the male himself and an assessment of the quality of the nesting site. In some cases, however, pairs are formed outside the nesting area, whereupon the pair establishes the nesting site together. Because the relationship between mates is so strong, it is not surprising that courtship is involved and occupies a great deal of time and energy. Male herring gulls attract females by means of a call and eventually by feeding the female. Once attracted, the male leads the female to his territory, and a new series of head-bobbing behaviors continues the process of mate attraction. In some terns, males catch fish and present them to prospective mates. Albatrosses have extremely elaborate displays, which involve rotating of male and female bills, flapping the lower mandible, occasional dances, and pronounced strutting about by the male. Nuptial coloration, communal displays by males, and elaborate facial adornments also aid in mate attraction by some species. Male king penguins, for example, raise their heads to expose the orange neck and bill patch.

* Nesting sites vary in substratum type, degree of slope, and degree of isolation. Many species may share a given area.

Many steep cliffs are sites for nests of many species of seabirds. They provide isolation from predators. A rich island may include nesting cormorants, gulls, guillemots, and many other species. Within these areas, however, individual species have more specialized requirements. Gannet colonies, for example, are best developed on the tops of cliffs, where there are broad horizontal ledges. By contrast, cormorant nests are often found on very narrow ledges in otherwise vertical cliffs. Gulls seem to occupy a broader variety of sites and may be found on cliffs but also on broader, flatter, and lower rocky areas.

* The breeding cycle includes periods of nest site establishment, egg laying and incubation, and fledging.

Seabirds become sexually mature rather late, usually at least in their second year. They then progress through a series of breeding stages, which in most cases are strongly seasonal. In temperate and higher latitudes, breeding must be timed to coincide with the plankton production cycle, when plankton and planktivorous fish are most abundant. This reaches an extreme in Antarctica, where birds must arrive in the spring (September–October in the Southern Hemisphere), dig through the ice and snow, establish nests, and fledge their young by the time of Antarctic summer, when the water is rich in fishes and euphausiid shrimps. Eggs are incubated sometimes for long periods, over 60 days in the case of albatrosses, and larger eggs are produced in areas where food is relatively scarce. Single-egg clutches are very common among seabird species, although females of some species may lay several eggs at a time. After hatching comes the fledging period lasts until the young birds are able to fly. The first flight is often a glide from a cliff, and a sudden downdraft may result in death. Fledging periods can be short, as in Antarctica, or as long as 6 months, especially if food is relatively scarce. A longer fledging period is associated with a chick's resistance to starvation, which can be crucial in birds that must travel a great distance to the first meal.

Large colonies of seabirds are probably limited by food in the surrounding area. In general, chick fledging weight is inversely related to breeding colony size, which suggests food limitation by the fish in the local surrounding area. Philip Ashmole and Myrtle Ashmole[11] suggested that seabird population size is regulated by food; in very large colonies, parental birds would have to travel very far from the nesting grounds to find food for their young. This would put an upper limit on the number of young that could be fledged in a given area.

* Seabirds migrate to maximize use of feeding and nesting areas.

Although as we have seen, nest sites are chosen for their suitability to avoid predators, for social interactions, and for proximity to local food, after the

[11] See Ashmole and Ashmole, 1967, in Further Reading, Birds and Reptiles.

breeding season birds may migrate to distant areas where food is more abundant. If breeding occurs on a mid-Pacific island, for example, this migration may involve moving hundreds to thousands of kilometers in any of several directions in the pursuit of food. Albatrosses and frigate birds fit this pattern. At the other extreme are many gull species, which move no more than a few tens of kilometers from the nesting grounds. In some species, however, the migration is much more directed, especially in high latitudes when the winter season is very harsh. Arctic terns breed during the Northern Hemisphere summer and then fly to Antarctica to take advantage of the austral summer. The short-tailed shearwater migrates between south Australia and the northern Pacific Ocean. Unlike many land birds, such long-distance migrants often touch down to feed during their journey.

FEEDING

✳ **Food gathering depends upon the alternative strategies of efficient long-distance flying and shallow diving.**

Seabirds employ two strategies for hunting food: flying and underwater swimming. To a large degree these strategies are mutually exclusive because efficient flying requires long and relatively inflexible wings, whereas underwater swimming requires short wings that are usable as flippers. Figure 8.25 shows the methods by which seabirds implement these strategies. Underwater swimmers may dive from the surface (e.g., penguins) or from the air (some shearwaters). Penguins have flightless wings that are highly specialized for underwater swimming and some can reach depths greater than 100 m. Cormorants use their feet as flippers to swim below the surface. Gulls and terns are good flyers but can dive to shallow depths by plunging into the water from the air. Some seabirds simply rest on the sea surface and pluck fish from the water; others dip their bills in the water and strain out zooplankton. Gulls also feed on land, either by scavanging or by taking prey such as mussels and urchins, and smashing them on the rocks or even the pavement of roads and parking lots.

The dichotomy of flying and underwater swimming influences the geographic location of seabirds. Efficient underwater swimmers cannot fly very far—or indeed at all, in the case of penguins. Thus one tends to find specialized underwater swimmers in coastal areas of very high productivity, such as upwelling and polar regions. By contrast, efficient flyers can be found throughout the relatively less productive tropical waters of the open Pacific.

Shorebirds

Shorebirds differ from seabirds in that they have a greater dependency upon terrestrial sites for nesting and often migrate between two sites on continents.

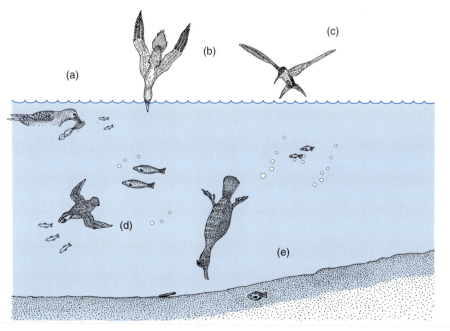

Fig. 8.25 Methods by which seabirds obtain prey: (a) feeding from surface (fulmar), (b) plunge diving (gannet), (c) diving from air (tern), (d) underwater pursuit diving using wings (puffin), and (e) use of feet in underwater propulsion (shag). (After Furness and Monaghan, 1987.)

Table 8.2   General classification of shorebirds

ORDER	FAMILY	NUMBER OF SPECIES	COMMON NAME
Charadriiformes	Haematopodidae	ca. 6	Oystercatchers
	Charadriidae	ca. 65	Plovers, turnstones, surfbirds
	Scolopacidae	ca. 85	Sandpipers
	Recurvirostridae	7	Avocets, stilts
	Phalaropodidae	3	Phalaropes

Like other migrating birds they usually migrate between winter feeding grounds and spring–summer nesting grounds. Shorebirds, including the sandpipers, plovers, and oystercatchers (Table 8.2, Figure 8.26), feed principally on intertidal soft-bottom and rocky-shore organisms and may exert profound effects on these ecosystems. Many species migrate in extremely large numbers and often stop for feeding at specific sites (e.g., Plymouth Bay, Massachusetts; Jamaica Bay, New York; Delaware Bay, Pennsylvania), where they devastate the local invertebrate populations. Sadly, hunting has reduced many of these species to numbers far below those that predated European colonization of North America.

* **Shorebirds migrate great distances between nesting and feeding grounds.**

Many shorebird species migrate truly great distances between summer nesting grounds and winter feeding grounds. Phalaropes, for example, nest in high northern latitudes, but they spend the winter in the Southern Hemisphere. The white-rumped sandpiper breeds on grassy or mossy tundra in north Alaska and on islands in the Canadian Arctic, migrates mainly within the continental interior, and winters east of the Andes Mountains, from Paraguay to the southern tip of South America. As mentioned earlier, many stop at major migratory feeding areas, where benthic

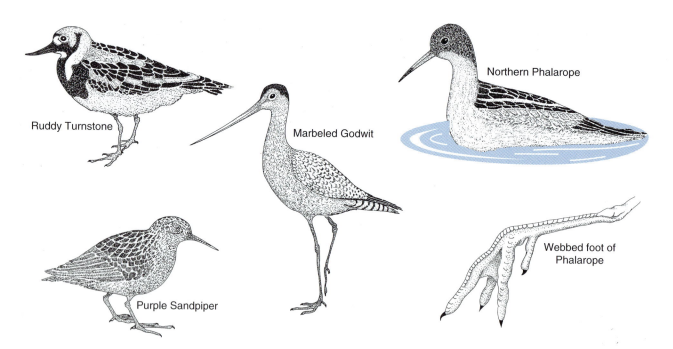

Ruddy Turnstone

Purple Sandpiper

Marbeled Godwit

Northern Phalarope

Webbed foot of Phalarope

Fig. 8.26   Some of the diversity of shorebirds: ruddy turnstone, purple sandpiper, marbled godwit, and northern Phalarope, with detail of webbed foot. (After Schneider, 1983.)

invertebrates are very abundant. Nests are often very simple, perhaps scrapes of the ground, adorned with just a few pebbles or shells.

✳ Feeding is diverse and is related to strong differences in beak morphology.

Shorebirds have a diversity of foraging behavior and beak types. Unfortunately, this diversity allows us to predict their prey types only partially. Nevertheless, there are several distinct types of feeding. **Running and stabbing** is done by plovers, which pursue a prey (e.g., a crab) and stab it with the beak. **Chiseling and hammering** is employed by turnstones and oystercatchers. Ruddy turnstones use their bills to excavate clams from the sand and chisel them open. Oystercatchers may chisel a hole directly in an oyster or mussel. They also can plunge the beak into a gaping bivalve and sever the adductor muscle, thus creating an opportunity to consume the rest of the bivalve's flesh. Sandpipers use **pecking and probing**, sometimes penetrating the sand to a depth of a few centimeters, to find infaunal prey. Probing is often accompanied by a shaking of the beak, which liquefies the sand. Godwits have heavily reinforced long bills, and the skull musculature is modified for vigorous probing into the sediment for razor clams and lugworms. Sanderlings simply pick out burrowing animals such as mole crabs that have been washed out by the waves on a beach. A number of species wade in the water and capture prey at the surface or perhaps by probing into the sediment. Phalaropes swim in the water and use a whirling motion to concentrate crustaceans, which are pecked out of the water.

Sea Snakes

Sea snakes are relatives of the cobras and are found commonly throughout the Indo-Pacific region, although a single species occurs along the west coast of North America. None are to be found in the Atlantic or the Mediterranean, although a now moribund proposal to build a sea-level canal across the Isthmus of Panamá raised concerns about an invasion from the Pacific Ocean into the Caribbean Sea. Sea snakes have fixed fangs and are largely venomous, although very few humans have died from sea snake bites and some species are surprisingly docile. Many snakes lay eggs on shore, but some complete their breeding cycle out at sea and young are born alive in the water.

Sea snakes usually prey upon fish, but since they are air breathers, they must periodically go to the surface to gather air. The lung is large and extends to the base of the tail. Some can dive to 150 m, but dives to 5 m are common. Some snakes can stay submerged for several hours and must be tolerant of anoxia. Some species actively trap fish prey in coral crevices and then grab them with the mouth, whereas others lie quite still, like sticks, and strike fish that approach them.

Sea Turtles

Sea turtles (Plate XIII) were once common throughout tropical waters, but are now nearly all endangered owing to direct hunting and indirect trapping in gear intended for fish and shrimp. The five species in United States and Caribbean waters (Figure 8.27) are all carnivores, with the exception of green tur-

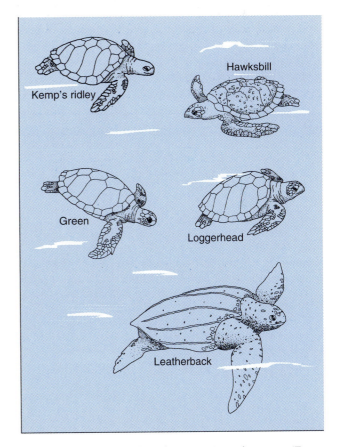

Fig. 8.27 Sea turtles found in U.S. coastal waters. (From National Research Council, 1990).

tles, which eat seaweeds and sea grasses. Kemp's ridley and loggerheads eat bottom-dwelling invertebrates, leatherbacks consume jellyfish from the water column, and hawksbills feed on sponges. The estuarine diamond-backed terrapin may walk along the bottom and seize a mussel by means of a rapid, downward protrusion of the head. Alternatively, the turtle may swim in a burst by means of a rapid backward movement of its forelegs, thrusting itself forward to allow for an attack on a more mobile prey such as a crab. Green turtles swim elegantly in shallow water, chewing on sea grasses. By contrast, leatherbacks suck in their gelatinous zooplankton prey.[12]

* Female sea turtles lay eggs on specific beaches, and migrate between the beach and seasonal feeding grounds great distances away.

Sea turtles are well known for their beach nesting and extensive migrations (see Chapter 5, Figure 5.10). Males and females mate near nesting beaches, and females come onto shore several times per breeding season. On each occasion, they lay approximately 100 eggs apiece, which are buried in the sand. Biologists and tourists alike flock to these beaches, which are to be found from Australia to Florida to Costa Rica to eastern Brazil. After about 2 months, young hatch from the eggs, crawl to the surface, and scramble to the sea. Their movements are guided by three distinct signals.[13] After hatching, they orient and move toward the horizon, which gets them to the strandline. They then move opposite to the direction of arriving waves, which moves them out to sea. These movements create an orientation that can be locked in by orienting to the earth's magnetic field; loggerhead turtles can respond to both the strength and the declination of a magnetic field. The magnetic field may provide information on geographic position and migration direction, but it is not known how turtles manage to migrate long distances to feeding grounds or locate their natal beaches.

They spend some time in open water and then move into bays and estuaries to feed as juveniles until sexual maturity, which usually takes more than ten years. Long-distance migration is common, and recent evidence demonstrates repeated homing by green turtles to the same beach (see Chapter 5).

[12] See Bels and others, 1998, in Further Reading, Birds and Reptiles.
[13] See Lohmann and Lohmann, 1998, in Further Reading, Birds and Reptiles.

Further Reading

CEPHALOPODS

Ambrose, R. F. 1982. Shelter utilization by the molluscan cephalopod *Octopus bimaculatus*. *Marine Biology*, v. 7, pp. 67–73.

Denton, E. J., and J. B. Gilpin-Brown. 1961. The distribution of gas and liquid within the cuttlebone. *Journal of the Marine Biological Association of the United Kingdom*, v. 41, pp. 365–381.

O'Dor, R. K., and D. M. Webber. 1986. The constraints on cephalopods: Why squid aren't fish. *Canadian Journal of Zoology*, v. 64, pp. 1591–1605.

Yang, W. T., R. F. Hixon, P. E. Turk, M. E. Krejci, and W. H. Hulet. 1986. Growth, behavior, and sexual maturation of the market squid, *Loligo opalescens*, cultured through the life cycle. *Fishery Bulletin*, v. 84, pp. 771–798.

FISHES

Block, B. A., J. R. Finnerty, A. F. R. Stewart, and J. Kidd. 1993. Evolution of endothermy in fish: Mapping physiological traits on a molecular phylogeny. *Science*, v. 260, pp. 210–214.

Bone, Q., N. B. Marshall, and J. S. Blaxter. 1994. *Biology of Fishes*, 2nd ed. Glasgow and London: Blackie Academic and Professional.

Carey, G. F., J. M. Teal, J. W. Kanwisher, K. D. Lawson, and J. S. Beckett. 1971. Warm-bodied fish. *American Zoologist*, v. 11, pp. 137–145.

Hastings, J. W. 1971. Light to hide by: Ventral luminescence to camouflage the silhouette. *Science*, v. 173, pp. 1016–1017.

Houde, E. D., and R. C. Schekter. 1980. Feeding by marine fish larvae: Developmental and functional responses. *Environmental Biology of Fishes*, v. 5, pp. 315–334.

Hughes, G. M., and A. V. Grimstone. 1965. The fine structure of the secondary lamellae of the gills of *Gaddus pollachius*. *Quarterly Journal of Microscopical Science*, v. 106, pp. 343–353.

Juanes, F., and D. O. Conover. 1994. Piscivory and prey size selection in young-of-the-year bluefish: Predator preference or size-dependent capture success? *Marine Ecology—Progress Series*, v. 114, pp. 59–69.

McAllister, D. E. 1967. The significance of ventral biolu-minescence in fishes. *Journal of the Fisheries Research Board of Canada*, v. 24, pp. 537–554.

Moyle, P. B., and J. J. Cech Jr. 1982. *Fishes: An Introduction to Ichthyology*. Englewood Cliffs, NJ: Prentice-Hall.

Norton, S. F. 1988. Role of the gastropod shell and oper-culum in inhibiting predation by fishes. *Science*, v. 241, pp. 91–94.

Rome, L. C., D. Swank, and D. Corda. 1993. How fish power swimming. *Science*, v. 261, pp. 340–343.

Sanderson, S. L., and R. Wassersug. 1990. Suspension-feed-ing vertebrates. *Scientific American*, v. 262(3), pp. 96–101.

Sanderson, S. L., J. J. Cech Jr., and M. R. Patterson. 1991. Fluid dynamics in suspension-feeding blackfish. *Science*, v. 251, pp. 1346–1348.

Schmidt-Nielsen, K. 1975. *Animal Physiology*. London: Cambridge University Press.

Webb, J. F. 1989. Gross morphology and evolution of the mechanoreceptive lateral line system in teleost fishes. *Brain, Behavior and Evolution*, v. 33, pp. 34–53.

Webb, P. W. 1984. Form and function in fish swimming. *Scientific American*, v. 256, pp. 58ff.

MAMMALS

Baird, R. W., P. A. Abrams, and L. M. Dill. 1992. Possi-ble indirect interactions between transient and resident killer whales: Implications for the evolution of foraging and speciation in the genus *Orcinus*. *Oecologia*, v. 89, pp. 125–132.

Beddington, J. R., R. J. H. Beverton, and D. M. Lavigne, eds. 1985. *Marine Mammals and Fisheries*. Boston: Allen & Unwin.

Berta, A. 1994. What is a whale? *Science*, v. 263, pp. 180–181.

Bonner, W. N. 1990. *The Natural History of Seals*. New York: Facts on File.

Brault, S., and H. Caswell. 1993. Pod-specific demography of killer whales (*Orcinus orca*). *Ecology*, v. 74, pp. 1444–1454.

DeLong, R. L., and B. S. Stewart. 1991. Diving patterns of northern elephant bulls. *Marine Mammal Science*, v. 7, pp. 369–384.

Ford, J. B. 1991. Vocal traditions among resident killer whales (*Orcinus orca*) in coastal waters of British Co-lumbia. *Canadian Journal of Zoology*, v. 69, pp. 1454–1483.

Gingerich, P. D., N. A. Wells, D. E. Russell, and S. M. Ibrahim Shah. 1983. Origin of whales in epicontinental remnant seas. *Science*, v. 220, pp. 403–406.

Harrison, R. J., ed. 1972. *Functional Anatomy of Marine Mammals*. New York: Academic Press.

Heyning, J. E. 1997. Sperm whale phylogeny revisited analysis of the morphological evidence. *Marine Mammal Research*, v. 13, pp. 596–613.

Highsmith, R. C., and K. O. Coyle. 1993. Production of Arctic amphipods relative to grey whale energy require-ments. *Marine Ecology—Progress Series*, v. 83, pp. 141–150.

Hochachka, P. W. 1992. Metabolic biochemistry and the making of mesopelagic mammals. *Experientia*, v. 48, pp. 570–575.

Hoelzel, A. R., ed. 1991. *Genetic Ecology of Whales and Dolphins*. Cambridge: International Whaling Commis-sion.

King, J. E. 1983. *Seals of the World*, 2nd ed. Ithaca, NY: Cornell University Press.

Milinkovitch, M. C., G. Orti, and A. Meyer. 1993. Revised phylogeny of whales suggested by mitochondrial DNA sequences. *Nature*, v. 361, pp. 346–348.

Nikaido, M., A. P. Rooney, and N. Okada. 1999. Phylo-genetic relationships among detartiodactyls based on in-sertions of short and long interspersed elements: Hippopotomuses are the closest extant relatives of whales. *Proceedings of the National Academy of Sci-ences*, v. 96, pp. 10261–10266.

O'Leary, M. A., and M. D. Uhen. 1999. The time of ori-gin of whales and the role of behavioral changes in the terrestrial-aquatic transition. *Paleobiology*, v. 25, pp. 534–556.

Payne, R., ed. 1989. *Communication and Behavior of Whales*. Boulder, CO: Westview Press.

Piatt, J. F., and D. A. Methuen. 1993. Threshold foraging behavior of baleen whales. *Marine Ecology—Progress Series*, v. 84, pp. 205–210.

Pivorunas, A. 1979. The feeding mechanisms of baleen whales. *The American Scientist*, v. 67, pp. 432–440.

Ridgway, S. H., and R. J. Harrison. 1981. *Handbook of Marine Mammals*. New York: Academic Press.

Schaeff, C. M., S. D. Kraus, M. W. Brown, and G. N. White. 1993. Assessment of the population structure of western North Atlantic right whales (*Eubalaena glacialis*). *Canadian Journal of Zoology*, v. 71, pp. 339–345.

Thewissen, J. G. M., S. T. Jissaom, and M. Arif. 1994. Fos-sil evidence for the origin of aquatic locomotion in ar-chaeocete whales. *Science*, v. 263, pp. 210–212.

Thomas, J. A., R. A. Kastelein, and A. Y. Supin, ed. 1992. *Marine Mammal Sensory Systems*. New York: Plenum Press.

Trillmich, F., and K. A. Ono. 1991. *Pinnipeds and El Niño*. Berlin: Springer-Verlag.

Wursig, B. 1979. Dolphins. *Scientific American*, v. 256, pp. 100–105.

BIRDS AND REPTILES

Ashmole, N. P., and M. J. Ashmole. 1967. Comparative feeding ecology of seabirds of a tropical oceanic island. *Peabody Museum of Natural History Bulletin*, v. 24, pp. 1–131.

Bels, V. L., J. Davenport, and S. Renous. 1998. Food ingestion in the estuarine turtle *Malaclemys terrapin*: Comparison with the marine leatherback turtle *Dermochelys coriacea*. *Journal of the Marine Biological Association of the United Kingdom*, v. 78, pp. 953–972.

Bowen, B. W., N. Kanezaki, C. J. Limpus, G. R. Hughes, A. B. Meylan, and J. C. Avise. 1994. Global phylogeography of the loggerhead turtle (*Caretta caretta*) as indicated by mitochondrial DNA haplotypes. *Evolution*, v. 48, pp. 1820–1828.

Carr, A. 1984. *The Sea Turtle. So Excellent a Fishe*. Austin: University of Texas Press.

Dunson, W. A., ed. 1975. *The Biology of Sea Snakes*. Baltimore: University Park Press.

Furness, R. W., and P. Monaghan. 1987. *Seabird Ecology*. Glasgow: Blackie.

Lohmann, K. J., and C. M. F. Lohmann. 1998. Migratory guidance mechanisms in marine turtles. *Journal of Avian Biology*, v. 29, pp. 585–596.

National Research Council. 1990. *Decline of the Sea Turtles: Causes and Prevention*. Washington, DC: National Academy Press.

Nelson, B. 1979. *Seabirds*. New York: A&W Publishers.

Nevitt, G. A. 2000. Olfactory foraging by Antarctic procellariiform seabirds: Life at high Reynolds numbers. *Biological Bulletin*, v. 198, pp. 245–253.

Pitelka, F. A., ed. 1979. *Shorebirds in Marine Environments: Studies in Avian Biology*, no. 2. Los Angeles: The Cooper Ornithological Society.

Schneider, D. C. 1983. Seabirds and shorebirds. *Oceanus*, v. 26, pp. 38–43.

Review Questions

1. How do cephalopods move to capture prey?

2. How do cuttlefish and the nautilus regulate their vertical position in the water column?

3. There are a few species of nautilus in the ocean, but these are the only cephalopods with external shells. Why were there many more externally shelled species in the geological past?

4. How does the fish swim bladder use a countercurrent exchange mechanism to regulate its vertical position in the water?

5. What is the difference in the pattern of tooth emplacement between bony and cartilaginous fishes?

6. How does a shark maintain a rigid body while performing undulatory swimming, despite the lack of a rigid bony skeleton?

7. What is the best body form for a fish that must maneuver among rocky crags in a coral reef?

8. Why is it that, in any one marine habitat, one finds fish species of many shapes rather than a single general shape?

9. Fishes often swim in large schools of a single species? Why might this be so?

10. Why is there a hydrodynamic conflict between feeding and rapid movement through the water in baleen whales?

11. Name two different mechanisms that marine mammals employ to insulate against heat loss.

12. Why may language tend to diverge among different groups of killer whales?

13. What might cause a diving carnivorous mammal to dive deeper for prey than before? What are the costs and benefits that might enter into such a decision?

14. In some whales, a third individual is required to "help" in copulation by leaning against the two participants. What evolutionary benefit might accrue to such a helper?

15. Males animals are about the same size as females when the species is monogamous. Male seals that maintain harems, however, are often much larger than females. Why should there be such sexual size difference when the species is polygamous?

IV

PROCESSES IN THE OPEN SEA

9

Critical Factors in Plankton Abundance

Plankton are extremely variable in abundance, both spatially and temporally. This is especially true of the phytoplankton, upon which zooplankton depend. Although many factors contribute to variation in the abundance of phytoplankton, water motion is in many ways the dominant factor. It regulates two factors that are vital for the growth of phytoplankton populations: the exchange of nutrients with deeper waters and the mixing of waters downward and therefore away from exposure to sunlight. Water motion is important for phytoplankton growth in all of the ocean, but its influence is best introduced in a discussion of the spring diatom increase, which occurs in nearly all areas where there is a seasonal change in day length. Of course, light and nutrients are crucial determinants of phytoplankton, which after all are photosynthetic and require these for growth. Finally, animal grazing affects phytoplankton abundance. It is the purpose of this chapter to discuss the importance and interaction of these various factors on plankton dynamics.

The Seasonal Pattern of Plankton Abundance

✳ **In midlatitudes, phytoplankton increase in the spring, decline in summer, and may increase to a lesser extent in fall.**

There is a predictable seasonal pattern to plankton abundance in temperate and boreal waters of depths of an approximate range of 10–100 m. Figure 9.1 traces the seasonal changes in phytoplankton, zooplankton, light, and nutrients during the year in a temperate–boreal coastal zone. Usually in the early spring, phytoplankton increase dramatically and are dominated by a few diatom species. This is known as the **spring diatom increase**. Although the exact time varies with latitude and year, the peak of phytoplankton abundance in waters of southern New England is usually in March. During this period, the surface waters are murky brown with phytoplankton. The phytoplankton begin to decline as summer approaches, when the early dominants give way to other phytoplankton species. In northeastern United States coastal waters and estuaries, microflagellates and dinoflagellates become abundant, but in higher latitudes diatoms continue to dominate. In some locations, the phytoplankton increase again in the fall, followed by a decline to very low abundance again in the winter.

✳ **Zooplankton start to increase as the phytoplankton bloom reaches its peak, attaining a maximum in the late spring or early summer.**

Zooplankton reach their yearly maximum after the spring phytoplankton increase begins to decline.

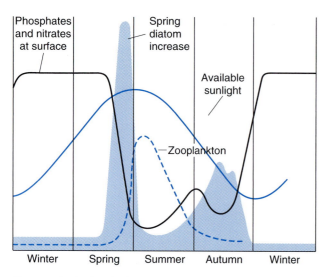

Fig. 9.1 Idealized diagram tracing changes in phytoplankton, zooplankton, light, and nutrients during the year in a temperate–boreal inshore body of water. (Modified after Russell-Hunter, 1970.)

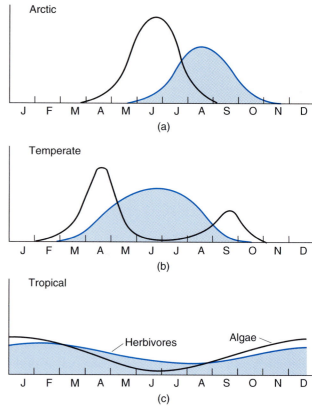

Fig. 9.2 Phytoplankton and zooplankton in a seasonal cycle: (a) Arctic, (b) temperate–boreal, and (c) tropical. (After Cushing, 1975.)

Calanoid copepods dominate this burst of abundance and are the principal grazers of the diatoms. Depending on the year and conditions, the genera *Acartia* and *Temora* may dominate the coastal New England zooplankton. At any one time, several different developmental stages of copepods may be found in the plankton. Planktonic larvae of benthic invertebrates are also common in the late spring and in early summer. Following this burst, zooplanktivorous fish and invertebrates become abundant. Comb jellies and jellyfish become especially abundant and are effective predators on copepods and planktonic larvae. These gelatinous creatures clog plankton nets in summer, when arrow worms and tunicates may also become abundant. Zooplanktivorous fishes, such as menhaden in the eastern United States, also are important predators in the plankton.

* The spring phytoplankton peak and the later zooplankton peak are shortest and sharpest in high latitudes, becoming indistinct in the tropics.

The temperate–boreal pattern is a good starting point for discussion, but this pattern is not universal. The strength of the spring phytoplankton bloom varies with latitude. In the Arctic, a single summer peak of phytoplankton abundance is followed by a zooplankton maximum (Figure 9.2a). The phytoplank-

ton production lasts as little as 2–4 weeks. In temperate–boreal waters, as we have discussed, a spring phytoplankton increase is followed by a decrease, coinciding with a zooplankton increase (Figure 9.2b). In late spring and summer, the zooplankton decline and a peak of phytoplankton may follow in the fall. In the tropics there is no clear alternating pattern of phytoplankton and zooplankton abundance (Figure 9.2c). This correlates with the relative lack of seasonality in tropical waters.

Water Column Parameters and the Spring Diatom Increase

Water Column Stability and Light

* Because light irradiance decreases exponentially with increasing depth, there is a compensation depth below which respiration for a given phytoplankton cell exceeds photosynthetic output.

Light irradiance decreases exponentially with increasing depth (Figure 9.3) and becomes a limiting factor to photosynthesis. The **compensation depth** is that depth at which the amount of oxygen produced by a phytoplankton cell in photosynthesis equals the oxygen consumed in respiration. We can estimate the compensation depth by placing phytoplankton cells in a clear bottle. At depths shallower than the compensation depth, there is a net increase of oxygen over time, whereas at depths greater than the compensation depth, there is a net decrease of oxygen over time. The compensation depth is thus an indicator of the potential of a photosynthesizing cell to be a net producer. These relationships would hold if there were no mixing with waters of differing oxygen content. The light intensity corresponding to the compensation depth is **the compensation light intensity**.

The compensation depth is controlled by season, latitude, and transparency of the water column. As the temperate–boreal spring progresses, the increasing photoperiod tends to increase the compensation depth to an eventual maximum. The Arctic winter photoperiod is zero and therefore the compensation depth is at the surface. Suspended matter in coastal waters reduces the compensation depth relative to the open sea. Similarly, as a phytoplankton bloom develops and as suspended matter (seston) becomes trapped in the water column, the compensation depth

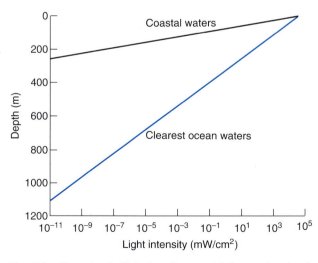

Fig. 9.3 Decrease in light irradiance with increasing depth in coastal water and clear ocean water. Note that the light intensity is plotted on a logarithmic scale; light is decreasing exponentially with depth.

decreases due to light absorption and shading by particles. A yellow pigment originating from rivers and other terrestrial sources is also important in the extinction of light with depth.

✳ In winter, the water density is about the same throughout the water column, and phytoplankton cells are stirred on average to depths that prevent average photosynthesis from permitting the phytoplankton population to increase.

Consider the state of the plankton and of the water column in the winter before the spring diatom bloom occurs. At this time the water column is isothermal, with little or no density variation with depth. Near the shore, temperate–boreal winters are times of high wind stress, resulting in extensive overturn of the water column. Because there are no density differences, the water column is unstable, and winds cause extensive vertical mixing. The **mixing depth** is the depth above which all water is thoroughly mixed under the wind's influence. Because the winter mixing depth is great due to storms, phytoplankton cells can easily be swept down to great depths, where there is not enough light for photosynthesis. A phytoplankton bloom cannot be initiated because any potential profit in photosynthesis would be lost through mixing to greater depths. So even though the photoperiod may increase as winter progresses, the instability of the water column may preclude the development of a phytoplankton bloom. Along with the increase of photoperiod, **water column stability** is thus an essential part of the development of the phytoplankton bloom.

Water is usually well mixed in winter, and plankton are uniformly distributed. Hence, respiration is about the same at all depths. Because a population increase of phytoplankton requires that total production exceed total respiration, a phytoplankton bloom can occur only when the volume of water in which photosynthesis occurs has a net excess of production over consumption (respiration, in this case). The depth above which total production in the water column equals total consumption (respiration) is known as the **critical depth**. If the mixing depth is less than the critical depth, phytoplankton should increase, but, if the mixing depth is greater than the critical depth, phytoplankton will be swept down below the area of optimum light, preventing a phytoplankton bloom (Figure 9.4).

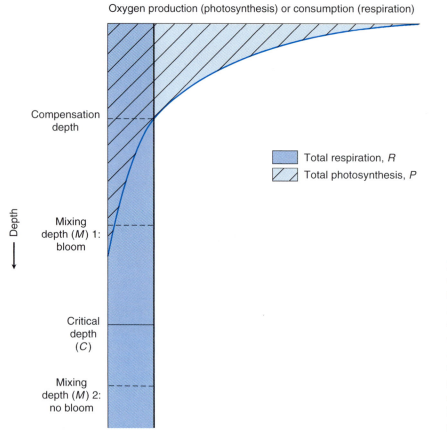

Oxygen production (photosynthesis) or consumption (respiration)

Depth

Compensation depth

Total respiration, *R*

Total photosynthesis, *P*

Mixing depth (*M*) 1: bloom

Critical depth (*C*)

Mixing depth (*M*) 2: no bloom

Fig. 9.4 The relation of critical depth (*C*) and mixing depth (*M*). In the absence of vertical mixing, *P* = *R* at the critical depth. If the mixing depth is less than the critical depth, *C* (e.g., at mixing depth 1), then *P* > *R* and a bloom will develop. If *M* > *C* (e.g., at mixing depth 2), a bloom fails because some phytoplankton cells are swept below to waters of light intensity low enough to yield the condition *P* < *R*.

✳ **As the spring temperature increases, the surface waters warm up, and the water column stabilizes. Nutrients are lost to deeper waters and the phytoplankton bloom is depressed.**

A spring thermocline generated by solar heating reduces the water density in the surface waters, which retards wind mixing and overturn. The water column then stabilizes, decreasing the mixing depth. The onset of relatively quiet summer conditions also further enhances the decrease of mixing depth. In inshore waters, the water column may be further stabilized by lowered salinity due to freshwater influxes from terrestrial sources. The late spring and early summer runoff from the Fraser River in British Columbia, for instance, causes a shallow-water salinity minimum and typically high production throughout the northern Puget Sound and the Strait of Georgia region. The earlier timing of inshore Norwegian fjord phytoplankton blooms, which appear earlier than open-water phytoplankton blooms, may be related to the greater stability of the water column inshore.

The stabilization of the water column causes both the birth and the eventual demise of the spring diatom bloom. The spring stabilization of the water column maintains phytoplankton in the upper layer, thus precluding its removal from the zone of active photosynthesis. As the water column stabilizes, however, and the thermocline is established, phytoplankton die or are ingested and egested by zooplankton, sinking below the compensation depth. Because of the stabilization of the water column, these materials and other nutrients are not returned to the surface from greater depths and from the bottom. In a shallow area, like Long Island Sound, New York, there is extensive exchange between the bottom and the overlying water in terms of resuspension of detritus and dissolved nutrients. Once the thermocline has been established in the spring and summer, however, this exchange is greatly diminished, trapping large amounts of detritus in the water column because of its stability. Toward the end of summer, with the advent of fall storms, the thermocline may be disrupted, bringing nutrients toward the surface from the bottom in shallow water. This may result in a fall increase of phytoplankton.

As the water column stabilizes in late spring and summer, phytoplankton organisms denser than seawater, such as diatoms, start to sink from the water column. Such considerations do not hold, however, for phytoplankton (e.g., dinoflagellates) whose swimming abilities circumvent this tendency to sink.

If these ideas are correct, then we may conclude that the hydrographic conditions tied to seasonal variation play the primary role in the birth, development, and demise of the spring phytoplankton increase. The stabilization of the water column in spring initially permits the development of the spring increase. The stability of the water column, however, prevents nutrients lost from the surface waters from returning to the surface where light is available for photosynthesis. Furthermore, dense phytoplankton organisms sink out of the water column when spring–summer stability sets in. The poor nutrient situation prevails until the fall and winter overturn of the water column.

Water Column Exchange in Shallow Waters and in Estuaries

✳ In very shallow water estuaries, nutrient exchange, or benthic–planktonic coupling, occurs between the bottom and the water column, fueling more phytoplankton growth.

The importance of water column stability may change with the depth of the basin and with season. In estuaries such as Long Island Sound, between New York and Connecticut, the spring phytoplankton bloom ends when the water column is strongly strat-ified, because nutrients are not regenerated from the bottom to the surface. Regeneration from the bottom occurs the following winter, as the water column homogenizes in density and is turned over by winter storms. In shallower waters, however, the nutrient exchanges between bottom and surface waters called **benthic–planktonic coupling** are much stronger and are not shut off by stratification. Many bays and estuaries are very shallow, and even modest wind mixing mixes the water column completely. Examples in the United States include Narragansett Bay, Rhode Island, and Great South Bay, New York. In late spring and summer, decomposition of organic matter in shallow bottoms, combined with vertical mixing, brings nutrients back to the surface, and a summer phytoplankton bloom occurs that can exhibit higher productivity than the spring event.

Benthic–planktonic coupling figures importantly in a high energy nearshore phenomenon known as **beach blooms**. These are phytoplankton blooms of a few species of very large diatoms and are especially well developed off the sandy coasts of Oregon and South Africa. Even though the coast appears to be very energetic with strong waves, an offshore submerged bar confines the vertically recirculating water to the nearshore region (Figure 9.5). Dense populations of benthic animals excrete into the water column nutrients that are thoroughly mixed and fuel enormous phytoplankton growth, enough to color the water a deep brown. The diatoms sink to the bottom and are eaten by the invertebrates, although some planktonic consumption occurs as well. Input from waters offshore and from the exposed beaches adds nutrients to the system.

In deeper water, especially seaward of the shelf–slope break, benthic–planktonic coupling is indirect

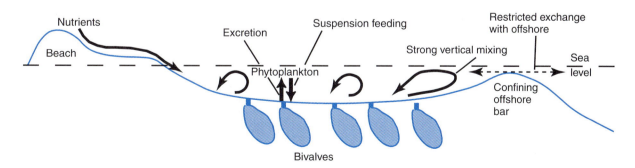

Fig. 9.5 A benthic–planktonic coupling system. Bivalve excretion provides nutrients for phytoplankton uptake, as do inputs from the beach. Bivalves in the bottom feed on the phytoplankton. A large confined circulation landward of an offshore bar recirculates the water. Water exchange with offshore waters is relatively restricted.

at best. In some parts of the ocean a surprising amount of material sinks from the surface waters to the bottom, but the return route to the surface is very indirect and depends upon bottom currents. Such currents could not restore organic material to the surface waters for many years.

* In estuaries, the spring freshet combines with net water flow to the sea and mixing to determine the nutrient regime.

Nutrients recycle extensively in the ocean, but the extraordinarily high primary production of estuaries owes itself to large inputs of nutrients from the tributaries of the watershed, with a special seasonal peak during the freshet. To appreciate the pattern of input and loss in an estuary, consider Figure 9.6, which

shows a budget of carbon input and loss in the Hudson River estuary. Carbon comes principally from tributaries and runoff from the land, although sewage input also occurs to a minor degree. Carbon then enters the main part of the estuary and is taken up by bacteria and phytoplankton, to be later passed on to consumers. Some of the carbon, however, is respired as carbon dioxide and leaves the estuary as gas. Another fraction is lost to the sediment. There is general downstream movement of the remainder, and finally about half the original carbon reaches the saline portion of the estuary. Similar budgets could be made for nitrogen and phosphorus, but the proportions lost, and even the processes of loss and gain, would differ. For example, nitrogen can be lost as nitrogen gas to the atmosphere via denitrification, a bacterially mediated process.

TEXT BOX 9.1 The Basics of Photosynthesis

Photosynthesis is the transformation of light energy into chemical energy for use in living cells. The overall process can be described by means of the following relation:

$$CO_2 + 2H_2O \rightarrow (CH_2O)_n + H_2O + O_2$$

With the aid of plants and light, water is split into hydrogen and oxygen, and carbohydrates (CH_2O), such as sug-

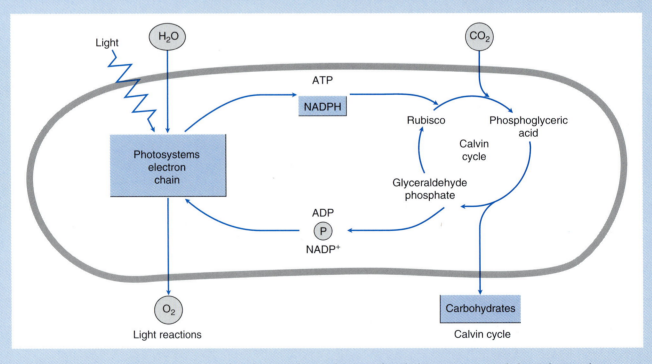

Box Fig. 9.1   A general scheme for photosynthesis, showing the "light" reactions and the Calvin cycle.

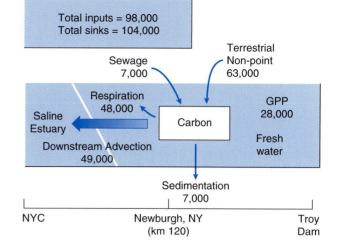

Fig. 9.6 A carbon budget for the Hudson River estuary, showing the inputs and losses (in tons) of carbon into the freshwater tidal part of the estuary and the degree of downriver loss of carbon to the saltwater part of the estuary: GPP, gross primary productivity; point sources, carbon from sewage treatment plants. (The total input does not match the total losses and sinks of carbon because these were measured independently and there is some error.) (Courtesy of Robert Howarth.)

ars and starches, are produced. In the equation, $n = 6$ in the case of glucose. In plants, photosynthesis occurs in the chloroplasts, which are cell organelles bounded by a double membrane. Within the chloroplast is a membrane system where light is captured and transformed into chemical energy, through the reaction of carbon dioxide and water. The membranes are surrounded by spaces where this energy is used to produce the carbohydrates.

Light energy is captured by substances known as photosynthetic pigments. All plants have chlorophyll a, which absorbs light energy most efficiently in the blue and green regions of the spectrum. Other pigments, known as accessory pigments, also absorb light, but in the parts of the light spectrum where chlorophyll a is relatively inefficient.

The overall process of photosynthesis can be divided into the light reactions and the Calvin cycle (Box Figure 9.1). As one would expect from the name, the light reactions involve the capture of light energy. They involve two processes. First, light energy is used in conjunction with a complex series of reactions to oxidize water and to split it to produce molecular oxygen, protons (H^+), and electrons (e^-). This is actually accomplished in two distinct biochemical photosystems, also with a chain of reactions to transport electrons, but the details are not of concern here. The electrons then enable ADP to be converted to ATP, which is the major source of energy for the Calvin cycle. Also, NADP is reduced by the electrons to NADPH, which

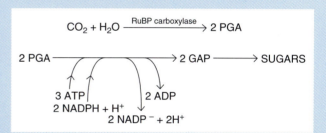

Box Fig. 9.2 The Calvin cycle of photosynthesis: PGA, a three-carbon molecule, 3-phosphoglyceric acid; GAP, another three-carbon molecule, 3-phosphoglyceraldehyde; RuBP carboxylase is an enzyme, catalyzing the first reaction.

is a major source of chemical reducing power in the Calvin cycle.

Carbon is "fixed" to carbohydrates in the Calvin cycle. The reaction is shown in Box Figure 9.2. The enzyme ribulose biphosphate carboxylase (RuBP carboxylase, or Rubisco) enables the reaction of carbon dioxide, water, and the five-carbon sugar ribulose biphosphate (RuBP) to produce a three-carbon molecule that is a precursor of the six-carbon sugars. Energy from ATP and the chemical reducing power of NADPH are both required to make the reaction go. Most of the reaction series (five out of every six three-carbon molecules) involves the regeneration of RuBP.

Nutrient input to estuaries occurs as a pulse during the spring freshet, but some estuaries have rather short residence times for water, which is rapidly lost to the sea. The Hudson estuary loses water over its main channel in a few tens of days, whereas the Chesapeake Bay estuarine system has a water residence time of hundreds of days. Carbon input into the Hudson system is about 30 times that into the Chesapeake Bay estuary. The longer residence time in the Chesapeake allows for more extensive recycling of nutrients within the estuary. The shorter water retention time in the Hudson River estuary results in more export of nutrients downriver into the saline part of the estuary and into the open sea. In most estuaries, primary productivity increases as the nutrient input of the freshet increases, and decreases as the degree of flow from the estuary increases in late spring and summer. In North Carolina estuaries, strong spring precipitation and nearly windless summers combine to trap larger concentrations of nutrients within the estuaries, which fuels phytoplankton blooms.

Light

✳ **Light is often inhibitory to photosynthesis near the surface, but a series of photosynthetic pigments capture light over much of the visible spectrum.**

Energy from solar sources is expressed in terms of energy units (such as langleys per minute: g cal cm^{-2} min^{-1}).[1] The angle of the sun at different times of day, the latitude, and other factors contribute to the spectral distribution of light that strikes the sea surface and the amount of back-scattering. The spectral distribution of light striking the sea surface includes a large part of the ultraviolet–infrared spectrum; however, only visible parts of the spectrum penetrate to great depths. At temperate latitudes in clear weather during the summer, the maximum energy striking the sea surface is about 1.4 langleys per minute. About one-half the total radiant energy is in the infrared region of the spectrum and so is not available to marine photosynthetic organisms.

Light is attenuated in the water column through **absorption** and **scattering**. Scattering can be accomplished by water molecules, dissolved organic matter, particulate organic and inorganic material, and living plankton themselves. Figure 9.7 shows attenuation values for different wavelengths of light. In the clear open ocean, the attenuation spectrum of light transmission maximizes transmission at about 480 nanometers (nm). In turbid inshore waters, however, a more pronounced maximum occurs at longer wavelengths, approximately 500–550 nm. Because ultraviolet light has detrimental effects on DNA, its

[1] One langley per minute = 0.0698 W cm^{-2}.

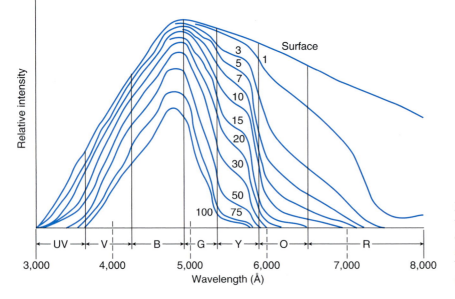

Fig. 9.7 Attenuation of different wavelengths of light with increasing depth (each curve is labeled by depth in meters) below the sea surface. Wavelength is in angstroms (1 Å × 0.1 = 1 nm). (From Clarke, 1939.)

penetration is of great interest. In moderately turbid coastal waters, incident light with a wavelength of 380 nm or less is almost attenuated at a depth of 1–2 m, but in very clear parts of the ocean, 20 m may be required to remove 90 percent of the radiation entering the surface.

Incident light near the surface is intense enough to inhibit photosynthesis through bleaching of photosynthetic pigments, such as chlorophyll *a*, or the arresting of pigment production. With increasing depth, light energy is absorbed to the extent that the inhibitory effect disappears.

Photosynthetic phytoplankton use chlorophyll *a*, chlorophyll *c*, and a variety of "accessory" pigments, such as protein-bonded fucoxanthin and peridinin, to utilize light energy from most of the visible spectrum. The **action spectrum** is the extent of utilization of different wavelengths of light, and can be determined by using different monochromatic sources of light and measuring the amount of photosynthesis. Within the usable wavelengths of 400–700 nm, the light absorbed by phytoplankton pigments can be divided into (a) light of greater than 600 nm, which is mainly absorbed by chlorophyll, and (b) light of less than 600 nm, which is mainly absorbed by accessory pigments. The combined absorption of chlorophyll and accessory pigments allows the yield of photosynthesis to be constant over a large portion of the visible light spectrum.

✳ **Photosynthesis increases with increasing light intensity, up to a plateau, and then is inhibited by high light intensity.**

Figure 9.8 illustrates a theoretical photosynthesis–light curve showing photosynthetic rate as a function of light intensity. Photosynthesis increases with increasing light intensity and then reaches a plateau at a maximal value P_{max}. At the compensation light intensity, the photosynthetic rate (in this case, measured in terms of oxygen evolution) equals the amount of oxygen consumed in respiration. Because the amount of light reaching a phytoplankton cell varies over a day, the compensation light intensity is usually expressed on a 24-hour basis. It is assumed here, for simplicity, that respiration occurs at the same rate in the light and the dark (a shaky assumption because of light-accelerated respiration, known as photorespiration). An average 24-hour compensation light intensity is in the range of 3–13 langleys per day in

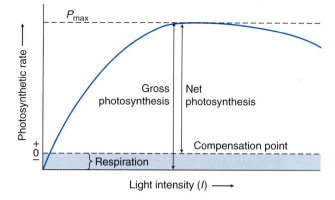

Fig. 9.8 The relationship between photosynthesis and light intensity.

temperate seas. Figure 9.8 shows photoinhibition at high light intensities.

The physiological adjustment to surrounding light conditions involves changes of some of the following morphological and biochemical factors: total photosynthetic pigment content, pigment proportions, morphology of the chloroplast, chloroplast arrangement, and availability of Calvin cycle enzymes. For example, under strong light conditions, diatom chloroplasts shrink and aggregate. Adaptations to changes in light intensity usually occur within 1 day. Deep-water phytoplankton can respond partially to low light intensities by increasing the total pigment content.

Nutrients Required by Phytoplankton

✳ **Nutrients are required by plants. They may occur in dissolved and particulate form.**

Nutrients are required by photosynthetic protists and plants. We can speak of required nutrient elements (e.g., nitrogen) that occur in dissolved inorganic form (e.g., ammonium, nitrate) or in organic form (e.g., amino acids). Major nutrient elements are required in great amounts and include carbon, nitrogen, phosphorus, oxygen, silicon, magnesium, potassium, and calcium. Trace nutrients elements are required in far smaller amounts and include iron, copper, and vanadium. Organic nutrients include vitamins. **Autotrophic** and **auxotrophic** uptake are the uptake of inorganic and organic nutrients, respectively, in association with photosynthesis. **Heterotrophic** uptake refers to uptake of organic substances for nutrition in the absence of photosynthesis. Many phytoplankton can absorb peptides and even engulf particles.

Many elements essential for phytoplankton nutrition can be found in both particulate and dissolved forms. The influxes from rivers and seaweed beds contribute to particulate organic matter. Carbon, nitrogen, and phosphorus may be found in carcasses of phytoplankton and zooplankton sinking in the water column. Microturbulent water motion may also create particles from collision of dissolved organic molecules. Toward the end of rich phytoplankton blooms, some species become sticky and exhibit negative buoyancy. As they sink, they aggregate and form particles. This resulting fragile organic aggregate material, known as **marine snow** (Figure 9.9), is often found to be enriched by a variety of planktonic organisms and detrital products of plankton.

The work of Mary Silver of the University of California at Santa Cruz, Alice Alldredge[2] of the University of California at Santa Barbara, and their colleagues has greatly illuminated our understanding of the global extent and importance of marine snow in transport of organic material in the ocean. Marine snow persists in relatively quiet water and can be a major mechanism of transport of material to deeper waters. Marine snow aggregates are rich in organic matter, derived from various sources including decaying phytoplankton cells, mucus from various gelatinous zooplankton, and marine bacteria. In the eastern Pacific, the gelatinous houses of larvacean tunicates are used to filter material but are shed several times daily. These houses often entrap particles and are a major part of the larger marine snow particle aggregates. Sometimes the aggregates are rich enough in organic matter and bacterial activity to cause the material to become anoxic.

Dissolved organics, particulate organics, and living biomass occur, on average, in the approximate proportions 100:10:2. The ocean is a solution of nutrients in the dissolved state, with a depleted surface layer due to uptake by living forms in the photic zone. Exchange processes, such as upwelling and wind mixing, may balance the loss of surface nutrients to greater depths via sinking carcasses and zooplankton fecal pellets. The turnover of the ocean's dissolved organic carbon is estimated to be 30–300 years, but the rate is much higher for some molecules (e.g., glucose) and lower for others.

Nitrogen

❋ **Nitrogen is required for protein synthesis and is taken up in the form of ammonium, nitrate, and nitrite.**

Nitrogen (N) is required for the synthesis of proteins and occurs in three principal inorganic dissolved forms: **ammonium** (NH_4^+), **nitrate** (NO_3), and **nitrite** (NO_2). Nitrogen also occurs in dissolved organic forms, such as urea, amino acids, and peptides.

Ammonium is usually the preferred form of nitrogen from a nutritional perspective because no chemical reduction is required to be used in protein synthesis. Nitrate and nitrite must be reduced by the enzymes nitrate reductase and nitrite reductase, respectively, making their uptake a chemically slower process. Dissolved ammonium can inhibit the uptake of nitrate. Ammonium can also be taken up more efficiently at low light levels than is the case for nitrate.

The highest concentration of dissolved nitrogen that occurs in the ocean is that of nitrate (roughly 100 micromolar: 100 μM), and nitrate is often the most abundant form of nitrogen in eutrophic coastal waters. Upwelling and storm-induced turbulence

Fig. 9.9 Marine snow in open-ocean surface waters off the coast of California. Fragments of gelatinous zooplankton have been colonized by bacteria, and other fragments have adhered to the original larger particle. Marine show derives from collisions of macromolecules and from the degradation of dead plankton. (Courtesy of Alice Alldredge.)

[2] See Alldredge and Silver, 1988, in Further Reading, Nutrients, Water Motion, and Phytoplankton Dynamics.

carry nitrate to the euphotic zone. Under certain circumstances, however, ammonium can surpass nitrate in abundance (usually when the nitrate is used up). Nitrite is generally the rarest of the three nitrogen-bearing nutrients and behaves similarly to nitrate in phytoplankton nutrient uptake. The dissolved concentrations of all three forms of nitrogen increase in the temperate–boreal winter and decrease in the spring and summer when phytoplankton populations build up.

* Nitrogen supplied to phytoplankton can be divided between that provided from new production and that provided from regenerated production.

A population of phytoplankton in the surface waters depends completely upon the supply of nutrients. Our earlier discussion demonstrated that this supply depends strongly upon mixing of water from a nutrient-rich pool of deeper water. In the open sea, wind mixes deeper, nutrient-rich water to the surface and supplies the phytoplankton with a rich supply of nitrates and other nutrients. As long as phytoplankton cells are not mixed downward too far, this supply, combined with light, allows phytoplankton to grow. The amount of primary production attributable to nutrient supply from deeper waters is known as **new production**.

By contrast, the surface waters also contain zooplankton and bacteria, which excrete nitrogen, usually in the form of ammonia. This, too, is available to the phytoplankton, and may be used in primary production. This amount of primary production attributable to nutrient supply from these sources is known as **regeneration production**. Because regeneration production comes from a recycling process that occurs within the system, the new production is often of greater interest in regional estimates of primary productivity because it represents the outcome of nutrient supply from outside the system.

* Nitrogen recycles between phytoplankton and the bottom in shallow-water environments. Zooplankton excretion is another major source of recycling.

In shallow coastal bays and estuaries, coupling with the benthic system may influence phytoplankton nutrient dynamics. During the process of decomposition, dissolved forms of nitrogen are released from the bottom into the overlying water. In very shallow

bays, much of the return of nitrogen from the bottom to the water column is probably in the form of organic nitrogen, such as urea. Some phytoplankton species are capable of taking up urea, uric acid, and amino acids.

Recycling of different forms of nitrogen depends on the habitat and the nature of the nutrient regeneration cycle. In coastal areas of high upwelling (e.g., off the coast of Peru), nitrate is the main form of nitrogen regenerated from the bottom. Tracer studies employing the ^{15}N isotope of nitrogen show that over half the nitrogen uptake in upwelling areas is in the form of nitrate. The remainder is in the form of ammonia that recycles from zooplankton excretion and decomposition, and back to the phytoplankton. Excretion of the anchovy may be the principal source of regenerated nitrogen in the Peru upwelling region. In contrast, in the nutrient-poor gyres, less than 10 percent of the measured nitrogen uptake is in the form of nitrate. Most nitrogen uptake must therefore involve efficient recycling of ammonia and organic nitrogen between the zooplankton and phytoplankton.

* Nitrogen cycling is intimately involved with microbial transformations.

Figure 9.10 shows the pathways of nitrogen exchange in the sea. Nitrogen cycling involves a large pool of gaseous nitrogen in the atmosphere. Several distinct groups of bacteria transform nitrogen from one form to another. Some **nitrifying bacteria** convert ammonium ion to nitrite, whereas other species of nitrifying bacteria oxidize nitrite to nitrate. Both processes require the presence of oxygen. Under anaerobic conditions, **denitrifying bacteria** reduce nitrate to ammonium ion. **Nitrate-reducing bacteria** can return nitrate to the atmosphere, in the form of nitrogen gas.

Nitrogen may be incorporated into marine food chains through the process of nitrogen fixation (accomplished by some bacteria, cyanobacteria, and yeasts). Gaseous nitrogen is converted eventually to nitrogen in proteins. Many diatoms living in nutrient-poor waters form symbiotic associations with a cyanobacterium suspected of nitrogen fixation. Small cyanobacterium cells are found ubiquitously in the ocean. A nitrogen-fixing cyanobacterial species, *Trichodesmium thiebautii*, is found in tropical gyre centers and may possibly contribute nitrogen to the phytoplankton. Such filamentous cyanobacteria sometimes bloom, and the ocean surface becomes thick

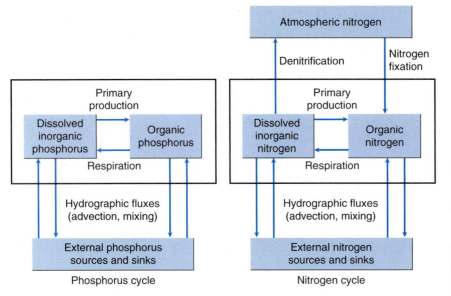

Fig. 9.10 Major transformations and movement of nitrogen and phosphorus in the water column.

with their filamentous colonies. The Red Sea is reputed to owe its name to a filamentous cyanobacterium that was dominated by reddish photosynthetic pigments. Nitrogen fixation is an anaerobic process, and *Trichodesmium* has local anoxic zones within groups of nitrogen-fixing cells. In some seas, nitrogen fixation is responsible for as much as 20 percent of the input of nitrogen into phytoplankton, but elsewhere the role of nitrogen fixation is trivial.

In the global ocean, nitrogen is gained by nitrogen fixation as already mentioned. It is also lost, however, because of the process of **denitrification**, which is most common in anaerobic environments such as inshore water columns and sediments. Denitrification and nitrogen fixation are not coupled processes, so there is a possibility that one will dominate the other. Current evidence suggests that there is a slight excess of denitrification, which implies that nitrogen is a more limiting nutrient element than phosphorus, at least as a world average.

Because of large uncertainties over the role of microbial transformations of nitrogen, the global cycle and the limitation of nitrogen are often unclear.

Phosphorus

✳ Phosphorus occurs in seawater mainly as inorganic phosphate, is required for the synthesis of ATP, and is a crucial energy source in enzymatic reactions.

The biochemical role of phosphorus (P) is different from that of nitrogen because phosphorus is used primarily in the energy cycle of the cell. Adenosine triphosphate (ATP) is a crucial energy source in all enzymatic reactions.

Phosphorus occurs in the ocean as inorganic phosphate, dissolved organic phosphorus, and particulate phosphorus. Phosphate is the form preferred by phytoplankton and exchanges rapidly between phytoplankton and seawater (Figure 9.10). Phosphate is taken up very rapidly by phytoplankton, and the concentration in surface waters is usually quite low. Phosphorus is recycled rapidly between the water and phytoplankton and is often a rate-limiting step in primary productivity as a result. Grazing and excretion by the zooplankton allow rapid regeneration in the plankton. As phytoplankton detritus settles from the water column, the sediments accumulate phosphorus. Benthic decomposition results in the diffusion of phosphorus from the bottom. The remixing of the water column in fall and winter returns phosphorus to the surface waters.

The absolute amount of phosphate in seawater probably limits standing crop, although there is some controversy over the minimum concentration sufficient to limit growth rate. Usually a shortage of nitrogen is believed to be responsible for stopping the growth of phytoplankton populations. A shortage of phosphorus in a cell must be extreme before it affects energy-related processes, such as enzymatic reactions, and photochemical processes in photosynthesis.

Nitrogen/Phosphorus Ratio in the Sea

* The nitrogen-to-phosphorus ratio in the sea is generally about 15:1 and is regulated by uptake and decomposition of phytoplankton.

The growth of phytoplankton results in the simultaneous depletion of both nitrogen and phosphorus. Those elements are available in ocean water (N:P = 14.7:1) in very nearly the proportions usually required by phytoplankton. Phytoplankton particles are more enriched in nitrogen (N:P = 16:1), which suggests that inasmuch as N and P are taken up in photosynthesis, nitrogen is a limiting factor in primary production. These proportions are known as Redfield ratios, after their discoverer. The near coincidence of nitrogen-to-phosphorus ratios in the sea and in phytoplankton requirements has evoked the hypothesis that growth of phytoplankton cells, followed by their sinking and decomposition, controls the N:P ratio in both the phytoplankton and seawater. A balance of nitrogen fixation and denitrification would fix the overall value of nitrogen relative to phosphorus. As we mentioned in the discussion of nitrogen, there appears to be a slight excess of denitrification, leading to the apparent difference in N:P ratios in phytoplankton and in water. Despite the probable correctness of this hypothesis, numerous examples exist of phytoplanktonic species that deviate from the "typical" N:P uptake ratio.

Silicon

Silicic acid, which contains the nutrient element silicon, is a constituent of seawater and is essential for the skeletons of diatoms. Depletion of silicon inhibits cell division and eventually suppresses the metabolic activity of the cell. In natural waters depletion of silicon can limit diatom populations and may direct the course of subtropical succession toward phytoplankton lacking a siliceous test. As diatoms sink from surface waters, they remove silicon, which may partially cause the demise of the spring diatom increase found in so many temperate and higher-latitude waters.

Trace Substances and the Iron Hypothesis

Metals, such as iron, manganese, and zinc, have important functions in oxidase systems (iron is the cofactor in the oxygen evolution step of photosynthesis) and serve as cofactors for enzymes essential for plant growth (e.g., molybdenum, zinc, cobalt, copper, and vanadium play such a role). Iron limits the photosynthetic potential of phytoplankton. Chelators may be important in affecting the utilization of trace metals. Phytoplankton may synthesize and release chelating substances into the water to alter the availability of trace metals such as iron. Oceanographer John Martin[3] first suggested that iron may be a strongly limiting nutrient in the North Pacific Ocean, and more recently several other investigators have suggested that this may also be true in parts of the Antarctic Ocean. These suggestions have stemmed from experimental findings that phytoplankton growth was strongly increased in water in bottles to which iron had been added. Such experiments were conducted in 1994 on a larger scale in the equatorial Pacific, and phytoplankton production was found to increase.[4] This would suggest that phytoplankton is controlled by a bottom-up process such as nutrient control, as opposed to a top-down ecosystem process such as grazing by zooplankton. Areas remote from windborne iron-bearing dust, such as the South Pacific, are more likely to be limited by iron than areas near such supplies (e.g., as the Atlantic is to the deserts and winds of northern Africa). It appears that during glacial maxima in the Pleistocene epoch, iron was eroded from the land and supplied at a much higher rate to the ocean.

Organic trace nutrients, particularly vitamins, may also be of great significance in the sea. Almost all marine phytoplankton species are auxotrophic and require cobalamine, thiamine, or biotin. In mixed cultures, vitamin production and release by one species may stimulate the growth of another species, although most vitamin production is probably by bacteria. In the Sargasso Sea, small diatoms requiring vitamin B_{12} increase in abundance relative to coccolithophores (which need only thiamine) at the time of year when the vitamin B_{12} concentration is great.

[3] See Martin, 1992, in Further Reading, Nutrients, Water Motion, and Phytoplankton Dynamics.
[4] See Kolber and others, 1994, in Further Reading, Nutrients, Water Motion, and Phytoplanckton Dynamics.

Phytoplankton Succession

✻ During the production season, there is a successional sequence of phytoplankton species whose general properties correspond to the seasonal trend of nutrient availability. Differential dependence on substances excreted by phytoplankton and production of toxic substances may influence the succession of phytoplankton species during the production season.

Seasonal changes in dominance are common in the phytoplankton. These changes are known as phytoplankton succession, and the mechanisms are complex. Groups such as dinoflagellates may depend on exudates and nutrients produced in the excretion and decomposition of species earlier in the successional sequence. Diatoms early in the successional sequence are autotrophic, requiring only inorganic nutrients for their survival, whereas species later in the successional sequence may be auxotrophic, requiring nutrients such as vitamins that they cannot produce themselves. Earlier diatom species often have large cell sizes, which allow them to store nutrients. Later species often are smaller in cell size, and thus the surface area (relative to cell volume) that is exposed to nutrient uptake is increased. This adds up to a shift of advantage for different species as the season progresses. But the auxotrophy of later species suggests that later species cannot reach great abundance until the flowering of earlier species. Dinoflagellates are known typically to require more nutrients that they cannot manufacture themselves than is the case for diatoms, which perhaps explains the successional sequence in the plankton. The demise of the diatoms in late spring may also be explained in terms of sinking from the surface waters of silica, which is a crucial nutrient for diatom growth. The depletion of silica at this time may favor phytoplankton groups that do not have a siliceous test.

Allelopathy—the production of toxic compounds by one organism to inhibit another—may play a role in succession. Cyanobacteria may dominate eutrophic lakes and can inhibit the development of diatom populations. Cell-free filtrates of cyanobacteria cultures inhibit the growth of diatoms isolated from the same lake. Cyanobacteria blooms may thus alternate with diatom outbursts in lakes and some polluted estuaries.

During succession, changes in relative abundance of phytoplankton species occur, but it is important to remember that all species are present at all times of the year. Otherwise there would be no seed population from which a population increase of a given species could develop. In some cases, a population of cysts in the bottom sediment may help initiate blooms of some species.

The diversity of phytoplankton also tends to increase as succession progresses. This may reflect an increasing diversity of nutrient sources, especially an increase in organic substances in the water. Later phytoplankton successional dominants also are often ornate and adorned with spines. This may be a response to a temporal increase in the presence of predators, or to an increase in surface area/volume ratio, which would increase nutrient uptake efficiency. It may also be a mechanism to induce rotation as the phytoplankton cell is sinking, which may break up the boundary layer around the phytoplankton cells and increase nutrient access.

The pattern of seasonal succession seems generally to reflect the geographic distribution of phytoplankton. Phytoplankton species that bloom early in succession are typical of nutrient-rich (eutrophic) coastal waters, whereas those that occur later dominate nutrient-poor (oligotrophic) offshore environments. Stages of diatom dominance are most pronounced in mid- to high latitudes and are indistinct in tropical locations.

The Microbial Loop: Nutrient Cycling by Heterotrophs and Chemoautotrophs

✻ Phytoplankton may take up organic molecules, but bacteria are the major heterotrophic consumers in the water column.

Certain phytoplankton organisms can grow heterotrophically on dissolved organic carbon sources, such as sugars, alcohol, lactic acid, and various amino acids. Generally, many substances can be taken up by a given planktonic species but will not support its growth in darkness. Species that cannot grow in the dark may nevertheless accumulate organic molecules of one type or another. For a given species, certain organic molecules taken up may inhibit growth, although other, quite similar compounds (e.g., a different amino acid) may be used heterotrophically. In some cases, light inhibits heterotrophic uptake. Heterotrophic capabilities are widespread among the pennate diatoms and rare among centric diatoms. Pennate diatoms are predominantly benthic, but heterotrophic pennate and centric diatoms are found in planktonic habitats with high concentrations of dissolved organic matter. A few dinoflagellates can engulf particulate matter, and phagocytosis may complement photosynthesis.

The presence of heterotrophic uptake in phytoplankton suggests an overlap of nutrition with saprophytic bacteria (Figure 9.11). In shallow bays, bacteria may also compete with phytoplankton for nutrients regenerated from the bottom.

Phytoplankton consume the majority of dissolved inorganic nutrients in well-lit surface waters. Phytoplankton are also responsible, especially near shore, for heterotrophic consumption of dissolved organic nutrients. Bacteria, however, are probably the principal heterotrophic consumers in the water column; they efficiently utilize both dissolved and particulate material including chitin and cellulose. In some areas, bacterial consumption of dissolved nutrients can even exceed the uptake by phytoplankton. Jonathan Cole and others[5] have shown that bacterial heterotrophic production overall averages about 20 percent of primary production.

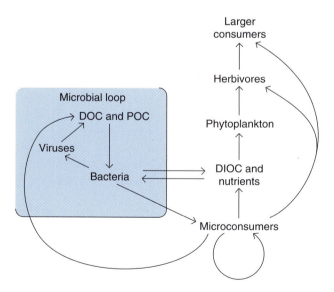

Fig. 9.11 Diagram of the cycling of organic material and nutrients through the phytoplankton and through the microbial loop (shaded box): DOC, dissolved organic carbon; POC, particulate organic carbon; DIOC, dissolved inorganic carbon.

Although ubiquitous, bacteria are most abundant in association with suspended particulate matter. In shallow basins during spring and summer, they often coincide with concentrations of particulates at the base of the thermocline. In the open ocean, they are similarly associated with the oxygen layer below the thermocline, which is itself associated with microbial consumption of organic matter. Resuspended particulates near the bottom also harbor large bacterial populations. These bacteria are consumed by protozoans and this part of the planktonic food web is known as the **microbial loop** (Figure 9.11).

Bacteria in the water column are responsible for the rapid uptake of such organic substrates as carbohydrates, amino acids, and peptides. Therefore, these dissolved substances can be converted to a particulate form available to consumers, such as protozoans and salps. Bacteria (and algae that are partially heterotrophic) attached to particulate organic matter may be a major source of food for smaller zooplankton.

A wide variety of bacteria are specialized to utilize specific inorganic dissolved substrates to generate the energy (ATP) required for synthesis. Such chemoautotrophs oxidize substances such as ammonium (by nitrifying bacteria) and H_2S (by means of sulfur bacteria). These bacterial groups tend to be inviable or very inefficient in oxygenated waters. Consequently, chemosynthesis is largely restricted to anoxic or poorly oxygenated waters, and some microbial transformations occur only at the border of oxygenated and anoxic water.

✳ **Protistans are the major consumers of water column bacteria and are themselves a major component of the food of zooplankton.**

Planktonic protistans are diverse and include such groups as heterotrophic nannoflagellates, heterotrophic dinoflagellates, ciliates, naked amoebae, foraminiferans, and radiolarians. Experimental studies show that ciliates promote growth of many suspension-feeding invertebrate zooplankton and fish larvae. Most of these studies have been done only in the last few years, but they show that protistans are a major component of the lower part of planktonic food webs. In the past, most oceanographers believed

[5] Cole and others, 1988, in Further Reading, Microbial Loop and Protozoan Feeding.

that the bacterial-feeding protistans were not consumed and therefore constituted a sort of trophic dead end in the plankton. It is now known that bacterial production is efficiently transferred to the zooplankton, via these protistan intermediate trophic levels. The microbial loop is therefore of considerable importance in planktonic food webs.

Rate of Nutrient Uptake

✻ **Nutrient uptake increases with increasing nutrient concentration, eventually leveling off to a plateau.**

Given that nutrients are usually superabundant at the beginning of a phytoplankton bloom and very low in concentration toward the end, it is desirable to determine the relationship between nutrient concentration and uptake rate by phytoplankton. Uptake rate may be measured directly (e.g., in terms of ammonium taken up per cell per unit time), or indirectly (in terms of the cell doubling rate). We assume here, for simplicity, that the faster the nutrient uptake, the faster the cell doubling rate. (This is not always a good measure on the time scale of a short laboratory

experiment, because some phytoplankton are known to take up nutrients at night and use them in photosynthesis the next day.)

Nutrient uptake usually follows the general pattern illustrated in Figure 9.12, which shows the relationship of cell doublings per day, D (which is an estimate of nutrient uptake, as mentioned earlier), as a function of nutrient concentration, C. The cell doubling rate increases with increasing nutrient concentration but then reaches a plateau, at a value of D_{max}. The nutrient concentration at which half the maximum cell doubling rate occurs is known as the half-saturation concentration,[6] or K. This is a useful measure of nutrient uptake, which we shall use later.

✻ **Inshore phytoplankton live at higher nutrient concentrations and would be expected to be able to take up nutrients at higher nutrient concentrations in the environment, relative to open-ocean phytoplankton, which would be expected to be more efficient at low concentrations.**

Open-ocean environments usually have far lower nutrient concentrations than do inner-shelf waters. As a result, we may expect phytoplankton living in these two different habitats to evolve differing patterns of nutrient uptake. Open-ocean phytoplankton should be able to take up nutrients efficiently from low environmental concentrations. They would be expected, however, to have low maximum uptake rates, because they never encounter high nutrient concentrations. By contrast, inner-shelf phytoplankton should be relatively inefficient at low nutrient concentrations but able to take up nutrients at far higher concentrations than the open-ocean phytoplankton. Expected curves for both types are shown in Figure 9.13.

The difference leads to a prediction, which turns out to be true, that the half-saturation concentration K should be greater for inner-shelf phytoplankton. For coastal phytoplankton, K is usually greater than 1 μM for nitrate uptake, whereas oceanic phytoplankton have values of about 0.1–0.2 μM. Clones of the same diatom species show high and low K values, depending on whether the clones are isolated from near-shore or oceanic waters, respectively. Oceanic phytoplankton are more efficient at taking

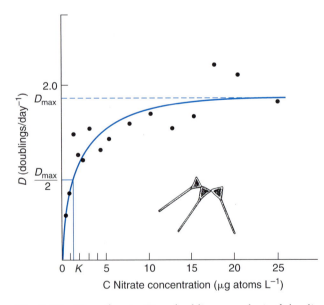

Fig. 9.12 Growth rate (D = doublings per day) of the diatom *Asterionella japonica* (circles) as a function of C (which, in this case, is the nitrate concentration in μg atoms L^{-1}). The half-saturation constant K is estimated at 1.5 by finding the value of C on the green curve that corresponds to $D_{max}/2$. (Modified from Eppley and Thomas, 1969.)

[6] K is formally estimated by plotting C/D as a function of C. K is the negative value of the point at which this line intersects the abscissa.

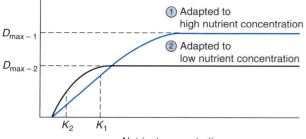

Fig. 9.13 A theoretical representation of the differences in nutrient uptake expected for a phytoplankton clone (1) adapted to high nutrient levels near-shore versus another clone (2) adapted to offshore low nutrient levels. D_{max-1} is the maximum doubling rate for clone 1, and K_1 is the estimated corresponding value of K (same procedure for clone 2).

up nutrients at low nutrient concentration. Thus, they may be competitively superior to coastal forms in the low nutrient concentrations of the open sea but unable to take up high concentrations.

Differences in nutrient limitation among competing species may also contribute to coexistence. For example, species 1 may be superior at phosphate uptake whereas species 2 may be superior at silica uptake. Thus, under conditions of varying nutrient concentration (spatial patchiness) different species may be favored. Such coexistence has been shown for lake phytoplankton.

Can interspecific competition for nutrients determine the success of some phytoplankton species relative to others? Many phytoplankton species seem to coexist with no competitive exclusion and similar nutrient preferences. This coexistence has been characterized by the ecologist G. Evelyn Hutchinson as "the paradox of the plankton." Conditions in the ocean may be too variable and may shift too rapidly for one species to drive others to extinction by means of greater efficiency of nutrient uptake. Many species may be relatively more efficient under conditions that change periodically.

✳ **A stable water column, input of nutrients, and sometimes an initial input of resting stages all combine to promote dense harmful phytoplankton blooms, principally some dinoflagellates and cyanobacteria.**

In the Book of Exodus we read that "All the waters that were in the river turned to blood. And the fish that was in the river died; and the river stank, and the Egyptians could not drink of the water in the river." Today, many interpret this passage as a description of a **red tide**, an intensely colored surface scum caused by dense populations of some species of dinoflagellates (Figure 9.14). Both dinoflagellates and cyanobacteria (only a few species of each) are responsible for such **harmful algal blooms**, which are found mainly in inshore waters and estuaries. Usually, the blooms are dominated by one species of phytoplankton. The blooms often result in subsequent population crashes, and bacterial decomposition may reduce the oxygen in the water, helping to reduce water quality and conditions for marine animals. Red tides are often accompanied by release of toxins by dinoflagellate species, which may poison higher levels of the food chain. This toxicity is sporadic. For example, shellfish often consume red tide dinoflagellates and sequester the toxin class known as saxitoxins. Evidence is now developing that these toxins affect the shellfish (e.g., bivalve mollusks), can cause reductions in metabolic rate and feeding rate, and may even cause death. These substances, however, are also virulently neurotoxic and strongly affect marine vertebrates and humans (we discuss this later in Chapter 18). In some dinoflagellate species, toxins also suppress other phytoplankton species, thus allowing the nuisance species to dominate the phytoplankton completely.

The principal causes of harmful algal blooms—dinoflagellates and cyanobacteria—are both motile and can migrate through the water column. They are

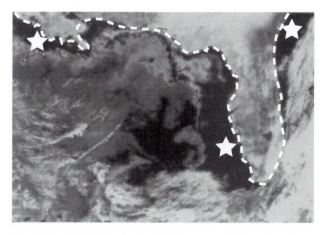

Fig. 9.14 Red tides in the Gulf coast and Florida region, detected by satellite imagery (starred areas) colored falsely as black.

often able to move several meters per day and to cross a strong thermocline. This gives them a tremendous advantage, since they can spend the daylight hours near the well-lit surface but can then dive at night to more nutrient-rich waters below. Dinoflagellates accomplish this by flagellar movement, whereas cyanobacteria adjust their depth by changing cell buoyancy. This advantage can be enjoyed only if the water column is stable. Harmful algal blooms therefore do not usually develop unless the surface part of the water column is stable or the phytoplankton population is trapped near a discontinuity of water density.

Although a stable water column is usually necessary for the generation of a harmful algal bloom, a couple of other factors are also necessary. First, the blooms are usually associated with some major input of nutrients, including those related to sewage. In the specific case of cyanobacterial blooms, the nitrogen-fixing species do better under conditions of low oxygen, and organic matter that promotes bacterial reduction of oxygen in the water column will promote cyanobacterial growth. In some cases, nutrient inputs from runoff include metals such as iron, which may be limiting to phytoplankton growth. Runoff may also include necessary organic nutrients, such as vitamins. Second, phytoplankton species in these blooms often appear to have **resting stages** (e.g., cysts of some dinoflagellates, cyanobacterial resting cells in estuaries), which may be abundant in the sediment and in the water column. The resting stages are often produced under unfavorable environmental conditions, but may be mobilized by, for example, a storm that erodes cysts from the sediment into the water column, and thus ends their dormancy. If, following this erosional event, the water column is stable and nutrient input is strong, then favorable conditions exist for a bloom.

Harmful algal blooms are becoming more and more frequent in coastal waters. This may be related to increased nutrient input, owing to increasing coastal populations of humans and increased sewage input. The blooms often have very negative effects because toxic substances are often produced. The nuisance phytoplankton choke or poison suspension-feeding benthic invertebrates, and the phytoplankton population crashes are followed by death and decomposition, which in turn may reduce the oxygen content of the water. The surface scums of phytoplankton are also unappealing aesthetically.

Zooplankton Grazing

Grazing in the Sea

✱ **Zooplankton abundance usually follows phytoplankton abundance. Although zooplankton growth depends upon phytoplankton growth, it is not always clear that zooplankton grazing controls phytoplankton abundance.**

Zooplankton usually increase after the peak of the phytoplankton bloom has passed, owing to the time lag in reproduction of the zooplankton; and the grazing down of the phytoplankton standing crop as the zooplankton population increases (see Text Box 9.2). Zooplankton such as copepods have a much longer generation time (weeks) than phytoplankton (hours to days), and a large population of zooplankton could graze a phytoplankton population to near-extinction. This would cause a collapse of the zooplankton until the phytoplankton recovered. Such overgrazing would thus result in strong oscillations in the zooplankton population. When the phytoplankton organisms reach very low densities, however, the zooplankton may not be able to find them. A low-density refuge for phytoplankton allows a subsequent increase when zooplankton decrease in abundance, and the severity of zooplankton oscillations may diminish.

The impact of zooplankton grazing is quite variable. In the coastal waters of California, Nova Scotia, and Great Britain, zooplankton grazing can

---

TEXT BOX 9.2 Quantification of the Effect of Grazing

Phytoplankton (e.g., diatom) population growth can be described in terms of the equation:

$$C_2 = C_1 e^{rt}$$

where cell concentration C_2 is produced by a previous concentration C_1 growing after t time units with a rate r. With a grazing rate g, this equation becomes

$$C_2 = C_1 e^{r-g}$$

If g exceeds r, the zooplankton will graze the phytoplankton population to extinction.

exceed 50 percent of the phytoplankton per day. However, in near-shore and estuarine waters of high production, much of the early spring phytoplankton production is not grazed by the zooplankton, if only because of a lag time for buildup of the zooplankton. For a brief time, the advent of zooplankton may even stimulate phytoplankton growth, owing to animal excretion. Eventually, zooplankton may cause significant reductions in the phytoplankton, but this does not seem to occur in estuaries and inshore bays. Rather, the stabilization of the water column and loss of nutrients to deeper water in summer seems to halt phytoplankton production there.

Spatial alternations of abundance of phytoplankton and zooplankton occur in some regions. This would suggest that an area currently rich in zooplankton but poor in phytoplankton had been rich in phytoplankton until the zooplankton grazed them down. Spatial patterns of alternation between phytoplankton and zooplankton are therefore indications of temporal patterns of phytoplankton growth and subsequent zooplankton population growth and strong grazing. For example, in the waters near Plymouth, England, there is a sudden drop in the phytoplankton population as the zooplankton first appear. This finding suggests that the zooplankton have an important influence on the decline of phytoplankton. In the North Sea, alternating areas of phytoplankton and zooplankton dominance occur, suggesting cycles of dominance by phytoplankton production and then by grazing (Figure 9.15). Phytoplankton in the Antarctic are grazed down by the zooplankton, resulting in large numbers of broken diatom tests being deposited on the seafloor.

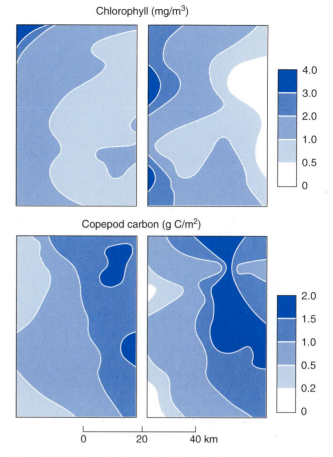

Chlorophyll (mg/m^3)

Copepod carbon (g C/m^2)

Fig. 9.15 Distribution of chlorophyll *a* and copepod carbon on a survey in the North Sea, showing an inverse relationship between phytoplankton and zooplankton standing stock. Phytoplankton are most abundant toward the left, whereas zooplankton are most abundant toward the right. (Modified from Steele, 1974.)

Zooplankton Feeding

❋ Feeding behavior varies with phytoplankton cell size and cell concentration (cells per unit volume).

Figure 9.16 shows the relationship between phytoplankton cell concentration and ingestion rate I, which is the cells ingested per animal per hour, I increases linearly with cell concentration until a maximum is reached, whereupon ingestion increases no further. Below the maximum, the feeding response of the copepod is indicative of an animal that searches, encounters, and feeds on particles in direct proportion to their concentration. Then there appears to be a saturation level, and ingestion increases no further. Feeding may cease because the feeding appendages

are saturated with food, or because digestion efficiency would decrease if material were moved through the gut any faster. There is also some evidence for a lower threshold, below which the copepod does not expend energy to search for food and to feed. Studies of spatial variation in the water column show that there are areas where phytoplankton concentrations are too low to support copepod growth. These areas may be within meters of dense patches of phytoplankton. Copepods are probably able to regulate their energy expenditure to the degree that feeding efforts are reduced in water parcels with very low phytoplankton concentrations.

Cell size and species type also influence feeding in

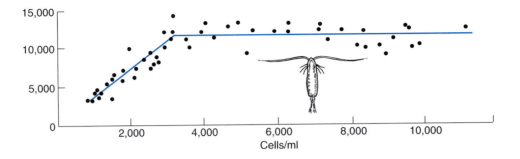

Fig. 9.16   Ingestion rate I (cells copepod^{-1} h^{-1}) for the copepod *Calanus pacificus* as a function of diatom cell concentration. (After Frost, 1972.)

copepods and other zooplankton. Copepods feed more rapidly on two-celled chains of diatoms than on single cells, and also feed at higher rates on diatom species with larger cell size, relative to smaller cells. Many zooplankton avoid toxic phytoplankton, but tintinnids actually feed preferentially on these forms.

✳ Zooplankton can select particles by size. Owing to the low Reynolds number for copepods, this involves direct plucking of preferred particles, rather than straining of particles on a feeding sieve.

In natural waters copepods are presented with a great variety of particle sizes that continually change in relative abundance with season. Adaptation to the particle size spectrum would maximize the efficiency of particle uptake. There is some evidence that copepods can shift the preferred particle size, depending upon the abundance of various food particles in the environment. As mentioned earlier, larger particles are usually preferred. As mentioned in Chapter 7, copepod feeding involves movement of feeding appendages and particles at low Reynolds number. Under these conditions, the water is viscous and it is not possible to strain phytoplankton cells out of the water on a sieve. Instead, feeding appendages actively

select different phytoplankton cells, but the exact mechanisms of search and sensory detection are poorly understood. In copepods, selection of particles involves chemical detection, and the animals tend to avoid inert particles, relative to live phytoplankton cells.

If copepods and other zooplankton can track changes in the size spectrum of the phytoplankton and graze the most abundant size classes, then selective grazing could favor the less abundant species and the less common size classes. Copepod grazing thus could affect the species composition of the phytoplankton and prevent one species from competitively excluding all others. If there is a threshold below which feeding is suppressed, then the phytoplankton always have a refuge, thereby preventing grazing to extinction. Recovery is possible because zooplankton will decrease in abundance and the lag time for zooplankton reproduction will allow the phytoplankton to recover if nutrients are available. Because the phytoplankton generation time is, in general, shorter than that of the zooplankton, the phytoplankton can recover quickly. This factor suppresses strong oscillations in both phytoplankton and zooplankton abundance. Shifts of grazing may also offer an explanation for the so-called paradox of the plankton.

Further Reading

NUTRIENTS, WATER MOTION, AND
PHYTOPLANKTON DYNAMICS

Alldredge, A. L., and M. Silver. 1988. Characteristics, dynamics and significance of marine snow. *Progress in Oceanography*, v. 20, pp. 41–82.

Banse, K. 1990. Does iron really limit phytoplankton production in the offshore subarctic Pacific? *Limnology and Oceanography*, v. 35, pp. 772–775.

Barber, R. T. 1992. Geological and climatic time scales of nutrient variability. In P. G. Falkowski, and A. D. Woodhead, eds., *Primary Productivity and Biogeochemical Cycles in the Sea*. New York: Plenum, pp. 89–106.

Barber, R. T., and R. L. Smith. 1981. Coastal upwelling systems. In A. R. Longhurst, ed., *Analysis of Marine Ecosystems*. New York: Academic Press, pp. 31–68.

Bruland, K. W., J. R. Donat, and D. A. Hutchins. 1991. Interactive influences of bioactive trace metals on biological production in oceanic waters. *Limnology and Oceanography*, v. 36, pp. 1555–1567.

Carpenter, E. J. 1973. Nitrogen fixation by *Oscillatoria* (*Trichodesmium*) *thiebautii* in the southwestern Sargasso Sea. *Deep-Sea Research*, v. 20, pp. 285–288.

Carpenter, E. J., and K. Romans. 1991. Major role of the cyanobacterium *Trichodesmium* in nutrient cycling in the North Atlantic Ocean. *Science*, v. 254, pp. 1356–1358.

Chisholm, S. W. 1992. What limits phytoplankton growth. *Oceanus*, v. 35, pp. 36–46.

Clarke, G. L. 1939. The utilization of solar energy by aquatic organisms. Problems in lake biology. *American Association for the Advancement of Science Publication*, v. 10, pp. 27–38.

Cushing, D. H. 1975. *Marine Ecology and Fisheries*. Cambridge: Cambridge University Press.

Dame, R. F. 1993. The role of bivalve filter feeder material fluxes in estuarine ecosystems. In R. F. Dame, ed., *Bivalve Filter Feeders in Estuarine and Coastal Ecosystem Processes*. NATO ASI Series, v. G33. Berlin: Springer-Verlag, pp. 245–269.

Dugdale, R. C., and J. J. Goering. 1967. Uptake of new and regenerated forms of nitrogen in primary productivity. *Limnology and Oceanography*, v. 12, pp. 196–206.

Dugdale, R. C., and F. Wilkerson. 1992. Nutrient limitation of new production in the sea. In P. G. Falkowski and A. D. Woodhead, eds., *Primary Productivity and Biogeochemical Cycles in the Sea*. New York: Plenum, pp. 107–122.

Eppley, R. W., J. N. Rogers, and J. J. McCarthy. 1969. Half-saturation constants for uptake of nitrate and ammonium by various phytoplankton. *Limnology and Oceanography*, v. 14, pp. 912–920.

Falkowski, P. G. 1984. Physiological responses of phytoplankton to natural light regimes. *Limnology and Oceanography*, v. 6, pp. 295–307.

Falkowski, P. G., R. T. Barber, and V. Smetacek. 1998. Biogeochemical controls and feedbacks on ocean primary production. *Science*, v. 281, pp. 200–206.

Harvey, H. W. 1926. Nitrate in the sea. *Journal of the Marine Biological Association of the United Kingdom*, v. 14, pp. 71–88.

Honjo, T. 1994. The biology and prediction of representative red tides associated with fish kills in Japan. *Review of Fisheries Science*, v. 2, pp. 225–253.

Kolber, Z. S., R. T. Barber, K. H. Conle, S. E. Fitzwater, R. M. Greene, K. S. Johnson, S. Lindley, and P. G. Falkowski. 1994. Iron limitation of phytoplankton photosynthesis in the equatorial Pacific Ocean. *Nature*, v. 371, pp. 145–149.

Malone, T. C., C. Garside, and P. J. Neale. 1980. Effects of silicate depletion on photosynthesis by diatoms in the plume of the Hudson River. *Marine Biology*, v. 58, pp. 197–204.

Mann, K. H. 1982. *Ecology of Coastal Waters. A Systems Approach*. Berkeley: University of California Press.

Martin, J. H. 1992. Iron as a limiting factor in oceanic productivity. In P. G. Falkowski and A. D. Woodhead, eds., *Primary Productivity and Biogeochemical Cycles in the Sea*. New York: Plenum, pp. 123–137.

Martin, J. H., and S. E. Fitzwater. 1988. Iron deficiency limits phytoplankton growth in the northeast Pacific subarctic. *Nature*, v. 331, pp. 341–343.

Morel, F. M. M., R. J. M. Hudson, and N. M. Price. 1991. Limitation of productivity by trace metals in the sea. *Limnology and Oceanography*, v. 36, pp. 1742–1755.

Morris, I, ed. 1980. *The Physiological Ecology of Phytoplankton*. Berkeley: University of California Press.

Nakatani, H. Y. 1988. *Photosynthesis*. (Series editor, J. J. Head, Carolina Biology Readers). Burlington, NC: Carolina Biological Supply Company, pp. 1–16.

Parsons, T. R., M. Takahashi, and B. T. Hargrave. 1977. *Biological Oceanographic Processes*. Oxford: Pergamon Press.

Paerl, H. W. 1988. Nuisance phytoplankton blooms in coastal, estuarine, and inland waters. *Limnology and Oceanography*, v. 33 (no. 4, part 2), pp. 823–847.

Platt, T. 1981. Physiological Bases of Phytoplankton Ecology. *Canadian Bulletin of Fisheries and Aquatic Sciences*, no. 210.

Riemann, R. 1989. Gelatinous phytoplankton detritus aggregates on the Atlantic deep-sea bed. Structure and mode of formation. *Marine Biology*, v. 100, p. 533–539.

Riley, G. A., and D. F. Bumpus. 1946. Quantitative ecology of the plankton of the western North Atlantic. *Journal of Marine Research*, v. 6, pp. 33–47.

Russell-Hunter, W. D. 1970. *Aquatic Productivity*. London: Macmillan.

Steele, J. H. 1974. *The Structure of Marine Ecosystems*. Cambridge, MA: Harvard University Press.

Strickland, J. D. H. 1965. Production of organic matter in the primary stages of the marine food chain. In J. P. Riley and G. Skirrow, eds., *Chemical Oceanography*, v. 1. London: Academic Press, pp. 477–610.

MICROBIAL LOOP AND PROTOZOAN FEEDING

Azam, F., T. Fenchel, J. G. Gray, L. A. Meyer-Reil, and T. Thingstad. 1983. The ecological role of water column microbes in the sea. *Marine Ecology—Progress Series*, v. 10, pp. 257–265.

Cole, J. J., S. Findlay, and M. L. Pace. 1988. Bacterial production in fresh and saltwater ecosystems: A cross-system overview. *Marine Ecology—Progress Series*, v. 43, pp. 1–10.

Conover, R. J., Jr. 1982. Interrelations between microzooplankton and other plankton organisms. *Ann. Inst. Oceanogr. Paris* v. 59S, pp. 31–46.

Eppley, R. W., and W. H. Thomas. 1969. Comparison of half-saturation constants for growth and nitrate uptake of marine phytoplankton. *Journal of Phycology*, v. 5, pp. 375–379.

Fenchel, T. 1988. Marine plankton food chains. *Annual Review of Ecology and Systematics*, v. 19, pp. 19–38.

Fuhrman, J. 1992. Bacterioplankton roles in cycling of organic matter. In P. G. Falkowski and A. D. Woodhead, eds., *Primary Productivity and Biogeochemical Cycles in the Sea*. New York: Plenum, pp. 361–383.

Hobbie, J. E., and P. J. LeB. Williams. 1984. *Heterotrophic Activity in the Sea*. New York: Plenum.

Shanks, A. L., and M. L. Reeder. 1993. Reducing microzones and sulfide production in marine snow. *Marine Ecology—Progress Series*, v. 96, pp. 43–47.

Stoecker, D. K., and J. M. Capuzzo. 1990. Predation on protozoa: Its importance to zooplankton. *Journal of Plankton Research*, v. 12, pp. 891–908.

GRAZING

Banse, K. 1992. Grazing, temporal changes of phytoplankton concentrations, and the microbial loop in the open sea. In P. G. Falkowski and A. D. Woodhead, eds., *Primary Productivity and Biogeochemical Cycles in the Sea*. New York: Plenum, pp. 409–440.

Enright, J. T. 1969. Zooplankton grazing rates estimated under field conditions. *Ecology*, v. 50, pp. 1070–1078.

Frost, B. W. 1972. Effects of size and concentration of food particles on the feeding behavior of the marine planktonic copepod *Calanus pacificus*. *Limnology and Oceanography*, v. 17, p. 805–815.

Mullin, M. M. 1963. Some factors affecting the feeding of marine copepods of the genus *Calanus*. *Limnology and Oceanography*, v. 8, pp. 239–250.

Richman, S., and J. N. Rogers. 1969. The feeding of *Calanus helgolandicus* on synchronously growing populations of the diatom *Ditylum brightwelli*. *Limnology and Oceanography*, v. 14, pp. 701–709.

Steele, J. H. 1974. *The Structure of Marine Ecosystems*. Cambridge, MA: Harvard University Press.

Valiela, I. 1983. *Marine Ecological Processes*. New York, Springer-Verlag. See Chapter 8.

Walsh, J. J. 1988. *On the Nature of Continental Shelves*. San Diego, CA: Academic Press.

Review Questions

1. In winter, which factors prevent a phytoplankton bloom from occurring? What factors are not limiting at that time?

2. Why does a spring phytoplankton bloom finally collapse?

3. Why does the peak of phytoplankton production not necessarily coincide with the maximum day length in the summer solstice, even if light is important for photosynthesis?

4. In very shallow estuaries, from where do nutrients come in summer to fuel phytoplankton growth?

5. Describe the relationship of photosynthesis to light intensity.

6. Why do phytoplankton blooms occur occasionally in the fall?

7. Why is nitrogen an essential nutrient element for phytoplankton, and in what major forms does it occur chemically in the water column?

8. Why is the spring freshet important in nutrient supply to an estuary?

9. Distinguish between new production and regeneration production.

10. How may microbial organisms influence the availability and forms of nitrogen for phytoplankton?

11. What is the microbial loop and what is its importance in nutrient cycling in the water column?

12. Why may different forms of phytoplankton dominate during different times of the year?

13. How may resting stages contribute to the eruption of nuisance blooms?

14. Why do you think that nuisance blooms are becoming more and more common in coastal waters of the world?

15. What would be good evidence, in your opinion, of the importance of the effect of marine grazers on phytoplankton abundance?

10

Productivity and Food Webs in the Sea

Productivity, the rate of production of living tissue, is a process that reflects the interaction of many factors, many of which were discussed in Chapter 9. In this chapter, we shall discuss how productivity is measured, how it is controlled, and how consumption links the marine biota into a web of food interactions. We will concentrate on productivity and food webs in the water column. Production in the water column accounts for most of the biological production in the sea, even though certain more locally restricted benthic environments, such as sea grass beds and coral reefs, are far more productive per unit area. Water column plant production, moreover, is the basis of the food web of most of our most important fisheries.

Productivity is the amount of living tissue produced per unit time. It is often estimated in terms of carbon contained in living material and expressed as grams of carbon (g C) produced per day, in a column of water intersecting one square meter of sea surface (g C m^{-2} d^{-1}), from the surface to the seabed. The amount of living material present in the water column at any one time is the **biomass** (expressed as g C m^{-2}). **Primary productivity** is that part of the production ascribed to photosynthesis. **Secondary productivity** refers to production by the organisms that consume the phytoplankton; and those that consume the organisms responsible for secondary production are said to be engaged in tertiary productivity. As was

explained earlier, **plant nutrients** are those constituents required by photosynthetic organisms. **Limiting nutrients** are those that are in potentially short supply and may limit phytoplankton growth. **Autotrophic** phytoplankton can photosynthesize and produce all necessary constituents of the cell by simply using light and inorganic nutrients. **Auxotrophic** phytoplankton require some sort of organic nutrient, such as a fatty acid or vitamin, synthesized by another organism. **Heterotrophic** organisms consume organic matter exclusively; they include animals and some plants that are partly heterotrophic. **Saprophytic** organisms decompose organic matter and include the bacteria and fungi.

The Food Chain Abstraction

✳ **A food chain is a linear sequence that reveals which organisms consume which other organisms in an environment. A food web is a more complicated diagram of feeding interactions.**

A **food web** is a diagram that shows the overall pattern of feeding among organisms, or who eats whom. Figure 10.1 shows such a diagram for the North Sea, leading finally to feeding by the herring *Clupea harengus*. The feeding relationships are a bit complex, but they can be simplified to the linear **food**

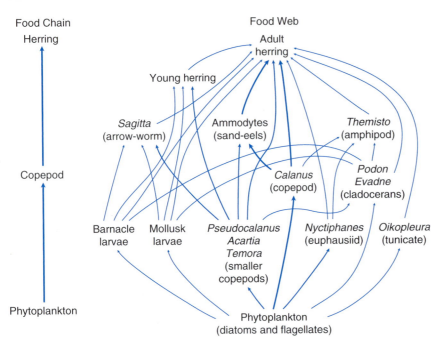

Fig. 10.1 Food web constructed from the feeding relationships of the North Sea herring *Clupea harengus*, during different life history stages. A simplified food chain leading to the adult herring is diagrammed at left. (Modified after Russell-Hunter, 1970.)

chain on the left-hand side of the figure, which takes into account the main species through which organic matter cycles. Each species represents a **trophic level,** which is defined as a species or group of species that all feed on one or more other species (which can be grouped into a lower trophic level). Species in one trophic level may also be consumed by species in a *higher* trophic level. As Mark Graham once wrote in a song: "You are what you eat—and you're also what eats you!" Nearly all oceanographic studies of productivity must focus on the abundant species and can produce only a simplified representation of nature. However, it is important to recognize the actual complexity of the situation. For example, species may change levels in the food chain at different stages of the life cycle. Fully adult codfish are predators on herring, but when they are small (<50 cm long), they feed on copepods and other planktonic crustaceans. Consumers may feed at several "levels" in the food chain. Copepods, for example, feed on phytoplankton, but they also feed on smaller zooplankton, fish eggs, and particulate organic matter.

Transfer Between Trophic Levels

* **Transfer from one trophic level to the next is not complete.**

Not all the production from one trophic level is transferred perfectly to the next. To estimate the potential production at the top of a food chain, such as fish production, we must tabulate the losses. They result from essentially two factors.

1. **Not eaten.** Some proportion of a given trophic level evades consumption through escape, unpalatability, or unavailability. Phytoplankton with large spines or toxins are avoided by zooplankton. Phytoplankton cell size may be too small, or too large, to permit ingestion.

2. **Inefficient conversion.** Some proportion of the food that is ingested is not converted into growth.

We can construct a budget for ingested food as follows:

$$I = E + R + G$$

where I is the amount ingested, E is the amounted egested, R is the amount consumed in respiration, and G is the amount used in growth; G can be partitioned between somatic (body) growth and reproduction. Such a budget is usually constructed in terms of energy units, such as calories.

Growth is usually a minority of the ingestion budget. Not all food can be digested and assimilated, and some, therefore, is egested (e.g., skeletal material). Some of the energy obtained in food is lost as respiration, hence is not available to the next trophic level.

The incompleteness of transfer up a food chain can be estimated in terms of **food chain efficiency**, E, defined as the amount of energy extracted from a trophic level divided by the amount of energy supplied to that trophic level. Growth efficiency, which is the proportion of assimilated food used in growth can be in a range as high as 30–45 percent, but food chain efficiency is usually far lower, often in the range of 10 percent. Higher-latitude planktonic systems may have much higher food chain efficiency.

❋ Food chain efficiency can be used to calculate the potential fish production at the top of a food chain.

Food chain efficiency can be used to estimate the potential fish production at the top of the food chain. If B is the biomass of the phytoplankton, and n is the number of links between trophic levels, then the production P of fish is:

$$P = BE^n$$

A change in E from 0.1 to 0.2 would magnify by 16 the estimate of fish production at the fifth trophic level!

Keeping the potential errors in mind, J. A. Gulland[1] estimated the potential annual yield of common commercially exploited fishes at 100 million tons. This figure is disturbing, because the 1970 yield was 55.7 million tons, and the world catch increased at the time about 7 percent per year. Gulland's estimate suggests that the world's fisheries may soon reach the limits of exploitation. The recent worldwide decline in fisheries may be a harbinger of Gulland's prophecy (see Chapter 18).

Patterns of Food Chain Variation

❋ Food chains can be classified on the basis of oceanographic conditions, which helps determine the number of trophic levels.

According to a classification devised by John Ryther,[2] marine planktonic food chains can be classified into three basic types (Figure 10.2). The **oceanic system type** has five trophic levels, with a low annual primary production of about 50 g C m^{-2} y^{-1}. The **coastal type** has three trophic levels, and primary production is about 100 g C m^{-2} y^{-1}. The **upwelling**

[1] See Gulland, 1972, in Further Reading.
[2] See Ryther, 1969, in Further Reading.

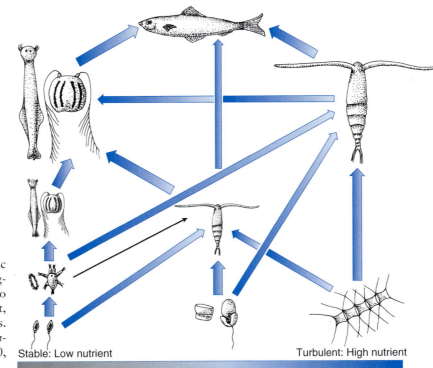

Fig. 10.2 Variation in the trophic levels of planktonic ecosystems, ranging from oceanic environments to coastal environments to turbulent, upwelled high-nutrient environments. (After Landry, *Helgoländer wissenschaffen Meeresuntersuchungen*, v. 30, pp. 8–17, 1977.)

Stable: Low nutrient   Turbulent: High nutrient

Table 10.1 Some characteristics of the three principal types of marine planktonic food chains

FOOD CHAIN TYPE	PRIMARY PRODUCTIVITY (g C m^{-2} y^{-1})	NUMBER OF TROPHIC LEVELS	FOOD CHAIN EFFICIENCY (PERCENT)	POTENTIAL FISH PRODUCTION (mg C m^{-2} y^{-1})
Oceanic	50	5	10	0.5
Shelf	100	3	15	340
Upwelling	300	1.2	20	36,000

Source: After Ryther, 1969.

type occurs in areas such as the Peru current and the Antarctic, and has only two trophic levels. Upwelling provides higher and more continuous nutrient supply, leading to a primary productivity of about 300 g C m^{-2} y^{-1}.

Table 10.1 shows the potential fish production of these three types. By far the greatest potential for fish production lies in the upwelling system, which in fact fulfills the prediction. The high potential of upwelling systems is enhanced by a greater food chain efficiency, which is related to the ease of ingestion and assimilation of large diatoms by planktivorous fishes. The low primary productivity and the large number of food chain transfers greatly reduce the fishery potential of the oceanic system.

Why do some food chains have many levels, whereas others are so short and simple? Temporal environmental stability and a stable water column may promote the survival of complex oceanic food webs. In near-shore and upwelling systems, strong temporal changes in temperature and circulation would, on the other hand, tend to collapse a complex multilevel food chain. In the Peru upwelling system, for example, periodic El Niño events cause great increases in temperature and sudden reductions in nutrient supply. These changes may prevent the addition of more trophic levels.

Measuring Primary Productivity

Primary productivity must be known before we can estimate the potential production of fish at the top of a food chain. Marine ecologists have attempted to develop accurate and relatively simple field-oriented techniques for measuring photosynthesis in the water column. **Gross primary productivity** is the total carbon fixed during photosynthesis. We are more interested, however, in **net primary production**, which

is the carbon that is available to higher trophic levels. Some of the primary production is respired away, leading to a somewhat lower net availability to higher trophic levels.

Oxygen Technique

✳ **The oxygen technique relies upon the fact that oxygen is released in proportion to the amount of photosynthesis.**

Because oxygen is released during photosynthesis, changes in oxygen concentration can be used to estimate primary productivity. A water sample is first collected and the zooplankton are strained from it, using a 150–300 μm plankton net. The remaining water is placed in an oxygen-tight biological oxygen demand (BOD) bottle and incubated in the light for several hours. Oxygen is measured at the start and finish of the experiment. Dissolved oxygen is measured either by the **Winkler method**, which is based upon chemical titration, or by a **polarographic oxygen electrode**, which measures an electric current that is proportional to the dissolved oxygen concentration.

During the course of the experiment, phytoplankton and bacteria are respiring in the bottle, reducing the oxygen concentration. The change in oxygen concentration is explained, therefore, in terms of **an addition from photosynthesis, and a subtraction from respiration**. We can estimate the loss from respiration by incubating a bottle that is completely covered in black tape so that no photosynthesis occurs. This **light–dark bottle technique** can be applied by incubating in various levels of artificial sunlight, or by actually dropping a string of bottles over the side and incubating them in natural light (Figure 10.3). With the artificial light technique, it is necessary to measure the light intensity as a function of depth, so that the shipboard laboratory measurements can be used to calculate the amount of photosynthesis in the nat-

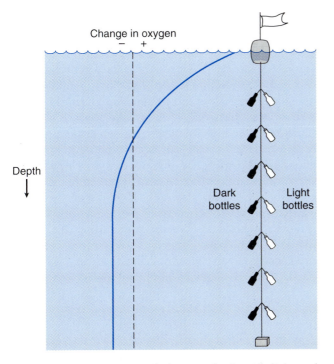

Change in oxygen
−   +

Depth

Dark
bottles

Light
bottles

Fig. 10.3 Measurement of photosynthesis with light and dark bottles. Light and dark bottles are suspended on a line in a series, at various depths. The change of oxygen in the light bottle provides data on photosynthesis plus respiration, whereas the change in the dark bottle provides data on respiration only.

ural environment. Practically speaking, respiration is often not actually measured, but is usually taken to be a value that corresponds to 10 percent of the total oxygen increase (which is based upon measurements in rapidly growing laboratory cultures).

Radiocarbon Technique

✳ **The radiocarbon technique uses the radioactive isotype ^{14}C as a tracer in uptake of bicarbonate during the process of photosynthesis.**

In this method, bicarbonate ion is labeled with the radioactive isotope of carbon, ^{14}C. (The common nonradioactive isotope is ^{12}C.) The uptake of carbon is measured directly, using ^{14}C-labeled bicarbonate as a tracer. First, a sample of seawater is inoculated with a small amount of radioactive bicarbonate solution, and the phytoplankton are allowed to photosynthesize for a specified number of hours. After this, the phytoplankton are trapped on a filter, and the ra-

---

TEXT BOX 10.1 How to Calculate Productivity, Using the Oxygen Technique

If I is the initial amount of oxygen in both a light bottle and a dark bottle (see Figure 10.3), P is the oxygen produced in photosynthesis, and R the amount consumed in respiration, we can show that after a specified amount of time, the amount of oxygen in the light bottle, L, is $L + I + P − R$. The amount in the dark bottle is $D = I − R$, because there are no additions from photosynthesis. Therefore, the amount of oxygen in the light bottle, minus the amount in the dark bottle, is

$$L − D = (I + P − R) − (I − R) = P$$

It is desirable to convert photosynthesis from units of oxygen to units of carbon in the photosynthetic product (e.g., sugars). To do this, we must know the ratio of the molecules of oxygen liberated during photosynthesis to the molecules of carbon dioxide assimilated. This **photosynthetic quotient** depends on the type of photosynthetic product. The photosynthetic quotient is 1 if hexose sugars are produced, 1.4 for lipids, 1.05 for protein with ammonia as the nitrogen source, and 1.6 for protein with nitrate as the nitrogen source. We must also know the **respiratory quotient**: the ratio of molecules of carbon dioxide liberated during respiration, to the molecules of oxygen used in respiration. The following relationships hold for gross primary production (GPP), respiration (R), and net primary production (NPP):

$$GPP = \frac{375(L − D)X}{PQ}$$

$$R = 375(I − D) \times RQ \times X$$

$$NPP = GPP − R$$

where I equals the initial oxygen content of the water added to the light and dark bottles, D equals oxygen in the dark bottle after a selected time period, L equals oxygen in the light bottle after the same time period, X equals the depth of a one-square-meter water column, RQ is the respiratory quotient, PQ is the photosynthetic quotient, and 375 is a conversion factor (mg oxygen to mg carbon) when the photosynthetic quotient is 1.

dioactivity is then measured in an instrument known as a scintillation counter. Productivity is proportional to the percentage of the radioactive label taken up by the phytoplankton.

The radiocarbon measurements of photosynthesis are usually lower than oxygen estimates. Phytoplankton may excrete some of the photosynthate as it is produced, and the radiocarbon technique therefore provides a good estimate of the production of particulate matter from photosynthesis. The radiocarbon technique is preferable in waters of low productivity, because of the very low changes in oxygen concentration during photosynthesis.

It is impractical to suspend bottles of radioactively labeled carbon at different depths. Instead, phytoplankton are usually incubated at optimal light intensity for the geographic location, thus revealing the maximum production for that area. If the relation of production versus light intensity is known, as well as the light intensity as a function of water depth, then one can calculate the primary productivity, integrating over the whole depth of the water column.

Satellite Color Scanning

✳ Satellite color scanners can crudely estimate relative standing stocks of phytoplankton, which can in turn be used with ocean surface data to estimate changes in primary production.

Our data on primary production of the world's oceans come from literally tens of thousands of measurements made from ships, using either seawater incubated on ship or samples directly measured within the water column. These efforts are very expensive, however, and we need a means of monitoring primary production more rapidly over large geographic scales. The only hope of achieving this is through the use of satellite data. Satellites can be outfitted with color scanners, and these have already been used to survey the ocean (Figure 10.4). The *Coastal Zone Color Scanner*, when it was in operation, measured the visible part of the spectrum of light that left the water (radiance).

With measurements of incident solar irradiance, the vertical structure of the water column, and sea temperature, it is beginning to be possible to estimate net primary production on a global scale. But there are several problems with using satellite data. First, one is attempting to gather data on light received over

a spectrum of wavelengths and to relate this radiance to the concentration of light-harvesting pigments in seawater. A set of equations that relate the concentration of photosynthetic pigments to the light received by the satellite is required. Phytoplankton and associated particles tend to absorb light in the green part of the spectrum, so relatively more blue light tends to reach the color scanner when phytoplankton are denser in the surface waters. An index of chlorophyll density can therefore be developed from the ratio of reflectance of blue to green light. However, there is a saturation effect, which prohibits the estimation of light-harvesting pigments above a few milligrams per cubic meter; thus water in phytoplankton-rich coastal seas and bays cannot be measured accurately in this way. A satellite can "see" only partly into the water column; in some parts of the ocean, much of the chlorophyll is beyond the reach of detection. Given the overall errors, this problem is probably usually a secondary one.

With the aid of the satellite's color scanner, we obtain an estimate of photosynthetic pigment concen-

---

TEXT BOX 10.2
Using the Radiocarbon Technique to Estimate Productivity

Gross productivity, *GPP*, can be estimated as follows:

$$GPP = \frac{(R_L - R_D) \times W \times 1.05}{R \times N}$$

where W is the weight of bicarbonate in water (mg C m^{-3}), N is the number of hours of the experiment, R is the counting rate expected from the entire amount of ^{14}C added to the sample, R_L is the counting rate of a light-bottle sample, R_D is the counting rate of the dark-bottle sample, and 1.05 is a constant allowing for the different rate of uptake of ^{14}C relative to the more common isotope ^{12}C.

The radiocarbon technique estimates gross productivity, but only if measurements are made quickly. If measurements are made over several hours or more, some dissolved organic carbon, including the radiolabeled ^{14}C, leaks from the phytoplankton cells.

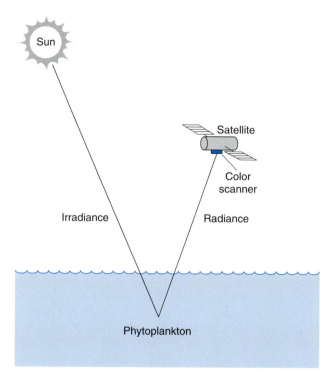

Fig. 10.4 Color detectors in a satellite measure the radiance of the ocean, or the light that is reradiated after the sun's light encounters and partially penetrates the ocean (the light that reaches the ocean is called the irradiance).

tration. However, the relationship between pigment concentration and primary production varies considerably throughout the ocean. We therefore need to find a relationship that ties together pigment concentration and primary production and is specific to a geographic region. Usually in a given region, there is a relationship between (a) the ratio of primary production to phytoplankton biomass and (b) irradiance, or light that hits the ocean surface. It is the dream of proponents of the satellite approach that a measure independent of local on-the-ground adjustments will some day be developed.

There is an overall inverse relationship between water temperature and nitrate concentration in the open ocean. In warmer seas, vertical water exchange is more limited and nutrients are relatively depleted; primary production is therefore reduced. Because satellites can also estimate sea-surface temperature, this relationship can be used in a local calibration. Even though photosynthesis increases with increasing temperature, the relationship of sea-surface temperature to nutrient concentration is more important, so, overall, primary production decreases with increasing temperature.

Oceanographers are now involved intensively with developing an algorithm to relate ocean color to primary productivity. The SeaWiFS program (**sea**-viewing **wi**de **f**ield-of-view **s**ensor), run by the National Aeronautics and Space Agency, has a new satellite in space, recording large-scale pictorial images and ocean color. Its principal purpose is to provide bio-optical data to the scientific community (see Plate VIII).

The Pump and Probe Fluorometer

* The amount of photosynthesis can be estimated by measuring the fluorescence obtained from phytoplankton that have been exposed to a sequence of light flashes.

Most estimates of photosynthesis are tedious and often involve incubations on shipboard. A device known as the **pump-and-probe fluorometer**[3] allows in situ measures of photosynthesis. The technique depends upon an important aspect of photosynthesis, which involves the transfer of electrons between different parts of the photosynthetic apparatus. It also depends upon the fact that photosynthetic pigments fluoresce when exposed to ultraviolet light. When a total phytoplankton sample is exposed to UV light it fluoresces, but the total amount of fluorescence corresponds to the total amount of pigment, not necessarily to the active photosynthesis. To actually probe the dynamic process of photosynthesis, the phytoplankton sample is exposed sequentially to three flashes of light. The details are beyond the scope of this text, but the response of the sample to the strength and the timing of the flashes allows calculation of the target area used by the phytoplankton cell for the absorption of light energy and the total photosynthetic rate.

This technique is complex in principle but extremely simple in practice. Indeed it allows the design of an instrument that can be deployed in the water to measure photosynthesis directly. The instrument can be gradually lowered through the water column to get an estimate of photosynthesis.

[3] See Falkowski and others, 1992, in Further Reading.

Geographic Distribution of Primary Productivity

Regional Variation in Productivity

* Continental-shelf and open-ocean upwelling areas are among the most productive, owing to winds that move surface water offshore and bring nutrient-rich water from below.

Continental-shelf upwelling regions have high productivity because of a consistent wind parallel or at a slight angle to the coast. Owing to the Coriolis effect, surface waters are deflected to the right in the Northern Hemisphere and to the left in the Southern Hemisphere. As the spring and summer seasons progress, the axes of the trade winds shift, resulting in the movement of the principal location of upwelling toward the pole.

The movement of water offshore results in replacement by cooler nutrient-rich water from the bottom. The bottom nutrients originate from sedimentation of phytoplankton to the seafloor. The regeneration of nutrients to the surface fuels high primary production, and the great fisheries of the world are located in upwelling regions. The California, Peru, Benguela, and Canary currents are all eastern boundary oceanic currents.

Upwelling can occur from as deep as 200 m but usually occurs from 100 m depth or less (e.g., off the Peruvian coast). In this region, dense standing crops of phytoplankton and anchovy occur. The anchovy further serve as food for tuna, and these species form two of the world's great fisheries. Anchovy populations also feed large populations of birds, whose guano production covers the famous guano islands off the coast of Peru.

Upwelling in Antarctic seas and in the eastern equatorial Pacific also fuels high phytoplankton production. In some areas upwelling is seasonal. In the monsoon regions of southern Asia periods of high production alternate with low production seasons. In upwelling areas, production can exceed 1.5–2 g C m^{-2} d^{-1} for protracted periods. Strong upwelling can also occur adjacent to submarine ridges, as well as in areas of strong currents, such as the Faroe–Iceland Ridge, where production equals 2.5 g C m^{-2} d^{-1}.

* Coastal areas are nutrient rich and productive.

Because of the shallowness of the water and the regeneration of nutrients from the bottom, waters close to shore are generally highly productive. At the shelf–slope break, intrusions of nutrient-rich slope water can fuel production in shelf waters. Much of the inner-shelf phytoplankton is not consumed by the zooplankton, and they sink to the bottom. On productive shelves, much of the food of the benthos comes from such sedimentation of uneaten phytoplankton. Some of the surface phytoplankton over the shelf–slope break may be exported to deep-sea bottoms, but probably far less material reaches the abyss.

Estuaries in Georgia are normally rich sources of nutrients, but the water is usually turbid and the depth of active photosynthesis is relatively shallow. Outer-shelf waters are clear, but nutrients are low and production is relatively low. Inner-shelf surface waters combine the best of both worlds: the water is relatively clear and nutrient rich. The combination causes higher primary productivity than is the case for either estuarine or outer-shelf surface waters (Figure 10.5).

In shallow coastal waters, sea grasses and seaweeds may be the dominant forms of primary production. In the tropics, turtle grass beds (*Thalassia testudinum*) are very common in waters of a couple of meters depth or less; eelgrass (*Zostera* spp.) dominates in higher latitudes. Such beds have primary production often greater than 1,000 g C m^{-2} y^{-1}. Kelp forests dominate the shallow coastal waters of wave-swept, high-latitude hard bottoms and match sea grass beds in production (see Chapter 15).

* At convergences and fronts, nutrients are concentrated and primary productivity is high.

A **front** is any rapid geographic change of seawater properties. Horizontal variation in hydrography has a significant effect on patterns of production. Surface convergences concentrate nutrients and plankton. Seabirds, fishes, and fishing boats congregate at convergences to take advantage of the food supply. Such fronts may be temporary (as are intrusions of slope water moving onto the shelf) or long-lived (as are some fronts off the shelf–slope breaks near Nova Scotia and near the southern Bering Sea). In general, fronts increase primary productivity when nutrient-rich water is upwelled from a deep source and is transported into a relatively stable shallow part of

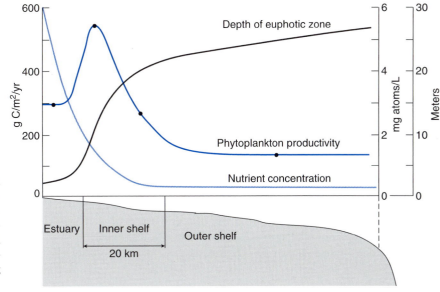

Fig. 10.5 Geographic variation in euphotic zone depth, nutrient concentration, and primary productivity, on a transect from the coast of Georgia, USA to the outer continental shelf. (After Haines, 1979, with permission of the Plenum Publishing Corporation.)

the water column, such as a warm surface layer. This is commonly the case in the surface waters over the shelf–slope regions, where deep nutrient-rich water seaward of the continental slope moves upward over the shelf–slope boundary region to fuel primary production in the surface waters.

In the southern Bering Sea, a series of fronts divides the ocean into distinctive food webs. A middle-shelf zone supports the production of large diatoms that are not grazed completely and settle to the bottom and provide food for benthic invertebrates. The outer-shelf phytoplankton are grazed in the water column efficiently, and the high fish production supports large populations of seabirds and mammals.

✱ Central oceans and gyre centers are nutrient poor and relatively barren of primary productivity.

Productivity is extremely low between latitudes 10 and 40 within tropical gyres, such as the Sargasso Sea and the North Pacific Gyre. The nutrient regime is poor, but is compensated by the year-round growing season and the great euphotic depth. Primary production is about 50 g C m^{-2} y^{-1}. An important feature of such regions is the permanence of the thermocline. The density stratification of the water column reduces the potential for nutrients to be regenerated from deeper water. Near the Bermuda islands, the thermocline breaks down during the winter

and early spring, resulting in upwelling and higher rates of primary productivity than in surrounding deep-water areas.

Productivity by Region

✱ Different oceans have different overall levels of primary productivity, which are determined by latitude, ocean basin shape, wind-driven surface currents, and the influence of surrounding continents.

Atlantic Ocean

The North Atlantic and the South Atlantic are both characterized by large circulating gyre current systems. **Eastern boundary currents**, such as the Benguela current of South Africa, produce a large upwelling system in the southern Atlantic. The Benguela current runs along the west coast of South Africa, and then the south equatorial current crosses the southern Atlantic Ocean westerly toward South America, just south of the equator. Finally, the Brazil current moves southerly along the coast of South America. The center of the South Atlantic Ocean at midlatitude thus consists of a large counterclockwise eddy. A similar clockwise eddy is found in the North Atlantic, in the Sargasso Sea, where daily primary production is less than 0.1 g C m^{-2} d^{-1}.

PACIFIC OCEAN

The Pacific Ocean is divided into the North Pacific and the South Pacific, and its pattern of water circulation and production is similar to that of the Atlantic, with low production in the ocean centers and higher production on shelf waters. The northern part of the North Pacific differs from the Atlantic in that there is no noticeable burst of phytoplankton biomass in the spring, although primary productivity does increase. It is believed that zooplankton predation in this part of the Pacific prevents phytoplankton biomass from accumulating, but it has been suggested that iron may be a limiting nutrient. We have already discussed the great production in the Peru upwelling system. There is also an upwelling system off the west coast of North America. Overall, the Pacific is more productive than the Atlantic.

INDIAN OCEAN

The rate of primary production in the open waters of the Indian Ocean is fairly low (about 0.2 g C m^{-2} d^{-1}), as in the South Atlantic. North of the equator, however, very high rates are found off the coasts of Somalia and India, and in the Arabian Sea. Winds called monsoons shift seasonally. Production rates vary with alternating periods of upwelling and influx of nutrient-poor surface water from offshore. During periods of upwelling, organic matter settles slowly in the thermocline region, and bacterial respiration causes the development of strong oxygen minimum layers.

ANTARCTIC OCEAN

One of the most productive areas of the world, the Antarctic supports a dense diatom flora and pro-

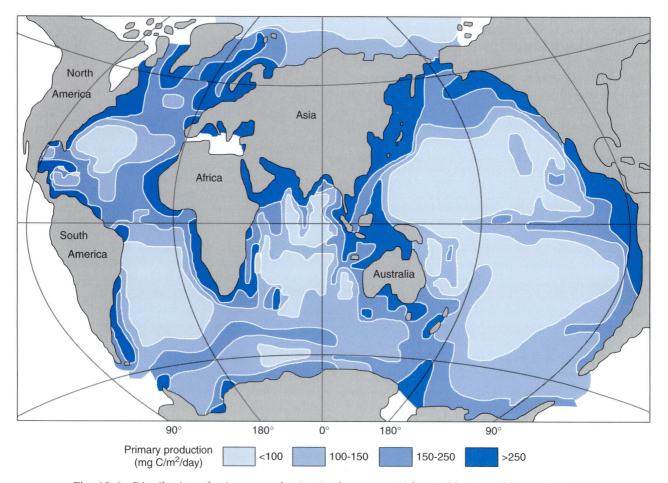

Fig. 10.6 Distribution of primary production in the oceans. (After Koblentz-Mishke et al., 1970.)

duction is probably 1 g C m^{-2} d^{-1}during the production season of about 100 days. A slow upwelling of nutrient-rich water fuels the high productivity. Phytoplankton production can proceed under the ice.

ARCTIC OCEAN

Production can be high in the Arctic Ocean in summer. Year-round figures are low, however, because of the short production season. Levels of 1 g C m^{-2} d^{-1} can be reached.

EQUATOR

Upwelling a few degrees on either side of the equator fertilizes tropical ocean waters in both the Pacific and Atlantic. In the Pacific Ocean, for example, there are two westward-flowing currents, one on either side of the equator. The Coriolis effect causes a net southerly transport of surface waters south of the equator and a net northerly transport north of the equator. The movement of surface waters away from the equator in the eastern Pacific causes upwelling of nitrogen-rich water from greater depths. Productivity varies around 0.3–0.4 g C m^{-2} d^{-1}. In the western Pacific, nutrients are far lower in concentration.

WORLD PRODUCTIVITY

Figure 10.6 shows the overall pattern of productivity in the ocean. The overall world production is computed on the basis of oxygen evolution measurements (gross production) to be 100–200 g C m^{-2} y^{-1}; on the basis of radiocarbon estimates (net production estimate), the computed range is 25–35 g C m^{-2} y^{-1}. John Ryther[4] has estimated that a realistic world average is about 50 g C m^{-2} y^{-1}. Overall, satellite-integrated scans of chlorophyll between 1978 and 1986 conform well to this regional map, which is based upon thousands of "ground-level" estimates. Satellite data suggest that world oceanic net primary productivity is in the area of 140 g C m^{-1} y^{-1}, which is a good deal larger than some previous estimates.[5] Any of these estimates can be used to show that all the nitrogen and phosphorus in the ocean must be used and recycled many times over. The supplies of nitrogen and phosphorus, therefore, limit overall world productivity.

[4] See Ryther, 1969, in Further Reading.
[5] See Field and others, 1998, in Further Reading.

Further Reading

Barber, R. T., and F. P. Chavez. 1991. Regulation of primary productivity rate in the equatorial Pacific. *Limnology and Oceanography*, v. 36, pp. 1803–1815.

Bargett, A. E. 1991. Physical processes and the maintenance of nutrient-rich euphotic zones. *Limnology and Oceanography*, v. 36, pp. 1527–1545.

Brown, O. B., R. H. Evans, J. W. Brown, H. R. Gordon, R. C. Smith, and K. S. Baker. 1985. Phytoplankton blooming off the U.S. east coast: A satellite description. *Science*, v. 229, pp. 163–167.

Cushing, D. H. 1975. *Marine Ecology and Fisheries.* Cambridge: Cambridge University Press.

Esaias, W., G. Feldman, and C. McClain, and R. Evans. 1986a. Global distribution of marine primary production derived from CZCS observations for 1969. *Eos*, v. 68, p. 1703.

Esaias, W., G. C. Feldman, C. R. McClain, and J. A. Elrod. 1986b. Monthly satellite-derived phytoplankton pigment distribution for the North Atlantic Basin. *Eos*, v. 68, pp. 835–837.

Falkowski, P. G., ed. 1980. *Primary Productivity in the Sea.* New York: Plenum.

Falkowski, P. G., and A. D. Woodhead, eds. 1992. *Primary Productivity and Biogeochemical Cycles in the Sea.* New York: Plenum.

Falkowski, P. G., R. M. Greene, and R. J. Geider. 1992. Physiological limitations of phytoplankton productivity in the ocean. *Oceanography*, v. 5, pp. 84–91.

Falkowski, P. G., R. T. Barber, and V. Smetacek. 1998. Biogeochemical controls and feedbacks on ocean primary production. *Science*, v. 281, pp. 200–206.

Field, C. B., M. J. Behrenfeld, J. T. Randerson, and P. Falkowski. 1998. Primary production of the biosphere: Integrating terrestrial and oceanic components. *Science*, v. 281, pp. 235–238.

Fuhrman, J. A., and D. G. Capone. 1991. Possible biogeochemical consequences of ocean fertilization. *Limnology and Oceanography*, v. 36, pp. 1951–1959.

Gulland, J. A. 1972. *The Fish Resources of the Ocean.* London: Fishing News Books.

Haines, E. B. 1979. Interactions between Georgia salt marshes and coastal waters: A changing paradigm. In R. J. Livingston, ed., *Ecological Processes in Coastal and Marine Systems.* New York: Plenum, pp. 35–46.

Hardy, A. C. 1954. *The Open Sea, Its Natural History. The World of Plankton.* London: Collins.

Koblentz-Mishke, I. J., V. V. Volkovinsky, and J. B. Kabanova. 1970. Plankton primary production in the world ocean. In W. S. Wooster, ed., *Scientific Exploration of the South Pacific*. Washington, DC: National Academy of Sciences, pp. 183–193.

Lalli, C. M., and T. R. Parsons. 1997. *Biological Oceanography: An Introduction*. Oxford: Butterworth Heinemann.

Landry, M. R. 1977. A review of important concepts in the trophic organization of pelagic ecosystems. *Helgoländer Wissenschaften Meersuntersuchungen*, v. 30, pp. 8–17.

Ryther, J. H. 1969. Photosynthesis and fish production in the sea. *Science* 166:72–76.

Russell-Hunter, W. D. 1970. *Aquatic Productivity*. London: Macmillan.

Steele, J. H. 1974. *The Structure of Marine Ecosystems*. Cambridge, MA: Harvard University Press.

Review Questions

1. Distinguish between autotrophic, auxotrophic, and heterotrophic.

2. What is the difference between a food chain and a food web?

3. What is the effect of increasing numbers of trophic levels on the productivity of the organisms in the top level?

4. Why is ecological efficiency in a food chain usually much less than 100 percent?

5. What gains and losses must be accounted for in the oxygen approach to measuring primary productivity?

6. Why are dark bottles used in the oxygen technique for measuring primary productivity?

7. In the radiocarbon technique of measuring primary production, why must a correction be made for the uptake of ^{14}C, relative to ^{12}C?

8. Over the years, the oxygen production method has lost out to the radiocarbon method of measuring primary productivity. Why might this have happened?

9. What contributes mainly to the high primary productivity of coastal areas?

10. Current estimates of primary productivity using satellite imagery are very poor, yet satellite data may still be very useful relative to ground-based measurements. Why?

11. Why is it important to have a global estimate of primary productivity? What may be the advantage of having total estimates of productivity on scales more regional than that of the global ocean?

V

ORGANISMS OF THE SEABED

11

The Diversity of Benthic Marine Invertebrates

There is a fantastic diversity of marine life, especially among the invertebrates, animals without backbones. If you go to the nearest tide pool, without much effort you will find snails (Phylum Mollusca, Class Gastropoda), barnacles (Phylum Arthropoda, Subphylum Crustacea), bivalve mollusks (Phylum Mollusca, Class Bivalvia), worms of various sorts (Phylum Annelida, among others), starfish (Phylum Echinodermata, Class Asteroidea), perhaps sea urchins (Phylum Echinodermata, Class Echinoidea), chitons (Phylum Mollusca, Class Polyplacophora), and more! Not to mention the plants! This diversity is bewildering, and you may some day want to take a course in invertebrate zoology to study these animals in greater detail. For now, this chapter will serve as an introduction to many of the invertebrate groups you may encounter on the seabed and shores.

In Chapter 3, we discussed the taxonomic hierarchy. This hierarchy, as you may recall, consists of successively inclusive sets of taxa: in ascending order, species, genus, family, order, class, subphylum, phylum, and kingdom. In this chapter, we will cover invertebrates, down to the level of class for the most part, which will give an adequate introduction to marine benthic animal diversity.

A number of evolutionary changes allowed the evolution of morphological and behavioral complexity. Grade of construction is especially important because it indicates the diversity of structure. Some animals merely consist of a single cell and therefore cannot be very complex or even very large. Others are just groups of cells. A few phyla possess a **tissue grade** of construction, wherein the body consists of layers of distinct cell types. This allows for some subdivision of labor among tissues. The most complex grade of construction is the **organ grade**, wherein tissues combine to form organs such as a liver or large intestine, which can also be combined into systems of organs. Here there is a large degree of flexibility in using different cell types (skin, secretory, nerve, etc.) in combination to serve a complex function.

Complexity and different degrees of evolutionary transformation can also be found in terms of symmetry and organ types. Many invertebrates are **radially symmetrical** (Figure 11.1), which simply means that they possess no true front or rear. A sea anemone is an example of a radially symmetrical invertebrate, which can capture food from all directions. By contrast, most animals are **bilaterally symmetrical**: they have a front and a rear, as well as a top and a bottom. This form of symmetry allows directional movement, and requires processing of information as the forward part moves along. Bilateral forms therefore usually have a head with anterior sense organs.

Another indication of complexity in invertebrates is the presence of a **coelom** or internal body cavity. A "true" coelom arises within the embryonic mesoderm, whereas a "false" coelom, or pseudocoel, has a different embryonic origin. The embryonic origin is important as a tool in inferring evolutionary rela-

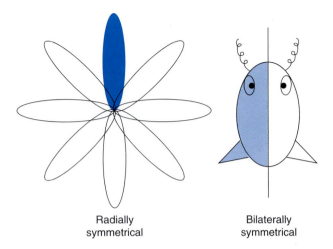

Radially symmetrical　　Bilaterally symmetrical

Fig. 11.1 A radially symmetrical organism looks the same in many directions relative to a center (one radial element is filled in green), whereas a bilaterally symmetrical animal has a front and a rear, an arrangement that promotes the evolution of sensory structures and a head.

tionships among groups. Coeloms allow compartmentalization of the body and specialization of function. They are also sometimes used as fluid-filled chambers to form a fluid skeleton, which becomes stiffened by pressure.

Before proceeding with this chapter, I would like to apologize for omitting many phyla that would only increase the student's appreciation of the diversity of the invertebrates. The worm-shaped Priapulida are relatively minor today, comprising only about 15 species. Yet they have a venerable fossil record, stretching back to the Cambrian period, nearly 600 million years ago. The Loricifera is the most recently described living phylum, and it has an eversible spiny head. Its relationships are poorly understood, but it has some resemblance to nematodes, priapulids, and kinorhynchs (yet another group we will not cover). The many phyla provide much information on the diversity of morphology and resemblances among the phyla. The interested student should consult one or more the invertebrate zoology texts cited in Further Reading, General Invertebrate Morphology.

Kingdom Protista: Single-Celled Organisms

BASIC FACTS

Taxonomic level: Kingdom Protista; grade of construction: in effect, single autonomous cells; symme-

try: variable; type of gut: none; type of body cavity other than gut: none; segmentation: none; circulatory system: none; nervous system: none; excretion: diffusion from cell surface. Other features: autonomous single cells, reproducing by fission. Number of species: over 30,000.

✻ **Protists are neither animals nor plants, but free-living, one-celled organisms.**

Protists include amoebas, ciliates, and a host of other organisms that consist of one cell living freely in the environment. All have a nucleus, cytoplasm, several types of cell organelle, and a cellular membrane. They are found abundantly in pore waters among sedimentary grains, and attached to the sediment and to hard surfaces. Most feed on very fine-grained particulate organic matter or bacteria, but some are carnivorous and feed on other Protista. Some ciliates have mouth openings lined with cilia that trap particles, whereas amoeboid forms surround and engulf food particles.

✻ **Amoeboid forms include naked amoebas, Foraminifera with calcium carbonate skeletons, and Radiolaria with silica skeletons.**

Amoebas and their allies all share the ability to stream cellular protoplasm and to form extensions, or pseudopodia, which can surround and engulf food items. Forms lacking an external skeleton can adopt many shapes (Figure 11.2) and can move in many di-

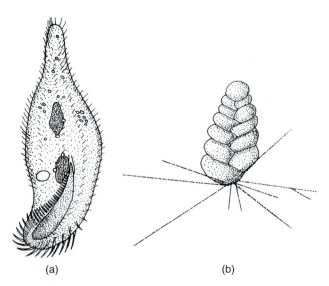

(a)　　　　　　　(b)

Fig. 11.2 Benthic protists: (a) ciliate and (b) foraminiferan.

Fig. 11.3 A xenophyophorid, a colonial foraminiferan from the deep sea that agglutinates sedimentary particles. This one can be seen in the core (7 cm diameter) from which it was extruded. (Photograph courtesy of Lisa Levin.)

rections. They usually live in pore waters or on the sediment surface. Foraminiferans construct beautiful chambers of calcium carbonate, which may be laid out in a row or in a spiral. The chambers may also be ornamented with spines, which deter predators from ingesting the organisms. Foraminiferans are usually quite small—less than 1 mm in length—but some such as the deep-sea xenophyophorids are colonial and quite large (Figure 11.3). Protoplasm streams out through holes in the hard-walled chambers, and food particles that stick to the pseudopodia are captured and engulfed. Some foraminiferans have associated with them symbiotic algae, known as zooxanthellae, which transfer sugars to their protistan hosts. Like the xenophyophorids, these foraminiferans also are larger than most, and some extinct forms reached several centimeters in length. Radiolarians, another group of amoeboid protists, have elaborate hard skeletons made of silica and long thin pseudopodia that extend outward from a central body.

✳ **Ciliates are elongate and their outer body is covered with cilia, which are used for locomotion.**

Ciliates, represented by the familiar freshwater genus *Paramecium*, are usually elongate, with a mouth at one end (Figure 11.2a). The cellular membrane is lined externally with cilia, whose coordinated beats propel the protist along. Depending upon the species, the mouth is of varying shape and may or may not be lined with cilia. Beating of the mouth cilia brings food particles such as bacteria toward the mouth. Cil-

iates are found in all marine habitats and are especially abundant as interstitial animals in sediment pore waters.

✳ **Flagellates also are elongate, but they are propelled by one or a few flagellae instead of by many cilia.**

Flagellates are similar to ciliates, except for their method of locomotion. Long flagellae, usually very few in number, are embedded externally in the cellular membrane, and these beat to move the organism. Like ciliates, flagellates usually have a mouth and consume bacteria or small organic particles. Some are carnivorous and feed on other protists. A host of photosynthetic phytoplankton are flagellates, and these are discussed in Chapter 7.

Phylum Porifera: Sponges, Simplest of Animals

BASIC FACTS

Taxonomic level: Phylum Porifera; grade of construction: cellular, with no distinct tissues or organs; symmetry: variable; type of gut: none; type of body cavity other than gut: none; segmentation: none; circulatory system: none; nervous system: none; excretion: diffusion from cell surface. Other features: flagellated cells called choanocytes drive water through pores and cavities. Number of species: 5,000.

✳ **Members of the phylum Porifera possess structures consisting of groups of flagellated cells, which move water and food particles into open chambers.**

Sponges, which comprise the phylum Porifera, are extremely simple animals that lack organs, do not have a gut, and have poorly developed tissue layers. They have several different specialized cells that are grouped to perform different functions. Sponges have an internal space, which is made up of a series of pores, canals, and chambers (Figure 11.4). Simple flagellated **collar cells**, or **choanocytes**, line the chambers and beat water through the entire system. The simplest of sponges is a two-layered animal built around one chamber, which is lined with collar cells. The outer part of the animal is a layer of epithelial cells interspersed with pore cells, through which water flows with food particles to the flagellae of the collar cells. Food particles are trapped on the flagellae and engulfed by the collar cells. Another cell type,

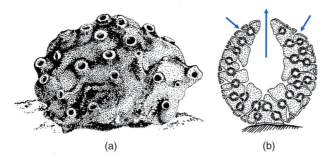

(a)   (b)

Fig. 11.4 Simplified diagrams of a sponge: (a) a whole animal and a cross section of a colony, showing chambers lined with collar cells.

the **amoebocyte**, is involved in transport of digested food to other cells in the sponge body. Amoebocytes also can transport waste products.

Although water enters through hundreds or thousands of pore cells, it leaves the central chamber through one exit, called the **osculum**, which may occur singly or as several openings. The action of thousands of collar cells moves water and concentrates it so that the wastewater leaves at a fairly high velocity through the osculum and is carried away by water currents (see Chapter 6 for a discussion of the water transport of sponges). All sponges (except one small carnivorous group) either have a simple one-chamber design or consist of hundreds to thousands of chambers in a complex array. Whereas some sponges consist of tiny filmy colonies, many are quite large and have distinct forms. Basket sponges, for example, which have a distinct vase shape, may be a meter high. For support of the colony, sponges secrete a fibrous protein known as **spongin** between the two main cell layers. Many sponges also secrete interlocking networks of needle-shaped **spicules**, which may be siliceous or calcareous. Sponges usually have several different specifically shaped spicule types, and these can be used to identify the sponge.

Sponges are capable of both asexual and sexual reproduction. Colonies can extend themselves over a surface by means of asexual reproduction. The relatively low level of organization allows extension of the colony in any direction. (To see how versatile a sponge can be, try straining the simplest of sponges through cheesecloth: the cells will reaggregate and form a new sponge. Many sponges have fairly complex chamber structure, however, and cannot perform this feat.) Sponges also produce gametes. The

eggs are usually retained within the body, to be fertilized by planktonic sperm. On coral reefs, whole sponge populations have been observed to produce a fog of sperm over the bottom, timed with the lunar cycle.

Sponges may seem to be easy targets for a predator, but many contain very toxic compounds that are distasteful and that can harm an attacker. Some of these compounds, especially those produced by some tropical sponges, are caustic to the touch.

Phylum Cnidaria: Hydrozoans, Jellyfish, Anemones, and Corals

BASIC FACTS

Taxonomic level: Phylum Cnidaria; grade of construction: two tissue layers; symmetry: radial; type of gut: blind gut; type of body cavity other than gut: none; segmentation: none; circulatory system: none; nervous system: network of nerve cells; excretion: diffusion from cell surface. Other features: two basic stages—sessile polyp and swimming medusa, both tentaculate; includes anemones, corals, sea whips. Number of species: 9,000.

✱ **The cnidarians are all built around a common cup-shaped polyp body plan, which has a ring of tentacles and a digestive tract with one opening.**

Although the cnidarians have a large range of morphologies, all are built around a basic individual. Although some are highly specialized, most are cup shaped and have a single mouth–anus opening, surrounded by a ring of tentacles. The tentacles are usually lined with several different types of **nematoblast**, which produce structures called **nematocysts**. Upon contact with prey, the nematocysts evert explosively, thrusting either a sticky extension or a sharp barb at the prey. Species may have more than one type of nematocyst. The nematocysts may be mucus coated and entrap the prey, but some have deadly poisons, which can stun and even kill prey animals. After the nematocysts have everted and made contact with prey, the tentacles draw the food item through the mouth and into the digestive cavity, where digestive enzymes are secreted. After a period of time, digested material crosses the gut cavity wall, and undigested remains are extruded back through the mouth.

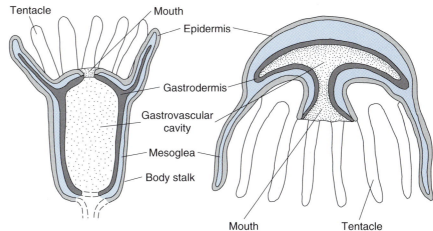

Fig. 11.5   Basic body plans and life cycle of a cnidarian, including the polyp and medusa stages.

This basic body plan (Figure 11.5), which is in the form of a **polyp**, can be used by attached benthic creatures, such as sea anemones, whose tentacles extend upward into the water column. The plan also serves, however, in free-swimming jellyfish in a **medusa** form; the tentacles of the animals hang down, and the body is free living; muscular, rhythmic contractions allow a medusa to expel water downward and thus swim upward. In groups like corals and sea anemones, the polyp is the main form in the life cycle, whereas in the true jellyfish, the medusa form is the main adult stage. In the hydrozoans like *Tubularia* (Figure 11.6), the life cycle consists of an alternation between small colonies of polyps and small medusa jellyfish stages (Figure 11.7). The jellyfish stage produces gametes, which fuse in the water column and develop into larvae, which settle and produce the polyp stage.

Many cnidarians are colonial. In the case of corals, some anemones, and the polyp stages of hydrozoans, all the polyps are essentially alike and have feeding tentacles. Many cnidarians, however, are polymorphic, and different polyp types perform functions that contribute to the functioning of the colony. Hydrozoans often have separate feeding and reproductive polyps. In the feathery *Plumularia*, the reproductive polyps are found along a central axis, and the feeding polyps are found on the projections from the central axis. The Portuguese man-of-war *Physalia* (Figure 7.7b) belongs to a group known as the siphonophores and has several distinctive types of polyp. One highly specialized polyp forms a float, from which the rest of the colony dangles into the water beneath. A disk underneath contains a common digestive cavity used by the whole colony.

✳ **Cnidarians are divided into Hydrozoa, Scyphozoa (true jellyfish), and Anthozoa (corals, anemones).**

The **Hydrozoa** are the simplest cnidarians, and most have a complex life cycle, with an alternation of a benthic colony and a planktonic small-jellyfish medusa stage. The benthic colonial stage may be polymorphic and may consist of both feeding and reproductive polyps, all connected by living tissue. The

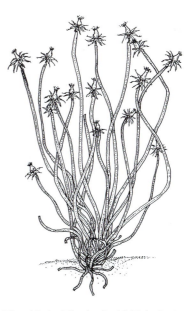

Fig. 11.6   The hydroid *Tubularia*.

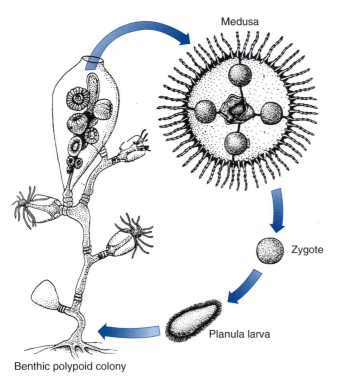

Fig. 11.7   Life cycle of a typical hydrozoan.

bush shaped, whereas others live as layers of polyps, attached to rocks or seaweeds. In a typical hydrozoan, reproductive polyps bud off small jellyfish-shaped individuals, which are known as medusae, and these medusae spawn gametes into the water column. The gametes form a larva called a **planula** (Figure 11.7), which swims for a period and settles to form the benthic colony stage.

The **Scyphozoa** are true jellyfish, which were discussed more completely in Chapter 7. They look like the medusa stage of a hydrozoan, although their digestive cavity is far more complex: they are much larger (sometimes a meter in diameter), and the periphery of the body is often covered with sensory structures. The jelly layer is also far thicker than in hydrozoan medusae. Unlike the Hydrozoa, planktonic scyphozoans have a life cycle that lacks a benthic polyp stage. Jellyfish shed gametes into the water and the embryo develops into a larva, which in turn metamorphoses into a jellyfish. Sometimes there is a very reduced polyp stage that produces the jellyfish asexually. Scyphozoans are all carnivores; they are propelled by rhythmic contractions of muscle rings that force water out of a bell-shaped structure known as a bell. After the contraction, the animal starts sinking, and tentacles draping from the bell can touch prey and fire off the nematocysts.

The **Anthozoa** include the anemones, corals, and sea fans (Figure 11.8). In these groups, the polyp stage dominates the life cycle, and there is no medusa stage. Gametes are shed by the animals, fertilization

polymorphism is developed especially well in **Hydractinia**, which lives as a coat on the snail shells occupied by hermit crabs. Polyps are specialized into feeding polyps, spines, reproductive polyps, and protective dactylozooid polyps. Some hydrozoans are

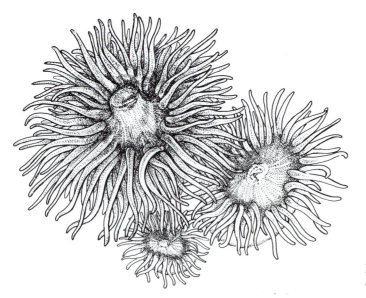

Fig. 11.8   A sea anemone, member of the Anthozoa.

occurs, and the resulting embryos develop into swimming planula larvae. These larvae settle and metamorphose into the polyp stage, which feeds on zooplankton. Anthozoans have an element of bilateral symmetry because a slitlike mouth is elongate. There is still a ring of tentacles, however, that catch differently sized prey depending upon tentacle morphology and overall polyp size. The tentacles surround the polyp and have radial symmetry. Many anemones are solitary, though some consist of many hundreds of individuals that arise by fission from one original animal. Stony corals and soft corals usually consist of hundreds to thousands of polyps, interconnected by soft tissue. Stony corals secrete calcium carbonate and may have quite large and branching skeletons, or may be mound shaped. The great diversity of stony-coral forms gives coral reefs their spectacular appearance (see Chapter 15).

Phylum Platyhelminthes: Flatworms

BASIC FACTS

Taxonomic level: Phylum Platyhelminthes; grade of construction: organs derived from three tissue layers; symmetry: bilateral; type of gut: blind; type of body cavity other than gut: none; segmentation: absent; circulatory system: none; nervous system: small bundles of nerves (ganglia), two ventral nerve cords; excretion: excretory organs in many species. Other features: flattened free-living worms, often with tubular pharynx to gather food; also, many parasitic species; representatives include the common freshwater planaria and the tapeworms. Number of species: 12,000.

* Flatworms are truly bilaterally symmetrical, with anterior–posterior differentiation. They also have distinct organs.

Flatworms represent a major transformation in evolution. Instead of having radial symmetry, they are bilaterally symmetrical, which means that their body exists in mirror images about a long anterior–posterior axis. They also have distinct anterior and posterior ends, a morphology that gives the opportunity for consistently directed forward motion. Such a motion requires the development of a battery of sensory structures for smell, sight, and touch to allow for interpretation of the environment as the flatworm

moves along. As the name indicates, the flatworms are wormlike, although the free-living forms are usually flattened. Movement is accomplished partially by means of ventral cilia in the small forms, but also by rhythmic contractions of an outer layer circular muscle and an inner layer of muscle that parallels the line of the body. Unlike the cnidarians, flatworms have distinct organs, or structures that consist of more than one kind of tissue. Examples are the excretory organs. Unlike the cnidarians, flatworms have tissues and organs that derive from three original cell layers (endoderm, mesoderm, and ectoderm) instead of from two. Flatworms have a central nervous system, with concentrations of nerve cells in the head that could be considered a brain.

Free-living flatworms (Figure 11.9) are flattened and usually have a muscular **pharynx**, which can be protruded to suck fluids from prey such as crustaceans and annelids. The animal has a blind digestive cavity, and undigested material must be egested from the same opening through which food enters. The larger forms have eyespots and sensory tentacles. Flatworms are found in crevices, under rocks, and sometimes on bare sediment surfaces.

The marine flukes parasitize many vertebrates and invertebrates. Many have complex life cycles with several hosts (see Chapter 3, Figure 3.14). Many flukes have one stage that infests a mollusk or annelid. This stage produces free-swimming individuals that enter the water column and may burrow into the legs of ducks or the skin of fishes. Eggs produced there develop into another free stage that reenters the invertebrate host. The fluke *Austrobilharzia variglandis* is an example of such a flatworm fluke that burrows into the skin of ducks or human bathers, causing "swimmer's itch." More serious are the flukes causing the debilitating disease schistosomiasis. The fluke *Schistosoma mansoni* moves between humans and a freshwater snail as hosts.

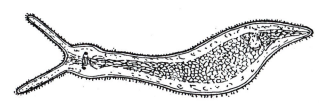

Fig. 11.9   A free-living flatworm.

Phylum Nemertea: Ribbon Worms

BASIC FACTS

Taxonomic level: Phylum Nemertea; grade of construction: organs derived from three tissue layers; symmetry: bilateral; type of gut: complete, with anus; type of body cavity other than gut: rhynchocoel surrounding proboscis; segmentation: absent; circulatory system: present; nervous system: small bundles of nerves (ganglia), two nerve cords; excretion: excretory organs in many species. Other features: elongate free-living worms, with complete gut; carnivorous, using barbed proboscis to kill prey; Number of species: greater than 800.

* **Ribbon worms have a proboscis and a complete gut and are mobile carnivores that burrow through the sand.**

Nemerteans are long, flat, carnivorous worms with remarkable powers of contraction. Some species can extend to over 10 m in length and can rapidly contract to just a few centimeters. They resemble flatworms to a degree and have an externally ciliated body; but they have a complete gut, with an anus, a circulatory system, and a pointed anterior with an elaborate proboscis for catching prey (Figure 11.10). Ribbon worms move with the aid of ciliary activity or by coordinated contractions of the body wall muscles.

When the animal is inactive, the **proboscis** is inverted and housed in a cavity separate from the mouth. The proboscis is surrounded by a fluid-filled cavity known as a rhynchocoel. When the ribbon worm detects a possible victim, the muscles contract the rhynchocoel and fluid pressure shoots out the proboscis, which is usually armed with a stylet that punctures the prey and injects a venom. The proboscis can also draw the prey to the worm's mouth.

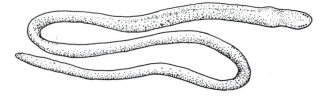

Fig. 11.10 A nemertean worm.

Phylum Nematoda: The Roundworms

BASIC FACTS

Taxonomic level: Phylum Nematoda; grade of construction: organs derived from three tissue layers; symmetry: bilateral; type of gut: complete; type of body cavity other than gut: pseudocoel; segmentation: absent; circulatory system: present; nervous system: small bundles of nerves (ganglia), two nerve cords; excretion: special excretory cells. Other features: small free-living and parasitic worms with only longitudinal muscles, circular in cross section; covered by cuticle; both free-living and parasitic. Number of species: 12,000.

* **Many nematodes live free in all marine environments, using longitudinal muscles that work antagonistically against a fluid-filled body with a rigid wall. Free-living nematodes may be carnivorous or plant eating; some even consume organic matter from sediment.**

Roundworms (Figure 11.11) are among the most widespread of all marine invertebrates and often have population densities of millions per square meter of mud. The free-living forms are generally small, usually less than 3 mm in length. They are cylindrical and have a rigid outer organic case known as a **cuticle**. Growth cannot be continuous because of this structure and the animal sheds, or molts, the cuticle four times in the adult life cycle. Unlike flatworms and ribbon worms, roundworms have longitudinal muscles only. These muscles work antagonistically against the body, which is made rigid by the cuticle and a fluid-filled cavity. The fluid-filled cavity is known as a pseudocoel because it is not located within the mesoderm, like a true coelom. The working of the longitudinal muscles pushes against sand grains and moves the animal along. Out of the sediment, the worms appear to thrash about wildly.

Depending upon the species, nematodes have a wide variety of feeding habits reflecting the different teeth and rods that project from the cuticular hard surface of the mouth. Some species can pierce algal cells and suck out the juices, and others can consume tiny invertebrates. The latter have the best developed

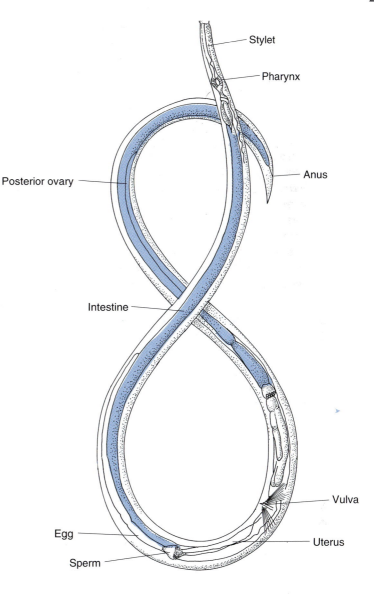

Stylet

Pharynx

Posterior ovary

Anus

Intestine

Vulva

Egg

Uterus

Sperm

Fig. 11.11 Cutaway diagram of a typical free-living nematode; such animals are found in densities of millions per square meter of mud. (After Barnes, 1987.)

teeth. Other species ingest mud and absorb organic matter from the sediment.

Phylum Annelida: Segmented Worms

BASIC FACTS

Taxonomic level: Phylum Annelida; grade of construction: organs derived from three tissue layers; symmetry: bilateral; type of gut: complete with anus; type of body cavity other than gut: coelom; segmentation: present; circulatory system: closed system; nervous system: brain, with nerve cords and bundles (ganglia); excretion: excretory organs in most segments. Other features: segmented worms, with great diversity of head and locomotory appendages; includes earthworm, sandworm, and lugworm. Number of species: 12,000.

✻ Annelids are worms divided into segments, with a tubular gut from mouth to anus and three distinct embryonic tissue layers.

Annelids include the earthworms, marine polychaetes, and leeches, and all share a wormlike form, with a mouth and anus connected by a tubular gut. The wormlike body is divided into distinct **segments**, and organs such as digestive glands, reproductive organs, and locomotory appendages are usually repeated in each segment. The segmentation of musculature allows localized contraction and the worm can crawl and even swim by sequentially moving appendages on individual segments. The body contains an interior fluid-filled space, or coelom, and the body wall muscles act against this fluid, which serves as a hydrostatic skeleton. In most of the annelids, only the head area and anal segments are differentiated (Figure 11.12). Annelids have a nervous system, which consists of a double nerve cord that extends from a brain through the length of the body. The nerve cord coordinates locomotion through the entire body.

* The annelids are divided into the Polychaeta, Oligochaeta, and Hirudinea.

The largest class of annelids is the **Polychaeta**; its members have distinct segments, each of which bears a pair of **parapodia**. The parapodia have bristles known as **setae** (Figure 11.13), which are used in burrowing, crawling, and swimming. Some polychaetes have elaborate head areas, which may have a specialized proboscis that everts and seizes prey; they may also bear external feathery gills. Some polychaetes have feathery extensions that have cilia and collect suspended food from the water column. The **Oligochaeta**, including the earthworms and other forms abundant in salt and fresh water, lack parapodia and usually have smaller setae and reduced heads. Marine forms usually feed by ingesting sedimentary organic matter. The **leeches** (Hirudinea) lack setae and use external suckers to attach so that the worm can move along by muscle contraction in the body wall. Leeches are parasites that use their suckers to feed on a host's body fluids.

* Annelid locomotion depends upon layers of longitudinal and circular muscles working against a rigid fluid, compartmentalized among many segments.

In most annelids, the body wall consists of an outer cuticle and inner skin. Beneath is a circular muscle layer and a longitudinal muscle layer, whose strands are parallel to the length of the body. By combining contractions, these two muscle layers work against the fluid-filled segment and can change its shape. In the polychaetes, this action moves the parapodia, whose setae press against the substrate and move the worm in a tube or burrow or along the surface. As the body wall muscles contract, the worm must also coordinate the protrusion and withdrawal of the parapodia and setae. Lacking parapodia, oligochaetes use setae and wriggling to burrow through the sediment. The segmentation allows localized movements, but this is absent in the leeches.

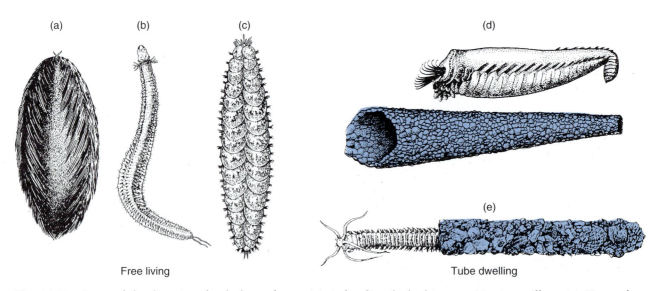

(a) (b) (c) (d)

Free living

(e)

Tube dwelling

Fig. 11.12 Some of the diversity of polychaete forms: (a) *Aphrodite*, (b) herbivorous *Nereis vexillosa*, (c) *Harmothoe*, (d) deposit-feeding *Pectinaria*, and (e) *Onuphis*.

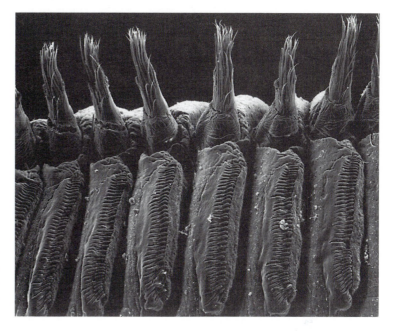

Fig. 11.13 Electron micrograph (ventral view: anterior is toward left of photo) of *Sabellastarte magnifica*, a tube-dwelling polychaete. The long capillary setae (protruding upward from the parapodia) aid the worm in moving along within its tube. The shorter hooked setae below used to hook into the tube wall to prevent removal by a predator. (Courtesy of S. A. Woodin.)

✱ Annelids adopt a wide range of living positions and include free burrowers, infaunal tube dwellers, and epifaunal tube constructors. They may be carnivores, deposit feeders, and suspension feeders.

The form and locomotory abilities of annelids have allowed them to evolve a wide range of living habits. Many annelids live freely in the sediment and may either be specialized carnivores or sediment-ingesting deposit feeders. Species of the sandworm genus *Nereis* have an eversible proboscis, which in some species is used for seizing prey and in others for tearing and ingesting fragments of seaweeds. The animal lives temporarily in mucus-lined tubes, but periodically burrows through the sediment by coordinated rhythmic muscular contractions, combined with extension and withdrawal of the parapodia and setae. A large number of species live in tubes or nearly permanent burrows. The western North Atlantic bamboo worm *Clymenella torquata* may be up to 25 cm in length and lives head down in a mud tube. The animal ingests sediment at depth and defecates on the sediment surface. Tube dwellers move rhythmically to irrigate the burrow and get rid of waste products and draw in fresh seawater.

A large number of species of annelids live in tubes but protrude a feeding organ into the water column to collect plankton. *Spirorbis borealis* secretes a small spiral tube that is cemented usually to the fronds of seaweeds. Ciliated tentacles protrude from the tube opening, and the cilia beat and drive water past the tentacles. Phytoplankton are collected and transported on a ciliated groove to the mouth. Species of the parchment worm *Chaetopterus* makes a U-shaped, paperlike tube and creates a current that is passed across a sheet of mucus stretched between a pair of specialized parapodia (see Figure 13.12a). This worm has extremely specialized body parts with very different appendages. It is so delicate that it cannot possibly live outside the tube.

Phylum Sipuncula: Peanut Worms

BASIC FACTS

Taxonomic level: Phylum Sipuncula; grade of construction: organs derived from three tissue layers; symmetry: bilateral; type of gut: complete with anus; type of body cavity other than gut: coelom; segmentation: absent; circulatory system: large coelomic cavity bathes most tissues; nervous system: brain with single ventral nerve cord, with branches; excretion: paired organs. Other features: worm shaped, living in sediment and among rocks, feeding on sediment with protrusible organ called an introvert. Number of species: about 300.

✱ Peanut worms live in burrows in soft sediment, and in rock crevices. They gather food by means of an introvert that has branched tentacles.

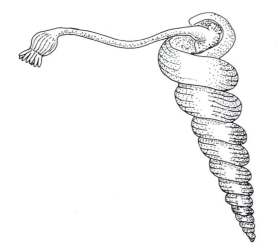

Fig. 11.14 A sipunculan, or peanut worm, protruding from the mud in a snail shell.

Sipunculans number only about 300 species; they are wormlike but are not segmented. They can be found in soft sediments and in crevices, but they are commonly seen living in the mud trapped in empty snail shells (Figure 11.14). They use fluid pressure in a large coelomic cavity to protrude the introvert, whose tip usually contains branched tentacles for feeding. Peanut worms feed on the organic matter in sediment. The introvert can be rapidly withdrawn by muscle contraction, giving the animal a bulging appearance.

Phylum Ponogonophora: Gutless Wonders

BASIC FACTS

Taxonomic level: Phylum Pogonophora; grade of construction: organs derived from three tissue layers; symmetry: bilateral; type of gut: none; type of body cavity other than gut: coelom in several sections; segmentation: present in one region; circulatory system: dorsal and ventral blood vessels, with part of the dorsal vessel muscularized into a heart; nervous system: brain with rudimentary nerve cords; excretion: paired organs. Other features: solitary worm-shaped individuals with no gut; rely upon symbiotic bacteria for nutrition. Number of species; about 100.

✱ The members of the phylum Pogonophora are generally deep-sea species that lack a gut and depend upon symbiotic bacteria.

If they did not exist, the Pogonophora would have to be invented to satisfy our imagination, because they surely contradict our intuition of what an animal should be. There are only about 100 species, and the phylum was only recently recognized. Generally, they are long and slender worms (Figure 11.15). For years, zoologists working in museums had only the anterior portions to examine. An anterior head bears sometimes as many as 200 tentacles that protrude into the water. This is set aside distinctly from a trunk region, whose anterior is ringed with hard plates, somewhat similar to the setae of annelids. A third section is segmented, and each segment has several stiff rods similar to setae. This arrangement allows the animal to anchor itself in a tube.

The most conspicuous feature is what is missing: a gut. For years, zoologists thought that pogonophorans could survive only by absorbing dissolved organic matter from seawater, and such matter may well be part of the nutrition of this animal. However, it has been recently discovered that a part of the trunk, the trophosome, contains large numbers of internal symbiotic bacteria. The animal may derive nutrition by digesting the bacteria or, alternatively, dissolved substances may leak from the bacteria.

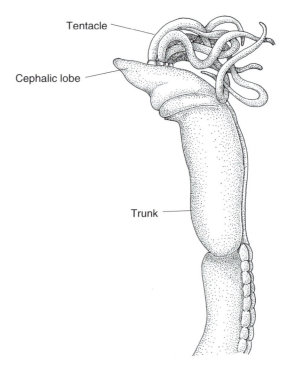

Fig. 11.15 A pogonophoran worm. (After Meglitsch and Schram, 1991.)

Recently, gigantic relatives of pogonophorans were discovered near hot vents adjacent to volcanically active parts of the seafloor (see Chapter 16). These **Vestimentifera** have been elevated to phylum status, and individuals can be as much as a meter in length and a centimeter in diameter. They secrete tubes, often 2 m long. They, too, have symbiotic bacteria living within the body. The bacteria are chemosynthetic and depend upon sulfide delivered by the worm's circulatory system, which contains a special hemoglobin with a binding site for sulfide. The worms have a ciliated free-living trochophore larva, whose morphology suggests a possible evolutionary relationship with the annelids.

Phylum Mollusca: Shelled Invertebrates (Mostly)

BASIC FACTS

Taxonomic level: Phylum Mollusca; grade of construction: organs derived from three tissue layers; symmetry: bilateral; type of gut: complete, with anus; type of body cavity other than gut: coelom; segmentation: absent; circulatory system: usually open to large coelomic cavity; nervous system: brain, with nerve cords and bundles (ganglia), brain very well developed in squids and relatives; excretion: excretory organs. Other features: typically externally shelled, mantle secretes shell; respires with ctenidium; includes clams, snails, squids, octopus. Number of species: more than 100,000.

✳ **Mollusks have a head–foot complex, a mantle that usually secretes a calcium carbonate shell, and a gill, suspended in a mantle cavity, which is used for respiration and commonly for suspension feeding.**

The mollusks are one of the most successful phyla, and there are over 100,000 living species. All are believed to have evolved from a primitive form with a head–foot complex, a cap-shaped shell, and a posterior gill that could be used both for respiration and to collect phytoplankton on ciliated tracts. The mollusks have a coelom, or fluid-filled body cavity, used in clams as a hydraulic device to burrow into the sediment. Most have a shell, secreted by the **mantle**. From this basic form sprang the supposedly ancestral **Monoplacophora**; the **Gastropoda** (snails), with their usually coiled shell and varied feeding types; the

Bivalvia (clams, oysters, and mussels), with two symmetrical shells and a variety of burrowing and epifaunal forms; the **Cephalopoda** (squids and octopus), with arms, and the ability to move rapidly; the **Polyplacophora** (chitons), which have a flattened foot; and the **Scaphopoda**, which have a tusk-shaped single external shell.

✳ **Members of the class Bivalvia are distinguished by two symmetrical shells connected in a hinge region. The mantle secretes the shell and a gill ctenidium usually helps in respiration and collects phytoplankton on ciliated tracts.**

The **Bivalvia** (Figure 11.16) are distinguished by two shells, or valves, usually mirror images of each other, hinged together by calcified teeth and by a tough or-

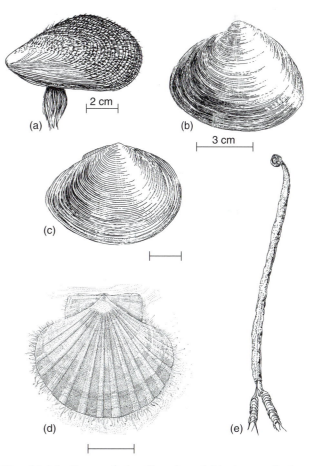

Fig. 11.16 Some of the diversity of bivalve mollusks: (a) mussel *Crenimytilus*, (b) suspension-feeding clam *Mactra*, (c) deposit-feeding clam *Macoma*, (d) scallop *Pecten*, and (e) boring bivalve *Bankia*.

ganic **ligament**, which tends to spring the shell open. In life, muscles counter this force and draw the valves closed. The mantle lines the valves and secretes new shell. Most bivalves have a powerful **foot**, which can be filled with fluid. The foot probes into the sediment and expands, whereupon the shell is drawn in behind. In some epifaunal forms such as mussels, the foot is strongly reduced and merely secretes byssal thread that glues the animal to the rock substrate.

Most bivalves have a **mantle cavity**, and water is drawn through siphons across a gill or ctenidium (Figure 13.12c). Phytoplankton are collected on the gill, and ciliated tracts pass the food to a **palp**, which also is ciliated and passes particles into the mouth and digestive system. The siphons connect the infaunal bivalves to the sediment surface. Some bivalves feed directly on the organic matter in sediment. In clams of the superfamily Tellinacea, the siphons are separate, the ventral siphon vacuums the surface sediment, and particles are sorted on the gill (Figure 13.7c). In nut clams (*Nucula, Acila*), palp tentacles collect sediment and the particles are passed directly to the mouth. Burrowing bivalves vary in shape, depending upon their speed of burrowing. Rapid burrowers tend to be elongate and smooth, whereas poor burrowers are stubbier and may be ornamented. Although most bivalves are burrowers, some (e.g., mussels) live on the surface, and the shipworm *Teredo*

bores into wood by means of mechanical abrasion and use of the wood-digesting enzyme cellulase (Figure 13.19). These bivalves have symbiotic bacteria that can convert nitrogen into useful forms for the bivalve. The borer *Lithophaga* can penetrate limestone and has special glands that secrete acid.

✳ Members of the class Gastropoda have a flattened foot, usually a cap-shaped or coiled shell, and a mouth apparatus known as a radula. They are characterized by a twisting of the body, known as torsion.

Gastropods (Figure 11.17) are usually distinguished by a flattened foot and a coiled shell, which can range from the high-spired (tall and pointy) type to low and flattened forms. The shell has one opening, and the animal often has an oval and stiff organic door, known as an **operculum**, which can seal in the animal, protecting it from predators and from drying out. Within the shell, the soft body is twisted, so that the anus can exit through the single shell opening; this condition is known as **torsion** and is unique to the gastropods. In the 1920s the distinguished embryologist Garstang explained torsion as a larval defense, because it allows a larva first to pull its head into the shell, and then to seal off the aperture with the foot and a hard operculum. This idea has been

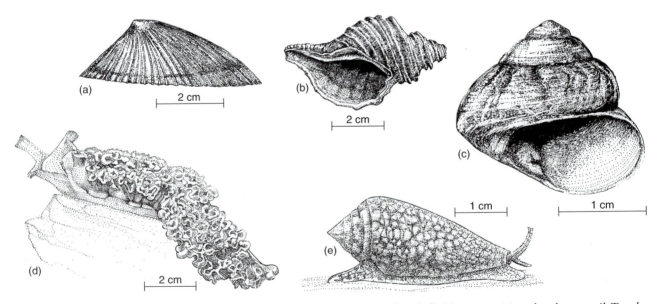

Fig. 11.17 Some of the diversity of gastropods: (a) limpet *Notoacmea*, (b) whelk *Neptunea*, (c) rocky-shore snail *Tegula*, (d) nudibranch *Tridachia*, a shell-less gastropod, and (e) carnivorous *Conus*.

questioned,[1] since experiments have failed to show an advantage against predators for larvae that had undergone torsion, as opposed to those that have not. Most larvae are swallowed whole by planktivores in any event, so torsion is still a bit of a mystery. Gastropods have a mouth apparatus that is distinguished by a **radula,** or tooth row, which moves back and forth over the food. The teeth differ according to the food source. Thus in herbivorous snails such as periwinkles and limpets, the tooth row scrapes attached microalgae from the rock surface (Figures 11.18 and 13.18a); in carnivores, the teeth are often fewer and stronger. Some species of the cone shell *Conus* have special modified radular teeth that inject a deadly poison, which is immediately fatal to the prey (and sometimes to humans!). Some snails (e.g., moon snails and dog whelks) have a mouth apparatus that can drill holes into calcium carbonate shells.

* Members of the class Polyplacophora (chitons) have a flattened foot, a radula, and eight dorsal articulated plates.

Chitons (Class **Polyplacophora**) (Figure 11.19) are found commonly on hard substrates; they are oval shaped, flattened, and eight articulated dorsal plates cover the dorsal mantle. Chitons have a flattened foot that can adhere to the rock by suction, and a mouth with a radula that is generally similar to that of gastropods. All chitons are herbivores that feed on microalgae and seaweeds attached to rocks.

* Members of the class Monoplacophora include simple animals with cap-shaped shell, posterior gill, flattened foot, and radula.

The **Monoplacophora** are represented by just a few species, but they are believed to have traits very similar to those of the ancestor of the mollusks. The animals have a cap-shaped shell and a flattened foot, with a radula in the mouth apparatus. The taxonomic class was once thought to have only fossil representatives, but the group was "rediscovered" in a dredge sample taken in the Danish *Galathea* expedition of 1952. Many believe that the presence in the "living fossil" *Neopilina galathea* of several examples of repeated structures (gills, muscles, auricles, nerves) links the Monoplacophora to other segmented groups, such as annelids.

* Members of the class Cephalopoda are distinguished by elaborate nervous and muscular coordination, the presence of grasping arms, and a carnivorous feeding mode.

[1] See Pennington and Chia, 1985, in Further Reading, Mollusca.

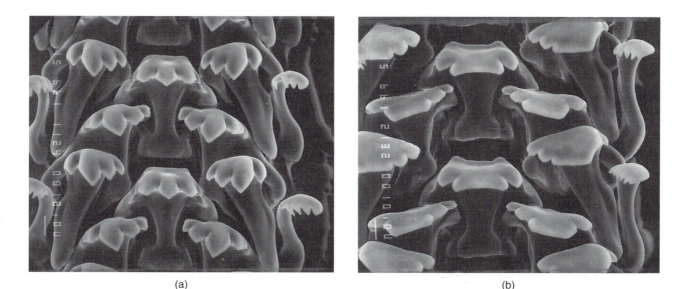

(a)   (b)

Fig. 11.18 Gastropods have a radula, which is a sliding belt of teeth. The snail *Lacuna variegata* is small and feeds on microalgae on sea grasses and hard surfaces, as well as on seaweeds. Tooth shape is determined by diet, and the snail develops pointier teeth (a) when feeding on seaweeds, but flatter, shovel-shaped teeth (b) when scraping microalgae from surfaces. Scale bars are 10 μm. (Photographs by Dianna Padilla.)

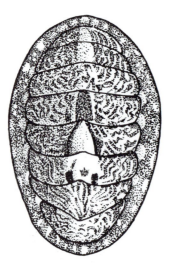

Fig. 11.19 Shell of the chiton *Tonicella*.

Fig. 11.20 *Octopus* is a major predator of benthic invertebrates, including crabs and mollusks. (Photograph by Paulette Brunner, with permission from Friday Harbor Laboratories.)

The **Cephalopoda**, which include squids, octopods, and the pearly *Nautilus*, represent the peak of invertebrate evolution in terms of nervous organization and behavioral complexity. They have the basic molluscan body plan, including a mantle, external shell or internal shell remnant, and a foot that is modified into a water-squirting funnel. They differ considerably from the other mollusks, especially because the head–foot complex is usually in line with the rest of the body (dominated by the mantle) and because they have arms, often with suckers, that can seize prey. All cephalopods have well-developed nervous systems, and, with the exception of *Nautilus*, have an eye with cornea, iris diaphragm, lens, and retina. Octopods (Figure 11.20) are locally important predators and usually live in crevices. Individuals often specialize on specific prey, such as individual clam or crab species. Because cephalopods live mainly in the water column, the subgroups are treated in more detail elsewhere (see Chapter 8).

✱ **Members of the class Scaphopoda (tusk shells) have an elongate conical shell and live buried within the sediment, feeding on foraminiferans and other small animals.**

Scaphopods (Figure 11.21) number only a couple of hundred species and live in sand and mud. They secrete a tusk-shaped shell, usually less than 10 cm in length. They have a foot resembling that of bivalve mollusks, and they burrow into the substratum. Food is collected by means of tentacular structures, which are extended and probe into the sediment searching for small-animal prey. The animals have a radula, which pushes food into the mouth.

Phylum Arthropoda: Jointed Appendages

BASIC FACTS

Taxonomic level: Phylum Arthropoda; grade of construction: organs derived from three tissue layers; symmetry: bilateral; type of gut: complete with anus; type of body cavity other than gut: coelom; segmentation: present; circulatory system: usually open to large coelomic cavity; nervous system: brain, with nerve cords and bundles (ganglia), compound eyes; excretion: excretory organs. Other features: external cuticle, jointed appendages; includes horseshoe crabs,

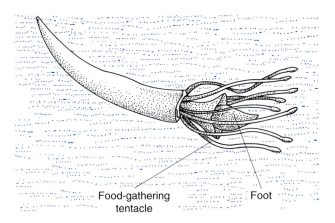

Food-gathering
tentacle   Foot

Fig. 11.21   A scaphopod in living position.

shrimp, crabs, sow bugs, insects. Number of species: over one million.

✱ Arthropods are characterized by an external cuticle of chitin, as well as by segmentation and jointed appendages.

The phylum **Arthropoda** contains the largest number of living species and includes the insects, arachnids (spiders), millipedes, and centipedes. In the marine environment, arthropods are represented mainly by the subphylum **Crustacea**, which includes shrimps, crabs, and lobsters. Also present in marine environments is the class **Merostomata**, or horseshoe crabs, and the class **Pycnogonida**, or sea spiders. An extinct class, the **Trilobita**, dominated ancient shallow seas, starting with their appearance in the early Cambrian period (roughly 500–550 million years ago) and ending with their extinction in the Permian period (roughly 225 millions years ago).

All arthropods have an external flexible skeleton composed mainly of the polysaccharide **chitin**, although it is calcified in many species. The skeleton is jointed, and thinner spots allow flexing, especially at the jointlike areas in the limbs. The body is segmented but usually consists of a main cephalothorax (head–trunk) region and a multisegmented abdomen. The head contains elaborate sensory organs and mouth parts that process food. The thorax usually contains a series of paired walking limbs, which are very differently formed in different groups, depending upon life habit.

Unlike some other phyla (e.g., annelids, mollusks), which use a hydrostatic coelom as a skeleton, the arthropods have a relatively rigid external skeleton.

Muscles are inserted into the interior of the skeleton, and limbs move and operate as a series of lever systems. The skeleton allows for precise movement. But since the animal has a rigid outer case, it must periodically molt the external skeleton in order to grow. After molting, the animal takes on water and increases body volume. It then forms a new cuticle on the larger "frame." In the case of the hermit crab, growth occasions the need of a larger shell house, and the animal's growth schedule is affected by shell availability. Successive molts in arthropods often are quite different. Different stages produced by the molts may vary in features such as swimming appendages and in terms of required foods. Most marine arthropods have a larval stage that is morphologically quite different from the adult stage, which in many species does not appear until after a major reorganization of the body, or metamorphosis.

Arthropods are sensory animals and usually have antennae that are covered with taste and smell receptors. A lobster has a set of antennae that is covered with tens of thousands of sensory nerves that can smell food at very low concentrations. Most have eyes of two types. **Compound eyes** consist of several units joined together and giving the appearance of a honeycomb. Each unit collects, focuses, and transmits light, and has separate nerve receptors. The animal pieces together the images, in mosaic fashion, to form a compound image. Some arthropods also have single eyes that are usually capable of detecting only the presence and strength of light. These eyes are usually used for orientation. Arthropods have well-developed nervous systems, and many species are capable of exchanging salts with the water, through gills and excretory organs.

✱ The Trilobita (Subphylum Trilobitomorpha) are an extinct class whose members had relatively unspecialized appendages and a compact body.

Trilobites come from an oceanic world no longer existing. They dominated the seas of the early Paleozoic era and survived until its end. Trilobite literally means "three lobes," signifying the three body sections (head, thorax, telson) that characterize the group (Figure 11.22). Paleontologists have found a few specimens with the appendages intact, and they report that the trilobites were far simpler in structure and more unspecialized than their living arthropod relatives. Over time, arthropods have evolved more specialized and different types of limbs. In the Pale-

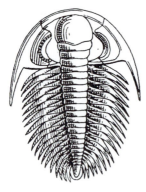

Fig. 11.22 The trilobites are now extinct, but they were dominant invertebrates in early Paleozoic times. Pictured is the Middle Cambrian *Paradoxides*.

ozoic era, trilobites dominated the marine arthropods, but became extinct at the era's end.

✳ **The subphylum Chelicerata includes the horseshoe crabs, spiders, terrestrial scorpions, and pycnogonids (sea spiders). These animals are characterized by a first pair of movable claws and a division of the body into two general sections.**

Chelicerates are more familiarly represented by the terrestrial spiders and scorpions than by the marine horseshoe crabs and pycnogonids (sea spiders). Chelicerates are distinguished by a division of the body into two major sections. The first section has six pairs of appendages, but the first pair consists of **chelicerae**, or movable claws, quite different from the first appendages of the other major marine arthropod group, the crustaceans.

In the **Merostomata**, which include the horseshoe crabs, the chelicerae and another pair of appendages handle and tear apart food before passing it to the mouth. The body is covered by a hard, calcified shield. There are two compound eyes on the dorsal surface. The second section of the body has several appendages, including highly modified, leafy book gills, which aid in swimming. The animal is often more than half a meter long and has a long tail spine, which can aid in righting the animal if it is flipped over (Figure 11.23). Horseshoe crabs prey on invertebrates living in sand- and mudflats. **Pycnogonids** (sea spiders) are far less conspicuous. They are spidery in appearance and have four pairs of walking legs. A sucking tube protrudes from the head. Sea spiders are often found perched upon colonial invertebrates and seaweeds, from which they suck fluids. Chelicerae are absent in many species of sea spiders.

✳ **The subphylum Crustacea is the largest marine arthropod group. Its members are characterized by a head with two pairs of antennae three pairs of mouthpart appendages, and a trunk with several specialized appendages. The trunk is sometimes divided into a thorax and a posterior abdomen.**

The crustaceans number about 50,000 species and include the lobsters, crabs, copepods, and shrimp. All have a head with two pairs of antennae and specialized mouthpart appendages, which process food before it enters the mouth. The antennae are so densely covered with taste and smell receptors that they may be able to locate live food in the dark while several tens of meters downstream from it. All crustaceans have feeding, walking, or swimming appendages on the trunk. In the **decapods** (crabs and lobsters) there are five pairs of appendages, but they are specialized differently in each group. In the crabs, the first pair consists of claws, and the remaining four pairs are walking legs. Lobsters (e.g., the New England lobster *Homarus americanus*) may have a large pair of claws and a couple of other pairs of legs armed with pincers that handle food, or they may have two or three anterior pairs of legs with small pincers (e.g., the spiny Pacific lobster *Panuliris interruptus*). In many species, the posterior part of the trunk contains a distinct area known as the abdomen. In the true crabs, this is bent over and more or less fused with the trunk, but in crayfish the abdomen has ap-

Fig. 11.23 A copulating pair of horseshoe crabs, *Limulus polyphemus*. (Photograph by the author.)

pendages used for respiration and egg incubation, and a couple of flattened appendages that can flex and move the animal rapidly backward. There is not enough space to describe crustaceans in detail, but Figure 11.24 shows the range of diversity of body shapes and appendages.

Crustaceans have a wide variety of life habits (Figure 11.24). For example, species of the burrowing shrimp *Callianassa* live in a series of interconnected burrows, sometimes a meter below the sediment surface. The ellipsoid *Emerita* burrows rapidly on wave-swept beaches and migrates up and down with the tides. The dorsoventrally flattened isopods live on hard surfaces and among debris, as do the laterally com-pressed amphipods. Crabs and lobsters move along the surface, walking on their legs and leaping by rapid flaps of their posterior appendages. There are also planktonic forms, as we discussed in Chapter 7.

Benthic crustaceans, such as many of the crabs, are voracious carnivores, using their claws to seize, crush, and tear apart prey. The Atlantic and Gulf coast blue crab *Callinectes sapidus* (Figure 13.17d) is abundant in western Atlantic estuaries; during its annual migration up the estuary, the Chesapeake Bay population devastates bivalve mollusk populations. Other species are scavengers and even eat sediment, digesting the organic debris contained in it. Fiddler crabs (genus *Uca*) (Figure 5.1) feed on sediment by using

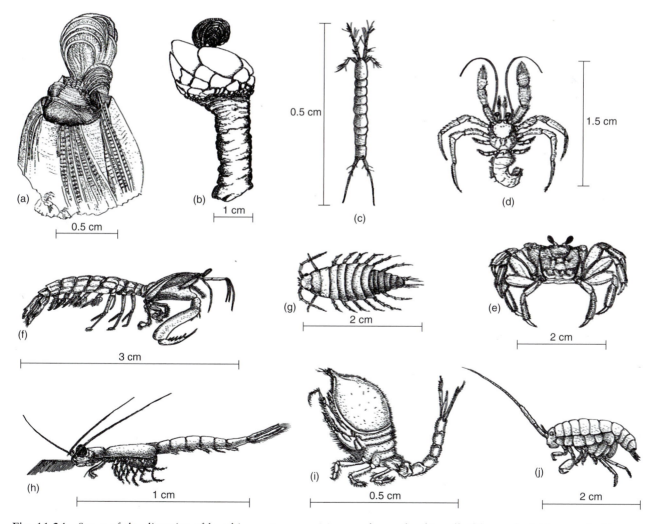

Fig. 11.24 Some of the diversity of benthic crustaceans: (a) acorn barnacle, (b) stalked barnacle, (c) harpacticoid copepod, (d) hermit crab (without shell), (e) decapod crab, (f) mantis shrimp or stomatopod, (g) isopod, (h) mysid shrimp, (i) cumacean, and (j) amphipod.

the claw to scoop sediment into the mouth cavity, where the mouth parts separate microalgae and bacteria from the sand grains, which are then rolled into sand balls and deposited on the sediment surface. Males have a large claw that is used in waving displays to attract females and plays no role in feeding. Females are at a feeding advantage because they have two feeding claws. Other crustaceans, such as barnacles and copepods, filter phytoplankton and bacteria from the water.

Acorn barnacles stand out as the most specialized of the crustaceans. Although the larvae are planktonic and typically crustacean, the adults are sessile and are enclosed by a series of calcium carbonate plates (Figure 11.24a). Midtrunk appendages, called cirri, collect food as the animal rapidly extrudes part of the body from the shell, and the feeding cirri comb phytoplankton from the water. Acorn barnacles and their relatives the stalked barnacles (Figure 11.24b) live on rocks, other invertebrate shells, and coral skeletons, and even on the skins of whales.

The Lophophorate Phyla

✳ The lophophorate phyla include the Bryozoa, Brachiopoda, and Phoronida. They are united by the presence of a looped feeding and respiring structure known as a lophophore.

Three invertebrate phyla—Bryozoa, Brachiopoda, and Phoronida—are believed to be closely related because they all have similar looped feeding structures known as **lophophores**. A lophophore is ciliated and gathers suspended food, mainly phytoplankton. The closeness of the phylogenetic relationship among the three phyla may be illusory, however, and the similarity may be more a result of convergent evolution of feeding structure.

Phylum Bryozoa: Moss Animals

BASIC FACTS

Taxonomic level: Phylum Bryozoa; grade of construction: organs derived from three tissue layers; symmetry: bilateral; type of gut: complete; type of body cavity other than gut: coelom, usually reduced; segmentation: absent; circulatory system: absent; nervous system: single ganglion with nerves branching throughout zooid; excretion: no special structures. Other features: colonial invertebrates, with small individuals (zooids) feeding with lophophore, growing in sheets or erect colonies; a colonial nervous system allows coordination of zooids. Number of species: 4,000.

✳ Bryozoans are abundant on hard surfaces and consist of colonies of small (<1 mm in width) individuals known as zooids.

There are about 4,000 species of living bryozoans, and they occur throughout the world on hard surfaces, mainly in shallow waters. Bryozoans are abundant on the undersides of stony corals, on rocky surfaces, and even on the fronds of seaweeds. They are all colonial, and individual **zooids** (Figure 11.25) are much less than 1 mm in diameter. The soft parts of the animal are encased in an organic box, which may also be calcified. The animal feeds on suspended matter using a ciliated lophophore, which can be thrust into the overlying water by fluid pressure. Ciliary movement creates a feeding current that draws suspended particles to the lophophore. Ciliary currents then direct food to the mouth. Interspersed among the feeding zooids are other beaklike specialized zooids called **avicularia**, which help keep the colony unfouled by pinching at settling organisms.

✳ Bryozoan colonies can occur as sheets, erect colonies, or units connected by runners, known as stolons.

Bryozoans have a wide variety of forms. *Electra*, *Membranipora*, and similar forms consist of a low-relief sheet (Figure 11.26). They usually live on flat surfaces. The peripheral zooids of some can produce spines to ward off predators. Many others, such as *Bugula*, grow as erect colonies (Figure 11.25) and could be confused with cnidarians but for the characteristic lophophore. The colony often grows by continuous dichotomous splitting. A minority of bryozoans consist of colonies connected by modified zooids that form runners or stolons. This form may have the advantage of allowing rapid colonization of new microsites, but it is not nearly as common as the other two growth forms.

Phylum Brachiopoda: Lingulas and Lampshells

BASIC FACTS

Taxonomic level: Phylum Brachiopoda; grade of construction: organs derived from three tissue layers; symmetry: bilateral; type of gut: complete; type of

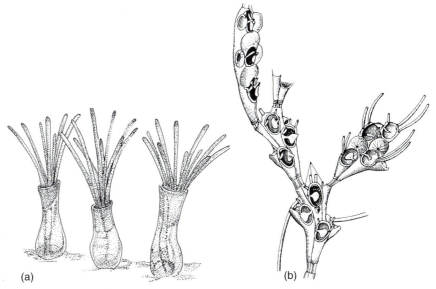

Fig. 11.25 (a) Close-up of zooids of the sheetlike bryozoan *Membranipora*. (b) The erect bryozoan *Tricellaria*.

(a)   (b)

body cavity other than gut: coelom, in two sections; segmentation: absent; circulatory system: heart with blood vessels; nervous system: rudimentary, but animals respond to light; excretion: no special structures. Other features: solitary individuals with two valves, attached to bottom with pedicle, feed with ciliated lophophore. Number of species: 300.

❋ Brachiopods have two shells and use a pedicle to attach to the bottom; a lophophore allows them to feed on suspended matter.

The living brachiopods represent a mere trace of their former abundance in the Paleozoic era, when they dominated shallow seas. Today, there are only about 300 species, which tend to live in cryptic coral reef habitats, some high-latitude shallow waters, and deeper waters. They are mainly sessile and live exposed on hard surfaces. They are quite distasteful to predators.

Brachiopods superficially resemble bivalve mollusks and have two calcareous valves, which enclose the animal's soft parts. In brachiopods (Figure 11.27), however, the valves are not the same: one usually has a perforation, through which a stalk, or pedicle, attaches to the bottom. Articulate brachiopods live on hard surfaces. The valves are articulated with a tooth-and-socket hinge, and the pedicle attaches to the rock surface. Inarticulates, such as *Lingula*, lack the articulate hinge system, however, and the valves can move more freely. The valves are connected by a complex musculature, which in burrowing species allows them to move like two sliding sheets of paper and permits the animal to dig into the sand, valves first (Figure 13.3). The animal assumes a living position with the valves up and pedicle down. When disturbed, the pedicle contracts and withdraws the valves below into the sand.

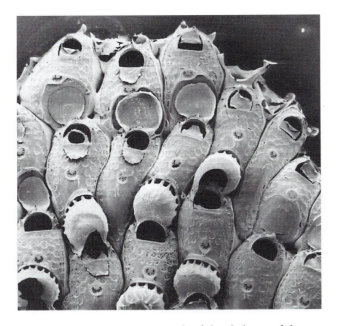

Fig. 11.26 Electron micrograph of the skeleton of the encrusting bryozoan *Fenestrulina*, showing pores through which feeding zooids emerge. Several egg-containing ovicells can be seen above the main zooid chambers. (Photograph by Sean Craig.)

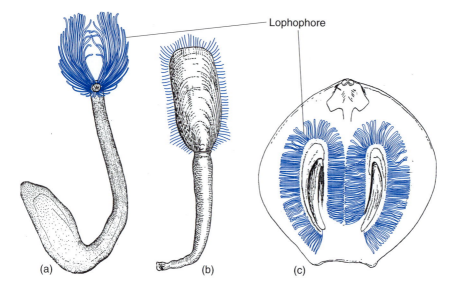

Lophophore

Fig. 11.27 Lophophorate phyla: (a) a phoronid worm, (b) an inarticulate brachiopod, and (c) opened articulated brachiopod, showing lophophore feeding organ.

All brachiopods have a coiled lophophore, which projects outward as two symmetrical sections, usually supported by a calcareous loop. In some fossil brachiopods, this support was often in the form of a lovely spiral. The cilia on the lophophore beat and create a current, which brings in particles through two lateral openings, across the two parts of the lophophore and toward a central area. Water and uningested particles exit through a central stream, between the valves.

Phylum Phoronida: Wormlike Animals with a Lophophore

BASIC FACTS

Taxonomic level: Phylum Phoronida; grade of construction: organs derived from three tissue layers; symmetry: bilateral; type of gut: complete; type of body cavity other than gut: coelom in several sections; segmentation: absent; circulatory system: blood vessels moving fluid with peristaltic action; nervous system: rudimentary, but giant neurons extend for the length of the animal to allow instant shortening; excretion: possible presence of organs of excretion. Other features: solitary individuals, wormlike, living in vertical tubes with lophophore protruding into water above. Number of species: approximately 10.

* Phoronids are wormlike, with a lophophore that protrudes above the substratum.

Phoronids would be hardly worth a mention except that they are occasionally abundant in sand flats of the Pacific coast of the United States. The soft bodies of *Phoronopsis californica* (Figure 11.27) and *P. viridis* are enclosed in a parchmentlike tube that is buried in the mud. A feathery lophophore protrudes above the sediment surface, and is a lovely orange in *P. californica* and green in *P. viridis*. As in the case of bryozoans and brachiopods, cilia on the lophophore circulate water across the lophophore, and mucus-laden cilia capture particles and transfer them in tracts, eventually to the mouth.

Phylum Echinodermata: Animals with Fivefold Symmetry

BASIC FACTS

Taxonomic level: Phylum Echinodermata; grade of construction: organs derived from three tissue layers; symmetry: radial, sometimes combined with bilateral; type of gut: blind sac with very reduced anus, or complete with anus; type of body cavity other than gut: coelom; segmentation: none; circulatory system: usually open to large coelomic cavity; nervous system: major nerves extending from nerve ring, no brain; excretion: diffusion. Other features: a spiny skin encloses an internal skeleton of interlocking calcium carbonate plates; feeding and locomotion on tube feet connected to water vascular system; includes starfish, sea cucumbers. Number of species: approximately 6,000.

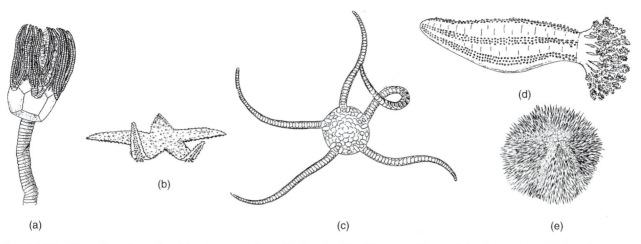

Fig. 11.28 The diversity of echinoderms: (a) stalked crinoid, (b) asteroid starfish, (c) ophiuroid brittle stars, (d) holothuroid (sea cucumber), and (e) echinoid (sea urchin). (After Meglitsch and Schram, 1991.)

* **Echinoderms are exclusively marine and have an outer skin that encloses a skeleton of interlocking ossicles.**

Members of the phylum Echinodermata (Figure 11.28) live only in the ocean; most are incapable of living even in estuaries. There are about 6,000 species, including the starfish (Class **Asteroidea**), sea urchins (Class **Echinoidea**), brittle stars (Class **Ophiuroidea**), sea cucumbers (Class **Holothuroidea**), and crinoids (Class **Crinoidea**). Among invertebrates possessing the organ grade of complexity, the echinoderms have a unique adult radial symmetry, and structures are often repeated in multiples of five (e.g., the arms of starfish). Echinoderms have an external leathery skin, which encloses an internal skeleton of interlocking calcium carbonate plates known as **ossicles**. In sea urchins, this system forms a calcareous ball, with openings for the mouth and other structures. The ball encloses the soft tissues, including digestive system and reproductive system. In starfish, the ossicles are less well fused, and the animal is flexible. On the outer surface of starfish and sea urchins, specialized groups of ossicles may form many **pedicillariae** (Figure 11.29), which can pinch and protect against predators or fouling organisms.

* **Both locomotion and feeding in echinoderms are based upon the water vascular system, which uses water pressure to operate many tube feet.**

Echinoderms all have a remarkable network of canals, pressure relief valves, and tubular suckers known as the **water vascular system** (Figure 11.30). Water is exchanged across a dorsal sieve plate, the **madreporite**. Combined with the action of relief valves, movement of water can create increases of pressure or suction in various parts of the canal sys-

Fig. 11.29 A tube foot of a starfish is held near a sea urchin, whose spines bend away and whose pedicillariae (in color) pinch at the tube foot in defense. (After Feder, 1972.)

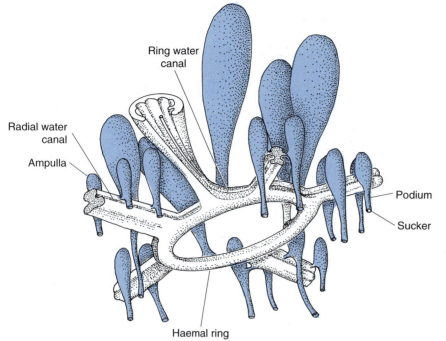

Ring water canal

Radial water canal

Ampulla

Podium

Sucker

Haemal ring

Fig. 11.30 The water vascular system and the operation of the tube feet in an asteroid.

tem. The animal connects with the outside world through thousands of **tube feet**, which can apply suction and also secrete mucus. On each tube foot is an **ampulla**, which looks like a rubber medicine-dropper bulb and functions by contracting and forcing fluid to expand the tube foot. When the tube foot contracts and withdraws, fluid is returned into the expanding ampulla. At the tip in some species is a sucker that sticks to the surface with the aid of mucus. Consider a starfish as an example (Figure 11.28b, Figure 3.2). The madreporite is located in the dorsal surface in the central disk area. The tube feet number in the thousands and are located on the ventral surface. Tube feet are expanded and withdrawn in coordination, and this allows the animal not only to pull itself along with thousands of tiny coordinated movements, but also to feed. In some groups, such as sea urchins and sea lilies, tube feet are used only to collect food and are not used for locomotion.

All echinoderms are built around a general body plan of nearly radial symmetry, with five (or multiples of five) ambulacral areas radiating from the central mouth. The **ambulacra** are lined with tube feet, which are reduced to varying degrees. In the starfish, the ambulacral areas are the ventral parts of the arms, but in sea cucumbers the areas are aligned parallel to the tubular body, with two on the dorsal surface and three on the ventral surface.

✳ Starfish (Asteroidea) are usually carnivorous, and many feed by extruding the stomach and digesting prey, mostly outside the body.

Starfish are ubiquitous throughout the ocean and nearly all are carnivorous. All have a central disk area, from which several arms radiate. Starfish have remarkable regenerative capabilities and can produce a complete new set of arms and disk as long as some major portion of the central disk survives an attack. The ventral surfaces are covered with tube feet, which can attach to a prey item. Depending upon the species, a starfish may attack anything it can handle or may be specialized to specific prey, some even to other starfish. The asteroid starfish *Luidia sarsi* uses its arms to leap onto brittle star prey. Leaping can be induced artificially by soaking cotton with extracts of brittle stars and placing the bait near the *Luidia*. Some species simply swallow their prey whole. Clam and scallop eaters attach to the valves of bivalves by means of the tube feet and use force, perhaps over a period of hours, to pull the shells apart a few millimeters. A starfish then pushes its stomach out through the mouth and into the bivalve and presses it against the soft tissues of the bivalve, secreting enzymes, and digesting the prey. The animal then withdraws the stomach and sucks in the digested material.

* Echinoids (Echinoidea) are usually covered with spines and may live on the surface, using the teeth of a structure called an Aristotle's lantern to scrape algae, or burrowing in the sand, feeding on sediment or suspended matter.

The sea urchins, sand dollars, and heart urchins (Class Echinoidea) have an internal test of interlocking ossicles. The outer skin is covered with pinching pedicillariae and spines, which also protect against predators. The spines of urchins are usually long and thin, but sand dollars and heart urchins are covered with a carpet of very short spines that are mobile and aid in burrowing. The Caribbean urchin *Diadema antillarum* has spines as long as 50 cm, which are very sharp, have thousands of tiny barbs, and produce great pain if touched.

Echinoids have a remarkable jawlike structure called an **Aristotle's lantern** (Figure 11.31), constructed from specialized ossicles connected by ligaments and moved by muscles. Five teeth come together at one point in the structure, allowing the animal to tear apart seaweeds or to scrape microscopic algae from a hard surface. Some urchins can actually scrape a depression in rock and nestle in it. The teeth are steadily worn down and replaced by secretion of calcium carbonate from dental sacs. Food is taken into the digestive system, and undigested remains are passed through the anus, on the dorsal surface. Nearly all sea urchins feed on attached algae, although some can catch drifting pieces of seaweed by means of long tube feet extending from the dorsal surface. Heart urchins and sand dollars are burrowers and feed on sedimentary organic matter, which they collect by means of modified tube feet on the ventral surface. Particles are passed from tube foot to tube foot and eventually to food grooves and to the mouth.

* Sea lilies and feather stars (Class Crinoidea) are characterized by a cup-shaped body, with upward-reaching arms that catch zooplankton on specialized tube feet.

Crinoids include the unstalked feather stars and stalked sea lilies. All have a basic cup-shaped body. The arms extend upward into the water column and are covered with pinnules, which are in turn covered by sticky tube feet that catch zooplankton from moving currents. Feather stars attach temporarily to the bottom with armlike **cirri**, but the animals have a stalk in early life. Feather stars can move into and out of crevices, and often are found at night feeding on the surface. They are sometimes common in coral reefs, but also occur in higher latitudes. Sea lilies have a long **stalk** that is embedded into the bottom, and often an umbrella-shaped array of arms. The stalk is composed of connected ossicles that resemble a stack of poker chips. Sea lilies tend to live in deep water today, but in the Paleozoic era they were a major dominant of shallow-water seas.

* Sea cucumbers (Class Holothuroidea) are tubular and have a crown of tentacles, which either feed on sediment or are directed upward to feed on zooplankton.

Sea cucumbers are shaped as their name suggests, and most have reduced ossicles, which makes them very soft and flexible. They move along by means of muscular contraction of the body wall and with the aid of the tube feet of three ambulacral areas that press against the bottom. The mouth has a crown of tentacles, consisting of highly modified tube feet. In cucumbers like the Pacific *Parastichopus californicus*, the tentacles press against the sediment, which is passed into the mouth and through the gut. Other species point the tentacles upward and collect suspended matter. *Leptosynapta* occurs on the Atlantic and Pacific coasts and lives burrowed in sand, feeding on the organic matter in it. The burrower *Molpadia oolitica* lives upside down in a vertical burrow, feeds on mud, and extrudes feces on the sediment surface.

Fig. 11.31 The Aristotle's lantern of a typical echinoid.

Sea cucumbers react to predators (or to marine biologists collecting them) by evisceration: rupturing the mouth or anal opening and expelling nearly all internal organs. The eviscerated animal can then escape and regenerate. Tropical sea cucumbers can also deter predators by means of highly toxic substances located in the surface skin.

Sea cucumbers have a unique pair of **respiratory trees**, which are highly branched outgrowths of the hind gut. The cucumber draws in water through the anal opening and contracts the hind gut, driving water through the branches of the respiratory trees. Then, the water is expelled through the anal opening. Species of crabs, protozoans, and fishes live as commensals among the respiratory trees of some tropical sea cucumbers. The fishes leave through the anal opening to forage outside.

✴ Brittle stars and basket stars have a central disk and distinct separate flexible arms that move the animal along the bottom without the use of tube feet.

Brittle stars and basket stars (Class Ophiuroidea) are generally small, although some basket stars are as large as a meter across. The disk is quite distinct from the arms, and the tube feet are not involved in locomotion. The arms are very flexible and muscular and move the animal along the sediment surface. The arms are far more flexible than those of regular sea stars, and the animals are much faster. Ophiuroids (Figure 11.28c) have only five arms. However, in basket stars the five arms are branched and subbranched, and the animals appear to have coiled tentacles (Figure 11.32). Basket stars often spend the day in

Fig. 11.32 An oral view of the eastern Pacific basket star *Gorgonocephalus eucnemis*. (Photograph by the author.)

crevices, with the arms wrapped in a compact ball. At night, they stretch out the arms and use their tube feet to trap zooplankton. Brittle stars may live within the sediment or on the surface, occasionally draped over erect colonial animals. They may be suspension feeders, sediment eaters, or carnivores. Some brittle stars live with the disk positioned below the sediment surface, the arms projecting into the water column. Particles are trapped on mucus suspended between modified spines, and tube feet remove the material and pass it to the mouth. Carnivores capture prey by looping the arms and bringing the captured victim to the mouth.

Phylum Urochordata: Sea Squirts

BASIC FACTS

Taxonomic level: Phylum Urochordata; grade of construction: organs derived from three tissue layers; symmetry: bilateral; type of gut: complete, with anus; type of body cavity other than gut: coelom; segmentation: none; circulatory system: heart with vessels; nervous system: brain with nerve cords; excretion: diffusion. Other features: benthic sea squirts have a barrel-shaped body, with incoming and outgoing siphons; tadpole larvae have features allying this group with vertebrates. Number of species: ca. 1,200.

✴ Sea squirts are barrel-shaped animals that filter water through a mucus sheet. A tadpole-shaped larva relates them to the vertebrates.

The phylum Urochordata includes the benthic sea squirts (Figure 11.33), but also the planktonic salps, which were discussed in Chapter 7. The sea squirts are found on hard surfaces nearly anywhere in the ocean and often are the dominant form of benthos. They have a barrel-shaped body, with an incoming and an outgoing siphon on top. The outer part of the body is sometimes covered with a tough tunic made of cellulose. To avoid predation and fouling, some have evolved the ability to concentrate toxic heavy metals and even to secrete sulfuric acid. The inner part of the barrel is lined with a lattice-shaped **pharynx**, which is covered with a ciliated layer, coated with mucus. Water currents bring plankton to the **mucus net**, which can filter particles as small as one micrometer in diameter.

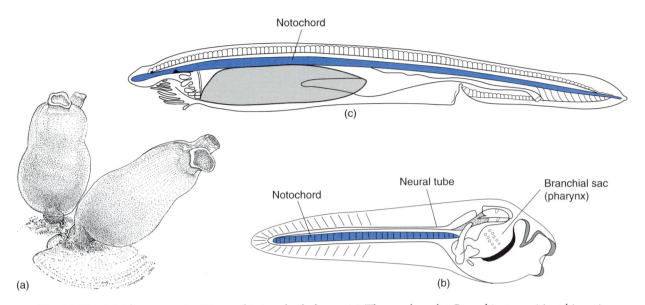

Fig. 11.33   (a) The sea squirt *Ciona*. (b) A tadpole larva. (c) The sea lancelet *Branchiostoma* (*Amphioxus*).

The larvae of sea squirts (Figure 11.33b) are especially exciting creatures, because they demonstrate an evolutionary relationship with our evolutionary branch, vertebrates. The larvae resemble tadpoles and have a trunk and tail. The tail is stiffened with a rod, called a notochord, and has a dorsal tubular nerve. The animal also has gill slits. These traits are characteristic of vertebrate embryos. They also ally sea squirts with another common marine animal, the sea lancelet, (Figure 11.33c.), which belongs to the phylum Cephalochordata. The sea squirt larva does not feed and cannot disperse very far.

Hot Topics in Marine Biology 11.1 From Where Did All This Invertebrate Diversity Come?

A rocky tide pool usually contains a fantastic diversity of creatures. Flatworms, anemones, hydroid cnidarians, crustaceans, snails, annelids, nematodes, starfish, small fishes, and more can be found with just a bit of effort. What are we to make of this hodgepodge? It is worth asking whether all these diverse creatures even had a common ancestor, and how we would be able to tell.

Many clues come from traits shared by different groups. These shared unique traits make it likely that the organisms share a common ancestor that had these traits. Embryology is a powerful tool for establishing such relationships. The mechanism of early cell cleavage allows most of the invertebrate phyla to be divided into two great groups, the **protostomes** (flatworms, annelids, mollusks, and arthropods) and the **deuterostomes** (echinoderms and chordates, which include human beings). Within each of these two groups,

members may look very different as adults, but their early embryology unites them. The protostomes, for example, are nearly all characterized by a peculiar pattern of cleavage that makes the embryo asymmetric (Box Figure 11.1). The early cleavage in deuterostomes is, by contrast, symmetric about a polar axis. Also, the deuterostomes are distinguished by a pattern of invagination of the early ball of cells, known as the blastula. The blastula invaginates, leading to the gastrula stage, and the opening formed to the outside eventually forms the anus. The mouth forms elsewhere on the embryo. In protostomes, the blastopore divides to form both mouth and anus. The sponges are the oddest and most remote of phyla. They seem to bear almost no relationships to the other phyla and probably arose from a colonial flagellate ancestor. Coelenterates also are difficult to place, though they have various tissues (e.g.,

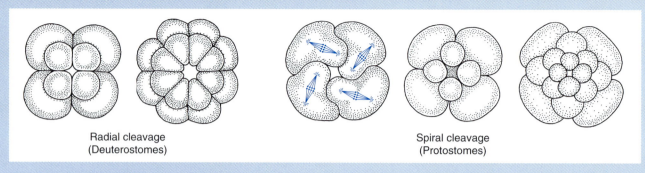

Radial cleavage
(Deuterostomes)

Spiral cleavage
(Protostomes)

Box Fig. 11.1   Protostomes and deuterostomes differ in the early cell cleavage of the embryo.

nerves) that can be related to those of most other invertebrate phyla.

The traits discussed in this chapter and many others lead to the evolutionary tree shown in Box Figure 11.2. The ancestor to all groups is unknown, but is some form of eukaryote, meaning that it had a true nucleus, nuclear membrane, and organized chromosomes. The protostomes and deuterostomes are united in having a fundamental bilateral body plan, which is only vaguely present in other phyla, such as Cnidaria. The Cnidaria and the other bilateral phyla are united by the presence of radial cleavage, which we guess to have been transformed evolutionarily into spiral cleav-

age with the rise of the protostomes. Given that the Cnidaria and deuterostomes both have radial cleavage, we assume that the deuterostomes retain the ancestral condition of cleavage. The Cnidaria can also be related to the protostomes and deuterostomes on the basis of molecular sequence similarities.

From what ancestral group did most of the invertebrates arise? Some believe that a ciliate organism became cellularized, but this seems unlikely because there is no living candidate for such a development among the vast diversity of ciliates. Others believe that a colonial flagellate might be a possibility in this regard. A flatworm-like organism named *Trichoplax* (Phylum Placozoa, Box Figure 11.3) may be a model for an early invertebrate ancestor. It is multicellular and reproduces by fission and by budding. It can change shape much like an amoeba, but is moved along on the substratum by an external layer of cilia. Unfortunately, there is too little to go on to say much about this most fundamental problem in biological diversity.

In recent years, modern molecular techniques have been used to give an independent perspective on the relationships among the phyla. DNA can now be sequenced, and simi-

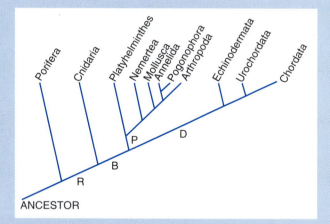

Box Fig. 11.2   An evolutionary tree of the animal phyla. Letters on the tree indicate traits that unify groups: R, radial cleavage; B, truly bilateral; P, traits unique to protostomes; D, traits unique to deuterostomes. The lophophorate phyla are omitted because molecular (indicating protostome affiliation) and morphological data (indicating deuterostome affiliation) are in conflict.

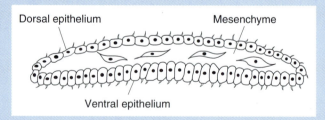

Dorsal epithelium

Mesenchyme

Ventral epithelium

Box Fig. 11.3   Cross section of the simple metazoan *Trichoplax*, showing ventral ciliated epithelium and free mesenchyme cells.

larities in certain sequences can be used to unite the phyla. Those with more sequences in common probably had a more recent common ancestor. By and large, some recent studies have confirmed the relationships established by more traditional methods of comparative morphology and comparative embryology, but some surprises may lie ahead. Many morphological traits could be similar simply because otherwise unrelated organisms evolved in similar environments. However, study of DNA sequences should sidestep this problem, because parts of most of the sequences studied are not functional. Shared sequences therefore reflect ancestry more than convergence of function. Thus far some interesting conflicts have been generated by DNA sequences. The velvet worm *Peripatus*, classified in the phylum Onycophora, possesses morphological traits linking the annelids and the arthropods. DNA sequencing, however, has allied the onycophorans far more closely with the arthropods, which has also been allied to the nematodes. DNA sequencing has also placed into question the alliance of brachiopods with other deuterostome phyla, which has been the point of view supported by embryological evidence.

We still cannot be sure that the invertebrates all arose at just one time, but the fossil record tells a very strange story indeed. Nearly all the living phyla arose in the Cambrian period, about 500–600 million years ago. Current radioactive dating evidence suggests that the entire animal explosion of life may have had its beginnings in a period of 12 million years or less. At this time, there was a fantastic diversity of creatures, perhaps much more so than any later time in the history of life. The great paleontologist C. D. Walcott discovered the Burgess Shale in British Columbia, Canada (Plate XXIV), in which were the remains of many fantastic and unknown creatures. We do not know the relationships of many of these fossils, but some may belong to phyla that have not survived to the present. The discovery of another spectacular locality in southern China has expanded our evidence on Burgess Shale–like creatures and appears to push the origin of most of the living phyla back to an early window of time at the beginning of the Cambrian. The fossil evidence, taken at first hand, suggests a "big bang theory" of metazoan evolution.

If one goes back a bit further in geological time, one encounters the Vendian faunas (Box Figure 11.4), first discovered in Australian rocks over 650 million years in age, and therefore somewhat older than those of the earliest Cambrian period. These have impressions of soft-bodied creatures that have little or no resemblance to anything in the younger Cambrian rocks. Were they an early experiment in evolution, unrelated to the living invertebrates?

Molecular evolution can also be used to estimate the time of divergence of the modern animal phyla. Consider a species that splits into two. Over time, DNA sequences

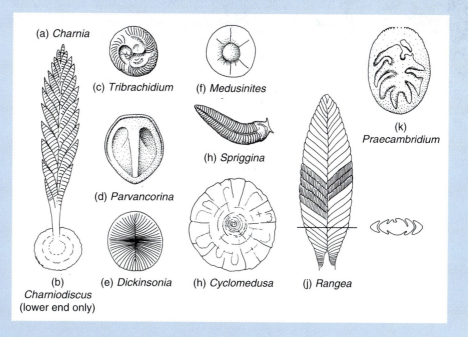

Box Fig. 11.4 Some of the peculiar soft-bodied fossils of the Vendian period. Could these be an early "experiment" in evolution, independent of invertebrates living today? (After Glaessner and Wade, 1966.)

(a) *Charnia*

(b) *Charniodiscus* (lower end only)

(c) *Tribrachidium*

(d) *Parvancorina*

(e) *Dickinsonia*

(f) *Medusinites*

(h) *Spriggina*

(h) *Cyclomedusa*

(j) *Rangea*

(k) *Praecambridium*

will evolve independently in the two daughter lineages. If the rate of change is the same in both lineages, then we can use the degree of sequence differentiation as a molecular clock, assuming that we can calibrate the differences on a time scale. We would conclude that the more molecular sequence divergence between any two species, the further back in time they diverged. This approach has been used to estimate the divergence of the protostomes and deuterostomes. A calibration for the clock can be established from fossil data documenting known divergences of vertebrate and other groups. A great and controversial surprise emerged in that nearly all estimates thus far place the protostome–deuterostome divergence considerably earlier than the so-called Cambrian explosion.[2] This may mean that there is a "lost" ancestral fauna that might consist of very small-bodied forms that not preserved by the fossil record. It may also mean that there is a bias in the molecular data, but it is too early to be sure. Even if there is such a lost fauna, it is still clear that there was a rapid appearance of large-bodied forms in the Cambrian period or just before. Paleontologists and DNA researchers now have more in common than ever before.

[2] See Wray and others, 1996, and Levinton, 2001, in Further Reading, Origin of the Invertebrates.

Further Reading

GENERAL INVERTEBRATE MORPHOLOGY

Barrington, E. J. W. 1979. *Invertebrate Structure and Function,* 2nd ed. New York: Wiley.

Brusca, R. C., and G. J. Brusca. 1990. *Invertebrates.* Sunderland, MA: Sinauer Associates.

Hyman, L. 1940–1966. *The Invertebrates*, v. 1–6. New York: McGraw-Hill. This is the definitive treatise on invertebrates that has been written in English.

Kozloff, E. N. 1990. *Invertebrates*. Philadelphia: Saunders.

Meglitsch, P. A., and F. R. Schram. 1991. *Invertebrate Zoology*, 3rd ed. London and New York: Oxford University Press.

Ruppert, E. E., and R. D. Barnes. 1994. *Invertebrate Zoology*. Philadelphia: Saunders.

Russell-Hunter, W. D. 1979. *A Life of Invertebrates*. New York: Macmillan.

Sherman, I. W., and V. G. Sherman. 1976. *The Invertebrates: Function and Form*, 2nd ed. New York: Macmillan.

Tasch, P. 1973. *Paleobiology of the Invertebrates*. New York: Wiley.

ANNELIDA

Dauer, D. M. 1983. Functional morphology and feeding behavior of *Scolelepis squamata* (Polychaeta: Spionidae). *Marine Biology*, v. 77, pp. 279–285.

Eckelbarger, K. J., and J. P. Grassle. 1987. Interspecific variation in genital spine, sperm, and larval morphology in six sibling species of *Capitella. Bulletin of the Biological Society of Washington*, v. 1987, pp. 62–76.

Fauchald, K. 1974. Polychaete phylogeny: A problem in protostome evolution. *Systematic Zoology*, v. 23, pp. 493–506.

Schroeder, P. C., and C. O. Herman. 1975. Annelida: Polychaeta. In A. C. Giese and J. S. Pearse, eds., *Reproduction of Marine Invertebrates*, v. 3. New York: Academic Press, pp. 1–214.

Woodin, S. A. 1987. External morphology of the Polychaeta: Design constraints by life habit. *Bulletin of the Biological Society of Washington*, v. 7, pp. 295–309.

Woodin, S. A., and R. A. Merz. 1987. Holding on by their hooks: Anchors for worms. *Evolution*, v. 41, pp. 427–432.

POGONOPHORA AND VESTIMENTIFERA

Jones, M. L., and S. L. Gardiner. 1989. On the early development of the Vestimentiferan tube worm *Ridgeia* sp. and observations of the nervous system and trophosome of *Ridgeia* sp. and *Riftia pachyptila. Biological Bulletin*, v. 177, pp. 254–276.

Meglitsch, P. A., and F. R. Schram. 1991. *Invertebrate Zoology*. New York: Oxford University Press.

Southward, E. C. 1988. Development of the gut and segmentation of newly settled stages of *Ridgeia* (Vestimentifera): Implications for relationship between Vestimentifera and Pogonophora. *Journal of the Marine Biological Association of the United Kingdom*, v. 68, pp. 465–487.

MOLLUSCA

Morton, J. E. 1979. *Molluscs*, 5th ed. London: Hutchinson.

Pennington, J. T., and F.-S. Chia. 1985. Gastropod torsion: A test of Garstang's hypothesis. *Biological Bulletin of Woods Hole*, v. 169, pp. 391–396.

Purchon, R. D. 1977. *The Biology of Mollusca*. Oxford: Pergamon Press.

Wilbur, K. M., ed. 1983–1986. *The Mollusca*. New York: Academic Press. 10 volumes.

ARTHROPODA

Bliss, D., ed. 1982–1986. *The Biology of Crustacea*. New York: Academic Press. 10 volumes.

Moody, K. E., and R. S. Steneck. 1993. Mechanisms of predation among large decapod crustaceans of the Gulf of Maine coast: Functional vs. phylogenetic patterns. *Journal of Experimental Marine Biology and Ecology*, v. 168, pp. 111–124.

Schram, F. R., ed. 1986. *Crustacea*. New York: Oxford University Press.

LOPHOPHORATE PHYLA

Emig, C. C. 1982. The biology of the Phoronida. *Advances in Marine Biology*, v. 19, pp. 1–89.

Ross, J. R. P., ed. 1987. *Bryozoa: Present and Past*. Bellingham, WA: Western Washington University Press.

Rowell, A. J. 1982. The monophyletic origin of the Brachiopoda. *Lethaia*, v. 15, pp. 299–307.

Rudwick, M. J. S. 1970. *Living and Fossil Brachiopods*. London: Hutchinson.

Ryland, J. S. 1970. *Bryozoans*. London: Hutchinson.

Taylor, P. D. 1988. Major radiation of cheilostome bryozoans: Triggered by the evolution of a new larval type? *Historical Biology*, v. 1, pp. 45–64.

Woollacott, R. M., and R. L. Zimmer. 1977. *Biology of Bryozoans*. New York: Academic Press.

ECHINODERMATA

Feder, H. M. 1972. Escape responses in marine invertebrates. *Scientific American*, v. 227, pp. 92–100.

Fenchel, T. 1965. Feeding biology of the sea-star *Luidia sarsi* Döben and Koren. *Ophelia*, v. 2, pp. 223–236.

Jangoux, M., and J. M. Lawrence. 1982. *Echinoderm Nutrition*. Rotterdam: Balkema.

Nichols, D. 1969. *Echinoderms*. London, Hutchinson.

Smith, A. B. 1984. Classification of the Echinodermata. *Palaeontology*, v. 27, pp. 431–459.

ORIGIN OF THE INVERTEBRATES

Anderson, D. T. 1982. Origins and relationships among the animal phyla. *Proceedings of the Linnaean Society of New South Wales*, v. 106, pp. 151–166.

Ballard, J. W. O., G. J. Olsen, D. P. Faith, W. A. Odgers, D. M. Rowell, and P. W. Atkinson. 1992. Evidence from 12S ribosomal RNA sequences that onychophorans are modified arthropods. *Science*, v. 258, pp. 1345–1348.

Barnes, R. D. 1987. *Invertebrate Zoology*, 5th ed. Philadelphia: Saunders. Chapter 3, "Introduction to the Metazoa," is especially recommended.

Conway Morris, S. 1998. *The Crucible of Creation*. Oxford: Oxford University Press.

Field, K. G., G. J. Olsen, D. J. Lane, S. J. Giovannoni, M. T. Ghiselin, E. C. Raff, N. R. Pace, and R. A. Raff. 1988. Molecular phylogeny of the animal kingdom. *Science*, v. 239, pp. 748–753.

Glaessner, M. F. 1984. *The Dawn of Animal Life: A Biohistorical Study*. Cambridge: Cambridge University Press.

Glaessner, M. F., and M. Wade. 1966. The late Precambrian fossils from Ediacara, south Australia. *Palaeontology*, v. 9, pp. 599–628.

Gould, S. J. 1989. *Wonderful Life. The Burgess Shale and the Meaning of History. New York*: Norton.

Lake, J. A. 1989. Origin of multicellular animals. In B. Fernholm, K. Bremer, and H. Jurnvall, eds., *The Hierarchy of Life*, Amsterdam: Elsevier, pp. 273–278.

Levinton, J. S. 2001. *Genetics, Paleontology, and Macroevolution*, 2nd ed. New York: Cambridge University Press.

Nielsen, C. 1995. *Animal Evolution: Interrelationships of the Living Phyla*. Oxford: Oxford University Press.

Vagvolgyi, J. 1967. On the origin of the molluscs, the coelom, and coelomic segmentation. *Systematic Zoology*, v. 16, pp. 153–168.

Valentine, J. W. 1989. Bilaterians of the Precambrian–Cambrian transition and the annelid–arthropod relationship. *Proceedings of the National Academy of Sciences USA*, v. 86, pp. 2272–2275.

Whittington, H. B. 1985. *The Burgess Shale*. New Haven, CT: Yale University Press.

Wray, G. A., J. S. Levinton, and L. H. Shapiro. 1996. Molecular evidence for deep Precambrian divergences among metazoan phyla. *Science*, v. 274, pp. 568–573.

Review Questions

1. What is the advantage of a radially symmetrical body plan? What is a disadvantage of such a body plan?

2. What is the function of collar cells in sponges?

3. What function does polymorphism serve in some cnidarians?

4. What is the major difference in overall body symmetry between flatworms and cnidarians?

5. What is the main feeding mode of nemerteans?

6. What is the distinction between locomotory mechanisms in polychaetes and oligochaetes?

7. What is the distinguishing characteristic of nutrition in the Pogonophora?

8. Describe three means by which different bivalve mollusks may feed.

9. What structure is used by gastropods to scrape, puncture, or tear apart prey?

10. What is the disadvantage of the stiff exoskeleton in arthropods?

11. Describe some of the different life habits of crustaceans.

12. What trait do the brachiopods, phoronids, and bryozoans share?

13. What general trait unites the members of the phylum Echinodermata?

14. How do echinoderms generally move about?

15. What characteristics unite sea squirts with the vertebrates?

16. Why is the Cambrian period special in the history of animal life?

17. What are the differences between protostomes and deuterostomes?

12

Seaweeds, Sea Grasses, and Benthic Microorganisms

Most marine naturalists tend to focus nearly all their attention on animals. Somehow, organisms such as whales and crabs tend to capture the imagination. However, the intertidal and shallow subtidal seabed is often covered with seaweeds and sea grasses. These plants are fascinating in their own right, but they also form the base of major food webs in the ocean. Kelps form large forests (see Chapter 15) in cold-water, wave-swept coasts and support enormous numbers of animals, both directly as food and indirectly as shelter. On rocky shores, seaweeds are abundant and are a major food source for grazing urchins and snails, among others. Microorganisms such as bacteria and fungi are also ubiquitous on the seabed. Some of the microorganisms are photosynthetic, but many, if not most, are capable of obtaining nutrition through the decomposition of organic matter. They are therefore extremely important in completing a cycle of decomposition, particularly when large amounts of dead plant matter enter the marine ecosystem.

The purpose of this chapter is to introduce the diversity of microorganisms and plants that live on the seabed. Unfortunately, there is often a simple confusion between true plants, all of which can manufacture food by photosynthesis, and microorganisms, such as bacteria, which may or may not be able to photosynthesize. We can distinguish among the organisms discussed in this chapter by the way they gather or manufacture food. Recall that autotrophic organisms are those that can manufacture food if provided only with inorganic nutrients (e.g., phosphate, nitrate). A variety of bacteria and all higher plants are autotrophic and photosynthetic: they need light energy to aid in the process of manufacturing sugars. Many other bacteria are **chemoautotrophic**: they oxidize or reduce some substance (e.g., sulfate) to obtain energy, which is required to reduce carbon dioxide in order to manufacture sugars. Finally, many bacteria are heterotrophic: they obtain their food from outside the body rather than by making it.

The distinction between **prokaryotes** and **eukaryotes** is also an important one to review. Prokaryotes occur either as single cells or as colonies of cells; they are distinguished by the lack of various features: a nucleus, cell organelles (e.g., mitochondria, chloroplasts), and cell reproduction by mitosis. Prokaryotes include bacteria. Eukaryotes have distinct nuclei and cell organelles, and all cells except gametes reproduce by mitosis. Eukaryotes include the animals, seaweeds, and terrestrial plants.

Seaweeds

General Morphology

✻ Seaweeds, like terrestrial plants, usually connect to a substratum, but they take up nutrients from the

surrounding water and need not have extensive support structures or other adaptations to life in air.

Plants gather light energy to fuel the process of photosynthesis. They also require nutrients, such as nitrogen and phosphorus, to build tissues. Light comes from above, and, on land, nutrients and water can be obtained only from the soil. This leads to the familiar terrestrial plant morphology of roots and shoots. The roots grow into the soil and take up nutrients, and the shoots protrude upward, so that the leaves can gather light. Air is thin (compared to water), and the shoots must therefore have strong support structures: stems, woody material, and so on. Excessive evaporation is also a problem, so leaves of higher plants are coated with a waxy material, and openings (stomata) control the entry and exit of water.

Seaweeds must gather nutrients, and they also require light, but they lack most of the other familiar terrestrial constraints. Water uptake and transpiration are not a problem except perhaps in higher intertidal species. Seaweeds gather nutrients in solution from the overlying water, and they therefore lack complex root systems. They also live in seawater, which is much denser than air. Seaweeds therefore do not have nearly the degree of support tissue found in terrestrial plants. They must grow upward toward the light, of course, but they can do this to some extent by making structures within them less dense than seawater. In many species, floats filled with gas suspend the plant in the water column.

✻ A seaweed usually occurs as an individual, or thallus, consisting of holdfast, stipe, and blade.

An individual seaweed, attached to the substratum, is known as a **thallus**, which can range in form (Figure 12.1, Plate XV.4) from a tarlike crust to a thin green sheet to an erect simple filamentous branching structure to a more elaborate plant, complete with highly differentiated structures for light gathering, reproduction, support, flotation, and attachment to the substratum.

The thallus may simply be attached to the surface, or it may be attached to the substratum by means of a **holdfast**. Holdfasts may range from a simple protuberance or a group of hairlike structures that are rooted in the sand (as is the case in *Penicillus*) or to an elaborate and strong branching structure that glues to a rocky surface (as in *Fucus* or in the sea palm *Postelsia*, Plate XIV.3).

Smaller seaweeds tend to be simple chains or sheets of cells, whereas larger forms are characterized by considerable differentiation. The bulk of the thallus usually consists of fused groups of chains of cells or a complex structure resulting from the growth of cells into many different planes. The **stipe** is a structure that is usually tubular and flexible; it connects to the holdfast and allows the thallus to bend over in a strong current. The stipe may be quite weak, or it may consist of a complex flexible network in species that live in strong wave surge. A flattened section connected to the stipe, known as the **blade**, is specialized for light capture, whereas other parts are specialized for reproduction or flotation. The leafy blades differ considerably from terrestrial leaves: they are symmetrical and do not have a top side (specialized for light capture) that is differentiated from a bottom side (specialized for gas uptake and release), as is the case for terrestrial plants. In the smallest seaweeds, growth can occur by cell division nearly anywhere in the thallus. Larger seaweeds, however, have specialized zones of cell division, known as **meristems**. Photosynthesis occurs mostly on the blade, but other parts of the thallus also have chlorophyll and can photosynthesize. In some seaweeds, specialized floats, known as **pneumatocysts**, keep the thallus suspended in seawater. Pneumatocysts occur commonly in kelps, which keep their blades near the surface atop an extremely long stipe, but they are also present in sargassum weed, and allow it to remain afloat in its pelagic habitat of the Sargasso Sea.

Means of Classification and Light Gathering in Photosynthesis

✻ Seaweeds are classified by the different pigments used in gathering light for photosynthesis, by their storage products, and by the types of flagellae in their spores.

Like most plants, seaweeds have a range of photosynthetic pigments that are specialized to capture light energy over a broad range of the visible spectrum. Different groups of seaweeds can be identified by their range of pigments and are commonly known as green, brown, or red algae. Also, carbohydrates are stored in different forms. Although all algae have cell walls composed primarily of cellulose, the main groups of seaweeds can be distinguished by other constituents, which often have commercial value. Finally, spores may be flagellated or not (e.g., red seaweed spores lack flagellae), and number and insertion

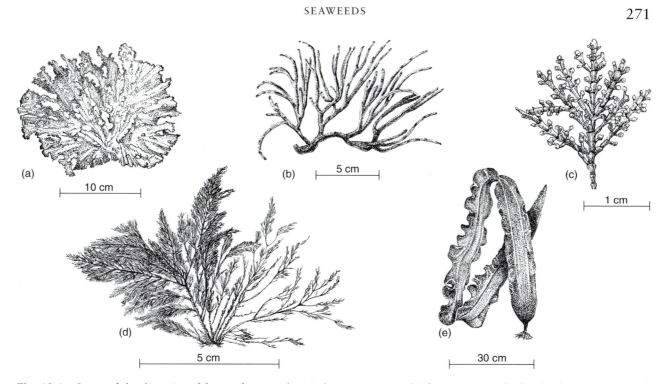

Fig. 12.1 Some of the diversity of form of seaweeds: (a) the green seaweed *Ulva* (ca. 25 cm high), (b) the green seaweed *Codium fragile* (ca. 30 cm high), (c) the red coralline alga *Corallina* (ca. 1 cm tall), (d) the red seaweed *Polysiphonia* (ca. 25 cm high); and (e) the brown seaweed *Laminaria* (ca. 2 m long).

also differ among the groups. The major differences are summarized in Table 12.1.

✱ Seaweeds depend upon light for photosynthesis and can adjust to the decrease in light with increasing depth.

Seaweeds differ in how they use absorbed light of different wavelengths. The relative utilization of different wavelengths is known as the **action spectrum**. Green algae use red and blue light the most and green light least, whereas red algae use green light the most. Browns are similar to greens in this respect, with a little more use of the red end of the spectrum.

In the ocean, light comes from the sun and light irradiance decreases exponentially with depth. Not only does the total light irradiance change, but blue-green light tends to penetrate to much greater depths than does light from the red part of the visible spectrum. It therefore stands to reason that algae might adjust to changing light in two ways. As total light irradiance decreases, algae could increase their total pigment concentration. In very shallow water, seaweeds living in sun and those that live in shade have lower and higher total pigment concentrations, respectively. The finger-shaped *Codium* is optically black, which means that it can absorb all incident light. It therefore can do rather well at great depth in low light[1] even though its action spectrum makes

[1] See Ramus and others, 1976, in Further Reading, Algae.

Table 12.1 Some general characteristics of the major seaweed groups

SEAWEED GROUP	PHOTOSYNTHETIC PIGMENTS	STORAGE PRODUCTS	CELL WALL
Green algae	Chlorophylls *a*, *b*	Starch	Cellulose (not all)
Brown algae	Chlorophylls *a*, *c*, fucoxanthin	Laminarin, mannitol	Alginate
Red algae	Chlorophylls *a*, *d*, phycoerythrin, phycocyanin	Floridian starch	Agar, carrageenan

it appear to be more efficient at higher light. In effect, the thallus has such high concentrations of pigment that it can capture all light.

A second possible means of adapting to changing light levels at different depths would be to alter the proportions of different light-harvesting pigments, to adapt to the spectral composition of light as it changes with depth. Seaweeds at depth might be expected to have pigment proportions that would maximize the capture of light under the differing spectral conditions there. At present, however, the evidence leans against this response as an adaptation to depth.

Life Cycles and Classification

❋ Seaweeds have a complex life cycle, with thallus stages alternating with dispersing stages.

Like many marine organisms, seaweeds have complex life cycles, often with more than one macroscopic form. The gametes may be motile in seawater, and many species have mobile spore stages, which facilitate dispersal to new sites.

Sexuality and life cycles of seaweeds are very complex. Often, a haploid (*N* chromosomes) generation alternates with a diploid (2*N*) generation. Attached large stages may be haploid or diploid. The **gametophyte** stage produces gametes, which may be motile (as in the brown algae) or nonmotile (as in red algae, whose gametes may still be planktonic, hence able to be carried by water currents). Gametes are usually formed and released from **gametangia**, which may be single cells or elaborate structures. The zygote, formed from the fusion of two gametes to make a diploid **sporophyte**, may be an attached plant or a motile flagellated form. Seaweeds may be monoecious (hermaphroditic) or dioecious (separate sexes).

There is a great deal of diversity in the relative size and form of the sporophyte and gametophyte stages. In some seaweeds, such as the sea lettuce *Ulva*, the stages are identical. In the Laminariales (kelps), the sporophyte is a large plant, whereas the gametophyte is a small filamentous form. The sporophyte and gametophytes are so different in some seaweeds that members of the same species were once erroneously assigned to different genera. It is not completely clear why such a condition of difference has evolved, but many of the contrasts in forms between sporophyte and gametophyte allow different life stages to dominate in different seasons, or to allow resistance against herbivores in one stage and an ability to grow rapidly in the other stage. These differences may allow greater exploitation of different ecological conditions and may therefore increase the ecological success of the given seaweeds.

Figure 12.2 is a diagram of some of the common life histories of benthic seaweeds. As can be seen, there are a great number of possibilities. Most notable is a lack of consistent relationship between haploid or diploid condition and morphological form of the vegetative stages.

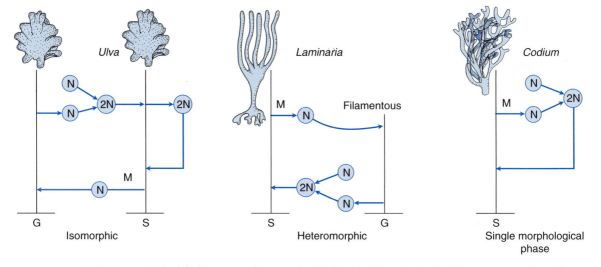

Fig. 12.2 Variation in the life histories of seaweeds. Diploid (2*N*) or haploid (*N*) status is indicated.

Types of Seaweed

GREEN SEAWEEDS (CHLOROPHYTA)

Green seaweeds have photosynthetic pigments similar to those of higher plants and they also have the same storage product, starch; indeed, members of the green seaweed group may have been the ancestors of higher plants. There is a fantastic array of morphologies, ranging from single-celled forms (e.g., the flagellated *Euglena*) to filmy seaweeds nearly a meter in length (e.g., *Ulva*, Figure 12.1a, and *Enteromorpha*. Along the shore, the sea lettuce *Ulva* is a familiar sight, and it occurs as mats on many sand- and mudflats. The genus *Ulva* has many species, but all share a delicate leafy morphology, which is very susceptible to grazing by snails and other herbivores. *Ulva* has relatively few compounds that deter feeding. Filmy green seaweeds are often associated with polluted environments and brackish water. They have a limited ability to store nutrients and so are associated with polluted environments where organic nutrients are in rich supply.

Codium fragile (Figure 12.1b) is a common green seaweed that has invaded coastal waters of the northeastern United States, destroying oyster beds by covering and sometimes smothering the oysters. On the large scale of oceans, *Codium fragile* was transported along with oysters transplanted for mariculture. Once arrived in a new area, the seaweed's dispersal is due mainly to the spread of flagellated spores that settle on rocks and oyster shells.

BROWN SEAWEEDS (PHAEOPHYTA)

Brown seaweeds dominate low intertidal and shallow subtidal environments in all latitudes. The largest seem to thrive in colder-water, nutrient-rich environments. They range from small filamentous forms to the largest seaweeds in the ocean (*Macrocystis*), which are often longer than 15 m. They include the large kelps (e.g., *Laminaria*, Figure 12.1e, Plate XXI.1 and *Macrocystis*), sargassum weed, and the intertidal *Fucus*, Plate XV.4, which has bulb-shaped reproductive structures. Brown seaweeds get their color from the pigments xanthophyll and carotene, which mask the green of chlorophyll in them. Brown seaweeds can be far more differentiated than green seaweeds and usually have distinct holdfast, stipe, blades, and reproductive structures.

Brown seaweeds contain phycocolloids, which are colloidal agents used in many foods and other products for human consumption. Many species of brown seaweed are harvested for the extraction of alginate, used in toothpaste, pills, and salad dressing. Kelp harvesting is especially well developed on the west coast of the United States, and grazing sea urchins are actively destroyed there to maximize kelp growth. During World War I, potash and acetone were extracted from brown seaweeds to manufacture a smokeless gunpowder known as cordite.

RED SEAWEEDS (RHODOPHYTA)

The pigment phycoerythrin gives red seaweeds their color, and nearly masks the green chlorophyll. Red seaweeds are found in shallow water and the intertidal zone, and the phycoerythrin in them is often bleached by the sun, causing the seaweeds to be any of a variety of colors.

The life cycle is complex and variable among species, but in the common *Porphyra*, the seaweed used in sushi, a sporophyte produces nonmotile gametes, which fuse with an undifferentiated thallus cell. The zygote thus formed grows, and it produces structures called carposporangia, which in turn form carpospores. The carpospore germinates to form an intermingled mesh of filaments, which may produce large spores, which germinate to form a filmy and large sporophyte.

Red seaweeds (Figure 12.1d) have a variety of forms, and their holdfasts may consist of a single cell, spread out into a pad, or of many cells. Some groups (e.g., *Gracilaria* and the Irish moss *Chondrus crispus*) are stringy or filamentous in appearance. *Porphyra* is thin and filmy. The **coralline algae** (e.g., Figure 12.1c) secrete calcium carbonate in the cell walls, and, depending upon the species, grow as upright branching forms or as crusts. In the tropics, some coralline algal species build up wave-resistant structures and a red algal ridge over a meter high is found on the windward side of many a Pacific coral reef.

Red algae often grow rapidly and some species contain substances of great value for food production. **Carrageenan**, obtained for centuries mainly from Irish moss, is used to thicken foodstuffs such as cream cheese, ice cream, and toothpaste. **Agar** is extracted from *Gracilaria* and other seaweeds and is used as a culture medium for the growth of bacterial strains. Agar is also employed in a variety of foods, including cake icing.

Sea Grasses

✳ Sea grasses are higher flowering plants, but the flowers are simplified and the pollen is transported by water.

Sea grasses occur in shallow temperate, subtropical, and tropical waters. Sea grasses, such as the Pacific *Phyllospadix*, may grow in wave-swept surf, but most others tend to grow in relatively quiet, shallow, subtidal soft bottoms. Their maximum depth is affected by light, so turbidity may greatly reduce the depth to which they extend. The eelgrass *Zostera marina* is widespread in shallow bottoms throughout the Atlantic, eastern Pacific, and Mediterranean. Other *Zostera* species can be found throughout coastal waters of the world. The turtle grass *Thalassia testudinum* (Figure 12.3) dominates shallow soft bottoms throughout the Caribbean.

Sea grasses are flowering plants and probably evolved from terrestrial ancestors. The flowers, however, are quite simple and reduced, and their pollen, which may be elongate and specialized to stick to the appropriate target, floats along until encountering a receptive stigma. Eelgrass seeds move a short distance before setting and germinating, but turtle grass produces a fruit that can move over relatively long distances.

✳ Sea grasses usually extend populations by asexual growth, via a subsurface rhizome system.

In many sea grasses, sexual reproduction and seed set represent a minor part of population increase. Sea grasses such as eelgrass and turtle grass have a complex **rhizome system** (a series of interconnected stems) that connects shoots beneath the sediment surface. Populations of sea grasses often extend mainly by rhizomal extension. Nutrients are obtained mainly from the pore waters of the substratum, although some sea grasses have symbiotic nitrogen-fixing bacteria in their rhizome systems, which aid in the uptake of nitrogen.

Sea grasses are tough, owing to the high cellulose content, and relatively few species can graze upon them. Eelgrass is especially indigestible, and few if any species actively consume it. Turtle grass is eaten by sea urchins and the green turtle. The latter hosts a symbiotic bacterial microflora, and residence time in the gut is very long; thus this animal is capable of digesting cellulose-rich sea grasses much as a cow digests grass.

Benthic Microorganisms

Diatoms

✳ Diatoms occur as single cells or chains of cells, are photosynthetic, and have a cell wall that is impregnated with silica.

Diatoms are ubiquitous on intertidal rocks and soft bottoms and are common on the shallow-water seabed. One can often see a golden-brown mat of diatoms on intertidal sand flats. These organisms have already been discussed as plankton, important especially in temperate and high latitudes. They will only be discussed briefly here. On the seabed, **pennate** diatoms tend to dominate. This group includes relatively elongate bilaterally symmetrical cells (Figure 12.4), as opposed to the more disk-shaped cells (**centric** diatoms) that are found in the plankton. Diatoms are eukaryotes and have a cell wall impregnated with silica. Reproduction is mainly asexual and involves cell division. Some diatoms are motile and can move several millimeters. While most occur singly, some exist as colonies, which are interconnected by a mucilaginous substance. Some intertidal diatom species move below the sediment surface in midday, to avoid desiccation.

Bacteria

✳ Bacteria occur as single cells and are the most numerous organisms in the sea. They are crucial in decomposition.

Fig. 12.3 A turtle grass (*Thalassia testudinum*) bed, St. Croix. (Courtesy of Thomas Suchanek.)

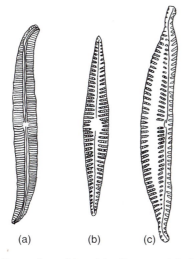

Fig. 12.4 Examples of benthic diatoms: (a) *Gyrosigma*, (b) *Navicula*, and (c) *Amphora*. (After Dawson, 1966.)

Bacteria are certainly the most numerous organisms in the ocean. They are prokaryotic and occur as single cells or as colonies. Most bacteria are tiny—only 0.5–10 μm in diameter. Although they are capable of genetic exchange, bacterial reproduction occurs through cell division. In laboratory culture, the capacity for increase is fantastic, and doubling may take place every 20 minutes or so. In the ocean, bacteria rarely achieve such high rates, and doubling can be as slow as once a day. Bacteria occur free in the water column, attached to particles both in the water column and in sediments, and also are found in the pore waters of sediments. Large numbers of bacteria

are also found in the guts of marine animals, such as the green turtle, as discussed earlier.

Most bacteria are heterotrophic and are important in the decomposition of organic matter in the sea. Such bacteria may absorb across the cell wall dissolved substances such as sugars. They may also produce enzymes externally. These break down organic material, which in turn is transported into the cell. Some bacteria can attach to particles and produce a network that allows enzymes to be trapped in an enclosed space so that they can work efficiently to break down organic matter. Bacteria involved in tooth decay use the same sort of mechanism to destroy human tooth enamel.

Blue-Green Bacteria (Cyanobacteria)

* **Blue-green bacteria are photosynthetic cells or chains of cells, but the benthic forms usually live in association with anoxic sediments.**

Cyanobacteria (members of the group Cyanophyceae, also known as blue-green algae) are widespread prokaryotic organisms that occur as free-living forms but also in symbiotic associations with marine benthic plants and animals. Cyanobacteria are photosynthetic, but many also are capable of nitrogen fixation, the conversion of gaseous nitrogen into ammonium ion, and synthesis of amino acids. In benthic habitats, cyanobacteria may consist of spherical cells that are 1–25 μm in diameter. Commonly, they are multicellular (Figure 12.5) and are arranged in rows of cells known as **trichomes**. **Filaments** are trichomes or groups of trichomes enclosed in sheaths, and many cyanobacteria occur as long or branched filaments.

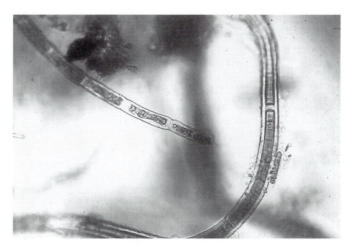

Fig. 12.5 The benthic cyanobacterium *Lyngbya* (right). (Photograph courtesy of Edward Carpenter.)

Many cyanobacteria have **akinetes**, resting spores that allow overwintering or sometimes survival of years of unfavorable conditions. Nitrogen fixation proceeds in enlarged cells known as **heterocysts**, and the enzyme nitrogenase enables the fixation of nitrogen. Cyanobacteria tend to favor environments where there is a local absence of oxygen, because nitrogen fixation is favored under these conditions. Low oxygen levels can sometimes be achieved merely by growth in the form of a bacterial mat, whose interstitial water is anoxic in the interior of the mat. Cyanobacteria are often found in black crusts on the upper parts of rocky shores and also on mudflats, where grazing animals are uncommon. Cyanobacteria also live in association with the rhizomes of many sea grasses, and the latter are benefited by the uptake of nitrogen into the rhizome system.

Fungi

✳ **Ubiquitous and usually filamentous chains of cells, fungi are very important in the decomposition of particulate organic matter.**

Fungi are either unicellular or multicellular, are eukaryotic, and occur ubiquitously in the ocean. They are relatively simple morphologically, lacking the true roots, stems, and leaves of true higher plants. They are usually heterotrophic and are important in the decomposition of plant matter in salt marshes and seaweed beds. Fungi are usually filamentous and can penetrate particulate organic matter, such as decaying sea grass. Fungi also live on the surface of seaweeds and sea grasses and probably survive by absorbing dissolved substances from the water.

Further Reading

ALGAE

Bold, H. C., and M. J. Wynne. 1978. *Introduction to the Algae. Structure and Function.* Englewood Cliffs, NJ: Prentice-Hall.

Chapman, A. R. O. 1979. *Biology of Seaweeds. Levels of Organization.* Baltimore: University Park Press.

Chapman, A. R. O. 1987. *Functional Diversity of Plants in the Sea and on Land.* Boston: Jones and Bartlett.

Dawson, E. Y. 1966. *Marine Botany, an Introduction.* New York: Holt, Rinehart and Winston.

Graham, L.E., and L.W. Wilcox. 2000. *Algae.* Englewood Cliffs: Prentice-Hall.

Lobban, C. S., and P. J. Harrison. 1994. *Seaweed Ecology and Physiology.* New York: Cambridge University Press.

Morris, I. 1968. *An Introduction to the Algae,* 2nd ed. London: Hutchinson

Ramus J., S. I. Beale., D. Mauzerali, and K. L. Howard. 1976. Changes in photosynthetic pigment concentration in seaweeds as a function of water depth. *Marine Biology,* v. 37, pp. 223–229.

Round, F. E., R. M. Crawford, and D. G. Mann. 1990. *The Diatoms. Biology and Morphology of the Genera.* Cambridge: Cambridge University Press.

South, G. R., and A. Whittick. 1987. *Introduction to Phycology.* Oxford: Blackwell Scientific.

Sze, P. 1986. *A Biology of the Algae.* Dubuque: William C. Brown.

MICROBES

Austin, B. 1988. *Marine Microbiology.* New York: Cambridge University Press.

Ford, T. E. 1993. *Aquatic Microbiology: An Ecological Approach.* Boston: Blackwell Scientific.

Moss, S. T., ed. 1986. *Biology of Marine Fungi.* New York: Cambridge University Press.

Sieburth, J. McN. 1979. *Sea Microbes.* New York: Oxford University Press.

Review Questions

1. Describe the major body parts of seaweeds.

2. What is the function of pneumatocysts in seaweeds?

3. What traits are used to classify seaweeds?

4. What are some economic uses for seaweeds?

5. By what principal means do sea grasses reproduce?

6. What special interaction do cyanobacteria have with nitrogen in seawater?

Plate XVII. *Spartina* Salt Marshes

XVII.1. *Left,* The salt marsh at West Meadow Creek, Long Island, New York. (Photograph by Jeff Levinton)

XVII.2. *Middle,* How it begins. Newly established seedlings of *Spartina alterniflora* on an open sand flat. (Photograph by Jeff Levinton)

XVII.3. *Bottom left,* Population of the semi-infaunal marsh mussel *Geukensia demissa,* among *Spartina* plants. (Photograph by Jeff Levinton)

XVII.4. *Bottom right,* The fiddler crab *Uca pugilator.* Its burrows enhance growth of salt marsh plants. (Photograph by Jeff Levinton)

Plate XVIII. Mangroves

XVIII.1. *Top left,* A recently established mangrove seedling. (Photograph by Robert Twilley)

XVIII.2. *Top right,* Mangroves colonize the intertidal zone of tropical tidal creeks, as shown here along Estero Pargo in Terminos Lagoon, Mexico. (Photograph by Robert Twilley)

XVIII.3. *Middle,* Mangrove islands also occur along inland coastal area of land, as shown here in the Everglades National Park in south Florida. These mangrove islands are known as hammocks and form tear-shape patterns parallel to the flow of water and to the coast. (Photograph by Robert Twilley)

XVIII.4. *Bottom,* Many species that find refuge in mangrove habitats are commercially important to local human populations that harvest them for food and revenue. Here an Ecuadorian carries a morning's harvest of *Ucides occidentalis* from a mangrove forest along the Churute River estuary for sale in the city of Guayaquil. (Photograph by Robert Twilley)

Plate XIX. The Soft-Bottom Intertidal Landscape

XIX.1. *Top left,* Burrow opening of the shrimp *Upogebia pugettensis.* This species lives in a deep Y-shaped burrow. (Photograph by Jeff Levinton)

XIX.2. *Top,* Surface mounds of burrowing shrimp and spiral fecal strands produced by the deposit-feeding polychaete *Abarenicola vagabunda* (upper right). (Photograph by Jeff Levinton)

XIX.3. *Middle,* The burrowing shrimp *Upogebia pugettensis,* ca. 10 cm long. (Photograph by Jeff Levinton)

XIX.4. *Bottom,* The deposit-feeding polychaete *Abarenicola vagabunda,* ca. 8 mm in diameter. (Photograph by Jeff Levinton)

Plate XX. Kelp Forests

XX.1. *Top,* Infrared-sensitive photograph of the shoreline, showing great abundance of seaweeds (red is false coloration representing chlorophyll). Bright red represents intertidal benches covered by the brown seaweed *Hedophyllum sessile.* Lighter red in foreground in *Alaria fistulosa.* (Photograph by Si Simonstad)

XX.2. *Middle,* A kelp forest in the Aleutian islands. *Cymathere triplicata* (foreground), *Alaria fistulosa* (rear). (Photograph by David Duggins)

XX.3. *Bottom left,* The Atlantic kelp *Laminara longicruris.* (Photograph by John Witman)

XX.4. *Bottom right,* The Pacific sea urchins *Strongylocentrotus purpuratus* (purple), *S, franciscanus* (red), and *S. droebachiensis* (white). (Photograph by Paul Banko)

**Plate XXI.
Kelp Forests**

XXI.1. A kelp forest bottom dominated by *Laminaria groenlandica.* (Photograph by David Duggins)

XXI.2. *Middle left,* An urchin barrens (Photograph by David Duggins)

XXI.3. *Middle right,* A kelp forest hard bottom, showing the diversity of benthic invertebrates. (Photograph by David Duggins)

XXI.4. *Bottom,* A sea otter, *Enhydra lutris,* with greenling. (Photograph by Si Simonstad)

Plate XXII. Subtidal Substrata, Solitary Invertebrates

XXII.1. *Top left,* The sea anemone *Metridium senile,* subtidal region of Washington. This anemone has bushy tentacles and feeds mainly on smaller zooplankton. (Photograph by Megan Dethier)

XXII.2. *Top right,* The subtidal carnivorous starfish *Pisaster giganteus,* near Bodega Head, California. (Photograph by Laura Rogers-Bennett)

XXII.3. *Middle,* The sea cucumber *Parastichopus californicus.* Specimens of this deposit feeder may reach lengths of 60 cm. (Photograph by Laura Rogers-Bennett)

XXII.4. *Bottom,* The aolid nudibranch *Hermissenda crassicornis* feeds on small colonial invertebrates. (Photograph by Sherry Tamone)

Plate XXIII. Subtidal Hard Substrata, Colonial Invertebrates

XXIII.1. A complex competitive interaction in the shallow subtidal region of the outer Washington coast: an anemone, *Epiactis prolifera,* competes with a red coralline alga (pink) and a sponge colony (orange). (Photograph by Dan Brumbaugh)

XXIII.2. Solitary coral–ascidian standoff, subtidal region near Pacific Grove, California. On the lower right, you can see the orange cup coral *Balanophyllia elegans.* Note the bare space between this solitary coral and the surrounding gray didemnid colonial ascidian. The standoff may be determined by the coral's stinging tentacles. (Photograph by Dan Brumbaugh)

XXIII.3. *Middle,* The hydrocoral *Allopora californica* is found commonly on the shallow subtidal and lower intertidal rocky surfaces. It can often be found growing in a sheetlike colony, but here it is growing vertically, which is perhaps a respone to the mode of feeding on zooplankton food that can be taken off the bottom in this habitat of relatively quiet water. (Photograph by Dan Brumbaugh)

XXIII.4. *Bottom,* The Pacific northwest nudibranch *Eubranchus rustyus* feeding on the hydroid *Plumularia lagenifera.* The nudibranch typically crawls along the central axis of each stalk, feeding on the gonophores, which are arrayed along the axis; it also reaches out onto the lateral branches to pick off the feeding polyps. The nudibranchs lay their eggs on the main axis of the hydroid. (Photograph by Michael McCartney)

Plate XXIV. The Strange Cambrian World

The shallow marine environment represented by the Cambrian Burgess Shale harbored a wide range of the earliest metazoan creatures on earth. (From Levinton, 1992, courtesy of *Scientific American*)

13

Benthic Life Habits

Life Habit Classification

Before going on to read about benthic life habits, it is important that you learn some basic terminology. **Benthos** are those organisms that are associated with the seabed, or **benthic habitats**. **Epibenthic** organisms (epifaunal, if they are animals) live attached to a hard substratum or rooted to a shallow depth below the surface, but most of an epibenthic organism projects into the water column. Seaweeds, limpets, crinoids, and corals all fit in this category. By contrast, infaunal organisms live below the sediment–water interface. They may be **burrowers**, such as clams and polychaetes, or they may be **borers** such as wood-boring shipworms and isopods. **Semi-infaunal** organisms live partially below the sediment–water interface but protrude above it. These include sea pens, which have a deeply rooted stalk but can protrude to varying degrees above the sediment–water interface. Some mussels live semi-infaunally, with the body partly in the sediment. Many **swimming** animals, such as scallops, sculpins, and shrimps, are essentially benthic. Finally, **interstitial** organisms live and move in the pore waters among sedimentary grains. These include foraminifera and harpacticoid copepods.

Because benthic animals are often collected and separated on sieves, a classification based upon overall size is useful. **Macrobenthos** include organisms whose shortest dimension is greater than or equal to 0.5 mm. **Meiobenthos** are smaller than 0.5 mm but larger than the **microbenthos**, which are less than 0.1 mm in size. Meiobenthos and microbenthos are often interstitial.

Feeding Classification

Suspension feeders feed by capturing particles from the water. They include feeders on bacteria (sponges, many ectoprocts), phytoplankton (most bivalve mollusks, many polychaetes), and zooplankton (corals, crinoids). **Passive suspension feeders** protrude a feeding organ into a current and collect particles as they are deposited. Crinoids are good examples of such organisms. By contrast, **active suspension feeders** draw particles toward the mouthparts by creating a feeding current. Siphonate bivalve mollusks are good examples. **Deposit feeders** ingest sediment and use organic matter and microbial organisms in the sediment as food. They include some bivalve mollusks and gastropods, many polychaetes, some sea cucumbers, and some crustaceans. **Herbivores** eat nonmicroscopic plants, such as seaweeds and sea grasses. They include sea urchins, some benthic fishes, and some polychaetes. **Microalgal grazers**, which feed on microalgae attached to surfaces such as rocks, include chitons, some snails, and some smaller polychaetes. **Carnivores** eat other animals and include asteroid starfish, many crabs, many fishes, anemones, and nemertean worms. However, there is no easy way to classify some organisms, such as the suspension feeders that ingest zooplankton. Finally, **scavengers** feed on carcasses and remains of other animals and plants. Many deposit feeders also scavenge. A good example of such species are the fiddler crabs, which are normally deposit feeders but can also tear apart dead fish.

Life in Mud and Sand

Mobility in Soft Sediments

Soft sediments are a mixture of inorganic particles, organic particles, and pore water. Benthic organisms are strongly affected by variations in all these constituents. **Particle size** is a good measure of current energy; large sedimentary grains are deposited by stronger currents, whereas fine-grained sediment indicates quieter water. Because currents carry suspended food to the bottom and also erode sediments, particle size reflects the current regime and helps to define the benthic environment. Burrowing organisms must push aside sedimentary grains, and the combination of grain size and pore-water conditions helps determine whether burrowing is possible. The range of adaptations to differing sediments and current regimes involves major differences in morphology, mode of feeding, and response to changes in water temperature, salinity, and pore-water chemistry.

✳ **Sediment grain size is an important determinant of the distribution of benthos and increases with increasing current strength.**

The size of soft-sediment particles affects the lifestyles of benthic organisms and also is a reflection of the hydrodynamic environment. Sediments consisting of only cobbles, for example, will lack the fine particles required of sediment-ingesting benthic animals. By contrast, sediments consisting of only very fine particles may be too unstable for a large and dense animal to maintain its necessary living position within the sediment. Grain size also reflects the current regime of the overlying water column. **The silt–clay fraction** is the percentage, by weight, of sediments finer than 62 μm in diameter. The **percent clay** (particles <4 μm) may also be useful in describing sediment properties relevant to benthic organisms.

Because stronger currents can transport larger particles, median grain diameter increases in areas of high current velocity (see Box Figure 13.1 for a discussion of the measure of grain size). Areas with high currents also experience extensive erosion and transportation of sediment. An animal in such an environment may be subjected to continual erosion from the substratum and must be capable of rapid reburial to reestablish its living position. Sediments from areas with low current strength have very small particle diameters. The abundance of sediment-ingesting animals may increase with the abundance of fine material, which contains more organic matter and small, ingestible inorganic particles with attached microorganisms.

✳ **Sediment sorting and grain size angularity also reflect the hydrodynamic regime.**

Sorting is an estimate of the spread of abundance of particles among the size classes. A sediment is poorly sorted when most of the sediment is spread over a large range of size classes, whereas in a well-sorted sediment, almost all the weight is confined to a few size classes, with a well-defined peak (Box Figure 13.1c).

A well-sorted sediment will be deposited in an environment with constant current strength. Sediment may also be well sorted at a given level on a beach, corresponding to a given amount of wave energy. Poorly sorted sediments usually reflect a heterogeneity of sedimentary processes. There may be variable depositional currents, or some historical reason for the sorting, such as the previous deposition of large cobbles by glacial streams, combined with present moderate currents. Some sediments may have more than one distinct size mode in the sediment (Box Figure 13.1c), and each different mode may indicate a different hydrodynamic regime.

✳ **In very shallow, sandy wave-swept bottoms, currents generate ripples and bars, which create strong microhabitat variation for benthic organisms.**

It is rare for the sediment surface to be completely flat. In areas of considerable current strength, surface sediment is transported continually, and a number of sedimentary structures may be established in equilibrium with this transport. On the largest scale, emergent and submergent bars may develop offshore. On a smaller scale, **ripple marks** commonly develop where sediment is in motion (Figure 13.1). In areas where currents are unidirectional, ripple marks are asymmetrical in cross section, with the steep slope facing downcurrent. In strong currents, such ripples

Text Box 13.1 Measuring Grain Size of Sediments

Median grain size is the simplest way to represent the overall particle size characteristics of soft sediments. By washing the sediment through a series of graded sieves, or by measuring the settling velocity of particles (larger particles settle faster in water), one can get the size class data to construct a **histogram of sizes** (Box Figure 13.1a). To accommodate a range of particle sizes of many orders of magnitude within one graph, we plot grain diameter in logarithmic form (log to the base 2, which means that a value of 1 in Box Figure 13.1 corresponds to 2 mm, a value of zero corresponds to 1 mm, and a value of −2 corresponds to a diameter of 0.5 mm). The diagram can be used to construct a **cumulative weight graph**, where the percent weights of the successive size classes are accumulated, and cumulative percent weight is plotted as a function of particle diameter (Box Figure 13.1b). The median diameter M, which corresponds to Q_{50}, is the particle diameter corresponding to 50 cumulative percent. Calculation of the 25 and 75 percent classes is also shown.

Sorting is a measure of spread among the grain sizes. This can be quantified by:

$$S = \frac{Q_{25}}{Q_{75}}$$

where Q_{25} is the grain size corresponding to the 25 percent cumulative weight (Box Figure 13.1b) and Q_{75} is the same value for 75 percent. As S approaches 1, the sediment is all the same size class and is perfectly sorted. Box Figure 13.1c shows examples sediments that have been poorly sorted and well sorted.

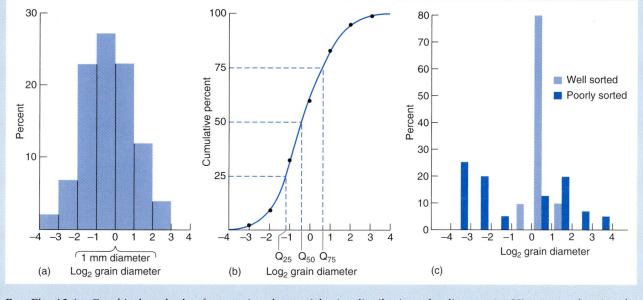

Box Fig. 13.1 Graphical methods of presenting the particle size distribution of sediments. (a) Histogram, showing the weight frequencies of each particle size class as a function of the log of particle diameter (we use the log of particle diameter to be able to plot an enormous range of particle sizes on a manageable scale). (b) Cumulative frequency distribution curve, showing Q_{25}, Q_{50} (the median particle diameter), and Q_{75}. (c) Examples of a poorly sorted and a well-sorted sediment.

may be reestablished to face in opposite directions with every change in tidal direction. By contrast, waves may produce sufficient oscillatory motion to generate symmetrical ripples.

The ripples create a local environment of their own, which strongly affects sediment stability and movement for organisms that are much smaller than the size scale of the ripples. For example, fine organic material tends to accumulate in the troughs, and deposit-feeding animals therefore are attracted to this

Fig. 13.1 Geometry of a sand ripple in a unidirectional current. Note direction of sand ripple movement from upper right to lower left, and the possibility of burial faced by invertebrates in its path. (Photography by the author.)

microenvironment. By contrast, the crests of the ripples are relatively bare of this material and are also localized sites of erosion. Because ripple crests migrate continually, benthic invertebrates may have to readjust their living position.

✳ Soft-sediment burrowers use hydromechanical and simple digging mechanisms to move through the substratum.

To penetrate the sediment, infaunal organisms must be able to displace sedimentary grains. Sediments may range from completely dry to plastic to very watery. Fine-grained sediments of moderate to high water content exhibit the phenomenon of **thixotropy**, wherein a force against the sediment is sufficient to permit further sediment displacement with a smaller force. In effect, the sediment becomes less resistant as you exert a concentrated shear force upon it.

The initial displacement of sedimentary grains requires that a firm structure be pushed into the sediment with a sufficient force. Many burrowing organisms, such as most worms and bivalve mollusks, have a soft and fluid-filled burrowing structure, which must be erected during the burrowing process. To accomplish this, these organisms have a **hydrostatic skeleton**, which is a flexible tube that can be stiffened by the injection of fluid. In the case of bivalve mollusks, the otherwise flaccid foot is filled with fluid and becomes a firm digging device. After the foot has pushed into the sediment, its distal end is engorged with fluid, creating an **anchor**. Contraction then carries the rest of the body along (Figure 13.2). Within the sediment, a part of the body can be dilated to form an anchor, so that another forward part of the body can be extended forward. A series of dilations and extensions allows the animal

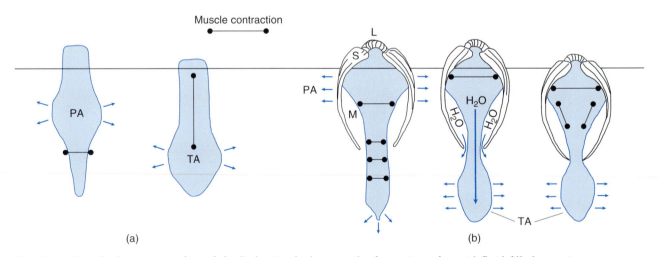

Fig. 13.2 (a) The burrowing of a soft-bodied animal, showing the formation of turgid fluid-filled mass into a penetration anchor (PA) and the dilation of a distal region, forming a terminal anchor (TA). Longitudinal muscles then drag the animal into the sediment. (b) How a clam uses its shell and fluid-filled foot to burrow. Left: clam is in sediment and presses shell outward, forming an anchor. At the same time, its fluid-filled foot thrusts into sediment. Middle: the foot fills with fluid at the tip, forming a new anchor. Right: muscle contraction draws the shell together and drags it downward. (After Trueman, 1975.)

to move within the sediment. This general principle applies to burrowing in mollusks, polychaetes, sipunculids, burrowing sea cucumbers, and other wormlike animals.

The other major mode of burrowing involves the use of **mechanical displacement**, based upon firm structures that act as spades and are moved by muscular action. A wide variety of crustaceans dig into the substrata by means of specialized digging limbs. For example, the mole crab, *Emerita talpoida*, has spadelike posterior appendages. Inarticulate brachiopods use a complex musculature to alternately push and rotate the two opposed valves through the sediment. The muscular rocking motion keeps the valves moving and constantly displacing sedimentary grains (Figure 13.3).

* Interstitial animals adapt to water flow and life in small spaces among particles by means of a simplified body plan, a wormlike shape, or by adhering to particles by means of mucus, suckers, and hooks.

Fig. 13.3 Burrowing inarticulate brachiopods have two symmetrical shells, which are connected by a complex musculature. They burrow in the substratum by scissoring the shells back and forth, which shovels the sediment aside and pushes the animal downward. (Courtesy of Charles W. Thayer.)

Interstitial animals[1] move among sedimentary grains but do not displace them in bulk, as do burrowing animals. Because they move through tight spaces, interstitial animals from many different phyla have evolved a wormlike shape and a simplified external body plan (Figure 13.4). Relative to their epibenthic relatives, for example, interstitial hydroids have reduced numbers of tentacles, for which capture of suspended prey from the water column is important. Smaller interstitial forms may be attached to sand grains by a variety of hooks and suckers.

The slender body form of some interstitial forms may be related to uptake of dissolved organic matter for food. Nematodes living in the low-oxygen parts of sediments (see the following) tend to be more slender than those living in the aerobic surface sediments. The slender form may be a design to increase surface area to facilitate uptake of dissolved organic matter.

The Soft-Sediment Microzone

* Sediments consist of an oxygenated layer overlying an anoxic zone.

If you dig into a protected sandy beach, you will first encounter light brown sediment but will soon reach a thin grayish zone, and then a black layer with an odor like that of rotten eggs. The changes in color and smell reflect a change of chemistry. The light brown zone contains pore water with dissolved oxygen, whereas the black smelly zone is devoid of oxygen, and the gray layer is a transition zone between the two. The smell in the black zone derives from hydrogen sulfide, H_2S, which is generated by sulfate-reducing bacteria. Overall, the oxic–anoxic zonation results from a shifting balance between addition and consumption of dissolved oxygen in the pore waters. The boundary between the oxygenated zone and the anoxic zone is known as the **redox potential discontinuity**, or RPD. It represents a sharp boundary between chemically oxidizing and reducing processes (Figure 13.5).

Near the sediment–water interface, oxygen diffuses into the pore water from the overlying water column, or it may be stirred in by means of current or wave action. As a result, the sediment in exposed beaches may be oxygenated down to depths of nearly a me-

---

[1] Interstitial animals are usually meiofauna, although benthos of meiofaunal size may also be epibenthic.

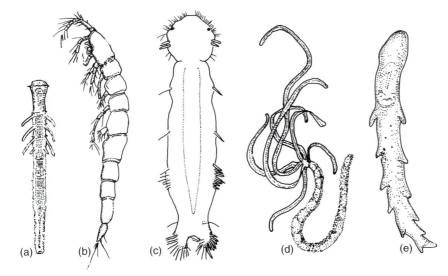

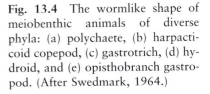

Fig. 13.4 The wormlike shape of meiobenthic animals of diverse phyla: (a) polychaete, (b) harpacticoid copepod, (c) gastrotrich, (d) hydroid, and (e) opisthobranch gastropod. (After Swedmark, 1964.)

ter. In quiet areas, especially in organic-rich, fine-grained sediments, the transition to the anoxic zone can occur only a few millimeters below the sediment surface. Infaunal organisms may carry in oxygen somewhat deeper by stirring the sediment or by irrigating their burrows. The combined action of infaunal organisms may bring oxygen down to depths of several centimeters in muddy sediments that would otherwise have their oxygen content controlled by diffusion and never surpassing a few millimeters depth. Because of vertical burrowers, the RPD may

not always be a horizontal surface, but may be vertical in places, parallel with tubes and burrows (see later: Figure 13.10).

* **In sediments in quiet water, there is usually a vertical zonation of species of microorganisms.**

As will be discussed later, microorganisms are important as food for soft-substratum benthos. Microorganisms are consumed as food directly, but they also help decompose particulate organic matter, which is another potential food source. The vertical gradient in oxygen, created by microorganisms, also affects the composition of the microorganism community. Near the surface, aerobic organisms can survive, but only anaerobic organisms can survive below the RPD. Below the RPD, nearly all animals must maintain contact with the sediment above the RPD by means of siphons, irrigated burrows, and tubes. It has been argued that a few metazoans, such as some nematodes, can survive without oxygen and that some macroinvertebrates can live for extended periods on the proceeds of anaerobic metabolism. The anoxic community, known as the **thiobios**, was first described by T. Fenchel and R. Riedl.[2] Some researchers, such as Riese and Ax,[3] have argued that this community does not really exist in truly anoxic sediments, but only in sediments of very low oxygen

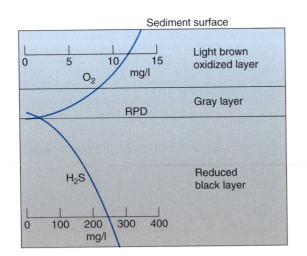

Fig. 13.5 Cross section of the sediment near the sediment–water interface, showing the redox potential discontinuity (RPD), which is a boundary between oxidative and reducing processes. The diagram shows the concentration of oxygen (above) and hydrogen sulfide (below).

[2] See Fenchel and Riedl, 1970, in Further Reading, Soft Sediments and Deposit Feeding.
[3] See Riese and Ax, 1979, in Further Reading, Soft Sediments and Deposit Feeding.

that are adjacent to anoxic microzones. Some protozoans are clearly anaerobic and contain symbiotic anaerobic bacteria.

Microbial organisms may be autotrophic or heterotrophic. Recall that autotrophic organisms produce their own carbohydrates or sugars by means of either photosynthesis or chemoautotrophy. Photosynthesis employs light as an energy source, whereas chemoautotrophy employs one of several chemical substrates (e.g., sulfate, hydrogen) to derive energy.

Figure 13.6 illustrates a generalized zonation of microbial communities in soft sediments. At the surface, aerobic photosynthetic microorganisms, such as diatoms and cyanobacteria, may predominate. These coexist with heterotrophic aerobic bacteria, which use oxygen and contribute to the oxygen depletion of the pore waters. Oxygen is the terminal hydrogen acceptor in the decomposition process. In the anaerobic zone, however, heterotrophic microorganisms use a variety of compounds as hydrogen acceptors. Most notable are **fermenting bacteria**, which use organic compounds and produce end products such as fatty acids and alcohols, and the deeper **sulfate-reducing** bacteria, which reduce SO_4 to H_2S. The reduced compounds diffuse upward and are used by chemoautotrophic **sulfur bacteria** (which oxidize H_2S) at the RPD region. Beneath the sulfate reducers are **methanogenic bacteria**, which break down organic substrates and produce methane as an end product.

Environmental constraints (e.g., the presence of oxygen) and energetic payoff combine to determine the successive dominance with depth of different heterotrophic bacteria groups (aerobic, fermentative, sulfate-reducing, and methanogenic). In the presence of oxygen, much more energy is obtained from the energy-efficient aerobic breakdown of organic matter by bacteria. Beneath this zone, however, there is little or no oxygen, and the energetically less efficient process of fermentation is performed by different bacteria. The processes of sulfate reduction and methanogenesis are still lower in energy efficiency. Ultimately, heterotrophic bacterial activity is limited with depth by the lack of a food source. This can be shown by the steady decrease of substrate use with depth into the sediment.

Feeding in Soft Sediments

✳ Deposit feeders ingest sediment and derive their nutrition mainly from microalgae and particulate organic matter, and to a smaller extent from the scarcer bacteria.

Deposit feeders are animals that ingest sedimentary material and derive their nutrition from some fraction of that material. As is obvious from preceding sections, sediment is a complex mixture of inorganic material, microorganisms, decomposing organic material, and pore water with dissolved constituents. Understanding the nutrition of these creatures therefore is a complex task, not at all like watching a caterpillar chew on a leaf! Deposit feeders tend to be more abundant in fine-grained sediments, but such sediments contain increased quantities of microorganisms, fine-grained particulate organic matter, *and* ingestible inorganic particles. Thus, simple correlations will not solve the question of nutrition.

Deposit feeders feed in a variety of ways that are associated with phylogenetic origins of the organisms and the environments within which they live (Figure 13.7). Representatives of many animal groups swallow sediment nonselectively, although there is an upper limit on the size of particle they can ingest. Many polychaetes have tentacles, which gather particles by means of a mucus-laden ciliated tract (Figure 13.7a). Sea cucumbers, such as the large northeast Pacific *Parastichopus californicus*, draw sediment into the mouth by means of a large crown of tentacles. Bivalves in the group Tellinacea use a separate inhalent vacuum hose siphon to suck up sedimentary grains (Figure 13.7c). In some other groups, the sediment is processed quite noticeably before a residue is ingested by the deposit feeders. Many amphipods tear particulate material apart and ingest considerably smaller particles (Figure 13.7e). Fiddler crabs handle sediment extensively and ingest only the fine particulate organic matter; they reject the inorganic sand grains. Although tellinacean bivalves may ingest particles by

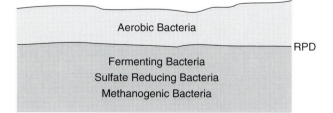

Fig. 13.6 Typical vertical zonation of bacterial components of quiet, muddy marine sediments.

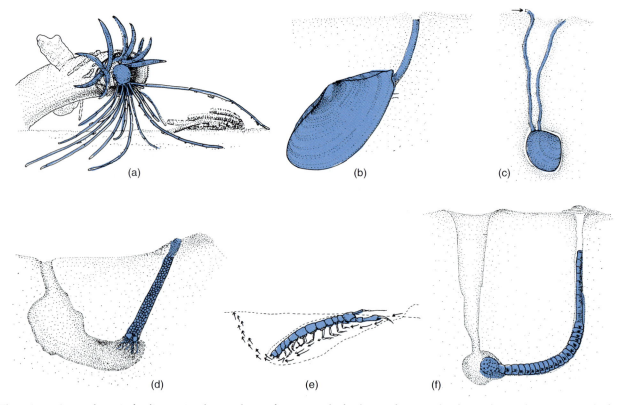

Fig. 13.7 Some deposit-feeding animals: (a) the surface tentacle-feeder *Hobsonia*, (b) the within-sediment tentacle-feeding bivalve *Yoldia limatula*, (c) the surface deposit-feeding siphonate bivalve *Macoma*, (d) the within-sediment feeding Atlantic polychaete *Pectinaria gouldii*, (e) the surface-feeding *Corophium volutator*, and (f) the deep-feeding *Arenicola marina*. (Drawing of *Hobsonia* copied with permission from an original by P. A. Jumars.)

means of a vacuum cleaner–like siphon, particle sorting occurs on palps and even in the digestive tract. Even in the so-called nonselective feeders, specialization is usually common. Feeding rate may increase if the sediment is richer in food.

Feeding structures used for deposit feeding are modified or used for other types of feeding in related groups; for example, the tentacle crown of deposit-feeding sea cucumbers is often used for suspension feeding in other sea cucumber species. When viewed from the perspective of the sediment column, there are a few distinct types of deposit feeder. **Surface browsers** use tentacles or siphons to collect surface sediment, which is rich in photosynthetic microbes such as diatoms. For example, spionid polychaetes have two tentacles that are pressed to the surface and thus collect both particles and attached photosynthetic forms such as diatoms. At the other end of the spectrum, **head-down deposit feeders** (e.g., many ver-

tical-tube-dwelling polychaete annelids) maintain their long axis vertical, consume particles at depth, and defecate at the surface (see later: Figure 13.10).

A series of experiments gave us some important insights on how deposit feeders deal nutritionally with the complex sediment to which they are exposed. In a classic series of experiments, B. T. Hargrave[4] fed decaying leaves to a freshwater amphipod and found that its ability to digest and assimilate the material was very low, in contrast to its high efficiency at digesting bacteria. These results were found to be similar for several marine invertebrate deposit feeders of widely varying origin (marine amphipods, gastropods, sea cucumbers), showing the results to be of broad application. The **microbial stripping hypothesis**, developed in the light of this finding, states that

[4] See Hargrave, 1970, in Further Reading, Soft Sediments and Deposit Feeding.

particulate organic matter is relatively indigestible and that microbial organisms are therefore the main source of nutrition for deposit feeders. To be nutritionally useful for deposit feeders, therefore, particulate organic matter (POM) must be decomposed and converted by microbes into digestible microbial tissue. Particulate organic matter is indigestible, particularly because deposit feeders usually lack sufficient cellulase enzyme activity to digest the complex carbohydrates in the POM, which is commonly derived from sea grasses that are rich in cellulose.

Particulate organic material is decomposed by three processes, which often act simultaneously. **Fragmentation** involves the breakdown of large particles into smaller ones. This may involve breakup by current action at inherent weak points; but animals, such as amphipods, may tear up material as they feed upon it. This reduces the grain size and increases the surface area available for microbial attack. **Leaching**, the loss of dissolved materials from once-living organisms, is accelerated by mechanical fragmentation. Finally, **microbial decay** is the active use of POM nutrients by surface-bound microbes. As microbes colonize, they enrich particulate organic matter with nitrogen (Figure 13.8). In intertidal environments, much decay can be attributed to marine fungi, especially in marsh grasses. Heterotrophic bacteria also contribute strongly to the microbial decomposition.

Grazing on the benthic microbial community stimulates microbial productivity and, by extension, detrital decomposition (Figure 13.9). Oxygen consumption by microbes increases while the organisms are being grazed by consumers. The mechanism behind this stimulation is not well understood. Grazing may reduce the standing crop of bacteria and select for metabolically active cells with higher cell division rates. Of course, with very high grazing rates, microbial biomass is cropped down faster than it can be renewed, but stimulation is the rule at intermediate levels of grazing.

The microbial stripping hypothesis is imperfect in several respects. First, although the digestion and assimilation of particulate organic matter may be inefficient, POM may be far more abundant than microbes. Sediments in sea grass meadows contain large amounts of decaying sea grass, and deposit feeders cannot help but ingest much of this material. Thus, a poor rate of uptake may be balanced by the sheer abundance of the poor food source. Many other sources of POM exist in marine habitats, particularly

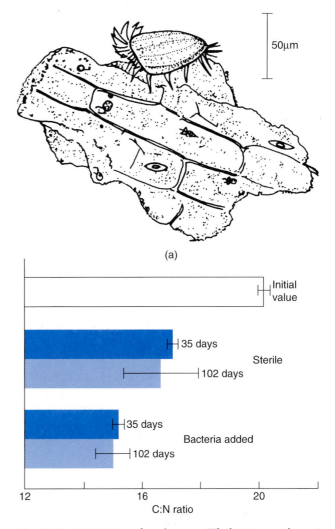

Fig. 13.8 (a) A piece of turtle grass (*Thalassia testudinum*) detritus and its microbiota. (Redrawn from Fenchel, 1972.) (b) Change in the carbon-to-nitrogen ratio in particulate organic matter over time, with and without the presence of bacteria. (Modified from Harrison and Mann, 1975, [c] Blackwell Scientific Publications, Ltd.)

the rain of dead phytoplankton in shallow embayments and on the continental shelf. Near shore, seaweeds may provide a significant input of POM. As it turns out, seaweed detritus is far more digestible to deposit feeders than is grass detritus, and seaweed detritus can fuel deposit feeder population growth. In kelp forests, the rain of decomposing seaweeds supports large populations of benthic suspension feeders.

Because both POM and some microbial organisms

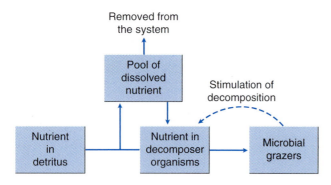

Fig. 13.9 The relationship between deposit-feeding microbial grazers and microbial decomposers, particulate organic matter, and the dissolved nutrients in the pore waters of sediments. Microbial grazers remove microbes, but they also stimulate decomposition by increasing microbial activity and by tearing apart particulate organic particles, which exposes more surface area to mechanical attack. (Modified after Barsdate et al., 1974.)

are important as food sources, certain habitats are probably dominated by only one type of source. For example, most of the available organic matter in sandy intertidal flats is in the form of living microbial organisms. Here the question of deposit feeder nutrition is relatively simple. By contrast, mud flats adjacent to salt marshes contain a complex mixture of POM derived from the cordgrass *Spartina* and seaweeds, combined with bacteria and diatoms and other microbial photosynthesizers. In subtidal shelf bottoms, the sediment is probably dominated by detritus derived from the rain of dead phytoplankton and indigenous bacteria, and even organic material adsorbed onto sedimentary grains.

Some quantitative estimates set some limits on the value of certain food sources. Although heterotrophic bacteria are usually readily digestible, they are too sparse in the sediment to be able to support macrofaunal deposit feeders. This can be shown by comparing the energetic content of bacteria in the sediment with the energetic requirements of a given deposit-feeding population. A number of studies suggest that bacteria can supply only a small fraction of the energy requirements of a macrofaunal deposit feeder such as a polychaete or a bivalve mollusk. These estimates apply only to bulk calculations of needs for carbon and nitrogen. It is likely that some fractions, such as sediment bacteria, are essential in the supply of important micronutrients such as specific fatty acids, amino acids, and vitamins. Rich diatom mats may be a more adequate food source, but

only in intertidal and very shallow subtidal sediments, and surely these cannot be very important in even the relatively shallow waters of bays and estuaries, below the light compensation depth. In estuaries and on the continental shelf, the spring diatom increase is often followed by a sinking of detritus, and this POM may be crucial in fueling the productivity of the deposit-feeding benthos (see Chapter 16).

In contrast to macrofauna, meiofauna probably depend mainly upon a combination of bacteria and fine-grained particulate organic matter. Because of their small size, they cannot feed on particles much larger than 10–30 μm.

We are still at a very immature stage in our understanding of the food value of the sediment for deposit feeders. On the one hand, we know that a simple measure of carbon or nitrogen of sediments does not necessarily tell us how much food is available for deposit feeders. Much of the carbon may be bound up as indigestible complex carbohydrates such as cellulose, and nitrogen may also be bound up as indigestible nitrogenous compounds. Lawrence M. Mayer and colleagues[5] have greatly advanced our perspective on this subject by developing methods to detect potential digestively available amino acids in sediments, which presumably come from the recent deposition of fresh organic matter such as phytodetritus and newly deposited fragments of seaweed fronds. A simple assay involves exposing the sediment to proteolytic (protein-digesting) enzymes (from bacteria, but the action is probably fairly similar to equivalent enzymes in deposit-feeding animals). Then the amount of amino acids released into the enzyme–sediment broth is measured. An interesting finding is that the rate of release of amino acids by the enzyme approach is sufficient to suggest that it could happen within the time span of a single passage through a deposit feeder's gut. This type of extraction could be used as an assay to examine the seasonal change in proteins and peptides in sediments that are relevant as food available to deposit feeders. For example, low-density particles in muddy sediments are strongly enriched in protein and peptides, which confirms the idea that fine low-density particles are a possible food source for deposit feeders.[6] Deposition of phytoplankton cells in the spring is a

[5] See Mayer, and others, 1995, in Further Reading, Soft Sediments and Deposit Feeding.
[6] See Mayer and others, 1993, in Further Reading, Soft Sediments and Deposit Feeding.

major source of sediment protein, but as time goes by, the material becomes more refractory and probably more difficult for deposit feeders to digest and assimilate.[7] It is of great interest that many larger deposit feeders employ surfactants (compounds with detergent properties) in high concentration, which probably enable the stripping of organic material from particle surfaces.[8] This activity may allow some deposit feeders to digestively attack relatively refractory organic material in sediments.

In conclusion, it appears that macrofaunal deposit feeders derive their nutrition from most of the conceivable sources within the sediment. Bacteria are probably not a significant component for any but interstitial and very young macrofaunal forms, and POM and benthic microbial algae combine as food for deposit feeders.

✳ Microbes and particles comprise a complex renewable resource system for deposit feeders.

As we have discussed, many sediments are dominated by POM, but some consist mainly of microbes and particles. In intertidal soft-bottom flats, microbes such as benthic diatoms are the main food source for surface feeders. In these cases, microbes may be (a) free-living among the sedimentary grains, (b) attached to sedimentary grains, or (c) living as a mat on the sediment surface. Because the microbes themselves seem to be limited by some resource, the abundance of microbes at any one time is a balance between the microbial population growth rate and the grazing rate. At high grazing rates, the steady-state abundance of diatoms is kept at a low standing stock.

A balance between production and renewal may also apply to particles that bear attached microbes. Some deposit-feeding invertebrates, such as polychaetes and gastropods, consume fine particles and bundle them into fecal pellets that are often not reingested. Deposit feeders may live in a mixture of fecal pellets and fine particles, ingesting only the latter. In some cases, the deposit feeder may try to get rid of the pellets. For example, the Pacific ampharetid polychaete *Amphicteis scaphobranchiata* has a specially modified branchium that flings fecal pellets out of the animal's feeding reach.

When such behavior is not possible, the deposit feeder must wait for the pellet to break down into its constituent particles before it can reingest the sediment. In crustaceans, pellets are often surrounded by a distinctive coating, and in mollusks the sediment is bound together by mucus. As the pellets break down, they are probably colonized by microbes, so there is a value to having the particles sequestered for a time. Presumably, the nutritive value of a new fecal pellet is far less than that of one that has had some time to simultaneously break down and be recolonized by microbes. In such a system, there will be an analogous equilibrium abundance of ingestible particles, which is determined by the competing rates of pelletization and pellet breakdown. Some sediments are completely pelletized, and, in these cases, deposit feeders may have reduced access to particles for ingestion. Mud snails of the genus *Hydrobia* slow down feeding and may emigrate from microsites with fully pelletized sediments.

✳ Many benthic animals do not feed directly on microorganisms but harbor symbiotic chemoautotrophic bacteria, which derive energy from dissolved ions in seawater.

Although many benthic animals feed actively on sediment, or on suspended organic matter (see section on suspension feeding), a large number of species depend upon **symbiotic bacteria**, which may live intracellularly or in chambers in various organs, depending upon the group. Many bivalve mollusks, for example, have bacteria living intracellularly in their gills. These bacteria oxidize reduced sulfur compounds. The oxidation processes provide energy, which is used by the bacteria to manufacture ATP, which, in turn, is used in bacterial cellular metabolism. Some species of the infaunal bivalve genus *Solemya* have a very small gut or lack one entirely. These forms rely exclusively on symbiotic sulfur-oxidizing bacteria. The animals are also tolerant to sulfide, which normally is quite toxic, especially to animals that use oxygen in metabolism. Mussels living near hydrocarbon seeps have intracellular bacteria in the gills. These bacteria rely upon methane from the seeps for nutrition and energy. The bivalves rely exclusively upon the bacteria for nutrition. This life habit is especially prominent in some deep-sea environments that are poor in organic matter but rich in sources of oxidizable sulfur compounds (see Chapter 16).

---

[7] See Mayer and Rice, 1992, in Further Reading, Soft Sediments and Deposit Feeding.

[8] See Mayer and others, 1997, in Further Reading, Soft Sediments and Deposit Feeding.

Burrowers and Sediment Structure

* Deep feeders cause overturn of the sediment and strongly affect the soft-sediment microzone.

Sediments with abundant burrowing animals may bear no resemblance to sediments that are relatively animal free. Donald C. Rhoads[9] investigated the properties of burrowed sediments and found their mechanical properties to be quite different. The production of fecal pellets may increase the grain size of the sediment. If a sediment with abundant deposit feeders is wet-sieved, it tends to be dominated by fecal pellets, which are often on the order of 50–150 μm in size. If the same sediment is placed in a blender and sieved, one finds that its constituent particles are more of the order of 50 μm or less. Burrowing, deposit feeding, and production of fecal pellets tends to make the sediment in the top few millimeters very watery, sometimes over 90 percent water.

* Head-down deposit feeders create biogenically graded beds.

As mentioned earlier, many deposit feeders feed in a head-down position and defecate sediment at the surface (Figure 13.10). Head-down deposit feeders tend to ingest particles that are smaller than the average size for the sediment. Such animals may select fine particles and transport them to the surface, leaving a lag deposit of coarser material at depth. For example, the bamboo worm *Clymenella torquata* usually does not ingest particles greater than 1 mm. In poorly sorted muddy sediments, dense populations produce a **biogenically graded bed**, with small particles concentrated at the surface (Figure 13.11).[10] Such biogenically graded beds can be easily detected by walkers who suddenly encounter squishy sediment.

* Deposit feeders can optimize their intake of food by adjustments of particle size to be ingested and adjustments of gut passage time.

[9] See Rhoads, 1967, in Further Reading, Soft Sediments and Deposit Feedings.
[10] See Rhoads, 1967, in Further Reading, Soft Sediments and Deposit Feeding.

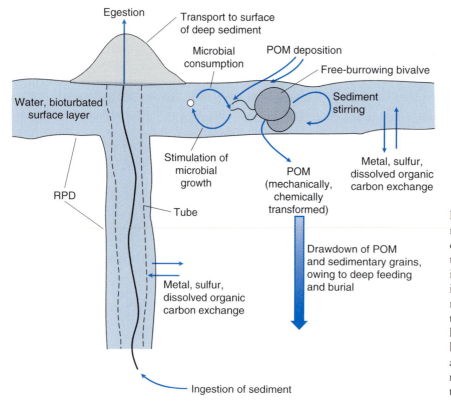

Fig. 13.10 General processes occurring within a sediment dominated by deposit feeders, including various transport processes. In deposit feeding, particles are taken up by a feeding organ, and some of them may be rejected before entering the gut. Particles may be packaged in fecal pellets, which are egested. As the pellets break down, the sedimentary grains are recolonized by microbes, which may be ingested and assimilated as the particles are ingested once again.

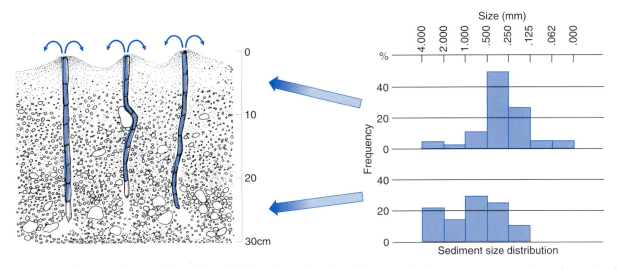

Fig. 13.11 Left: vertical reworking of intertidal sediments by the tube worm, *Clymenella torquata*; right: change in the vertical distribution of particle size as a result of vertical reworking of the sediment. (After Rhoads, 1967.)

Natural selection can be expected to optimize food choice and feeding rate to maximize fitness. In the case of deposit feeders, sediments with fine particles tend to be easier to ingest, but fine particles should be preferred in complex sediments. This is related partially to the ease of ingestion of smaller particles, but the food value may also have an influence. A unit volume of small particles may have more attached microbial organisms, owing to the greater surface area per unit volume of the particles. This has been shown to be the case for surface-bound bacteria, but some sediments may be characterized by very complex relationships between grain size and organic matter and microbes. If the expected relationship holds, then deposit feeders should select for fine particles, assuming that the cost of rejecting those particles is not too great.

Feeding rate and gut passage time may also be regulated according to food value. There may be an optimal feeding rate simply because feeding too quickly reduces the time available for digestion, whereas feeding too slowly may waste valuable time that could otherwise be applied to feeding on new material. This idea could be tested by consideration of foods of differing quality. If the cost–benefit approach is appropriate, deposit feeders should feed more rapidly on richer sediments. This has turned out to be true in several experiments on polychaetes, for which feeding and gut passage are steady.

* **Most infaunal suspension feeders in soft sediments generate a current into the burrow or siphon, and strain plankton on a feeding organ.**

Many suspension feeders live infaunally and semi-infaunally in soft sediments. For example, the siphonate infaunal bivalve mollusk *Mercenaria mercenaria* creates a current by means of a ciliated gill. Water is drawn into an inhalant siphon, and the cilia strain and sort particles (Figure 13.12). The polychaete *Chaetopterus* lives in a U-shaped burrow, and specialized parapodia drive an inhalant current into the tube. A sheet of mucus stretched between another pair of specialized parapodia captures particles, and this sheet is periodically rolled into a ball and passed to the mouth. Most soft-sediment suspension feeders rely on phytoplankton for food. In coastal waters, large numbers of detrital particles are in the water column, and these are digested poorly. Benthic algae, however, are often resuspended and these may be an important food for benthic suspension feeders.

* **Hydrodynamic forces at the sediment–water interface cause sediment transport, which often induces switches of feeding from deposit feeding to suspension feeding.**

Deposit feeders are usually found in fine sediments, which are deposited in relatively quiet water. In sand

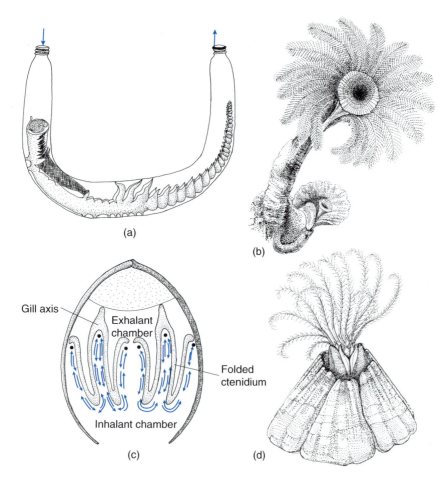

Gill axis

Exhalant chamber

Folded ctenidium

Inhalant chamber

(a)

(b)

(c)

(d)

Fig. 13.12 Some suspension-feeding invertebrates. (a) The active suspension-feeding parchment worm *Chaetopterus*. (b) The suspension-feeding polychaete *Serpula*, which uses ciliary currents to draw particles to tentacles. (c) Cross section of a bivalve mollusk, an active suspension feeder (arrows denote ciliated tracts transporting particles). (d) The acorn barnacle *Semibalanus balanoides* with cirri protruded like a basket, the concave side pointing into the flow, and particles trapped on feeding appendages, which are then withdrawn. If the flow increases beyond a point, the basket is reversed, to maintain stability in the flow.

flats, however, deposit feeders may dominate in areas where sediment is transported extensively, owing to tidal and wave action. In a sandy bay, sediment transport may far exceed recycling of the sediment in place by deposit feeders. In other words, in such habitats deposit feeders may look upstream for their next meal. During some periods, the bottom is quiescent, as evidenced by the presence of distinct fecal mounds that continually build up with each tidal cycle. In many cases, however, these mounds are eroded away, indicating that the animals have fresh sediment to deal with from tide to tide or even from wave to wave. In such cases, there is no need for the animals to deal with accumulations of fecal pellets because the waste is transported away. During quiescent periods, pellets build up and feeding may decrease, but the next food supply may be only a wave or a tide away.

If wave action is sufficiently strong, deposit feeders may change their behavior significantly. In mov-

ing waters, some of the normally deposit-feeding tellinacean bivalves switch to suspension feeding. This may be a reaction to particle saltation, which would be common during wave action. Spionid polychaetes have tentacles, which in quiet water pick up particles from the sediment surface. If current speed is increased sufficiently to transport particles above the bottom, these worms deploy their tentacles in an erect spiral, which then serves as a suspension-feeding organ. If prominent sedimentary structures, such as sediment ripples, develop, animals may locate themselves in hollows so that they may feed efficiently, perhaps even moving toward fine particles that may accumulate in the hollows.

✱ Suspension feeders and deposit feeders must be able to avoid clogging from heavy particle loads.

When water moves above a soft sediment, the erosive power of fluids eventually saltates particles into suspension. For suspension feeders, this process dilutes their plankton food source with unwanted inert particles such as sand grains. Higher particle loads usually clog suspension-feeding organs, such as ciliary tracts and siphons. At very high water velocity, sediment moves laterally, and ripples form. As crests and troughs alternately pass over a suspension-feeding animal, it becomes difficult for the animal to maintain a stable feeding position. Water eddies often form in the trough of a ripple, which creates a complex flow pattern.

Infaunal animals have a variety of means of dealing with increasing particle flux near the sediment–water interface. Some suspension-feeding siphonate bivalves have a ring of papillae at the siphon opening, which can act as a protective network against influxes of sand grains. The inhalant siphon of some tellinacean bivalves is lined with papillae, which can help in rejecting unwanted sand. Most eulamellibranch bivalves can "sneeze," or suddenly expel water and an overload of sand through the inhalant siphon. In cases of extreme sediment instability, many infaunal animals must continually readjust their burrowing position. Bivalves such as the surf clam *Spisula solidissima* are continually unburied in the surf zone and must be rapid burrowers. A remarkable series of species that have adapted to a tidal cycle of erosion and burial will be discussed in Chapter 14.

Benthic Life on Hard Surfaces and in Moving Waters

Hard surfaces include rock, clamshells, coral skeletons, or any other surface that does not consist of grains that can be pushed apart. Hard-surface organisms include forms cemented along a flat surface (e.g., acorn barnacles), forms attached by threads or knobby structures (e.g., mussels, seaweeds), and mobile forms (e.g., snails, sea urchins).

Flow and Orientation of Sessile Benthos

✳ **Organisms must adapt to strong laminar flow above the bottom, but near the surface they experience more complex flows of lower strength.**

Sessile benthos may live in widely different current regimes depending upon their living position relative to the sediment–water or rock–water interface. A directional current may be far more irregular near the interface. This has effects on species of different sizes and especially upon species that may experience complex flow as small juveniles but strong directional flow as adults. This can be seen in the sea fan *Gorgonia*. Small colonies have an irregular shape and no preferred orientation, in contrast to larger colonies, whose fan shape is oriented approximately normal to unidirectional currents. It may be that the near-bottom currents are erratic in direction, owing to irregularities of bottom topography and surrounding erect organisms. As the colony grows, it probably protrudes more directly into the main current. The fan-shaped colony probably "grows" into its orientation.

This difference in current direction can be seen among species of feather star crinoids. Feather star species found in crevices generally experience multidirectional currents and have their pinnules arranged in four rows at approximate right angles, which maximizes food capture from several possible directions. By contrast, the erect Caribbean feather star *Nemaster grandis* protrudes strongly into unidirectional currents and has its pinnules arranged in a plane, which maximizes capture under these circumstances. In the brittle star *Ophiothrix fragilis*, tube feet arise from either side of the tentacle and are also arranged in a two-row plane (Figure 13.13). Food particles are captured by the tube feet and are compacted into a mucus-clad bolus that is passed down the arm.

✳ **Sessile epibenthos may experience pressure drag and must have traits to minimize drag by changing orientation.**

Sessile benthos feel drag when they protrude from the bottom into a strong current. (Recall that drag is a force parallel to the direction of the current.) In some cases, the shear force on a nonflexible body may be sufficient to tear it from the bottom or snap a weak but erect skeleton. To reduce this problem, sessile organisms must be able to minimize drag by adjustments of behavior and by having shapes and orientations that minimize drag. There are two distinctly different types of situation. Passive animals are oriented by a strong unidirectional current until some equilibrium orientation is achieved. In many cases, this orientation is no different from that achieved by an inanimate object of the same shape and bulk density. By contrast, active animals may use muscular ac-

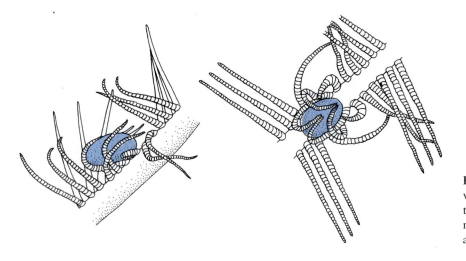

Fig. 13.13 The food-collecting wave of a suspension-feeding brittle star, showing the planar arrangement of the pinnules. (From Warner and Woodley, 1975.)

tion, behavior, or other means to orient actively in a flow. The active adjustment minimizes drag, while maintaining access to food.

Under conditions of strong flow it is hard to maintain an erect position. As explained in Chapter 6, pressure drag would be maximal upon a cylinder with its long axis maintained perpendicular to the flow. For example, the Caribbean elkhorn coral *Acropora palmata* (Figure 13.14) forms large erect branching colonies, sometimes greater than 2 m across. In the strong currents of exposed reefs, a branch growing perpendicular to the flow could easily snap off. As a result, colonies protruding into strong unidirectional flow tend to have their branches pointing nearly horizontally, to minimize the great potential shear. Obviously, this coral is stiff and cannot bend over in a current. Both sessile animals and plants can be flexible and thereby reduce drag. The anemone *Metridium senile* can grow to be quite tall, sometimes exceeding a meter in length. The animal is quite flexible, so in fast currents the body bends over and points downstream. In this posture, the bushlike crown of tentacles can collect food particles in the wake of the flow, although the crown may be withdrawn at very high current velocities. Some seaweeds are equally flexible, and the stipe can bend and point downstream to minimize drag in a strong current. The kelp *Nereocystis luetkeana* has a series of fibrils in the cortical cell walls that have an average angle of 60 degrees to the axis of the stipe. This increases the extensibility of the stipe, which prevents breakage in a strong current.

The eastern Pacific stalked sea squirt *Styela montereyensis* is remarkable for its occurrence in a wide array of environments on the California coast, ranging from wave-swept outer coasts to quiet bays. In quiet water, this species resembles typical solitary sea squirts, and the siphons orient upward. In outer coasts, most animals are attached by relatively slender stalks and sway with the wave surge (Figure 13.15). The incurrent siphon is bent approximately 180 degrees. When the animal is bent over by the current, the water can ram into the siphon, which facilitates flow of water and particles. It is not clear how these two morphologies are determined; they may be genetic variants.

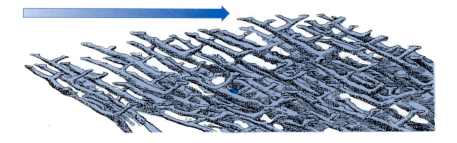

Fig. 13.14 The elkhorn coral *Acropora palmata* has a stiff skeleton. In a strong unidirectional or bidirectional flow, the colony grows with the branches subparallel to the current, which minimizes drag and the possibility of snapping off of coral branches.

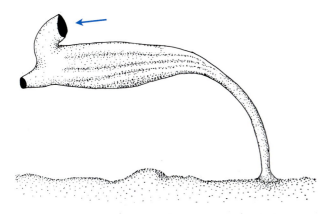

Fig. 13.15 Orientation of the outer-coast form of the sea squirt *Styela montereyensis*. Note that a current causes the individual to bend, but the incurrent siphon is bent, forcing its opening to face the current, which facilitates intake of water and particles for food. (After Young and Braithwaite, 1980).

✳ **Mobile benthos may passively orient to the position that minimizes drag.**

The rules of drag are not very different for nonsessile benthos. The orientation that minimizes drag is usually preferred. Mobile snails often find themselves in strong currents, and an orientation that places the axis of coiling parallel with the current will minimize pressure drag. The mud snail *Ilyanassa obsoleta* often finds itself in tidal creeks with flow exceeding 30 cm s^{-1}, and orients with the apex pointing upstream. Another possibility is to have a very low profile, to permit streamlines to flow smoothly over the shell. Many intertidal limpets live under conditions of severe wave shock and have such a low profile.

Animals that are not sessile have the luxury of relatively rapid movement, hence can remove themselves from limiting flow conditions. The Caribbean feather star *Nemaster grandis* lives on coral promontories and seems to prefer high flow conditions. Nevertheless, flow occasionally becomes so extreme that the feather star cannot protrude its arms and tube feet into the mainstream. In the case of strong currents, it actively moves to a protected crevice. Even snails may protrude so far into a stream that they are unable to resist the shear stress. The mud snail *Ilyanassa obsoleta* responds to strong flow by burrowing in the sand; this ability is especially important during a strong incoming tide, when shear stress

is maximized and erosion of sediment around the shell is very strong.

Suspension Feeding From Hard Surfaces

✳ **Particle capture may involve simple sieving, but particles may directly impact on tentacles and tube feet, or they may leave the flow by a variety of other mechanisms and impact on the feeding structure.**

We intuitively think of particle capture as a sieving process. An animal would therefore capture particles larger than the distance between tentacles (if the animal were a coral), or between tube feet (if it were a brittle star or a crinoid). However, velocity drops greatly near any closely spaced structures, and recent studies show that biological sieving is unusual. The Reynolds number in such structures is so low that the water between the fibers is more like a nearly impervious wall than a passage. Instead of considering these structures as simple sieves, we must consider the hydrodynamic features of particles as they approach the suspension-feeding capturing fibers. Figure 13.16 illustrates the possible means of capture by a passive set of fibers:

1. **Sieving.** The particle is trapped between fibers.
2. **Direct interception.** The particle follows water streamlines and comes within the distance of one particle radius.
3. **Inertial impaction.** The particle has inertia and crosses streamlines.
4. **Motile particle deposition.** The particle moves randomly and hits a fiber.
5. **Gravity impact.** The particle crosses streamlines because it has inertia and is pulled by the force of gravity to impact on the fiber.

In Chapter 6, we discussed the effect of flow when velocity is low and the structures are small. Under these conditions of low Reynolds number, inertia is not nearly as important as the viscous forces in the water. If sieving does not occur, direct interception is the most likely mechanism of particle capture at low Reynolds number.

✳ **Passive suspension feeders collect food by means of morphological structures that protrude into the flow and capture particles.**

Suspension feeders commonly protrude a feeding organ into a mainstream current and suffer the prob-

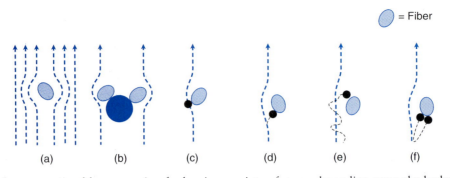

Fig. 13.16 Particles are captured by suspension feeders in a variety of ways, depending upon the hydrodynamic regime, which is characterized partially by the Reynolds number. (a) Streamlines around a fiber. (b) A particle is captured between fibers that act like a sieve (this rarely happens in suspension feeders because there is not enough inertia at the small size scale and low Reynolds number). (c) Direct interception, a common means of intercepting phytoplankton particles by cilia of bivalves and by polychaetes. (d) Inertial impaction. (e) Motile-particle deposition, which occurs because particles are not moving in perfectly smooth streamlines but move erratically and impact on the fiber. (f) Gravitational deposition, which occurs when particles are moving rapidly at high Reynolds number or are dense enough to cross streamlines. (Modified from Rubenstein and Koehl, 1977.)

lems already mentioned. In a moderate unidirectional current, the best strategy would be to deploy a network whose plane is perpendicular to the flow. This maximizes the opportunity for food particles to be intercepted. Gorgonians are branch- or stemlike colonies of feeding polyps. When small, the branches have no particular orientation because the current flow is complex near the bottom. However, as the colony gets larger and protrudes into a predictable and unidirectional flow, it "pays" for the colony to have a preferred orientation. The branches then often grow into a plane that is perpendicular to the current.

Not all benthic passive suspension feeders adopt a vertical planar form, nor are their feeding structures always oriented upstream. Many suspension feeders are colonial and are bushlike (e.g., the hydroid coelenterates), or simply form a thin layer over the substratum (e.g., many sponges and corals). Although this may seem to be an inefficient way of feeding in a unidirectional current, a multidirectional orientation serves well when the current flow is complex. In many benthic habitats, water motion is oscillatory; the water just sloshes back and forth over the bottom. In other cases, tidal currents cause a reversal of flow. Under such circumstances, a bushlike shape will gather more food and oxygen than will a planar shape with individuals pointing upstream.

It is still not very clear how particles are captured by passive suspension feeders. Overall, viscosity should be very important in particle capture, owing to the small scale of the capturing structures and the relatively low velocity. In animals like corals, crinoids, and suspension-feeding brittle stars, tentacles or tube feet probably capture particles upon impact. However, there may be some cases in which the Reynolds number is sufficiently high that particles have some inertia and actually fall out of the water, cross local streamlines, and impact on the feeding organ. In brittle stars, particles much smaller than the distance between the tube feet are captured, indicating that simple sieving is not the main mechanism of capture. Zooplankton landing on the tentacles of large anemones may be a case of suspension feeding by means of gravitational deposition.

✳ Active suspension feeders are under some constraints similar to those that apply to passive suspension feeders, but they also generate their own water currents to channel and ingest particles.

In contrast to passive suspension feeders, active suspension feeders create a current to take in planktonic food. In many polychaete annelids and bivalve mollusks, ciliary currents draw particles toward the cilia, which capture particles and transport them down ciliated tracts (Figure 13.12b,c). Processes near cilia are at low Reynolds number, and the cilia must directly reach out and capture particles. (See the discussion in Chapter 5 on planktonic larval ciliated feeding for more on this subject.) As transport occurs, the tracts reject unsuitable food particles. Many inter-tidal acorn barnacles use a different active strategy: the thoracic limbs move actively and capture particles that are drawn to and processed by appendages surrounding the mouth. Barnacles can adjust the ori-entation of the thoracic appendages at different flow velocities. At low velocity, the feeding cirri face into the current and capture particles. If the current passes a threshold velocity, the cirri are suddenly reversed and pointed downstream, to minimize drag.

HOT TOPICS IN MARINE BIOLOGY 13.1 Suspension Feeding by Bivalves: The Inside Story

The ocean is clouded with small particles of many types, defying a careful and complete description. Large rivers empty specks of degraded leaves, clay and silt particles, and detritus of many other types into an ocean already filled with phytoplankton, smaller zooplankton, and a plethora of animal and plant degradation products and exudates that collide in the water column, to eventually form what is often termed marine snow. A typical coastal water column has all this complexity, but even deep waters may be surprisingly rich in suspended particles.

Suspension-feeding animals must face this complexity and eek out a living by collecting particles, ingesting them, and absorbing nutrients. Many of the particles in the water, however, have no nutritive value. Of course one might imagine a suspension feeder ingesting all particles indiscriminately, digesting only those that are palatable, but this strategy would have two important disadvantages. First, the gut would be filled with large amounts of indigestible material. This would waste opportunities to digest perfectly good food. Second, materials might enter the gut that have negative physiological effects and even are poisonous. For example, the saxitoxin-bearing dinoflagellate *Alexandrium* has strong negative effects on bivalve function, so why would a bivalve want to ingest it? Why not reject the toxic substance beforehand?

Long ago, a simple observation demonstrated that bivalves could reject particles and even discriminate among different particle types. If you placed an oyster in a dish and fed it a high concentration of phytoplankton or a large amount of silt particles, you would soon see diffuse material at the bottom of the dish, emanating usually from the incurrent siphon or from between the shells. How did this happen? The internal space of a bivalve, its **mantle cavity**, contains a series of structures that transport and potentially reject particles before they enter the mouth. This process is known as **preingestive selection**. Particles enter a siphonlike entrance and then are drawn toward the surfaces of the gills. The water movement within the mantle is stimulated by a complex set of ciliated tracts that beat continuously, creating a pressure differential that moves water and particles through the incurrent siphon and toward the **gill surfaces**. It appears that particles hit ciliated surfaces by inertial impaction, whereupon they become subject to selectivity (see Box Figure 13.2b), are moved by ciliary beating to more centralized ciliary tracts, and may then be transported to another main structure known as the **palp** (Box Figure 13.2b). The bivalve palp varies in structure depending upon species, but it is typically highly ridged, and the ridges and valleys are also ciliated. The palp moves particles along its own ciliated systems and eventually transfers some to the mouth.

To investigate this type of feeding, particles could be placed upon both gills and palps that have been surgically removed and mounted on a slide. If the ciliated tracts were "smart," they could reject bad particles before the palp had transferred them to the mouth. The rejected material that is never ingested is known as **pseudofeces**; this is the stuff mentioned earlier that was found in a dish with a bivalve, as opposed to true feces, which is matter that is egested from the gut. Pseudofeces production may occur by the gill sloughing off material directly or passing it to the palp, which can also form and pass pseudofeces outside the mantle cavity.

Do bivalves manage to select against nutritionally poor or even toxic particles before ingestion? A simple experiment suggests that they do. One can make a mixture of silt and phytoplankton and feed the particles to the bivalves. Presumably the phytoplankton are suitable food and the silt is not. One can measure the organic content of the food by sampling the experimental water and compare the results to measurements made of pseudofeces. Such experiments generally show that the pseudofeces are proportionally depleted in organic material, suggesting that selective rejection of nonnutritious material is occurring.

But where is the site of selectivity? This has been a mystery for many years. It has been possible to surgically re-

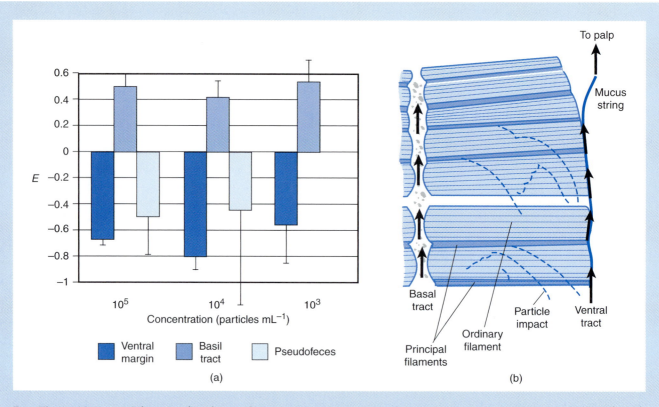

Box Figure 13.2 (a) Selectivity for algae, relative to nonnutritive particles, at total particle concentrations of 10^3, 10^4, and 10^5 particles per milliliter. Positive values indicate that there are relatively more algal particles than what was fed to the animal. Note the relative abundance of algal particles in the basal tract, which delivers particles eventually to the mouth. Algal particles are depleted in the ventral tract, where particles are eventually rejected as pseudofeces, where algal particles are also depleted. This means that algae are preferentially selected and ingested. (b) The gill surface of an oyster. Note the different directions in which particles are transported by ciliary tracts. Particles transferred to the basal tract usually are moved to the palp and are ingested. Particles transferred to the ventral tract are usually eventually rejected by the palps as pseudofeces and are not ingested.

move structures such as gills and palps to trace the directions of ciliary beating, but this precludes observing the entire mantle cavity organ system as an integrated unit and disrupts any special hydrodynamic conditions that might occur in the intact mantle cavity.

Enter (literally) the surgical endoscope, which is a tubular lens that may be made of a fiber-optic bundle or may be a single glass lens (Box Figure 13.3). For bivalves, endoscopes of as little as 1 mm in diameter are available, which allows for easy insertion into the mantle cavity, especially of larger bivalves. The endoscope can be attached to a video monitor and video tape recorder, allowing recordings of the movements of particles within the mantle cavity.[11]

Another instrument aids us greatly in studying particle selectivity. The flow cytometer is an instrument adapted

from medical uses to marine biology. The principle is fairly simple. A small volume of water is drawn into a tube and as particles pass through the tube, they are struck by a laser beam. Particles reflect the laser light and can be counted and even sized, because the larger the particle cross section, the greater the reflection. Phytoplankton cells have a variety of photosynthetic pigments (see Chapter 9), many of which fluoresce when bombarded by blue light. For example, when chlorophyll is bombarded with ultraviolet light it fluoresces a vivid red color.

The flow cytometer reads the light reflected and fluo-

[11] If you want a flavor of the movement of particles, I encourage you to visit the following Web site: http://www.sp.uconn.edu/~jeward/endoscopy/text2.html

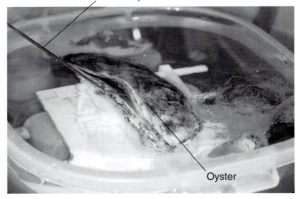

Endoscope insertion tube

Oyster

Box Figure 13.3 An endoscope inserted into a bivalve. This surgical telescope allows us to observe how internal mantle cavity organs of a bivalve function in processing particles without surgically altering the mantle organs, such as gills and palps. (Photograph by the author.)

resced from each particle by means of a series of detectors. The combined signal of each particle allows us to discriminate among a variety of different particle types (Box Figure 13.4). Flow cytometers are extremely useful in analyzing marine water samples, but they also can be used to tell if a suspension feeder is selecting among different particle types. For example, one could feed a series of phytoplankton species and tell if some are being selectively ingested merely by comparing counts in the water column with particles in the pseudofeces. If there are proportionally more particles in the suspended food than in the pseudofeces, then the bivalve must be selectively ingesting those particle types because they are depleted in pseudofeces (Box Figure 13.2a).

A simple experiment demonstrated an interesting result. Sandra Shumway and colleagues[12] pioneered in using the flow cytometer to study selective feeding by bivalves. They fed mixtures of algal cultures to see whether species of algae were preferentially ingested, which could be told by comparing available food particles in the water with the material rejected in the pseudofeces. On comparing samples of algal mixtures, it was clear that a variety of bivalves were selectively rejecting certain phytoplankton species and preferentially retaining others for ingestion.

But where did the selection occur? Roger Newell and Stephen Jordan[13] did an experiment that set the stage for the problem. After feeding the eastern American oyster

Crassostrea virginica a mixture of silt and the alga *Tetraselmis* sp., they found that the pseudofeces were strongly enriched with silt relative to the mixture in the water. This suggested that oysters are able to ingest organic rich particles and reject nonliving particles in the pseudofeces. Not only was the selectivity clear, but it worked in favor of excluding indigestible material (i.e., the silt) from ingestion. Because this study compared food in the water and in pseudofeces, it was not possible to demonstrate where selection had occurred in the mantle cavity. Newell and Jordan followed the popular wisdom that oysters selected particles on the palp, maintaining that the gills had no role.

The use of a surgical endoscope clarified this problem definitively. J. Evan Ward and colleagues[14] examined the feeding of *Crassostrea virginica* by placing the tip of an endoscope between the open valves. The structure of oyster gills (Box Figure 13.2b) is rather different from those of many other bivalves. The gill is folded, or plicate, and the valleys are lined with cilia that beat dorsally. The "hills" also are covered with cilia and can either transfer particles dorsally (refer to the figure or you will become confused!) or ventrally. If particles were transported dorsally, they arrived at a basal ciliated tract that transferred particles in a slurry toward the palp and were ingested. Particles transported ventrally, however, were moved to a so-called ventral groove and were either rejected from the gill or transported to the palp and rejected from there as pseudofeces. The bidirectional option of particle transport immediately suggested the gill as the possible site of particle selection, rather than the palp, which was the standard belief of bivalve researchers for many decades.

A simple experiment performed by J. Evan Ward and colleagues[15] combined observations using the endoscope with sampling directly from various ciliated tracts in the mantle cavity. While the endoscope was focused on various parts of the gill, a fine pipette was located to sample material from the dorsal and ventral ciliated tracts. It was already known that the dorsal tract carries particles that will be ingested and the ventral ciliated tract carries parti-

[12] See Shumway and others, 1985, in Further Reading, Suspension Feeding.

[13] See Newell and Jordan, 1983, in Further Reading, Suspension Feeding.

[14] See Ward and others, 1994, in Further Reading, Suspension Feeding.

[15] See Ward and others, 1998a, 1998b, in Further Reading, Suspension Feeding.

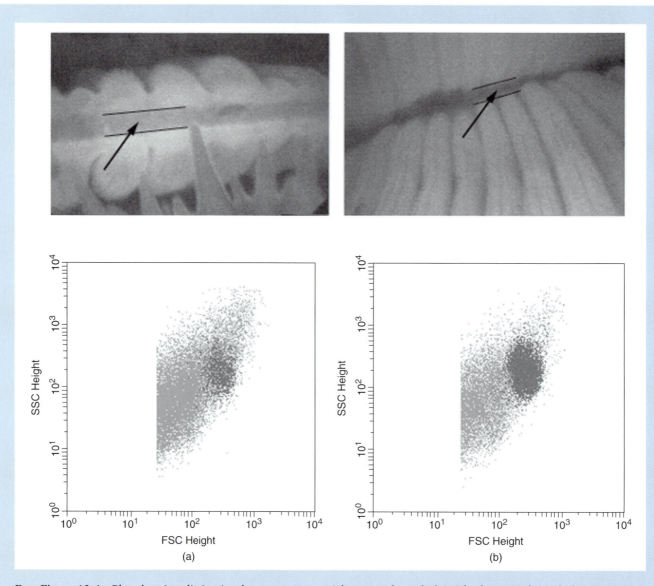

Box Figure 13.4 Plot showing distinction between two particle types, the red-alga *Rhodomonas lens* (darker gray areas of graphs) and cellulose-rich particles derived from the decomposition of the cord grass *Spartina alterniflora* (lighter gray areas of graphs): FSC, Forward Light Scatter (estimate of particle size); SSC, Side Scatter (estimate of index of refraction). (a) Particles sampled from a ciliated tract (arrow in accompanying photo) that collects rejected particles; note the enrichment of cord grass detritus. (b) Particles sampled from another ciliated tract (arrow in photo) that collects particles that will be passed to the palp and then ingested. Note the increased abundance of *Rhodomonas lens* and relatively smaller amount of cord grass detritus.

cles that are to be rejected. Therefore, one could sample these two tracts and compare them with the particles in the water, representing the available food. A flow cytometric analysis could then determine whether the gills were sorting the good from the bad particles.

Ward and colleagues used an approach similar to that of Newell and Jordan. They fed the oysters *Crassostrea virginica* and *C. gigas* with aged cord-grass (*Spartina alterniflora*) detritus, which they had ground to match an alga (*Rhodomonas*) in particle size. The detritus was very poor in nutrients, much like the silt used by Newell and Jordan. The results were immediately apparent, especially because *Rhodomonas* is a vivid red. The particles in the dorsal tract were clearly red and the ventral tract was tan, the color of the cord grass detritus. This demonstrated that it was the gill doing the sorting, not the palp.

Not all bivalves work this way. Mussels mainly transport particles ventrally and these particles are usually transported to the palps. Here the structure of the gill (much simpler) precludes sorting at this stage, and it is the palp that does the job. Indeed, the palp sorts among the cord grass detritus and *Rhodomonas* algal particles with the same degree of selectivity as the gills of oysters. There's more than one way to get the job done, apparently.

This research opens up the "black box" and allows us to see what is happening inside the bivalve mantle cavity. Many problems however, remain. While it is clear that the oyster gill is "deciding" to reject or reject particles instantaneously, the mechanism is completely unknown. No chemosensory cells have yet been found on bivalve gills. Yet the impact of good and bad particles is literally spread all over the gill, so there must be a general system of detection. We also have no idea why different bivalves have such different gill architectures. Mussels do not select particles on the gills, yet oysters do this routinely. Both types of bivalve often live in near-shore estuaries with complex arrays of particle types, nutritionally poor as well as appropriate. The results thus far therefore just scratch the surface.

Benthic Carnivores

✽ **Carnivory relies on mechanisms of prey search, location, seizure, and ingestion.**

Carnivorous animals hunt and eat other animals (Figure 13.17). Defining benthic carnivores is not entirely straightforward, because those that eat zooplankton are as much suspension feeders as they are carnivores. Of necessity, most carnivores are mobile and have a variety of means of prey detection. Many species are capable of detecting soluble substances emanating from the prey. The European sea star *Astropecten irregularis* moves along the sediment surface but can detect its prey within the sediment. Many carnivores orient to prey upstream. Specialized bivalve mollusks known as septibranchs detect prey by chemical means. A specialized pumping septum moves suddenly, expels water through the exhalant siphon, and draws water plus prey into the inhalant siphon (Figure 13.17b).

Vision is a common means of prey detection. Bottom-feeding birds, crabs, fishes, and cephalopods such as cuttlefish all detect prey visually. Visual detection is usually accompanied by sophisticated and rapid eye–motor coordination. The oystercatcher *Haematopus ostralegus* can dash onto an open mussel as a wave recedes and plunge its beak into the mussel, severing they prey's adductor muscles, thus making it helpless. More rarely, the oystercatcher hammers with its bill and crushes the shell. In either case, the bird assesses the size of the mussel and tends to take prey that are larger than average size. Lobsters and crabs use both chemical detection and vision in predation, and can rapidly attack and immobilize prey.

The several strategies for attacking and seizing prey are obviously related to the mode of prey detection. Many predators are essentially sessile and must wait for prey to arrive. Anemones usually remain fixed to a hard surface and have access only to prey that swims or falls in contact with the tentacles. One large eastern Pacific anemone, the intertidal *Anthopleura xanthogrammica*, lives in low intertidal pools, and depends for food upon mussels that fall from above. This mode of feeding is greatly aided by starfish, which are somewhat sloppy as they pry mussels from the bed and allow some to fall to the anemones below. More mobile carnivores have a variety of sophisticated search behaviors, aided by vision and olfaction. Some fishes and crustaceans are "sit-and-wait" predators, whereas many others cruise continuously until they detect a prey item.

The handling of prey varies with phylogenetic background because morphologies are so disparate. Seizing prey involves some sort of appendage, such as a crab claw or a starfish arm. Many crab species have large crushing claws with denticles that enable handling of prey. Some crabs, such as the stone crab *Menippe mercenaria*, have robust claws and musculature and can crush thick-shelled mollusk prey. Others, such as the shore crab *Carcinus maenas*, are not terribly strong and have trouble crushing mussels unless they discover a weak spot in the shell. Some crabs

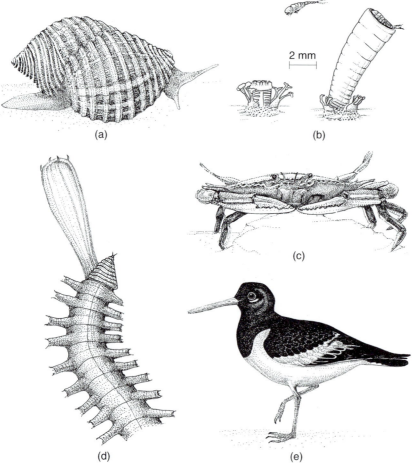

Fig. 13.17 Some marine benthic carnivores. (a) Gastropod *Nucella*, which uses a specialized radula and buccal mass to drill holes in barnacles and bivalve mollusks. (b) Bivalve mollusk *Cuspidaria*, which uses a pumping septum to suck up small prey. (c) Decapod crab *Callinectes sapidus*, whose strong claw can crush mollusks. (d) Polychaete *Glycera*, which has a proboscis armed with hooks, used in seizing and tearing prey. (e) The oystercatcher, *Haematopus ostralegus*, a predator on intertidal bivalve mollusks.

repeatedly apply a crushing load to bivalves. Eventually, after several applications of pressure, the shell fatigues and can be crushed. Some tropical crabs can easily peel the shell of a snail to expose the soft parts. Polychaetes such as some species of *Glycera* (Figure 13.7d) have a protrusible proboscis with hook-shaped teeth; other polychaetes have large chitinous jaws that can tear prey apart.

For some carnivores, success involves immobilizing the prey by means of a sting. Cone snails of the genus *Conus* have a highly movable proboscis and long, barbed, radular teeth. The proboscis is protruded very rapidly, one or a few teeth stab the prey, and a poison is injected along a groove. The speed of attack and the virulence of the poison allow some species to immobilize and kill small fishes, which are then swallowed whole. Some nemerteans can seize prey and pierce them with stylets, injecting a poison.

In the gastropods, drilling is a specialized way of penetrating prey that have exoskeletons. This occurs

in the prosobranch families Muricidae (*Urosalpinx*, *Murex*), Naticidae (*Polinices*), and Thaiidae (*Nucella*), and involves alternations of mechanical rasping and chemical secretions from an accessory boring organ.

Benthic Herbivores

✳ **Benthic herbivores are divided between microphages and macrophages.**

The food of benthic herbivores (Figure 13.18) can be divided by size class into two major categories. Benthic microalgae include a variety of groups, such as diatoms, cyanobacteria, and microscopic stages of seaweeds. These organisms may form a thin layer on a rock surface or on the surface of sediment. Consumers have a range of morphological features that allow them to graze efficiently on this layer. Chitons,

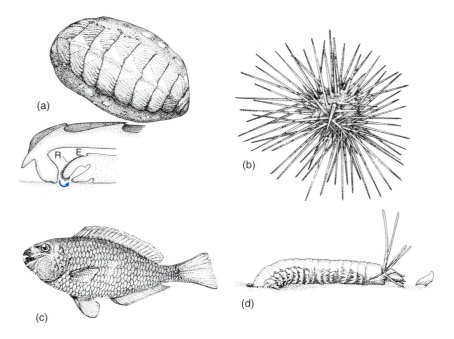

Fig. 13.18 Some benthic herbivores. (a) The chiton *Tonicella*, a scraper of microalgae; inset shows anterior sagittal cross section, indicating the action of the radular tooth belt in scraping algae from the substratum (R, radula; E, esophagus). (b) The sea urchin *Arbacia*, which uses a toothed Aristotle's lantern to scrape microalgae or to tear apart seaweeds. (c) A parrot fish, which uses a specialized mouth to scrape algae from coral surfaces. (d) The nereid polychaete *Nereis vexillosa*, which tears apart sea lettuce. (Copied from an original by K. Fauchald.)

limpets, and other grazing mollusks employ a radula, a belt of teeth that scrape along the surface. The movement of the subradular membrane over a cartilaginous portion of the buccal mass erects the teeth and scrapes them over the surface. The radula and buccal mass are retracted, and food trapped on the teeth is delivered to the buccal cavity. This feature can be used on rocks, and limpet grazing scars are common on rocky shores. However, radular scraping is also employed by gastropods feeding on soft-sediment surfaces. Some polychaetes can graze on sediment microalgae by pressing their tentacles onto the surface and collecting particles and microalgae, which are transported to the mouth by means of a ciliated tract.

A wide variety of herbivorous animals can tear apart and consume macroalgae and marine higher plants. The radular apparatus of many mollusks has been adapted in many cases to tearing apart seaweeds, and periwinkles, for example, can rasp and tear apart a large number of seaweeds. Their effectiveness, however, is often restricted to seaweeds that have rather delicate structures, such as the sea lettuce genera *Ulva* and *Enteromorpha*. Sea urchins possess an Aristotle's lantern, which is a complex of calcareous teeth, ligaments, and muscles. This device can tear apart a variety of seaweeds, and some urchins are even capable of devouring relatively less digestible

sea grasses, such as the tropical Caribbean *Thalassia testudinum*.

Many crustaceans are also herbivorous. Many smaller amphipods and isopods feed on relatively soft seaweeds, or on the microalgae growing on seaweed surfaces. A variety of fishes are also efficient herbivores, particularly on coral reefs. The jaw teeth of parrot fishes (Scaridae), which are fused into plates, are capable of cutting material from the surface of coral skeletons. Surgeon fishes (Acanthuridae) also can scrape algae from corals, and the two groups are major causes of erosion on coral reefs. Even smaller invertebrates, such as isopods and polychaetes, have sufficiently strong mouth parts to tear apart algae. The buccal hooks of some species of the sandworm *Nereis* are employed in tearing apart soft green algae.

Although herbivores are usually mobile, many rock scrapers are capable of homing. A home base may provide a reference location, allowing efficient exploitation of the renewable resource of microalgae living on hard surfaces. The eastern Pacific owl limpet *Lottia gigantea* and the limpet *Patella longicosta* both defend territories.

Although most benthic herbivores search for suitable food, some wait for the food to come to them. Many urchins capture drifting seaweed fragments on their dorsal spines and dorsal tube feet transfer them

Hot Topics in Marine Biology 13.2 Claws! The Crushing and Tearing Story of Mechanical Advantage

If you have ever handled a blue crab (*Callinectes sapidus*) or a Maine lobster (*Homarus americanus*), chances are your fingers have been pinched. These creatures, like many other crustaceans, have powerful claws capable of extraordinary closing forces. Lobsters and blue crabs feed on mollusks and easily crush even thick shells. How can such great force be exerted?

Like Popeye the Sailor with his muscular forearm, a lobster has most of its crushing force concentrated in muscles in the claw itself. The claw is a lever system, and muscles attached to one of two sides of the pivot can either open or close the claw (Box Figure 13.5a). From the outside, you can see a movable finger, the dactyl, which opposes a fixed finger, which is attached to the main part of the claw. Muscles contract and pull on a slender projection of the dactyl, known as the extensor apodeme. This pulls the lever system around the pivot point, and the dactyl is raised. Relatively little musculature is required to open the claw. Far more muscle is attached to a much broader flexor apodeme, whose pull closes the claw. The flexor apodeme attaches to the dactyl on the other side of the lever, so the dactyl is pulled down. Muscular force is proportional to cross-sectional area; there is much more area for muscle attachment on the flexor apodeme than on the extensor apodeme.

More muscle cross-sectional area can generate more force, but there is another means of changing force, simply by changing the proportions of the lever system. Think of a seesaw in a playground. If the seats are equidistant from the pivot point, then two people of equal weight will exactly balance each other. However, if we move the board so that one side has twice the length of the other, a person on the long end can exactly balance someone twice her weight on the short end. The weight, after all, is a force,

so one can see that the change of proportions has changed the mechanical properties of the machine. The **mechanical advantage** is found from the ratio of lengths of the sides of the seesaw (long to short).

The lobster claw obeys the same principles, though the orientations of this living machine are bent relative to a seesaw. (The pivot and appropriate lengths are shown in Box Figure 13.5a.) Force F_1 is exerted over length L_1, and this generates force F_2 over length L_2 which is the length of the dactyl. To increase the closing force, one must either increase the ratio of L_1 to L_2 or increase the area of muscle attachment to the flexor apodeme. Both can be accomplished by increasing the volume of the compartment that encloses the muscles.

Maine lobsters have two distinct claws. One is a crusher claw, and its proportions are such that its mechanical advantage is greater than the other, which is a slicing claw (Box Figure 13.5b,c). Amazingly enough, the side each claw type will be on cannot be predicted. Various random events cause one claw to develop into the crusher chela and the other into a slicer chela, so a population of lobsters will have approximately equal proportions of right-handed and left-handed crusher claws.

The mechanical advantage of the crusher claw is double that of the cutter claw, but the cutter claw has an advantage, even if it is a bit weaker. The complement of mechanical advantage is speed. To understand this point, remember the seesaw. If you move the tip of the short side of the seesaw a given distance, the longer side will move much more rapidly. Thus, the cutter claw of a lobster can move more rapidly and handle food much more efficiently than the crusher claw can. Thus the lobster uses the cutter claw to manipulate food and the crusher to perform the gruesome final act.

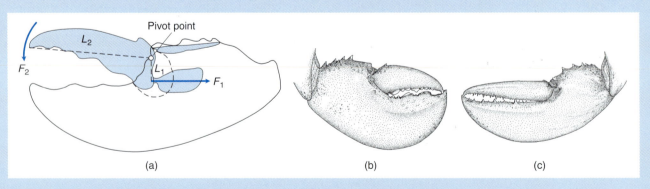

Box Figure 13.5 Claws of the lobster *Homarus americanus*. (a) Features of the claw, forces and pivot of the claw apparatus. (b) The crusher claw. (c) The cutter claw. Scale for (b) and (c) is 5 cm. (After Elner and Campbell, 1981.)

toward the mouth. Sand-flat polychaetes such as species of *Nereis* and *Lumbrinereis* can drag seaweed fragments down into their burrows. In some cases the downward dragging is incomplete and the seaweed actually can start to grow. Some polychaetes practice farming by attaching fragments of *Ulva* to their tubes and letting them grow.

✻ Some benthic herbivores can feed on highly indigestible plant material.

Most marine herbivores are restricted to relatively soft seaweeds and microalgae, with a minimum of relatively indigestible complex carbohydrates, such as cellulose. A small number of species, however, have adapted to such difficult food sources. Some invertebrates can bore into wood and digest it or may depend upon the marine microbiota living in the wood. The wood-boring bivalve *Teredo* (shipworm) and *Bankia* scrape the wood particles and use the digestive enzyme cellulase to attack the cellulose (Figure 13.19). The wood-boring isopod *Limnoria* can also digest cellulose, but it requires wood-boring fungi as a source of nitrogen. Wood-boring bivalves derive their nitrogen from symbiotic nitrogen-fixing bacteria, since nitrogen is not present in sufficient quantities in the wood.

Sea grasses, such as eelgrass (*Zostera*), salt marsh cord grass (*Spartina*), and Caribbean turtle grass (*Thalassia*), are relatively indigestible to most marine consumers because of the abundance of cellulose the grasses contain. Some small grazers consumer the microalgal surface layer, but relatively few species can consume, digest, and assimilate material from the grass itself. As mentioned earlier, a few species of

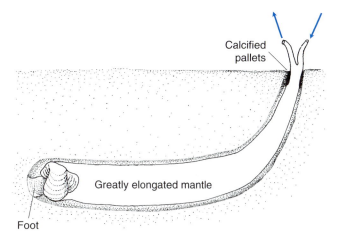

Fig. 13.19 Cross section showing the living position of the wood-boring bivalve mollusk *Teredo*. [After Trueman, 1975, *The Locomotion of Soft-Bodied Animals*, with permission of Edward Arnold (Publisher) Ltd.]

urchins can deal with turtle grass. Eelgrass and cord grass are remarkable for the minuscule amount of grazing they experience from marine herbivores. An interesting exceptional species is the green turtle, *Chelonia mydas*, which can digest cellulose derived from turtle grass.[16] It has a hindgut that bears a functional resemblance to the stomachs of ruminant mammals, such as cows and horses. The postgastric gut region is greatly elongated, and postgastric fermentation is facilitated by the presence of digestion-aiding symbiotic bacteria and protozoa.

[16] See Fenchel and others, 1979, in Further Reading, Herbivores.

Further Reading

SOFT SEDIMENTS AND DEPOSIT FEEDING

Aller, R. C. 1980. Relationship of tube-dwelling benthos with sediment and overlying water chemistry. In K. R. Tenore and B. C. Coull, eds., *Marine Benthic Dynamics*. Columbia: University of South Carolina Press, pp. 285–308.

Anderson, A. E., J. J. Childress, and J. A. Fanuzzi. 1987. Net uptake of CO_2 driven by sulfide and thiosulphate oxidation in the bacterial symbiont-containing clam *Solemya reidi*. *Journal of Experimental Biology*, v. 133, pp. 1–31.

Barsdate, R. J., R. T. Prentki, and T. Fenchel. 1974. Phosphorous cycle of model ecosystems: Significance for decomposer food chains and effect of bacterial grazers. *Oikos*, v. 25, p. 239–251.

Cammen, L. M. 1980. Ingestion rate: An empirical model for aquatic deposit feeders and detritivores. *Oecologia*, v. 20, pp. 33–49.

Carey, S. C., C. R. Fisher, and H. Felbeck. 1988. Mussel growth supported by methane as sole carbon and energy source. *Science*, v. 240, pp. 78–80.

Eckman, J. E., A. R. M. Nowell, and P. A. Jumars. 1981. Sediment destabilization by animal tubes. *Journal of Marine Research*, v. 39, pp. 361–374.

Fenchel, T. 1972. Aspects of decomposer food chains in marine benthos. *Verhandlungen der Deutschen Gesellschaft fur Zoologie* v. 65, pp. 14–22.

Fenchel, T. 1978. The ecology of micro- and meiobenthos. *Annual Review of Ecology and Systematics*, v. 9, pp. 99–121.

Fenchel, T., and H. Blackburn. 1979. *Bacteria and Mineral Cycling*. Berlin: Springer-Verlag.

Fenchel, T., and R.J. Riedl. 1970. The sulfide system: A new biotic community underneath the oxidized layer of marine sand bottoms. *Marine Biology*, v. 7, pp. 255–268.

Hargrave, B. T. 1970. The utilization of benthic microflora by *Hyalella azteca*. *Journal of Animal Ecology*, v. 39, pp. 427–437.

Harrison, P. G., and K. H. Mann. 1975. Detritus formation from eel grass (*Zostera marina* L.): The relative effects of fragmentation, leaching and decay. *Limnology and Oceanography*, v. 20, pp. 924–934.

Jensen, P. 1987. Differences in microhabitat, abundance, biomass and body size between oxybiotic and thiobiotic free-living marine nematodes. *Oecologia*, v. 71, pp. 564–567.

Kepkay, P. E., and J. A. Novitsky. 1980. Microbial control of organic carbon in marine sediments: Coupled chemoautotrophy and heterotrophy. *Marine Biology*, v. 55, pp. 261–266.

Lopez, G. R., and J. S. Levinton, 1987. Ecology of deposit-feeding animals in marine sediments. *Quarterly Review of Biology*, v. 62, pp. 235–260.

Lopez, G. R., G. L. Taghon, and J. S. Levinton, eds. 1989. *Ecology of Marine Deposit Feeders*. New York: Springer-Verlag.

Mayer, L. M., and D. L. Rice. 1992. Early diagenesis and protein: A seasonal study. *Limnology and Oceanography*, v. 37, pp. 280–295.

Mayer, L. M., P. A. Jumars, G. L. Taghon, and S. A. Macko. 1993. Low-density particles as potential nitrogenous foods for benthos. *Journal of Marine Research*, v. 51, pp. 373–389.

Mayer, L. M., L. L. Schick, T. Sawyer, C. J. Plante, P. A. Jumars, and R. F. L. Self. 1995. Bioavailable amino-acids in sediments—A biomimetic, kinetics-based approach. *Limnology and Oceanography*, v. 40, pp. 511–520.

Mayer, L. M., L. L. Schick, R. F. L. Self, P. A. Jumars, R. H. Findlay, Z. Chen, and S. Sampson. 1997. Digestive environments of benthic macroinvertebrate guts: Enzymes, surfactants and dissolved organic matter. *Journal of Marine Research*, v. 55, pp. 785–812.

Penry, D. L., and P. A. Jumars. 1987. Modeling animal guts as chemical reactors. *American Naturalist*, v. 129, pp. 69–96.

Penry, D. L., and P. A. Jumars. 1990. Gut architecture, digestive constraints and feeding ecology of deposit-feeding and carnivorous polychaetes. *Oecologia*, v. 82, pp. 1–11.

Rhoads, D. C. 1967. Biogenic reworking of intertidal and subtidal sediments in Barnstable Harbor and Buzzards Bay, Massachusetts. *Journal of Geology*, v. 75, pp. 461–474.

Riese, K., and P. Ax. 1979. A meiofauna "thiobios" limited to the anaerobic sulfide system of marine sand does not exist. *Marine Biology* 54:225–237.

Swedmark, B. 1964. The interstitial fauna of a marine sand. *Biological Reviews*, v. 39, pp. 1–42.

Taghon, G. L. 1981. Beyond selection: Optimal ingestion rate as a function of food value. *American Naturalist*, v. 118, pp. 202–214.

Trueman, E. R. 1975. *The Locomotion of Soft-Bodied Animals*. Bristol, U.K.: Arnold.

HARD SUBSTRATA

Chamberlain, J. A., Jr., and R. R. Graus. 1975. Water flow and hydromechanical adaptations of branched reef corals. *Bulletin of Marine Science, Gulf and Caribbean*, v. 25, pp. 112–125.

Denny, M. W. 1988. *Biology and the Mechanics of the Wave-Swept Environment*. Princeton, NJ: Princeton University Press.

Koehl, M. A. R. 1976. Mechanical design in sea anemones. In G. O. Mackie, ed., *Coelenterate Ecology and Behavior*. New York: Plenum, pp. 23–31.

Koehl, M. A. R., and S. A. Wainwright. 1977. Mechanical adaptations of a giant kelp. *Limnology and Oceanography*, v. 22, pp. 1067–1071.

Wainwright, S. A., and J. R. Dillon. 1969. On the orientation of seafans (genus *Gorgonia*). *Biological Bulletin*, v. 136, pp. 130–139.

Warner, J. F., and J. D. Woodley. 1975. Suspension-feeding in the brittle star *Ophriothrix fragilis*. *Journal of the Marine Biological Association of the United Kingdom*, v. 55, pp. 199–210.

Young, C. M., and L. F. Braithwaite. 1980. Orientation and current-induced flow in the stalked ascidian *Styela montereyensis*. *Biological Bulletin*, v. 159, pp. 428–440.

SUSPENSION FEEDING

Fréchette, M., C. A. Butman, and W. R. Geyer. 1989. The importance of boundary layer flows in supplying phytoplankton to the benthic suspension feeder, *Mytilus edulis* L. *Limnology and Oceanography*, v. 34, pp. 19–36.

Hunter, T. 1989. Suspension feeding in oscillating flow: the effect of colony morphology and flow regime on plankton capture by the hydroid *Obelia longissima*. *Biological Bulletin*, v. 176, p. 41–49.

Jørgensen, C. B. 1966. *Biology of Suspension Feeding*. Oxford: Pergamon Press.

Jørgensen, C. B. 1990. *Bivalve Filter-Feeding: Hydrodynamics, Bioenergetics, Physiology and Ecology.* Fredensborg, Denmark: Olsen and Olsen.

LaBarbera, M. L. 1978. Particle capture by a Pacific brittle star: Experimental test of the aerosol suspension feeding model. *Science,* vol. 201, pp. 1147–1149.

LaBarbera, M. L. 1984. Feeding currents and particle capture mechanisms in suspension feeding animals. *American Zoologist,* v. 24, pp. 71–84.

Leversee, G. J. 1976. Flow and feeding in fan-shaped colonies of a gorgonian coral, *Leptogorgia. Biological Bulletin,* v. 151, pp. 344–356.

Newell, R. I. E., and S. J. Jordan. 1983. Preferential ingestion of organic material by the American oyster *Crassostrea virginica. Marine Ecology Progress Series,* v. 13, pp. 47–53.

Okamura, B. 1987. Particle size and flow velocity induce an inferred switch in bryozoan suspension-feeding behavior. *Biological Bulletin,* v. 173, pp. 222–229.

Rubenstein, D. E., and M. A. R. Koehl. 1977. The mechanisms of filter feeding: Some theoretical considerations. *American Naturalist,* v. 111, pp. 981–994.

Shimeta, J. S., and P. A. Jumars. 1991. Physical mechanisms and rates of particle capture by suspension feeders. *Oceanography and Marine Biology Annual Review,* v. 29, pp. 191–257.

Shumway, S. E., T. L. Cucci, R. C. Newell, and C. M. Yentsch. 1985. Particle selection, ingestion, and absorption in filter-feeding bivalves. *Journal of Experimental Marine Biology and Ecology,* v. 91, pp. 77–92.

Walne, P. R. 1972. The influence of current speed, body size and water temperature on the filtration rate of five species of bivalves. *Journal of the Marine Biological Association of the United Kingdom,* v. 52, pp. 345–374.

Ward, J. E., P. G. Beninger, B. A. MacDonald, and R. J. Thompson. 1991. Direct observations of feeding structures and mechanisms in bivalve molluscs using endoscopic examination and video image analysis. *Marine Biology,* v. 111, pp. 287–291.

Ward, J. E., B. A. MacDonald, R. J. Thompson, and P. G. Beninger. 1993. Mechanisms of suspension feeding in bivalves: Resolution of current controversies by means of endoscopy. *Limnology and Oceanography,* v. 38, pp. 265–272.

Ward, J. E., J. S. Levinton, S. E. Shumway, and T. L. Cucci. 1998a. Site of particle selection in a bivalve mollusc. *Nature,* v. 390, pp. 131–132.

Ward, J. E., J. S. Levinton, S. E. Shumway, and T. L. Cucci. 1998b. Particle sorting in bivalves: *In vivo* determination of the pallial organs of selection. *Marine Biology,* v. 131, pp. 283–292.

Ward, J. E., R. J. Thompson, R. I. E. Newell, and B. A. MacDonald. 1994. In vivo studies of suspension-feeding processes in the eastern oyster *Crassostrea virginica* (Gmelin). *Biological Bulletin,* v. 186, pp. 221–240.

CARNIVORES

Alexander, R. M. 1983. *Animal Mechanics.* Oxford: Blackwell Scientific.

Barbeau, M. A., and R. E. Scheibling. 1994. Behavioral mechanisms of prey size selection by sea stars (*Asterias vulgaris* Verrill) and crabs (*Cancer irroratus* Say) preying on juvenile sea scallops (*Placopecten magellanicus* Gmelin). *Journal of Experimental Marine Biology and Ecology,* v. 180, pp. 103–136.

Birkeland, C., and S. Neudecker. 1981. Foraging behavior of two Caribbean chaetodontids: *chaetodon capistratus* and *C. aculeatus. Copeia,* v. 1981, pp. 169–178.

Boulding, E. G., and M. LaBarbera. 1986. Fatigue damage: Repeated loading enables crabs to open larger bivalves. *Biological Bulletin,* v. 171, pp. 538–547.

Elner, R. W. 1978. The mechanics of predation by the shore crab, *Carcinus maenas* (L.), on the edible mussel, *Mytilus edulis* (L.). *Oecologia,* v. 36, pp. 333–344.

Elner, R. W., and A. Campbell. 1981. Force, function and mechanical advantage in the American lobster *Homarus americanus* (Decapoda: Crustacea). *Journal of Zoology, London,* v. 193, pp. 269–286.

Fenchel, T. 1965. Feeding biology of the sea-star *Luidia sarsi* DÅben and Koren. *Ophelia,* v. 2, pp. 223–236.

Norton-Griffiths, M. 1967. Some ecological aspects of the feeding behaviour of the oystercatcher, *Haematopus ostralegus* on the edible mussel, *Mytilus edulis. Ibis,* v. 109, pp. 412–424.

Reid, R. G. B., and A. M. Reid. 1974. The carniverous habit of members of the septibranch genus *Cuspidaria* (Mollusca: Bivalvia). *Sarsia,* v. 56, pp. 47–56.

HERBIVORES

Fenchel, T. M., C. P. McRoy, J. C. Ogden, P. Parker, and W. E. Rainey. 1979. Symbiotic cellulose degradation in green turtles. *Applied Environmental Microbiology,* v. 37, pp. 348–350.

Hawkins, S. J., and R. G. Hartknoll. 1983. Grazing of intertidal algae by marine invertebrates. *Oceanography and Marine Biology Annual Review,* v. 21, pp. 195–282. This is a comprehensive review with extensive references.

Hay, M. E., J. E. Duffy, and W. Fenical. 1990. Host–plant specialization decreases predation in a marine amphipod: An herbivore in plant's clothing. *Ecology,* v. 71, pp. 733–743.

Hay, M. E., Q. E. Kappel, and W. Fenical. 1994. Synergisms in plant defenses against herbivores: Interactions of chemistry, calcification, and plant quality. *Ecology,* v. 75, pp. 1714–1726.

Pennings, S. C., and V. J. Paul. 1992. Effect of plant toughness, calcification, and chemistry on herbivory by *Dola-
bella auricularia. Ecology*, v. 73, p. 1606–1616.

Steinberg, P. D., and I. van Altena. 1992. Tolerance of marine invertebrate herbivores to brown algal phlorotan-
nins in temperate Australasia. *Ecological Monographs*,
v. 62, pp. 189–222.

Steneck, R. S., and L. Watling. 1982. Feeding capabilities
and limitations of herbivorous molluscs: A functional
group approach. *Marine Biology*, v. 68, pp. 299–319.

Review Questions

1. Distinguish between active and passive suspension
feeders.

2. What type of hydrodynamic condition does a well-
sorted sediment reflect?

3. What is a burrowing anchor, and why is it required in
a burrowing organism?

4. What do most interstitial marine animals have in common, in spite of being from quite different taxonomic
groups?

5. What factors help to determine the depth of the redox
potential discontinuity?

6. Why do bacteria of different types tend to dominate
at different depths below the sediment–water interface of
a muddy sediment?

7. What is the microbial stripping hypothesis?

8. Describe the components of decay of particulate organic matter in sediments.

9. How does a bivalve like *Solemya*, which lacks a gut,
manage to derive its nutrition?

10. Why and under what conditions do some benthic infaunal species switch between suspension feeding and deposit feeding?

11. How can sessile epibenthos reduce pressure drag?

12. Why are many suspension-feeding structures not simple sieves, whose interfiber distance can be used to predict
the diameter of particles that can be captured?

13. What is the advantage to carnivorous crustaceans of
having differentiated crusher and tearing claws?

14. Why is it possible for some marine animals to digest
cellulose, which is nearly indigestible for most organisms?

VI

COASTAL BENTHIC ENVIRONMENTS

14

The Tidelands: Rocky Shores, Soft-Substratum Shores, Marshes, Mangroves, and Estuaries

Coastal benthic habitats are among the most productive marine environments. They receive a rich nutrient supply, which is influenced both by terrestrial nutrient sources and by rich coastal phytoplankton production. The richness of these habitats also make them feeding and nursery grounds for migratory species, particularly fishes, crustaceans, and birds. The rich nutrient supply and the visitations by predators often lead to cycles of population increase of prey species, followed by population crashes because of predation. The great abundance of organisms also leads to rapid depletion of resources and strong interspecific competition. Because these processes are so strong, and because such coastal habitats are so accessible to ecologists, we know most about marine ecological processes from these habitats. This is especially true of the intertidal zone, where strong interactions occur within a gradient from fully terrestrial to fully marine habitats. Biological interactions thus occur against a backdrop of strong variation in the physical environment and the ability to survive nearly terrestrial conditions often has a strong influence on the relative success of different marine species.

The Intertidal Zone

Vertical Zonation

✳ **Vertical zonation, the occurrence of dominant species in distinct horizontal bands, is a nearly universal feature of the intertidal zone, but many localities do not "obey" the rules.**

The intertidal zone is the shoreward fringe of the seabed between the highest and lowest extent of the tides (see Chapter 2 for an explanation of tidal motion). The upper part of the intertidal zone is easy to define in areas of quiet water, but on open coasts waves splash many feet above the normal upper limit of tidal reach. This extends the vertical range of many intertidal species.

An important, but variably developed, feature of the intertidal zone is **vertical zonation,** the occurrence of dominant species in distinct horizontal bands (Figure 14.1). For example, a general pattern of zones is found throughout temperate and boreal rocky shores. From highest to lowest the zones are (a) a black lichen zone, (b) a periwinkle (littorine gastropod) zone, (c)

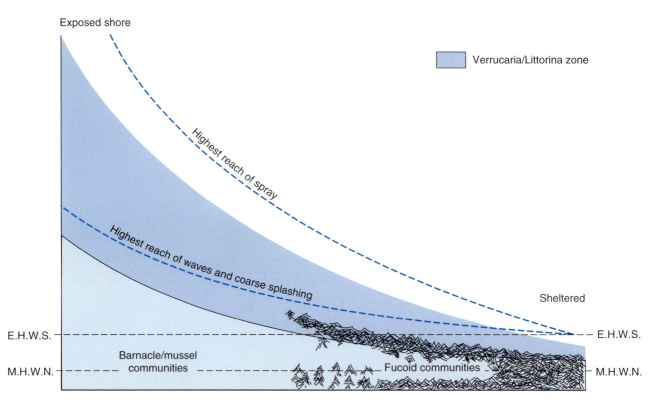

Fig. 14.1   The effect of wave exposure in broadening biotic zones of the rocky intertidal zone of the British Isles: EHWS, extreme high water spring tide; MHWN, meanhigh water neap tide. In quiet water habitats, represented on the right of the diagram, the mid-intertidal zone is dominated by seaweeds, but with more wave exposure, barnacles and mussels come to dominate. The upper intertidal is dominated by herbivorous snails in all cases. In strongly wave-swept habitats (left), the vertical zones are expanded. (After Lewis, 1964.)

a barnacle-dominated zone (Plate XIV.1), and (d) a zone dominated variously, depending upon locality. On North American shores, mussels (e.g., *Mytilus edulis*) dominate below the barnacle zone. In protected waters, zones are compressed between the highest and lowest extent of the tides. On open coasts, however, wave splash allows organisms to live higher, which expands the zones (Figure 14.1). Vertical zonation can also be found on sand- and mudflats, but it is rarely as distinct as on rocky surfaces. The lower limits of species zones are often determined by the presence of predators or competing species, whereas the upper limit is often controlled by physiological limits on the species tolerance of temperature and drying.

While overall patterns of zonation exist, on a gross scale, it is equally noticeable how often there is not a predictable pattern of dominance at different levels of the tide zone. Most commonly one sees patches dominated by different species at the same tidal lev-

els. Also, many areas consist simply of bare space. Explaining such departures from zonation is a major issue. Indeed, one can question whether a predictable zonation is the typical situation on many rocky-shore localities. Often as not, complexity of species occurrence obscures the zonation. We shall come back to this shortly.

Factors Affecting Intertidal Organisms

* **The intertidal zone is alternately a marine and a terrestrial habitat. The changes cause heat stress, desiccation, shortage of oxygen, and reduced opportunities for feeding.**

With every low tide, the intertidal zone is a nonmarine environment for all organisms not residing in tidal pools or in sediments deep enough to retain cool seawater. Even when water is retained, a long period of exposure changes the water properties significantly. After a strong rain, the salinity of a small tidal pool may drop nearly to that of fresh water. With increasing tidal height, more of the shore is exposed to air for a greater proportion of the day. The highest part of the intertidal zone is essentially a terrestrial environment, but organisms living at the low-tide level are only barely affected by aerial exposure.

At low tide, marine organisms face both **heat stress** and **desiccation stress**. On a hot day, invertebrates rapidly heat up, although they possess several adaptations to counteract thermal stress. If body fluids become too hot, physiological function and even the stability of proteins may diminish. In the worst case, intertidal organisms can literally dry up. In summer, this happens commonly to fragile species, such as thin green seaweeds. Organisms such as barnacles and mussels survive drier, more exposed areas because their shells can enclose water at low tide (Plate XIV.1).

Heat and desiccation stress vary on quite small spatial scales. As might be expected, sessile animals on upward-exposed flat rocks will gain far more heat than those in a moist crack or in the shade. The timing of the tide also can have profound effects on the heat inputs experienced by sessile intertidal organisms.[1] In the outer exposed coast of Washington, for example, spring low tides in summer come quite early in the morning, when air temperatures are often only 15°C. In contrast, summer spring low tide in Puget Sound comes at midday, when air temperatures are often above 20°C and can surpass 25°C. Small-scale spatial differences in microhabitat and timing of low tides may therefore have far greater effects than overall annual variation and even broad-scale latitudinal variation, at least on the west coast of North America where climatic gradients are slight between Washington and much of the California coast.

Body size and **body shape** both influence the degree of heat and water loss. As body size increases, the surface area, relative to body volume, tends to decrease, and this aids in reducing water loss. So, from the point of view of water loss, it is better to be big. However, the decrease in surface area, relative to body volume, that often comes with increasing size is a disadvantage with regard to heat loss. Small animals, with their higher relative surface area, tend to lose heat faster. The combination of these two factors must strike a balance. An intertidal animal cannot be too small or it will dry up in the sun. If it gets too large, however, it may not be able to dissipate heat fast enough through its body surface. Shape

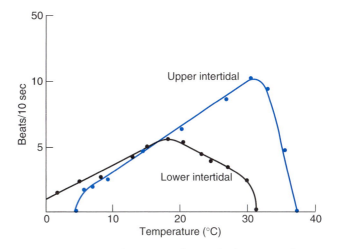

Fig. 14.2 Beating frequency of cirri (feeding appendages) in relation to increasing temperature in an upper intertidal (color curve) and lower intertidal (solid curve) barnacle species. (After Southward, 1964.)

has a similar effect. Long and thin organisms will dry up much more easily than spherical ones. This is one reason why a sea anemone contracts into a small equidimensional cylinder during low tide. The change in shape reduces surface area and water loss.

Intertidal invertebrates can avoid overheating by **evaporative cooling**, combined with **circulation of body fluids**. As a result of such processes, intertidal snails are usually cooler than inanimate objects of the same shape, size, and color. Higher-intertidal animals are better adapted to desiccation than lower-intertidal species, because the higher-intertidal forms experience more hours of the drying sun during the day. Movement of the cirri (feeding appendages) in intertidal acorn barnacles increases with increasing temperature but declines near an upper thermal limit (Figure 14.2). Upper-intertidal barnacles tolerate high temperatures better than do species found in the lower intertidal. Species living in the high-intertidal zone in the tropics tend to maintain coordinated ciliary motion at higher temperatures than do species living in the low-intertidal zone or subtidal zone.

Some genetic variation has been found in shell color, which may be related to reflection of sunlight. Along the east coast of the United States, for example, the mussel *Mytilus edulis* has light and dark shell color forms, which are genetically determined. On

[1] See Helmuth, 1999, in Further Reading, Intertidal Zone.

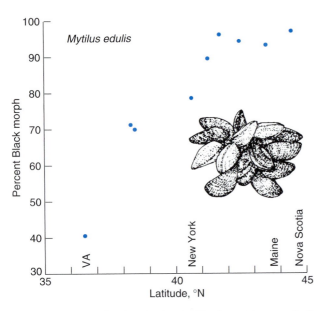

Fig. 14.3 Latitudinal variation in black and brown shell color forms of the blue mussel *Mytilus edulis*. The brown morph reflects solar radiation more efficiently and is favored in lower latitudes. (Data from Innes and Haley, 1977, and Mitton, 1977.)

the east coast of North America, the light-colored form is found more frequently toward the south (Figure 14.3). Dark mussels gain heat more rapidly and are superior in environments where cold water and freezing are common, whereas light mussels reflect heat in the sun and are superior in the south, where high-temperature stress is a problem.

Because high-intertidal habitats are more stressful, species confined to the high-intertidal zone tend to be more resistant to high temperature and desiccation stress at the cellular and biochemical levels. Species living in the high part of the shore retain cellular function at higher temperatures than those living in lower shore levels (Figure 14.4). The high-marsh mussel *Geukensia demissa* (Plate XVII.3) is very tolerant to desiccation and often suffers strong changes in the concentration of cellular constituents. It regulates free amino acids to maintain cell volume (see Chapter 4).

Animals living in intertidal sand- and mudflats face the same problems of temperature and desiccation stress. At low tide, the water level drops, draining the water from the sediment pore spaces. During low

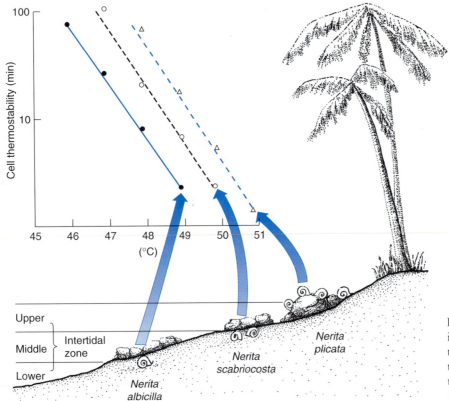

Fig. 14.4 Survival of ciliated epithelial cells in three species of the intertidal gastropod genus *Nerita* and the relation of temperature tolerance to position on shore. (After Ushakov, 1968.)

tide, animals can remain within cool and wet sediment by living in relatively deep burrows. As a result, many intertidal sand-flat invertebrates maintain burrows as much as 25–50 cm below the sediment surface (Figure 14.5). At the time of high tide, the animals either come to the surface or extend siphons, tentacles, or other feeding and respiratory organs to the surface. Shallow-burrowing eastern U.S. coast bivalves, such as the eastern hard clam *Mercenaria mercenaria*, deal with desiccation by tightly sealing the shell. Deep-burrowing clams, such as the razor clam, *Ensis directus*, and the soft-shell clam, *Mya arenaria*, are bathed in pore water and have shells containing gapes, which preclude the shell from being tightly sealed. In fine sediments, water does not drain quite as easily, and small soft-bodied organisms may live closer to the surface.

With increasing height above the low-water level, animals are exposed to air for a greater proportion of the day. This reduces access to food in nearly all species, although the specific mechanisms differ. Mobile carnivores, such as drilling gastropods and starfish, have to spend considerable time to seize and consume prey. Grazers such as periwinkles and limpets cease feeding at low tide on hot dry days, although they may feed when the air is cool and moist. Suspension feeders obviously have no access to phytoplankton at low tide, so higher-intertidal animals have reduced feeding periods and grow more slowly than do low-intertidal forms. Upper-intertidal herbivorous gastropods may be able to compensate for the reduced feeding time by feeding more rapidly. Nevertheless, reproductive output is usually far less for high-intertidal animals than for low-intertidal members of the same species.

At low tide, marine organisms are exposed directly to air, or they are confined in burrows or with small volumes of trapped water (as in barnacles or mussels). As a result, most intertidal organisms face shortages of oxygen and buildup of metabolic wastes. **Reduction of metabolic rate** is a common means of reducing the need for oxygen during low tide. Certain burrowing invertebrates, such as the lugworm *Arenicola marina*, have blood pigments, such as hemoglobin, that bind tightly to oxygen and release it for metabolic needs during low tide. Some species have developed the capacity to breathe air. Mussels may gape at low tide and consume oxygen from the air, as long as their gills are moist. The eastern North American salt marsh mussel *Geukensia demissa* lives in the upper intertidal zone and probably obtains more oxygen from air than from water. Many high-intertidal crabs, such as fiddler crabs and ghost crabs, spend much of their active time in air. Some high-intertidal crabs have membranous disks on their legs that are specifically designed for gas uptake from the air, rather than from water (Figure 14.6).

✴ Mobile intertidal animals may remained fixed at one tidal level, becoming active at the time of low tide, while others migrate up and down the shore with each tidal cycle in order to remain moist.

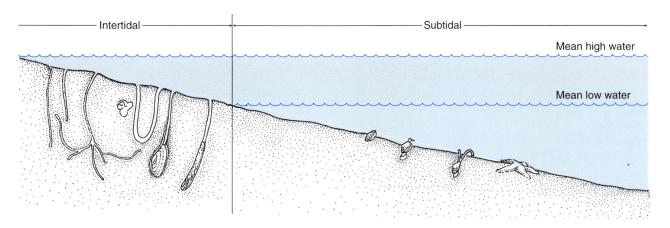

Fig. 14.5 Cross section of the bottom of intertidal and subtidal marine soft sediments. The depth of burrowing is deeper in the intertidal, where exposure to desiccation and temperature variation at the sediment surface is greatest. (After Rhoads, 1966.)

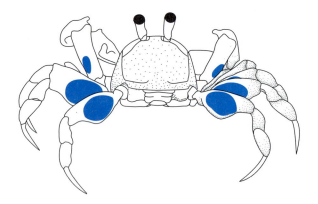

Fig. 14.6 The high-intertidal sand-bubbler crab *Scopimera inflata* has a membrane on each leg (shaded green) designed to exchange gas from the air for uptake into arterial blood. (After Maitland, 1986.)

Like other marine animals, intertidal forms often have adaptations to maintain position, which can be critical in a strong physical gradient. Many limpet species remain in one spot at low tide and graze on algae in the vicinity when the area is covered by water. This spot is often marked by an erosional scar in the rock. The west coast owl limpet *Lottia gigantea* is territorial and actively ejects other limpets. Other grazing limpets and snails spend low tide in moist cracks in the rock, or among the byssal threads of mussels, departing for open areas during high tide. Some intertidal species can use the sun as a light-compass to guide their movements both in sand flats and on rocky shores.

Many species have a series of responses designed to keep a position relative to the water level. In the case of rocky-shore mobile forms, this usually involves responses to light, gravity, and water that combine to keep the animal at the correct level. This can be illustrated by G. S. Fraenkel's work on the high-intertidal periwinkle *Littorina neritoides*. When submerged in seawater, *L. neritoides* is negatively geotactic. It is negatively phototactic when immersed and right side up, but positively phototactic when immersed and upside down. When the animal is moist but not submerged, it has no light response. Figure 14.7 illustrates how these responses combine to lead *L. neritoides* to its high-intertidal splash zone habitat after it has been dislodged by waves or after a feeding excursion in the lower intertidal zone.

Some ocean beach inhabitants continually live on the edge of doom, because they may find themselves in sediments too high and dry, or too low and wave swept. As the tide falls, an infaunal sandy-beach species may find itself in the equivalent of a desert sand. As the tide rises, waves may suddenly make it impossible to maintain a burrow. As a result, some species have evolved novel ways of migrating up and down the shore by periodically leaving the sediment, riding the swash, and reburrowing. Species of the mole crab *Emerita* (*E. analoga* on the west coast of the United States and *E. talpoida* on the east coast) and several species of the wedge clam *Donax* have adopted the strategy of behaving as **swash riders,** leaving the sediment as they feel the pressure of an approaching wave (Figure 14.8). They then ride the wave to a higher position in the intertidal zone and rapidly burrow. When the tide falls, they reverse the process, which guarantees that they will always be located in a moist but not excessively wave-swept level of the beach. As the waves wash back down the beach, the mole crab extends its feathery second antennae, which trap phytoplankton. The mole crab has a streamlined shape and special digging appendages (Figure 14.9).

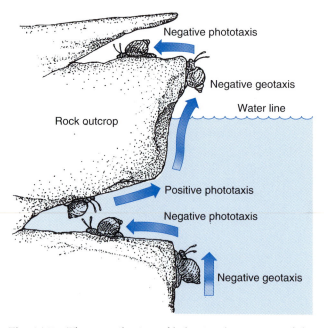

Fig. 14.7 The contribution of behavioral responses of the high-intertidal gastropod *Littorina neritoides* to the regulation of its vertical position on rocky shores. (After Newell, 1979.)

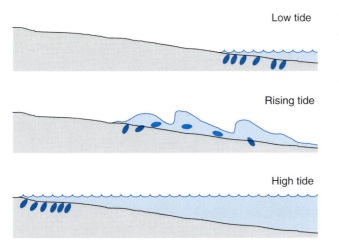

Fig. 14.8 Vertical beach migration of swash riders, such as the mole crab *Emerita* and the beach clam *Donax variabilis*. Note that animals seek sediments that are neither too dry nor too wave washed.

Wave Shock

✳ **Wave shock is a major factor determining the distribution and morphology of intertidal organisms.**

The impact of waves and the material that they carry (sand, pebbles, logs) are important in selecting morphological adaptations to intertidal life. We need only visit a rocky shore on a stormy day to witness the tremendous energy that is focused on the shoreline.

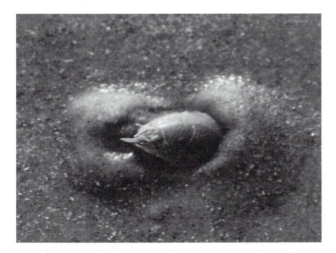

Fig. 14.9 The streamlined shape of the mole crab *Emerita talpoida* allows the animal to burrow into the sand rapidly. (Photograph by the author.)

Waves may rip organisms from the rocks, erode large volumes of sand, or propel a variety of projectiles to the shore.

Breaking waves can damage rocky-shore organisms in the following ways:

1. **Abrasion.** Particles in suspension or floating debris scrape delicate structures. Water turbulence may whip seaweeds and other erect organisms against rocks.
2. **Pressure.** The hydrostatic pressure exerted by breaking waves can crush or damage delicate and compressible structures, such as gas-filled bladders of seaweeds. Most intertidal organisms are liquid filled and therefore are relatively incompressible.
3. **Pressure drag.** Water exerts a directional force against intertidal organisms. The forces may rip apart support structures or dislodge holdfasts, such as byssal threads of mussels or holdfasts of seaweeds.

On rocky shores, open-coast biotas usually differ from those of protected coasts. This difference is especially prominent on the west coast of North America. Many species found in outer open coasts are absent from inshore protected waters. The mussel *Mytilus californianus*, for example, is an outer-coast form with thick shell, large size, and thick byssal threads for firm attachment to the wave-beaten rocks. In species that occur in both habitats, there are often strong morphological differences between open-coast and protected-coast forms. The integument of the starfish *Pisaster ochraceus* is much thicker and harder in the outer-coast individuals than in members of the same species living in protected water. Protected-coast individuals of the drilling snail *Nucella lamellosa* have a delicate shell with large finger-like projections, but open-coast forms have thicker shells with less ornament.

Wave action is also a strong limiting factor on exposed ocean beaches, owing to the tremendous erosive force. Open-ocean sandy beaches are among the most abiotic zones in the world, and very few species can burrow rapidly enough to counteract the continual uprooting process of erosion. We mentioned a few species of swash riders that maintain a migratory cycle of burrowing and wave riding. The eastern surf clam *Spisula solidissima* lives in such exposed areas, and, when dislodged, is capable of rapid reburrowing. In the winter, when much of a beach is eroded, intertidal open-beach invertebrates often migrate to subtidal parts of the beach because the erosional forces are too great to allow them to maintain position.

Causes of Zonation

✳ **Zonation results from preferential larval settlement and adult movement, differential physiological tolerance, and biological interactions such as competition and predation.**

Having examined the important factors in the intertidal gradient, we can ask the question: What causes zonation? The remarkably similar patterns of species's vertical ranges, combined with the sharp boundaries between zones, seem to suggest that the zonation is caused by simple factors. However, this is rarely the case. Several major factors combine to form zones.

1. **Physiological tolerance**. Species found higher on the shoreline are generally more tolerant of desiccation, reduced feeding time, reduced access to oxygen, and extreme temperature. For example, because of their sealable test, barnacles can live higher on the shore than sea anemones.
2. **Larval and adult preference**. To some degree, larvae of sessile animals are able to locate the tidal height suitable for adults of their own species. For example, larvae of the common Atlantic higher-intertidal barnacle *Chthamalus stellatus* settle in the high shore, although they also settle somewhat lower. **Gregariousness** (settling in groups, on adults of the same species) also causes preferential settlement on the "right" level of the shore. In eastern Australia, larvae of the barnacle *Hexaminius popeiana* settle preferentially on adults of their own species.[2] If adults are transplanted above the typical shore level at which the adults usually live, the larvae settle there preferentially as well. Survival of recruits, however, is much lower than within the normal tidal level where the adults usually live. Mobile animals can adjust their tidal height by a combination of responses to light, gravity, and moisture, as discussed earlier.
3. **Competition**. Intertidal habitats, particularly rocky shores, may be severely space limited (Figure 14.10). Species capable of overgrowing or undercutting others may come to dominate

Fig. 14.10 Top: intertidal rocks are often completely covered by organisms. This rock, on Tatoosh Island, Washington, is covered by barnacles (light) and mussels (dark). Bottom: close-up of a mussel-dominated rock. (Photographs by the author.)

that level of the shore in which they can do well physiologically. Mussels, for example, can usually move by forming and detaching byssal threads. They can climb on top of competitors and smother them. As a result, mussels often come to dominate the lower shore. Some barnacles are more resistant to drying and can displace other species on the upper shore.

4. **Predation**. Predators are often strongly limited by the time of immersion because carnivores such as starfish and snails must be moist as they move to locate, seize, and ingest prey. This requirement usually limits predation to the lower

[2] See Coates and McKillup, 1995, in Further Reading, Intertidal Zone.

part of the shore. In some cases, a **refuge** exists, above a certain shore height, where predators do not have enough time to capture prey. Figure 14.11 shows a rocky-shore mussel bed on the coast of England. Dog whelks (*Nucella lapillus*) come out of moist lower-intertidal cracks during the rising tide and prey upon barnacles (see Plate VI.2). They must return to the cracks when the tide withdraws. Mussel beds on the outer Pacific coast, dominated by *Mytilus californianus*, often have a similar sharp lower limit, which is controlled by the carnivorous starfish *Pisaster ochraceus*. Robert Paine re-

Fig. 14.11 Predation line on rocky shores of England. Below this line (adjacent to the crack in the rock), the gastropod *Nucella lapillus* can clear the rocks of its prey, the mussel *Mytilus edulis* (dark-colored areas). (Photograph courtesy of Raymond Seed.)

moved starfish for over a decade and found that the mussel bed gradually extended lower in the tide zone.

Interspecific Competition

✳ **Field experiments show the importance of interspecific competition.**

What determines the sharp zonation often found on rocky shores? **Field experiments** were crucial in understanding the important processes. One may expect differential competitive success to be common in a strong physiological gradient such as that of the rocky intertidal zone. On rocky shores of both the American and northern European sides the North Atlantic, one often sees the zonation pictured in Figure 14.12. In the highest part of the shore, the barnacle *Chthamalus stellatus* dominates available space down to the approximate level of mean high tide. Below this, the shore is usually monopolized by the white barnacle *Semibalanus balanoides*. At the level of mean low water, however, barnacles are usually rare. What explains the distribution?

Joseph Connell studied this problem by selective removals of hypothesized competitors and by caging selected areas to keep away the common predatory gastropod there, the dog whelk *Nucella lapillus*. First he considered larval recruitment. Cyprids of *Chthamalus* tended to settle high on the shore, but they also settled well within the range in which adults were rare. *Semibalanus* cyprids recruited in large numbers throughout the intertidal, but failed to do very well above mean high water. Hence, there was a substantial area of overlap, although differential larval recruitment explains some of the difference in adult vertical distribution.

Connell transplanted rocks downward with newly settled *Chthamalus* and found that these barnacles were rapidly overgrown and undercut by recruited *Semibalanus*, which are much faster growing. Settlement of *Semibalanus*, however, was very dense, and intraspecific competition also resulted in extensive *Semibalanus* mortality. In the upper reaches of the shore, *Chthamalus* survival was greater, but the *Semibalanus* there died, owing to their poor survival in the dry reaches of the upper intertidal (Figure 14.12).

These experimental results show that the lower limit of *Chthamalus* on the shore is regulated by competition with *Semibalanus*, whereas the upper limit is probably controlled by desiccation. The upper limit

Hot Topics in Marine Biology 14.1 Supply-Side Ecology:
Importance of Larval Dispersal in Community Structure

Marine biological communities are dynamic, and few places anywhere in the ocean are the same from one decade to the next. In the past, ecologists have concentrated upon interspecies interactions among adults as the source of much of this change. We know that predators can have devastating effects. Migratory birds, for example, can extinguish all the denizens of a huge mud-flat system. Ecologists have thought that in most environments the interactions of competition, predation, and occasional disturbance were the main determinants of species composition. We thought that we could extrapolate to large areas the interactions that could be studied experimentally on a small spatial scale.

In the last few years, the roles of larval supply and transport have become more obvious as strong controllers of regional differences in community composition.[3] On the west coast of the United States, kelp forests separate the rocky coasts from the open sea. Where kelps are scarce, barnacle larvae appear to recruit more heavily. Kelp forests seem to act as larval filters, partially because they slow down the advance of shoreward water movement, which would otherwise carry larvae to the coast. However, kelp forests are also dense with planktivorous fishes, and these pick off much of the larval population that enter the relatively quiet waters. On the Great Barrier Reef of Australia, coral larval recruitment is densest where natural eddies focus streams of water, with entrained larvae, toward the coast. If larvae do settle, competition is strong, predators grow, and interspecies interactions matter. However, if there is a failure of larval arrival and settlement, none of these local biological processes amount to much.

Coral reefs are regularly destroyed by hurricanes and other strong storms, and larval colonization is apparently crucial in their recovery. In recent years, the reefs on the north coast of Jamaica, once among the most beautiful in the world, have been nearly destroyed because of strong hurricanes that swept along the coast. Only recently, a good larval set of the staghorn coral *Acropora cervicornis* has brought the possibility of recovery, assuming that a common coral-eating snail species does not kill off the new colonies.

When will larvae fail to appear? Many larvae feed on phytoplankton and smaller zooplankton. If these are absent, a larval population will collapse. In many coastal areas, the net water flow is landward (see Chapter 5), but water flow is seaward in estuaries and also on some open coasts. Larvae released in a deep embayment may be trapped inside for weeks, and adjacent coastal embayments may never profit from the large flush of larvae in any one locale. In southeast Australia a persistent current tends to transport surface water offshore, thus larval settlement success depends upon local larval production and retention, not on regional supply of larvae from far down the coast. We are still very ignorant about the movement of coastal water in many regions and can say little about the general importance of larval supply, although some bays and estuaries are now well understood with regard to larval supply.

It is easier to examine recruitment, which is the appearance of newly settled larvae as metamorphosed juveniles, as opposed to settlement of the larvae themselves, which are small and hard to observe directly. The classic experiment done by R. T. Paine demonstrated that removal of the predatory starfish *Pisaster ochraceus* resulted in dominance by the mussel *Mytilus californianus*, but this effect depended upon a good recruitment season. As it turned out, many subsequent seasons were quite poor, and the experiment probably would not have "worked" in quite the same way had it been carried out later. Realistically, most communities consist of species whose respective temporal variations may depend upon very different factors. A recent study by Curtis Lively and colleagues, for example, demonstrated that, in the Gulf of California, barnacle abundance is mainly a function of microspatial factors, such as predation intensity by a carnivorous gastropod. The abundance of mussels, by contrast, seemed to be best related to temporal changes in the environment, which probably cause variation in larval recruitment. This study reveals the realism of natural communities; the dominance of different ecological processes (e.g., competition, predation, larval recruitment) will vary among environments and among ecological groupings within habitats.

[3] See Underwood and Fairweather, 1989, in Further Reading, Intertidal Zone.

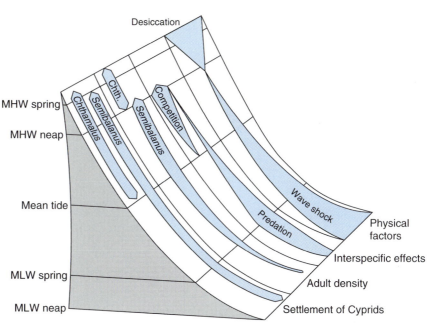

Fig. 14.12 Distribution on rocky shores of Scotland of adult and newly settled larvae of the barnacles *Semibalanus balanoides* and *Chthamalus stellatus*. Width of the bars indicates the relative effects of principal limiting factors; MHW, mean high water; MLW, mean low water. (Modified after Connell, 1961.)

of the *Semibalanus* zone is also controlled by desiccation. Predation by the drilling gastropod *Nucella lapillus* controls the lower limit of *Semibalanus*. These results have led to the generalization that the upper limit of an intertidal species is regulated by physical factors, whereas the lower limit is regulated by biological factors (e.g., competition). This generalization, however, has many exceptions. For example, the lower limit of some exposed coast species is largely determined by wave stress, a physical factor.

The same sort of process works in soft sediments. The common mud snail *Ilyanassa obsoleta* dominates eastern muddy shores of temperate North America. In the last few decades, it was introduced and spread rapidly into San Francisco Bay. It has displaced the local mud snail *Cerithidea californica*, which now occupies a refuge in the upper part of the intertidal zone. This seems to be due only to the poor desiccation tolerance of *I. obsoleta*. Margaret Race[4] built an enclosure spanning the intertidal zone, and removed *I. obsoleta* from the lower part of the shore. *Cerithidea* soon expanded its vertical range downward. Predation of *Cerithidea* eggs also contributed to the reduction of its ecological range.

In muddy subtidal bottoms, burrowing invertebrates compete for space within the sediment, and one often sees **vertical stratification** of dominant species within the sediment. Some species, for example, are free burrowers, and they disrupt the permanent burrows of other species (Figure 14.13). The limitation of space can be shown by a field experiment. If deep-dwelling, suspension-feeding clams are reduced in density, the remaining individuals grow faster and have greater survival. In Mugu Lagoon, California, the deep-dwelling clam *Sanguinolaria nuttali* is affected by the presence of other deep-dwelling species. Variation in density of a shallow-burrowing species, however, causes no effect. Because all the species feed by means of a siphon connected with the surface, this experiment suggests that space, and not food, is probably the limiting factor.

A limited number of benthic species, including some polychaetes, hemichordates, and a species of phoronid, produce high concentrations of **bromine-containing aromatic compounds**, which are toxic and are avoided by other infaunal organisms. These compounds are released into the sediment, so a halo of brominated aromatics surrounds the worms. The sediment is smelly to humans. Some species exist as isolated large animals, while others are only a few millimeters long and live in high densities, sometimes as a single-species patch of organisms. Presumably the chemical production is an adaptation to defend against incursions by competitors, but it may also de-

[4] See Race, 1982, in Further Reading, Intertidal Zone.

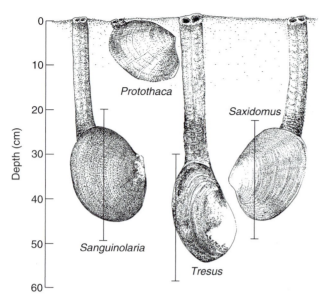

Fig. 14.13 Stratification of bivalve mollusks in intertidal sediments of Mugu Lagoon, southern California. *Prototothaca staminea*, a shallow-burrowing clam (at top), is unaffected by the presence of deep-burrowing clams. Experiments demonstrate that *Sanguinolaria nuttalli* (lower left) is strongly depressed by other deep-dwelling clams. Vertical bars show vertical burrowing position. (Drawn following descriptions in Peterson and Andre, 1980.)

ter predators. Newly recruited bivalves and polychaetes clearly avoid patches of benthic invertebrates with brominated aromatics. If these aromatics are injected into the sediment at the same concentrations found naturally next to chemical-producing animals, juvenile bivalves and nonbrominated polychaetes burrow much less than in aromatic-free sediment.[5]

Competition and Character Displacement

* Competition among similar species may lead to a separation of size and resource exploitation known as character displacement. Interpretations of this phenomenon in the marine environment are controversial.

The examples just cited involve short-term ecological interactions. However, one may predict that over the long term, reduced reproductive performance may result in the evolution of changed ecological preferences. If a species were in continual competition with a superior species, genotypes that specialized on other resources might win out in the evolutionary race. This line of reasoning would predict that assemblages of continually competing species would show niche partitioning, owing to a period of coevolution. The **theory of limiting similarity** predicts the minimum overlap that could be tolerated without competitive displacement. Such a theory works effectively only when the resource under consideration is simple and can preferably be quantified in a single dimension on a linear scale, such as prey size or location along a shore gradient. If there is a simple scale, one may expect **character displacement**, a pattern in which ecologically similar species differ more in resource exploitation and associated morphology (e.g., head size, which is related to prey size) when the species are co-occurring, but are nearly identical when they do not overlap geographically.

Unfortunately, very few good examples of character displacement have ever been discovered for marine organisms. The mud snails *Hydrobia ulvae* and *H. ventrosa* live in mudflats in the Limfjord, a body of water in Denmark that became saline in 1825 after a storm broke through a barrier bar. Although *H. ventrosa* prefers lower salinities, it overlaps broadly with *H. ulvae*. In the areas of overlap, *H. ulvae* tends to be large, whereas *H. ventrosa* tends to be small (Figure 14.14). In single-species localities, either species is intermediate in size. This may reflect competition pressure for benthic diatoms, which would be most intense when snails are of similar size. The divergence in body size may reflect a response that reduces competitive pressure for diatoms in the sediment. Large snails may be feeding on larger diatoms, whereas small snails may feed on smaller diatoms. One would have to conclude that evolution since 1825 has accomplished this divergence.

There are some complications to this story. Feeding in *Hydrobia* is somewhat more complicated than the large-snail–large-diatom hypothesis can account for. Snails spend a good deal of time scraping diatoms off sand grains, and the two species under consideration differ considerably in this regard. This discrepancy may require only a revision of the mechanism for displacement. An alternative and noncompetitive explanation may be called for, however. In the Limfjord, *Hydrobia ulvae* in single-species localities are curiously small, relative to single-species locales in the more open areas of the North Sea. It may be that these localities are somehow physiologically unsuit-

[5] See Woodin and others, 1997, in Further Reading, Intertidal Zone.

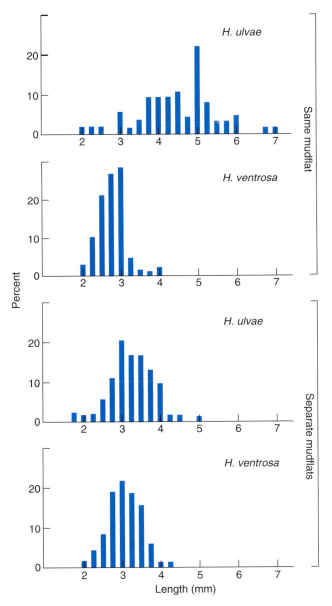

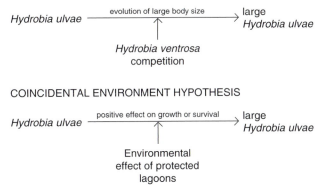

Fig. 14.15 Two hypotheses explaining the size differences between *Hydrobia ulvae* when it co-occurs with *Hydrobia ventrosa* and when it occurs alone.

Fig. 14.14 When the Danish deposit-feeding snails *Hydrobia ulvae* and *Hydrobia ventrosa* live apart, they have about the same size range (lower two diagrams), but when they live in the same mudflat, *H. ulvae* is larger and *H. ventrosa* is somewhat smaller. This may reflect evolved differences in feeding when the species coexist, but current evidence suggests that competition probably does not explain the difference. (After Fenchel, 1975.)

able for *H. ulvae*. It would be useful to work out the genetics of body size in this system.

I. Saloniemi[6] discovered that *Hydrobia ulvae* may be large when sympatric with *H. ventrosa*, but this relationship may be only coincidental. After exam-

ining size variation along the Baltic Sea coast of Finland, Saloniemi found that *Hydrobia ulvae* seems to do best in sheltered lagoons, which coincidentally are places where *H. ventrosa* is mainly found. Thus, the large size of *H. ulvae* may be due to the sheltered lagoons, not evolutionary competitive pressure from *H. ventrosa*. The smaller *H. ventrosa* are annuals (they die after reproducing, after one year of life), which explains perhaps why they are rather similar in maximum size everywhere. Instead, *Hydrobia ulvae* lives potentially for several years. In better habitats (i.e., the sheltered places where both species tend to co-occur), *Hydrobia ulvae* may either grow faster or live longer and therefore reach larger sizes. This leaves two competing hypotheses to explain size differences in mud snails: evolutionary character displacement and habitat effects (Figure 14.15).

Predation and Interspecific Competition

✳ Predation (or herbivory) may ameliorate the dominance achieved by competition and may strongly affect species composition.

The importance of predation has also been demonstrated through field experiments. Recall, for example, that Paine studied this effect on west coast American rocky shores by removing the carnivorous starfish *Pisaster ochraceus* (Figure 14.16, Plate XIV.4). The mussel *Mytilus californianus* soon came to dominate. Such experiments suggest a common effect: predation delays the competitive displacement of

[6] See Saloniemi, 1993, in Further Reading, Intertidal Zone.

Fig. 14.16 The starfish *Pisaster ochraceus* is a top predator on eastern Pacific rocky shores. Starfish of this species were turned over while consuming a barnacle (top) and while eating a mussel (bottom). (Photographs by the author.)

as **keystone species**. These species exert so-called top-down effects on ecosystems.

While the starfish *Pisaster ochraceus* is a keystone species with respect to species attached directly to rocks, one must remember that the interstices of the mussel bed may contain large numbers of smaller animal species, such as polychaetes, barnacles, and smaller gastropods. Starfish predation may remove the competitive dominant and increase the number of coexisting large-bodied species attached to rocks, but by removing the mussel bed it also acts in the same manner as clear-cutting a terrestrial forest, namely, causing the disappearance of many smaller dependent animal species.

Strong interactions come not only from animal predators but from herbivores. On rocky shores, limpets are especially important in the mid-intertidal zone because they are very effective grazers of diatoms attached to rocks, coralline algae, and seaweeds. Some limpets forage and return to a home scar, whereas others move about on the rocks. Simple removal experiments are very informative. Under normal circumstances, limpet-covered rocks are quite barren. However, Betty Nicotri[7] removed limpets from the intertidal zone of San Juan Island, Washington, and soon observed the development of a lush cover of diatoms. Limpet grazing has also been found to inhibit the growth of seaweeds on rocky shores in many regions of the world. Finally, limpets affect strongly the distribution of barnacles, because the limpets bulldoze aside and ingest newly settled cyprid larvae. On rocks dense with limpets, it is common to observe (Figure 14.17) only a few very large barnacles that managed to survive the limpets (and other sources of mortality).

Predation may be so intense that the competitive dominant is rather rare. For example, several species of turf-forming coralline algae compete for space in the lower intertidal zone of the outer coast of Washington. Experimental combinations of species produced a surprise: the dominant competitor was rather rare in the natural community. In a sense, the basis for its competitive superiority was the key to its population downfall. It grew rapidly enough to displace other competitively inferior species, but this also made it the favorite food of some common grazing chitons and limpets. As a result, the relatively slower

competitively inferior species by the competitive dominant. Recall that in North Atlantic rocky shores, the dog whelk *Nucella lapillus* preys in the lower intertidal upon the competitive dominant barnacle *Semibalanus balanoides*. This predation opens up space for colonization by the competitively inferior *Chthamalus stellatus*. This general effect has also been observed among seaweed competitors: that is, competition is reduced by the introduction of grazers such as urchins and snails. In New England shores, snail and starfish predation on mussels permits the Irish moss *Chondrus crispus* to dominate lower intertidal shores. Mussels tend to dominate available space on more exposed headlands where predators are eliminated owing to wave shock. Species, such as starfish, that prevent the monopolization of a habitat by preying on most potential competitors are known

[7] See Nicotri, 1977, in Further Reading, Intertidal Zone.

Fig. 14.17 Limpets on a mid-intertidal rocky shore, San Juan Island, Washington. Note the unoccupied space, caused by intense limpet grazing, which consumes the diatoms but also bulldozes aside newly settled barnacle larvae. (Photograph by the author.)

growing and less preferred algal species dominated the space, which was monopolized by coralline algae.[8] This result leads to some interesting conclusions. First, the complete monopolization of a resource does not necessarily indicate that predation is unimportant. Second, competition cannot be divorced from predation.

∗ Spatial heterogeneity may strongly affect the pattern and intensity of predation on rocky shores.

We have discussed the decreasing effectiveness of mobile carnivores in the higher part of the shore. Thus many upper intertidal species escape predation by snails and starfish. But even at the same level in the intertidal zone, spatial heterogeneity can strongly influence local patterns of predation. Cracks retain

moisture and many mobile carnivores spend the time of low tide there, with no need to retreat to the lowest part of the tide zone. It stands to reason, therefore, that prey items within easy reach of the cracks and pools will be taken with great frequency. In the intertidal zone of the west coast of the United States, species of the genus *Nucella* drill a variety of barnacle and mussel species. They are often found abundantly at the edges of mussel beds and can stay in these relatively moist areas during low tide. When the tide rises, they move from these protected edge habitats out on the rocks to prey upon barnacles and smaller mussels. If a large patch is opened by a storm, however, the snails tend not to move across a large area of open rock. Thus, barnacles and mussels settling in the middle of the patch may occupy a refuge from predation.

While rock cracks and pools are relatively permanent, mussel beds and other biological refuges from desiccation need not be permanent. Anthony J. Underwood[9] was surveying a rocky shore in New South Wales, Australia, when a severe storm hit in 1974. Before the storm, the rocks were covered by large patches of the seaweed *Hormosira banksii*, and the carnivorous drilling gastropod *Morula marginalba* survived desiccation at low tide by retreating to the wet shelter of the seaweed canopy (Figure 14.18). Barnacle mortality was very high near and within the seaweeds, which was explained mostly by snail predation. In some cases seaweeds whip back and forth, scraping barnacle-settling larvae from the rocks, but that was not the case in this study. The seaweeds indirectly affected the barnacles by positively enhancing the local abundance of the snail. The storm therefore set a new pattern for predation interactions by removing the shelter for a major predator. It took about 5 years for the strongly disturbed patches of *Hormosira* to recover fully. To this day, the effects of the great storm of 1974 can be registered on rocky shores of southeastern Australia in terms of abundance patterns. We explore this further later in Hot Topics in Marine Biology 14.2, The Challenge of Disturbance Scale and Recruitment to Rocky-Shore Community Ecology.

∗ Seasonal influxes of predators in the intertidal can devastate local communities.

[8] See Paine, 1990, in Further Reading, Intertidal Zone.
[9] See Underwood, 1999, in Further Reading, Intertidal Zone.

Hormosira banksii
(a)

Morula marginalba
(b)

Fig. 14.18   (a) The seaweed *Hormosira banksii* forms a canopy under which snails hide and keep moist at low tide. (b) The gastropod *Morula marginalba* is a major predator of invertebrates on rocky shores of southeastern Australia. It is in the process of consuming a barnacle. Snail is about 1.5 cm long. (Photograph by the author.)

Although many habitats have their indigenous permanent population of predators, many seasonal habitats experience periodic invasions of predators with devastating consequences for prey populations. In the intertidal zone, the most prominent example are migrating shorebirds, whose migrations may extend for thousands of kilometers. Shorebirds (Figure 14.19) often have favored feeding grounds, which they visit successively during their migration. Along the coast of eastern North America, sites such as Plymouth Bay (Massachusetts), Cape May (New Jersey), and Jamaica Bay (New York) are famous for their periodically dense populations of shorebirds. Sanderlings (*Calidrus alba*), semipalmated sandpipers (*Calidrus pusilla*), short-billed dowitchers (*Limnodromus griseus*), and black-bellied plovers (*Pluvialis squatarola*) all prey on mudflat invertebrates. In one study, David Schneider[10] "caged" areas of a Massachusetts mudflat simply by placing a rope suspended from four short sticks. Within the enclosed area, the benthos survived for 2 months at the same densities. Outside the enclosure, the soft-bodied benthos (e.g., polychaetes) were devastated. The birds concentrated their efforts on the most abundant benthic species (Table 14.1). Similarly, migrations up Chesapeake Bay of the blue crab *Callinectes sapidus* also obliterate the shallow-burrowing benthos. Spring migrations of fish and crustacea into shallow water embayments and mudflats cause great decreases in benthic population size. In many cases, however, smaller predators, including gastropods, nemertean worms, and polychaete annelids have much larger effects than the larger predators such as birds and flatfishes.

Fig. 14.19 Laughing gulls, *Larus atricilla*, feeding on horseshoe crab eggs on a sand flat at Cape May, New Jersey. (Courtesy of Joanna Burger.)

[10] See Schneider, 1978, in Further Reading, Intertidal Zone.

Table 14.1 Percent mortality of intertidal mudflat invertebrates fed on by migratory shorebirds in Plymouth Harbor and Kingston Harbor, Massachusetts: Mortality is listed as a function of the prey species' rank at the start of the study

SPECIES RANK IN JULY	PERCENT MORTALITY
1	84
2	78
3–5	67
5	36

Source: Data from Schneider, 1978.

Fig. 14.20 Aggregation of starfish, *Pisaster ochraceus*, at the periphery of a mussel bed dominated by *Mytilus californianus*. (Courtesy of Thomas H. Suchanek.)

✳ Mobile predators can often detect newly recruited sessile benthos and may form strong aggregations and feed on new recruits.

As we have stressed, larval recruitment is not homogeneous, either between sites or from one year to the next. Predators, such as starfish, drilling snails, and crabs would alternately be overwhelmed by food and starved if they did not attempt to find spatially concentrated new sources of food. Such concentrations are liable to develop in sites where larval recruitment was very high; perhaps the water flow was high and brought settling larvae to a given location. It is quite common to see abundant predators where recruitment is high. On rocky shores of Maine, where springtime is a wet, bone-chilling affair, large herds of the drilling snail *Nucella lapillus* move along at low tide, mowing down newly settled barnacle recruits. On the west coast, large aggregations of the starfish *Pisaster ochraceus* can often be seen on the lower shore periphery of dense mussel beds (Figure 14.20). Working on the outer coast of British Columbia, Canada, Carlos Robles and colleagues found that starfish aggregated around local areas of high mussel recruitment, consumed the mussels, and then dispersed.

✳ Many intertidal organisms have behavioral and structural defenses against predators.

Predation can be very intense, but many potential prey species have evolved effective defenses against predators. Hiding in cracks, living high in the intertidal zone, and simply growing too large all provide effective refuges against predation. Many species, however, cannot escape predator contact. Sessile species and slow-moving mobile species cannot escape predators unless they are cryptic or have some sort of defense. Mussels and barnacles can deter predators with their shell, and large mollusks often have shells too thick to be penetrated by boring snails. The mussel *Mytilus edulis* is especially vulnerable to drilling gastropods, such as the eastern Atlantic dog whelk *Nucella lapillus*, but it has an effective defense: when a whelk wanders onto a group of mussels, the prey animals react by secreting byssal threads, which can entwine and trap the snail.

✳ Strongly interacting species often cause indirect effects in rocky-shore food webs.

Rocky-shore food webs often consist of several strongly interacting predator and prey species; the latter also compete intensely for space. Such a strongly interacting web can lead to unexpected interactions. If these interactions are diagrammed by means of arrows, one can sometimes see indirect effects. Consider Figure 14.21, which partially depicts the interactions among sessile species and mobile carnivores in the northern Gulf of California, studied by Curtis Lively and his colleagues. As can be seen, mussels compete with barnacles and the seaweed *Ralfsia*. The snail *Acanthina* consumes barnacles, which, in turn, allows *Ralfsia* to expand. In turn, this provides more food for grazing limpets. So, indirectly, *Acanthina* benefits *Ralfsia* and limpets. By grazing on *Ralfsia*, limpets allow barnacles to settle, which indirectly benefits *Acanthina*. Because the interactions among

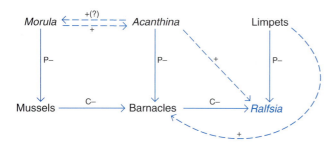

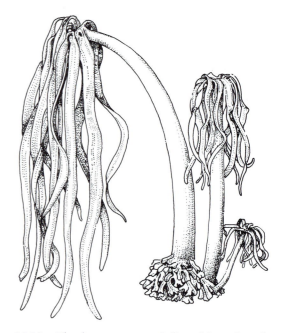

Fig. 14.21 The direct (solid arrows) and indirect (dashed arrows) interactions in an intertidal rocky-shore community in the northern Gulf of California. For example, limpets graze the seaweed *Ralfsia*, whose removal allows greater settlement by barnacles. Key: (P–, a negative effect by a predator; C–, a negative effect by a competitor; + symbol alone specifies a positive indirect effect. (Drawn from descriptions in Lively et al., 1993.)

Fig. 14.22 The brown seaweed *Postelsia palmaeformis* colonizes eastern Pacific rocky shores that have been recently disturbed.

species are so strong, such indirect effects are common. It is even possible to imagine mutualisms. For example, the gastropod *Morula* consumes mussels, which can overgrow barnacles. Consumption of mussels will benefit the barnacle-consuming *Acanthina*, but consumption of barnacles by *Acanthina* may accelerate the competitive dominance by mussels, a development that benefits *Morula*. Therefore, *Morula* and *Acanthina* are mutualists, via these indirect effects. Although such examples may seem clear, the actual details of the natural history muddy the waters substantially. For example, mussels settle a bit more frequently on barnacles than on bare rock.

Disturbance and Interspecific Competition

* Physical and biological disturbance often determines the species composition of intertidal communities.

The intertidal zone stands in the way of some of the most intense storms in the ocean. This is especially true of rocky headlands exposed to the open sea, where waves, boulders, and logs all crash on rocky shores. On the outer Pacific coast of North America, such storms commonly strip off the dominant mussels and seaweeds and thereby open space for colonization. The palm seaweed *Postelsia palmaeformis* (Figure 14.22, Plate XIV.3) colonizes such newly opened spaces, mainly by dispersal of adult plants. Once it has colonized a newly opened area, the seaweed produces spores that move down the fronds and increase the patch size of the plant population.

Disturbance by storms on rocky shores has much the same effect as predation. It liberates space and allows inferior competitors to persist. Moderate disturbance should allow more species to coexist than strong storm disturbance, which would strip off most species. Storms may indirectly benefit anemones in tidal pools, which can eat mussels stripped from the rocks.

Soft-bottom areas, especially exposed beaches, are also disturbed continually. Open-ocean beaches are continuously eroded by waves and currents, and very few animal species can survive the instability. In some locations, biological disturbance is an important factor. In eastern American protected beaches, the horseshoe crab *Limulus polyphemus* comes into shore to breed, and its burrowing activities can destroy the burrows of many species of invertebrates.

Succession in the Intertidal Zone

* Succession of intertidal seaweeds may be irregular, but an overall spectrum of life histories begins with early successional good colonizers, which are prone to grazing, yielding in many cases in late succession to good space holders that are resistant to grazing and to competitors.

Seaweed succession on rocky shores cannot be divorced entirely from animals because several animal grazers can affect succession, and some rocky-shore animals compete with seaweeds for space. These facts make it difficult to discuss seaweed succession without some oversimplification. Consider the colonization of a recently constructed rocky jetty on a North American coast. Succession is initiated by a disturbance, and continual disturbance maintains the earliest stage of succession. At first, nothing colonizes the boulders, but later a surface slime develops, consisting of bacteria and other microorganisms. After this, species of the green seaweed genus, *Ulva* or *Enteromorpha*, will most likely colonize. Limpets or periwinkles may graze the seaweed down to bare rock.

After several cycles of colonization and grazing, the brown seaweed *Fucus* (*Gigartina* or others on the west coast) may appear slowly and come to dominate the lower shore. Its later appearance does not depend upon the previous appearance of any seaweed species. It recruits more slowly than the early successional greens, and it is not likely to dominate unless they have been grazed away. In continually disturbed areas, *Ulva* or another green seaweed will persist, and the later successional dominants will never appear. Once *Fucus* has appeared, however, greens will not likely dominate again unless and until *Fucus* is removed by a storm disturbance or by grazers. The periwinkles *Littorina* spp. are not very good at eating *Fucus*, owing to its relatively tough stipe and holdfast and, possibly, its poisonous compounds. The combined operation of these factors makes succession quite irregular and far from a predictable sequence of changes.

Despite a considerable degree of irregularity and the absence of a guarantee that macroalgal succession will move toward completion, certain properties of seaweeds reflect their position in the successional sequence. In early succession, seaweeds place a premium on rapid growth. Later in succession, however, resistance to grazing and the ability to combat competitors are more important, but the investment of seaweeds in tough tissues and various poisonous compounds imposes a cost that reduces growth rate. In California, the sea lettuce *Ulva californica* dominates early colonization, whereas the fucoid brown seaweed *Pelvetia fastigiata* dominates the late successional stages. Algal productivity and nutritive value (in kilocalories per ash-free unit dry weight) decrease from early to late successional species. Late successional seaweeds allocate more energy to the synthesis of support structures, attachment structures, and structures that fend off urchin grazers. Later successional forms are tougher and more resistant to wave shear. These features all contribute to persistence, the hallmark of late successional species.

✳ Succession in the intertidal depends strongly on the spatial scale of the disturbance that begins the process.

Storms, logs, and starfish may all cause disturbances that initiate the process of succession. The size of the disturbance is very important in directing the course of events. Remember, succession is a fairly sloppy process and is not inevitable. If disturbances are on a relatively small scale, then colonization will be from the area surrounding the disturbed patch. The first colonists may be species characteristic of later stages of succession. If the disturbance is on a much larger scale, then only the fastest colonizers from a distance will have access. These will most likely be the early successional species.

These principles can be applied to successional patterns on rocky shores. Robert Paine and Simon Levin[11] studied patch dynamics on intertidal rocks of the northwest coast of the United States. Patches that were artificially opened within a bed of the mussel *Mytilus californianus* gradually recovered after a number of months. An interesting pattern of colonization occurs within relatively small patches that are opened by storms and logs crashing against the rocks. In extremely wave-swept microsites, the sea palm *Postelsia palmaeformis* takes over, and spores with very reduced dispersal potential will permit the seaweed to hold space for several years. In less wave-swept microsites, the rapidly colonizing mussel *Mytilus trossulus* often appears and takes over the space. Carnivorous snails of the genus *Nucella* are common within the surrounding *M. californianus* beds and soon move into an opened patch and typically kill off all the *Mytilus trossulus*. As this proceeds, the surrounding mussel bed congeals and covers up the original patch.

If a storm strips off a mussel bed over an area of several square meters, however, such a reestablishment pattern is far less likely. Instead, the community dominants will be determined by recruitment mainly from the planktonic larval and spore pool. If the patch is very small (<0.1 m² or so), grazing limpets and

---

[11] See Paine and Levin, 1981, in Further Reading, Intertidal Zone.

Hot Topics in Marine Biology 14.2 The Challenge of Disturbance Scale and Recruitment to Rocky-Shore Community Ecology

The pioneering experiments of Joseph Connell and Robert Paine and their students in the 1960s and 1970s brought us a new language for the study of controls of community structure. We could visualize a community as the product of a number of processes—competition, predation, and disturbance. These worked within the context of a steep intertidal physiological gradient that strongly influenced predator behavior and competitive abilities. Most importantly, these processes could be studied by field manipulative experiments at relatively small spatial scales, usually much less than a square meter. Remove predators and one could expect intensification of competition, followed by dominance of a superior competitor for space. Disturb an area and space might be opened up, creating space for inferior competitors. The intertidal landscape was taken to be dominated by interactions among recruits or adults and at relatively small spatial scales.

This point of view is a classic example of a paradigm, or a model for all to follow, if you wish to understand rocky shores throughout the world. It is fair to say that an intellectual assault of about 15 years is now complete and the paradigm has been cracked.

The beginning of the turnaround starts in another marine system. Peter Sale attempted to explain the distribution of coral reef fishes (see Chapter 15) and concluded that recruitment to open sites might matter much more than predictable competitive bouts between benthic fish, even though these are territorial. A particularly telling study demonstrated that two species of damsel-fish on an Australian coral reef took nearly all the space, dividing it up into individual territories. If an individual of one species was removed, the species that would recruit to take its place could not be predicted. Thus Sale made us wonder whether recruitment was really the driving force behind the distribution and abundance of marine species.

Enter Tony Underwood,[12] who also argued that great variation in larval recruitment might alter the outcome of interactions in the intertidal zone. If larval supply was paltry, then how could a sessile species come to displace all others? After all, a mussel could grow to be only so large, and it could not displace another sessile organism even a few centimeters distant. As mentioned in the text, Robert Paine's classic study showing the competitive domination by the mussel *Mytilus californianus* was based to a degree on an unusually high recruitment season. In most subse-

quent years recruitment of these mussels would be too low to produce the expected result of competitive displacement.

The issue of recruitment raises a fundamental assumption behind rocky-shore ecology and marine ecology in general. We often perform small-scale experiments and extrapolate the outcome to larger time scales and spatial scales. Is this valid? Not if processes cannot be extrapolated from within a small cage or patch of rocky shore.

What if rare events were a driving force behind the pattern of community structure? That is the message conveyed by Tony Underwood's analysis of intertidal recovery (discussed in the text) following an extremely unusual storm in the region of Sydney, Australia, in 1974. The storm stripped away a seaweed canopy, and recovery did not occur for years. Such large-scale disturbance may completely alter community structure because of dependencies of predators upon shelter, as is the case in the Australian example. The drilling gastropod *Morula marginalba* no longer could be sheltered under the seaweed canopy, which resulted in reduced predation because *Morula* had less immediate access to prey (Figure 14.18). Low recruitment (there is a persistent current that moves larvae offshore in this region) also contributed to the failure of large numbers of mussels to recolonize Sydney rocky shores. For someone accustomed to North American and northern European rocky shores, I was surprised to see the nearly nonexistent mussel population.

The role of rare events can strongly affect intertidal landscapes, especially because patches are continually being opened up by storms on exposed coasts. In the U.S. Pacific Northwest, such patches are opened continuously, and the sea palm *Postelsia palmaeformis* can colonize open spaces and maintain very localized populations by means of very short distance dispersal of spores. This observation, made by Robert T. Paine and students, is the first key we have to the potential of disturbance to produce structure to an intertidal landscape.

A suggestion by Thomas Suchanek[13] placed a strong focus on the role of spatial scale in patterning intertidal landscapes of exposed coasts of the western United States (Box Figure 14.1). If disturbance was not too frequent, then mid-intertidal areas were dominated by the California sea

[12] See Underwood and Denley, 1984, in Further Reading, Intertidal Zone.
[13] See Suchanek, 1981, in Further Reading, Intertidal Zone.

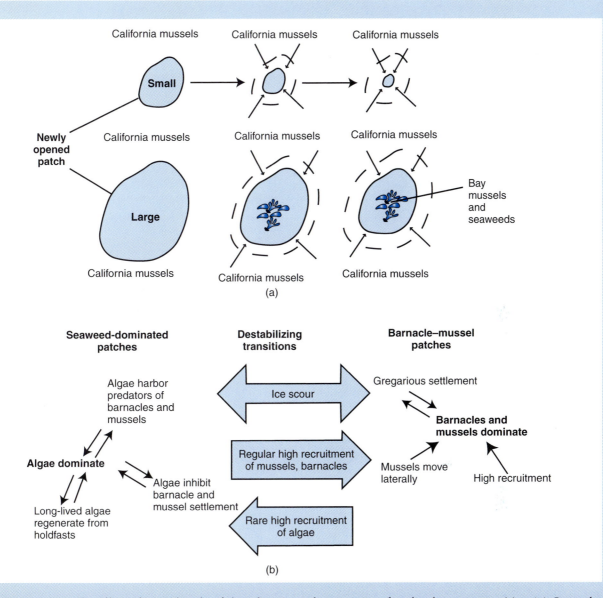

Box Figure 14.1 The effect of spatial scale of disturbance on the structure of rocky-shore communities. (a) On rocky shores of the western United States, the size of a patch opened by disturbance strongly affects the future of the patch. If small patches are opened, mussels move in toward the interior by means of crawling, and the patch is closed. If the patch is large, however, the ratio of perimeter to interior area is smaller, which reduces access by the surrounding mussels and inhibits incursions of predatory snails. This gives enough time for colonization of the patch center by seaweeds and another species of mussel, which may persist. (b) Alternative stable states on the rocky shores of Maine, and the factors that perpetuate the states and the strong events that effect a transition from one state (seaweed dominated) to the other (barnacle and mussel dominated). (b after Petraitis and Latham, 1999.)

mussel, *Mytilus californianus*. Simon Levin and Robert Paine[14] found that if patches were relatively small, on the order of a square meter or less, the mussels would gradually move into the open space. But what if rare events

opened large patches? If storms opened large patches then the immigration by adult mussels was too slow to seal up

[14] See Paine and Levin, 1981, in Further Reading, Intertidal Zone.

the patch (Box Figure 14.1a). It was then that open space toward the patch centers was colonized by seaweeds and another mussel, *Mytilus trossulus*. In larger patches the perimeter is smaller relative to patch area, and this reduces access of California mussels to the patch interior. Predatory snails might have been abundant in the surrounding California mussel bed, but they would not move across open bare rock space. This situation allowed rapid growth of *M. trossulus*, and it was able to reach sexual maturity in a few months. Suchanek accumulated data on recruitment in California and discovered that *M. trossulus* recruited in the winter, which might have been an adaptation to the opportunity to settle in open spaces in the mussel bed created by winter storms. To conclude, Suchanek's message was that, above a certain size threshold, disturbance patches were ecologically distinct from their smaller cousins.

We now have a fascinating possibility of a landscape that is not only patterned by disturbance but even maintained as a series of alternative stable states. This may make sense of one of our commonest observations about the marine environment: the coexistence of large patches with completely different dominating species, even when all of the space is occupied by organisms. How can this be, when intertidal experimentation tells us that space shortage should be a sure-fire indication of a predictable process of competition?

On protected rocky shores of Maine, one is struck by how different species may dominate different sites that appear to be similar in physical features. Some areas are covered with barnacles and mussels, while others are covered exclusively by brown seaweeds, most commonly stands of *Ascophyllum* or *Fucus*. Peter Petraitis[15] and students examined these assemblages and could not see any reason for competition or predation alone to lead to this sort of alternative dominance. As in rocky shores near Sydney, Australia, and on the exposed coast of the western United States, the key was thought to be in rare events that could flip a site from one stable state to another. Peter Petraitis and Steven R. Dudgeon[16] demonstrated this by clearing patches at several spatial scales (1, 2, 4, and 8 m in diameter). They planted a small patch of mussels in the center of each cleared patch. A discontinuity was discovered between patch sizes of 2 and 4 m diameter: mussel mortality dropped in a distinct step in the 4 and 8 m patches, relative to the 0, 1, and 2 m patches (Box Figure 14.2). It is possible that at this threshold of disturbance, perhaps due to rare ice scour, mussels can colonize and resist predation by the drilling snail *Nucella lapillus*. This explana-

tion is similar to Suchanek's explanation of the role of large open patches in discouraging the movement of carnivorous snails and therefore reducing the mortality of colonists at the patch centers.

On the other hand, it is not likely that *Ascophyllum* will rapidly recruit to large opened spaces. The periwinkle *Littorina littorea* appears to be capable of consuming new seaweed recruits. It would take an extraordinary year of *Ascophyllum* recruitment, correlated with the recent establishment of large open-spaced patches, to result in seaweed-dominated patches. Predation by the drilling snail *Nucella lapillus* might also prevent the establishment of barnacle–mussel patches, because large herds of snails often arrive the moment barnacle larvae settle and clear off the rocks. The moist, cool weather in Maine allows drilling snails to be very active on air-exposed rocks during the time of low tide.

Another study followed a major ice-scouring event on rocky shores of Alaska and demonstrated that abundant local patches of drilling *Nucella lima* could readily consume all settling mussels, but in specific areas.[17] Recovery from disturbance was strongly affected by spatial variation in predator density at the time of mussel settling. Thus patch differentiation could be set in motion by spatial variation in predators. But typical barnacle–mussel recruitment in Maine is so high that open spaces will inexorably be covered and dominated by these sessile forms. Petraitis speculates that rare icy winters scour the intertidal, which opens up space for colonization.

Box Figure 14.1b summarizes the processes that initiate the formation and tend to maintain alternative stable states on rocky shores of Maine. Once seaweeds have colonized a patch, they provide moist shelter for predatory gastropods, which rapidly consume recruiting barnacles and mussels. To some degree, the complete rock cover by seaweeds prevents successful settlement of barnacles and mussels in the first place, and whiplashing by seaweed fronds also tends to damage newly settled barnacle larvae. On the other hand, barnacle–mussel patches are maintained by high recruitment and to a degree by lateral movement of mussels, which can fill in empty spaces (the eastern mus-

---

[15] See Petraitis and Latham, 1999, in Further Reading, Intertidal Zone.
[16] See Petraitis and Dudgeon, 1999, in Further Reading, Intertidal Zone.
[17] See Carroll and Highsmith, 1996, in Further Reading, Intertidal Zone.

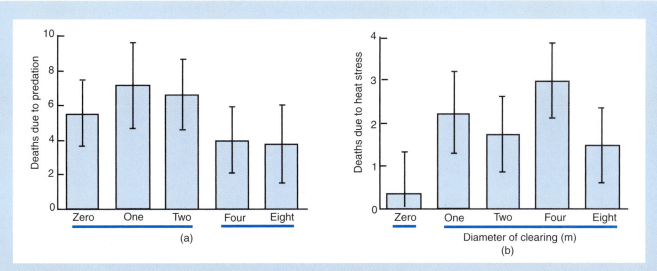

Box Figure 14.2   An experiment on rocky shores of Maine. Patches of diameters of 1, 2, 4, and 8 m were scraped clear and planted with a clump of the mussel *Mytilus edulis*, including areas where no patch was opened up (0 m diameter). (a) Predation success by the gastropod *Nucella lapillus* is distinctly higher on 0, 1, 2 m patches than on 4 and 8 m patches. The green bars unite patch size treatments that are statistically indistinguishable. (b) Mortality of mussels owing to physiological stress. Note that mortality in all cleared areas is indistinguishable statistically, but all cleared areas have greater stress-induced mortality than the uncleared areas. (After Petraitis and Dudgeon, 1999.)

sel *Mytilus edulis* is far more mobile than the western mussel *Mytilus californianus*). Most importantly, barnacles and mussels settle gregariously, which tends to perpetuate patches. Mussels also can resist predation when in a dense bed. Predatory *Nucella lapillus* snails can attack only from the periphery of the mussel bed, because mussels can enmesh snails that venture closer with byssal threads.

It is essential for a major perturbation to occur to cause a shift from one patch type to the other. That is the important lesson of this research. Small-scale processes may be inferred by use of manipulative field experiments, but it is the longer-term and sometimes unpredictable disturbance events at larger spatial scales that have such an impact on community structure.

chitons will move from the surrounding mussel bed and remove all potential colonizing algae and animal larvae. The scale of disturbance may be a major determinant of intertidal landscapes, and infrequent but severe events may overprint the intertidal with a pattern of species distribution and abundance that effectively cancels out all we have discussed thus far as important interactive effects (see Hot Topics 14.2).

✱ **In soft sediments, deposit feeders appear to be food limited, whereas suspension-feeding populations are more variable and are not affected as much by population density.**

Deposit feeders depend upon microbial organisms and particulate organic matter for food. By contrast,

suspension feeders feed upon phytoplankton or resuspended particles. In reality, the difference is not so great because particulate organic matter may be deposited on the bottom from the water column, which is an especially important process in subtidal habitats. In the intertidal, however, the food source of deposit feeders may be dominated by microalgae such as benthic diatoms. I once suggested[18] that deposit-feeding populations may be living with a more stable food source, whereas suspension feeders may be exposed to much more variable phytoplankton and may have an excess food supply or at least a food supply that may be quite variable, owing to

[18] See Levinton, 1972, in Further Reading, Intertidal Zone.

the large spatial and temporal changes that occur in the water column.

The hypothesis of deposit feeder stability versus suspension feeder variability has now been tested several times and has been confirmed for soft sediments in intertidal and very shallow subtidal areas. Einar Olafsson[19] planted varying densities of the bivalve *Macoma balthica*, which suspension-feeds in sand and deposit-feeds in mud. He found that higher population densities reduce growth only in the deposit-feeding populations. This approach has also been extended to the community level in a study of tidal mudflats of the Dutch Wadden Sea.[20] Deposit-feeding species were both spatially and temporally more homogeneous than suspension-feeding species.

✻ Feeding by deposit-feeding populations may involve overexploitation of renewable resources or seasonal decline of food in the sediment, causing severe food limitation.

Many intertidal deposit-feeding species graze diatoms from the surface of the sediment. If you add several surface/deposit-feeding clams to a tray of sand, you will find that the surface remains clean. Remove the clams and, as long as there is light, diatoms will grow profusely. Deposit feeders therefore are exploiting a renewable resource. A number of studies show that grazing by bivalves and gastropods reduces sediment microalgae to low levels and that animal growth is inhibited. This is a mechanism for population regulation of surface-grazing deposit feeders.

Many deposit feeders, such as small polychaetes and oligochaetes, appear in great numbers in the spring in temperate, protected tidal soft sediments of southern New England. Their generation times are short and populations usually build up, eventually declining by end of summer. A set of studies on the marine oligochaete *Paranais litoralis* have demonstrated that spring sediments probably have a considerable amount of food-rich material deposited on the sediment surface. As the spring progresses, this material is eaten by deposit feeders and fuels population growth.[21] By the middle or end of summer, however, this particulate material either is all eaten or is degraded microbially into material that is relatively indigestible to deposit feeders. In response to this cycle, *P. litoralis* at first reproduces by budding, and the population increases, reaching a peak in early summer (Figure 14.23). Then food quality declines

and the population crashes. If the organic matter collected over the season is labeled with ^{14}C one can observe a steady decline in absorption of the material when fed to the worms. This suggests that the material is becoming steadily less digestible. A caging experiment suggested that declines in *Paranais* were not accelerated by predators; the decline of numbers within cages protected from fish predators occurred at the same rate as outside those cages.

Outside Forces: Productivity and Flow

✻ Water flow and primary productivity interact strongly to determine intertidal community dynamics.

For many years, marine investigators have known that community interactions are influenced by larger regional forces. We shall discuss estuaries more later, but an important example emerges from studies of estuaries and bays, which have spatially and temporally varying degrees of water exchange with open coastal areas. J. D. Andrews[22] found that a range of estuarine tributaries feeds into Chesapeake Bay. The James River has relatively high flow and tends to have very poor larval sets of oysters, which increase a bit downstream as the river widens. On the other hand, a number of smaller rivers have very poor exchange with the open parts of the bay, and in those tributaries there are commonly very dense sets of oyster larvae. These observations can be embraced under the concept of **flushing time**, which is the mean time that water and entrained larvae and phytoplankton are maintained in a bay (Figure 14.24). Apparently the flow regime, especially the flushing time of water in an area, is a good predictor of whether larvae will be kept in a system until settlement or will be flushed out. If flushing time is short, then larvae and phytoplankton will be mainly flushed out and larval sets will be poor. But if flushing time is great, then larvae will build up and settle in great densities. Both outcomes depend upon a source of larvae from adult populations within the bay.

In Narragansett Bay, Rhode Island, some years are characterized by a great deal of river flow into the bay, which results in stronger flushing of the bay and mixing with coastal waters, where larval density is

[19] See Olafsson, 1986, in Further Reading, Intertidal Zone.
[20] See Kamermans and others, 1994, in Further Reading, Intertidal Zone.
[21] See Cheng and others, 1993, in Further Reading, Intertidal Zone.
[22] See Andrews, 1984, in Further Reading, Intertidal Zone.

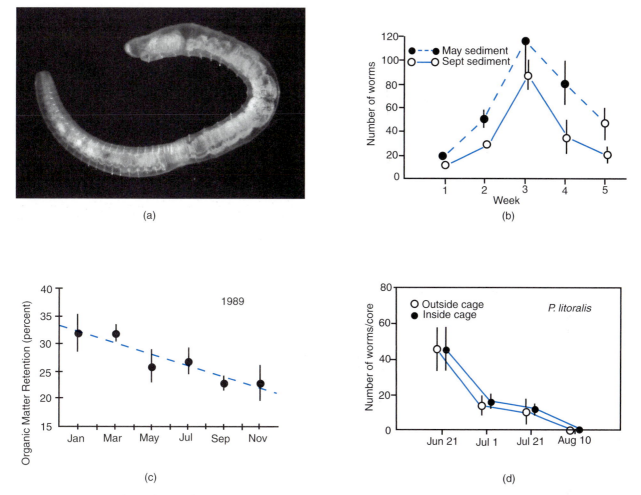

Fig. 14.23 Testing a hypothesis of the regulation of deposit feeder populations by food limitation in Flax Pond, New York (a) *Paranais litoralis*, which reproduces mainly by budding asexually. (Courtesy of Alexa Bely.) (b) If food has entered the mudflat in a pulse, populations should increase and then crash, and sediment from the beginning of the cycle (May) should be capable of growing more worms than sediment taken from late in the population cycle (September). (c) If food is becoming less abundant, carbon in the sediment should become more and more difficult to assimilate during the spring–summer season. (d) If predation is not significant in causing population decline, then *Paranais litoralis* populations should decline at the same rate on the open mudflat and within predator-exclusion cages. (After Cheng et al., 1993.)

usually lower. In such years, barnacle sets of *Semibalanus balanoides* are quite low in Narragansett Bay. In years with low flow, the retention time is much greater and barnacle recruitment is greater as well[23] (Figure 14.25).

On more open rocky shores, there is the question of larval supply. In such an open system, water flowing from elsewhere is the only delivery mechanism for planktonic larvae and phytoplankton that could be consumed by sessile suspension feeders. As flow increases, one might expect higher larval supplies and more suspension feeder food. In low flow situations the larval supply would be low, and the suspension feeders might be able to filter all the phytoplankton, thereby putting a cap on their food supply.

These predictions were explored by George Leonard and colleagues[24] in a tidal estuary in Maine, and some of the results are summarized in Table 14.2.

---

[23] See Gaines and Bertness, 1992, in Further Reading, Intertidal Zone.

[24] See Leonard and others, 1998, Further Reading, Intertidal Zone.

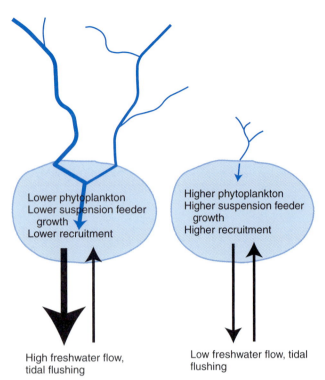

Fig. 14.24 The expected relationship between retention time of water in an estuary, water entering from the watershed, and recruitment rate of an estuarine invertebrate.

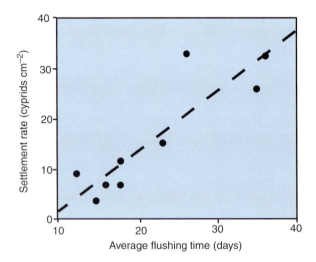

Fig. 14.25 Settlement rate of the barnacle *Semibalanus balanoides* in different years as a function of flushing time of Narragansett Bay, Rhode Island. When rainfall is high, fresh water enters the bay more frequently and flushing time decreases. (Modified from Gaines and Bertness, 1992.)

In high-flow localities, larval recruitment was high for mussels and barnacles, as expected. In effect, this type of habitat was driven by **bottom-up processes**, since flow supplied both food and recruits. The recruitment was so high that it could not be affected very much by a common carnivore, the drilling snail *Nucella lapillus*, which nevertheless benefited from the large prey abundance. In low-flow sites, larvae and phytoplankton food were in short supply and there was much open rocky-shore space. Predation was far more effective here in pruning sessile suspension-feeding adult populations. In other words, top-down processes were at work. This study demonstrated that community structure may be controlled by very different forces, depending upon the delivery of water to the site.

Regional differences in primary productivity also should influence the pattern of rocky shore species abundances. Bruce Menge had observed striking differences in rocky-shore animal abundances at different sites along the coast of Oregon. In one striking case, abundance was very different on the two sides of a headland. As it turned out, the side with abundant rocky-shore benthos was a site of strong upwelling, which delivered to the local site large abundances of nutrients, which stimulated phytoplankton growth. Subsequent investigations using satellite photos demonstrated that shores with abundant benthos are also characterized by abundant nearshore phytoplankton.[25] As we have discussed, the bottom-up process of phytoplankton delivery can strongly influence community structure.

Estuaries

✳ Estuaries are geologically ephemeral but biologically rich.

Because rather small changes of sea level can completely fill or empty out an estuarine basin, estuaries are geologically impermanent features of the coastline. As sea level rises, river valleys may be flooded by the sea; but such basins will also empty as sea level drops, such as during a worldwide increase in glaciers, which lock up the water of the ocean into ice. It is believed that estuaries therefore rarely tend to exist for more than 10,000 years or so.

[25] See Menge and others, 1997, in Further Reading, Intertidal Zone.

Table 14.2 The effects of low flow and high flow on rocky-shore community structure

FACTOR AFFECTED	HIGH FLOW	LOW FLOW
Nutrient supply	High	Low
Planktonic larval supply	High	Low
Individual growth	High for barnacles but not for mussels; higher for predatory snails feeding on barnacles and mussels	Lower for barnacles
Competition	Strong; influenced by recruits	Perhaps strong, but direction not determined as much by recruits
Predation	Low predation rates	Higher predation rates

Despite the ephemeral nature of estuaries, they are among the biologically richest habitats in the world. Supplies of nutrients from freshwater sources and recycling of nutrients from the seabed (see Chapter 9) combine to support large numbers of estuarine benthic invertebrates, fishes, plants, and birds. In eastern North America, estuaries such as the Hudson River and Chesapeake Bay support major fisheries. The Chesapeake Bay system had enormous oyster beds (they are now depleted owing to overfishing and disease) and populations of blue claw crabs, although pollution seems to be taking its toll. The Atlantic menhaden *Brevoortia tyrannus* spawns offshore, but larvae and juveniles live and feed in estuaries such as Chesapeake Bay, which act as nurseries. The Hudson River has large populations of fishes such as shad and striped bass. The bottom is alive with annelids, mollusks, crustaceans, and insect larvae (in the freshwater parts of the estuary).

✳ **The decreased salinity at the headwaters of estuaries can reduce the number of marine species.**

The most noticeable gradient in an estuary is that of decreasing salinity as one goes upstream. Marine species, as discussed in Chapter 3, generally can tolerate fluctuations in salinity, but their tolerance is often exceeded when salinity falls below 10–15 o/oo. Some major invertebrate groups, such as echinoderms (sea urchins, starfish), tend to drop off in estuaries, owing to a general incapacity to evolve resistance to lowered salinity. Others, such as crustaceans, are capable of good regulation in the face of osmotic stress and are often quite abundant in estuaries. In estuaries with some degree of vertical density stratification, the salinity is greater on the bottom than at the surface, where freshwater flow moves lower-salinity water downstream (see Chapter 2). As a result, bottom animals can often penetrate an estuary farther upstream than surface planktonic organisms can. Infaunal species experience less salinity variation than do epifaunal species over a tidal cycle in a very well mixed estuary because of the buffering effect caused by sediment pore waters that exchange water slowly with the overlying water column (Figure 14.26).

The second major estuarine salinity transition is the critical salinity range of 3–8 o/oo. Many marine groups apparently find it hard to survive in this salinity range, even though many more species are capable of living either in fully fresh water or in waters of higher salinity. Along the estuarine gradient, species numbers are at a pronounced minimum in the critical salinity range (Figure 14.27). Mollusks may be incapable of cell volume regulation at salinities this low. Freshwater species, however, can regulate ionic concentrations and maintain a hyperosmotic state. They have lost the ability to regulate cell volume, however, and therefore cannot penetrate even the low salinities of the critical salinity range. The critical salinity is thus a no-man's land, hospitable to neither marine nor freshwater species.

Although salinity change is often a critical factor in limiting the range of marine species, many are capable of rapid regulation and adjustment to the changing osmotic stress of varying salinity. Many fish species are capable of extensive regulation of tissue fluids (see Chapter 3) and can swim across strong salinity gradients. Striped bass, salmon, and killifish are just a few examples of fishes that migrate from

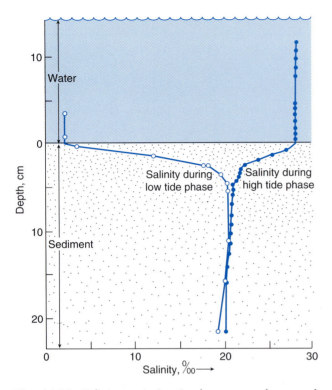

Fig. 14.26 Salinity variation in the water column and within the sediment in the small tidal Pocasset River estuary, Massachusetts. Note that the salinity varies a great deal in the water column, owing to tidal motion and freshwater flow. Within the sediment, however, the salinity is relatively constant and intermediate in value. (After Sanders et al., 1965.)

completely saline water to fresh water in a few weeks or even days. Some crab species are also quite adaptable to changing salinity and perform migrations over nearly the same range of salinity traveled by fishes. In the Hudson River, males of the blue crab *Callinectes sapidus* can migrate from full-strength seawater in the lower New York Bay to fresh water in the upper reaches of the estuary.

✳ Some estuarine species are adapted to counteract the estuarine flow to the sea, in order to be retained within the estuary.

Estuarine flow is usually seaward in the low-salinity waters at the surface. Small estuaries with extensive tidal flushing cannot support large nurseries of juvenile fishes; and retention adaptations probably would be insufficient to counteract the tidal exchange with the open sea. Larger estuaries with longer flushing times and vertical density stratification can support

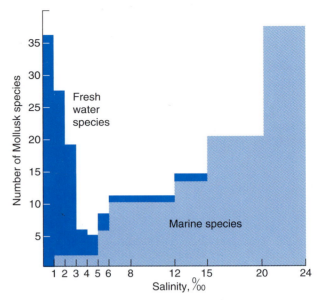

Fig. 14.27 Species richness along the estuarine gradient of the Randersfjord, Denmark. (After Remane and Schlieper, 1971.)

fisheries because of the reduced loss of larvae to sea. Within large estuarine systems, such as Chesapeake Bay, tributaries like the St. Mary's River have relatively low tidal exchange rates with the rest of the estuary. The reduced exchange tends to trap larvae and may be the reason for heavy larval sets of the oyster *Crassostrea virginica*. To counteract the estuarine flow that does exist, estuarine fish and invertebrate larvae have been observed to keep to the bottom during the ebbing tide, swimming actively at the surface with the incoming tide (Figure 14.28).

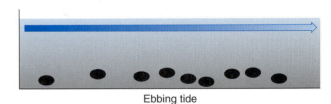
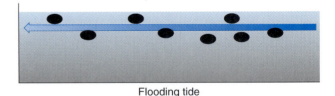

Fig. 14.28 Adaptation to prevent washout from estuaries involves moving into the water column at flood tide and keeping near the bottom during the ebb.

Menhaden are more easily netted during flood tides, indicating their adaptations for retention within the estuarine system. Week by week, yolk-sac larvae of the striped bass in the Hudson River are found at sites progressively upstream. This may suggest an active swimming at flood tide only. The mud crab *Rhithropanopeus harrisii* is also found in greater abundance upstream as the larval life period progresses. As mentioned in Chapter 5, some species export larvae to adjacent coastal waters and depend on a variety of tidal and wind sources to enable them to reinvade the estuary.

Spartina Salt Marshes

✴ *Spartina* **salt marshes are dominated by cord grasses, which bind fine sediment and cause the buildup of meadows above low water.**

Spartina salt marshes (Plate XVII) develop in tidal areas of quiet water, where a variety of salt-tolerant grasses colonize the sediment and then trap fine sediment. Characteristic of these grasses is an extensive **rhizome system** beneath the sediment surface. *Spartina alterniflora*, the eastern and Gulf Coast American species found lowest in the intertidal zone, must put up with long periods of immersion in salt water, to which it is more tolerant than other grass species. The sediment pore water has little or no oxygen, and the root system of *S. alterniflora* connects to air by air pockets in the midcortex. Much of the leaf and stem section of the plant is highly vascularized. A cross section of the plant shows the large amount of open space near the surface, which is devoted to air and oxygen transport (Figure 14.29).

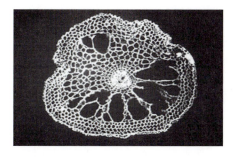

Fig. 14.29 The aerenchymal tissue allows *Spartina* to exchange gases, even when surrounded by an anoxic soil. The tissue in this photograph is visible as a series of circular passageways around the periphery. (Photograph courtesy of Mark Bertness.)

Plants cannot use nutrients efficiently without oxygen, so this tissue allows a connection between the aerobic leaves to the stems, which are surrounded usually by anoxic water.

The rhizomes take up nutrients, but they are also the means by which the plants extend their coverage. Rhizomes extend laterally, and shoots grow above the surface. Although these grasses develop flowers and set seed, the asexual spread of the rhizome system is usually the major form of local spread. Genetic analysis using enzymes demonstrates that marshes often consist of a few clones, or groups of genetically identical individuals, each of which has spread by asexual growth.

A single plant may colonize an open area of sediment (Plate XVII.2), either by rafting or by setting of seed. After the clone has developed, the grass blades will develop a density sufficient to slow current speeds and accelerate the deposition of fine-grained sediment. This will gradually cause the development of a rising sediment level. Thus, a salt marsh begins to spread and evolve into a meadow of sediment held together by dense grass stands. The meadow usually encloses a system of creeks, which are often rich nursery grounds for juvenile crustacea and fishes (Figure 14.30). If there is no major change of local sea level, then the level of the sediment surface will rise, and the dominant plants will change gradually from low-intertidal grasses to terrestrial grasses. Thus a "mature" marsh has passed through the stages of (1) bare intertidal sediment, (2) early colonization of patches of grass, (3) extension of the grass patches and trapping of sediment, (4) gradual rise of the sediment surface, and (5) development of a higher marsh, dominated by terrestrial plants.

In North America, salt marshes are best developed on the east and Gulf coasts, and marshes of hundreds to thousands of acres can be found. The most spectacular marshes are found in the southeastern United States, especially in South Carolina, Georgia, and Louisiana. The cord grass *Spartina alterniflora* dominates from mean low water to mean high water. Its blades may be as tall as 1.5–2 m, especially near rapidly flowing marsh creeks. Further from the creek edge is a shorter form of this species, which grows more slowly owing to nutrient limitation. *Spartina alterniflora* is very salt tolerant and contains siliceous deposits, presumably to deter grazing by birds and mammals. The cellulose composition and the mechanically tough leaves seems to prevent much successful grazing, although recent evidence demon-

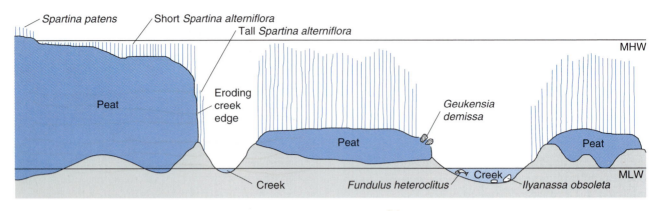

Fig. 14.30 Subhabitats of the *Spartina* salt marsh environment. A tall form of *Spartina alterniflora* is associated with the high nutrient supply of flowing creeks. (After Redfield, 1972.)

strates that the southeastern periwinkle *Littorina ir-rorata* does significant damage to leaves.[26] Usually much less than 10 percent of the leaf production is consumed by herbivores. The great majority decomposes and may support large populations of decomposing bacteria and fungi (see Chapter 13). Predation on flowers, however, is often intense, and *Spartina* seed production is therefore often very limited. An overall tidal zonation exists, and *Spartina patens* and other grasses dominate above high water.

Vegetational Zonation and Plant Interactions in Salt Marshes

✳ **Vegetational zones in salt marshes develop from the interaction of competition and physiological ability to survive salt and drowning.**

If one moves from the low-water mark to the terrestrial environment, one encounters a series of distinct vegetational zones, each dominated by a different grass. In most marshes, grasses occur in the following order, from low to high intertidal: *Spartina alterniflora, Spartina patens, Distichlis spicata, Juncus gerardi*. The border between zones is often quite sharp and at a predictable tidal height. For example, the zone boundary between *S. alterniflora* and the higher-intertidal *S. patens*, which lies approximately at spring high tide, is used in legal disputes to define the marine–terrestrial border. Why are such zones present? Research by Mark Bertness[27] and colleagues showed that all these grass species grow more rapidly alone on the terrestrial border of the marsh, in the highest zone. In other words, no physiological factor limits any of these species from invading up-

per levels. When species are placed in combination, however, it quickly becomes apparent that *Distichlis* can outcompete all other species in its own zone when plants are placed side by side. In its own zone, *Spartina patens* outcompetes *S. alterniflora*. If one transplants a higher-zone plant to a lower zone, however, the former does badly physiologically, owing to a relative intolerance to immersion in salty water and to drowning for more of the day at the lower-tide level. *Spartina patens* lacks an efficient mechanism to survive in lower-intertidal anoxic sediments and therefore does badly within the lower-level *Spartina alterniflora* zone, where oxygen supply to roots is a major limiting factor. Successful dominance in a zone is therefore a combination of physiological tolerance and relative competitive ability.

✳ **Floating wrack often smothers plants and creates patches of bare sediment, which become salty and inhibit colonization for a time.**

High-marsh zones often consist of acres of continuous tracts of one species of grass. Bare patches are common, however, and sometimes are several meters across. Considering the lushness of a salt marsh, the patches are surprising. Some are covered by mats of cyanobacteria, and others are nearly abiotic, with layers of salt on the surface. The high-marsh patches are often a remnant of floating rafts of decaying *Spartina* shoots, which are concentrated by currents and then float up to rest on top of grass in the high marsh.

[26] Silliman, 2000, in Further Reading, Salt Marshes.
[27] See Bertness, 1991a, in Further Reading, Salt Marshes.

The grass is smothered, and a bare zone is created. Once the area has become bare, strong sunlight causes evaporation, which in turn greatly increases the salt content of the sediment pore waters, and sometimes a layer of salt develops on the surface. Mark Bertness[28] demonstrated that the salty water prevents seed germination and the bare patch is self-sustaining. If a plant can colonize, the shading reduces evaporation immediately, and the saltiness of the water decreases. Many of the grasses can extend from the edge of the patch through vegetative growth, but a strong rain may reduce the salt content, and seeds may then be able to germinate. By shading and helping to reduce the salt content, an initial colonizing plant facilitates further colonization by other plants and the patch will be eventually covered by vegetation.

Salt Marsh Creeks

✳ **Salt marshes enclose creeks, which are often biologically rich and are nurseries for many marine fish species.**

As mentioned earlier, salt marsh habitats are usually a series of broad *Spartina* meadows, alternating with salt marsh creeks (Figure 14.31, Plate XVII.1). At a creek edge, a marsh may be at a standstill, eroding, or accreting in size. As the grass stands trap sediment, the marsh can extend over bare sediment and into a creek. Several species are often found in the high-intertidal zone, often at the creek edge. The marsh mussel *Geukensia demissa* lives semi-infaunally in the sediment and apparently aids marsh accretion by trapping sediment (Plate XVII.3). It also enhances grass growth through the addition of organic-rich fecal material on the sediment surface. Fiddler crabs of the genus *Uca* (Plate XVII.4) burrow in the upper-intertidal zone and also are found on creek edges in marshes. The burrowing apparently enhances the growth of *Spartina*, perhaps by aerating the sediment (Plate XVII.4). The creeks themselves often have strong tidal flow, with bottoms of well-sorted sand, or fine mud, depending upon the degree of current strength. Marsh creeks often have rich soft-sediment faunas, including polychaete annelids and mollusks. In tidal creeks of northeastern North America, the mud snail *Ilyanassa obsoleta* occurs in densities of hundreds per square meter. In the south, the marsh periwinkle *Littorina irrorata* feeds on the muddy surface, but climbs grass blades at high tide to avoid in-

coming predators. Many species of smaller crabs and shrimp are also abundant, especially the mud crab *Rhithropanopeus harrisii*. Smaller fish, such as killifish (species of *Fundulus*) and silversides (*Menidia menidia*) are also common and may attract predatory bluefish into the creeks. A variety of wading birds such as night herons and snowy egrets are also common. These birds stalk mobile forms such as crabs and shrimp. Diving birds such as terns and kingfishers also frequent the creeks, where they dive for smaller fish.

Spartina Marshes as Sources of Organic Matter

✳ ***Spartina* salt marshes produce large amounts of particulate and dissolved organic matter, which may influence the food webs of salt marsh benthos and perhaps the food webs of coastal marine systems.**

In the late fall, leaves of the dominant lower-intertidal *Spartina alterniflora* senesce, turn a lovely yellow brown, and eventually sever from the main plant. Large amounts of floating material enters the marsh system, although the material is relatively rich in indigestible cellulose and takes some time to be decomposed by physical fragmentation, tearing apart and ingestion by detritivores, and bacterial decomposition. This material is probably a source of nutrition for a large fraction of the deposit feeders in marsh soft sediments. Minimally, the *Spartina* fragments are substrates for microbes, which are consumed by mussels, deposit-feeding polychaetes, and gastropods. However, these animals can also inefficiently digest particulate organic matter and probably derive some of their nutrition from the *Spartina* detritus itself. Experiments show that additions of such particulate matter can stimulate somatic and population growth of salt marsh oligochaetes.

Very little of the *Spartina* production is consumed by herbivores. The salt-marsh ecologist John Teal showed that in Georgia salt marshes, herbivores were ineffective in consuming the plants, the majority of which entered into the detritus food chain (Figure 14.32). Many believed that this material floated from salt marsh creeks into coastal waters and was a major source of nutrition both for zooplankton and benthos. Further studies, however, showed that salt

---

[28] See Bertness, 1991a, 1991b, in Further Reading, Salt Marshes.

(a)

(b)

(c)

Fig. 14.31 The *Spartina* salt marsh at West Meadow Creek, north shore of Long Island, New York: (a) West Meadow Creek and marsh, (b) An eroding marsh bank, with *Spartina alterniflora*, and (c) close-up of eroding bank showing individuals of the ribbed mussel, *Geukensia demissa*. (Photographs by the author.)

marsh plant detritus does not significantly contribute to coastal production (see Chapter 10). This can be shown by examining the isotopic composition of detritus in the coastal zone, which resembles phytoplankton, not salt marsh detritus.

Although we have few studies to bolster this idea, it appears that export of particulate organic matter may be slight and the export of dissolved nitrogen sources may be considerable. A study of the enclosed marsh–creek system in Sippewissett marsh on Cape Cod, Massachusetts, shows that considerable amounts of dissolved nitrogen are exported to the adjacent Buzzards Bay, which can fuel a considerable amount of the primary productivity by phytoplankton.[29]

Invasion of Species

✳ *Spartina* **species have been introduced, accidentally and purposefully, and have greatly modified shoreline environments throughout the world.**

[29] See Valiela and Teal, 1979, in Further Reading, Salt Marshes.

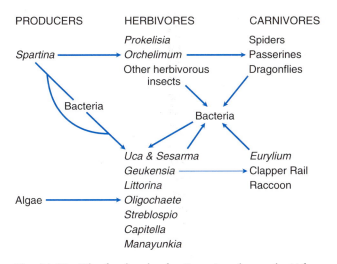

PRODUCERS   HERBIVORES   CARNIVORES

Fig. 14.32 The food web of a Georgia salt marsh. (After Teal, 1962.)

Spartina species have been introduced in many parts of the world, and, like the brooms in the story of the sorcerer's apprentice, have spread rapidly by vegetative growth while altering greatly the nature of shoreline habitats throughout the world. The spread has been both purposeful and accidental. The eastern American *Spartina alterniflora* was introduced accidentally into England in the early 1800s and into the state of Washington probably in the 1940s. However, other introductions were done to stabilize the shoreline (recall the earlier discussion of sediment trapping and spread of *Spartina* marshes).

The English introduction is particularly fascinating.[30] *Spartina alterniflora* was introduced accidentally and hybridized with the English native *S. maritima*, which produced a form known as *Spartina townsendii*, which was believed to be sterile. Somehow, a chromosome doubling in this form produced the perfectly fertile *Spartina anglica*, which spread rapidly throughout the protected coastal regions of Great Britain. This new form displaced the native marsh species in many locations. It is exceptionally efficient at spread by rhizomes. Because *Spartina* marshes change the local topography greatly and grow over bare mudflats, often rich with burrowing invertebrates, there were many changes in shore habitats, including a reduction of usable habitats for some birds and spawning fishes. Some ecologists have argued that a decline in numbers of the dunlin (*Calidris alpina alpina*), a shorebird, was related directly to the expansion of *Spartina anglica*, but this is in dispute.

The supposedly sterile *Spartina townsendii* was imported to many localities around the world for shoreline stabilization, creation of duck habitat, and so on, and somehow the reproducing *Spartina anglica* was brought along, or arose independently in many localities. Now, *Spartina anglica* and other *Spartina* species are being regarded as a scourge in the Pacific from New Zealand to the state of Washington, where they are spreading rapidly to protected shores along the west coast of the United States. *Spartina anglica* is spreading rapidly in the Puget Sound region of Washington and may not be stoppable. Its major means of spread appears to be by floating seeds. In Willapa Bay, Washington, *Spartina alterniflora* is displacing all viable oyster grounds and intertidal mudflat spawning sites for fishes, and may be creating habitat unsuitable for the local migrating birds. The spread of marshes in this bay is also displacing eelgrass beds, which are believed to be important feeding and nursery grounds for commercially important fishes. It is not clear how this species was introduced from the eastern United States to the western coast, but it may have arrived with oysters that were shipped west. Once established in Willapa Bay, it has spread inexorably by means of vegetative growth.

Control measures against *Spartina* infestations has been attempted on a small scale, and include (1) spread of plastic, which eliminates light and access to oxygen, (2) mowing, (3) spraying of herbicides, and (4) trampling of marshes. No technique has worked, however, and history tells us that it is nearly impossible to stop the invasion of a new weed. Perhaps some form of biological control is possible, but remember that very few insects or marine grazers eat *Spartina*. Perhaps a specific pathogen may be used in the future to control the spread of this prolific marsh plant.

Filling and Erosion

✳ **Filling and erosion have caused the loss of marshlands, but restoration efforts are now under way.**

Although *Spartina* invasions are a worrisome factor in coastline management, marshes are being lost at an alarming rate within their natural geographic ranges. Losses derive principally from filling and from coastal erosion. Until the passage of a series of state and federal protection laws, it was customary

[30] See Thompson, 1991, in Further Reading, Salt Marshes.

to fill in *Spartina* and other marshes to allow shoreline development of housing tracts, airports, and industrial parks. Coastal erosion has resulted from problematic water management practices. In the Mississippi Delta region, the main channel of the Mississippi River was walled off by levees, which prevented the deposition of sediment in marsh areas. Combined with land subsidence, this has led to the loss of millions of acres of marshland. Finally, because marshes trap fine sediment, many organic pollutants and metals that associate with fine particles tend to accumulate in marsh sediments. Some sites, principally those near industrial outfalls, have become very contaminated and are now being dredged. Active efforts are under way to preserve marshlands. We have come to realize their value as nurseries for juvenile fish, as buffers to coastal erosion, and simply as places of great beauty. Legislation now largely protects marshes from destruction throughout the United States. Restoration projects are being actively pursued where destruction has occurred. In Chesapeake Bay, an early effort by a company called Environmental Concern, Inc. involved the movement of large amounts of sediment and the transplantation of *Spartina* seedlings, which were raised from seed in greenhouses.[31] Some efforts have also been made at planting marshes in newly established housing developments, and a new type of environmental consultant who specializes in restoration is now in demand.

Mangrove Forests

Mangrove forests are intertidal and emergent plant communities dominated by **mangroves**, which are rooted in marine soft sediment. They are tropical and subtropical in distribution and can be found in Australia, the Americas, Asia, and Africa. Growth in lower latitudes is continuous, and mangrove trees tend to grow to greater heights than trees in higher-latitude mangrove forests. Mangroves are rich in marine and terrestrial animal life. Large numbers of falling leaves provide a continual source of detrital material.

Mangrove forests are found along quiet water tropical and subtropical marine coasts (Figure 14.33). They are dominated by shrub- or treelike mangroves, which are rooted in anoxic sediment that is waterlogged with seawater. Waterlogging is a very profound

Fig. 14.33   A mangrove forest on the Caribbean coast of Panama. (Photograph courtesy of Harris Lessios.)

physiological problem, especially because the sediment pore water is often anoxic. Mangrove below-ground tissue is therefore subjected to long periods of exposure to anaerobic conditions, which slows nutrient uptake and allows the accumulation of toxins such as hydrogen sulfide, methane, carbon dioxide, and reducing metals. Exposure to decomposing bacteria is also a problem, which may explain the high concentrations in mangrove tissues of tannins that function to protect against bacterial invasion.

Mangroves are usually broadly rooted but only to a shallow depth. This may be a response aimed at avoiding exposure to deeper-lying anoxic sediments. Above the water level, mangroves are in many ways typical terrestrial shrubs, with trunks, stems, leaves, and flowers. Their root system, however, is adapted to the anoxic sediment, and all mangrove species have root extensions that project into the air so that the underground parts of the plant root system can obtain oxygen.

The variety of root morphologies maintained by a single tree allows differentiation of function. Mangroves can have prop roots, structures that extend midway from the trunk and arch downward for support, roots that direct upward into the air (knee roots or larger pneumatophores, depending upon the species), and finer roots for gathering nutrients (Figure 14.34). Oxygen is gathered and directed into the highly chambered upward-directing roots, which transport oxygen to below-ground tissues. This as-

---

[31] See Berger, 1985, in Further Reading, Salt Marshes.

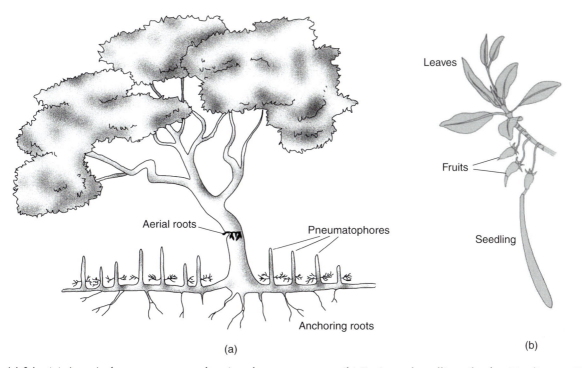

Fig. 14.34 (a) A typical mangrove tree, showing the root systems. (b) Fruits and seedling. (b after Tomlinson, 1986.)

sures aerobic metabolism of the plants within the anaerobic environment of the sediment. The formation of pneumatophores in the genus *Avicennia* is induced by anoxia in the sediment.

Mangroves are also quite salt tolerant. They have a variety of mechanisms for excluding salt, including salt glands that secrete salt from the leaves (Figure 14.35). In the morning it is common to see small dots of moisture where the salty drops have been excreted. One often sees leaves covered with salt crystals after a few hours in the sun. Roots are capable of avoiding salt uptake to a degree, but because the mangrove circulatory system fluids contain less salt than in the sediment pore waters, there is an osmotic gradient to maintain. This is quite costly, energetically speaking.

Mangrove forests can be divided into a series of zones, with different tree species dominating with increasing distance from the shoreline. Landward of the mangrove forest one tends to find typical terrestrial trees, which differ depending upon location. In the Caribbean, the red mangrove *Rhizophora mangle* dominates the seaward part of mangrove forests. It has prop roots, which extend into the water.

Red mangrove seeds germinate while still attached to the parent plant. Seedlings develop and dangle

from the parent, until they either drop into the mud, like darts, or float away in the water (Figure 14.34b). Those that float away are finally carried by winds to another muddy shore, where they root in the sediment. A seedling coat is shed, giving the seedling negative buoyancy, which causes it to drop to the bottom.

Although subtropical forests may have only one

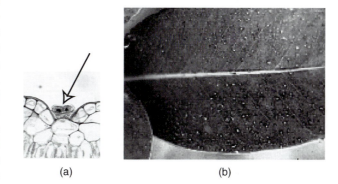

Fig. 14.35 (a) A vertical section through a mangrove leaf, showing the salt gland (arrow). (Photograph by Peter Saenger.) (b) A mangrove leaf with numerous excreted salt crystals. (Photograph by Robert Twilley.)

mangrove tree species, many tropical forests have several dominants, which are usually zoned in abundance with increasing distance from the shoreline. In the Caribbean, landward of the red mangrove zone is a black mangrove zone and a white mangrove zone. Each is dominated by a different mangrove species and associated flora and fauna.

The prop roots of mangroves usually support a rich invertebrate and seaweed community. The flat tree oyster (*Isognomon alatus*) attaches to the roots (Figure 14.36), as do a variety of crabs, shrimp, and barnacles. A number of species also live on the trunks and leaves, including barnacles and snails. A number of crab species move cyclically between the trees and mudflats, where they feed at low tide when predators are absent.

The roots and sediments in the seaward part of the mangrove forest are dominated by detritus feeders. In the inner part of the forest, mudflats are rarely inundated, and many of the dominant species live in air for extended periods of time. The mudskipper, *Periophthalmus*, is a small Indo-Pacific fish with excellent eyesight, capable of slithering on the surface of the intertidal mudflat and even climbing low branches. Fiddler crabs in these habitats release larvae during the rare times of tidal inundation. The

Fig. 14.36 Wood tree oysters attached to a Caribbean mangrove. (Photograph by Robert Twilley.)

soft-sediment fauna often contains crustacean species capable of deposit feeding and scavenging. This abundance is probably related to frequent leaf falls, which enhance the supply of particulate organic matter into the ecosystem. Stable isotopes demonstrate that a large amount of this leaf matter is consumed and used by benthic mangrove mudflat animals. Unlike other marine habitats, mangrove forests are also terrestrial habitats, and southwest Pacific mangrove forests may have large populations of herbivorous insects, monkeys, and bats.

Further Reading

INTERTIDAL ZONE

Andrews, J. D. 1984. Transport of bivalve larvae in the James River, Virginia USA. *Journal of Shellfish Research*, v. 3, pp. 29–40.

Bertness, M. D. 1989. Intraspecific competition and facilitation in a northern acorn barnacle population. *Ecology*, v. 70, pp. 257–268.

Branch, M., and G. Branch, 1981. *The Living Shores of South Africa*. Capetown: C. Struik Publishers.

Brown, A. C., and A. McLachlan. 1990. *Ecology of Sandy Shores*. New York: Elsevier.

Carefoot, T. 1977. *Pacific Seashores*. Seattle: University of Washington Press.

Carroll, M. L., and R. C. Highsmith. 1996. Role of catastrophic disturbance in mediating *Nucella–Mytilus* interactions in the Alaskan rocky intertidal. *Marine Ecology—Progress Series*, v. 138, pp. 125–133.

Carter, A. R., and R. J. Anderson. 1991. The intertidal alga *Gelidium pristoides* in the eastern Cape, South Africa. *Journal of the Marine Biological Association of the United Kingdom*, v. 71, pp. 555–568.

Cheng, I.-J., J. S. Levinton, M. M. McCartney, and D. E. Martinez. 1993. A bioassay approach to seasonal variation in the nutritional value of sediment. *Marine Ecology—Progress Series*, v. 94, pp. 275–285.

Coates, M. and S. C. McKillup. 1995. Role of recruitment and growth in determining the upper limit of distribution of the intertidal barnacle *Hexaminius popeiana*. *Marine and Freshwater Research*, v. 46, pp. 1065–1070.

Connell, J. H. 1961. The influence of interspecific competition and other factors on the distribution of the barnacle *Chthamalus stellatus*. *Ecology*, v. 42, pp. 710–723.

Dayton, P. K. 1971. Competition, disturbance and community organization: The provision and subsequent utilization of space in a rocky intertidal community. *Ecological Monographs*, v. 41, pp. 351–389.

Fenchel, T. 1975. Factors determining the distribution patterns of mud snails. *Oecologia*, v. 20, pp. 1–17.

Gaines, S. D., and M. D. Bertness. 1992. Dispersal of juveniles and variable recruitment in sessile marine species. *Nature*, v. 360, pp. 579–580.

Helmuth, B. 1999. Thermal biology of rocky intertidal mussels: Quantifying body temperatures using climatological data. *Ecology*, v. 80, pp. 15–34.

Innes, D. J., and L. E. Haley. 1977. Inheritance of a shell-color polymorphism in the mussel. *Journal of Heredity*, v. 68, pp. 203–204.

Kamermans, P., H. W. van der Veer, L. Karcsmarski, and G. W. Douglas. 1994. Competition in deposit- and suspension-feeding bivalves: Experiments in controlled outdoor environments. *Journal of Experimental Marine Biology and Ecology* v. 162, pp. 113–135.

Leonard, G. H., J. M. Levine, P. R. Schmidt, and M. D. Bertness. 1998. Flow-driven variation in intertidal community structure in a Marine estuary. *Ecology*, v. 79, pp. 1395–1411.

Levinton, J. S. 1972. Stability and trophic structure in deposit feeding and suspension feeding communities. *American Naturalist*, v. 106, pp. 472–486.

Lewis, J. R. 1964. *The Ecology of Rocky Shores*. London: English University Press.

Liveley, C. M., P. T. Raimondi, and L. F. Delph. 1993. Intertidal community structure: Space–time interactions in the northern Gulf of California. *Ecology*, v. 74, pp. 162–173.

Maitland, D. P. 1986. Crabs that breathe air with their legs—*Scopimera* and *Dotilla*. *Nature*, v. 319, pp. 493–495.

Menge, B. A. 1976. Organization of the New England rocky intertidal community: Role of predation, competition, and environmental heterogeneity. *Ecological Monographs*, v. 46, p.. 355–393.

Menge, B. A. 1991a. Relative importance of recruitment and other causes of variation in rocky intertidal community structure. *Journal of Experimental Marine Biology and Ecology*, v. 146, pp. 69–100.

Menge, B. A. 1991b. Generalizing from experiments: Is predation strong or weak in the New England rocky intertidal? *Oecologia*, v. 88, pp. 1–8.

Menge, B. A., B. A. Daley, P. A. Wheeler, E. Dahlhof, E. Sanford, and P. T. Strub. 1997. Benthic–pelagic links and rocky intertidal communities: Bottom-up effects on top-down control? *Proceedings of the National Academy of Science USA*, v. 94, pp. 14530–14535.

Mitton, J. B. 1977. Shell color and pattern variation in *Mytilus edulis* and its adaptive significance. *Chesapeake Science*, v. 18, pp. 387–389.

Moore, P. G., and R. Seed. 1985. *The Ecology of Rocky Shore Coasts*. London: Hodder and Stoughton Press.

Nicotri, M. E. 1977. Grazing effects of four marine intertidal herbivores on the microflora. *Ecology*, v. 58, pp. 1020–1032.

Newell, R. C. 1979. *Biology of Intertidal Animals*. Faversham (Kent, U.K.): Marine Ecological Surveys.

Olafsson, E. B. 1986. Density dependence in suspension-feeding and deposit-feeding populations of the bivalve *Macoma balthica*: A field experiment. *Journal of Animal Ecology*, v. 55, pp. 517–526.

Paine, R. T. 1966. Food web complexity and species diversity. *American Naturalist*, v. 100, pp. 65–75.

Paine, R. T. 1977. Controlled manipulations in the marine intertidal zone, and their contributions to ecological theory. *The Changing Scenes in Natural Sciences, 1776–1976*. Academy of Natural Sciences (Philadelphia), Special Publication 12, pp. 245–270.

Paine, R. T. 1988. Habitat suitability and local population persistence of the sea palm *Postelsia palmaeformis*. *Ecology*, v. 69, pp. 1787–1794.

Paine, R. T. 1990. Benthic macroalgal competition: Complications and consequences. *Journal of Phycology*, v. 26, pp. 12–17.

Paine, R. T., and S. A. Levin. 1981. Intertidal landscapes: Distribution and dynamics of pattern. *Ecological Monographs*, v. 51, pp. 145–178.

Peterson, C. H. 1991. Intertidal zonation of marine invertebrates in sand and mud. *American Scientist*, v. 79, pp. 236–249.

Peterson, C. H., and S. V. Andre. 1980. An experimental analysis of interspecific competition among marine filter feeders in a soft-sediment environment. *Ecology*, v. 61, pp. 129–139.

Peterson, C. H., and R. Black. 1987. Resource depletion by active suspension feeders on tidal flats: Influence of local density and tidal elevation. *Limnology and Oceanography*, v. 32, pp. 143–166.

Petraitis, P. S. 1990. Direct and indirect effects of predation, herbivory and surface rugosity on mussel recruitment. *Oecologia*, v. 83, pp. 405–413.

Petraitis, P. S. 1995. The role of growth in maintaining spatial dominance by mussels (*Mytilus edulis*). *Ecology*, v. 76, pp. 1337–1346.

Petraitis, P. S., and S. R. Dudgeon. 1999. Experimental evidence for the origin of alternative communities on rocky intertidal shores. *Oikos*, v. 84, pp. 39–45.

Petraitis, P. S., and R. E. Latham. 1999. The importance of scale in testing the origins of alternative community states. *Ecology*, v. 80, pp. 429–442.

Race, M. 1982. Competitive displacement and predation between introduced and native mud snails. *Oecologia*, v. 54, pp. 337–347.

Rhoads, D. C. 1966. Missing fossils and paleoecology. *Discovery* (Yale Peabody Museum of Natural History), v. 2(1), pp. 19–22.

Robles, C., D. Sweetnam, and J. Eminike. 1990. Lobster predation on mussels: Shore-level differences in prey vulnerability and predator preference. *Ecology*, v. 71, pp. 1564–1577.

Robles, C., R. Sherwood-Stephens, and M. Alvarado. 1995. Responses of a key intertidal predator to varying recruitment of its prey. *Ecology*, v. 76, pp. 565–579.

Saloniemi, I. 1993. An environmental explanation for the character displacement pattern in *Hydrobia* snails. *Oikos*, v. 67, pp. 75–80.

Schneider, D. C. 1978. Equalisation of prey number by migratory shorebirds. *Nature*, v. 271, pp. 353–354.

Sousa, W. P. 1979a. Disturbance in marine intertidal boulder fields: The nonequilibrium maintenance of species diversity. *Ecology*, v. 60, pp. 1225–1239.

Sousa, W. P. 1979b. Experimental investigations of disturbance and ecological succession in a rocky intertidal algal community. *Ecological Monographs*, v. 49, pp. 227–254.

Southward, A. J. 1964. The relationship of temperature and rhythmic cirral activity in some Cirripedia considered in connection with their geographic distribution. *Helgolaender Wissenschaftliche Meeresuntersuchungen* v. 10, pp. 391–403.

Stephenson, T. A., and A. Stephenson. 1972. *Life Between Tidemarks on Rocky Shores*. San Francisco: W. H. Freeman.

Suchanek, T. H. 1981. The role of disturbance in the evolution of life histories in the intertidal mussels *Mytilus edulis* and *Mytilus californianus*. *Oecologia*, v. 50, pp. 143–152.

Underwood, A. J. 1999. Physical disturbances and their direct effect on an indirect effect: Responses of an intertidal assemblage to a severe storm. *Journal of Experimental Marine Biology and Ecology*, v. 232, pp. 125–140.

Underwood, A. J., and E. J. Denley. 1984. Paradigms, explanations, and generalizations in models for the structure of intertidal communities on rocky shores. In D. Simberloff, D. R. Strong, L. Abele, and A. R. Thistle, eds., *Ecological Communities: Conceptual Issues and the Evidence*. Princeton, NJ: Princeton University Press, pp. 151–180.

Underwood, A. J., and P. G. Fairweather. 1989. Supply-side ecology and benthic marine assemblages. *Trends in Ecology and Evolution*, v. 4, pp. 16–20.

Ushakov, B. P. 1968. Cellular resistance adaptation to temperature and thermostability of somatic cells with special reference to marine animals. *Marine Biology*, v. 1, pp. 153–160.

Vadas, R. L., W. A. Wright, and S. L. Miller. 1990. Recruitment of *Ascophyllum nodosum*: Wave action as a source of mortality. *Marine Ecology—Progress Series*, v. 61, pp. 263–272.

Wertheim, A. 1984. *The Intertidal Wilderness*. San Francisco: Sierra Club.

Wilson, W. H. 1991. Competition and predation in marine soft-sediment communities. *Annual Review of Ecology and Systematics*, v. 21, pp. 221–241.

Woodin, S. A. 1974. Polychaete abundance patterns in a marine soft-sediment environment: The importance of biological interactions. *Ecological Monographs*, v. 44, pp. 171–187.

Woodin, S. A. 1977. Algal "gardening" behavior by nereid polychaetes: Effects on soft bottom community structure. *Marine Biology*, v. 44, pp. 39–42.

Woodin, S. A., S. M. Lindsay, and D. E. Lincoln. 1997. Biogenic bromophenols as negative recruitment cues. *Marine Ecology—Progress Series*, v. 157, pp. 303–306.

ESTUARIES

Day, J. W., Jr., C. A. S. Hall, W. M. Kemp, and A. Yanez-Arancibia. 1989. *Estuarine Ecology*. New York: Wiley.

Dyer, K. R. 1997. *Estuaries: A Physical Introduction*. Chichester: Wiley.

Hobbie, J. E., ed. 2000. *Estuarine Science: A Synthetic Approach to Research and Practice*. Washington, D.C.: Island.

McLusky, D. S. 1989. *The Estuarine Ecosystem*. Glasgow and London: Blackie.

Remane, A., and C. Schlieper. 1971. *Biology of Brackish Water*. New York: Wiley.

Russell, R. J. 1967. Origins of estuaries. In G. H. Lauff, ed., *Estuaries*. Washington, DC: American Association for the Advancement of Science, Publication 83, pp. 63–70.

Sanders, H. L., P. C. Mangelsdorf, and G. R. Hampson. 1965. Salinity and faunal distribution in the Pocasset River, Massachusetts. *Limnology and Oceanography*, v. 10, pp. R216–R229.

SALT MARSHES

Berger, J. J. 1985. *Restoring the Earth*. New York: Knopf. See Chapter 4 for discussion on salt marsh restoration.

Bertness, M. D. 1991a. Interspecific interactions among high marsh perennials in a New England salt marsh. *Ecology*, v. 72, pp. 125–137.

Bertness, M. D. 1991b. Zonation of *Spartina patens* and *Spartina alterniflora* in a New England salt marsh. *Ecology*, v. 72, pp. 138–148.

Bertness, M. D. 1999. *The Ecology of Atlantic Shorelines*, Chapter 7. Sunderland, MA: Sinauer Associates.

Ellison, A. M. 1987. Effects of competition, disturbance, and herbivory on *Salicornia europaea*. *Ecology*, v. 68, pp. 576–586.

Nixon, S. W. 1980. Between coastal marshes and coastal waters—A review of twenty years of speculation and research on the role of salt marshes in estuarine productivity and water chemistry. In P. Hamilton and K. McDonald, eds., *Estuarine and Wetland Processes*. New York: Plenum, pp. 437–525.

Nixon, S. W., and C. A. Oviatt. 1973. Ecology of a New England salt marsh. *Ecological Monographs*, v. 43, pp. 463–498.

Raybould, A. F., A. J. Gray, M. J. Lawrence, and D. F. Marshall. 1991. The evolution of *Spartina anglica* C. E. Hubbard (Graminae): Origin and genetic variability. *Biological Journal of the Linnaean Society*, v. 43, pp. 111–126.

Redfield, A. C. 1972. Development of a New England salt marsh. *Ecological Monographs*, v. 42, pp. 201–237.

Silliman, B. R. 2000. Top-down control of *Spartina alterniflora* growth by periwinkle grazing. Abstract, Benthic Ecology Meeting, Wilmington, NC, March 2000.

Teal, J. M. 1962. Energy flow in the salt marsh ecosystem of Georgia. *Ecology*, v. 43, pp. 614–624.

Teal, J. M., and J. W. Kanwisher. 1966. Gas transport in the marsh grass, *Spartina alterniflora*. *Journal of Experimental Botany*, v. 17, pp. 355–361.

Teal, J. M., and M. Teal. 1969. *The Life and Death of a Salt Marsh*. Boston: Little, Brown.

Thompson, J. D. 1991. The biology of an invasive plant. *BioScience*, v. 41, pp. 393–401. This is a discussion of *Spartina* invasion.

Valiela, I., and J. Teal. 1979. The nitrogen budget of a salt marsh ecosystem. *Nature*, v. 280, pp. 652—656.

Valiela, I. J. M. Teal, and W. G. Deuser. 1978. The nature of growth forms in the salt marsh grass, *Spartina alterniflora*. *American Naturalist*, v. 112, pp. 461–470.

MANGROVE FORESTS

Hutchings, P., and P. Saenger. 1987. *Ecology of Mangroves*. St. Lucia: University of Queensland Press.

Micheli, F. 1993. Feeding ecology of mangrove crabs in northeastern Australia: Mangrove litter consumption by *Sesarma messa* and *Sesarma smithii*. *Journal of Experimental Marine Biology and Ecology*, v. 171, pp. 165–186.

Robertson, A. I., and D. M. Alongi, eds. 1992. *Tropical Mangrove Ecosystems*. Washington, DC: American Geophysical Union.

Saenger, P. 1998. Mangrove vegetation: An evolutionary perspective. *Marine and Freshwater Research*, v. 49, pp. 277–286.

Saenger, P., and S. C. Snedaker. 1993. Pantropical trends in mangrove above-ground biomass and annual litterfall. *Oecologia*, v. 96, pp. 293–299.

Scholander, P. F., L. Van Dam, and S. I. Scholander. 1955. Gas exchange in the roots of mangroves. *American Journal of Botany*, v. 42, pp. 92–98.

Teas, H. J. 1983. *The Biology and Ecology of Mangroves*. The Hague: W. Junk.

Tomlinson, P. B. 1986. *The Botany of Mangroves*. New York: Cambridge University Press.

Review Questions

1. Why are vertical zones expanded on wave-swept coasts, relative to protected shores?

2. How may an intertidal animal prevent overheating upon exposure to air and sunlight?

3. How do swash riders manage to survive the tide and waves of an exposed beach?

4. What factors contribute to determining zonation on a rocky shore?

5. Why is zonation often not as clear-cut on a soft-sediment intertidal environment as on rocky shores?

6. How do vertical burrowing differences affect coexistence of potentially competing burrowing species?

7. What is character displacement, and what evolutionary process may it reflect?

8. How does predation alter the course of interactions among major space keepers on a rocky shore?

9. How does the size of a disturbed patch influence the subsequent course of colonization and dominance by sessile marine invertebrates?

10. Why are so few species able to live in the transition between estuarine and freshwater habitats?

11. How does invasion by *Spartina* lead to a rise of the sediment level in a protected coastal flat?

12. Why do *Spartina* marsh plants violate the general intertidal rule that an upper limit to a species distribution is determined by physiological problems?

13. Why has *Spartina* had such an important impact on shorelines to which it has been introduced?

14. What is the main means by which *Spartina* extends its population on a tidal flat?

15. Why are mangrove roots important to the invertebrates of mangrove forests?

15

Sea Grass Beds, Kelp Forests,
Rocky Reefs, and Coral Reefs

In this chapter we discuss a series of shallow, coastal subtidal environments that have several features in common. Grass beds, kelp forests, and coral reefs are the most productive subtidal benthic environments. They are all dominated by highly active benthic primary producers, even though the biology of the specific groups differs considerably among the environment types. The primary producers are important not only in fixing carbon, but also in contributing to the structural habitat and providing a substrate for other organisms. They are even important in strongly altering water flow. The grass bed, kelp forest, and coral reef environments are also biologically diverse, owing to the diversity of microhabitats created by the effects of the dominant primary producers.

Sea Grass Beds

✳ **Sea grasses are marine higher (flowering) plants that are confined to very shallow water and extend mainly by undersediment rhizome systems.**

In soft sediments, from the low-water mark to depths of about 3–5 m, a number of broad flats are inhabited by sea grasses and a rich associated biota. There only are 50–100 species of sea grasses, as opposed to the far more diverse submerged flowering-plant flora in fresh water. Shallow subtidal sea grass beds are found throughout the tropical and temperate oceans. In the temperate zone, the eelgrass *Zostera marina* forms thick beds in sediments ranging from sand to mud (Figure 15.1). In the southeast American tropics, the turtle grass *Thalassia testudinum* dominates and covers very shallow flats of carbonate sediment. Like *Spartina* marshes, both grasses are typified by a large complex of rhizomes, a root network within the sediment. As the rhizome system grows and extends laterally, shoots may be sent up. Dissolved nutrients are probably taken up by the rhizomes, mainly from pore waters in sediments. Sea grasses extend their coverage principally by this growth mechanism. These grasses, however, are flowering plants, and pollen moves between plants in the water currents. Fancy flowers are associated usually with attraction of pollinators, so sea grass flowers are rather dull and usually quite inconspicuous. Eventually, these grasses set seed, and the seeds are also borne by water currents. Flowering is rather rare in sea grasses, which appear to reproduce more by asexual spread, through the rhizome system.

Sea Grass Growth Conditions, Succession, and Production

✳ **Sea grasses grow best in conditions of relatively high light and modest current flow.**

Fig. 15.1 A bed of *Zostera marina* in Padilla Bay, Washington. Blades of this sea grass are 50–100 cm high. (Photograph courtesy of Charles Simenstad.)

Sea grasses grow densely in shallow water only; rarely are they found in depths greater than 3 m. At the deeper end of sea grass meadows, light is strongly limiting to photosynthesis, and the lower depth limit is usually strongly related to light irradiance. Sea grasses generally form linear leaves, with the growing meristem region at the base. These leaves extend from the rhizome system toward the surface. Epiphytes, such as bryozoa and microalgae, grow on the leaves and may strongly reduce light capture and photosynthesis. Sea grass meadows appear to develop more extensively under modest current conditions, which may reflect the delivery of nutrients. Sea grasses obtain a portion of their nitrogen by means of nitrogen-fixing bacteria on the rhizome, but most is probably taken up from the water. Under high flows of 0.5 m s^{-1}, sea grasses are affected adversely, perhaps owing to erosion and shear stress on the leaves.

* Sea grass beds most easily colonize sediment after a successional sequence featuring a previous colonization by seaweeds.

A well-developed sea grass bed may extend laterally into bare sediment by means of the rhizome system. Bare sediment, however, is generally low in nutrients. The rhizome system aids in transport of nutrients from the already established bed. Colonization of new areas by seedlings is difficult unless the sediment is already physically stable and rich in dissolved nu-

trients. This can be accomplished by the presence of other plants, such as seaweeds, which stabilize the sediments and add nutrients in forms such as ammonium. Thus, bare sand may change, by succession, to a sea grass bed (Figure 15.2).

* Sea grass beds have high primary production and support a diverse group of animal species.

Along with intertidal *Spartina* salt marshes (see Chapter 14), shallow subtidal sea grass beds are among the most productive of all marine communities. The grasses themselves grow rapidly, but there are a large variety of associated animal and algal species. Sea grasses form dense meadows and are often colonized by a wide variety of fouling organisms, such as hydroids, sponges, bryozoans, and seaweeds. Many invertebrate species depend upon sea grasses and would become extinct or greatly reduced in abundance if the grasses were to disappear. The common Atlantic bay scallop *Argopecten irradians*, for example, has a planktotrophic larva that recruits to sea grasses. The early juvenile stage attaches by byssal threads to the grass blades of *Zostera marina* and remains attached for several months before living free on the bottom. The attachment places the juveniles out of the reach of predators.

Water currents are greatly reduced within sea grass meadows, and a number of species of burrowing invertebrates increase greatly in abundance, relative to nearby bare bottoms. The explanation for this increase is unclear. The reduced flow may encourage settlement of swimming larvae. Some experiments show that larval settlement is greater when simulated sea grass bottoms of medium grass density are constructed from soda straws (Figure 15.3).

* Eelgrass beds reduce current flow, deter the entry of some predators, and may enhance the growth and abundance of infaunal suspension feeders.

Eelgrass beds are often quite dense, with thousands of shoots per square meter. The density of blades, combined with the density of the root system, tends to deter some carnivores from entering the beds. Large carnivorous snails and flatfish have some trouble entering eelgrass. The exact mechanism is not clear, but larger suspension-feeding invertebrates, such as the clam *Mercenaria mercenaria*, tend to be larger and grow faster within the eelgrass than out-

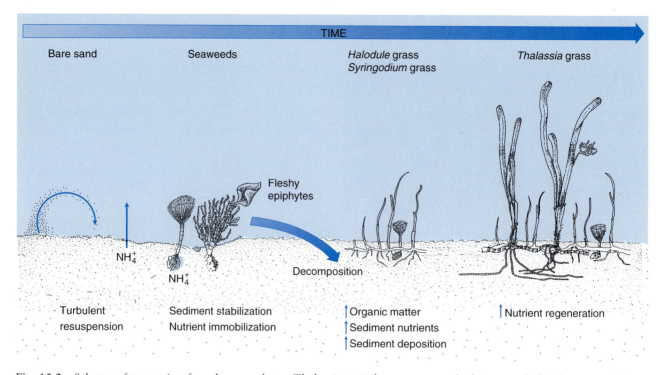

Fig. 15.2   Scheme of succession from bare sand to a *Thalassia testudinum* sea grass bed in coastal Florida waters. (Courtesy of Susan Williams.)

side the bed in bare sand or muddy sand. There are two popular explanations. First, suspended food may be more abundant within the eelgrass bed. Because of the reduced current speed, resuspended material may be retained and ingested by the suspension feeders. Much of the suspended food of clams comes from resuspension of benthic diatoms. Second, it is very likely that predation is reduced within the eelgrass beds for a reason mentioned earlier: prevention of predator entry.

* **Sea grasses are grazed to variable degrees.**

One of the most interesting aspects of sea grasses is their apparent unpalatability to grazers. By and large, the sea grass *Zostera marina* is not consumed by any major herbivores. The situation differs for the tropical turtle grass *Thalassia testudinum*. A few sea urchins and the green turtle *Chelonia mydas* consume this grass, which is rich in cellulose.

In the tropics, sea grasses are often associated with coral reefs, and smaller, so-called patch reefs usually stand as islands in the middle of sea grass meadows. Patch reefs are often surrounded by areas, known as

halos, that are bare of turtle grass. The urchin *Diadema antillarum* is responsible for the halos. By day, the urchins hide in crevices on the patch reef, but at night they move out on to the surrounding sand flats and consume turtle grass. To avoid predators, they must return to the patch by morning, so usually only a limited area around the reef is denuded.

Decline of Sea Grasses

* **Disease was a major cause of eelgrass decline. Now many sea grasses are declining because of pollution and water turbidity.**

Sea grasses have changed greatly in abundance, and a disease nearly collapsed one species in the Atlantic. The great eelgrass (*Zostera marina*) epidemic of the 1930s is an excellent example of a disease-induced catastrophe, even if we can only make an educated guess as to the pathogen. In the early 1930s eelgrass meadows declined rapidly in the bays of Europe and the eastern United States, though for some reason the Mediterranean was spared. During this period, many species dependent upon eelgrass declined precipi-

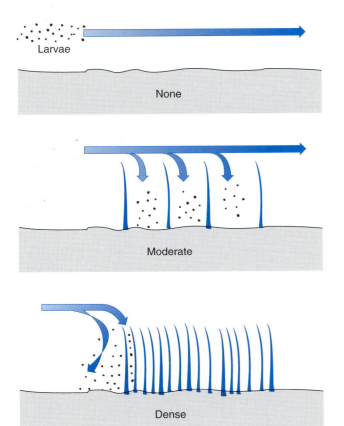

Fig. 15.3 An experiment showing the possible effect of sea grass beds on larval settlement. A larval swarm is carried over a bottom with no impediments to flow (top). Then a low density of soda straws reduces flow and causes dense larval settlement throughout the imitation grass bed. Finally, a dense array of straws cuts flow so much that only a little settlement at the edge of the "bed" is permitted. (Figure after work described by Eckman, 1983.)

tously. For example, settling larvae of the bay scallop *Argopecten irradians* must attach their byssal threads to grass blades, and these were suddenly absent. Thus, a related important commercial shellfish population went into a tailspin. Many other benthic species declined in abundance, and certain soft-bottom habitats that were protected by eelgrass beds, now exposed to the brunt of the sea, were eroded away. It was only in the late 1940s and early 1950s that extensive recovery occurred.

Unfortunately, the cause of this disaster will never be known with certainty, but a recent eelgrass epidemic has been studied properly and was attributed to a fungal pathogen. Because the pathogen does

poorly in low salinity, eelgrass survived in refuges at the heads of estuaries.

Eelgrass is otherwise declining at this time along the northeast coast of North America. Nutrients from sewage tends to increase phytoplankton density, and this decreases the amount of light that can reach the bottom. Sea grasses lose out and have declined greatly. Once abundant, eelgrass is nearly gone from Chesapeake Bay. However, there is an alternative hypothesis for the decline. Increased fishing may have reduced grazing on the epiphytes that live on the sea grass blades. This may have resulted in an explosion of epiphytes and a choking off of light to the sea grass. In basins where grasses have been shaded to extinction, scallops have also decreased greatly.

Kelp Forests

✳ **Kelp forests are dominated by a few species of brown seaweeds with fantastic growth rates.**

Kelp forests are dominated by brown seaweeds of the group Laminariales and are among the most beautiful marine habitats (Figure 15.4, Plates XX, XXI). They are found throughout the world in shallow open coastal waters, and the larger forests are restricted to temperatures less than 20°C, extending to both the Arctic and Antarctic Circles. A dependence upon light for photosynthesis restricts them to clear shallow water, and they are rarely found much deeper than 5–15 m. The kelps have in common a capacity for some

Fig. 15.4 A kelp forest in the Aleutian Islands: *Cymathere triplicata* (foreground); *Alaria fistulosa* (rear). (Courtesy of David Duggins.)

of the most remarkable growth rates in the plant kingdom. In southern California, the kelp *Macrocystis* can grow 30 cm per day, and a plant may grow 25 m, from the bottom to the surface, in 120 days. Species of *Macrocystis* are widespread and occur on the west coasts of North America, Australia, New Zealand, on both coasts of South America, and in South Africa. Somewhat smaller laminarians such as species of *Laminaria* and other large kelps extend the range of kelp forests to Alaska and the northwestern Atlantic. Kelp forests are therefore among the most widespread and productive of coastal marine habitats.

Kelps differ widely in size and growth rate, so the "forests" actually range from beds of seaweeds of a few meters depth, as in the Atlantic *Laminaria* beds of Nova Scotia, to vast stands of plants that extend from holdfasts on the deep (25 m) bottoms to stipes and blades floating at the surface (as in the *Macrocystis* and *Nereocystis* beds of the Pacific Coast from California to Alaska). Kelps usually have a complex life cycle (Figure 15.5), which alternates between a large asexual sporophyte and a tiny filamentous gametophyte. The sporophyte is the "kelp plant," which consists of a holdfast for attachment to the bottom, a stipe, which looks much like a stem and is strong and flexible, and a leaflike blade, the main site of photosynthesis and growth. The blade tips are often eroded rapidly, but growth at the base replaces the tissue. In the smaller Atlantic *Laminaria*, the stipe is the only support for the plant, aside from water currents. In larger Pacific, South Atlantic, and Indian Ocean kelps (e.g., *Nereocystis*, *Macrocystis*), the stipe is quite long, and large floating air bulbs, or pneumatocysts, support long blades, forming a canopy above the seabed.

✳ Kelp forests are biologically diverse and support many plant and animal species.

The seabed beneath kelp forest canopies often supports rich communities of seaweeds and animals. To use the forest analogy, the kelp forest floor often has an understory consisting of a wide variety of seaweed species. To some degree, certain kelp forests have several distinct canopy layers, each dominated by a different kelp. In rock crevices and in open areas, a number of grazing species are very common. In the *Laminaria* kelp beds of Nova Scotia, urchins are quite abundant, as are suspension feeders such as the blue mussel *Mytilus edulis*. The lobster *Homarus americanus* is abundant and preys mainly on mussels, but it may also consume other mollusks and urchins. In Pacific kelp forests, grazers include a number of species of abalones, limpets, and sea urchins. The urchins have a particularly interesting behavior of trapping waterborne kelp and other seaweed fragments on the dorsal spines and tube feet. This material is then transported to the oral surface and ingested. Urchins feeding in this way are usually rather sedentary and wait for the material come to them. In dense kelp forests, kelp erodes and fragments into minute particles, which support dense populations of suspension feeders such as mussels.

Although it is now extinct as a result of hunting in the eighteenth century, Steller's sea cow (*Hydrodamalis gigas*) was once a major browser on the upper kelp canopy throughout the north Pacific rim.

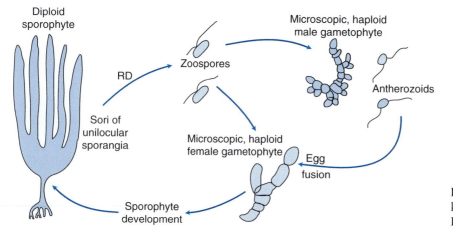

Fig. 15.5 The life cycle of a typical kelp of the genus *Laminaria*. (After Dawson, 1966.)

Fig. 15.6 The sea otter *Enhydra lutris* bringing a sea urchin to the surface to bash and eat. (Photograph by James A. Mattison III.)

Another animal associated with kelp, the sea otter *Enhydra lutris*, is a major predator on sea urchins, mollusks, and fish (Figure 15.6, Plate XXI.4). It dives to the bottom, removes an animal, and brings it to the surface. It often smashes the prey open with a rock and then floats on its back at the surface while eating. The otter was hunted for fur and was reduced to just a few populations, mainly in the northern Gulf of Alaska. It was nearly driven to extinction, but was protected after 1911 and is now increasing rapidly through natural population increase and reintroductions.

Factors Affecting Kelp Communities

✳ **Kelp communities are often strongly affected by a combination of storms, presence or absence of sea otters, and behavioral changes in herbivores.**

Kelps are attached to the bottom by means of sturdy holdfasts, but they project upward into very rough surface seas. In the winter on the Pacific coast of North America, storms coming from the open sea can rip kelps from their holdfasts, and those with floats and blades at the surface are the most vulnerable. In quiet waters, the upper-canopy kelps are competitively superior because they have first access to the sunlight. However, lower-canopy kelps survive better after a storm, and it may be difficult for the larger kelps to become established and grow from beneath the dense canopy of smaller kelps. The periodic El Niño events (see Chapter 2) that affect the eastern

Pacific may thus cause a kelp forest to shift in character toward domination by lower-canopy species. In the extreme case, all kelps may disappear from a region, especially because El Niño also brings water that is too warm for the seaweeds to survive very well.

Storms can initiate changes in urchin behavior that can strongly affect the later history of a kelp forest. When the bottom is stripped of kelps, large aggregations of urchins move along the bottom, and newly recruiting seaweeds are the preferred food, because drift algae are no longer available. As a result, coralline algal turfs take over, and **barrens**, or bare bottoms, develop (Figure 15.7, Plate XXI.2). Harrold and Reed[1] found that the barrens remain devoid of dense kelp until there is a good year for macroalgal

[1] See Harrold and Reed, 1985, in Further Reading, Kelp Forests.

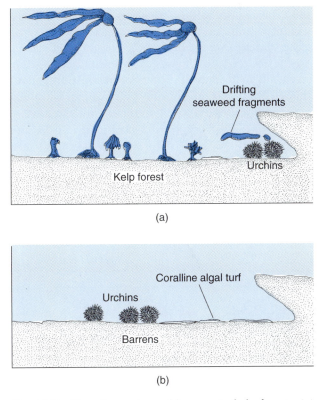

Fig. 15.7 Two alternative stable states in kelp forests. (a) Luxuriant kelp growth, with urchins that are sedentary and trap drift algae on the dorsal surface. (b) A storm has stripped the kelps, and urchins rove about the bed, denuding the bottom of all potential new recruiting kelps.

recruitment. Then, some seaweeds grow sufficiently to escape urchin devastation. Once this threshold has been breached, dense kelp can develop, and drift algae become abundant once more. As a result, the urchins adopt a more sedentary behavior and cease to devastate new seaweed recruits by marauding over the rock pavement. This only reinforces the redevelopment of the kelp community. Trophic structure, therefore, is strongly affected by extrinsic factors, such as storms, and intrinsic factors, such as switches of sea urchin behavior. Sometimes storms can strip the bottom of the roving urchins themselves and allow the colonization of kelp. Figure 15.8 summarizes these interactions, also see Hot Topics 15.1, p. 359.

Kelp systems can be simplified to a few main trophic levels. Although sea otters (*Enhydra lutris*) were once actively hunted to extinction in many east Pacific kelp beds, they still have large populations in some parts of Alaska, and they have recolonized some California beds. They prey upon urchins and abalones, the principal grazers in the system. In turn, these species, as well as others, graze upon attached seaweeds and crustose coralline algae. Urchins can graze attached seaweeds, but the common red urchin *Strongylocentrotus franciscanus* also feeds extensively on drift algae, a complex of algal fragments that either fall to the bottom or are caught on spines and conveyed by tube feet to the urchin's mouth.

Sea otters can completely change the structure of a kelp forest. In the 1970s,[2] Amchitka Island in the Aleutian Islands of Alaska had a dense (20–30 animals km^{-2}) otter population, low urchin density, and a lush cover of kelp. By contrast, Shemya Island lacked otters, had a high density of sea urchins, and lacked macroalgae except in the lower intertidal zone. Otters therefore controlled the pattern of macroalgal productivity, by mediating the herbivore trophic level, which initiated a trophic cascade to the seaweeds. On a smaller spatial scale, the carnivorous starfish *Pycnopodia helianthoides* preys on patches of urchins, a practice that may allow locally dense patches of kelp to develop.

As sea otters have recovered in the 1970s and 1980s in many sites in Alaska from their former low abundance, the resulting effects on kelp communities have been striking. Sea urchin population densities declined by 50 percent in the Aleutian Islands, and some species were nearly eliminated in southeast Alaska. In southeast Alaska kelp growth and increase in numbers was very rapid, but the removal of only about half of the urchins in the Aleutians resulted in a smaller increase of kelps. Why the smaller response in the Aleutians? Estes and Duggins[3] collected urchin fragments in sea otter feces and found that the otters rarely took urchins 15–20 mm in test diameter. High recruitment at the Aleutian sites almost guaranteed the presence of large numbers of such small urchins, however, which kept up the grazing rate on kelps despite abundant otters.

* In kelp forests, succession depends upon the interplay of grazing pressure, disturbance, and competition for light.

The strong interaction between carnivorous sea otters, grazing sea urchins, and kelps is now well known. Sea otters are effective predators and reduce urchins rapidly, allowing kelps to grow and become abundant. Although there are many alternative paths, consider the case in which an urchin population collapses because of otter predation or a disease affecting the urchins. Such a collapse can be simulated by an experimental removal of urchins, as performed in a southern Alaska area rich in urchins and barren (at first) of kelps (Figure 15.9). After only one year, the experimental plots were dominated by several species of kelps living at several levels above the bottom. An upper-canopy annual species, *Nereocystis luetkeana*,

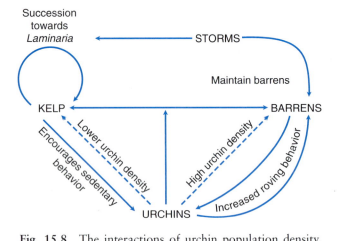

Fig. 15.8 The interactions of urchin population density, urchin behavior, and storms as they affect the character of a kelp forest.

[2] See Estes and Palmisano, 1974, in Further Reading, Kelp Forests.
[3] See Estes and Duggins, 1995, in Further Reading, Kelp Forests.

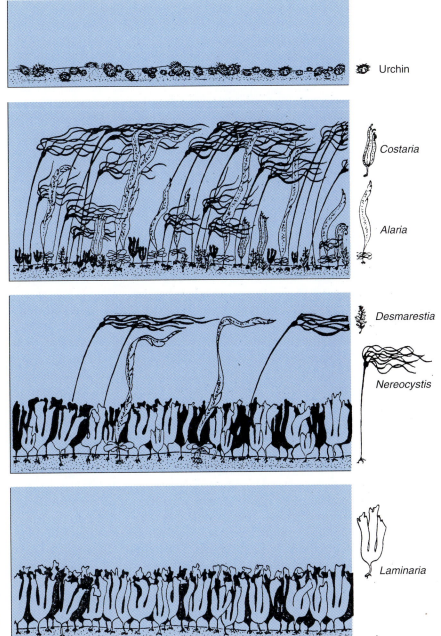

Fig. 15.9 Succession in an Alaskan kelp forest. Eventually (bottom) the kelp *Laminaria groenlandica* dominated the forest and prevented taller species from reinvading by shading out juveniles. (Courtesy of David Duggins.)

was abundant at first. Eventually the lower-story perennial *Laminaria groenlandica* dominated the forest. By covering the bottom, it shaded out juvenile seaweeds of other species, and prevented annual species from reinvading and growing above the lower story. In the initial stages of succession, occupation of many layers above the bottom and rapid colonization tended to increase greatly the diversity of algal species. However, in the end *Laminaria* came to dominate and reduce kelp species diversity.

Subtidal Rocky Reefs

✽ **Rocky subtidal reefs harbor abundant communities of algae and invertebrates and are often dominated by colonial invertebrates.**

Rocky shallow subtidal environments are common off rocky coasts. Our discussion here is confined mainly to temperate and higher-latitude environments. Subtidal rocky outcrops occur in the tropics, but they are usually dominated by corals and other large calcareous organisms. In the temperate zone and in higher latitudes, the rocks are colonized by a variety of plants and animals that do not secrete calcium carbonate at the rate that can be achieved in the tropics, although calcareous algae and animals with calcareous skeletons are usually very abundant.

Subtidal rocky reefs differ in an interesting way from intertidal rocky shores in that they tend to be dominated in many sites by colonial animals, whereas rocky shores are dominated usually by individuals, such as barnacles and mussels. Subtidal reefs are often covered by colonial bryozoans, hydroids, sponges, and colonial sea squirts, as well as by a variety of forms that are individuals, such as anemones, bivalves, and barnacles. In the rocky subtidal coast of Maine, many areas are dominated by mussels and anemones, so the colonial dominance pattern is not always found. The occurrence and often dominance of colonial organisms, however, does deserve comparison with the rocky intertidal and an explanation.

The preponderance of colonial animals on subtidal rock reefs may relate to the physiological limitations of colonial animals in the intertidal zone. Most are not capable of closing off soft tissues from the extreme temperature and salinity changes and desiccation of the intertidal. On exposed rocky shores of the Pacific northwestern United States, the calcareous hydrocoral *Allopora* can be found abundantly, but only on the lowest level of the shore. It grows lushly in subtidal rocky areas. Sessile animals of upper shore levels tend to be dominated by barnacles and mussels. Salinity variation may also be an important factor, as many colonial invertebrates are not well adapted to strong salinity variation. In waters off of Vladivostok on the Russian Pacific coast, only mussels are found abundantly in the upper meter or two. Below this depth one finds exceptionally rich epibenthic invertebrate assemblages, including bryozoans and ascidians. In the upper meter or so of depth the water temperature and salinity vary a great deal seasonally. Salinity can be below 20 o/oo owing to freshwater sources, and the surface water temperature can surpass 15–20°C. In contrast, water below this surface layer is predictably cool and of open marine salinity all year round. A similar layer can be found in New Zealand fjords, and mussels similarly dominate in the shallow layer.[4]

If colonial animals are excluded from the intertidal owing to physiological stress, then why aren't species that occur as individuals more often dominant on subtidal rocky surfaces? We can only speculate, but the answer may relate to the difference between **individuals** and **modular** organisms.

Modular organisms consist of interconnected and usually identical units, such as polyps of a colonial anemone, or zooids of a bryozoan. The ability of a modular organism to spread along a surface makes it more impervious to predation and perhaps to competition as well. A predator, such as a grazing snail, may take a portion of a colony, but the penetration of a mussel shell means death to the individual. One can imagine a group of predators consuming some of all the colonial prey species, which in turn grow back to cover the space. This would not be true for removal of individuals. Predation might move mixed populations of individual and modular species toward dominance by modular species. Competition is not nearly as clear, although the ability of a single settling larva to metamorphose and then to spread and occupy space might be an effective means of competitively displacing others, or at least arriving and dominating by virtue of their early arrival, a priority effect. This is not as simple as it sounds, since individuals may grow substantially more when they are living at lower population densities. A barnacle growing alone might do much better than in a crowd, where the size of the basal attachment it can form will be limited. A lower density of barnacles or mussels might result in greater individual growth, and the same proportion of space occupation as at high densities. We therefore can make no strong conclusion over the possible inferiority of individuals in space occupation. Since, moreover, mussels can readily smother other species, it is not self-evident that individuals are inherently competitively inferior. This question needs further investigation.

[4] See Smith and Witman, 1999, in Further Reading, Subtidal Rocky Reefs.

✳ Rocky reefs often are very patchy, with alternations of rocks dominated by rich invertebrate assemblages and turf-forming calcareous red algae.

Epibenthic organisms of rocky subtidal reefs are very
patchy, at scales from centimeters to hundreds of meters. This seems to characterize such reefs, whether
from New Zealand or from New England in the
United States. On the largest scale, it is common
to see large patches with rich assemblages of sessile
marine invertebrates, especially on vertical or near-
vertical rock walls. Within these patches space is
clearly very limited, suggesting the possible importance of interspecific and intraspecific competition.
This is especially interesting because species with extremely different morphologies and life histories are
competing for the same space (Figure 15.10). We will

(a)

(b)

Fig. 15.10 Biota of subtidal rock walls of Maine. (a) Patch
dominated by calcareous turf-building red algae. (b) Patch
dominated by a number of species of colonial invertebrates.
(Photographs by Kenneth Sebens.)

return to this later. The other common patch type is
a turf that is dominated by red calcareous algae. This
turf is quite distinctive because of the reddish tint to
the crust on the bottom. While sessile invertebrates
are found in such patches, they are decidedly rarer
than in the other patch type. On more horizontal surfaces, seaweeds predominate in shallow waters, except when they are kept in check by large populations
of urchins.[5]

The alternation of patches appears to maintained
by a number of processes. Disturbance, both biological and physical, appears to select for the calcareous
algal patches. In rocky reefs of New England, urchins
are associated with patches of calcareous algal turfs,
which are more resistant to the urchin herds than the
more delicate hydroids, ascidians, and other colonial
invertebrates. This is especially true when you consider newly settled recruits, which are well within the
range of foods susceptible to scraping by an urchin's
Aristotle's lantern. In subtidal rock walls of New
Zealand, algal turfs appear to be facilitated by landslides, which do more damage to the epibenthic invertebrates than the turf-forming calcareous algae.
Where landslides have disturbed the bottom, large
patches of rich sessile invertebrates are interspersed
with algal turfs.

The distribution of subtidal rock wall faunas in
New Zealand integrates nicely a set of physiological
effects, disturbances, and the dispersal biology of the
dominants of the animal-rich patches (Figure 15.11).
Mussels dominate a surface low-salinity layer of 0–5
m depth, possibly because they are physiologically
more tolerant. Colonial invertebrates may not be able
to penetrate, owing to an inability to survive the low
salinity. It is also possible that the strong pycnocline
presents a barrier for dispersal of larvae from deeper
water in the surface layer. In deeper waters, one finds
patches of dimensions of hundreds of meters. High-
diversity (28–32 species) patches, dominated by bryozoans, sponges, and ascidians, are interspersed with
low-diversity patches dominated by encrusting calcareous algae.

The deeper low-diversity patches were maintained
by disturbance. Landslides are quite localized but
when they occur they destroy delicate invertebrates,
favoring the encrusting calcareous algae. What is less

[5] Harvesting of urchins in recent years by people has led to enormous increases of seaweed abundance and major reorganizations
of subtidal hard-surface communities throughout the world.

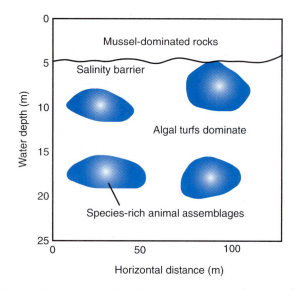

Fig. 15.11 Scheme of biotic patches on subtidal rock walls of New Zealand. The upper water layer is dominated by mussels, which are physiologically more tolerant than deeper-water species. Patches with abundant sessile invertebrates are interspersed by those dominated by calcareous turf algae. The algal turfs are sites of disturbance by landslides. Short-term dispersal, within the diameter of a typical animal patch, provides a reinforcing mechanism to perpetuate an animal patch. (Drawn from descriptions in Smith and Witman, 1999.)

clear is whether the colonial animal–dominated patches are on a path headed for extinction. Most species have planktonic larvae, even if many are lecithotrophic. Do they escape to the water column, or are the animal patches perpetuated by localized return of short-lived planktonic larvae?

Franz Smith and Jon Witman[6] placed settling plates within and at various distances from the diverse patches and found that settlement was much higher within the patches and trailed off with distance from the patch. This was especially true in depths shallower than the animal-rich patches; there was a strong gradient, with species diversity reduction into shallow water, where the low-salinity surface layer was located. Therefore, larval recruitment was sufficient to perpetuate the local patches. On Massachusetts subtidal rock surfaces, many of the animal species with short larval lives were found most abundantly within a few centimeters of the rock walls.[7] This provides a continuous supply of larvae. In effect, the rock wall communities are relatively closed systems.

One must keep in mind that the animal-rich patches are dominated by species with short-dispersal larvae, such as ascidian tadpole larvae, bryozoan short-lived larvae, and hydroid *Actinula* larvae. It may well be that gregarious settlement also affects recruitment. It is apparent that the patches may be largely self-seeding. Less frequent larval dispersal connections between patches establishes a metapopulation. Short-lived dispersal and gregarious settlement would prolong the survival of stable animal-rich patches, as long as there is no major disturbance.

✳ Subtidal rock wall patches of animals often are short on space, suggesting the importance of competition.

Competitive interactions on subtidal rocky reefs are poorly understood, and investigations are hampered by the need to understand the competitive abilities of a wide variety of species with different ecologies. While animal patch edges allow growth into empty space (or at least space dominated by turf algae), the middle of patches are characterized by crowding. Alice Kay and Alan Butler[8] reported on a photographic survey of sessile invertebrates on pilings in South Australia that followed quadrats for over 2 years. Over this period nearly all of the space was occupied. Overall, the dominant species did not change in total area covered. If you looked at individual quadrats, however, a considerable proportion of the total space was lost by some species, to be taken by others. It was as if the community was following the Red Queen's advice in *Alice in Wonderland*: moving just to stay in the same place.

Competitive success must be related to the ability to overgrow neighbors and to resist overgrowth by other species. As mentioned earlier, many if not most species in such communities spread asexually, which makes spread at the expense of other species crucial in competitive success. Kenneth Sebens[9] followed rock walls in northern Massachusetts by photographic plots for two years and was able to determine competitive interactions in a complex fauna.

[6] See Smith and Witman, 1999, in Further Reading, Subtidal Rocky Reefs.
[7] See Graham and Sebens, 1996, in Further Reading, Subtidal Rocky Reefs.
[8] See Kay and Butler, 1983, in Further Reading, Subtidal Rocky Reefs.
[9] See Sebens, 1986, in Further Reading, Subtidal Rocky Reefs.

The rock wall was dominated by a species of ascidian, a sponge, an octocoral, a red crustose alga, and a mat of amphipod tubes. Space occupation was generally positively correlated with competitive ability, either in displacing neighbors or by holding out against incursions by neighbors. Rapidly growing species were good at overgrowing but were not good at resisting overgrowth themselves. Larger and thicker colonies, somewhat slower growing, were usually at the top of the competitive hierarchy.

The variety of competitive mechanisms may make the outcome of competition unpredictable. Sebens observed a large number of reversals of expected competitive dominance of one species by another. In all likelihood a number of important variables, including secretion of poisons and context-dependent changes of morphology, influence reversals in the outcome of competition. In pilings of South Australia, many interspecies interactions did not result in consistent competitive outcomes, although sponges and ascidians tended to do better in larger, more stable patches.[10]

In Chapter 3, we discussed **alternative stable states**, which John Sutherland used to explain subtidal hard surfaces in coastal North Carolina. Often certain species would colonize a bare surface and then no other species could invade. Such **priority effects** led to strikingly different patterns of dominance in different experimental settling tiles, depending on the time of year the tile was placed. Colonization of bare space, combined with the ability to stand off many newly entering species, may be the key to much of the coexistence.

While space competition appears to be important on rocky reefs, most sessile species are suspension feeders, and the question of whether food limitation influences colony growth rates arises. Beth Okamura[11] demonstrated that lateral movement of water across a bryozoan colony resulted in depletion of food. Downstream zooids of a bryozoan received less food than upstream zooids. This result suggests food might be quite limiting in colonies that are very far downstream from the colony edge or from an animal-dominated patch border. Food limitation would be especially strong in cases where phytoplankton are confined to a surface layer of lower density, with limited exchange with the bottom. Many of these colonial species are passive suspension feeders, and the flux of particles is crucial for feeding. In New England, the subtidal mussel *Modiolus modiolus* has a lower scope for growth offshore, which may reflect lower food concentrations.[12]

[10] See Kay and Keough, 1981, in Further Reading, Subtidal Reefs.
[11] See Okamura, 1992, in Further Reading, Subtidal Rocky Reefs.
[12] See Lesser and others, 1994, in Further Reading, Subtidal Rocky Reefs.

HOT TOPICS IN MARINE BIOLOGY 15.1 Connecting a Kelp Ecosystem to the Open Sea

Kelp ecosystems are surprising in their tight linkage among different trophic levels. It is quite simple to describe eastern North Pacific kelp community trophic linkages: kelps are eaten by sea urchins, which are eaten by sea otters. The presence of the otters at the apex of this food chain causes a cascading effect to lower trophic levels. Predation on urchins reduces herbivory, resulting in kelp growth.

Hunting in the nineteenth century nearly drove the sea otters to extinction, a circumstance that must have allowed sea urchins to reach much greater densities than ever before. As a result, kelp forests would be few and far between except in places where urchin abundance was quite low, owing to poor larval recruitment, disease, or reduction perhaps by alternative predators (e.g., starfish). As the otters have returned in the 1970s and 1980s, urchins have declined and kelps have increased in abundance. Some urchin species are known to cease moving very much when kelp are very abundant; they just wait in cracks and capture drift algae. This behavior allows the urchins to avoid some of their predators such as the otters. This simple response results in even more luxuriant kelp growth.

This exquisite linkage has been twisted in recent years by an unexpected turn of events. James Estes and his colleagues[13] have been following for several years the precipitous decline in sea otters in Alaskan coastal waters (Box Figure 15.1), where they formerly were increasing in population size and spreading rapidly to new coastal localities. In the 1990s they were increasing about 25 percent a year.

[13] See Estes and others, 1998, in Further Reading, Kelp Forests.

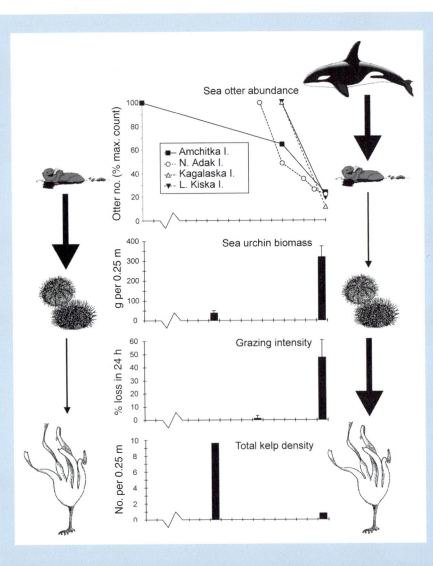

Box Figure 15.1 Changes in abundance of kelp, urchins, and sea otters. (After Estes et al., 1998.)

To give one perspective on this, the International Conservation Union uses a 30 percent annual decline as a dynamic gauge of endangerment for a species. Through the decade, the otter population declined by a factor of 10. Most importantly, these declines were not confined to specific kelp forests but were widespread throughout the Aleutian Islands region of western Alaska.

What caused the sudden decline? Hunting was now illegal, and no widespread pollution or disease could be detected. During this time a large number of otters were tagged, and there was no apparent decline in reproductive activity or survival of pups, nor was any massive emigration observed. All the evidence pointed to killer whales,

even though attacks on sea otters had not been observed by investigators before 1991. After this time, attacks were observed regularly. Most convincing was a local Aleutian bay known as Clam Lagoon. Here sea otters were inaccessible to cruising killer whales and their numbers were unchanged between 1993 and 1997. In the nearby Kuluk Bay, otter numbers declined 76 percent over the same period; these otters were exposed to open-coastal killer whales.

It is not entirely clear how many whales killed how many otters, or even whether the killer whale population increased in western Alaska during this period. Killer whales often hunt in pods and commonly hunt large teleost fish, such as salmon. They are extremely social and intelligent, and hunt in groups. They are quite capable of maintaining communication, trapping schools of salmon into coastal coves. However, there have been many observations of individual killer whales that hunt marine mammals. Energetic calculations suggest that a single killer whale would have to kill 1,825 sea otters each year to maintain its metabolic requirement. Estes and colleagues estimated that 6,788 otters a year were dying in Kuluk Bay. The mortality could be explained therefore by 6,788/1,825, or 3.7 killer whales, maybe even fewer when energetic requirements for reproduction are factored in.

Killer whales have shifted their feeding preferences for sea otters, but why? The reason is not easy to pin down, but it appears to be related to a precipitous decline in alternative prey-pinnipeds, including harbor seals and Steller's sea lions, which began in the 1970s. Pinniped declines can be traced to declines in North Pacific fish stocks, which have also caused a precipitous decline in fish-eating seabirds. The apparent reorganization of the North Pacific ecosystem is still more difficult to explain, but an increase in sea-surface temperature over the past few decades or perhaps strong fishery pressure in the North Pacific may be at least partly responsible. Whatever the explanation, there is a clear connection between open-ocean ecosystems and those near shore (Box Figure 15.2).

The otter decline has had predictable effects on Alaskan kelp forests (Box Figure 15.1): sea urchin density and grazing of kelp declined over 90 percent, and kelp biomass correspondingly increased.

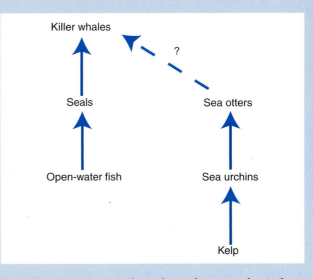

Box Figure 15.2 Interdependency between the inshore kelp food web and the offshore food web. (After Estes et al., 1998.)

More and more examples connecting near-shore ecosystems to the open-ocean change completely our perspective on marine ecology. Coral reefs, be they located on oceanic islands or near the coasts, respond strongly to transoceanic phenomena such as El Niño. Anoxic events in Swedish fjords resonate to the rhythms of the open North Atlantic oscillations. This degree of connectivity is only beginning to be fathomed.

Coral Reefs

✴ **Coral reefs are constructional wave-resistant features, which are built by a variety of species and are often cemented together. The growth of these structures is aided by zooxanthellae, algae that are symbiotic with the reef-building corals.**

Coral reefs are the most diverse and beautiful of all marine habitats. Large wave-resistant structures have accumulated from the slow growth of corals. The development of these structures is aided by algae that are symbiotic with reef-building corals, known as **zooxanthellae** (which are discussed in detail later). Coralline algae, sponges, and other organisms, combined with a number of **cementation** processes, also contribute to reef growth. The dominant organisms are known as **framework builders** because they provide the matrix for the growing reef. Corals and coralline algae precipitate calcium carbonate, whereas the framework-building sponges may also precipitate silica. Most of these organisms are colonial, and the slow process of precipitation moves the living surface layer of the reef upward and seaward.

The reef is topographically complex (Plate XXV.1). Much like a rain forest, it has many strata, as well as areas of strong shade due to the overtowering coral colonies. Because of the complexity, thousands of species of fish and invertebrates live in association with reefs, which are by far our richest marine habitats. In Caribbean reefs, for example, several hundred species of colonial invertebrates can be found living on the undersides of platy corals. It is not unusual for a reef to have several hundred species of snails, 60 species of corals, and several hundred species of fish. Of all ocean habitats, reefs seem to have the greatest development of complex biological associations.

✴ **Reef-building corals belong to the calcium carbonate secreting Scleractinia. Hermatypic corals contribute most to reef growth and have abundant endosymbiotic zooxanthellae.**

While many members of the phylum Cnidaria (see Chapter 11) occur on coral reefs, reef building is due mainly to some members of the calcium carbonate secreting cnidarian group Scleractinia, which mainly consists of colonial species. In all cases, individuals, known as **polyps** (Figure 15.12), feed with tentacles and secrete a calcium carbonate skeleton. While some species are solitary and consist of a single polyp and skeleton, most are colonial and consist of hundreds to thousands of polyps that are interconnected by living tissue. In all cases, the polyps collect zooplankton by means of nematocysts on their tentacles and digest their small prey in a blind gut. As we discuss shortly, the symbiotic zooxanthellae contribute to nutrition. Corals that have large numbers of zooxanthellae and calcify at high rates are important con-

tributors to reef growth. These are known as **hermatypic corals**.

Reef-building corals can be divided into **massive and branching forms**, although there are intermediates between these two types. Massive corals are mound shaped and often irregular. They tend to grow slowly, not usually increasing in any linear dimension much more than one centimeter per year. Branching coral colonies usually have either tree-branch forms, or elkhorn shapes (Plate XII.2). They tend to grow rapidly, on the order of 10 cm y^{-1}. The more rapid linear growth sometimes allows branching corals to spread rapidly on the reef. Storms sometimes break up branching corals, but the fragments may be able to start new colonies.

The Crucial Role of the Coral–Zooxanthellae Symbiosis

✴ **Zooxanthellae are symbiotic with many invertebrates, and they are crucial as a source of nutrition for reef-building corals.**

The remarkable mutualism between hermatypic corals and the photosynthetic zooxanthellae is the driving force behind the growth of coral reefs. Because they have an intimate interrelationship with zooxanthellae, hermatypic corals respond to the environment in many ways that are reminiscent of plants. As a result, the growth of the corals is strongly light dependent, as is the overall growth of reefs (see

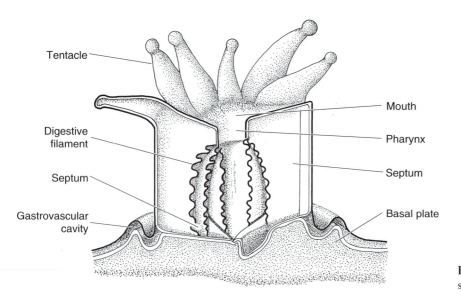

Fig. 15.12 Diagram of a polyp of a scleractinian coral.

next section). Coral reefs are among the most productive marine benthic habitats, mainly owing to the photosynthetic activities of the coral zooxanthellae.

Zooxanthellae are specialized single-celled algae, derived from dinoflagellates, that live intracellularly within the endodermal tissues of scleractinian corals. They are concentrated in the tentacles. Zooxanthellae are also found in a wide variety of other marine organisms, including many sea anemones and the so-called killer clams of the genus *Tridacna*.

Zooxanthellae benefit from the coral association by obtaining protection from grazing and perhaps by access to nutrients derived from coral excretion. The benefits of zooxanthellae to corals are more complex and have been the subject of great controversy over the years. The possible benefits of zooxanthellae to hermatypic corals include the following: (a) zooxanthellae aid in the removal of dissolved excretory products; (b) they provide oxygen via photosynthesis; (c) they manufacture carbohydrates that can be used for coral nutrition; (d) they enhance coral calcification; and (e) they aid in the synthesis of lipids. The first two hypotheses relating to waste removal and oxygen supply are unlikely, since hermatypic corals live in wave-washed environments, where moving waters carry away excretory products and bring ample dissolved oxygen. We therefore focus on the remaining factors.

Zooxanthellae are photosynthetic, and the carbohydrates they manufacture are transferred to the corals. This can be shown by labeling bicarbonate with radioactive ^{14}C, which is taken up by the zooxanthellae during photosynthesis. Eventually, a considerable amount of this carbon can be detected in the animal tissues by a technique known as autoradiography. Slices of corals are placed on radiation-sensitive film, and coral tissues that take up radio-labeled carbon from the zooxanthellae release radiation, which reacts with the film. Corals therefore benefit from the association by getting carbohydrate. (Incidentally, most of the carbon produced by the zooxanthellae is not used by the corals.)

All hermatypic species depend upon the zooxanthellae for food. All hermatypic corals, however, capture small animals by use of their tentacles and nematocysts (Figure 15.13). In the Pacific coral *Stylophora pistillata*, the carbon received from the symbiotic zooxanthellae must be supplemented by nitrogen, which probably comes from feeding on zooplankton. The nitrogen is essential for amino acids, the building blocks of proteins. The mixture of dependence upon zooxanthellae has also been demonstrated for species of the bivalve *Tridacna* (Plate XXVII.3). The dependence of these clams upon the symbiotic zooxanthellae is supplemented by suspension feeding upon phytoplankton. The contribution of suspension feeding to overall nutrition, however, appears to decline in larger-sized species of *Tridacna*. There are groups of soft corals, such as the family Xenidae, that have lost the ability to feed and appear to depend entirely on zooxanthellae.

The presence of zooxanthellae also enhances calcification in hermatypic corals. By using a radioactive isotope of calcium, ^{45}Ca, Thomas and Nora Goreau2[14] demonstrated that calcium uptake is greater when hermatypic corals are exposed to light. Shade or photosynthetic inhibitors reduce the rate of skeletal growth. By removal of carbon dioxide, zooxanthellae may shift the carbonate–bicarbonate–carbon dioxide interactions toward conditions favorable for calcium carbonate secretion.

[14] See Goreau and Goreau, 1959, in Further Reading, Coral Reefs.

Fig. 15.13 Close-up view of the coral *Montastrea cavernosa*, showing expanded individual polyps and tentacles used to capture smaller zooplankton. (Photograph courtesy of James W. Porter.)

It is also believed that zooxanthellae influence lipid production in corals. Lipids constitute about a third of the dry weight of corals and serve both as structural support and perhaps as a source of energy. Lipogenesis is strongly accelerated in the light, which is likely due to the uptake of acetate by zooxanthellae, followed by the synthesis of lipids.

Factors Limiting Reef Growth

* Reef development is limited by the presence of relatively high temperature, open marine salinity, available light, and low turbidity.

Coral reefs are confined to the tropical and subtropical waters, generally between 25° north and 25° south latitude. The high calcification rates required for vigorous coral growth are limited to warm waters. In general, coral reefs are also confined to regions of open marine salinities. Reefs are rarely found in tropical estuaries.

Next to temperature, light is probably the most important limiting factor to well-developed coral reefs because of the symbiosis between reef-building corals and zooxanthellae. Because light intensity decreases exponentially with increasing depth, active reef building ceases below depths of 25–50 m, at least as far as coral growth is concerned. Deeper than this, cementation and growth of sponges permit some reef accretion. Basically, however, coral reef growth is a light-limited process.

Although light is essential for the growth of hermatypic corals, the ultraviolet part of the spectrum is potentially dangerous, especially in clear tropical ocean water, where light penetrates to great depths. Because of the dependence of the zooxanthellae symbiosis on light, reef corals thrive in shallow well-lit waters where ultraviolet radiation is intense. A variety of UV-absorbing materials have been discovered to be common in corals. When corals are exposed to photosynthetically active radiation combined with UV, corals produce more of the UV-absorbing materials. Some corals have conspicuously dark pigments that also are believed to absorb ultraviolet radiation.

Turbidity (number of particles per unit water volume) and sedimentation both have adverse effects on reef-building corals. Turbid waters intercept light and reduce the photosynthesis possible. Sedimentation tends to smother coral colonies and inhibit feeding and the extension of the polyps' crowns of tentacles. Blankets of sediment also may encourage the growth

of disease-causing bacteria. Corals with relatively flat surfaces (e.g., brain corals) tend to produce large amounts of mucus, which traps the sedimented particles. The polyps can then transport the material off the colony. Erect corals intercept less material and usually have a lower capacity to produce mucus. The overall effect of turbidity and sedimentation reduces the development of corals. For example, the coast of Venezuela is very turbid, and coral development there is quite poor. In waters greater than 20 km offshore, coral growth is better developed.

* Coral reefs live exposed to high wave energy, but strong waves can break coral colonies and limit reef growth.

Because coral reefs require clear water and are constructional topographic features, they tend to be located in areas of high wave energy. Moving water brings nutrients and zooplankton to the corals, and is therefore beneficial. The exposure to wave also has its disadvantages. Erect branching forms, such as the Caribbean elkhorn coral *Acropora palmata* (Figure 15.14), are often greater than 2 m across and live in reef crest zones that must withstand severe wave shock. Storms, such as cyclones in the Pacific and hurricanes in the Atlantic, often topple coral colonies and may exert massive destruction of coral reefs.

* Reef development is a balance between growth and bioerosion.

As was just mentioned, reef organisms precipitate calcium carbonate, which enables reef growth. In Jamaica, the growth rate is about one centimeter per thousand years, both upward and seaward. Cementation is the chemical process that solidifies the reef by precipitation of calcium carbonate in crevices and cracks.

Coral reefs are continually under attack by bioerosion. Many species of invertebrates bore into coral skeletons. Parrot fish and surgeon fish rasp away at coral surfaces, and urchins enlarge crevices and bore into colony bases. A group of sponges (Family Clionidae) penetrate coral rock by chemical attack. Although destructive in the long run, the erosion provides living and hiding crevices for many reef species. The collective effect of bioerosion of reefs produces tons of fine-grained sediments, which may be transported from the reef to deeper water, sometimes to depths of 1,000 m or more.

Fig. 15.14 Colonies of the elkhorn coral *Acropora palmata*, a dominant of the reef crest of Caribbean coral reefs. Note the preferred orientation of the branches. In center foreground is the branching coral *Acropora cervicornis*. (Photograph courtesy of P. Dustan.)

Geographic Distribution of Reefs and Types of Reef

✳ There are two distinct biogeographic realms, Pacific and Atlantic.

As reef corals and their associated faunas evolved, most of the world's tropical oceans were interconnected. The Atlantic and Pacific were connected by a broad seaway known as Tethys, or the Tethyan Sea, which dried up toward the end of the Miocene epoch, roughly 10 million years ago. Even at the beginning of the Miocene, divergence began between the Pacific and Atlantic, but by the end of the Miocene, faunas in the Atlantic had diverged from those of the Pacific, although there still was a connection through what is now Panama, because the Isthmus of Panama did not exist. More than 3 million years ago, the Isthmus of Panama did not exist, and many groups with Pacific affinities lived in the Caribbean Sea. After the rise of the isthmus, the Pacific forms in the Caribbean largely became extinct. In general, the number of living species is about twice as great in Pacific reefs as in the Atlantic reefs, with a maximum level in the southwest Pacific (Figure 15.15).[15] Overall there is a great deal of difference in the species composition of Pacific and Atlantic reefs, now that there are essentially no connections.

Maximum diversity seems to be associated with areas that have been historically subject to the least elimination of major regions of appropriate tropical open marine habitat, owing to climatic changes. The southwest Pacific has the least such environmental variation and has had the most continuous occupation of coral reefs over geological history. Paleogeographic maps demonstrate that fluctuations of sea level during the Pleistocene epoch caused major changes in the distribution of land and sea in the Indo–West Pacific. At times of glacial maxima, sea level was low and many areas between Australia and mainland southeast Asia were mostly dry land. At times of high sea level such as the present, most of this region is under water. Times of low sea level might have isolated coral populations and contributed to speciation.[16]

✳ Coral reefs can be divided into atolls and coastal reefs.

Coral reefs often have complex histories, but they can be divided into atolls and coastal reefs. Atolls are horseshoe- or ring-shaped island chains that cap an oceanic island of volcanic origin (Plate XXVI.1). You can think of an atoll as resembling a truncated cone capped by a necklace of islands. They are open-ocean structures, not usually found near a continental coast. Although most are in the Pacific, a few have been

---

[15] More information can be found in Veron, 1995, Chapters 8 and 9, in Further Reading, Coral Reefs.
[16] The interested student should consult Potts, 1983, and Veron, 1995, in Further Reading, Coral Reefs.

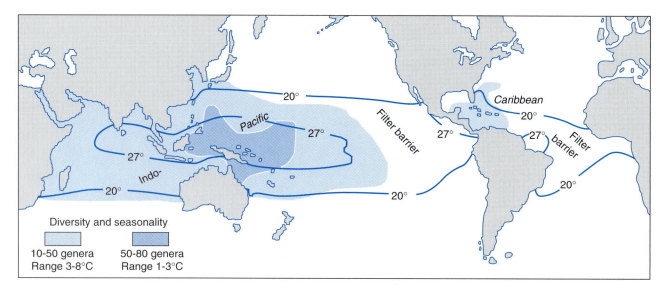

Fig. 15.15   Approximate limits of the tropical Indo-Pacific and Atlantic coral reef provinces as compared with minimum average sea temperatures. (After Newell, 1971.)

found in the Caribbean. **Coastal reefs** are elongate structures that border a continental coast. They may be enormous structures, like the 2,000 km long Great Barrier Reef off the east coast of Australia, or a **fringing reef**, which is a thin veneer of corals on a subtidal coast, as on the coast of Israel and Egypt, in the Red Sea. **Platform reefs** are complexes of coral reefs found on continental shelves—for example, the reefs on the Bahama Banks.

Atolls have a unique origin, which we discussed briefly in Chapter 1. Charles Darwin reasoned that atolls developed when coral grew upward from the top of a sinking volcano on the seafloor. At first, a volcano grows and reaches the surface. Coral then grows around the fringe of the volcanic island. As the volcanic island submerges, corals reefs continue to develop and grow upward. This forms the island ring, leaving a lagoon in the center (Figure 15.16a). After many millions of years of upward growth, there should be a great thickness of coral rock capping the sunken volcano. In the 1950s Harry Ladd and colleagues drilled a bore hole in Enewetak Atoll and had to penetrate 1,400 m of reef rock before they struck volcanic rock. The reef dates back to Eocene times, or 40–60 million years before the present. Thus Darwin was correct about the origin of these reefs. He erred, however, in believing that the crust beneath all coral reefs is subsiding; this is not true for many coastal reefs.

Atolls usually reside in a stable wind system, and

Figure 15.16b shows a typical cross section. The side facing the wind is usually strongly affected by wave action. Corals facing the sea do not grow very well, and **a red algal ridge** usually accumulates on the seaward side of a broad reef flat. The ridge is caused by algal precipitation of calcium carbonate and trapping of sediment by coralline red algae. The ridge protects a broad **reef flat**, which has abundant small colonies of corals. Large coral colonies grow somewhat better at depth, but coral growth is much more lush on the leeward side of the open ocean part of the reef and within the lagoons.

Coastal reefs parallel shorelines and have diverse origins. Some of the Great Barrier Reef develops by means of the growth of corals on subsiding rock, but this is not universal. Because of the combination of sea-level rise and fall, the topography of coastal reefs is often the net result of the interaction upward and seaward growth of coral as sea level rises, and erosion, as sea level falls. The erosion often creates hollows and caverns, through which seawater may gush as the waves beat on shore.

✳ **Both atolls and coastal reefs have prominent depth zones, each which has its own dominant framework-building corals.**

Depth zonation is a major feature of coral reefs, whether they are atolls or coastal reefs. If you swim from shallow to deep reefs, you will cross a succes-

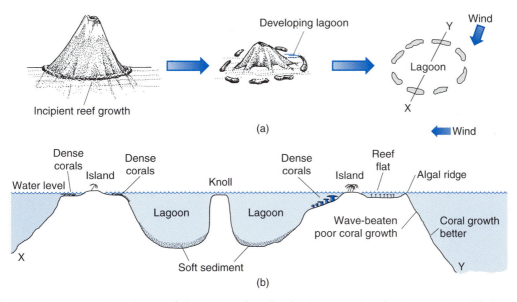

Fig. 15.16 (a) Darwin's theory of the origin of atolls. (b) Cross section showing major subhabitats.

sion of distinct bottoms, or zones, each dominated by a different species or group of coral species. Although the exact explanation for dominance by certain species is often unknown, the following factors must be important: (1) wave and current strength, (2) light, and (3) suspended sediment. Competition among species often leads to dominance by single competitively superior species. Because of the strong dependence of reef-building corals on light and space for attachment, one species may displace another by overgrowth. Figure 15.17 shows a mound-building coral being overgrown by a branching coral. On Pacific reefs, species of the genus *Pocillopora* are superior at overgrowing other species and come to dominate shallow reef zones (Figure 15.18).

Coral reefs usually protect an inner lagoon (Figure 15.19). The lagoon has a soft-sediment bottom and is dominated by sea grasses, urchins, sea cucumbers, and sparse corals and sea whips. As one moves seaward in Caribbean reefs, different corals dominate. The lagoon is protected by an intertidal reef flat, which is dominated by small corals and colonial anemones. Seaward is the reef crest, which is dominated by large colonies of the elkhorn coral, *Acropora palmata* (Figure 15.14). In strong unidirectional currents, the branches grow to point into the current, to reduce tensile stress.

At slightly greater depths is a **buttress zone**, dominated by the massive coral *Montastrea annularis*. These corals form large mounds, which form a **spur**

Fig. 15.17 The shallow-water mound-building coral *Montastrea annularis* being overtopped by the elkhorn coral *Acropora palmata*. (Photograph courtesy of James W. Porter.)

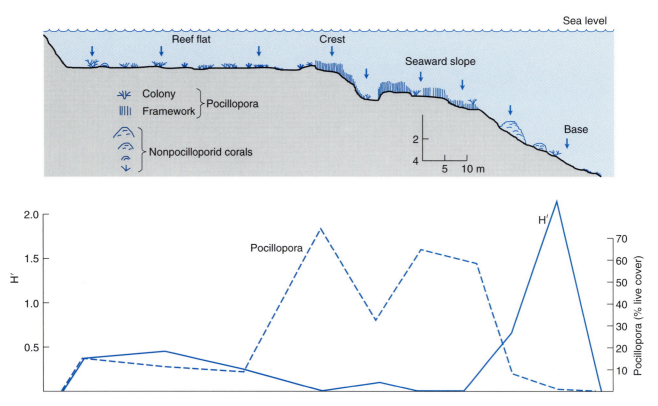

Fig. 15.18   Percentage of cover of the coral *Pocillopora damicornis* on a coral reef on the Pacific coast of Panama versus diversity of all corals. Diversity, *H′*, increases with increasing number of species and with increasing evenness of distribution of numbers among the species. (After Glynn, 1976.) Hurricanes have strongly altered these reefs in recent years.

and groove topography, or an alternation of mounds and channels (Figure 15.19). The channels may be grown over by coral, forming long caves. Below the buttress zone, a broad low-relief bottom is covered with thickets of the staghorn coral, *Acropora cervicornis* (Figure 15.20). Deeper than this zone, corals are much more sparse, although there is a zone of platy corals of a species that is closely related to the shallow-water *Montastrea annularis*. This coral looks like a tree fungus, and its flat aspect may serve to capture whatever little light exists at these depths. Seaward of this zone in Jamaica, the reef drops nearly vertically, and only sponges and the cementation of downward-sliding sediment contribute to reef growth.

Mass Spawning on Coral Reefs

✳ Most corals spawn gametes in the water column, often in synchronized multispecies mass spawnings. This is also true of species of other taxonomic groups.

In Chapter 5, we discussed larval dispersal and mentioned that some Australian reefs were found to have large populations of planktonic larvae, which tended to settle on the patch reef from which they were derived. Self-seeding of the reef was promoted by currents, which tended to create eddies, which in turn brought larvae back to the reef. There has been very little information until recently concerning the important question of whether corals spawn gametes into the water column or whether gametes are brooded. Also, the relative constancy of the tropical climate suggests to many that corals and other coral reef invertebrates probably reproduce at low levels continuously throughout the year, as opposed to the highly seasonal reproductive cycles found among invertebrate species in higher latitudes, which are more obviously seasonal.

It was thought that gamete brooding was the rule until some remarkable observations were made by both Australian and American divers. In Australia,

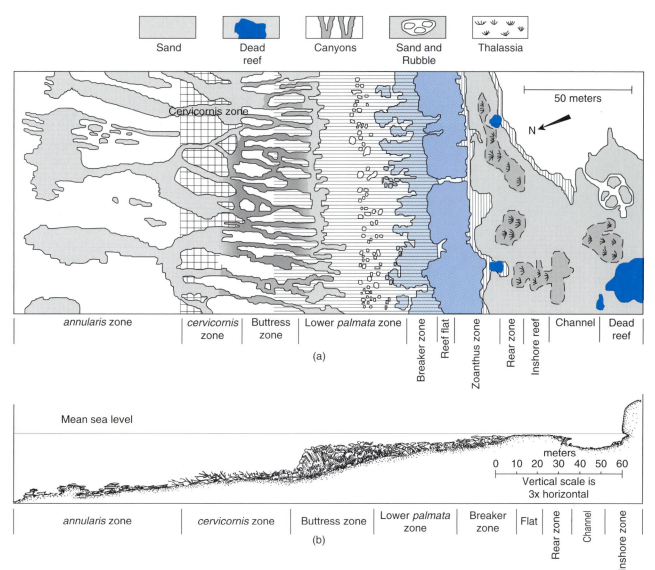

Fig. 15.19 (a) Map view of coral reef environments of the north coast of Jamaica. (After Goreau and Land, 1974.) (b) Cross-sectional view of depth zonation of the coral reef at Discovery Bay, Jamaica. (After Goreau, 1959.) Hurricanes have strongly altered these reefs in recent years.

divers observed massive spawning by over 30 species a few nights after full moons in spring. (Keep in mind that this discovery could be made only by doing large numbers of SCUBA dives at night!) Nearly all species were simultaneous hermaphrodites, but most coral species released both eggs and sperm in the water column. Fertilization was followed by the development of swimming planula larvae.

Despite the tropical habitat, coral mass spawning is highly seasonal, although the time of spawning seems to differ by locale (e.g., in the Great Barrier Reef of eastern Australia, spawning is in the Southern Hemisphere spring, but mass spawning occurs in late summer or fall in western Australia and also on the Flower Garden Banks of the northern Gulf of Mexico). Mass spawning is also not universal, but there are not yet enough data to describe regional variations in this phenomenon.

Fig. 15.20 Thickets of the branching coral *Acropora cervicornis* on a reef in Jamaica. Colonies of the massive coral *Montastrea annularis* in background. (Photograph courtesy of Philip Dustan.)

There are two questions raised by mass spawning. First, why would one species population spawn simultaneously? Second, given the multispecies phenomenon, why should many species spawn in the same evening? In Chapter 5, we discussed shedding of gametes and simultaneous spawning. It makes sense that population-level mass spawning would increase the probability of fertilization. Starfish and coral eggs have been placed in the field when spawning was at its peak and when spawning was off peak. Fertilization frequency increased dramatically[17] on major spawning nights, which suggests that sperm concentration in the water column regulates fertilization success to a strong degree. Corals spawn in evenings during slack neap tides at times of low wave action. This timing might minimize the spread of gametes by waves and turbulence, thus increasing fertilization rates. It might also overwhelm potential predators with too many gametes to possibly eat. The multispecies mass spawning is coincidental, because there is no adaptation to cross-breed with other species—in fact, quite the opposite. Crosses in the laboratory between species shows very limited fertilization success for the most part, but occasional high fertilization rates.[18] Most likely, all species independently evolved a spawning time to coincide with some predictable time marker, such as the full moon. It may be that such timing may also relate to favor-able current conditions, which maximize the meeting of gametes.

* **Known strong dispersal potential of coral larvae presents a conundrum. Is the great potential for long-distance dispersal realized in widespread mixed populations?**

The enormous production of larvae during a mass spawning event might be the beginning of a long journey of dispersal. In Chapter 5 we discussed two studies whose implications are inherently contradictory. On the one hand, a study of larval recruitment in a patch reef in the Great Barrier Reef tract demonstrated the unexpected result that larvae settling on or near the reef probably came from the very same reef. But despite this conclusion, it has become clear that coral larvae have the potential for long-distance dispersal. An example of this is the larval stage of the common Pacific coral *Pocillopora damicornis*, which can last 105 days or more before settling.[19] *P. damicornis* broods larvae, which are endowed with more energy reserves and therefore a longer period of larval settling competency. Larvae derived from broadcasted planktonic gametes have been studied less, but they still have considerable larval competency periods ranging from 26 to 56 days. This would still allow considerable dispersal. Such long-lived larvae suggest that extensive dispersal may connect widely separated island populations across the Pacific.[20] This possibility is certainly conflicting to a degree with the results mentioned in Chapter 5 on a patch reef. It may well be that the potential for long-distance dispersal is not realized under some conditions.

Some independent evidence on genetic differentiation of sea urchins over long distances in the Pacific demonstrates the presence of regional differences in gene frequencies. This would suggest that there is isolation by distance across the Pacific, which in turn means that the potentially long dispersal distances of the planktonic larvae of many reef-associated species do not prevent local genetic differentiation. However, we cannot yet be sure whether the local differentiation is realized as the result of regional differences in natural selection or whether the regional differentia-

[17] See Babcock and Mundy, 1992, and Oliver and Babcock, 1992, in Further Reading, Coral Reefs.
[18] See Willis and others, 1997, in Further Reading, Coral Reefs.
[19] See Richmond, 1987, Chapter 5, Further Reading.
[20] See Wilson and Harrison, 1998, in Further Reading, Coral Reefs.

tion arose as the product of genetic drift. If it is the latter, the degree of isolation would likely be very great, since a single individual dispersed per generation suffices to prevent differentiation. Natural selection more successfully enforces geographic differentiation because there is strong localized selection and often for different alleles. Such local selection can counter the dispersal of inappropriate genotypes into a local population.

Biological Interactions and Community Structure of Coral Reefs

✳ Herbivores exert strong effects on the species composition of reefs.

Shallow soft sediments associated with reefs are often covered with turtle grass and seaweeds (see the sea grass section earlier in this chapter), and hard surfaces free of coral are also potential sites for seaweed growth. Large numbers of herbivores are common on reefs and can exert strong controls on the abundance of both plant and animal species. Many species of fish are herbivorous, including the parrot fishes (Scaridae, Plate XXXI.2) and surgeon fishes (Acanthuridae). Both groups have specialized mouthparts that are capable of scraping algae from coral rock. Urchins are especially efficient and can denude a reef of its plants. Experimental removals of the common Caribbean urchin *Diadema antillarum* are followed by colonization and heavy growth of algae. In the early 1980s, a disease devastated this urchin in the Caribbean. This die-off was followed by extensive seaweed growth in shallow parts of the reef, where the urchins were abundant. Deeper-living herbivorous fishes moved into shallow water to take the place of the urchins.

✳ Space is limiting on coral reefs, and the great variety of colonial species compete for space by overgrowth, shading, and aggressive interactions.

Large numbers of species coexist on coral reefs, and space is often fully occupied. Most of the dominant invertebrates on reefs are colonial invertebrates. Many species spread by overgrowing their neighbors. Because of the symbiosis with zooxanthellae, reef corals are strongly light limited, and erect corals can reduce the growth rate of neighbors by shading them out.

Coral reef ecologists Thomas and Nora Goreau

Fig. 15.21 Interspecific digestion of the coral *Scolymia cubensis* (left) by the coral *Scolymia lacera*. Note bare skeleton to the right of *S. cubensis*, whose soft tissues have been digested away. (Photograph courtesy of Judith Lang.)

discovered a great paradox on the coral reefs of Jamaica. After measuring the growth rates of many species of corals, they found that the most rapidly growing species did not necessarily dominate the reef. The relatively slow-growing mound-building corals were often dominant. Why?

Judith Lang[21] made a pioneering discovery that helped account for this phenomenon. Having observed that some reef coral colonies often have bare zones between them, she placed two species of the solitary and slow-growing coral genus *Scolymia* side by side. After several hours, one of the two forms extruded mesenterial (digestive) filaments through openings in the polyp wall and extended them toward the other form. Within 12 hours, the mesenterial filaments had completely digested part of the second individual, leaving the underlying skeleton exposed. Areas of the victimized coral that were beyond reach seemed to be quite healthy. Interspecific digestion was an effective means of competition (Figure 15.21).

There is a hierarchy of ability to digest other species. The most aggressive species at digesting their neighbors are in general the slowest-growing forms. Some slow-growing species also have long **sweeper tentacles** that damage neighbors within reach of the stinging cells. Rapid growth is therefore not necessarily a guarantee of competitive success among corals. The race is not always to the swift.

[21] See Lang, 1973, in Further Reading, Coral Reefs.

On coming into contact, members of the same species of coral usually fuse. The interaction of competition with fusion between neighbors is an important aspect of studying the success of colonial organisms. Many colonies of scleractinian corals are probably chimeras of genotypes that have met and fused. Fusion may enhance survival by reducing the perimeter open to attack by competitors and predators.

In Caribbean coral reefs, hundreds of species of invertebrates coexist on the undersides of the skeleton of the platy coral genus *Agaricia* and in coral reef caves, even though nearly all the space is occupied. By searching for instances of one species overgrowing another, it is possible to develop a table showing which species routinely outcompete others for space. These successful species consist mainly of colonial ectoprocts and sponges, which come eventually to dominate experimental plates. Why did one or a few species not win? Predation is not as important in this environment, as in our example of coexisting rocky-shore coralline algae. It is possible that there is no clearly competitively dominant species in the first place. The species with the lowest growth rate, for example, may have another means to allow competitive success, such as production of a poison. As we discussed earlier, slower growing corals can outcompete faster ones by interspecific digestion or by deploying specialized sweeper tentacles. This leads to standoffs, and to coexistence of competitors. In some cases, even different conditions may reverse the outcome of competition. In the cryptic invertebrate communities under platy corals, cheilostome bryozoans may or may not be able to overgrow other species, depending upon the orientation or the degree to which a competing colony is fouled by other organisms (Figure 15.22). Larval recruitment in this habitat is minimal, making unlikely the rapid colonization and rise to dominance by single species.

Because of the great diversity of competitors, the complexity of interspecific interactions, and frequent small-scale disturbances, it is unlikely that coral reefs are controlled by any single keystone species, as is the case on some rocky shores and in kelp forests, except under normal circumstances. While urchins such as the Caribbean *Diadema antillarum* and the Pacific crown-of-thorns starfish *Acanthaster planci* (see later) can occasionally exert such control, most predatory effects are more modest, and owing to the complexity of interspecific interactions, there is less commonly a predictable path of succession.

✳ Coral reefs are often subject to extensive disturbance from tropical storms. In the eastern Pacific, El Niño events are also major sources of large-scale disturbance.

In the popular literature, reefs are often pictured as completely benign environments, where exquisite biological interactions have evolved over eons of constancy. Although some of the most remarkable symbiotic relationships are to be found in coral reefs, the environment is far from benign and has experienced many catastrophic changes. First, rises and falls

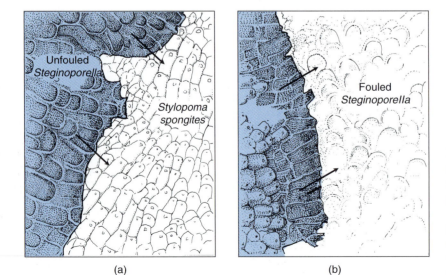

(a)   (b)

Fig. 15.22 Overgrowth of cryptic cheilostome bryozoans, showing different possible outcomes of overgrowth by one species over another: (a) unfouled *Steginoporella* sp. overgrows *Stylopoma spongites* and (b) *Stylopoma spongites* overgrows fouled *Steginoporella* sp. (After Jackson, 1979.)

Plate XXV. Coral Reefs

XXV.1. The reef crest at Discovery Bay, Jamaica (1973), dominated by the elkhorn coral *Acropora palmata*. (Photograph by Philip Dustan)

XXV.2. The elkhorn coral *Acropora palmata* in a strong current. Note the preferred orientation of branches. In the foreground is the massive coral *Montastrea annularis*. (Photograph by Philip Dustan)

Plate XXVI. Coral Reefs

XXVI.1. *Top,* Air photograph of an atoll in the Marshall Islands, central Pacific. (Photograph by Robert Richmond)

XXVI.2. *Middle,* A patch reef in Guam. (Photograph by Robert Richmond)

XXVI.3. *Bottom left,* A Pacific *Acropora* sp. preparing to spawn. (Photograph by Robert Richmond)

XXVI.4. *Bottom right,* The unstalked crioid *Comaster* sp. on a promontory at Enewetak Atoll, Pacific. (Photograph by Robert Richmond)

Plate XXVII. Coral Reefs

XXVII.1. *Top,* Close-up of polyps of an octocoral. (Photograph by Robert Richmond)

XXVII.2. *Middle,* Close-up of polyps of a scleractinian coral. (Photograph by Robert Richmond)

XXVII.3. *Bottom right,* The "killer clam" *Tridacna gigas* on the reef at Enewetak Atoll. (Photograph by Robert Richmond)

XXVII.4. *Bottom left,* The crown-of-thorns starfish *Acanthaster planci.* (Photograph by Robert Richmond)

Plate XXVIII. Coral Reefs

XXVIII.1. *Right,* Jacks at the reef in the Grenadine Islands, Caribbean. (Photograph by Robert Richmond)

XXVIII.2. *Below,* Mutualistic association between an anemone and a clown fish. (Photograph by Robert Richmond)

XXVIII.3. *Above,* The reef-dwelling gastropod *Cyphoma* sp., grazing on a gorgonian. (Photograph by Robert Richmond)

XXVIII.4. *Left,* A serpulid polychaete, growing amid the polyps of a coral colony. (Photograph by Robert Richmond)

Plate XXIX. Coral Reefs

XXIX.1. *Top left,* Close-up showing the pinnules of the feather star *Comaster bella,* Lizard Island, Great Barrier Reef of Australia. (Photograph by David Meyer)

XXIX.2. *Top right,* The giant tube sponge *Aplysina lacunosa,* at 5 m depth, Bonair, Caribbean. (Photograph by Jeff Levinton)

XXIX.3. *Middle,* The sclerosponge *Ceratoporella nicholsoni,* taken from a submarine at a depth of 80–90 m, north coast of Jamaica. (Photograph by David L. Meyer)

XXIX.4. *Bottom right,* A stalked crinoid, photographed from a submarine at a depth of about 230 m, north coast of Jamaica. (Photograph by David L. Meyer)

Plate XXX. Coral Reefs and Reef Flats

XXX.1. *Top left,* Annual banding can be seen in this table-shaped coral, *Acropora sp.* (Photograph by Robert Richmond)

XXX.2. *Top right,* Spur-and-groove topography often occurs in coral reefs, apparently as a growth response to wave fronts hitting the reef (here at Enewetak Atoll). (Photograph by Robert Richmond)

XXX.3. *Middle,* The Caribbean urchin *Diadema antillarum* is a major grazer on Caribbean reefs and patch reefs. (Photograph by Jeff Levinton)

XXX.4. *Bottom,* Air photo of an area with patch reefs in St. Croix, U.S. Virgin Islands. Halos around patch reefs are created by grazing *Diadema antillarum* that hide in the patch reef during the day, but depart and graze on the surrounding sea grasses at night, when visual predators cannot spot them. (Photograph by Thomas H. Suchanek)

Plate XXXI. Coral Reef Fishes

XXXI.1. *Top left,* The foureye butterfly *Chaetodon capistratus* on the reef at Bonaire, Caribbean. It preys upon polyps of corals. (Photograph by Jeff Levinton)

XXXI.2. *Middle,* A male of the stoplight parrotfish *Sparisoma viride* in Bonaire, Caribbean, which feeds on algae attached to hard calcareous surfaces. (Photograph by Jeff Levinton)

XXXI.3. *Bottom left,* A squirrelfish on the reef of Bonaire. (Photograph by Jeff Levinton)

XXXI.4. *Bottom right,* Blue tangs, *Acanthurus coeruleus,* patch reef in Bonaire. (Photograph by Jeff Levinton)

Plate XXXII. The Deep Sea

XXXII.1. A hot-vent fauna, dominated by vestimentiferan tube worms. (Photograph by Ruth Turner)

XXXII.2. A bed of the deep-sea mussel *Bathymodiolus thermophilus,* which has symbiotic sulfur bacteria. The chamber in the background was deployed by the deep-sea submersible *Alvin.* (Photograph by Richard Lutz)

of sea level over the last several hundred thousand years have been a source of profound disturbance; many reefs have failed to keep up with rising sea level and have been drowned. At the other extreme, falling sea level has caused major changes in sedimentary regimes and extensive erosion. On smaller time scales, reefs are often located within very strong tropical storm belts. In the Caribbean, many reefs have been pounded by hurricanes, and cyclones do similar damage in the Indo–West Pacific region.

The north coast of Jamaica was hit by Hurricane Allen in the 1970s and by Hurricane Gilbert in 1988. A formerly exquisite coral reef near Discovery Bay has been devastated and thus far has been recolonized by very few large colonies; recovery will take many years. Staghorn corals are beginning to grow, but the coral-eating snail *Coralliophilia* may prevent the small colonies from enlarging. Also in the 1980s, a disease nearly eliminated the grazing urchin *Diadema antillarum*, resulting in algal overgrowths and inhibition of coral settlement (Plate XXX.4).

In the Pacific, cyclones are often devastating, and enormous brain corals are often toppled easily. Outbursts of predators can also be a major source of disturbance. In the 1960s, the crown-of-thorns starfish *Acanthaster planci* (Plate XXVII.4) increased in numbers in many Pacific reefs, and corals were devastated. Strong outbreaks also occurred in the 1980s, and many areas of the Great Barrier Reef still show the effect. It has never been clear why this species increased so rapidly, and then declined. Endean[22] noted the removal by shell collectors of its major predator, the giant triton *Charonia tritonus*. *C. tritonus* does not seem to be a major source of starfish predation. Charles Birkeland[23] claimed that storms have caused washouts of nutrients, which in turn have increased larval recruitment by the starfish, owing to an increased food supply. This would suggest that larvae are typically limited by starvation, but this appears not to be the case.[24] In either case, the devastation was real, and it will take many years for corals to recover completely from the algal-dominated bottoms that the starfish outbursts left behind.

In the eastern Pacific, El Niño events can have drastic effects. As we discussed in Chapter 3, these periodic episodes bring warm surface water poor in nutrients to the shallow coastal zones of the eastern equatorial Pacific. In 1982–1983 El Niño was particularly severe, and the increase in water temperature caused massive mortality among the dominant corals of Costa Rica, Panama, and, especially, the Galapagos Islands. Many dependent species also declined, and the lovely reefs were replaced by turfs dominated by filamentous algae and eroded dead coral rock. Such strong changes probably occur only once a century.

✳ Disease can devastate the dominant species of a coral reef, which can result in major changes in reef community structure.

To further weaken the image of stable coral reefs, we now know that disease can rapidly destroy important species in coral reef systems. The sea urchin *Diadema antillarum* is a major grazer on Caribbean coral reefs, but it declined catastrophically in the early 1980s. This species is normally very abundant and conspicuous, so its rapid disappearance over the span of one year was noticed and studied intensively.[25] Although the pathogen was not conclusively identified, it was clear that the pattern of initial spread followed the path of surface water currents, so the pathogen was waterborne. Since other urchins did not suffer, the pathogen is likely to have been host specific. Mortality was nearly 100 percent in many locales, starting with coastal Panama. A catastrophic decline of several species of sponges occurred in the 1930s in the Caribbean.[26] A fungal parasite probably caused the disease, though increased salinity may have exacerbated mortality. Here too, the sequence of local mortality could be predicted by water current patterns.

The geographically large scale of these declines suggests that complete extinction of a widespread species can occur, simply by waterborne transmission. Some of these species, like the Caribbean sea urchin *Diadema antillarum*, exert strong effects on communities, and their disappearance may change completely the relative abundance of most other species. The characteristically tremendous surge in urchin population growth is usually followed by a decline of seaweeds. Similarly, sudden mortality is usually followed by lush seaweed growth. Considering the structural complexity that a seaweed population can bring to

[22] See Endean, 1977, in Further Reading, Coral Reefs.
[23] See Birkeland, 1982, in Further Reading, Coral Reefs.
[24] See Olson, 1987, in Further Reading, Coral Reefs.
[25] See Lessios, 1988, in Further Reading, Coral Reefs.
[26] See Smith, 1941, in Further Reading, Coral Reefs.

an otherwise featureless seabed, one can well imagine the importance of disease in the economy of marine communities.

Corals themselves are strongly affected by diseases and physiological weakening, but often the source of the decline cannot be fathomed. Many corals are **bleached** of their zooxanthellae, a change that may have involved the expulsion of the zooxanthellae by corals under some form of physiological stress such as warming. After strong oceanic warming owing to El Niño, corals on the Pacific coast of Panama were bleached. Many coral reef workers have observed an apparent increase of bleaching in recent years and many wonder whether coral reefs are under some sort of physiological stress, owing to regional warming of the sea, pollution or some other factor. Bleaching also seems to be correlated with times when the water column is very clear and stable. This suggests that ultraviolet radiation is a possible source of bleaching.

Many corals have a so-called **black band disease**, which is caused by the cyanobacterium *Phormidium corallyticum*. A densely interwoven black mat of filaments develops, which eventually separates the coral tissue from its underlying calcium carbonate skeleton. Other bacteria then invade, and the coral tissue dies. Coral species differ strongly in susceptibility to black band disease, but the condition is widespread and can kill corals rapidly. Some believe that the disease is becoming more widespread, as corals become physiologically weaker and more susceptible to disease. Bleaching and black band disease have both been noticed more frequently in Atlantic corals, and many believe that this reflects the weakening condition of corals in many areas. Corals have been declining especially in the Florida Keys, and both conditions apparently are sources of the coral mortality. Other diseases have also been identified, sometimes in areas remote from obvious localized sources of human impacts.

✳ Owing to strong regional disturbances and other factors, coral reefs may exist as alternative stable states.

If the water is oxygen rich, warm, and free of very strong storms, coral reefs will grow vigorously in waters of moderate to strong wave action. The diversity of corals creates a landscape within which hundreds of species of fish and invertebrates can reside as well.

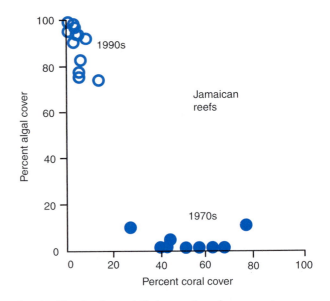

Fig. 15.23 A phase shift in coral reef community structure. Reef areas have either high coral cover combined with low algal cover or high algal cover combined with high coral cover. Solid circles correspond to sites on the north coast of Jamaica in the 1970s, whereas open circles refer to sites in the 1990s. (After Hughes, 1994.)

Such a reef is, in the large sense, a stable state, supporting further growth of corals and allowing continued residence of hundreds to thousands of dependent species. But what if there are significant disturbances? The case of Jamaica, mentioned earlier, suggests that it has been propelled into a new alternative stable state. Hurricanes, the disease-driven collapse of the herbivorous urchin *Diadema antillarum*, and previous overfishing of algal-grazing fish by humans have combined to convert a coral world to a seaweed world. Shading by seaweeds inhibits coral settlement, and a lack of grazing encourages algal growth. If we examine coral cover in the 1970s before the storms, it was usually in the range of 20–80 percent, and algal cover was usually a few percent. Now, coral cover is just a few percent and algal cover is always greater than 75 percent (Figure 15.23). Hughes[27] has concluded that coral reefs of the north coast of Jamaica have been forced into a completely different phase that is a stable state.

[27] See Hughes, 1994, in Further Reading, Coral Reefs.

HOT TOPICS IN MARINE BIOLOGY 15.2 Billions and Billions of Coral Reef Species?

Carl Sagan looked into the eye of a television camera and told us of the "billions and billions of stars" that spread across the universe. The numbers were daunting, but the potential for individuality of stars opened an exciting door to the quirks of stellar history. Some stars were at death's door, while others burned brightly. With this abundance of different stars, was it not inevitable that life somewhat like ours existed in some planetary system somewhere?

This sort of thinking lends itself well to our current quest to understand the earth's biodiversity. Is every species a different individual, capable of diverse interactions? Obviously not, because we need at least two individuals of different sexes for most species to survive. But what if our current species inventory missed thousands and thousands of cryptic species, all nearly identical in external form but nevertheless reproductively isolated? If there are so many species, is it likely that many are ecologically identical, having the same habitat and food requirements? Does it matter that we have 10 million species as opposed to 1 million?

Our estimates of species diversity are poor at best. It is believed that there are between 3 and 30 million species on the earth,[28] but this range is likely to underestimate the true total number, perhaps by no more than a factor of 10. The underestimate is unlikely to be of the same magnitude across all taxonomic groups. Especially in the marine environment, many species with broad geographic ranges experience strong gene flow, and therefore speciation by means of isolation by distance is much less prevalent than in species with very reduced dispersal. Despite this, sibling species have been discovered in groups that are entirely conspicuous and with considerable dispersal potential. The mussel genus *Mytilus*, found on temperate and higher-latitude shores throughout the world has only recently been discovered to be a larger complex of species than previously thought. Each newly recognized species, however, still has a coastal range of up to thousands of kilometers.

Coral reefs are probably the most species-rich marine environments on the earth. On the Great Barrier Reef of Australia, one can often find over 500 species of fish in a relatively small reef area. Many of these species have planktonic larvae, ensuring a good deal of dispersal and wide biogeographic ranges. Their morphologies are distinctive enough to suggest that we may be close in estimating the total number of fish species. But what about corals, sponges, and bryozoans? These are modular species whose morphology appears to change greatly in response to different degrees of wave action and water depth. In the Caribbean,

Box Fig. 15.3 Phenotypic plasticity in the Caribbean elkhorn coral *Acropora palmata*, ranging from bidirectional flow environments (top) to more multidirectional flow (bottom). (Photographs by Phil Dustan (top and bottom) and Jeff Levinton (middle).

[28] See May, 1990, and Gaston, 1991, in Further References, Coral Reefs.

the elkhorn coral *Acropora palmata* may obtain extremely different forms even over a short distance (Box Figure 15.3).

We are conditioned by experience to expect a large degree of phenotypic plasticity in corals. When coral planula larvae settle on the bottom, they must grow to maximize their exposure to oxygenated and zooplankton-bearing water, but they may not be able to choose the best place for a specific, rigidly determined growth form. As the coral grows larger, phenotypic plasticity allows the coral to "grow into" a local hydrodynamic regime that could not have been detected at the time of larval settlement. After all, the hydrodynamic regime where a larva metamorphoses is far different from that when it reaches a much larger size as an adult coral colony. Because we expect this plasticity, our sights are tuned toward minimizing the perceived number of coral species and maximizing cases of phenotypic plasticity.

Despite our expectations, it is clear that more and more cases of **sibling species** are being discovered in the ocean. As we discussed in Chapter 3, sibling species are closely related species that often cannot be distinguished easily on the basis of morphology alone. Because they are true species, however, they are reproductively isolated from each other. Nancy Knowlton[29] compiled over 130 records of occurrences of sibling species in the sea. It is fair to say that sibling species have only recently been thought to matter in the ocean, so this compilation is probably only the beginning of a long journey of discovery of the true magnitude of marine biodiversity.

This brings us back to coral reefs, where colonial organisms and their varied form makes it especially difficult to identify species. A good example is the common mound-building coral named *Montastrea annularis*. This coral often dominates shallow Caribbean reefs, forming so-called buttresses that may be as much as 2–3 m high, and colonies often protrude into the waves. In the 1970s, a deep-water platy form was also classified *M. annularis*, owing to the similarity of its corallite (the skeletal unit enclosing each polyp, which contains important skeletal characters used in coral species identification) to that in the shallow mound-building form. But even in shallower water it was clear that there were several distinct morphs, only they were thought to result from phenotypic plasticity. This was ironic, because in a previous era of taxonomic splitting, three species of the group *M. annularis* had been named, only to be merged into one species, in the belief that the

forms were not genetically distinct, but phenotypically plastic forms of the same species.

Nancy Knowlton and colleagues[30] took another look at this problem and found a number of characters that distinguished three common species, including molecular sequences, enzyme polymorphism, corallite morphology, and life histories. The reproductive separation between the three species is complex. *M. faveolata*, the most molecularly distinct species, hybridizes poorly with the other two, *M. annularis* and *M. franksi*, which hybridize rather easily with each other. On the other hand, the two genetically similar species spawn at different times and may thus preserve their reproductive isolation and minimize hybridization.

Studies of DNA polymorphisms support the separate species status of *M. faveolata*, which has fixed genetic variants that are not found in the other two putative species (Box Figure 15.4). Because with interbreeding the "species" would exchange genetic variants freely, these differences can be readily explained as a complete reproductive isolation between *M. faveolata* and the other two. But such differences were not found between *M. annularis* and *M. franksi*.[31] It may therefore be that some of the variation we see involves sibling species differences while other instances do indeed involve phenotypic plasticity. It is too early to tell right now, because more DNA markers might reveal fixed differences between the two problematic taxa. A lot of difficult research has gained us just one more species. Cross sections of the polyp skeletons of the three species do show distinct differences, however (Box Figure 15.5).

What is the significance of underestimating the number of species on a coral reef, or any other habitat? We certainly may have underestimated the degree of isolation leading to speciation in the sea. Free spawners with planktonic larvae may not be as broadly distributed as we once thought. In the case of corals and other free spawners, we must also acknowledge that the mechanisms of isolation among gametes of different species may be very fine-tuned to this high diversity. We discuss mass coral spawning in

[29] See Knowlton, 1993, in Further Reading, Coral Reefs.
[30] See Knowlton and others, 1997, and Lopez and others, 1999, in Further Reading, Coral Reefs.
[31] See Lopez and others, 1999, in Further Reading, Coral Reefs.

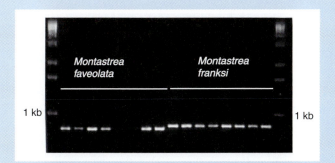

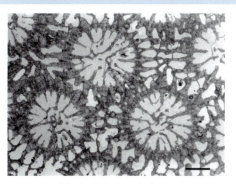

Montastrea annularis

Montastrea faveolata

Box Fig. 15.4 A gel in which nuclear DNA fragments have been run, showing a distinct difference between two hypothesized species of the coral genus *Montastrea*. (Photograph provided by J. V. Lopez.)

this chapter. These events bring gametes of a large number of species in potential contact, and the potential exists for failed fertilization or perhaps successful hybridization. We know little yet of the consequences of such gamete encounters, but we do know of specialized molecular isolating mechanisms between gametes of species of urchins and abalones (see Chapter 5), and strong gamete incompatibility has also been found among species of hydroids, which are at least in the same phylum as corals. The encounters between gametes also invite cases of interspecies hybridization, which would introduce even more complexity into this story.

There are also ecological implications of underrecognizing the number of marine species. Many coexisting sibling species were formerly thought to be phenotypically plastic forms of the same species. This would suggest that plasticity is not as common as we think. We associate plasticity with situations in which an arriving individual (e.g., a planktotrophic larva) is just as likely to encounter any one of many distinct environments. Plasticity thus has a value in permitting responses to fine-tune the organism to the local microenvironment. But if the many species are not only reproductively isolated from each other but ecologically distinct, this may mean that there is much more potential for ecological specialization in diverse communities. It may mean that the degree of local adaptation is much stronger than those who believe in the power of long-distance dispersal to blunt local adaptation.

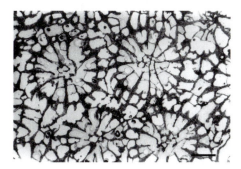

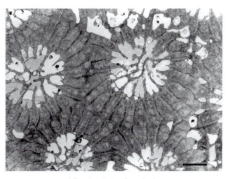

Montastrea franksi

Box Fig. 15.5 Cross sections of the polyp skeletons of three newly recognized sibling species of the *Montastrea annularis* complex of the Caribbean. (Photographs by Nancy Budd.)

Mutualisms Involving Corals

* **Coral reefs and nearby environments harbor some of the most remarkable mutualisms in the sea.**

Coral reefs are fantastically diverse, and many interdependencies have evolved between species. Many such dependencies begin with the protection that the crevices of a large sessile species can give to smaller mobile animals. In many sponges, the crevices are inhabited by a large number of species. In some cases, such commensal relationships have apparently evolved into mutualisms. For example, the Pacific branching coral *Pocillopora damicornis* harbors a group of species of crabs, shrimp, and fishes. One of the coral's principal predators is the crown-of-thorns starfish, *Acanthaster planci* (Plate XIV.4). When the starfish mounts the coral, shrimps and crabs emerge from crevices within the coral colony and actively attack the starfish, usually causing a retreat.

In many marine habitats, several animal species may share the same burrow, constructed by one of the species. On coral reefs, one of these relationships has apparently evolved into a mutualism. While diving on reefs, one can often see the head of a goby poking upward from a burrow in the sand. Beneath, a shrimp of the genus *Alpheus* can be found burrowing. In the reefs of Eilat, Israel, each of four shrimp species has its own companion fish species that lives in the burrow. The shrimp has poor vision and relies upon the fish to warn against predators. In three of the four cases, the shrimp alone excavates the burrow; but in one case the digging is shared by the two species. When digging sand, the shrimp exits from the burrow and communicates by pressing its antennae to the goby's tail. When an intruder approaches, the fish flicks its tail and the shrimp become motionless or retreats into the burrow. The fish then follows if the danger is severe.

Certain species of fish are found among the tentacles of highly virulent coelenterates. The Portuguese man-of-war, *Physalia physalis*, often has several species of fish swimming among the tentacles. These fishes are immune from predators. It is also possible that these fish lure others into the clutches of the man-of-war, but this relationship may be commensal. Clown fish (Family Pomacentridae) often live among the tentacles of large tropical anemones (Plate XXVIII.8). This relationship is more clearly mutualistic. The fish do not induce the firing of the anemone's nematocysts. To avoid attack, a clown fish slowly rubs the anemone's mucus onto its body wall, so that the anemone no longer perceives the fish to be a prey item. The fish then lies among the tentacles, which would otherwise be fatal. The fish benefits by protection from predators and by getting food items such as anemone tissue, wastes, and some prey collected by the anemone. The anemone benefits by the fish's removal of necrotic tissue, and by the strong territorial behavior of the fish, which tends to scare off some potential anemone predators.

One of the most remarkable mutualisms in coral reefs is the cleaning mutualism between cleaner shrimp or cleaner fishes and a large number of fish species. Cleaning organisms are strongly aggregated on reefs in "cleaning stations." Cleaner shrimp and fishes feed by picking ectoparasites off the fishes, which visit regularly. There is limited evidence that the absence of such cleaning fish increases the parasite load of fishes. The cleaning fish *Labroides dimidiatus* maintains cleaning stations that are visited by about 50 species of fishes each day. "Customers" are attracted to the undulating movements of the cleaning fish. During cleaning, the black-colored surgeon fish *Acanthurus achilles* turns a bright blue. Such movements and color change suggest a complex set of recognition signals that have evolved to reinforce the interspecific relationship between cleaners and customer. The movements of cleaner fish are so stereotyped that it probably not surprising that there might be deceptive interlopers. The fish *Aspidonotus taeniatus* mimics the undulation of *L. dimidiatus*, but, instead of picking parasites, it attacks an approaching fish and bites its fins.

Defenses Against Predation

* **Predation is intense on coral reefs, and many species have evolved strong defenses.**

Coral reefs are dominated by many species of voracious predators. Although many species of invertebrates live part of the day in the safe crevices and cracks of corals and sponges, many species live exposed to predatory fishes and crabs. Corals and other calcified species are protected to a degree, but soft tissues are still exposed to predators. Corals and sea whips, for example, can be attacked by starfish, snails, and polychaetes.

Many reef species have acquired strong chemical defenses against predators and fouling organisms. The tropical Caribbean tunicate *Phallusia nigra* has acid vacuoles filled with sulfuric acid. Fishes usually refuse to eat these sea squirts, which are also loaded with other unpalatable substances, including the toxic metal vanadium. Saponins (triterpene glycosides) are produced by many invertebrates and discourage predators. Many sea cucumber species produce saponins, and fish usually die if they feed upon them. The sea cucumbers are also unpalatable, but this is apparently due to some other substance. One of the most spectacular examples of toxicity is the Caribbean gorgonian *Plexaura homomalla*, which is comprised of about 5 percent by wet weight of the hormone prostaglandin. This hormone is highly active and is about five orders of magnitude lower in concentration in other species. Fishes that try to eat extracts of the sea whip soon vomit their meal. Prostaglandin is not distasteful, but the fish learn to avoid ingesting pellets that contain the hormone because they associate vomiting with the pellets. (This is analogous to getting sick after eating at a bad restaurant and deciding not to return.) Surprisingly, the flamingo tongue *Cyphoma gibbosum* (gastropod) and the fire worm *Hermodice carunculata* (polychaete) have somehow managed to circumvent the toxic effects of prostaglandin to eat the gorgonian. *Cyphoma* species are commonly found on gorgonians that are well defended with chemicals (Plate XXVIII.3). How the predators detoxify their prey remains a fascinating and unsolved problem. There are some very interesting inconsistencies, however. In Panama, the butterfly fish *Chaetodon capistratus* (Plate XXXI.1) preys extensively on *P. homomalla*, although predation on this sea whip is otherwise generally rare.

Many species with chemical deterrents smell bad, and predators can probably detect them at a distance. The most poisonous marine organisms are often conspicuous, as opposed to being cryptic. The poisonous tunicate *Phallusia nigra* is black and conspicuous against a usually white coral reef or sand background. Many toxic tropical species are bright red or yellow. The Panamanian tunicate *Rhopalea birklandii* is acidic and a bright electric blue.

The extent to which chemical defense influences herbivory in the sea is still not very clear. Many herbivores, such as parrot fish and sea urchins in coral reefs, are generalists and can consume most or all of the plant production. This would argue for strong natural selection for the production of deterrent compounds. It is hard to separate the effects of mechanical from chemical defense without experimentation. For example, tropical calcified seaweeds also commonly have very toxic chemical defense compounds. Figure 15.24 shows the results of an experiment in which various toxic compounds occurring in seaweeds were painted onto blades of the turtle grass *Thalassia testudinum* and exposed to coral reef fishes. The effect of chemistry was isolated because the grass remained the same in mechanical properties. As can be seen, various compounds had widely varying effects, but fishes that bit a blade of grass usually detected the toxic compound on it and then were reluctant to continue feeding. Smaller invertebrates may live on toxic seaweeds and evolve a tolerance for these same compounds that deter more mobile grazers.

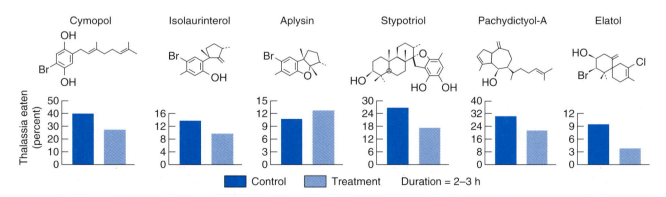

Fig. 15.24   The effect of six different seaweed compounds on feeding by herbivorous reef fishes. Compounds were painted onto blades of turtle grass, *Thalassia testudinum* and the amount eaten was compared to control unpainted blades. In five of the six compounds, fish preferred the unpainted blades. (After Hay et al., 1987.)

Further Reading

SEA GRASS BEDS

Coen, L. D., K. L. Heck Jr., and L. G. Abele. 1981. Experiments on competition and predation among shrimps of seagrass meadows. *Ecology*, v. 62, pp. 1484–1493.

Dennison, W. C., and R. S. Alberte. 1982. Photosynthetic responses of *Zostera marina* L. (eel-grass) to in situ manipulations of light intensity. *Oecologia*, v. 55, pp. 137–144.

Dennison, W. C., R. C. Aller, and R. S. Alberte. 1987. Sediment ammonium availability and eelgrass (*Zostera marina*) growth. *Marine Biology*, v. 94, pp. 469–477.

Eckman, J. 1983. Hydrodynamic processes affecting benthic recruitment. *Limnology and Oceanography*, v. 28, pp. 241–257.

Fonseca, M. S., J. C. Zieman, G. W. Thayer, and J. S. Fisher. 1983. The role of current velocity in structuring eelgrass (*Zostera marina* L.) meadows. *Estuarine and Coastal Shelf Science*, v. 17, pp. 367–380.

Hemmings, M. A., P. G. Harrison, and F. van Lent. 1991. The balance of nutrient losses and gains in seagrass meadows. *Marine Ecology—Progress Series*, v. 71, pp. 85–96.

Iizumi, H., A. Hattori, and C. P. McRoy. 1980, Nitrate and nitrite in interstitial waters of eelgrass beds in relation to the rhizosphere. *Journal of Experimental Marine Biology and Ecology*, v. 47, pp. 191–201.

Irlandi, E. A., and C. H. Peterson. 1991. Modification of animal habitat by large plants: Mechanisms by which seagrasses influence clam growth. *Oecologia*, v. 87, pp. 307–318.

Leber, K. M. 1985. The influence of predatory decapods, refuge, and microhabitat selection on seagrass communities. *Ecology*, v. 66, pp. 1951–1964.

McRoy, C. P., R. J. Barsdate, and M. Nebert. 1972. Phosphorus cycling in an eelgrass (*Zostera marina* L.) ecosystem. *Limnology and Oceanography*, v. 17, pp. 58–67.

Orth, R. J. 1977. The importance of sediment stability in seagrass communities. In B. Coull, ed., *Ecology of Marine Benthos*. Columbia: University of South Carolina Press, pp. 281–300.

Peterson, C. H., H. C. Summerson, and P. B. Duncan. 1984. The influence of seagrass cover on population structure and individual growth rate of a suspension-feeding bivalve. *Journal of Marine Research*, v. 42, pp. 123–138.

Stevenson, J. C. 1988. Comparative ecology of submerged grass beds in freshwater, estuarine, and marine environments. *Limnology and Oceanography*, v. 33 (no. 4, part 2), pp. 867–893.

KELP FORESTS

Dawson, E. Y. 1966. *Marine Biology: An Introduction.* New York: Holt, Rinehart and Winston.

Dayton, P. K. 1985. The ecology of kelp communities. *Annual Review of Ecology and Systematics*, v. 16, pp. 215–245.

Dayton, P. K., and M. J. Tegner. 1984. Catastrophic storms, El Niño, and patch stability in a southern California kelp community. *Science*, v. 224, p. 283–285.

Duggins, D. O. 1983. Starfish predation and the creation of mosaic patterns in a kelp-dominated community. *Ecology*, v. 64, pp. 1610–1619.

Duggins, D. O., C. A. Simenstad, and J. A. Estes. 1989. Magnification of secondary production by kelp detritus in coastal marine ecosystems. *Science*, v. 245, pp. 170–173.

Ebeling, A. W., D. R. Laur, and R. J. Rowley. 1985. Severe storm disturbances and reversal of community structure in a southern California kelp forest. *Marine Biology*, v. 84, pp. 287–294.

Eckman, J. E., and D. O. Duggins. 1991. Life and death beneath macrophyte canopies: Effects of understory kelps on growth rates and survival of marine, benthic suspension feeders. *Oecologia*, v. 87, pp. 473–487.

Elner, R. W., and R. L. Vadas. 1990. Inference in ecology: The sea urchin phenomenon in the northwestern Atlantic. *American Naturalist*, v. 136, pp. 108–125.

Estes, J. A., and D. O. Duggins. 1995. Sea otters and kelp forests in Alaska: Generality and variation in a community ecological paradigm. *Ecological Monographs*, v. 65, pp. 75–100.

Estes, J. A., and J. F. Palmisano. 1974. Sea otters: Their role in structuring nearshore communities. *Science*, v. 185, pp. 1058–1060.

Estes, J. A., N. S. Smith, and J. F. Palmisano. 1978. Sea otter predation and community organization in the western Aleutian islands, Alaska. *Ecology*, v. 59, p. 822–833.

Estes, J. A., M. T. Tinker, T. M. Williams, and D. F. Doak. 1998. Killer whale predation on sea otters linking oceanic and nearshore ecosystems. *Science*, v. 282, pp. 473–476.

Harrold, C., and D. C. Reed. 1985. Food availability, sea urchin grazing, and kelp forest community structure. *Ecology*, v. 66, pp. 1160–1169.

Miller, R. J. 1985. Succession in sea urchin and seaweed abundance in Nova Scotia, Canada. *Marine Biology*, v. 84, pp. 275–286.

Reed, D., C. Amsler, and A. Ebeling. 1992. Dispersal in kelps: Factors affecting spore swimming and competency. *Ecology*, v. 73, pp. 1577–1585.

Van Blaricom, G. R., and J. A. Estes. 1988. *The Community Ecology of Sea Otters*. Berlin: Springer-Verlag.

SUBTIDAL ROCKY REEFS

Butler, A. J. 1986. Recruitment of sessile invertebrates at five sites in Gulf St. Vincent, South Australia. *Journal of*

Experimental Marine Biology and Ecology, v. 97, pp. 13–36.

Butler, A. 1995. Subtidal rocky reefs. In A. J. Underwood and M. G. Chapman, eds., *Coastal Marine Ecology of Temperate Australia*. Sydney: University of New South Wales Press, pp. 83–105.

Butler, A. J., and P. L. Chesson. 1990. Ecology of sessile animals on sublittoral hard substrata: The need to measure variation. *Australian Journal of Ecology*, v. 15, pp. 521–531.

Graham, K. R., and K. P. Sebens. 1996. The distribution of marine invertebrate larvae near vertical surfaces in the rocky subtidal zone. *Ecology*, v. 77, pp. 933–949.

Kay, A. M., and A. J. Butler. 1983. 'Stability' of the fouling communities on the pilings of two piers in South Australia. *Oecologia*, v. 56, pp. 58–66.

Kay, A. M., and M. J. Keough. 1981. Occupations of patches in the epifaunal communities on pier pilings and the bivalve *Pinna bicolor* at Edithburgh, South Australia. *Oecologia*, v. 48, pp. 123–130.

Keough, M. J. 1984. Effects of patch size on the abundance of sessile marine invertebrates. *Ecology*, v. 65, pp. 423–437.

Lesser, M. P., J. D. Witman, and K. P. Sebens. 1994. Effects of flow and seston availability on scope for growth of benthic suspension-feeding invertebrates from the Gulf of Maine. *Ecology*, v. 187, pp. 319–335.

Okamura, B. 1992. Microhabitat variation and patterns of colony growth and feeding in a marine bryozoan. *Ecology*, v. 73, pp. 1502–1513.

Sebens, K. P. 1986. Spatial relationships among encrusting marine organisms in the New England subtidal zone. *Ecological Monographs*, v. 56, pp. 73–96.

Smith, F. and J. D. Witman. 1999. Species diversity in subtidal landscapes: Maintenance by physical processes and larval recruitment. *Ecology*, v. 80, pp. 51–69.

CORAL REEFS

Babcock, R. C., and C. N. Mundy. 1992. Reproductive biology, spawning and field fertilization rates of *Acanthaster planci*. *Australian Journal of Marine and Freshwater Research*, v. 43, pp. 525–534.

Babcock, R. C., C. N. Mundy, and D. Whitehead. 1994. Sperm diffusion models and in situ confirmation of long-distance fertilization in the free-spawning asteroid *Acanthaster planci*. *Biological Bulletin*, v. 186, pp. 17–28.

Birkeland, C. 1982. Terrestrial runoff as a cause of outbreaks of *Acanthaster planci* (Echinodermata: Asteroidea). *Marine Biology*, v. 69, pp. 175–185.

Birkeland, C. 1989. The influence of echinoderms on coral-reef communities. In M. Jangoux and J. M. Lawrence, eds., *Echinoderm Studies*. Rotterdam: Balkema, pp. 1–79.

Buss, L. W., and J. B. C. Jackson. 1979. Competitive networks: Nontransitive competitive relationships in cryp-

tic coral reef environments. *American Naturalist*, v. 113, pp. 223–234.

Coll, J. C. 1992. The chemistry and chemical ecology of octocorals (Coelenterata, Anthozoa, Octocorallia). *Chemical Reviews*, v. 92, pp. 613–631.

Dollar, S. J. 1982. Wave stress and coral community structure in Hawaii. *Coral Reefs*, v. 1, pp. 71–81.

Dubinsky, Z., ed. 1990. *Ecosystems of the World, v. 25: Coral Reefs*. Amsterdam: Elsevier Science Publishers.

Endean, R. 1977. *Acanthaster planci* infestations of reefs of the Great Barrier Reef. *Proceedings of the Third Coral Reef Symposium*, Miami, v. 1, pp. 185–191.

Falkowski, P. G., Z. Dubinsky, L. Muscatine, and J. W. Porter. 1984. Light and the bioenergetics of a symbiotic coral. *BioScience*, December, pp. 705–709.

Feder, H. M. 1966. Cleaning symbioses in the marine environment." In S. M. Henry, ed., *Symbiosis*. London: Academic Press, pp. 327–380.

Gaston, K. J. 1991. The magnitude of global insect species richness. *Conservation Biology*, v. 5, pp. 283–296.

Gerhart, D. J. 1984. Prostaglandin A2: An agent of chemical defense in the gorgonian *Plexaura homomalla*. *Marine Ecology—Progress Series*, v. 19, pp. 181–187.

Gerhart, D. J. 1991. Emesis, learned aversion, and chemical defense in octocorals: A central role for prostaglandins? *American Journal of Physiology* 260 (Regulatory Integrative Comparative Physiology 29), pp. R839–R843.

Gittings, S. R., G. S. Boland, K. J. P. Deslarzes, C. L. Combs, B. S. Holland, and T. S. Bright. 1992. Mass spawning and reproductive viability of reef corals at the east Flower Garden Bank, northwest Gulf of Mexico. *Bulletin of Marine Science*, v. 51, pp. 420–428.

Glynn, P. W. 1976. Some physical and biological determinants of coral community structure in the eastern Pacific. *Ecological Monographs*, v. 46, pp. 431–436.

Glynn, P. W. 1988. El Niño–Southern Oscillation 1982–1983: Nearshore population, community, and ecosystem responses. *Annual Revue of Ecology and Systematics*, v. 19, pp. 309–345.

Glynn, P. W. 1990. Feeding ecology of selected coral-reef macroconsumers: Patterns and effect on coral community structure. In Z. Dubinsky, ed., *Coral Reefs*. Amsterdam: Elsevier, pp. 365–400.

Glynn, P. W., and L. D'Croz. 1990. Experimental evidence for high temperature stress as the cause of El Niño–coincident coral mortality. *Coral Reefs*, v. 8, pp. 181–191.

Goreau, T. F. 1959. The ecology of Jamaican coral reefs. I. Species composition and zonation. *Ecology*, v. 40, pp. 67–90.

Goreau, T. F., and N. I. Goreau. 1959. The physiology of skeleton formation in corals. II. Calcium deposition by hermatypic corals under various conditions in the reef. *Biological Bulletin*, v. 117, pp. 239–250. The classic paper on the role of zooxanthellae in coral calcification.

Goreau, T. F., and L. Land. 1974. Fore-reef morphology and depositional processes, north Jamaica. In L. F. Laporte, ed., *Reefs in Time and Space*, Society of Economic Paleontology and Mineralogy, Special Paper 18, pp. 77–89.

Harrison, P., and C. C. Wallace. 1990. Reproduction, dispersal and recruitment of scleractinian corals. In Z. Dubinsky, ed., *Coral Reefs*. Amsterdam: Elsevier, pp. 133–207.

Harrison, P., R. Babcock, G. D. Bull, J. Oliver, C. Wallace, and B. Willis. 1984. Mass spawning in tropical reef corals. *Science*, v. 223, pp. 1186–1189.

Hay, M.E., W. Fenical, and K. Gustafson. 1987. Chemical defense against diverse coral-reef herbivores. *Ecology*, v. 68, pp. 1581–1591.

Hughes, T. P. 1994. Catastrophes, phase shifts, and large-scale degradation of a Caribbean coral reef. *Science*, v. 265, pp. 1547–1550.

Huston, M. A. 1985. Patterns of species diversity on coral reefs. *Annual Review of Ecology and Systematics*, v. 16, pp. 149–177.

Jackson, J. B. C. 1979. Overgrowth competition between encrusting Cheilostome ectoprocts in a Jamaican cryptic reef environment. *Journal of Animal Ecology*, v. 48, pp. 805–823.

Jones, O. A., and R. Endean, eds. 1973. *The Biology and Geology of Coral Reefs*. 4 Volumes. New York: Academic Press.

Kinzie, R. A., III. 1993. Effects of ambient levels of solar ultraviolet radiation on zooxanthellae and photosynthesis of the reef coral *Montipora verrucosa*. *Marine Biology*, v. 116, pp. 319–327.

Klumpp, D. W., B. L. Bayne, and A. J. S. Hawkins. 1992. Nutrition of the giant clam *Tridacna gigas* (L.). I. Contribution of filter feeding and photosynthesis to respiration and growth. *Journal of Experimental Marine Biology and Ecology*, v. 155, pp. 105–122.

Knowlton, N. 1993. Sibling species in the sea. *Annual Review of Ecology and Systematics*, v. 24, pp. 189–216.

Knowlton, N., J. L. Maté, H.M. Guzmán, and R. Rowan. 1997. Direct evidence for reproductive isolation among the three species of the *Montastrea annularis* complex in Central America (Panama and Honduras). *Marine Biology*, v. 127, pp. 705–711.

Lang, J. C. 1973. Interspecific aggression by scleractinian corals. 2. Why the race is not only to the swift. *Bulletin of Marine Science, Gulf and Caribbean*, v. 23 pp. 260–279.

Lasker, H. R. 1985. Prey preference and browsing pressure of the butterfly fish *Chaetodon capistratus* on Caribbean gorgonians. *Marine Ecology—Progress Series*, v. 21, pp. 213–220.

Lessios, H. A. 1988. Mass mortality of *Diadema antillarum* in the Caribbean: What have we learned? *Annual Review of Ecology and Systematics*, v. 19, pp. 371–393.

Lessios, H. A., D. R. Robertson, and J. D. Cubit. 1984. Spread of *Diadema* mass mortality through the Caribbean. *Science*, v. 226 pp. 335–337.

Lewis, S. M., J. N. Norris, and R. B. Searles. 1987. The regulation of morphological plasticity in tropical reef algae by herbivory. *Ecology*, v. 68, pp. 636–641.

Lopez, J. V., R. Kersanach, S. A. Rehner, and N. Knowlton. 1999. Molecular determination of species boundaries in corals: Genetic analysis of the *Montastrea annularis* complex using amplified fragment length polymorphisms and a microsatellite marker. *Biological Bulletin*, v. 196, pp. 80–93.

May, R. M. 1990. How many species? *Philosophical Transactions of the Royal Society B*, v. 330, pp. 293–304.

Muscatine, L. 1980. Productivity of zooxanthellae. In P. G. Falkowski, ed., *Primary Productivity in the Sea*. New York: Plenum, pp. 381–402.

Newell, N. D. 1971. An outline history of tropical organic reefs. *Novitates*, no. 2465, pp. 1–37.

Ogden, J. C., R. A. Brown, and N. Salesky. 1983. Grazing by the echinoid *Diadema antillarum* Philippi: Formation of halos around West Indian Patch reefs. *Science*, v. 182, pp. 715–717.

Oliver, J., and R. Babcock. 1992. Aspects of the fertilization ecology of broadcast spawning corals: Sperm dilution effects and in situ measurements of fertilization. *Biological Bulletin* (Woods Hole), v. 183, pp. 409–417.

Olson, R. R. 1987. In situ culturing as a test of the larval starvation hypothesis for the crown-of-thorns starfish, *Acanthaster planci*. *Limnology and Oceanography*, v. 32, pp. 896–904.

Pawlik, J. R., and W. Fenical. 1989. A re-evaluation of the ichthyodeterrent role of prostaglandins in the Caribbean gorgonian coral *Plexaura homomalla*. *Marine Ecology—Progress Series*, v. 52, pp. 95–98.

Porter, J. W., and O. W. Meier. 1992. Quantification of loss and change in Floridian reef coral populations. *American Zoologist*, v. 32, pp. 625–640.

Potts, D. C. 1983. Evolutionary disequilibrium among Indo-Pacific corals. *Bulletin of Marine Science*, v. 33, pp. 619–632.

Rogers, C. S. 1983. Sublethal and lethal effects of sediments applied to common Caribbean reef corals in the field. *Marine Pollution Bulletin*, v. 14, pp. 378–382.

Rützler, K., D. L. Santavy, and A. Antonius. 1983. The black band disease of Atlantic reef corals. III. Distribution, ecology, and development. *Marine Ecology*, v. 4, pp. 329–358.

Sale, P. F., ed. 1991. *The Ecology of Fishes on Coral Reefs*. San Diego, CA: Academic Press.

Sammarco, P. W. 1982a. Echinoid grazing as a structuring force in coral communities: Whole reef manipulations. *Journal of Experimental Marine Biology and Ecology*, v. 61, pp. 31–55.

Sammarco, P. W. 1982b. Effects of grazing by *Diadema antillarum* Philippi (Echinodermata: Echinoidea) on algal diversity and community structure. *Journal of Experimental Marine Biology and Ecology*, v. 65, pp. 83–105.

Sapp, J. 1999. *What Is Natural? Coral Reef Crisis*. New York: Oxford University Press.

Smith, F. G. W. 1941. Sponge disease in British Honduras, and its transmission by water currents. *Ecology*, v. 22, p. 415–421.

Stafford-Smith, M. G., and R. F. G. Ormond. 1992. Sediment-rejection methods of 42 species of Australian scleractinian corals. *Australian Journal of Marine and Freshwater Research*, v. 43, pp. 683–705.

Stoddart, D. R. 1969. Ecology and morphology of recent coral reefs. *Biological Reviews*, v. 44, pp. 433–498.

Tanner, J. E. and T. P. Hughes. 1994. Species coexistence, keystone species and succession: A sensitivity analysis. *Ecology*, v. 75, pp. 2204–2219.

Veron, J. E. N. 1995. *Corals in Space and Time*. Sydney: University of New South Wales Press.

Willis, B. L., R. C. Babcock, P. L. Harrison, and C. C. Wallace. 1997. Experimental hybridization and breeding incompatibilities within the mating systems of mass spawning reef coral. *Coral Reefs*, v. 16 (suppl.), pp. S53–S65.

Wilson, J. R., and P. L. Harrison. 1998. Settlement-competency periods of larvae of three species of scleractinians. *Marine Biology*, v. 131, pp. 339–345.

Review Questions

1. How do sea grasses spread to occupy more space?

2. Which would be better for a sea grass and under what circumstances: spreading seed or growing vegetatively?

3. Why are sea grasses important for the existence of other marine species?

4. Why did eelgrass decline so rapidly in the 1930s? What was the consequence for near-shore marine animal populations?

5. Why is sea grass grazed relatively little?

6. Why are kelps important in the survival of many benthic invertebrate species?

7. What would happen if a sea otter population were removed from a kelp forest rich in otters, kelps, urchins, and abalones?

8. How do kelp forests affect the local hydrodynamic regime, and how may this affect benthic invertebrate larval recruitment?

9. What is a barrens, and why might it be a stable state in a kelp habitat?

10. What characteristics determine whether a given coral species is likely to be a hermatypic coral?

11. What processes determine whether and at what rate an entire coral reef will grow upward and seaward?

12. What evidence is there that zooxanthellae provide nutrition to hermatypic corals?

13. What evidence is there that space is a limiting factor in interactions among sessile species in coral reefs?

14. How may corals protect against incursions by other corals into their space?

15. Why may it be adaptive for many species of corals to have synchronized mass spawnings? What problems are created by multispecies mass spawning?

16. Many people once thought that coral reefs were rather stable environments and used this stability to explain the high species diversity found in reefs. Is this explanation valid? Why or why not?

VII

PROCESSES ON THE SEABED, FROM THE SHELF TO THE DEEP SEA

16

From the Continental Shelf to the Deep Sea

Most of the ocean is beyond the reach of divers. About 84 percent lies deeper than 2,000 m and is inaccessible to all but a few submarines, remote samplers, and highly specialized remote underwater vehicles. The average depth of the ocean is 4,000 m, and the trenches can be as deep as 10,000 m. It is no wonder then that we know so little about the seabed and have made major new discoveries in only the past few years. Consider the sampling problem. The abyssal seabed in any one ocean has an area of millions of square kilometers. Ridges extend for thousands of kilometers and cover depth ranges of thousands of meters. Until the 1950s, the only access to this enormity of space was a motley assortment of dredges, corers, and grab samplers. These gave some idea of the types of benthic creatures, but nothing of their spatial distribution, and many important habitats were missed. A vast new world of life has been discovered by the use of submarines, remotely controlled vehicles, and underwater photography and video. A sense of the spatial and temporal change of the deep-sea benthos has been achieved, and wholly new habitats, such as the hot vents (to be discussed later), have been discovered. With remote video techniques we have the means to scan efficiently for new habitats and creatures over large areas of the seabed.

It is the purpose of this chapter to give a broad overview of the changes in bottom life over the enormous depth gradient of the sea, from the inner continental shelf to the deep-sea bed. Most of this habitat is covered by soft sediments. As you will see when we discuss the depth gradient of the sea, increasing depth is accompanied by increasing environmental stability, coupled with a steady decline of food input. The deep sea is dark, physically stable, and utterly dependent upon the import of organic matter from surface waters. Yet, it has a rich indigenous biota. There are some rare exceptions, however, in which food is produced within the deep sea.

Sediment Type and Spatial Distribution

Suspension Feeders Versus Deposit Feeders

∗ Suspension-feeding benthic animals dominate sandy sediments, whereas deposit feeders dominate muds.

Most of the subtidal seabed consists of soft sediments. As was discussed in Chapter 13, clean sands generally occur in areas where currents are relatively strong; fine muds dominate bottoms where currents are sluggish and only fine particles settle from the water column. Although feeding types such as carnivores occur in sediments of both types, *muds are generally dominated by deposit feeders, whereas sands are dominated by suspension feeders*. These two feed-ing groups tend not to co-occur, and the phenome-non has stimulated a great deal of research.

As was discussed in Chapter 13, small sedimentary particles are indicative of a quiet water environment, and it is here that fine-grained organic matter tends to settle from the water column. After the time of the spring diatom increase (see Chapter 9), many diatoms often settle to the bottom in estuaries and inner con-tinental shelves. These relatively small and low-density particles often come to reside in quiet bot-toms. Thus, fine-grained sediments have particle sizes and relatively high organic contents that are suitable for animals that eat sediment and digest and assimi-late the organic matter and microbes. Howard Sanders found that the abundance of deposit feeders correlates best with variation in the clay size (parti-cles <4 μm in diameter) fraction of the sediment. This fraction is a good indicator of the settling of fine-grained organic particles.

Suspension feeders appear to do best in well-sorted sandy substrata. To some degree, suspension feeders must depend upon currents to deliver planktonic food (mainly phytoplankton). Sandy bottoms have faster currents, and therefore probably have greater access to phytoplankton.

✻ Suspension feeders function poorly in muds, ow-ing to the clogging effect of resuspended particles and to the destabilizing effect of deposit feeders on the sediment.

Why are suspension feeders relatively rare in muddy substrata? The answer has to do with particles sus-pended near the sediment–water interface. Water is often turbid just above muddy bottoms. Bottom cur-rents over the muddy seafloor erode and suspend fine particles especially easily. The feeding and burrow-ing activities of deposit feeders further increase the water content of the sediment and convert the sedi-ment to mainly fecal pellets, which can be eroded rel-atively easily by bottom currents. If a vane is moved over the bottom, it generates a current and shear stress at the sediment–water interface. In bottoms oc-cupied by deposit feeders, the vane causes more re-suspension of sedimentary particles than is observed when deposit feeders are absent. This effect is due mainly to the higher water content and relatively low stability of the deposit-feeder-dominated sediment (Figure 16.1).

Both the instability of the muddy sediment and the near-bottom turbidity generated by the action of bot-tom currents over soupy muds have a strongly neg-

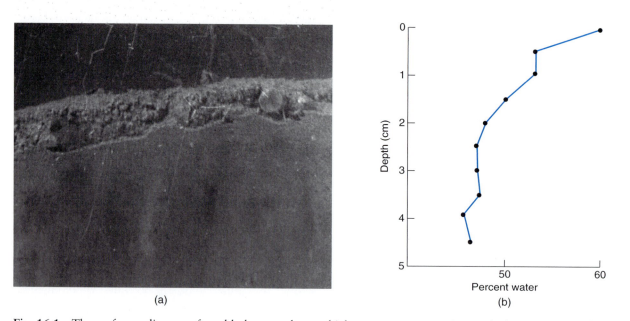

(a)   (b)

Fig. 16.1   The surface sediments of muddy bottoms have a high water content, owing to the burrowing and feeding ac-tivities of invertebrates. (a) A cross section of the bottom. (b) Diagram showing the high water content of the surface sed-iments. (Photograph by the author.)

ative effect on the feeding efficiency, growth, and survival of benthic suspension feeders. Because deposit feeders make the sediment more watery and unstable, they only further deteriorate the environment of suspension feeders. The negative effect exerted by deposit feeders on suspension feeders is an example of **trophic group amensalism** (amensalism is a negative effect of one organism on another; a trophic group is a group of organisms that feed in the same way). Suspension-feeding organs tend to be clogged at high particle concentrations in the water. Figure 16.2 illustrates this effect on the suspension-feeding estuar-

ine clam *Rangia cuneata*. The condition index (the ratio of dry tissue weight in grams to shell cavity volume in cubic centimeters) is a general measure of nutritional state. Over a 2-year period, the condition index was greater for clams living in sands than for those living in muds. Suspended solids were greater in bottom waters over muds, relative to sands. Suspension feeders living near the bottom grow more slowly. If they are suspended in trays somewhat above the bottom, however, the suspended particles are not too concentrated and the animals can feed on them and grow. This finding can be used to design a mariculture system in which oysters and mussels are suspended in racks above muddy bottoms.

Over muddy bottoms, the weak bottom currents resuspend particles continually, and well over 90 percent of the fine particles spend some time resuspended in the water column. This probably enhances bacterial decomposition of the organic material associated with the resuspended matter (Figure 16.3). In late spring and summer, the water column becomes stable, owing to the presence of low-density warm water above and cooler water below (see Chapter 9). During that time, suspended fine particles are often trapped in a layer over muddy bottoms. Divers often call such layer a "false bottom" because it appears to be the seabed as one descends. Such near-bottom, high-turbidity layers can often be traced over tens of square kilometers of estuarine and shallow shelf bot-

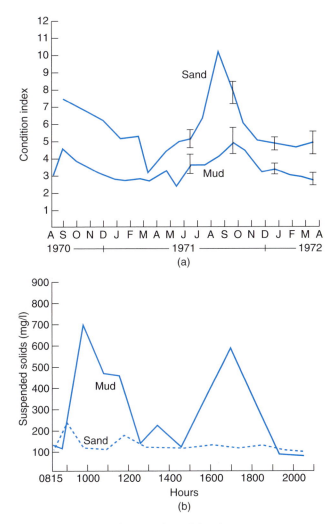

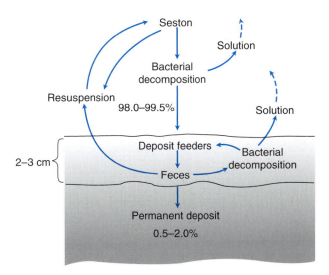

Fig. 16.2   (a) Condition index of the clam *Rangia cuneata* in sand and mud (vertical bars indicate 95 percent statistical confidence limits). (b) Suspended solids in the water column during a day over sandy and muddy bottoms. (After Peddicord, 1977.)

Fig. 16.3   The movement of fine particles over muddy bottoms in shallow-water marine environments. (After Young, 1971.)

toms. In these areas, the near-bottom turbidity is too great for suspension feeder survival.

Patchiness of Organisms on the Seabed

✱ Shelf and deep-sea bottoms are mainly sand and mud, but organismal densities are patchy, owing to animal modification of the sediment, bottom currents, and patchy larval settlement.

In Chapters 14 and 15 we examined a variety of coastal habitats that are notable for their diversity of species and high biological productivity. A traverse along the coastal rim of the "oceanic bathtub" leads from coral reefs to mangroves to mudflats to rocky shores. In the deeper parts of the ocean (the continental shelf, slope, and the deep sea), the seabed appears at first glance to be far more homogeneous, even dull, relative to the shallow parts. Most of the seafloor is covered with soft sediments, ranging from clean sand to fine muds. Such bottom habitats are dominated by deposit feeders and suspension feeders, living mainly within the sediment.

✱ Modification of the sediment by burrowers and digging predators causes local patches.

The impression of homogeneity is only reinforced by visual inspections of the seabed. When the first photographs (Figure 16.4) and, later, videos, were taken

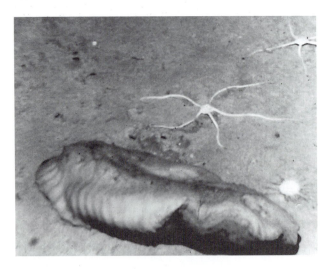

Fig. 16.4 Photograph of the soft-bottom seabed at 1,800 m depth. A sea cucumber is in the foreground, and brittle stars are to the right rear. (Picture taken by J. F. Grassle, courtesy of Woods Hole Oceanographic Institution.)

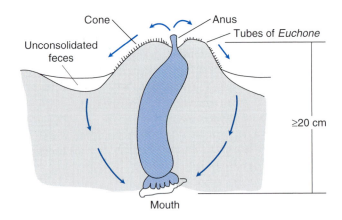

Fig. 16.5 Cross section of the sediment showing the feeding position, surface cone, and overall microtopography generated by the burrowing sea cucumber *Molpadia oolitica*, in Cape Cod Bay, Massachusetts. (From Rhoads and Young, 1971.)

of the bottom, one could see traverses over endless expanses of muds and sands, sometimes covered by mobile organisms such as brittle stars, bottom-living shrimps, and benthic fishes. The homogeneity is a bit of an illusion, because much activity may occur beneath the sediment surface. Animals live and die, hunt and are hunted, much the same as on a coral reef or on a rocky shore. Soft-sediment organisms strongly modify the sediment and can convert an otherwise spatially homogeneous sedimentary environment into one that is discontinuous.

Figure 16.5 shows a cross-sectional view of a mound made by the burrowing sea cucumber, *Molpadia oolitica*, which is common on the subtidal muddy seafloor of New England, north of Cape Cod. The animal creates fecal mounds protruding a few centimeters above the seabed. These attract smaller suspension-feeding bivalves, amphipods, and polychaetes, which live on the flanks of the mound. Between the mounds, deposit-feeding animals decrease the sediment stability and make the bottom inhospitable to suspension feeders. In effect, a population of *Molpadia oolitica* creates a small mountain range, with distinct "mountain villages," as opposed to the "valley communities." In the deep sea, similar mounds are created by echiurid worms and several species of polychaetes, which generate a landscape of hills and valleys in the seabed.

Larger mobile animals feed on benthic organisms and thereby increase the patchiness of the seafloor.

Benthic fishes plow through the bottom in their search for prey. Skates and rays, for example, leave large depressions in the sediment as they feed, and create a series of hills and valleys. Many smaller whales dive to the bottom and strongly disrupt the seabed as they feed. The mysticate gray whale, *Eschrichtius robustus,* for example, is a major predator of amphipods in the Bering Sea, and it turns over the sediment extensively as it feeds on the bottom. It is likely that the majority of the seafloor is disturbed by such bottom-feeding predators. As was discussed in Chapter 13, most infaunal benthic animals must maintain a certain orientation to feed and to maintain a respiratory connection with the surface. Such mobile predators not only kill some of the benthos, but also completely disrupt the local sedimentary regime and the living positions of the survivors.

* Bottom currents disturb the sediment and generate a series of microenvironments.

Bottom currents create a second source of spatial patterning. As was discussed in Chapters 6 and 13, water exerts a shear stress on the seabed, and sedimentary particles are bounced and dragged along the bottom, or propelled into the mainstream. The movement of sedimentary particles often results in a variety of sedimentary structures, including ripple marks if the current moves in a single direction for any length of time. The local environment on the crests and the troughs of the ripples can be quite different when compared with the needs of differing benthic creatures. Fine-grained organic detritus often accumulates in the ripple troughs and is of use to detritus feeders. The crests are a bit more exposed to the currents, which is of advantage to suspension feeders. If the current is particularly strong, then the crests migrate rapidly, sometimes on the scale of centimeters per hour. Under these circumstances, animals requiring a stable seabed to maintain a permanent burrow do not survive very well.

Until recently, most marine ecologists believed that the deep sea was very quiet. Marine benthic animals probably lived in a dull and stable sedimentary environment with little change over decades, if not centuries. We are now beginning to realize that the deep-sea bed can be very unstable. The new perspective comes from deep-sea photography and video, which have advanced greatly in recent years.

To study this environment more directly, the Office of Naval Research established the High Energy Benthic Boundary Layer Experiment (HEBBLE), which made direct measurements and observations of a deep-sea site on the continental rise of the western North Atlantic Ocean. HEBBLE's researchers found that the sediment surface was not quiet at all. Erosion and deposition were fast and furious, and sediment was eroded suddenly, with large-scale declines of benthic populations.[1] Bottom currents were strong, and surface features such as ripple marks developed frequently. The bottom "storm" events often produced bottom currents as strong as 15 cm s^{-1}, which alternated with very slow currents where biological activity dominated the structures of the sediment–water interface. No longer could one say that the deep sea was an unchanging world.

Recent studies have followed the abundance of benthic animals, and certain groups, such as clams and polychaete annelids, tend to be continuously abundant at this site. Following storm events, however, members of meiofaunal groups, such as harpacticoid copepods, are often eroded away in this high-energy environment. To experimentally test the hypothesis that bottom currents altered meiofaunal and macrofaunal communities, David Thistle and Lisa Levin placed a flume on the seabed at about 583 m depth in a Pacific locality.[2] The flume altered local bottom currents by guiding water through a narrow passage, which increased the velocity at the seabed surface (remember the principle of continuity, discussed in Chapter 6). As in observations made following storms, meiofauna were depleted relative to control sites, where currents were not altered experimentally.

* Organic material is distributed discontinuously, generating local sites of high food value.

Certain parts of the seafloor comprise a sort of trash heap. When a large fish dies, it often falls to the seabed, where it commences to decompose. Soon thereafter, the decomposition consumes local oxygen, and the surrounding sediment becomes anoxic. Animals unlucky enough to be directly beneath the fishfall will inevitably die if they cannot burrow laterally to escape the anoxia. Near the coastal zone, a sur-

[1] See Aller, 1998, in Further Reading, Sediment Type and Benthic Distributions.
[2] See Thistle and Levin, 1998, in Further Reading, Sediment Type and Benthic Distribution.

prising amount of plant matter is carried by currents, often to deep-sea bottoms. For example, large fragments of kelp have been collected or photographed at depths of 2,000–3,000 m on the seafloor adjacent to the southern California coastline. Here, the continental shelf is only a few kilometers wide and occasionally intersected by canyons, so it is quite easy for such material to be swept out to the deep sea. North of St. Croix, in the U.S. Virgin Islands, large fragments of the turtle grass *Thalassia testudinum* can be found in deep-sea sediments. Again the seaward extent of the shallow seabed is quite small, and only a few kilometers from the shoreline there is a rapid plunge to the deep-sea bottom.

Although the fall of such animals and plants to the seabed can kill organisms unfortunate enough to be in the wrong place at the wrong time, there is also a potential for a positive effect. As will be discussed further, the deep sea is a world of very low food supply. These falls, therefore, may be the source of quite patchy organic matter inputs, which can be eaten by scavengers, or may gradually be broken down into particles of a size suitable for deposit feeders. Alternatively, bacteria may decompose the material, and deposit feeders may derive nutrition by eating the bacteria in the sediment.

The impact of food falls was discovered by the use of a "monster camera" that was placed over a bucket of fish scraps placed on the deep-sea bed. Very rapidly, large populations of fishes and a very large amphipod were attracted to the scraps and consumed them (Figure 16.6). This experiment demonstrated that there are deep-sea organisms adapted to moving toward such fishfalls, which are rare riches in an otherwise poor environment.

✳ Patchy dispersal and recruitment also create spatial change in benthos on the seabed.

The nature of dispersal in the marine environment is another source of patchiness. As was discussed in Chapter 5, many marine invertebrates have planktonic larvae, often with dispersal periods of several weeks. Owing to currents, predation, and other factors such as bottom substratum, temperature, and salinity, larval settlement is very patchy. Thus a series of samples taken along a transect of seabed may have quite different assemblages of animals, mainly as a result of patchy settlement.

There often are very good and very bad years for larval survival. Thus, every few years, a bottom may

Fig. 16.6 A shark attracted to a bait bucket placed on the deep-sea floor. (Photograph by John Isaacs, courtesy of Scripps Institution of Oceanography.)

be colonized by an extremely large larval population in one local area. This colonization may survive for many years, leaving a strong impact on the local animal community. This is probably a widespread phenomenon, but it is most noticeable in the echinoderms, which are relatively large and therefore conspicuous organisms. Brittle stars and sea urchins often arrive suddenly and in great numbers, lasting for many years. Through bottom sampling, they have been identified often as a series of patches of differing age, depending upon the time at which a larval swarm settled successfully in a given spot. The seabed is therefore a mosaic of settlement of larval swarms. This picture is more appropriate to the continental shelf, where planktonic larvae are the major dispersal type. As one goes to deeper continental rise and abyssal seabed environments, most of the invertebrates have much shorter dispersal times, usually producing lecithotrophic larvae of only a few hours duration. In these cases, larvae disperse a much shorter distance, and one does not see quite the same level of larger-scale patchiness, owing to the variation of larval settlement success.

✳ Succession of subtidal soft sediments involves colonization of a disturbed and abiotic substrate by rapidly colonizing surface dwellers, followed by colonization and dominance of deeper-burrowing species.

Marine soft sediments are well known to have a continual cycle of disturbance and succession. In Chapter 13, we discussed the physical and chemical properties of sediments that are important to benthic organisms. Lack of sediment aeration and continual erosion are two major sources of stress, and a disturbance required to initiate succession usually involves these two factors. In Long Island Sound, New York, winter storms commonly erode the bottom down to water depths of approximately 10 m. The top few millimeters or centimeters of surface sediments are removed, exposing a sulfide-rich anoxic layer. The removal of burrowing organisms and the increase of sulfide make the sediment inhospitable for a wide variety of benthic infauna. Anoxia also affects intertidal salt marsh flats in the winter and spring, when decaying green seaweeds of the genus *Ulva* colonize and decay on the mudflat surface.

In subtidal muds, newly disturbed bottoms are rapidly colonized by surface-dwelling polychaetes with high reproductive rate and strong colonization potential. In strongly polluted habitats, one of several species of the small red polychaete *Capitella* can rapidly invade. In less polluted bottoms, the invading species feed either on suspended matter or on surface detritus. Deeper feeding is precluded, owing to the anoxic hydrogen sulfide-rich pore waters. Some early successional surface dwellers have been hypothesized to have planktonic larvae that cue onto hydrogen sulfide, which suggests that they are adapted to finding disturbed areas. Surface dwellers, being more accessible to surface-feeding fish and crustaceans, are therefore vulnerable to rapid removal. The colonists, however, condition the sediment for further colonization by burrowing and aerating it (Figure 16.7).

The small Atlantic North American bivalve *Mulinia lateralis* is an example of a successful pioneer colonist. It is well adapted to anoxic conditions and can sustain normal activity after several days of anoxia. It feeds on suspended matter. Unlike most bivalves, it has a generation time of only 2 months and, as a consequence, can colonize rapidly and produce a set of offspring. Populations of this bivalve colonize in great numbers in disturbed habitats, such as a subtidal area of Buzzards Bay, Massachusetts, that was covered by oil residues in 1969. Although most recruitment is erratic, the species clearly recruits mainly in the spring, a fact that imposes a seasonal pattern on the nature of succession in disturbed soft sediments.

Later colonists are able to burrow deeper, but they also continue the process of sediment detoxification by burrowing and by circulating water through tubes and siphons that extend to the sediment–water interface. These species, which include gastropods, bivalves, and polychaetes, tend to be characterized by lower colonization potential, smaller investment in reproduction, and lowered mortality rates, probably stemming from their burrowing below the sediment–water interface.

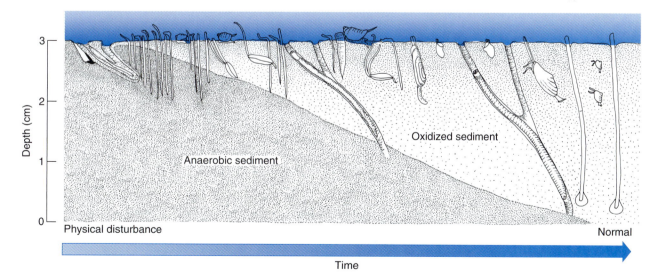

Fig. 16.7 Dominants of early (left) and late succession (toward right) in soft sediments of southern New England. Note the transition from dominance by surface forms to deeper-burrowing species. (After Rhoads et al., 1978.)

Sampling the Subtidal Soft-Bottom Benthos

* Most sampling of continental shelf and deep-sea bottoms in the past involved remote sampling, which often failed to give a complete picture of sea-bottom biology.

Understanding of the subtidal seabed is hampered by the very remoteness of the target area. In the 1870s, the sailing ship H.M.S. *Challenger* circumnavigated the globe and provided the first images of the seabed on a large scale. To do this, a dredge was cast over the side, attached to thousands of meters of piano wire, and dropped to and dragged along the bottom. It was then winched to the surface, bringing a bag of sediment plus animals to be sifted and picked. This was very much the pattern of the nineteenth-century investigation, as we mentioned in Chapter 1. It took hours to get a sample from the deep-sea floor. In doing this sort of sampling, one could not see the animals themselves, and it was impossible to know the exact point from which the sample was taken. This is especially problematic, considering the spatial patterning we discussed above. How could one know whether the sample hit an area of ripple marks or a level seabed, a fishfall or a biogenic sediment mound? The sample, moreover, came up to the surface completely scrambled.

Our knowledge of the shelf and deeper seabed was greatly advanced with the development of better bottom samplers. An ideal sampler should (1) sample a large area of bottom, (2) sample a defined area and uniform depth below the sediment–water interface, (3) sample uniformly in differing bottom substrata, (4) have a closing device, to avoid washout of specimens as the sampler is brought to the surface, sometimes for a distance of thousands of meters, and (5) bring up a sediment sample intact, so that the living positions of the animals may be examined directly. It is also important for bottom samplers to hit the bottom gently. Most devices push water ahead as they hit the bottom, creating a **bow wave** that erodes the sediment and scares the organisms.

Most sampling of the seabed has been done with dredges and sleds. **Dredges** are heavy metal frames with cutting edges designed to dig into the sediment. An attached burlap or chain bag is attached at the rear and collects sediment as the device is dragged along the bottom. The **anchor dredge** has a control plane that constrains the dredge to bite to a defined depth (Figure 16.8). The area sampled can be calculated from the volume of sediment collected, divided by the biting depth. **Sleds** are dredges with ski-like runners that permit the device to dig in only a few centimeters. They are usually not quantitative, but they can collect large amounts of material. This is a virtue in the deep sea, where animals are often sparse.

Gravity and spring-loaded **grabs** are designed to sample a precise area of bottom with two or more sharp digging sections. As the **Peterson grab** hits the bottom and the supporting wire has some slack, the hook, whose support depends upon the wire's tension, releases and allows a chain to pull the two sections closed (Figure 16.9a). If the wire is suddenly slackened upon lowering, the device will fire prematurely. The **Smith–McIntyre grab** is a heavy, spring-loaded device that digs efficiently in both muds and sands. Non-spring-loaded devices often fail to dig efficiently in hard sandy bottoms.

Shipboard-deployed **gravity corers** take a sample of uniform depth and exact area. Small-diameter (<10 cm) cylindrical devices, such as the **Phleger corer**, are useful for meiofauna, sediment, and microbial samples. A weight drives the corer into the bottom. The **box corer** is a rectangular gravity corer that is guided into the bottom by a movable plunger mounted on a frame. When the frame hits the bottom, a spade is released that digs into the sediment and closes the bottom of the corer, as the frame is lifted by a wire (Figure 16.9b). This device is the one

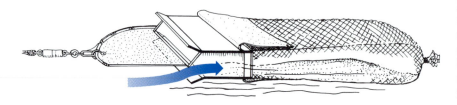

Fig. 16.8 A deep-sea anchor dredge used to sample soft bottoms. The metal plane forces the dredge to bite at a defined depth. The sediment is collected in the bag to the right, brought to the surface, and sieved for organisms. (From Sanders et al., 1965.)

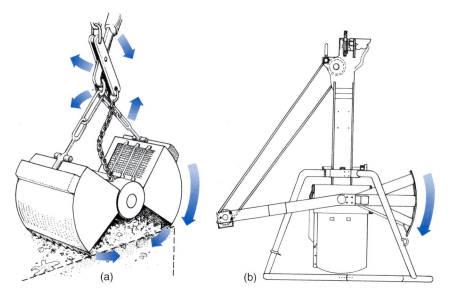

Fig. 16.9 Some benthic sampling devices. (a) The Petersen grab taking a sample from the seabed. (b) A box corer developed to take samples from the deep-sea bottom. (After Hessler and Jumars, 1974.)

(a)   (b)

of choice in the deep sea; it is lowered slowly to the bottom, where it takes a sample of defined area and depth.

✳ Submersibles, remote underwater vehicles, underwater video, and other remote devices are crucial tools in investigating the seafloor.

Remote, shipboard-deployed samplers are by far the least expensive means of exploring the seabed. There are, however, a number of disadvantages connected with their use. Samples cannot be located exactly as the sampler hits the seabed. Also, it is difficult to control the spatial distribution of a series of repeated samples at the same general location. Drift of the vessel and currents make the exact sampling locations very indefinite. One literally cannot see what one is getting until the sampler is brought onboard ship. Finally, most samplers can take only a small amount of bottom.

The use of several remote manned and unmanned vehicles has greatly aided the sampling of the deep sea. The device most similar to a shipboard sampler was the pioneering remote underwater manipulator (RUM), first used in the 1960s. This device was attached to the ship by means of cables and moved on treads along the muddy seafloor. It had a mechanical arm, which could be used to take a sample similar to that taken by a box cores. The RUM could be programmed to move along specific paths, to ensure an exact spatial array of samples. In recent years, a

number of **remotely operated vehicles** (ROVs) have been developed. These can sample and move along the seabed without the need for a cable to the surface. Some ROVs have manipulator arms with samplers attached; others are equipped with lights and cameras for roving photography. The vehicles are relatively inexpensive to operate, and they can navigate with great precision (Figure 16.10). Vehicles such as these were used to seek out and explore the sunken ocean liner *Titanic*, which had been lost for many years.

Manned submersibles are among the most valuable vehicles for underwater exploration, although their operation is quite expensive. The *Alvin*, operated by the Woods Hole Oceanographic Institution, is capable of diving to depths of 4,000 m. The submarine usually has a pilot and a scientist onboard and is equipped with cameras, manipulator arms, and observation lights. The manipulator arms have permitted the establishment of many experiments at great depths. It is possible, in conjunction with accurate satellite-based navigation, for *Alvin* to establish an experiment, and to return later to the same site. The *Johnson-Sea Link* submarine, located at the Harbor Branch Oceanographic Institution in Florida, is also capable of deep diving and has been used in a number of experiments (Figure 16.11). A number of smaller submarines have been developed for shallow-water work, especially under the auspices of the National Undersea Research Program.

Ship-deployed devices have been used in recent

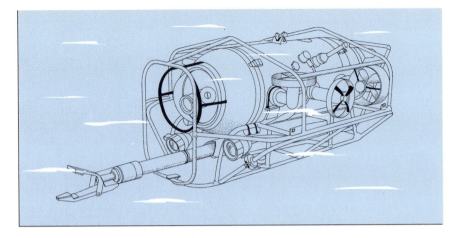

Fig. 16.10 A remotely operated vehicle, or ROV, capable of precise deep-sea navigation, underwater photography, and sampling with a manipulator arm.

Fig. 16.11 The *Johnson-Sea Link*, a deep-diving submersible operated by the Harbor Branch Oceanographic Institution, in Fort Pierce, Florida. (Courtesy of Harbor Branch Foundation.)

years to rapidly construct a map of the seabed, and changes in surface properties of the bottom can be related to benthic communities that differentially affect sediment structure. **Side-scan sonar** is especially useful. A sonic signal emanates from a device that scans the bottom as the ship is moving, allowing a signal to be gathered from a spreading beam 10–20 degrees from the ship's keel. A map of the seabed can be constructed and benthic samples can be collected to understand the changes in seabed surface features detected by the side scan.

The Shelf–Deep-Sea Gradient

✳ **A transect from the shelf to the deep sea defines a shift from shallow-water bottoms with high biomass and productivity to deep-sea bottoms with low biomass and productivity.**

Once the problem of quantitative sampling was solved, it was possible to understand the degree of change of animal density from the shelf to the deep-sea abyssal plain. The standing stock of animals, measured in terms of either numbers or biomass, declines extensively with increasing depth. Howard Sanders and colleagues established a transect from Cape Cod to Bermuda and showed that numbers declined substantially as one moved seaward of the continental slope. Abyssal densities of polychaetes are 100-fold less than those on the upper continental slope (Figure 16.12). The same pattern applies for meiobenthos, whose biomass decreases toward the

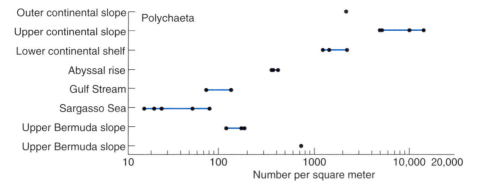

Fig. 16.12 The number of polychaetes found in different depth zones, along the Gay Head–Bermuda transect. (From Sanders et al., 1965.)

deep sea. The question raised is: Why is benthic biomass so reduced in the deep sea?

Input of Organic Matter

❋ **The input of organic matter to the deep-sea floor is low because of the distance from the coast and the great depth through which organic particles must travel.**

The supply of food to the seabed depends greatly on production in the water column. In the well-lit surface waters, primary production is great, and some of the production inevitably leaves the surface and reaches the bottom. In the deeper-water parts of the seabed, animals depend upon this deposition, because light there is too low for indigenous plant production. Near shore on the inner continental shelf, primary production is especially high, and bottom deposition is also great. There is also a peak of production over the continental shelf–slope break, where nutrient-rich waters intrude onto the shallow waters of the shelf. In the open ocean, seaward of the shelf–slope break, however, primary production is much reduced, which in turn reduces the supply of organic matter to the bottom. In the inner continental shelf, for example, 30–50 percent of the phytoplankton may reach the bottom. However only about 2–7 percent of the surface production reaches the bottom beneath the gyre center of the North Atlantic. Because the primary production in the open-ocean surface waters is less than that of the inner shelf, the supply of organic matter to the deep-sea bed is much less than the supply to bottoms of the inner continental shelf.

Along with production at the surface, the depth of the water column influences the amount and quality of the organic matter that reaches the bottom. The deep sea averages about 4,000 m in depth and organic particles may take weeks or even months to reach the bottom. During this period of descent, bacteria attack the particles, and only a very refractory material eventually reaches the bottom. This not universally true, however, because some organic material falls rather rapidly. There is evidence that benthic animals in the deep sea, especially in the North Atlantic where oceanic primary production is relatively high, respond to seasonal deposition of organic matter by consuming sediment more rapidly when the organic matter reaches the bottom. Some material, moreover, is consumed by midwater creatures, which, in turn transfer organic matter deeper by defecating relatively nutritious material. Such rapid transfer has been found over a wide variety of latitudes and therefore may be general for bottoms that lie beneath surface waters with strong seasonal pulses of primary production.[3]

In bays and estuaries, and in continental shelf environments, the fall of phytoplankton detritus is important in fueling benthic animal production. One therefore should expect that the reduced supply to the deep sea would strongly affect deep-sea benthic

[3] See Billet and others, 1983, and Graf, 1989, in Further Reading, The Deep Sea.

animal production as well. In the Antarctic, meiobenthic biomass declines tenfold from deep-sea bottoms with relatively high food input relative to bottoms with low food supply.

✳ The decline of input of organic matter to the deep-sea bottom is accompanied by strong changes in feeding types of benthic animals there.

The reduced organic input to the deep sea also affects the relative abundance of the major feeding types of benthic animal. Carnivores become less abundant as one samples from the shelf to the open deep-sea abyssal plain. This is probably due to the general reduction of animal biomass, which makes it difficult for the carnivores to find prey. Being at the top of the food web, hence relatively scarce numerically, may also create problems in finding mates. With increasing depth, suspension feeders become relatively less abundant, whereas deposit feeders become the dominant feeding type. The lack of phytoplankton makes the deep sea inhospitable for suspension feeders, but deposit feeders can still ingest the sediment and extract the organic matter.

✳ The deep-sea bed consists of sediment with little organic matter, and microbial activity is very low relative to upper-slope and shelf bottoms.

The input of organic matter from the overlying water column to the seabed declines steadily as we go seaward of the continental shelf–slope break. In inner continental shelf bottoms, the organic content of the sediment is in the range of 2–5 percent. In the open ocean, the seabed is very low in organic content, between 0.5 and 1.5 percent. The situation reaches an extreme in open-ocean abyssal bottoms beneath water columns of very low primary production. Here, the sediment is so low in organic matter (< 0.25 percent) that it is oxygenated and reddish. Such sediments are often called red clays. One may expect that such bottoms would be very low in microbial and animal biomass and activity.

The morphology and activity patterns of some deep-sea animals reflect the low food supply. Many deep-sea bottom fish are quite sluggish and have low muscle mass relative to animals of similar size in shallow water. Metabolic rates of deep-water fish are very low. Many abyssal fishes and fishes living in the deep water column seem adapted for the consumption of the rare meal that comes along. Unfortunately, we know far too little about the actual activity patterns of deep-sea fishes. About a decade ago, I. G. Priede and colleagues[4] put acoustic transmitters in baits placed on the seabed of the North Pacific and North Atlantic Oceans, at depths of 4,000–6,000 m. Grenadier fish consumed the baits and moved at speeds of approximately 0.11 m s^{-1}. The fish seemed to be quite active and probably fairly rapidly moved toward the bait, rather than sluggishly waiting until they smelled the rotting fishfall. Even though meals may be rare in the deep sea, animals may of necessity move briskly to take advantage of the events.

Many predatory fish species have the capacity to consume a victim that is much the same size as the predator. Midwater deep-sea fishes often have enormous mouths and stomach capacities. In the deep-sea benthos, one bivalve has been found to have an extraordinarily long gut, relative to its close relatives of the same size in shallow water. This may be a reflection of the very low food availability, as well as the need to digest as much as possible from the food as it passes through the gut.

The impoverished nature of the deep-sea bed was highlighted no better than by an event that occurred in 1968. A shipside accident caused the loss of the deep-sea submarine *Alvin* to the seafloor at a depth of 1,540 m. Luckily, no one was hurt, but some Woods Hole Oceanographic Institution scientists were unlucky enough to have left their lunches in the submarine that went over the side. In a remarkable salvage operation, the *Alvin* was recovered about a year later, and the food (thermos with bouillon, apple, and bologna sandwich) showed little decomposition. Dr. Holger Jannasch tasted the bread, and it was edible. The soup, initially prepared from canned meat extract, was not decomposed very much after 1 year. When such food was kept in a refrigerator at 3°C, bacterial attack was immediate and starch and protein fractions spoiled in just a few weeks. Therefore, the low temperature of the deep sea (2–4°C) was not the factor that retarded decomposition.

The impression of a low rate of decomposition in the deep-sea floor is only strengthened by other investigations. It is possible, for example, to measure the oxygen consumption of the seafloor by placing a

---

[4] See Priede and others, 1991, in Further Reading, The Deep Sea.

bell jar on the seabed and inserting a polarographic oxygen electrode in the water enclosed by the jar. Such measurements show that oxygen consumption of the abyssal seabed is 100-fold less than on continental shelf bottoms. The lowered oxygen consumption reflects the severely reduced animal and microbial activity in deep-sea sediments. It has been found also that bacterial substrates labeled with radioactive carbon are taken up by bacteria at a rate usually less than 2 percent of the rate measured in shallow waters.

The mechanisms accounting for the lowered rate of decomposition are not understood completely. The low temperature of the deep sea cannot explain the difference because, at the same low temperature, the shallow-water measurements of decomposition rate are much higher. The deep-sea bacteria may be adapted to the low food input by placing their metabolism in slow motion. There may also be a direct effect of the great pressures on the deep-sea bottom (the pressure at 4,000 m, for example, is about 400 surface atmospheres). Deep-sea bacteria are known to be barophilic—that is, they function best under high pressure—but still, their maximum rates of decomposition may be less than those of shallow-water bacteria.

Environmental Stability in the Deep Sea

* As we go from the shelf to the deep sea, we pass from physically variable environments to those that are stable.

The continental shelf is a physically unstable environment. In the midlatitudes, especially, there are strong seasonal changes in temperature, salinity, dissolved oxygen, and light. In North Carolina, for example, coastal water temperatures vary from winter minima near 3°C to 30°C in summer. At these latitudes, light also varies extensively, owing to changes in day length. At higher latitudes, temperature is a bit more constant through the year, though low. Day length, however, is even more variable; consider the midnight sun of the Arctic versus the sunless days of the Arctic winter. The only real exception to the general rule of shallow-water variation is in the tropics, where the water is rather warm and stable all year round. The same holds for day length. In the summer in midlatitude estuaries and inner continental shelves, oxygen may become a limiting factor because

much of the phytoplankton is consumed by bacteria, which also consume the dissolved oxygen.

On longer time scales, the shelves and estuaries of the world are even less stable than deeper water habitats. Because of the advance and retreat of the glaciers over the last 2 million years or so, large amounts of seawater alternately have been locked up in glacial ice or have been in the ocean. This difference is sufficient to cause sea-level fluctuations of 100–200 m. During the last glacial maximum, about 11,000 years ago, global sea level was more than 100 m lower than it is today, making the continental shelves far more narrow in geographic extent than they are today. Some of the major estuarine systems, such as Chesapeake Bay, San Francisco Bay, and the Baltic Sea, did not even exist.

The variability on the inner continental shelves and estuaries is often increased by the dominating effect of continental processes. Seasonal weather change in the continental interior is often extreme, and weather systems often strongly affect the near-shore ocean. In North America, weather systems move from west to east, so continental effects are stronger on the eastern coast.

Thus, seasonal temperature change is more extreme on the east coast of the United States than on the west coast. Off the coast of Oregon and California, there is very little difference in sea temperature between winter and summer; but in coastal waters of New York, water temperatures range from near freezing in winter to greater than 20°C in summer. Even on the west coast, however, the spring period of increased river runoff can decrease the salinity and increase both temperature and nutrient input to estuaries. These effects make west coast estuarine regions, such as Puget Sound and San Francisco Bay, far more variable than nearby open-coast waters.

As we move from the inner to the outer shelves, the variability in the physical environment is dampened considerably. Figure 16.13 shows seasonal variation in bottom temperatures at varying depths on the continental shelf and upper continental slope off the coast of New Jersey. Note how much more constant the 200 m bottom is, relative to the 30 m bottom. This trend continues as we move to deep-sea bottoms. At the base of the continental slope, toward the abyss of 4,000 m, the temperature and salinity variations are minuscule. Over the year, temperature in the abyss varies less than 1°C. The deep sea is thus a physically constant environment.

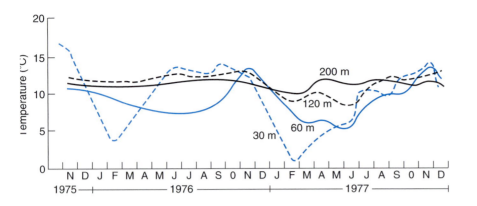

Fig. 16.13 Seasonal variation in bottom-water temperatures decreases with increasing depth. (Courtesy of D. F. Boesch.)

Deep-Sea Biodiversity

✳ Despite the poor food conditions, the deep sea in the vicinity of the continental slope and continental rise actually has more soft-bottom species than do corresponding environments on the inner continental shelf.

As we have seen, the deep sea is food poor. Remember that light in the deep sea is generated only by bioluminescence; there is not enough light to support photosynthesis and primary production. The remoteness from inner-shelf sources of organic matter, combined with the great depth of water through which organic matter passes and is decomposed, makes the deep sea an impoverished environment. Animal density is very low. Until the late 1950s and early 1960s it was generally believed that the deep sea also had very few species. This belief was based mainly on deep-sea dredge samples, which pulled up very few individuals and very few species. The impression was an illusion. Howard Sanders, Robert Hessler, and their colleagues[5] began a sampling program in the deep North Atlantic, but they used much higher-quality samplers than had been used previously. The large deep-sea sled they used could collect a great deal of sediment and was equipped with a closing device, which tripped just as the sampler was raised from the seafloor, to prevent washout. As a result, they collected far more small animals than had ever been collected before, with the exception of an earlier Soviet expedition. Previous samplers had no closing device, and most small animals were washed out of the sampler during its ascent through several thousand meters of water column.

The sampling showed that the deep sea, despite its low animal density, has large numbers of species relative to similar muddy-bottom habitats on the continental shelf.[6] Grassle and Maciolek[7] more recently reported that North Atlantic muddy bottoms at depths of 1,500–2,500 m had at least 1,500 species!

It is important to realize that estimates of deep-sea diversity are not straightforward. First of all, samples may have large differences in numbers of specimens. This makes it difficult to know the number of species that come from the locality from which the sample is extracted. We might expect that the more specimens we have, the more species we collect. This positive trend between specimen number and number of collected species would continue with increased sample size up to a plateau; eventually more individuals would reveal no more new species. But such a plateau has yet to be reached in benthic samples from the deep sea. Even with hundreds of samples, one still finds many species with only one individual, suggesting that there are many rarer species that have not yet been detected.

✳ Biodiversity of soft bottoms increases with increasing depth, but then declines as depths surpass about 2,000 m.

Before estimating species diversity from a site, deep-sea biologists use a large number of samples to construct a **rarefaction curve**, which relates the number of species collected to the number of specimens. Then a sample can be standardized to a standard number

[5] See Sanders and others, 1965, and Sanders and Hessler, 1969, in Further Reading, The Deep Sea.
[6] See Sanders and Hessler, 1969, in Further Reading, The Deep Sea.
[7] See Grassle and Maciolek, 1992, in Further Reading, The Deep Sea.

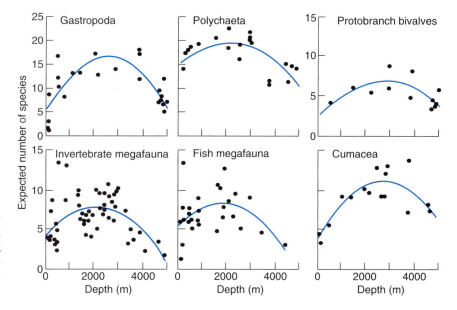

Fig. 16.14 Variation in species diversity of various benthic animal groups as a function of increasing water depth. Species number is an estimate for a sample of 50 individuals. (Data compiled by Rex, 1981.)

of specimens collected. This allows one to compare samples with thousands of specimens to others with just a few and get an estimate of species richness. It is important to remember that this assumes that there is such a continuous relationship between sample size and species number. Most studies now use the so-called **expected number of species**, which is the expected number of species to be found in a sample of n individuals that are selected at random from a larger collection containing N individuals, S total species, and a defined distribution of abundance among the species.[8]

Figure 16.14 shows trends in numbers of species as a function of water depth, with samples standardized to a sample of 50 specimens. For a wide variety of animal groups, diversity first increases to a maximum at a depth of about 2,000 m; at greater depths the diversity decreases.[9] The increased biodiversity at intermediate depths has been found in temperate and tropical latitudes, although areas with strong upwelling can depart from this pattern.

It is possible that in depths much greater than 2,000 m the scarcity of food has its effect on the biota and outweighs the effects of increased environmental stability. Population sizes may be so low that extinction is more likely, thus lowering diversity. This hypothesis is consistent with the especially strong decline in carnivores. In most communities, carnivores are less common in numbers and biomass because they depend upon other animals for their food, and

they cannot be completely efficient at converting their food to growth. As overall food declines, it would stand to reason that carnivores would be the first to disappear. Carnivorous gastropods decline more rapidly than deposit-feeding snails, as one samples with increasing seaward distance from the shelf–slope break. By contrast, the more abundant microscopic Foraminifera increase in diversity over the same gradient. This may be explained by their greater population sizes, which lower the probability of extinction. Ron Etter and Fred Grassle[10] discovered that high-diversity sites at intermediate depths also have greater sedimentary particle diversity. The particle diversity may reflect a diversity of ingestible particles or even a diversity of hydrodynamic regimes, both of which may support a greater diversity of ecologically different species. The only problem with this hypothesis is that the same diversity gradient occurs in many groups that are not deposit feeders and are not particularly dependent upon sediment grain size. Perhaps there is some form of trophic cascade effect on these other groups.

✳ Despite the apparent similarities in sediment properties, there is a prominent latitudinal diversity gradient in deep-sea benthos.

[8] The interested student should consult Hurlbert, 1971, in Further Reading, The Deep Sea.
[9] See Rex, 1981, in Further Reading, The Deep Sea.
[10] See Etter and Grassle, 1992, in Further Reading, The Deep Sea.

There is no obvious difference in sediment properties among latitudes. Indeed, the differences are much more pronounced with changes of depth, since both coarse and fine sediments and mixtures are found on the continental shelf. Given the latitudinal similarities in sediment and temperature, one might expect that there would be no latitudinal gradient in species diversity, but this is not so. Michael Rex and colleagues[11] compiled data for deep-sea bivalves, gastropods, and isopods and found clear latitudinal diversity gradients, especially in the North Atlantic (Figure 16.15). South Atlantic species diversity declined with increasing latitude but the pattern was much less clear.

These results suggest that there must be something more global driving deep-sea diversity, because the latitudinal pattern parallels a shallow-water latitudinal gradient (see Chapter 17). We have discussed earlier that the deep sea is not disconnected from the surface, as we once thought. Seasonal primary production at the surface is often exported to the deep-sea bed in rapid pulses. It may therefore be that patterns of longer-term change at the surface affect the deep sea. This does not answer the important questions of how and why, however. We discuss this problem further in Chapter 17.

✳ Deep-sea biodiversity may have been promoted by a stable environment, although disturbance and microhabitat heterogeneity also may have promoted diversity.

The high biodiversity of the deep sea was a great surprise to ecologists. Despite the absence of light, very low food supplies, and very low microbial activity, fantastic numbers of species have been recovered. At first, one tends to conceive of the deep sea as a monotonous muddy bottom, which causes us to ask: How can so many species coexist in what appears to be a continuous environment with no variation? One might think that the reduced food would cause strong interspecific competition, resulting in competitive extinctions and low diversity. The deep sea is not homogeneous, however. We have discussed the strong effects of currents, which disturb the sediment and reduce populations, much in the same way that disturbance affects rocky shores. This might promote the coexistence of species to some degree. But there is no reason to believe that such disturbance is any more prevalent than on the continental shelves, where biodiversity is less. Muddy-seabed animals also produce tubes, burrows, and mounds. This would surely increase local environmental heterogeneity and promote an increase of microhabitats that would allow more species to coexist. But, again, one cannot see why bioturbation, and bioturbation-promoted sediment structures, would be less on the shelf.

To understand the high biodiversity of the deep sea, we have to think about the process of speciation. Deep-sea species mostly have reduced dispersal mechanisms relative to their closest shelf relatives. This

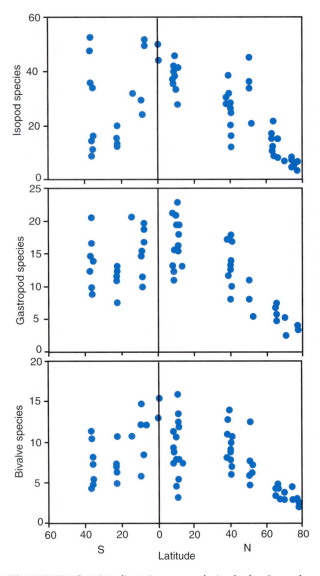

Fig. 16.15 Species diversity versus latitude for Isopoda, Gastropoda, and Bivalvia in the Atlantic. (Data from Rex et al., 1997.)

[11] See Rex and others, 1993, in Further Reading, The Deep Sea.

might help in causing isolation, which is necessary for speciation. Sanders and colleagues explained the high diversity in terms of the constancy of the deep sea. Because the deep sea was environmentally stable and perhaps an older habitat, species may have lower extinction rates and may accumulate to higher diversities than on shelf habitats. On the shelf, the habitat is less stable, and large-scale fluctuations in sea level may have contributed to increased extinction rates and lower speciation rates. If we go back just 8,000–10,000 years, the world was much more covered by continental glaciers, sea level was lower by about 150 m, and most of the continental shelves of the world were exposed to air. Such disturbance over millions of years may have caused extinctions on the shelves, but the deep sea of course has never been subjected to such massive disturbances. The solution to the high diversity of the deep sea still eludes us because its explanation must be found in historical factors that are poorly understood. We know that in the last few million years there have been significant extinctions of the shelf macrofauna, but we do not have a good record of deep-sea macrofauna on the same time scales.

Hot-Vent Environments

* **Hot vents, located near areas of submarine volcanic activity, spew out hot, sulfide-rich water and support unique biotic assemblages.**

The deep-sea bed gives the impression of monotony and poor productivity. Most deep-sea bottoms are low in production when contrasted to the upper continental slope and the continental shelf. In the 1970s, the submarine *Alvin* descended to a midoceanic ridge to examine the generally steep and rocky bottom environments. The expedition, led by Robert Ballard, came upon a series of fissures from which spewed hot water of partial volcanic origin.

The midoceanic ridges are volcanically active, and the molten rock heats seawater in crack systems, often to a superheated state of several hundred degrees. The underwater landscape often consisted of a series of "smoking chimneys" emanating hot water that is loaded with high concentrations of various metals and dissolved sulfide. However, these discoveries paled in comparison with the discovery of a fantastic assemblage of animals that were often associated with these hot vents, although in not-so-hot waters (the animals live in temperatures ranging from cold ambient to over 16°C). In contrast to adjacent cold-

Fig. 16.16 Dense stand of a vestimentiferan tube worm at the Galapagos Rift. (Photograph by Kathleen Crane, courtesy of Woods Hole Oceanographic Institution.)

water deep bottoms, the rocky surfaces surrounding the hot vents are covered with animals, including large limpets, clams, and mussels. The most curious and spectacular organism is a large type of tube worm, *Riftia pachyptila* (Figure 16.16, Plates XXXII.1 and XXXII.2). The worm is often 1 m in length and secretes tubes up to 3 m long, which are rich in chitin and protein. Several species of tube worms have now been discovered and classified in the phylum Vestimentifera, which has affinities to both the Pogonophora and the Annelida. The growth rate of these worms is very rapid. Sequential dives on a newly established volcanic terrain demonstrated[12] that the worms could grow tubes over 1.5 m long in just 18 months!

The community presents a paradox. How can such a productive biota occur in the midst of general poverty? The supply from surface waters is clearly insufficient to maintain these rich animal communities. The dissolved sulfide emanating from the vents seems to be the answer. Certain bacteria derive energy from the oxidation of sulfide. As the sulfide is released from the vent, bacterial populations grow and provide a food source for animals such as bivalve mollusks. Some animals may feed directly on sulfur bacteria from the water column, or they may graze on bacteria on surfaces. In the vicinity of the vents, where sulfide-rich water is spewed out, bacteria that derive energy from the oxidation of sulfide live as mats on rocky surfaces and on the surface of soft sediments. These bacteria are probably grazed

---

[12] See Lutz and others, 1994, in Further Reading, The Deep Sea.

Hot Topics in Marine Biology 16.1 Bacterial–Animal Symbiosis in the Sea

We usually think of life as a struggle for existence. Organisms compete for resources, avoid predators and parasites, and generally are adversaries with their surrounding biological world. Symbiosis, although admired and fairly commonplace, is usually thought of as the minority option for organisms. However, in recent years, marine biologists have come to find mutualisms between microorganisms and animals to be a commonplace and perhaps dominant theme on the soft seafloor.

Oddly enough, the discovery of the bizarre vestimentiferan worms near deep-sea hot vents has been the gateway to further discoveries. Often as much as a meter in length, the worms live near the hot vents, which spew out sulfide, and near so-called cold seeps, which spew out methane. The animals have a spongy tissue, filled with cavities that are packed with bacteria. In the case of the hot vents, the bacteria use sulfide as an energy source, and the animals have a special hemoglobin, which is adapted to ensure safe transport of the otherwise poisonous sulfide through the hosts' bodies en route to the bacteria.[13] Two hemoglobin sites with the amino acid cysteine are used for binding of sulfide onto the hemoglobin.[14] While not as well characterized, a group of polychaetes, living in this environment, the Alvinellids, also apparently have a metal-binding protein, which probably is also a hemoglobin.[15] The worms probably digest the bacteria; they have no mouth or digestive tract to deal with food from the outside world. However, dissolved nitrogen and carbon may be transferred from the bacteria to the worm. A truly remarkable interdependence has evolved. Presumably the bacteria gain a new habitat, safe from grazers. The worms do not have very high concentrations of sulfide, which may reflect a metabolic purifying process. It is also possible that the high concentrations of sulfide near the vents consists mostly of solid-phase sulfides, which are not taken up by the worms.[16]

This discovery was made in one of the most inaccessible environments on earth, and certainly one of the most newly discovered. However, many other benthic animals seem to have symbioses with bacteria. In shallow subtidal and intertidal muddy sediments, bivalves of the genus *Solemya* live in Y-shaped burrows. Most are small, less than 2 cm long, and several species have no gut at all. As it turns out, these species also have symbiotic bacteria. Bacteria have been found living on the gills of a wide variety of bivalve mollusks, including species living adjacent to deep-sea hot vents. Some of the bacteria use sulfide and others use methane, which emanates from cracks in the seafloor, perhaps from hydrocarbon pools below. Methane is also a product of decomposition of organic matter in sediments, and certain bacteria can use the methane as a source of energy to manufacture sugars. The possible mechanisms of food collection of the bacteria are not well understood, but more study will probably show that the kinds of symbiosis just described are widespread in the ocean.

The significance of bacteria and sulfide does not stop there. Bacteria near the hot vents gain energy by exploiting strong chemical gradients. Be it via methane or sulfide, a variety of different bacterial species manage to survive in lightless worlds, often at very high temperatures. These organisms have reopened some of the most fundamental aspects of the origin of life. Molecular sequences have been used to construct a "Universal Tree of Life," and the bacteria deriving energy at the hot vents and cold seeps have been found to be among the most primitive organisms known, the Archaebacteria. This has changed our conception of the possible locations where life may have arose. Instead of the usual conception of solar or lightning energy being the catalyst for the origin of life, it is now possible that life began in the chemical furnaces of the ocean ridges.

[13] See Childress and others, 1987, in Further Reading, The Deep Sea.
[14] See Zal and others, 1998, in Further Reading, The Deep Sea.
[15] See Martineu and others, 1997, in Further Reading, The Deep Sea.
[16] See Truchet and others, 1998, in Further Reading, The Deep Sea.

by a variety of consumers, including crabs, limpets, and bivalves. Sulfide-dependent bacteria also are found in "bacterial snow" that has been observed in the water column coming from vent waters. This material is also a source of nutrition for consumers.

However, many (and perhaps most) vent animals do not seem to feed directly from the external medium. The Vestimentifera, for example, have no obvious gut. Instead, they have a specialized organ, the trophosome, which has dense concentrations of chambers that contain sulfide-oxidizing bacteria. The animal, in turn, can digest the bacteria and derive nutrition. *Riftia* has a specialized hemoglobin, which binds to both sulfide and oxygen. The sulfide is transported to the trophosome, where it can be used by the symbiotic bacteria. The bivalve mollusks *Bathymodiolus thermophilus* and *Calyptogena magnifica* (Figure 16.17, Plate XXXII.2) have sulfur-oxidizing bacteria as symbionts in the gills. The sulfur source is not unique to the deep-sea vent animals. Shallow-water animals in sulfide-rich sediments also appear to be festooned with sulfide-oxidizing bacteria and also derive nutrition from them. There is some evidence that all the bacteria in both deep-sea and shallow-water bivalves belong to the same group of purple bacteria.

The vent faunas are also fascinating because they are clearly ephemeral, yet widespread throughout the ocean. Although there are local differences, the vent biota can be found near hot vents throughout the Pacific and Atlantic. Some "dead" vents have been encountered, with piles of shells nearby. Hot vents may last no more than hundreds of years or a few thousand years at most. The question then arises of how vent animals disperse from one vent system to the next. Although there is not yet an accurate map of the distribution of the vents, they are probably often hundreds and even thousands of kilometers apart. Despite the distance, genetic differentiation among marine invertebrates is very slight along thousands of kilometers of the same ridge system. There must be extensive dispersal to homogenize the distant populations. The degree of genetic differentiation of populations of the same species between separated ridge systems is quite high, however, indicating that connections between populations occur only along the axes of ridges.

Most of the vent species probably have lecithotrophic larvae, which in shallow waters would not be expected to disperse very far. Deep-sea biologist Richard Lutz has suggested that the coldness of the deep sea may greatly slow down larval development rate, increase larval life, and therefore allow more time for dispersal. Larvae may be entrained in the plumes of water that rise from hot vents, and these plumes may disperse laterally for distances of kilometers. As was discussed in Chapter 5, some of the vent animals probably have planktotrophic larvae, a circumstance that raises the enigma of how a larva can find the surface and then find a highly localized vent biota of only a few hundred square meters at most. Some have suggested that certain vent organisms may be able to disperse to nonvent habitats, such as fishfalls or whale carcasses. At present, knowledge of these habitats is sketchy. A group of organisms similar to the hot-vent biota has been found near a cold-water seep on the continental slope off southern Florida. Here, it is likely that methane emanating from cracks in the bottom is used by methanogenic bacteria as a carbon source. The larger animals are not of the same species as the hot-vent groups but are clearly related to them, and probably depend upon the bacteria for food.

There are other habitats that violate the general picture of a deep sea in slow motion. As we discussed earlier, experimental introductions of barrels of rotting fish are usually immediately followed by recruitment of various scavengers, such as amphipods and fishes, to the rotting fish. Ruth Turner[17] has found rapid colonization and rapid growth by boring bivalves on pieces of wood she placed in deep-sea habitats. The bivalves derived their carbon from the wood and obtained nitrogen from symbiotic nitrogen-fixing bacteria. A few bottoms have been found that are rich in hydrocarbons, and these are dominated by rich faunas of invertebrates, including an abundant type of polychaete. Thus depth or pressure per se cannot explain slow rates of bacterial and animal activity in more typical deep-sea bottoms.

Fig. 16.17 The hot-vent bivalve *Calyptogena magnifica*. (Courtesy of Richard Lutz.)

[17] See Turner and Lutz, 1984, in Further Reading, The Deep Sea.

Pressure Change

✳ Pressure increases of approximately 1 atmosphere in every 10 m depth strongly affect the biochemistry of deep-sea organisms.

The deep sea is beyond the reach of sunlight and, as has been discussed, is poor in organic matter input. However, the great pressure of the water column also exerts strong effects on the functioning of deep-sea organisms. The average depth of the ocean is about 4,000 m and the pressure is 400 atmospheres. Some species live over depth ranges of greater than 1,000 m, and therefore members of the same species can experience pressure differences of as much as 100 atmospheres. What is the effect of such differences?

Aside from living at high pressure, most deep-sea organisms live at low temperature, ranging from 1 to 4°C. Temperature has a strong effect on enzymes, which catalyze all important reactions in cells. An enzyme binds to a substrate and other molecules and catalyzes a reaction in which the substrate is converted to a product. The efficiency of the enzyme depends on its ability to bind and then detach rapidly. At low temperature, this process slows greatly, and the time of binding to substrate and other molecules increases. This decreases enzyme function.

Pressure affects the volume of biologically active molecules such as enzymes. The simplest effect is in changing the volume of the voids in enzymes, in which substrates and enzyme cofactors attach briefly during the reaction process. This effect tends to decrease the efficiency of catalysis. In deep-water fishes, enzymes that are less sensitive to pressure change have evolved. Increased pressure tends to harm the functioning of enzymes derived from shallow-water fish, whereas enzymes taken from deep-water fish manage to function at much higher pressures. Enzymes of deep-sea fishes are structurally far more stable than those of shallow-water species. This may be an adaptation to prevent high rates of protein turnover, which would be maladaptive in the food-poor deep-sea environment. These properties may also apply to proteins of deep-sea bacteria, that, as discussed earlier, seem to metabolize very slowly as well. Pressure tends also to compress membranes, and cell membranes taken from deep-water fish species are designed to remain fluid at pressures much higher than membranes from shallow-water related species.

Pressure does not seem to cause the low metabolic rates observed in deep-sea species. This seems to be more closely related to a set of adaptations to reduce activity, since food is so rare and often of such poor quality in the deep sea. Muscle-related enzymes, such as lactate dehydrogenase, are much lower in activity in deep-sea fish species relative to shallow-water relatives. Muscle-related metabolic power is very low, much lower than one would predict from the low temperature of the deep sea. The lowered metabolism is therefore an adaptation to low food conditions, not a response to pressure, temperature, or other physical conditions.

Further Reading

SEDIMENT TYPE AND BENTHIC DISTRIBUTIONS

Aller, J. 1998. Benthic community response to temporal and spatial gradients in physical disturbance within a deep-sea western boundary region. *Deep-Sea Research, Part I, Oceanographic Research Papers*, v. 44, pp. 39 ff.

Peddicord, R. K. 1977. Salinity and substratum effects on condition index of the bivalve *Rangia cuneata*. *Marine Biology*, v. 39, pp. 343–359.

Rhoads, D. C., and D. K. Young. 1970. The influence of deposit-feeding organisms on sediment stability and community trophic structure. *Journal of Marine Research*, v. 28, pp. 150–178.

Rhoads, D. C., and D. K. Young. 1971. Animal–sediment relationships in Cape Cod Bay. II. Reworking by *Mol-*

padia oolitica (Holothuroidea). *Marine Biology*, v. 11, pp. 255–261.

Rhoads, D. C., P. L. McCall, and J. Y. Yingst. 1978. Disturbance and production on the estuarine seafloor. *American Scientist*, v. 66, pp. 577–586.

Sanders, H. L. 1958. Benthic studies in Buzzards Bay. I. Animal–sediment relationships. *Limnology and Oceanography*, v. 3, pp. 245–258.

Thistle, D. and L. A. Levin. 1998. The effect of experimentally increased near-bottom flow on metazoan meiofauna at a deep-sea site, with comparison data on macrofauna. *Deep-Sea Research, Part I, Oceanographic Research Papers*, v. 45, pp. 625ff.

Young, D. K. 1971. Effects of infauna on the sediment and seston of a subtidal environment. *Vie Milieu*, v. 22 (suppl.), pp. 557–571.

THE DEEP SEA

Arp, A. J., and J. J. Childress. 1983. Sulfide binding by the blood of the hydrothermal vent tube worm *Riftia pachyptila*. *Science*, v. 219, pp. 295–297.

Billett, D. S. M., R. S. Lampitt, A. L. Rice, and R. F. C. Mantoura. 1983. Seasonal sedimentation of phytoplankton to the deep-sea benthos. *Nature*, v. 302, pp. 520–522.

Cavanaugh, C. M. 1985. Symbiosis of chemolithotrophic bacteria and marine invertebrates from hydrothermal vents and reducing sediments. *Bulletin of the Biological Society of Washington*, v. 6, pp. 373–388.

Childress, J. J., ed. 1988. Hydrothermal vents, a case study of the biology and chemistry of a deep-sea hydrothermal vent of the Galapagos rift. *Deep-Sea Research*, v. 35, pp. 1677–1849 (a series of papers on a vent system).

Childress, J. J., H. Felbeck, and G. N. Somero. 1987. Symbiosis in the deep sea. *Scientific American*, v. 256, pp. 106–112.

Cosson-Sarradin, N., M. Sibuet, G. L. J. Paterson, and A. Vangriesheim. 1998. Polychaete diversity at tropical Atlantic deep-sea sites: Environmental effects. *Marine Ecology—Progress Series*, v. 165, pp. 173–185.

Coull, B. C., and others. 1977. Quantitative estimates of the meiofauna from the deep sea off North Carolina USA. *Marine Biology*, v. 39, pp. 233–240.

Cronin, T. M. and M. E. Raymo. 1997. Orbital forcing of deep-sea benthic species diversity. *Nature*, v. 385, pp. 624–627.

Dayton, P. K., and R. R. Hessler. 1972. Role of biological disturbance in maintaining diversity in the deep sea. *Deep-Sea Research*, v. 19, pp. 199–208.

Dayton, P. K., and J. S. Oliver. 1977. Antarctic soft-bottom benthos in oligotrophic and eutrophic environments. *Science*, v. 197, 55–58.

Etter, R. J., and J. F. Grassle. 1992. Patterns of species diversity in the deep sea as a function of sediment particle size diversity. *Nature*, v. 360, pp. 576–578.

Felbeck H., and J. Jarchow. 1998. Carbon release from purified chemoautotrophic bacterial symbionts of the hydrothermal vent tubeworm *Riftia pachytpila*. *Physiological Zoology*, v. 71, pp. 294–302.

France, S. C., R. R. Hessler, and R. C. Vrijenhoek. 1992. Genetic differentiation between spatially-disjunct populations of the deep-sea, hydrothermal vent–endemic amphipod *Ventiella sulfuris*. *Marine Biology*, v. 114, pp. 551–559.

Gage, J. D. 1996. Why are there so many species in deep-sea sediments? *Journal of Experimental Marine Biology & Ecology*, v. 200, pp. 257–286.

Gage, J. D., and P. A. Tyler, 1991. *Deep-Sea Biology: A Natural History of Organisms at the Deep-Sea Floor.* Cambridge: Cambridge University Press.

Graf, G. 1989. Benthic–pelagic coupling in a deep-sea benthic community. *Nature*, v. 341, pp. 437–439.

Grassle, J. F. 1987. The ecology of deep-sea hydrothermal vent communities. *Advances in Marine Biology*, v. 23, pp. 301–362.

Grassle, J. F. 1989. Species diversity in deep sea communities. *Trends in Ecology and Evolution*, v. 4, pp. 12–15.

Grassle, J. F., and N. J. Maciolek. 1992. Deep-sea species richness regional and local estimates from quantitative bottom samples. *American Naturalist*, v. 139, pp. 313–341.

Grassle, J. F., and L. S. Morse-Porteus. 1987. Macrofaunal colonization of disturbed deep-sea environments and the structure of deep-sea benthic communities. *Deep-Sea Research*, v. 34, pp. 1911–1950.

Grassle, J. F., and H. L. Sanders. 1973. Life histories and the role of disturbance. *Deep-Sea Research*, v. 20, pp. 643–659.

Hecker, B. 1985. Fauna from a cold sulfur-seep in the Gulf of Mexico: Comparison with hydrothermal vent communities and evolutionary implications. *Bulletin of the Biological Society of Washington*, v. 6, pp. 465–473.

Hessler, R. R., and P. A. Jumars. 1974. Abyssal community analysis from replicate box cores in the central north Pacific. *Deep-Sea Research*, v. 21, pp. 185–209.

Hessler, R. R., C. L. Ingram, A. A. Yayanos, and B. R. Burnett. 1978. Scavenging amphipods from the floor of the Philippine Trench. *Deep-Sea Research*, v. 25, pp. 1029–1047.

Hurlbert, S. H. 1971. The nonconcept of species diversity: A critique and alternative parameters. *Ecology*, v. 52, pp. 577–586.

Isaacs, L. D., and R. A. Schwartzlose. 1975. Active animals of the deep-sea floor. *Scientific American*, v. 233, pp. 85–91.

Jannasch, H. W. 1987. Effects of hydrostatic pressure on growth of marine bacteria. In H. W. Jannasch, R. E. Marquis, and A. M. Zimmerman, eds., *Current Perspectives in High Pressure Biology*. London: Academic Press, pp. 1–14.

Jannasch, H. W. 1989. Serendipity in deep-sea microbiology: Lessons from the *Alvin* lunch. *Oceanus*, v. 31, pp. 28–33.

Jannasch, H. W., K. Eimhjellen, C. Wirsen, and A. Farmanfarmaian. 1971. Microbial degradation of organic matter in the deep sea. *Science*, v. 171, pp. 672–675.

Karl, D. M. 1987. Bacterial production at deep-sea hydrothermal vents and cold seeps: Evidence for chemosynthetic primary production. In M. Fletcher, T. R. Gary, and J. G. Jones, eds., *Symposium of the Society of General Microbiology*, v. 41. Cambridge: Cambridge University Press, pp. 319–359.

Karl, D. M., C. O. Wirsen, and H. W. Jannasch. 1980. Deep-sea primary production at the Galapagos hydrothermal vents. *Science*, v. 207, pp. 1345–1347.

Lutz, R. A., and R. R. Hessler. 1983. Life without sunlight—Biological communities of deep-sea hydrothermal vents. *Science Teacher*, v. 50, pp. 22–29.

Lutz, R. A., D. Jablonski, and R. D. Turner. 1984. Larval development and dispersal at deep-sea hydrothermal vents. *Science*, v. 226, pp. 1451–1454.

Lutz, R. A., L. W. Fritz, and R. M. Cerrato. 1988. A comparison of bivalve (*Calyptogena magnifica*) growth at two deep-sea hydrothermal vents in the eastern Pacific. *Deep-Sea Research*, v. 35, pp. 1793–1810.

Lutz, R. A., T. M. Shank, D. J. Fornari, R. M. Haymon, M. D. Lilley, K. L. Von Damm, and D. Desbruyeres. 1994. Rapid growth at deep-sea vents. *Nature*, v. 371, pp. 663–664.

McCave, I. N., ed. 1976. *The Benthic Boundary Layer*. New York: Plenum.

Martineu, P., S. K. Juniper, C. R. Fisher, and G. J. Massoth. 1997. Sulfide binding in the body fluids of hydrothermal vent Alvinellid polychaetes. *Physiological Zoology*, v. 70, pp. 578–588.

Menzies, R. J., R. Y. George, and G. T. Rowe. 1973. *Abyssal Environment and Ecology of the World Oceans*. New York: Wiley-Interscience.

Mullineaux, L. S., P. H. Wiebe, and E. T. Baker. 1991. Hydrothermal vent plumes: Larval highways in the deep sea? *Oceanus*, v. 34, pp. 64–68.

Priede, I. G., P. M. Bagley, J. D. Armstrong, K. L. Smith Jr., and N. R. Merrett. 1991. Direct measurement of active dispersal of food-falls by deep-sea demersal fishes. *Nature*, v. 351, pp. 647–649.

Rex, M. A. 1981. Community structure in the deep-sea benthos. *Annual Review of Ecology and Systematics*, v. 12, pp. 331–353.

Rex, M. A., C. T. Stuart, R. R. Hessler, J. A. Allen, H. L. Sanders, and G. D. F. Wilson. 1993. Global-scale latitudinal patterns of species diversity in the deep-sea benthos. *Nature*, v. 365, pp. 636–639.

Rex, M. A., R. J. Etter, and C. T. Stuart. 1997. Large-scale patterns of biodiversity in the deep-sea benthos. In R. F. G. Ormond, J. D. Gage, and M. V. Angel, eds., *Marine Biodiversity: Patterns and Processes*. Cambridge: Cambridge University Press, pp. 94–121.

Rhoads, D. C., and D. K. Young. 1970. The influence of deposit-feeding organisms on sediment stability and community trophic structure. *Journal of Marine Research*, v. 28, pp. 150–178.

Rowe, G. T., ed. 1983. *The Sea, v. 8: Deep Sea Biology*. New York: Wiley-Interscience.

Sanders, H. L., and R. R. Hessler. 1969. Ecology of the deep sea benthos. *Science*, v. 163, pp. 1419–1424.

Sanders, H. L., R. R. Hessler, and G. R. Hampson. 1965. An introduction to the study of deep-sea benthic faunal assemblages along the Gay Head–Bermuda transect. *Deep-Sea Research*, v. 12, pp. 845–867.

Siebenaller, J. F. 1987. Pressure adaptation in deep-sea animals. In H. W. Jannasch, R. E. Marquis, and A. M. Zimmerman, eds., *Current Perspectives in High Pressure Biology*. London: Academic Press, pp. 33–48.

Smith, C. R. 1985. Food for the deep-sea: Utilization, dispersal and flux of nekton falls at the Santa Catalina Basin floor. *Deep-Sea Research*, v. 32, pp. 417–422.

Truchet, M., C. Ballandufrancais, A. Y. Jeantet, J. P. Lechaire, and R. Cosson. 1998. The trophosome of the Vestimentifera *Riftia pachyptila* and *Tevnia jerichoana*—Metal bioaccumulations and sulfur metabolism. *Cahiers de Biologie Marine*, v. 39, pp. 129–141.

Tunnicliffe, V. 1991. The biology of hydrothermal vents: Ecology and evolution. *Oceanography and Marine Biology Annual Reviews*, v. 29, pp. 319–407.

Turner, R. D., and R. A. Lutz. 1984. Growth and distribution of molluscs at deep-sea vents and seeps. *Oceanus*, v. 27, pp. 54–62.

Tyler, P. A. 1988. Seasonality in the deep-sea. *Oceanography and Marine Biology Annual Reviews*, v. 26, pp. 227–258.

Van Dover, C. L. 2000. *The Ecology of Deep-Sea Hydrothermal Vents*. Princeton, NJ: Princeton University Press.

Wilson, R. R., and K. L. Smith. 1984. Effect of near-bottom currents on the detection of bait by the abyssal grenadier fishes *Coryphaenoides* spp. *Marine Biology*, v. 84, pp. 83–91.

Zal F., E. Leize, F. H. Lallier, A. Toulmond, A. Vandorsselaer, and J. J. Childress. 1998. S-Sulfohemoglobin and disulfide exchange—the mechanisms of sulfide binding by *Riftia pachytptila* hemoglobins. *Proceedings of the National Academy of Sciences, USA*, v. 95, pp. 8997–9002.

Review Questions

1. What general sediment type is dominated by deposit feeders? By suspension feeders?

2. How do deposit feeders contribute to making their benthic environment relatively inhospitable to suspension feeders?

3. Name three mechanisms that might create patchiness of benthic organisms on the seafloor.

4. What is an advantage of using remotely operated submersibles, relative to more standard bottom samplers such as dredges?

5. What feeding type of benthic organism dominates the deep sea? Why?

6. How does benthic biomass change as one goes from the continental shelf to the deep-sea floor of the open sea?

7. What evidence is there for a low rate of organic matter decomposition in the deep sea?

8. Describe the change in variability of the physical environment as one passes from shallow shelf habitats to those of the abyssal deep-sea floor?

9. Why are hot vents places of rather high benthic bio-mass and secondary production, despite their location in the deep sea?

10. How does benthic diversity change over a gradient from the inner continental shelf to the abyssal deep-sea floor?

11. Why do hot vents present a paradox in terms of some species with planktotrophic larval dispersal?

12. Why is the activity of deep-sea fishes so low?

17

Biodiversity and Conservation of the Ocean

In this chapter, we shall consider patterns of biological diversity from the broader perspectives of evolution, extinction, and biogeography. The number of species in a region is controlled by short-term ecological processes, but in the long term the relative rates of speciation and extinction explain the number of species. We therefore need to consider the causes of speciation and extinction. Regional variation in geography and climate influences the origin of species, and also the degree of structuring of the world into subdivided biotas, or provinces. We must therefore also consider the conditions that enhance the presence of such geographic or climatic barriers to dispersal among provinces. Finally, there has been an ever-increasing recognition of the importance of conserving the earth's biodiversity,[1] and we will discuss in this chapter some of the important factors in understanding biodiversity conservation issues.

Patterns of Species Diversity

The Speciation–Extinction Balance

✶ Although local patterns of species diversity are often explained in terms of short-term dynamic interactions, regional patterns are probably as much explained by the balance of speciation and extinction.

In Chapter 3 we saw how a series of factors interacted to control the number of species in a community. In a simple case, according to the intermediate predation hypothesis, diversity is the result of an interaction between short-term effects of competition and predation. Although this hypothesis may explain events in the short haul, we must take into account the processes of species origins and extinctions to explain variations in diversity on larger geographic and temporal scales. A predator may depress interspecific competitive exclusion and increase local diversity, but this process does not create species. Similarly, intense predation rarely drives a species to extinction over its entire geographic range. Rather, extinction usually follows larger-scale regional changes, such as a deterioration of climate. Thus, an understanding of the factors that regulate the number of species over the long term leads to the need for an understanding of processes controlling speciation and the regional changes that cause extinction.

✶ Speciation requires some degree of isolation of populations, which results eventually in reproductive incompatibility between them.

It is generally believed that **geographic isolation** promotes speciation. This idea is known as the **allopatric**

[1] See Further Reading, Conserving Biodiversity.

410

model of speciation. Isolation permits the separated populations to evolve in different directions and achieve incompatibility. It is possible that two populations might differentiate from each other even with some contact. The **parapatric model** of speciation emphasizes the possibility of the origin of differentiation despite contact, as long as there are different natural selection pressures in the semi-isolated populations, allowing divergence even in the face of some gene flow from other populations. In effect, according to this model, natural selection is sufficiently strong to balance the influx of genes from another population, and reproductive incompatibility eventually develops between the two semi-isolated populations.

Imagine a species divided by the rising of the Isthmus of Panama above sea level a few million years ago. Populations of the single species on either side of the isthmus became isolated at that time, and now a large number of recently derived pairs of species exist on either side of the land barrier. Natural selection, or evolutionary change involving a shift in genetic variants, may have been promoted by differing environmental conditions on either side of the isthmus. Such differences in selection on either side of the isthmus would enhance the differences among the species and might even accelerate the degree of reproductive isolation, owing to genetic incompatibility. Random changes in populations might also eventually result in increased genetic differentiation between the two populations. Nancy Knowlton and colleagues[2] have demonstrated that snapping shrimp species pairs on either side of the isthmus have concomitantly evolved differences in mating preference and overall genetic difference, as measured by genetically controlled enzyme variation.

The sum total of such differentiation tends to increase the total number of species in the world. If the geographic barriers then break down and shift rapidly, formerly isolated species will be combined, and the number of species in a given region will increase, unless interspecific competition results in the loss of some species.

✳ Extinction may be caused by habitat change or destruction, widespread diseases, biological interactions, or random fluctuations of population size.

Extinction, or the loss of species, may be caused by shifts in the environment, such as sudden changes of temperature. Destruction of major habitats may also cause extinction. For example, most coastal lagoons are geologically unstable habitats and are destroyed after the passage of several thousand years. Any species restricted to the lagoons would be in similar danger of extinction unless they had sufficient geographic range over a large number of lagoons, leading to a high probability that some would survive in the long run. Another such example of habitat destruction would be marginal seas, such as the Mediterranean, that have experienced massive anoxia, nearly complete evaporation, and other effects so major as occasionally to make them nonmarine habitats.

Biological interactions may also be the cause of sudden extinctions. In Chapter 3, we discussed the rapid spread of disease in the ocean and the potential of disease to destroy even a widespread dominant species. The immigration of competitors and predators may have a similar effect. The introduction by human beings of many species has resulted in major shifts of habitat use by formerly dominant species, although no extinctions of marine species seem to have been induced. Dispersal into new areas can result in rapid expansions. For example, when the common Atlantic mud snail *Ilyanassa obsoleta* was introduced into San Francisco Bay, the local dominant was subsequently restricted to high intertidal sites. After the arrival of the periwinkle *Littorina littorea* from Europe to North America in the late 1800s, the snail spread south and reduced the range of habitats used by the snail *I. obsoleta*.

The total number of species in a region, therefore, is the result of a net balance between the rate of species production and the rate of extinction. In unstable areas, where fluctuation between environmental extremes is the norm, newly isolated species are liable to become extinct, or are never established at all, owing to rapid shifts of climate and habitats. In such areas, the rate of extinction is likely to be high as well. Thus the speciation–extinction balance leans toward low standing diversity. In more stable habitats, species might have a higher probability of survival when newly formed, thus increasing diversity.

[2] See Knowlton and others, 1993, in Further Reading, Patterns of Species Diversity.

Biogeographic Factors

* Geographic isolation and major geographic gradients in temperature and salinity combine to determine provinces of statistically distinct groupings of species.

The spatial arrangement of the continents and oceans, combined with the influence of the latitudinal gradient of temperature, organizes the world oceans into a series of distinct areas, characterized by geography, local circulation patterns, and water properties. Owing to seafloor spreading and continental drift, most coastlines are oriented approximately north–south. Thus, the geographic ranges of many shallow-water species are limited by temperature, which varies greatly along a coast with changing latitude. Major differences in food regime usually limit a species's range of distance away from shore. Most inshore assemblages of both plankton and benthos are distinct from offshore assemblages. This leaves blocks of habitats that are distinct in their combined hydrographic and trophic characteristics.

Because of the set of geographic and environmental barriers, the ocean can be divided into a series of **provinces**, or biogeographic regions with characteristic assemblages of species. Boundaries between provinces can be water-mass borders, major thermal discontinuities, points of land coinciding with thermal discontinuities, or boundaries between water bodies of differing salinity. Along the west coast of North America, for example, there is a series of provinces (Figure 17.1) whose boundaries coincide with major thermal breaks. For example, Point Conception in southern California marks a major shift from southern warm water to northern cold water. Although most boundaries coincide with major environmental shifts, currents may isolate one region from another, which is true through the summer at Point Conception. Provinces are usually recognized statistically, not by unique assemblages of species.

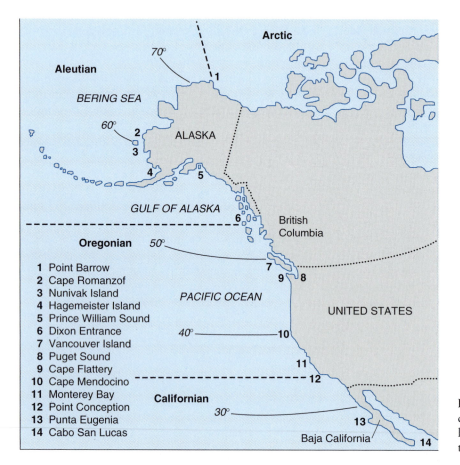

Fig. 17.1 Biogeographic provinces defined by the continental shelf mollusks of the northeastern Pacific. (After Valentine, 1966.)

While provincial boundaries can be recognized by the ends of the geographic ranges of marine species, one would expect that broader-ranging species might still be isolated to some degree across the same boundaries. Therefore, one might expect some degree of genetic difference between populations of the same species on either side of the boundary. Figure 17.2 shows differences in frequencies of major variants in mitochondrial DNA length polymorphisms along the east and Gulf coasts of North America. The Atlantic coast of southern Florida is known to be an important provincial boundary. Species with ranges that transcend this boundary nevertheless are strongly differentiated genetically on either side of it.

Major Gradients of Species Diversity

* We must distinguish between within-habitat and between-habitat comparisons of diversity.

In discussing geographic variation in species diversity, it becomes difficult to make comparisons between very different regions. An increase of species numbers, from one place to the next may be explained simply in terms of an increase in the number of habitats. The shallow-water tropics, for example, have coral reefs that provide a large number of available microhabitats that are nonexistent in temperate-zone bottoms. The difference in diversity explained by multiplication of habitats is the **between-habitat component of species diversity**. Any changes in species diversity between regions in a single defined habitat comprise the **within-habitat component of species diversity**. For example, comparison of the number of species living in muddy-bottom shelf subtidal sediments in the Atlantic Ocean versus the Pacific Ocean is a within-habitat comparison of species diversity.

Effect of Latitude

* Species diversity tends to increase with decreasing latitude.

The best-known diversity gradient is an increase of species diversity from high to low latitudes in continental shelf benthos, in the plankton in continental shelf regions, and in the open ocean. Figure 17.3 shows the latitudinal gradient for bivalve mollusks, but the trend applies to other shelf invertebrate groups and to planktonic groups such as copepods. There are some exceptions, and a couple of groups

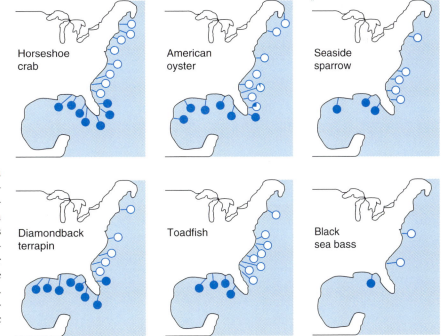

Fig. 17.2 Geographic distribution of mitochondrial DNA length variants in six coastal species. The pie diagram for a given species at each locality shows the relative frequencies of the two most common monophyletic groups of length variants (or groups of length variants with the same ancestor) for the given species. As can be seen, the south Florida region marks a major biogeographic discontinuity. (After Avise, 1992.)

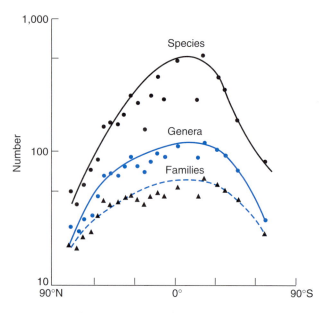

Fig. 17.3 The relationship of bivalve mollusk taxon diversity versus latitude. Points are average number of species, genera, and taxonomic families. (After Stehli et al., 1967, courtesy of the Geological Society of America.)

Comprehensive comparisons of latitudinal gradients in continental shelf species richness are far too scarce throughout the world to permit us to draw general conclusions. A recent compilation comparing the Atlantic and Pacific coasts of the Americas gives us the first insight into interoceanic comparisons. A compilation of nearly 4,000 species of marine prosobranch gastropods shows a latitudinal gradient in species richness on both coasts (Figure 17.4). It is a bit surprising to those who think of the Pacific as more diverse (which it is, overall) to find that there is not much difference in the number of species between coasts in the tropics. In higher north latitudes, the Pacific coast has about double the number of prosobranch species found in the same latitudes on the Atlantic coast, which may possibly be explained by a more extreme climatic history during the past few million years in the North Atlantic.

In Chapter 16, the diversity gradient of benthos along a depth gradient from the continental shelf to the deep sea was discussed. It has been a recent surprise that there is a latitudinal diversity gradient in the deep sea of the North Atlantic Ocean (see Chapter 16). Formerly, marine biologists thought that there was no environmental latitudinal gradient in the deep sea at all. It has been becoming more obvious, however, that high-latitude deep-sea environments differ from low-latitude bottoms. This is especially apparent in terms of the fall of detritus from the plankton, which is greater in abundance and also more erratic in higher latitudes. It will be very interesting to see if there is a latitudinal gradient in the deep Pacific.

Differences Between and Within Oceans

* There are differences in species diversity between ocean basins.

Even when latitude is held constant, the Pacific Ocean has far more species than the Atlantic. This fact has been documented for a wide variety of invertebrate groups and fishes, especially in coral reef habitats. The contrast is also obvious for many groups on the Atlantic and Pacific sides of North America (Table 17.1). In the northwest United States, one can find 19 species of shallow-shelf asteroid starfish, in contrast to only 6 species in the southern New England region. An interesting exception is the polychaete annelids, which are slightly more species rich in the New England area. Most of the diversity difference, how-

(e.g., nuculid bivalve mollusks) even increase in diversity toward higher latitudes. The generalization of increased diversity with decreasing latitude also applies to higher taxonomic categories, such as genera and families. This generalization applies to species lists, but not necessarily to the number of species living within a small area of, say, a square meter. The regional species list for Costa Rica contains five times as many species as the list for coastal Washington State, yet a square meter of typical tropical beach contains no more species than typical temperate-beach samples. The overall difference in species diversity seems to be due to the relative richness in Costa Rica of the low-shore cobble-beach habitat.

The present-day latitudinal species diversity gradients correlate with the strong climatic latitudinal gradient. One should keep in mind some of the conditions that are nearly unique to our time in geological history. At present, there is a strong latitudinal gradient in climate, owing to the recent (last few million years) cooling of world climate and glaciation from the poles. Also, seafloor spreading and continental drift have produced extensive north–south trending coastlines, a development that in turn places shallow-water biotas along a lengthy climatic–geographic gradient.

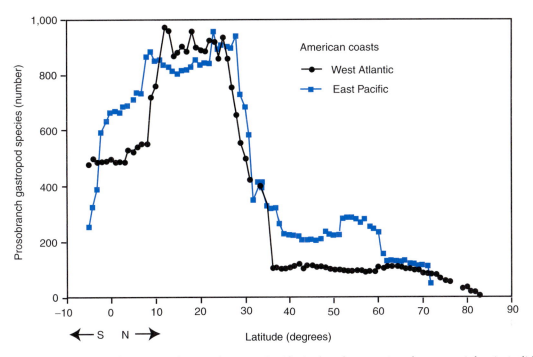

Fig. 17.4 Latitudinal pattern of species richness of eastern Pacific (colored squares) and western Atlantic (solid squares) marine prosobranch gastropods, divided into units of one degree of latitude. (After Roy et al., 1998.)

ever, is apparent in comparing the tropical Atlantic with the tropical Indo-West Pacific, the latter being far more diverse. The inner-shelf mollusks on the east and west coasts of the tropical Americas appear to be rather similar in species richness.

✳ Within the Pacific Ocean, species diversity in coral reefs declines in all directions from an Indo-Pacific diversity maximum.

Although the latitudinal gradient is most prominent in all oceans, the Pacific has a prominent peak of diversity, especially in the species associated with coral reef habitats. Diversity reaches a maximum in the southwest Pacific, in the region of the Philippines and Indonesia (Figure 17.5). From this center, diversity declines in all directions, although the latitudinal gradient is steeper than the longitudinal gradient.

Table 17.1 Species diversity of various invertebrate groups in the vicinity of Woods Hole, Massachusetts, and Friday Harbor, Washington

	NUMBER OF SPECIES	
TAXONOMIC GROUP	FRIDAY HARBOR, WASHINGTON	WOODS HOLE, MASSACHUSETTS
Shell-less opisthobranch gastropods	61	23
Shelled gastropods	88	51
Bivalve mollusks	114	55
Asteroid starfish	19	6
Polychaetes	174	218
Isopods	42	27
Amphipods	76	47

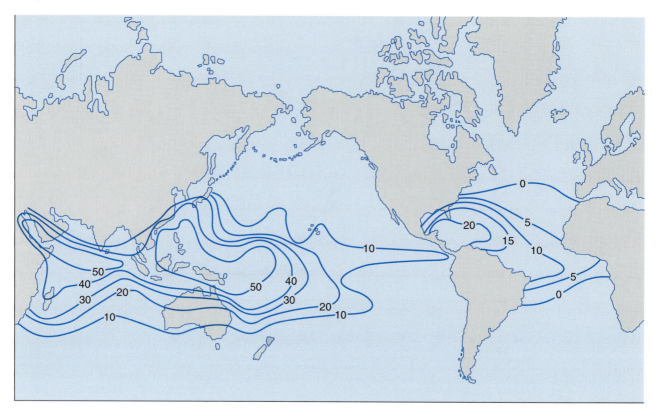

Fig. 17.5 Variation in numbers of genera of reef-building corals in the Indo-Pacific and Atlantic provinces. (After Stehli and Wells, 1971.)

Lower Diversity in Inshore and Estuarine Habitats

✻ **Inshore and estuarine habitats are poorer in species than comparable habitats in the open sea.**

In both benthic and water column assemblages, the open sea tends to have more species than inshore habitats. Estuaries are the extreme cases, where one finds a small number of species, dominated by very few species with high abundance. By contrast, open continental shelf communities usually have more species, and the species abundances are more evenly divided. In the plankton, the increasing complexity offshore is correlated with an increase in the number of trophic levels, as was discussed in Chapter 10.

In Chapter 16 the high diversity of the deep sea benthos was discussed. If one considers the muddy-bottom habitat alone, there is a regular change in benthic diversity from the coast to the abyssal plain. Species diversity of macroinvertebrates and fishes increases with depth to a maximum just seaward of the continental rise, then decreases with increasing dis-

tance toward the open abyssal plain. By contrast, the species diversity of meiofaunal and microfaunal groups seems to increase steadily toward the abyss.

Some Explanations of Regional Diversity Differences

Ecological Interactions

✻ **Ecological interactions may determine regional differences in species diversity.**

The processes of competition, predation, and disturbance interact to determine species diversity in a local area. Predation and disturbance tend to reduce population sizes of prey species, which relaxes competition among species. This suggests that predation, for example, might allow more species to coexist. Dayton and Hessler[3] have cited this as an explana-

---

[3] See Dayton and Hessler, 1972, in Further Reading, Patterns of Species Diversity.

tion for high deep-sea diversity. In effect, a short-term ecological process may control long-term species co-occurrence. There is a problem with this hypothesis, because the life histories of deep-sea benthic species are characteristic of organisms that have suffered little predation. Species suffering high predation would be expected to have high reproductive output, rapid growth, and early reproductive maturity. Deep-sea benthic species seem generally to have the opposite of these traits. However, there seems to be no strong evidence of ecological differences among coexisting species that would be indicative of the influence of competition in the evolution of the deep-sea benthos. Disturbances are probably small in scale, which may preserve unpredictable combinations of species, with little extinction. Almost everything we might say about the deep-sea benthos with regard to competition and disturbance is very limited by a lack of direct observations or experimental studies of community structure.

Effects of Recent Events

✳ Complex recent historical events may explain current regional differences in species diversity.

As mentioned earlier, some of the large-scale geographic difference in diversity can be ascribed to recent events. The gradients may not be due to a balance of extinction and speciation, as was postulated. For example, the present-day species pattern along the eastern U.S. coast is due primarily to a large-scale extinction in the southeast that probably was followed by the appearance of large numbers of new species. An extinction, followed by an invasion from the northeastern Pacific, strongly influenced the diversity of higher-latitude, shallow-water marine faunas in the northwestern Atlantic. The diversity of the Caribbean has also been shaped by relatively recent events. The large-scale diversity gradient in the tropical American coral reef biota may be due in part to extinctions around the periphery of the province, as the earth's climate became cooler and the latitudinal oceanic temperature gradient steepened strongly in the Atlantic basin during the last few million years. The rise of the Isthmus of Panama a few million years ago was followed by an extinction of many Caribbean species that had close relatives in the Pacific. Despite this extinction, many new species were produced, so the total effect on diversity was not very great. Molluscan diversity apparently even increased

after the emergence of the isthmus.[4] The change in diversity in the last few million years, in sum, was very complex and probably consisted of a series of expansions and contractions of species richness.

Factors Causing High-Diversity Regions

✳ Areas of high diversity may be centers of origin of new species, or they may be regions where speciation is steady and extinction is low, resulting in geographically separate (vicariant) regions.

There is still controversy over why some regions are rich in species. This is particularly true of tropical species-rich regions, such as the southwest Pacific in the vicinity of Indonesia. There are two problems. First, why do more species seem to accumulate in species-rich regions? Second, are the species-rich regions sources of new species that migrate out to regions of lower diversity? Many favor the **center-of-origin theory**, which states that the tropical high-diversity regions are the source of species, which disperse to regions of lower diversity. For this to be true, either extinction must be lower in the tropical regions of higher diversity or speciation rates must be higher. This hypothesis is difficult to test without reference to a fossil record. A study by Stehli and Wells[5] demonstrated that geologically younger reef coral genera are found most abundantly in the southwest Pacific, where diversity today is the greatest. This would seem to support the center-of-origin theory. A more recent study of western Pacific sea grasses by Mukai[6] has also supported the center-of-origin idea. Mukai hypothesized that species should be able to migrate most easily along major current systems. He then examined the number of species of sea grasses from the Indonesian southwest Pacific center of diversity along the Pacific equatorial current and several other current systems and found that they decreased steadily downcurrent (Figure 17.6).

In contrast to the center-of-origin hypothesis, the **vicariance hypothesis** argues that species arise in situ on either side of a geographic barrier. Such a hypothesis requires that an ancestral species (or a group of species, all with the same geographic range) be

---

[4] See Jackson and others, 1993, in Further Reading, Patterns of Species Diversity.

[5] See Stehli and Wells, 1971, in Further Reading, Patterns of Species Diversity.

[6] See Mukai, 1993, in Further Reading, Patterns of Species Diversity.

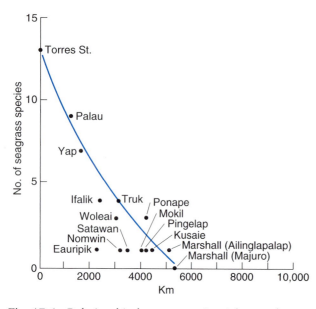

Fig. 17.6 Relationship between species richness of tropical sea grasses on various Indo-Pacific islands and distance from the Torres Strait (northern Australia), eastward along the equatorial countercurrent. (After Mukai, 1993.)

* Species diversity increases with increasing habitat area.

It has long been known that there is a quantitative correlation between geographic area and the number of species that area contains (Figure 17.7). This is best described in the form of a simple equation:

$$S = CA^z$$

where S is the number of species, A is the area, z is an exponent used to fit the equation, and C is a constant that partly accommodates the use of different metrics for area (e.g., km^2 vs. $miles^2$). The exponent z is usually less than 1, so a proportionate increase of area results in a smaller proportionate increase in species; that is, the curve is concave down. On a qualitative basis, this means that the differences in species numbers between the Atlantic and Pacific, for example, can be related to the fact that the Pacific Ocean has a larger area. Similarly, the Caribbean coral reef province is far smaller in area than that of the Indo-Pacific, and the latter contains far more species of most invertebrate and vertebrate groups.

A simple theory may explain why larger numbers of species are associated with larger areas. Remember that the number of species in a large region is probably a result of a balance between speciation and extinction. In larger areas, the speciation per unit area may be greater, owing to the larger diversity of habitats associated with an increased area. Similarly, the rate of extinction may be lower, because more refuge habitats are available, and also because larger population sizes and geographic extent in the larger area may make species less prone to extinction. Areas that are richer in food might also support larger populations, which would retard the extinction rate and raise the equilibrium number of species. Thus, food-rich bodies of water might support more species than food-poor ones (Figure 17.7).

There is a real problem with the area hypothesis when diversity in the deep sea is considered. It is true that deep-sea diversity is greater in the near-slope abyss. However, diversity increases with greater distance toward the midoceanic bottoms beneath the gyres. These bottoms are greater in areal extent, yet harbor fewer species than are found in the near-shelf abyssal bottoms. Clearly, reduced food supply has a major influence on diversity in these remote bottoms,

widespread and be subsequently isolated into separate populations by a barrier such as an isthmus. This model may be appropriate for the division of the Atlantic and Pacific marine biotas. Before the Oligocene epoch, approximately 40 million years ago, when the broad shallow sea, known as Tethys connected the Atlantic and Pacific, there was greater homogeneity of the marine faunas. Following the rise of landmasses, the Tethyan Sea disappeared and divergence soon followed. The barrier to dispersal not only allowed more local speciation in the two oceans, but also was accompanied by strong climatic differentiation, so that species also became adapted to the more variable Atlantic, for example, as opposed to the relatively more equable Pacific. The vicariance model also fits well with our understanding of the many species pairs on either side of the Isthmus of Panama. In a few other cases, the appearance of peninsulas appears to have caused separation and speciation on either side of these barriers. Within the central and eastern Pacific a great deal of evidence shows that longitudinal distance is accompanied by large differences in species occurrences and also genetic differences within widespread species, which reflects genetic isolation by distance.[7]

[7] See Palumbi and others, 1997, in Further Reading, Patterns in Species Diversity.

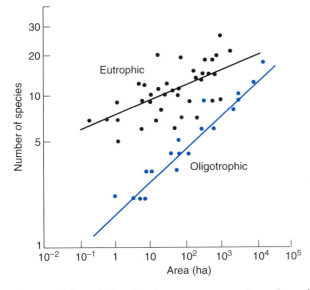

Fig. 17.7 The relationship between area and number of species of freshwater snails in nutrient-rich (eutrophic) and nutrient-poor (oligotrophic) ponds in Denmark. (After Lassen, 1975.)

perhaps by increasing extinction. Area cannot be the direct and only factor that regulates diversity.

* Increasing long-term habitat stability may tip the speciation–extinction balance toward higher species diversity.

Physically stable environments may accumulate more species than variable environments. Howard Sanders and Lawrence B. Slobodkin proposed that environmental stability influences the flexibility required of the resident species. By environmental stability, we mean the short-term range of environmental variation: for example, the degree of seasonality, seasonal changes in salinity, and frequency of storm events. If the environment varies often from state a to state b, then a species will become extinct unless it can survive and reproduce in both states. Given that resources may be limiting, the environment may be able to support only relatively few species of such broad adaptability. Unpredictably variable environments may have even more severe effects, because species must then have enough flexibility to deal with a wide range of environmental changes. J. W. Valentine[8] suggested that fluctuations in primary production may be a major influence on the degree of specialization of consumers. Stable environments may also accu-

mulate more species owing to reduced extinction rates. In stable environments, major habitat alterations may be less frequent and extinction may therefore be less important. This hypothesis also fits the recently discovered latitudinal gradient of benthic diversity in the deep sea of the North Atlantic. Sites in very high latitudes are characterized by greater deposition of detritus from the plankton, and the variation of this deposition is greater than at lower latitudes.

Several large-scale patterns of diversity might support the stability hypothesis. The Pacific coral reef province is demonstrably more seasonally constant in temperature variation than the Atlantic. The high diversity of the Pacific coast versus the Atlantic coast of the United States may be explained in the context of the more constant maritime climate of the Pacific as opposed to the continentally dominated climate of the North American Atlantic coast. The deep-sea increase in diversity may be explained by the extreme constancy of temperature and salinity, whereas the dropoff in diversity toward the open-sea abyssal plain may be due to severe food stress. In stressful environments, only a few species capable of evolving tolerance will come to dominate.

While the deep sea may appear to be more stable today, the difference may be more apparent than real. We have encountered a number of examples in this text of large-scale climatic changes that have ocean-wide if not global effects on the shallow-water marine biota, as occur during El Niño events. One might expect such climatic oscillations to cause changes in global circulation patterns. As we discussed in Chapter 2, the hydrography of the deep sea is very intimately dependent upon the surface.

The deep-sea fossil record of the Pliocene epoch (2.85–2.40 million years ago) demonstrates that the deep sea may not have been rock stable in diversity at all. Deep-sea ostracodes apparently fluctuate in diversity by a factor of 3, and fluctuations obey a 41,000-year cycle that correlates with changes in the earth's tilt.[9] Diversity tends to be high during warm periods and very low during glacial maxima. This suggests that some form of surface climatic oscilla-

[8] See Valentine, 1966, in Further Reading, Patterns of Species Diversity.
[9] See Cronin and Raymo, 1997, in Further Reading, Patterns of Species Diversity.

tion is affecting deep-sea benthic diversity, perhaps by forcing of climate fluctuation.

Sanders[10] also suggested that ancient environments may accumulate more species than do young environments. This conclusion was based mainly on observations of the spectacular diversity of many of the rift valley lakes in east Africa and of Lake Baikal in Siberia. All those lakes are millions of years old, unlike the majority of the world's lakes, which are only a few thousand years old at most. The concept is harder to apply to the ocean. The deep-sea biota does not seem to be more ancient on the whole than that of shallow water, despite the presence of a number of famous "living fossils" such as the stalked crinoids and the monoplacophoran *Neopolina*.

Expansion and Extinction in the Geological Past

* **The fossil record allows us to distinguish periods of origin and extinction. During the Phanerozoic era, there have been periods of rapid expansion and several episodes of major extinction.**

Unfortunately, present-day species distributions provide only a kind of snapshot of the current situation, often offering little possibility of insight into the historical events that may have led to current diversity conditions. Although the fossil record is known to be characterized by poor preservation and by marked gaps where there is no record at all, we nevertheless can get an idea of how diversity changed over longer periods of time, and of whether current conditions might be explained in terms of major changes in recent geological history. Only a brief sketch of some important patterns can be provided here, but such patterns demonstrate that many major changes have occurred, embedded in a history of past climate change and lateral and vertical movements in the earth's crust, which have reoriented coasts and changed current systems.

Figure 17.8 shows the geological time scale, which is broken up into eras, periods, and epochs. (We are living in the Phanerozoic era, the Neogene Period, and the Pleistocene epoch.) Although absolute time boundaries have been determined by the use of various radioactive isotopes, most of the time scale comes originally from the rock record and the location of fossils in relative positions (older fossils are lower down in the rock record, generally). Figure 17.9 shows that there was an explosive appearance of new phyla near the beginning of the Paleozoic Era,

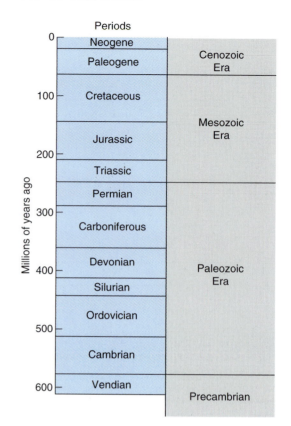

Fig. 17.8 The geological time scale, showing eras, periods, and epochs, and the absolute number of years, determined from radiometric dating.

but that there were two more explosive periods, which led to our current high level of diversity. The general increase, however, was punctuated by so-called mass extinctions. The most dramatic occurred at the end of the Paleozoic Era, and paleontologist David Raup[11] has estimated that over 95 percent of marine species became extinct at that time. More famous is the somewhat less dramatic extinction at the end of the Cretaceous period, when many marine species and, of course, the dinosaurs became extinct. There is still no general theory that satisfactorily and conclusively explains these mass extinctions, but the end-Cretaceous event coincided with a rather large asteroid fall, for which there is evidence in the form of an extraordinary iridium anomaly and a large craterlike structure in Yucatan (Mexico), which sug-

[10] See Sanders, 1968, in Further Reading, Patterns in Species Diversity.

[11] See Raup, 1979, in Further Reading, Patterns of Species Diversity.

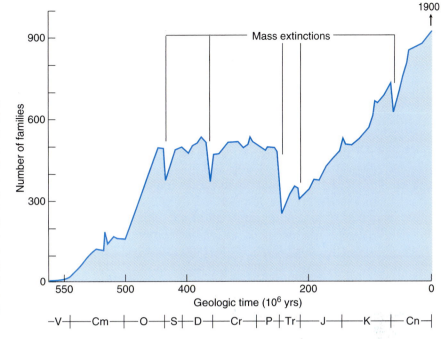

Fig. 17.9 Changes in the numbers of readily fossilizable marine fossil families throughout the Phanerozoic era (Cambrian times to the present). Major episodes of mass extinction are indicated by downward arrows. Key: V, Vendian; Cm, Cambrian; O, Ordovician; S, Silurian; D, Devonian; Cr, Carboniferous; P, Permian; Tr, Triassic; J, Jurassic; K, Cretaceous; Cn, Cenozoic. Because of subsequent adjustments to the dates since this work was published, the absolute time scale is not linear before 400 million years ago. (After Sepkoski, 1984.)

gests a major impact in the Caribbean basin. In the rocks at the end of the Cretaceous there also are in some places minerals indicative of high pressure (i.e., an impact). It is still possible, however, that some of the major extinctions were caused by major climate change or even changes in sea level. The Permian mass extinction coincided with what was probably the greatest fall in sea level in geological history. On the other hand, the large-scale Pleistocene changes in sea level, which occurred because of expansion and contraction of glacial ice, appear to have had little impact on extinctions.

The fossil record also allows examination of the origin of major evolutionary novelties. The Cambrian explosion was discussed in Chapter 11, but throughout the fossil record we can see the rise of new morphologies, which often are identified by the rise of new higher taxa (e.g., orders). Recent evidence compiled by David Jablonski suggests[12] that the tropics are a major source of new evolutionary novelties in the marine biota. The reason for this is not entirely clear, but it may be related to the presence of tropical reefs, which are complex habitats that could allow for more morphological evolution as adaptation to a larger number of habitats. More new taxonomic groups also seem to arise near shore, relative to offshore habitats. This apparent trend may be related to the greater degree of environmental change near

shore, which may cause more directional selection and evolution of new adaptive types.

Conserving Marine Biodiversity

Estimating Diversity

* **A great deal of the total diversity of marine life is as yet unknown.**

Although many important conservation efforts are underway, most ecologists and systematists agree that there is not yet an adequate understanding of the biodiversity of many marine habitats, particularly in the diverse tropics and in the deep sea. Coral reefs, for example, have hundreds of known species of epibenthic invertebrates, but many species remain undescribed (see Hot Topics Box 15.2), and probably many more remain undiscovered. Seaweeds are in particular very poorly understood. Application of molecular techniques, such as sequencing of DNA and evaluation of length polymorphisms of mitochondrial DNA may provide us with important tools for distinguishing unambiguously among marine species that are otherwise very difficult to identify.

[12] See Jablonski, 1993, in Further Reading, Patterns of Species Diversity.

Value of Biodiversity

* Diversity may increase the potential for change in an ecosystem and also may have the value of providing more sources of drugs and other products.

There of course is an aesthetic and ethical issue behind the notion that biodiversity should be conserved. We cannot deny our biological heritage, and we benefit in many aesthetic and even emotional ways by conserving the diversity of life. Most of us would agree that a world with only mussels and blue-green algae would be very dull. Some would even argue that it is immoral to allow the destruction of species. A new awareness of the continuity of all of life with humankind has led to a new concern for our fellow species. On the other hand, many see preservation as antithetical to the idea that humanity should have dominion over the earth. This argument is valid only if dominion involves acting purposefully to cause extinctions. Dominion can also involve behaviors to maintain our biological heritage.

There are also practical reasons to be concerned about the loss of diversity. Much of our knowledge of community ecology suggests that many species play crucial roles in elemental cycling and in regulating the distribution and abundance of marine organisms. Some of these important species are particularly vulnerable to human destruction. Top carnivores in food webs, for example, are likely to be lowest in population size, and any negative effects of the environment may drive them to extinction. While being most vulnerable, they also may play an important regulatory role in keeping prey populations down. Accidental effects of environmental degradation could eliminate species, such as sea grasses and reef-building corals, that create structural habitats for large numbers of other species, many of which are commercially important (e.g., scallops depend upon sea grasses in eastern U.S. coastal habitats). Marine biodiversity may also be a source of innumerable drugs and other products. Unfortunately, it is not predictable where the next important drug will emerge.

Reduction of Biodiversity

* Marine biodiversity may be reduced by habitat destruction, habitat fragmentation, and habitat degradation.

Biodiversity within a given habitat type can be reduced by three human activities. **Habitat destruction,** such as filling of marshes, human-induced erosion, and destruction of shorelines by the building of seawalls is one of the major sources of biodiversity reduction because it removes the habitats upon which species depend. Removal of one component location of a species's life cycle, such as disruption of a spawning area, may also be very potent. A more subtle, but equally important, source of reduction is **habitat fragmentation**, in which a previously continuous marine habitat is broken up into smaller isolated parcels. Shoreline development in the tropics, for example, tends to have a negative effect upon coastal coral reefs, which are broken up into smaller tracts. As a large habitat, such as a sea grass bed, is broken into fragments, colonization of planktonic larvae becomes more difficult. Also, larger predators and other foragers will no longer have a continuous range of habitat over which to feed. A habitat fragment may still contain members of a species, but the population size may be too small to sustain the species over long periods. On the other hand, habitat fragments may have a greater length of habitat edges, which may create new habitats for some species. Finally, **habitat degradation** is probably the most important potential source of loss of biodiversity (see Chapter 19). Nutrient input into estuaries of North America has caused increased phytoplankton density, which in turn has choked off light from sea grasses. Because many fishes and invertebrate species depend upon the sea grass as a structural habitat, many species are lost when sea grasses disappear. Toxic substances tend to eliminate sensitive species and leave a hardy residue of fewer species. Degraded marine environments are typically species-poor environments.

* The species–area effect might be used to predict the loss of species, but an understanding of habitat and biogeographic effects is crucial as well.

Earlier in this chapter we introduced the species–area effect, whereby the total number of species in a region can be predicted approximately through a knowledge of habitat area. Robert H. MacArthur and Edward O. Wilson[13] suggested that the number of species on an island (this can generalized to larger areas of land or ocean) is a dynamic balance between arrival of species and extinction. Larger areas receive

[13] See MacArthur and Wilson, 1967, in Further Reading, Patterns of Species Diversity.

more species per unit time and have lower extinction rates; they therefore maintain more species than do smaller islands. The arrival of species may stem from colonization, or "arrival" may mean speciation within the area.

To some degree, therefore, the degree of habitat destruction might be used to predict the loss of species. Destruction of coral reefs, for example, might be used to predict loss of coral-dependent species if one knew the degree of loss of total reef areal extent. Such an approach, however, would have to be modulated by knowledge of effects on habitat loss. Species may not be lost in proportion to area, for example, if some species can be supported on very small habitat fragments. Alternatively, larger top carnivorous fishes—which require large foraging areas and are more sensitive to extinction owing to smaller population size—may be lost even faster than the general species–area relationship would predict. Biogeographic provincial structure may also have an important influence on loss of species. In marine habitats comprising many biogeographic zones, loss of habitat may have far stronger effects than it does in habitat types found in provinces of greater geographic extent.

Marine Invasions

✳ **Many invasions are facilitated by human transport, and introductions of ecologically potent species may cause local extinctions and a homogenization of the world marine biota.**

Increasingly, marine commerce is delivering exotic species over great distances, causing major changes in marine communities around the world. The few excellent colonists have the potential of homogenizing the between-habitat diversity of our coastal environments. In Chapter 14, the tremendous colonization potential of species of the marsh grass *Spartina* was discussed, as was the disruption of a great many coastal habitats. The arrival in the 1980s of the ctenophore *Mnemiopsis leidyi* in the Black Sea resulted in the loss of a number of zooplankton and fish species. Owing to these transoceanic movements, the world's coastal biota is losing its individual differences. For example, all the dominant hard-bottom intertidal species of San Francisco Bay are now exotics; local species have become rare or extinct.

✳ **Invaders probably rarely establish successful populations and usually become extinct but vectors, high invasion frequency, and ecological suitability of the target habitat all contribute to invasion success.**

Any community is continually invaded by species from other biogeographic provinces, or from different habitat types. Most invasions are failures. Imagine a snail larva arriving on a coast after crossing an ocean. Even if the animal is a pregnant female, the chances are good that she and her offspring will not survive. Normal population fluctuations would cause such extinction, but rarity of mates and the likelihood that the habitat is unsuitable make for the poor prospects for such colonists.

Successful invaders must have the following properties:

1. **Vector.** An invading species must have a means of crossing a long distance, such as an ocean. Many Pacific islands have colonist species with quite limited dispersal, which suggests rafting on floating materials. In recent years, more and more ships use large amounts of ballast water, which is rapidly transported across oceans. Long-lived planktonic larvae might survive the journey, as might microbial organisms such as bacteria, protozoa, and phytoplankton species.
2. **Invasion frequency.** It stands to reason that invasion success must be related to frequency of transport. Rudolph Scheltema[14] found that when closely related species lived on either side of the tropical Atlantic large numbers of larvae were present in the open-ocean plankton. Frequency of ship transport must also matter in human-aided invasions.
3. **Ecological compatibility.** When a species arrives, it will not survive or reproduce unless it can survive and exploit resources in the target location. Thus one does not expect tropical species to successfully invade a polar sea. No such newly arrived species would be able to evolve new thermal tolerances within a few generations—assuming that members of a tropical species could survive a polar environment even that long. It would make sense for the most successful invaders to be ecological generalists,

---

[14] See Scheltema, 1971, in Further Reading, Patterns of Species Diversity.

capable of using whatever resource is available. A specialist is likely to fail to find an appropriate local habitat, even if it did manage a long journey across the ocean. The Asian clam, which tolerates wide ranges of salinity, is a particularly good example of a successful generalist; it probably invaded San Francisco Bay in an area where very low salinity had knocked out a resident biota. When salinity increased, however, the former residents could not reinvade.

4. **Survival of initial population variation**. A newly arrived population is likely to be very small, and random changes in survival and resources will likely drive it to extinction. The population size therefore must increase beyond this initial roadblock. There is an interesting trade-off between dispersal ability and the ability to survive the initial population variability at low population size. A poor disperser, once arrived at an invasion site, might be able to build up a large and local population, since larvae would not spend a long time in the water column. This would allow buildup of a local population, which would then disperse. By contrast, a species with a long-distance-dispersing larva might become extinct after arriving at a spot because the next generation of larvae, being widely dispersed, were not able to find mates at the time of reproduction.

An invader may hang on for many years in low abundance until an environmental change allows a local population expansion. For example, the Chinese mitten crab, *Eriocheir sinensis*, probably invaded the United Kingdom from Europe (which was invaded originally from the Far East) at least as far back as the 1930s. Careful surveys in the Thames estuary of southern England reported its presence in low numbers through the 1980s, but it suddenly expanded to a large population with an estuarine migration cycle in 1992.[15] The population flush coincided with a drought period of very low river flow, which probably facilitated retention of larvae and enhanced recruitment dramatically, which allowed the development of an indigenous reproducing population. Before this, it is possible that crabs found in the Thames arrived continuously from other areas in continental Europe, where the crab was abundant. This case and others suggest that it is not the vector alone that matters in a successful invasion; a disturbance may be required to allow the population to pass the threshold from an invading propagule to a locally sustaining population.

※ Marine invasions are common, and invaders often come to dominate their newly adopted homes.

Despite the odds, invaders do often come to dominate and even cause major changes in the communities to which they live disperse. The periwinkle *Littorina littorea*, which arrived in Nova Scotia about one hundred years ago, is now the most abundant snail on New England rocky shores and exerts major effects on other species, most clearly seaweeds. (It is interesting that the periwinkle had arrived before, as witnessed by its presence in a thousand-year-old Viking boat in Newfoundland, but it did not "take" that time.) Brenchley[16] showed that the invasion of the periwinkle also reduced the ecological distribution of the local eastern mud snail *Ilyanassa obsoleta*, by grazing shore cobble habitats, which removed fine material that might be eaten by the mud snails. The periwinkles also preyed upon egg cases laid by mud snails on hard surfaces.

Because invasions are now noticed by inquisitive naturalists, the rate of spread and the ecological effects of an invader are becoming well understood. Because shipping is often the vector, invaders can be traced to a starting point. For example, the feather duster worm *Sabella spallanzanii*, a native to the Mediterranean Sea, established itself in Port Phillip Bay in Victoria, southern Australia, in the late 1980s (Figure 17.10). It likely came from an invasion in the 1960s to western Australia. It was found at first only in a small bay in the western part of the larger bay but then spread throughout the bay over the next few years (Figure 17.11). Short-distance larval dispersal reduces the rate of spread, but, on the other hand, gregarious settlement allows locally large populations to build up, assuring the probable survival of the population. Their tubes may be 50 cm in height and suspension feeding must be affecting the local biota, but this has not yet been evaluated.

Another ecologically significant and broadly occurring invader is the shore crab *Carcinus maenas*, which arrived on the east coast of the United States from northern Europe, probably in the early nineteenth century, and has become a major predator in

[15] See Attrill and Thomas, 1996, in Further Reading, Invasions.
[16] See Brenchley, 1982, in Further Reading, Invasions.

Fig. 17.10 The polychaete *Sabella spallanzanii*, a world-wide invader from the Mediteranean. Breadth of crown is about 10 cm. (Photo courtesy of David Paul.)

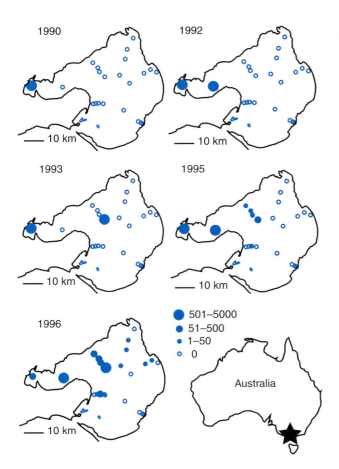

Fig. 17.11 The spread of the Mediterranean polychaete *Sabella spallanzanii* into Port Phillip Bay in Australia. (After Parry et al. 1996.)

the northeastern United States.[17] It is now the most abundant carnivorous crab in shallow waters of southern Australia, where it also seems to have arrived in the early nineteenth century. It is found in South Africa and Japan, as well. But this species's original geographic range was confined to northern Europe. Chances are it was brought both as larvae and as adults on ships and perhaps as juveniles with oysters and mussels, which have been transported widely for culture around the world.

We are now in the middle of an invasion by the shore crab on the west coast of the United States. The crab appeared in San Francisco Bay in 1989 or 1990. By 1993 it had spread 80 km northward to Bodega Bay, California. In 1997 it had been spotted in Coos Bay, Oregon and in the summer of 1999 it was spotted off southern Vancouver Island in Canada. It also has spread south from San Francisco. So rapid a spread is likely due only to dispersal via planktotrophic larvae. During the 1997–1998 El Niño, coastal water flow was northward along the Pacific coast of North America, which probably brought larvae-bearing waters to the outer coast of Vancouver Island.

Invasions sometimes cause major alterations of a local biota. The predatory nature of *Carcinus maenas* makes it a logical candidate for strong effects on near-shore benthic communities, especially because U.S. west coast populations are of far larger body size than their probable east coast progenitors. The invasion of the green crabs has resulted in a 90 percent decline in populations of the native shore crab *Hemigrapsus oregonensis* and two clam species.

San Francisco Bay is an extreme case of alteration by invasion of marine exotic species. The eastern mud snail *Ilyanassa obsoleta* was introduced into San Francisco Bay and has driven the local mud snail to near extinction. More recently, the Asian clam *Potamocorbula amurensis* has invaded the bay and has displaced nearly all other invertebrates.[18] Because San Francisco Bay is surveyed regularly, we can be fairly certain that the clam arrived in the northern end in 1986, probably as planktonic larvae in the ballast tanks of ships. It is now the dominant species there and has also spread to the rest of the bay. The Asian clam is very tolerant of salinity variation and probably invaded during a flood period, when the

[17] See Grosholz and Ruiz, 1996, in Further Reading, Invasions.
[18] See Carlton and others, 1990, in Further Reading, Invasions.

HOT TOPICS IN MARINE BIOLOGY 17.1 The Molecular Sleuth: Where Did the Invaders Come From?

Invasions from distant waters seem to be more and more frequent in recent years. Sometimes the source of an invasion is obvious. Some species are introduced purposefully, usually because they are useful food species and cultured easily. Until recently such introductions were rather frequent, with no attention paid to the possible disastrous side effects. The most dramatic of such changes in terrestrial faunas have occurred on oceanic islands, where introductions of pigs (done on purpose) and rats (by accident) have caused extinctions of many endemic species of snails, birds, small reptiles, and other groups all over the world.

When an invasion occurs, the source of the new population may not be evident. This is of great interest from the points of view of both population biology and management. If a species invaded Europe from eastern North American waters, has a single shipping source been the invasion route, or are propagules coming from the entire eastern coast? From the point of view of management, a single source might suggest some decisions about regulating shipping.

Even if you arrive at a site within weeks of an invasion, you are probably unlikely to witness the vector, such as a shipload of larvae. More often, your objective will be to trace the invasion to its source, sometimes months or years after the new species has arrived. But what is the source? Populations of the green crab *Carcinus maenas* are found throughout the world and are now spreading rapidly up from California to the Pacific Northwest, even to British Columbia. If a crab appears suddenly in Japan, for example, how will we know whether it came by shipping from South Africa, California, Europe, or Australia? Or could a few individuals have arrived on some floating material?

Morphological features are usually not very informative in tracing an invasion to its source. However, the rate of molecular evolution is so rapid in some parts of the genome that populations may diverge sufficiently to permit the fixation of specific DNA sequence differences among separated populations of a species. The question is, To what degree can these differences be used to trace invasion routes?

Toward the end of the twentieth century, a small polychaete, *Marenzellaria viridis*, appeared in the North (1979) and Baltic Seas (1985). *M. viridis* came from somewhere in eastern North American waters, but several alternative hy-

potheses can explain the observations. Perhaps the invasion of the North and Baltic Seas happened from one source simultaneously, but the Baltic invasion was not noticed for a few years. Or did worms invade the North Sea and then spread to the Baltic? Or, finally, were there at least two different invasions from eastern North America into the North Sea and the Baltic, respectively? The salinity of the Baltic is typically only a few parts per thousand, whereas the North Sea has full open-ocean salinity. It is therefore of great interest that two such different bodies of water could be invaded by a single species, which apparently is very broadly adapted to salinity change.

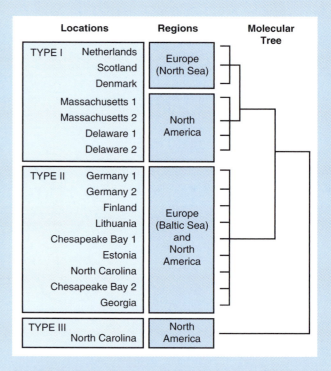

Box Figure 17.1   Genetic relationships among samples of the polychaete *Marenzellaria viridis* from eastern North American and European localities. Note that the North Sea samples cluster with Delaware and Massachusetts sites, but Baltic Sea samples cluster with more southern sites, many with low salinity. The clustering suggests separate invasions into the North Sea and Baltic Sea from two different source populations. (After Bastrop et al., 1998.)

Ralf Bastrop, Karl Jürss, and Christian Sturmbauer[19] attempted to solve the invasion problem by finding a molecular marker for source populations. They took samples from a number of localities along the east coast of North America and also from the North Sea and the Baltic Sea, and sequenced 16S rDNA, a sequence found in animal mitochondrial genomes. The results were striking and supported the separate invasion hypothesis (Box Figure 17.1). North Sea samples could be clustered with American samples taken in Delaware and northward. Baltic samples, however, clustered with American worms taken from Chesapeake Bay and nearby localities. The Baltic affinities were quite interesting because Chesapeake Bay has very low salinities and could have been the source of a population of immigrants that were adapted to low salinity and could better invade the Baltic.

The complications of invasions today make genetic markers essential in tracing the many sources and targets of invasion. The green crab (also known as the shore crab)

[19] See Bastrop and others, 1998, in Further Reading, Invasions.

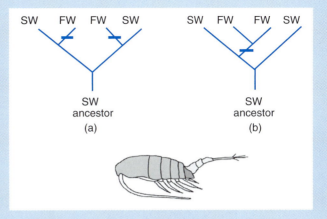

Box Figure 17.3 Two hypotheses of invasion of freshwater (FW) and marine (SW) populations of the copepod *Eurytemora affinis*. (a) A marine ancestor gave rise to two lineages, and the freshwater populations arose independently. (b) All freshwater lineages derive from one evolutionary event, with a marine lineage as an ancestor. Subsequently the freshwater lineage invaded many widespread freshwater localities directly, resulting in divergence. Molecular evidence supports alternative (a).

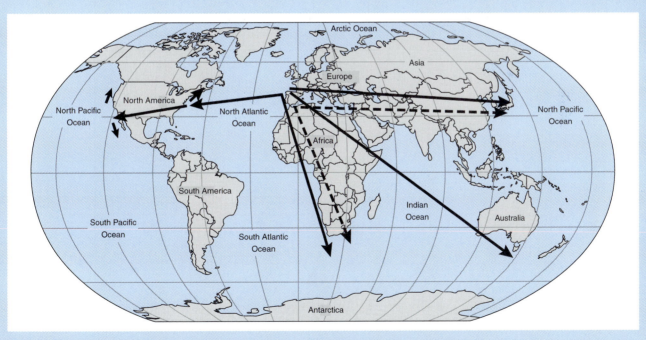

Box Figure 17.2 Presumed invasion routes of the crabs *Carcinus maenas* and *Carcinus aestuarii*. (Information for map mainly from Geller et al., 1997, and Grosholz and Ruiz, 1996.)

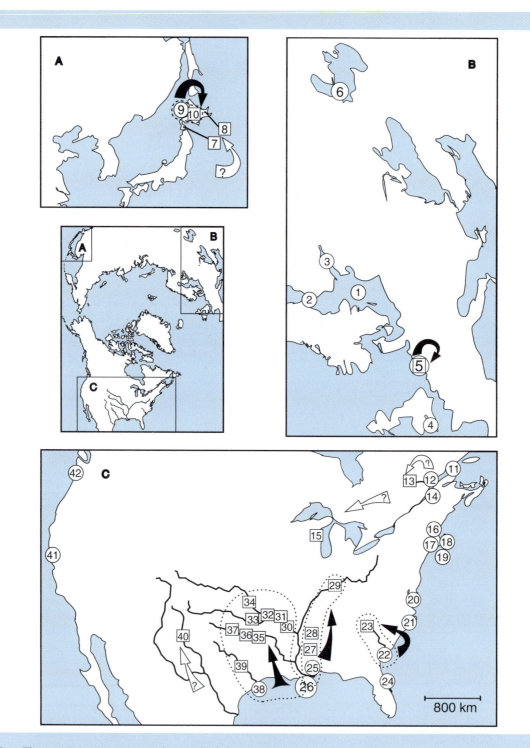

Box Figure 17.4   Invasions and spread of the copepod *Eurytemora affinis*. (Courtesy of Carol E. Lee.)

is an excellent example (Box Figure 17.2). First of all, there are two closely related sources of the invasions. *Carcinus maenas* from the North Atlantic and *C. aestuarii* from the Mediterranean Sea. Both species apparently have invaded South Africa and Japan, but only *C. maenas* has made it to Australia and North America.[20]

While invasions facilitated by human transport are now commonplace, one must remember that before people, chance was a major source of long-distance transport. Populations of coastal benthic species could hopscotch from continent to continent if their larval life was long enough or if they could travel on floating materials. Once a population has arrived on a coast, it is far easier to spread along that coast than to cross an ocean.

The *Marenzellaria viridis* invasion example raises an interesting question that can be applied directly to a natural distribution problem. The copepod *Eurytemora affinis* is found widely distributed throughout the world in open marine coastal and freshwater populations. Two alternative hypotheses might explain the distributions (Box Figure 17.3). First, the same freshwater species might have invaded throughout the world, along with an independent invasion of an open marine species. Alternatively, invasions of open marine species onto a number of coasts might have been followed by subsequent local invasions of nearby freshwater rivers. The latter hypothesis implies multiple origins of the freshwater populations and therefore multiple instances of adaptation to fresh water by marine populations. Carol Lee[21] used sequences of the gene for mitochondrial cytochrome oxidase I and obtained evidence for at least eight invasions of coastal waters throughout the world. At least five of these freshwater invasions most likely arose independently in different river drainages, but each was derived from saltwater sources on the coast near each river drainage (Box Figure 17.4).

[20] See Geller and others, 1997, in Further Reading, Invasions.
[21] See Lee, 1999, in Further Reading, Invasions.

salinity was very low. Now it seems to be able to resist the invasion by more typical benthic invertebrates that formerly dominated. It can filter all the phytoplankton from the water column and probably will greatly affect the entire structure of the San Francisco Bay biota as it continues to spread. Currently, San Francisco Bay is dominated by exotic benthic species.

* In recent years, the ballast water of ships has greatly accelerated transport of exotic species across oceans. While transoceanic dispersal of marine larvae certainly occurs occasionally, shipping is a far more frequent source of movement, currently.

Until the last 25 years, invasions facilitated by people probably were due to transport of associated material. The mud snail *Ilyanassa obsoleta* may have arrived in San Francisco Bay with oysters for culture. In the early mid-nineteenth century, invasions with larvae were less likely, given that most larvae could not last long enough to cross an ocean. But now ships are much faster and use seawater for ballast, which may provide a suitable habitat for surviving the journey.

Although it is a freshwater case, the zebra mussel *Dreissena polymorpha* is particularly spectacular. It was introduced from Eurasia, probably in a ship's ballast water, and has spread throughout the Great Lakes. It was first sighted in 1988 but now has been seen in all the Great Lakes, sometimes in densities as high as 700,000 per square meter! The mussels coat and ruin valuable fish spawning beds and are displacing the local clam species. In 1991, *Dreissena* invaded the Hudson River estuary and has taken over shallow water bottoms in the freshwater parts of the estuary. In 1993, Nina Caraco and colleagues[22] found that the typical strong phytoplankton bloom in the midreaches of the estuary failed to develop. The mussels are apparently clearing out all the phytoplankton in the river!

This is not the first time a major invader has caused such havoc. About 150 species have invaded the Hudson River estuary, and several have had major ecological effects. In the 1830s, the sea lamprey invaded the Hudson River and the Great Lakes, causing tremendous destruction of fishes. More recently, the invasion of water chestnut has choked off tidal lagoons in the Hudson and has greatly altered benthic

[22] See Caraco and others, 1997, in Further Reading, Invasions.

habitats by changing the current regimes, substratum type, and input of nutrients.

* Habitats vulnerable to invasion might include those that are disturbed.

It is difficult to make rules about the success of invasions, although marginal and disturbed habitats seem to be the most vulnerable. This may be because the best invaders are those adapted to such habitats and therefore they succeed best under those circum-

stances. It may be no accident that so many invasions have occurred in estuaries such as the Hudson River and San Francisco Bay. While these areas are obviously perfect targets for delivery of propagules by ships, they also may change radically, owing to strong environmental changes from one year to the next.

* Canals may be a source of large-scale invasions.

The possibility of such tremendous ecological disruption has always worried biologists contemplating the

HOT TOPICS IN MARINE BIOLOGY 17.2 Does Biodiversity Beget Stability?

If one principle has emerged from marine systems, it is this: There is little evidence of any community that is static, with species numbers and abundance unchanging. Indeed, all environments from wave-bashed rocky shores to coral reefs are dynamic, with many incidences of disturbance, changes of species abundance, predation, and competitive displacement. The dynamism of living communities begs a fascinating question that also has practical applications.

1. The theoretical question: Just what is the significance of having more species in a community?
2. The practical question: Given the large number of invasions of exotic species facilitated by human transport, can one assemblage of species better resist invasions than other assemblages?

Theory helps us to develop some expectations of the properties of diverse communities, as opposed to those with very few species. Suppose we have an environment with many potential microhabitats. With a large number of species, it is probable that more and more will exist that will occur in the different microhabitats. The net result will be an occupation of all available food and space resources.[23] The complete occupation of resources leads to an obvious prediction: Invasive species should have trouble invading high-diversity communities.

High-diversity communities would surely be invaded if something about their internal organization made them unstable. If large fluctuations of population sizes occurred in all species simultaneously, invading species might successfully colonize at times when populations were at their minima. It is therefore of great importance to think through the issues of stability and diversity. Remember, the model of Slobodkin and Sanders claims that stability begets higher

diversity. But does the reverse happen? Does high diversity increase stability?

Theory suggests that high diversity should beget stability. The more species there are, the more chances there are for one species to fill in a gap created by another species.[24] High diversity should also increase total productivity because more species will allow more resources to be used. A stable high-diversity community will consist of species that, in total, consume a large fraction of the available resources, suggesting that the possibility for successful invasion is less in high-diversity communities. At any one time, a given species might be declining in population density, but if there are many species, then some other species will take its place on the resource occupied by the first species. Thus the total resource use will remain high, making it hard for additional species to invade from without.

These statements can be true only in a community of species that does actually utilize all available resources. This is a relatively easy thing to assess when the resource is simple, such as space. Therefore, space occupation is an excellent resource base to test theories of diversity and resistance to invasion.

John Stachowicz, Robert Whitlatch, and Richard Osman[25] have completed a study using a lovely system of sessile invertebrate species that compete for space on subtidal hard surfaces, such as rockwalls. In southern New England, the Pacific Ocean ascidian *Botrylloides diegensis*

[23] See Tilman, 1999, in Further Reading, Conserving Biodiversity.
[24] See McGrady-Steede, 1997, in Further Reading, Conserving Biodiversity.
[25] See Stachowicz and others, 1999, in Further Reading, Conserving Biodiversity.

completion of canals that connect two bodies of water that had been isolated. The Suez Canal, connecting the Red Sea with the Mediterranean, has been a source of major invasion and change, especially in the Mediterranean. In the past, plans for a sea-level canal across Panama worried many. Would sea snakes move from the Pacific into the Caribbean? Would the crown-of-thorns starfish, famous for decimating corals throughout the Pacific, make an even bigger splash in the Caribbean? The tremendous ecological importance of invasions is now being appreciated.

Conservation Genetics

✳ Genetic approaches can be used to identify species and to identify genetically distinct populations within species.

We have come to appreciate that many marine species have been overlooked completely, partially because morphological similarities often mask complete reproductive separation and independent evolutionary histories. In recent years a number of molecular tech-

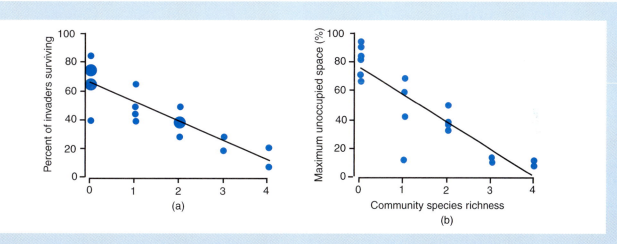

Box Figure 17.5   A study of the effect of species diversity of sessile suspension-feeding invertebrates of southern New England on invasion by the Pacific exotic colonial ascidian *Botrylloides diegensis*. (a) Survival of *Botrylloides*. (b) Percent of total space occupied by sessile invertebrates as a function of the number of species established experimentally. (Courtesy of John J. Stachowicz.)

has invaded and spread in recent years. The native community in any one small area of rock usually consists of three or four different sessile species, and the question is whether they can resist invasion by virtue of their diversity.

Stachowicz and others hit upon an ingenious and simple experimental design (see Box Figure 17.5). They allowed species of native sessile suspension-feeding invertebrates to settle and dominate small 2 × 2 cm² tiles, which were then combined into assemblages of one, two, three, and four species. The "community" consisted of 25 tiles with the different species combinations and 5 tiles covered by the potential invader *Botrylloides diegensis*. The results were impressive: *Botrylloides* mortality was higher in the tile assemblages with higher species diversity.

The mechanism for resistance to invasion by more ses-

sile species may lie in the compensating effect of having more species that can utilize a resource. Stachowicz and others found that if the tiles had only one or two species, population fluctuations often resulted in large areas of empty space. But with more species, one species might increase in numbers and take the space given up by species in decline. Under these circumstances, an assemblage with low species richness might be more likely to have available resources for an invading population, whereas higher diversity might result in more complete and continuous resource use, which would repel invasions. It is possible for one species to dominate space and repel invasions, but the seasonal nature of the environment and the eventual occurrences of disturbances will kill off some of the species and open space for invasion.

niques have greatly improved our ability to discriminate among species. DNA sequencing is now becoming the routine means of identifying new species.

It is also important to be able to identify genetically distinct populations that belong to the same species. Consider baleen whales, whose geographic range is often worldwide. If a humpback whale is hunted in Australia, will that affect the New Zealand population? If summering whales in one part of Antarctica are killed, will that affect the population of whales throughout the world, or just a section of the population that migrates to a specific place? For example, eastern Pacific humpback whales appear to have a population that migrates between Hawaii and Alaska, and another population that migrates between Mexico and California. DNA differences demonstrate that these two populations are, in fact, genetically distinct. The populations must therefore be managed separately.

Genetic markers also have been very useful in proving that the hunting of loggerhead turtles in the western Mediterranean was the cause of decline of the nesting population in the southeastern United States. It is now possible to trace the effects of hunting on populations in distant places. This is quite useful for cetaceans and turtles, both of which have species with long-distance migration. Management of the individual populations for the purposes of conservation can thus be based on a detailed knowledge of inter-change of individuals among separately migrating populations. A United Nations convention has given the nation in whose waters a species nests or spawns the right to file complaints against those who hunt such species in places removed from the territorial waters of the reproductive site.

Protective Legislation

✱ Laws are now being used to protect both biodiversity and habitats.

In recent decades, a new concern for conservation has led to a variety of legislative acts that attempt to conserve biodiversity itself and to protect the habitats upon which marine species depend. Table 17.2 provides a list of major programs or legislation. Much of the legal effort has been directed towards the identification and protection of endangered species. In the United States, the Endangered Species Act has been the basis for protection of many species. The problem lies with a definition of "endangered." This is a difficult issue, since one cannot always state how much habitat loss or population decline constitutes endangerment, especially because population recovery is difficult to predict. When population density of free spawners becomes very low, as it has for a few abalone species in California, the probability of gamete encounter, let alone larval survival, is very

Table 17.2   Some major programs and legislation now being used to conserve marine biodiversity

PROGRAM OR LEGISLATION	MAJOR OBJECTIVES
National Marine Sanctuaries Program	Identifies and provides management programs to protect important and potentially endangered marine habitats (e.g., Channel Islands off California, Flower Garden Banks in the Gulf of Mexico)
National Estuarine Research Reserve System	Identifies estuarine sites important in long-term ecological research (e.g., Hudson River estuarine reserve)
National Wildlife Refuge System	Identifies areas where wildlife is especially valuable, especially those in which species or migratory-bird sites are threatened by extinction
Endangered Species Act	Identifies species that are in danger of extinction (marine species identification is not well developed as yet)
Marine Mammal Protection Act	Intended to halt the decline of marine mammal species and to restore populations to healthy levels
Fisheries Conservation and Management Act	Intended to prevent decline of fisheries within 200 miles of U.S. coast, especially with regard to foreign fisheries (and did not work; see Chapter 18)

low. On the other hand, one can imagine bringing a vegetatively reproducing species to very low numbers and yet, because recovery is easy, not risking extinction. As a rule of thumb, the International Conservation Union has devised a criterion according to which populations that have experienced 80 percent decline three years in a row are considered to be endangered. While this rule has no real teeth except by general agreement, it at least provides a means of quantifying trends that may be cause for alarm.

Further Reading

PATTERNS OF SPECIES DIVERSITY

Abele, L. G., and K. Walters. 1979. The stability–time hypothesis: Reevaluation of the data. *American Naturalist*, v. 114, pp. 559–568.

Avise, J. C. 1992. Molecular population structure and the biogeographic history of a regional fauna: A case history with lessons for conservation. *Oikos*, v. 63, pp. 62–76.

Bakus, G. J. 1974. Toxicity in holothurians: A geographic pattern. *Biotropica*, v. 6, pp. 229–236.

Cronin, T. M., and M. E. Raymo. 1997. Orbital forcing of deep-sea benthic species diversity. *Nature*, v. 385, pp. 624–627.

Dayton, P. K., and R. R. Hessler. 1972. Role of biological disturbance in maintaining diversity in the deep sea. *Deep-Sea Research*, v. 19, pp. 199–208.

Harland, W. B., R. L Armstrong, A. V. Cox, L. E. Craig, A. G. Smith, and D. G. Smith. 1989. *A Geologic Time Scale*. Cambridge: Cambridge University Press.

Huston, M. 1979. A general hypothesis of species diversity. *American Naturalist*, v. 113, pp. 81–101.

Hutchinson, G. E. 1961. The paradox of the plankton. *American Naturalist*, v. 95, pp. 137–145.

Jablonski, D. 1993. The tropics as a source of evolutionary novelty through geological time. *Nature*, v. 364, pp. 142–144.

Jackson, J. B. C., P. Jung, A. G. Coates, and L. S. Collins. 1993. Diversity and extinction of tropical American mollusks and emergence of the Isthmus of Panama. *Science*, v. 260, pp. 1624–1626.

Knowlton, N., L. A. Weigt, L. A. Solarzano, E. K. Mills, and E. Bermingham. 1993. Divergence in proteins, mitochondrial DNA, and reproductive compatibility across the Isthmus of Panama. *Science*, v. 260, pp. 1629–1632.

Lassen, H. H. 1975. The diversity of freshwater snails in view of the equilibrium theory of biogeography. *Oecologia*, v. 19, pp. 1–8.

Levinton, J. S. 1979. A theory of diversity equilibrium and morphological evolution. *Science* v. 204, pp. 335–336.

MacArthur, R. H., and E. O. Wilson. 1967. *The Theory of Island Biogeography*. Princeton, NJ: Princeton University Press.

May, R. M. 1973. *Stability and Complexity in Model Ecosystems*. Princeton, NJ: Princeton University Press.

Mukai, H. 1993. Biogeography of the tropical seagrasses in the western Pacific. *Australian Journal of Marine and Freshwater Research*, v. 44, pp. 1–17.

Palumbi, S. R., G. Grabowsky, T. Duda, L. Geyer, and N. Tachino. 1997. Speciation and population genetic structure in tropical Pacific sea urchins. *Evolution* 51, pp. 1506–1517.

Pianka, E. R. 1966, Latitudinal gradients in species diversity: A review of concepts. *American Naturalist*, v. 100, pp. 33–46.

Raup, D. M. 1979. Size of the Permo-Triassic bottleneck and its evolutionary implications. *Science*, v. 206, p. 217–218.

Rex, M. A. 1981. Community structure in the deep sea benthos. *Annual Review of Ecology and Systematics*, v. 12, pp. 331–353.

Roy, K., D. Jablonski, J. W. Valentine, and G. Rosenberg. 1998. Marine latitudinal diversity gradients—tests of causal hypotheses. *Proceedings of the National Academy of Sciences*, v. 95, pp. 3699–3702.

Sanders, H. L. 1968. Marine benthic diversity: A comparative study. *American Naturalist*, v. 102, pp. 243–282.

Scheltema, R. S. 1971. Larual dispersal as a means of genetic exchange between geographically separated populations of benthic marine gastropods. *Biological Bulletin*, v. 140, pp. 284–322.

Sepkoski, J. J., Jr. 1984. A kinetic model of Phanerozoic taxonomic diversity. III. Post-Paleozoic families and mass extinctions. *Paleobiology*, v. 10, pp. 246–267.

Stehli, F. G., and J. W. Wells. 1971. Diversity and age patterns in hermatypic corals. *Systematic Zoology*, v. 20, pp. 115–126.

Stehli, F. G., A. L. McAlester, and C. E. Helsley. 1967. Taxonomic diversity of Recent bivalves and some implications for geology. *Geological Society of America Bulletin*, v. 78, pp. 455–466.

Valentine, J. W. 1966. Numerical analysis of marine molluscan ranges on the extratropical northeastern Pacific shelf. *Limnology and Oceanography*, v. 11, pp. 198–211.

INVASIONS

Attrill, M. J., and R. M. Thomas. 1996. Long-term distribution patterns of mobile estuarine invertebrates

(Ctenophora, Cnidaria, Crustacea: Decapoda) in relation to hydrological parameters. *Marine Ecology—Progress Series*, v. 143, pp. 25–36.

Bastrop, R., K. Jürs, and C. Sturmbauer. 1998. Cryptic species in a marine polychaete and their independent introduction from North America to Europe. *Molecular Biology and Evolution*, v. 15, pp. 97–103.

Brenchley, G. A. 1982. Predation on encapsulated larvae by adults: Effects of introduced species on the gastropod *Ilyanassa obsoleta*. *Marine Ecology—Progress Series*, v. 9, pp. 255–262.

Caraco, N., J. J. Cole, P. A. Raymond, D. L. Strayer, M. L. Pace, S. E. G. Findlay, and D. T. Fischer. 1997. Zebra mussel invasion in a large, turbid, river: Phytoplankton response to increased grazing. *Ecology*, v. 78, p. 588–602.

Carlton, J. T., and J. B. Geller. 1993. Ecological roulette: The global transport of non-indigenous marine organisms. *Science*, v. 261, pp. 78–82.

Carlton, J. T., J. K. Thompson, L. E. Schemel, and F. H. Nichols. 1990. Remarkable invasion of San Francisco Bay (California, USA) by the Asian clam *Potamocorbula amurensis*. I. Introduction and dispersal. *Marine Ecology—Progress Series*, v. 66, pp. 81–94.

Geller, J. B., E. D. Walton, E. D. Grosholz, and G. M. Ruiz. 1997. Cryptic invasions of the crab *Carcinus* detected by molecular phylogeography. *Molecular Ecology*, v. 6, pp. 101–106.

Grosholz, E. D., and G. M. Ruiz. 1996. Predicting the impact of introduced marine species: Lessons from the multiple invasions of the European green crab *Carcinus maenas*. *Biological Conservation*, v. 78, pp. 59–66.

Lee, C. E. 1999. Rapid and repeated invasions of fresh water by the copepod *Eurytemora affinis*. *Evolution*, v. 53, pp. 1423–1434.

Nichols, F. H., J. K. Thompson, and L. E. Schemel. 1990. Remarkable invasion of San Francisco Bay (California, USA) by the Asian clam *Potamocorbula amurensis*. II. Displacement of a former community. *Marine Ecology—Progress Series*, v. 66, pp. 95–101.

Parry, G. D., M. Lockett, D. B. Crookes, N. Coleman, and M. Sinclair. 1996. Mapping and distribution of *Sabella spallanzanii* in Port Phillip Bay. Melbourne, Australia: Fisheries Research and Development Corporation, pp. 1–14.

Strayer, D. L. 1991. Projected distribution of the zebra mussel, *Dreissena polymorpha*, in North America. *Canadian Journal of Fisheries and Aquatic Sciences*, v. 48, pp. 1389–1395.

CONSERVING BIODIVERSITY

Eldredge, N. 1992. *Systematics, Ecology, and the Biodiversity Crisis*. New York: Columbia University Press.

Franklin, J. F. 1993. Preserving biodiversity: Species, ecosystems, or landscapes. *Ecological Applications*, v. 2, pp. 202–205. See also other articles in the journal issue of *Ecological Applications*, v. 3, no. 2.

Groombridge, B. 1992. *Global Biodiversity: A Status of the Earth's Living Resources*. London: Chapman & Hall.

Hayden, B. P., G. C. Ray, and R. Dolan. 1984. Classification of coastal and marine environments. *Environmental Conservation*, v. 11, pp. 199–207.

Lawton, J. H., and R. M. May, eds. 1995. *Extinction Rates*. Oxford: Oxford University Press.

May, R. M. 1992. How many species inhabit the Earth? *Scientific American*, v. 267, pp. 42–48.

McGrady-Steed, J., P. M. Harris, and P. J. Morin. 1997. Biodiversity regulates ecosystem predictability. *Nature*, v. 390, pp. 162–165.

Norse, E., ed. 1993. *Global Marine Biological Diversity: A Strategy for Building Conservation into Decision Making*. Washington, DC: Island Press.

Primack, R. B. 1993. *Essentials of Conservation Biology*. Sunderland, MA: Sinauer Associates.

Quinn, J. F., and S. P. Harrison. 1988. Effects of habitat fragmentation and isolation on species richness: Evidence from biogeographic patterns. *Oecologia*, v. 75, pp. 132–140.

Stachowicz, J. J., R. B. Whitlatch, and R. W. Osman. 1999. Species diversity and invasion resistance in a marine ecosystem. *Science*, v. 286, pp. 1577–1579.

Thorne-Miller, B. L. 1991. *The Living Ocean: Understanding and Protecting Marine Biodiversity*. Washington, DC: Island Press.

Tilman, D. 1999. The ecological consequences of changes in biodiversity: A search for general principles. *Ecology*, v. 80, pp. 1455–1474.

Wilson, E. O. 1992. *The Diversity of Life*. Cambridge, MA: Belknap Press, Harvard University Press.

Wilson, E. O., and F. M. Peter. 1988. *Biodiversity*. Washington, DC: National Academy Press.

Zaitsev, Y. P. 1992. Recent changes in the trophic structure of the Black Sea. *Fisheries Oceanography*, v. 1, pp. 180–189.

Review Questions

1. What major factors enhance speciation in the sea?

2. What has caused groups of coastal marine species to have rather similar biogeographic ranges?

3. Describe four consistent gradients or regional differences in species diversity.

4. How might long-term historical factors and shorter-term ecological interactions combine to determine species diversity in a given area?

5. How might environmental stability have contributed to high species diversity?

6. Describe the difference between the centers-of-origin theory of diversity differences and the vicariance hypothesis of the development of diversity. Is it possible to test between these two?

7. Why might regions of greater areal extent tend to harbor greater numbers of species?

8. What does the fossil record reveal about the extent of extinction over long periods of geological time?

9. What are the major processes that contribute to human influences on the loss of biodiversity?

10. How might the species–area relationship allow us to predict the reduction in biodiversity by means of estimates of the extent of habitat destruction? Why might such estimates be inaccurate?

VIII

HUMAN IMPACT ON THE SEA

18

Food from the Sea

Human exploitation of the sea predates written history, but we can get a glimpse of ancient fishing practices by examining the reports of anthropologists and explorers who have observed aboriginal societies. When Europeans first encountered them, the aborigines of Australia were using bark boats, from which they threw snail-shell hooks attached to a line fashioned from beaten bark. When a fish was pulled in close to the canoe, it was speared and brought onboard. Native Americans of Pacific coastal Canada dropped lures to the bottom and bobbed them to the surface. The fishes followed the lures and were speared. These natives also used baited hooks, crab traps, and comblike devices with small bone spears to impale fishes.

By the usual standards of human activity, fishing is a form of hunting and gathering, the most primitive of food-collecting methods. Often, a fish population is discovered and then exploited to near-extinction. After overexploitation, fishers move on to new grounds, whose fish subsequently meet the same fate. Open-ocean fisheries have also been depleted. Many whale species have been hunted nearly to extinction, and some species of open-ocean whales are now endangered. Along with the collapse of a fishery, there is much human suffering, because families have invested heavily in boats and gear, which may become a source of debt instead of income. Fishing communities are also often torn apart, never to re-cover. The coastal fisheries of the United States are now mostly in dangerous decline.

Overexploitation first occurred in shellfish beds and in estuaries, because it was possible to clean out the stock by means of efficient hunting. In the twelfth century, a new type of English trawl was so efficient King Edward II banned its use in the Thames River. Open-ocean fisheries were essentially limitless until the development of power-driven vessels and gear. The harpoon cannon sounded what was nearly a death knell for many whale species. The internal combustion engine permitted fishing vessels to move far and wide, and many of the open-sea fisheries have since been devastated as a result. The worldwide search for edible protein has further compounded the problem, and fish landings have doubled every 10 years in recent decades. Some modern nets are 35 miles long and can quickly clear the sea of squid, salmon, and tuna. Fishes are not only consumed directly by people, but are also used as meal for fowl and pigs, and for pet food (which, ironically, is then used as a favored bait for shrimp fishing in the Puget Sound region). These practices have led to many shortages and international disputes. The establishment of 200-mile national fishing zones is a reflection of the common exploitation of fishing nations of the coastal zones of other countries. But, in the United States, this protection only led to severe overfishing by domestic fishing fleets (see later).

For nearly a hundred years, many scientific organizations and governmental agencies have been learning about the life histories of important fishery species and have been attempting to monitor the populations and devise management schemes. This began in 1902 with the establishment of the International Council for the Exploration of the Sea, to address fishery problems in the North Sea. The International Whaling Commission, established in 1946, is another of the many forums for international research and cooperation. Also, fisheries are now being harnessed as farms, and a few species have been domesticated and cultivated with remarkable success. In this chapter, the principles of fishery science and the strengths and weaknesses of various population assessment, management, and farming techniques will be discussed.

The Fishery Stock and Its Variability

* Fish populations are renewable resources.

If coal continued to be mined at a rapid rate, the supplies would eventually be exhausted. The coal resource is therefore **nonrenewable.** By contrast, fishery populations can grow, even if they are exploited. Such resources are **renewable.** This concept of renewable resources serves as the entire basis of fishing and fishery management. In some cases, even modest fishing depletes the stock much more rapidly than it can be renewed. Such overexploitation can lead to cycles of near-extinction of the fishery population, followed by eventual recovery. This is the case with many marine mammal populations, which have been hunted nearly to extinction. The California sea otter, for example, was hunted until only a few populations survived, in the Aleutian Islands and in a few isolated spots on the west coast of North America. The otters were prized for their skins. It is only now, after many years in which there has been no hunting, that population growth, migration, and restocking have resulted in a dramatic recovery of the *Enhydra lutris* population.

Proper fishery management requires the setting of fishing limits that prevent the collapse of the fished population. The fact that the resource is renewable does not guarantee the immortality of a fishery. It simply means that maintaining a sensibly limited fishing level may allow the fish stock to survive to be fished again.

Stocks and Markers

* Fishery species are divided into stocks. Various tags and markers can be used to monitor them.

In Chapter 5, we discussed fish migration, which may occur between spawning, nursery, and adult feeding grounds. A fishery species may have a broad geographic range, but, for management purposes, it makes sense to divide its range into populations, or **stocks,** that are relatively independent of other stocks. Each stock has nearly complete reproductive separation from all others. This usually involves separation of the spawning grounds, but it may also involve separation of nursery and feeding areas. **Tags** are commonly inserted into fishes (Figure 18.1). This allows the recording of time and location of release and capture. Some tags are visible outside the fish, but completely internal tags can be inserted simply and with little mortality. We can detect them by means of magnetic inserts or even by small radiotransmitters that are induced by a scanning device. Radiotagging and acoustic reflection are also used, but they are more efficient in confined freshwater areas.

Biochemical and molecular markers of various types can also be used to diagnose differences among fish stocks that have become genetically isolated.

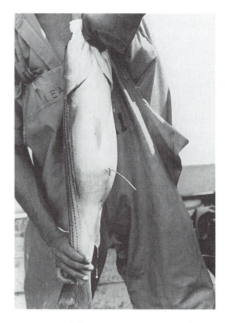

Fig. 18.1   A striped bass, *Morone saxatilis*, tagged ventrally with an internal tag. (Courtesy of John Waldman, Hudson River Foundation.

Presumably, natural selection or random processes change the genetic composition of stocks after they have been separated for a time. The cod is an example of a species divided into a number of geographically isolated populations, which can be diagnosed by a number of biochemical genetic markers. Transferrins, for example, can be used to diagnose different cod stocks in the North Atlantic. Other markers, including enzyme polymorphisms (genetically distinct variants of enzymes), morphology, and unique parasites can be used to distinguish among stocks. In recent years, DNA markers and sequences have been used to delineate stocks. For example, mitochondrial DNA can be fragmented by different types of restriction enzymes, each of which breaks up the DNA at specific points. Such enzyme digestion produces a series of length variants, which can be identified by their migration on a stained gel. Different populations tend to exhibit different length variants, and therefore DNA markers are useful in stock identification (Figure 18.2).

Spawning separation may be temporal as well as spatial. In the pink salmon, spawning occurs in streams. The juveniles move out to sea for 2 years and then return to spawn. There are odd- and even-year populations, which are reproductively independent, even though the populations may share the same streams. In this case, the odd- and even-year stocks associated with a given stream share a crucial resource and may also overlap in the open-ocean feeding grounds.

The delineation of stocks is useful ecologically because population dynamics may be understood in the context of local environmental change. The genetic isolation among stocks may also permit the study of evolutionary change in response to local environments. Such studies, however, are not the main motivations of fishery biologists. Fishermen often have a proprietary interest in their own local fishes, but the fishes also may range over broad areas. For example, many native American tribes have rights, negotiated by treaty, to specific stocks of Pacific salmon. This requires the ability to distinguish among salmon stocks. To negotiate border disputes between Canadian and U.S. fishers, for example, it is useful to have an accurate means of distinguishing among local stocks and to be able to identify the geographic extents of spawning and feeding grounds. Even when armed with such information, fishery experts may still be frustrated in their efforts to reach an agreement because fishes often swim across international

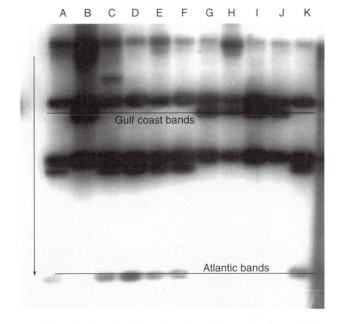

Fig. 18.2 Identification of stocks of the striped bass *Morone saxatilis* by use of DNA markers. An agarose gel, in which restriction-enzyme-digested fragments of mitochondrial DNA (labeled with radioactive carbon) have been separated by size. (Fragments migrate in the direction of the arrow at left, within lanes marked by letters.) Marked fragments in lanes B, G, H, I, and J exhibit a mitrochondrial genotype that is seen only in striped bass from the Gulf of Mexico coast, whereas marked fragments in lanes A, C, D, E, F, and K display a mitochondrial DNA genotype diagnostic of striped bass of Atlantic coast ancestry. (Courtesy of Isaac Wirgin.)

borders during migrations. As a result, there have been disputes between the United States and Canada over border fisheries.

Life History and Stock Size

❋ **The life history of a fishery species and the size of the stock must be understood before sensible planning can be done regarding fishery management.**

Because fishing is so important a livelihood, government agencies are interested in any means that can be found to best manage fisheries. To do this, it is crucial to understand the life history of the fishery species. To make sensible decisions about fisheries management, the following must be known: (1) range of temperatures and salinities for maximum growth

HOT TOPICS IN MARINE BIOLOGY 18.1 The Return of the Molecular Sleuth: Keeping Whalers Honest

As we have been discussing, one of the most important problems in fisheries biology is the identification of stocks, which are a series of populations of a species that respond to different factors, usually because they are somewhat isolated from each other in specific geographic areas. If stocks can be identified, it is possible to study the separate populations and perhaps to develop a fisheries management strategy to manage the stocks.

Compliance with fisheries management plans may have the teeth of the law on its side, but fishers may purposefully or inadvertently hunt stocks that are prohibited from exploitation. Morphological identification of exploited species in a given stock is usually unlikely to be definitive. This problem has motivated research on more specific markers, and molecular markers have been developed to identify stocks. This is not as easy as it seems. Because the evolutionary separation of most stocks is probably geologically recent, one needs a genetic marker that can evolve quickly to detect genetic differences among the stocks.

Whaling is surely among the most controversial fishery management problems. As mentioned, whales have been hunted to very low numbers, almost to the brink of extinction in the case of the blue whale. In 1982, the International Whaling Commission (IWC) voted a moratorium on whaling, with the exception of so-called scientific whaling (done by Japanese whalers) and subsistence whaling done by a few native peoples. Whaling thus continues, albeit at a substantially lower level than in the past. Such hunting is especially a problem if endangered stocks of some whale species are hunted. There also is good evidence for illegal whaling, aside from the limited International Whaling Commission allowances. Whale meat has been found as attempts to export several tons from Norway to South Korea and from Russia to Japan.

What can be done to enforce the IWC's present and future mandates? This is especially important inasmuch as whaling will eventually be resumed, as stocks of various species recover thanks to the moratorium. Markers are needed to trace whale meat in a can or in a freezer to its stock of origin.

C. S. Baker and Stephen Palumbi[1] took a molecular genetic approach to the tracing of whale meat to its stock source. There was some delicacy in this investigation because the investigators had to enter Japan, find cans of whale meat for sale, and study the meat without taking it out of the country, which would have violated export laws. It was not hard to find whale meat, which came frozen, cured in sesame oil and soy sauce, and dried and cured with salt. A portable laboratory was set up in a hotel room to extract DNA, and copy it many times by means of an amplification technique known as the polymerase chain reaction (PCR). One stretch of whale mitochondrial DNA,

[1] See Baker and Palumbi, 1994, in Further Reading, Fisheries.

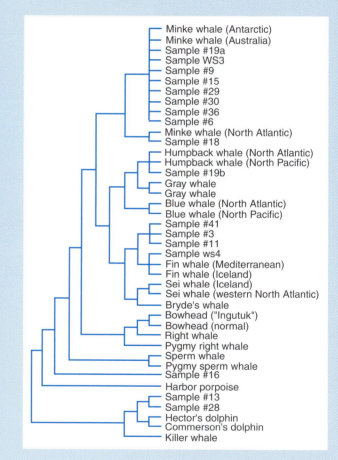

Box Fig. 18.1 Identification of whale stocks and origins by means of a molecular marker, the mitochondrial control region. A tree is established based upon similarities of sequences between different stocks. (After Baker and Palumbi, 1999.)

the so-called control region, was extremely variable and therefore a good candidate to distinguish among stocks.

The sequences could be compared with those taken from various whale populations around the world, and the results were striking (Box Figure 18.1). By examining sequence differences it was possible to construct a tree of genetic resemblance. This is not entirely a straightforward procedure but we will gloss over the details. If you look at the top of the tree you can clearly see that a group of whale meat samples cluster closely with sequences from minke whales found in Antarctica and Australia. Other samples could be associated with North Pacific humpback whales. All in all, different whale species could be discerned, as could stocks of individual species, at least to a limited extent. The presence of meat from a North Atlantic minke whale was especially interesting, since it suggested that the meat had been imported into Japan in violation of the IWC understanding. The prospects for detection of illegal fishing were now greatly improved.

A recent discovery showed the power of genetic analysis. One can only call this a lucky accident, but Frank Cipriano and Stephen Palumbi[2] managed to match the control region sequence of mitochondrial DNA in a Japanese meat sample to a specific whale that had been harpooned in the region of Iceland in the North Atlantic in 1989. The whale turned out to be a hybrid between the blue whale and the fin whale. Subsequent investigation confirmed this match with the sequencing of other genetic loci.

So how did whale meat get from Icelandic waters to Japan? Was it legal? All indications are that, technically, there was no illegality in the killing of whales and the export of whale meat to Japan (some thousand tons in 1990). Iceland also had permission from the IWC to do "scientific whaling," and its citizens were not restricted from shipping the meat to Japan. The Convention on International Trade in Endangered Species might have been a regulatory instrument to prevent Iceland's export of whale meat to Japan, but Japan was entitled to file an objection to the classification as endangered of the whales caught. This permitted Japanese importers to bring the whale meat into their country legally.

The technology of DNA sequencing is now sufficiently fine-tuned that a collection of samples can be taken from all whales killed, which will allow scientists to follow the trail to processed meats, even if they have been transported for thousands of kilometers. It is remarkable that DNA markers can be used to identify not only stocks but even specific whales that have been taken. Illegal whalers can run, but they can no longer hide.

[2] See Cipriano and Palumbi, 1999, in Further Reading, Fisheries.

---

and survival, (2) location of spawning habitat, (3) location of migration routes, (4) location of feeding grounds, and (5) biological information that minimizes unintended mortality during fishing.

To make proper management decisions, it is necessary to estimate the size of a stock. It is also desirable to know the abundance of the different year and size classes, because only fishes of certain sizes are economically important. In some cases, fish stocks are assessed by fishery agencies through a carefully designed sampling program that takes into account migration patterns and spatial distribution. Migration over wide geographic regions makes it very difficult to assess a stock size, because one cannot easily design a sampling program that adequately follows a fish population over hundreds or thousands of kilometers. The spatial distribution is important because fishes are rarely distributed evenly, but usually occur in distinct patches; they may also school. In such cases, small numbers of samples are liable either to miss the population altogether or to hit a dense patch accidentally. In either case, one cannot simply assume that the density estimated from a very few trawls adequately measures the population size.

It is desirable to use the same types of sampler in different studies, because samplers vary in their ability to catch fish. In many cases, population estimates are made with different types of sampling gear, and different results are obtained. When this happens, we cannot be sure whether the difference is due to the gear employed or to actual differences in population size. For purposes of cross-checking, it would be desirable to employ different methods of assessing abundance in the same area at the same time, but this is rarely done. A good example of disparities in stock size estimates is the northern right whale dolphin, *Lisodelphis borealis*, which is often a victim of large drift nets.[3] Estimates from line transects yield far

[3] See Mangel, 1993, in Further Reading, Fisheries.

lower population size estimates than do estimates made from accidental catches in drift nets. These differences are crucial because they lead to very different conclusions about the degree of endangerment of this dolphin species in the Pacific Ocean.

Moreover, it is usually not possible to sample eggs, larvae, and adult fishes with the same type of sampling gear. Because each gear type has a different efficiency of catching fish (or eggs), and different life stages have differing spatial distributions, it is often impossible to compare estimates of abundance between very different life stages.

* The size of a stock is mainly assessed by quantity of landings, which is principally a function of the population size, the spatial variability of the fish, and the amount of fishing effort.

Most fishery data do not come from scientific surveys, but from data on **landings** by fishing boats. For most important U.S. fisheries, federal and state agencies collect data presented by the fishermen. To some degree, inspectors may check the data, principally through onboard checks during the fishing. By the end of a year, a fishery agency may know the total number and poundage of fishes, and may have a breakdown by size, and a record of the number of boats and the time each one spent fishing.

Although the quantity of landings is to some extent a function of population size, landings are difficult to interpret for one of the same reasons applicable to more systematic surveys: the spatial distribution is rarely even, and sampling therefore is biased away from the true density. Moreover, fishermen work in areas that customarily have the highest fish densities, but landings data may ignore other important areas, such as spawning and juvenile feeding grounds. There is also a complication generated by variations in the numbers of boats deployed, fishermen working, and hours spent fishing. These three factors are collectively known as **fishing effort**. If effort increases, the catch will increase, but not necessarily because there are more fishes in the ocean. If the fishing effort doubles and the catch does not double, however, one may assume that the population is declining. All stock estimates must therefore take into account the catch per unit effort. Figure 18.3 shows the declining trend in catch per unit effort for the blue whale fishery.

Stock Health and Production

* The health of a stock is assayed by its production, which is explained in terms of growth of previous year classes and recruitment into the new year class.

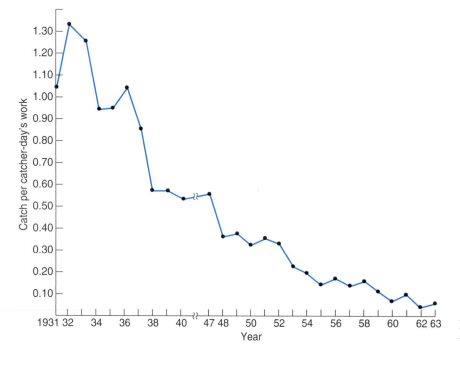

Fig. 18.3 Landings of the blue whale, as compared with effort.

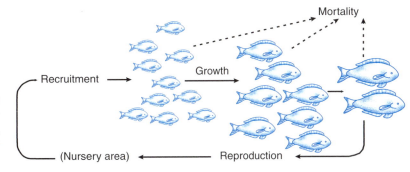

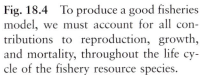

Fig. 18.4 To produce a good fisheries model, we must account for all contributions to reproduction, growth, and mortality, throughout the life cycle of the fishery resource species.

Because a stock has been defined to be a population that is relatively isolated from others of the same species, one can try to evaluate the potential yield of a stock to a fisherman. We are interested in the change in the potential **yield**, measured in pounds of fish per year, from the previous year. To make such an estimate, we have to have a general model for gains and losses in a stock of resource organisms (Figure 18.4), which includes recruitment, somatic growth, mortality, and reproduction.

The relationships among these factors can be expressed in an equation that relates them all to the change in population size:

$$\Delta W = W_{t-1} - MW_{t-1} + RW_{t-1} + GW_{t-1}$$

where ΔW is the change in mass of the total fish population, W_{t-1} is the mass of the population at the same time in the previous year, M is a mortality fraction (varies from 0.0 to 1.0) giving the fraction of the weight of fish that was lost owing to death, R is the fraction by weight that is added from recruitment, and G is the fraction by weight that is added owing to growth of individuals that lived from the previous year to the present. We can neglect immigration and emigration because the stock has been defined to be independent of other populations.

This equation represents an oversimplification from the perspective of both fishing exploitation and population dynamics. Although one can describe all changes in terms of fish mass, one unit of mass is not equivalent to another. A bag of yolk-sac larval striped bass is not really the equivalent of one adult striped bass of the same mass. First, in terms of population biology, each has a different potential for growth, survival, and reproduction. Second, in terms of exploitation, those who would buy and cook the latter may not have the same appetite for the former. Thus, for example, it may be of more interest to follow the mass of the year class(es) or corresponding body lengths that are marketable.

In some cases, age can be identified by size. This is true in seasonal environments, where birth and rapid growth are often confined to short periods of the year when food is abundant. After several years, the population will consist of a series of **year classes**, which are distinguishable by size (Figure 18.5). Although this may be sufficient, it is often not possible to age fishes in this way because all year classes may

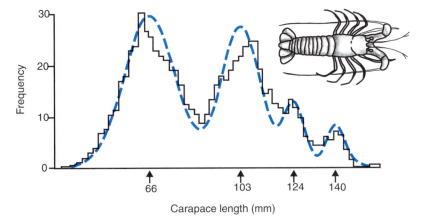

Fig. 18.5 Age classes of the lobster *Panuliris ornatus*. Curved line estimates age classes from the more discontinuous distribution of the histogram. Note the older age classes to the right, which are more indistinct. (Modified from King, 1994.)

not be distinguishable. Growth slows in later years, and the variation in growth tends to blend the later age classes. Finally, fishes growing at different rates may be mixed together in a feeding ground, and the age–size relationship may therefore be invalid. Because of this, fishery biologists have resorted to other techniques. Both otoliths (small precipitated spheres used as part of a balancing organ) and scales have rings that record seasonal growth. **Otoliths** can be used to age larval fish to the day, which is quite important in the understanding of early life histories (Figure 18.6). For shellfish, growth rings can also be used, and cross sections of shells often give an accurate age and growth history.

The equation just shown may lead one to believe that mortality rates, recruitment, and growth rates are constant from year to year. Such constancy is rarely the case, however. Variability and its causes are major problems of fisheries assessment and management. In a seasonal environment, reproduction is also seasonal, and recruitment of juveniles usually occurs as a pulse. The reproductive season usually corresponds to the time of year when food for juvenile fishes is likely to be most abundant. The success of the reproductive season determines the initial size of that year class. Because fishes and shellfish have the capacity to produce thousands to hundreds of thousands of eggs per female, the initial population size

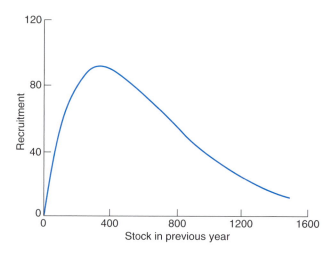

Fig. 18.7   An expected relationship between the size of the stock and recruitment in the following year.

of eggs, larvae, and juveniles is astoundingly large. A plankton net can bring in millions of fish eggs and thousands of yolk-sac larvae. As these young of the year (0+ year class) feed and grow, one can trace a peak of body size as it increases with time. There is also a noticeable decline in the numbers in the year class as it ages. This is a result of mortality. It is no coincidence that egg production per female is so prodigious. Most eggs and larvae are eaten, starve, or are lost to inappropriate habitats. Few survivors make it to adult age and size, to contribute the next generation of young.

If resources (e.g., food, space for spawning) are limiting, we might expect a relationship between the size of the stock and the recruitment as depicted in Figure 18.7. As may be expected, as the stock increases, the number of juvenile recruits should increase as well. Past a critical stock size, however, food per adult may start to decrease to the point that eggs per female will decrease correspondingly. As the stock increases past this critical threshold, we may therefore expect a decline in recruits per reproductive female. In practice, often there is hardly a relationship between the size of the stock and the number of recruits. This suggests that, in the main, postreproductive processes determine population size, and these processes (e.g., predation) are independent of the parent stock.

A simple example may illustrate why such a decoupling seems to exist. Let us take 1,000 eggs as the average produced by a female and assume that all

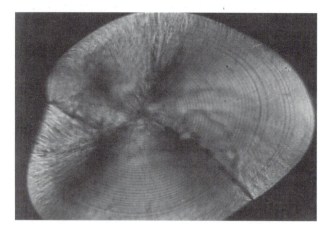

Fig. 18.6   A cross section of the otolith of the bluehead wrasse *Thalassoma bifasciatium* that settled from the plankton at a length of 13 mm. The daily growth record reveals the 41-day mark after hatching, when settlement on the bottom occurred (transition between closely spaced lines and relatively broad band). Otolith is about 300 mm long. (Courtesy of Robert Cowan.)

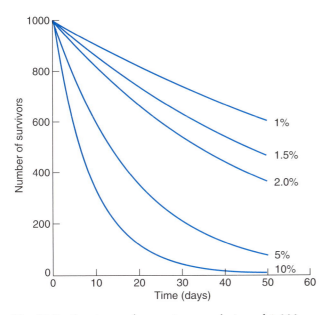

Fig. 18.8 Survivors of a starting population of 1,000, as a function of a differing daily mortality rate.

these eggs are fertilized. Figure 18.8 shows the different adult population sizes obtained when the mortality per day is 1, 1.5, 2, 5, and 10 percent. Note the striking differences in adult population size, even with very slight differences in mortality. It seems quite possible that actual mortality could fluctuate to this degree, so it should be no surprise that there usually is virtually no systematic relationship between stock size and the recruits of the next generation.

Fishing Techniques

✳ Fin fishes are mainly caught by (1) hooking fishes individually, (2) entangling fishes in nets or, (3) catching fishes in nets or traps.

Although the methods may have improved slightly, the basic ways of catching fishes have not changed for centuries. Perhaps spearing is the one technique that has been essentially abandoned by all but aboriginal peoples. It is still quite useful for hunting fishes living on rock and coral reefs. Harpoons are used widely for whales and swordfish. Currently, hooked lines and nets are the two most popular techniques (Figure 18.9). Hook and line may appear to be equally primitive, but variants are used extensively. Most popular in the hooking method are longlines,

which are main lines from which short leaders with hooks extend at regular intervals. The line may be paid out and suspended by floats, or it may be trolled behind a boat. Trolling is a common means of fishing for albacore and some species of Pacific salmon. Of course many sports fisherman use single-hook and longline arrangements for catching many species.

Many different types of shoreline net staked into the bottom have been developed that entrap fishes swimming in schools along the shoreline. Such nets are especially effective in trapping migratory fish that have strongly constrained routes, such as Hudson River shad, which move upriver to spawn. **Pound nets** often have some sort of leader wire that guides the fish into a blind-ended net. If currents are strong enough, there is no need for a long leader, and the fish can be trapped directly. These net arrangements can also be suspended from rigid floating frames and are used to catch some species of salmon. **Fyke nets** are long nets that are usually staked to the bottom. They have leading wings that guide fishes moving with the current into the main bag, from which escape is difficult.

Gill nets differ in that the mesh size is chosen to entrap the fishes by their gills, fins, and jaws as they attempt to swim through. These nets are towed from one end through the water, while the other end is attached to a smaller stationary boat or is anchored to the bottom. From both shore and boat, a variety of fine-mesh nets are hauled to entrap large numbers of fishes. One of the laziest techniques is to drop a net stretched over a horizontal frame. The net is raised periodically and may haul in a large school of small fishes. This is a common technique used in the Mediterranean to trap small fishes and squid. Many fishing peoples haul **seine nets** across beach areas, sometimes with the aid of four-wheel-drive trucks on the beach. The top of the net is attached to floats, and the bottom is weighted down. This method catches a large variety of fishes, and has been used for hundreds of years to catch striped bass on the south shore of Long Island, New York. In other areas, a boat offshore hauls the net, which is fixed to the bottom at the beach. The boat gradually brings its end of the net toward the beach, and the fishes are entrapped in the closed portion.

Seine nets are also deployed entirely from boats. **Purse seines** are particularly effective. The end of the net is attached to a small boat, and the net is paid out as a larger boat moves away. The larger boat at-

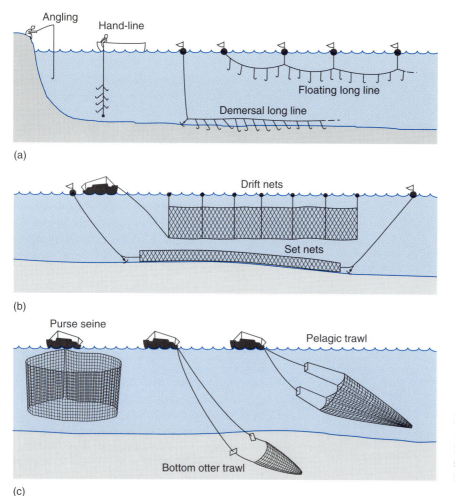

Fig. 18.9 Some fishing techniques: (a) hook-and-line and longline fishing, (b) gill net fishing, and (c) purse seine and trawl netting. (From Cushing, 1988.)

tempts to encircle a school of fishes with the net, and a rope acts like a purse string as the net is drawn aboard and closes about the fishes. This method is used for a wide variety of fish species, and is used from inshore waters to the blue ocean. Environmentalists have complained in recent years about the damage to porpoises, which are entrapped in purse seines with the tuna that are sought by fishermen. For reasons that are not well understood, schools of tuna congregate beneath groups of porpoises, and fishing boats therefore seek out the porpoises. As the net is hauled in, the porpoises are often trapped underwater and drown. Many fishing boats now include divers to help save the porpoises, and there has been strong pressure to eliminate the fishing technique. Many large fishing fleets in the Pacific have changed fishing practices to avoid porpoises, which enjoy popular support among the public.

Shrimp nets used in the Gulf of Mexico are known to entrap and drown turtles; some of the nets are now equipped with trap doors to allow the turtles to escape drowning, but fishermen complain of a loss of shrimp and would prefer instead to employ shorter tow times, in the hope that the nets will be raised before any turtles drown.

Many other nets that are employed are essentially bags towed by a harness of wire. **Otter trawls** are especially efficient because the harness is attached to two or more panels called otter boards that spread the opening of the net efficiently. The bottoms of the boards are weighted in order to drag the net over the bottom. This technique is used widely to bring in bottom fishes such as plaice and flounder.

In recent years, extremely large open-ocean nets have been employed to catch squid and other abundant nekton. Nets over 30 miles long are being used by fishers from a number of nations. These nets kill off large numbers of squid and fish and threaten to wipe out a number of different species. A treaty has been ratified to ban the nets.

✳ Bottom-associated animals are trapped in baited mesh traps.

A variety of mobile animals, but principally large crustacea such as crabs and lobsters, are taken from baited traps that are usually marked with surface buoys. Of necessity, these are generally coastal and estuarine fisheries. Traps often have inverted cones at the entrance, making it difficult for the animals to leave the trap. Unlike most net fisheries, trapping is associated with specific areas of bottom, and this has led to a variety of informal and formal divisions of fishing grounds among fishers.

Human Impact on Fisheries and Fish Habitats

The Concept of Maximum Sustainable Yield

✳ Fishery management strategies seek to maximize the sustainable yield.

Fishing increases the rate of mortality beyond that due to natural sources. Too much fishing will reduce the stock and may drive it to extinction. Effective management therefore must set a fishing rate that permits removal at a sustainable yield but does not eliminate the population. By continually modifying a fishing level to maximize production, fishers attempt to obtain the **maximum sustainable yield**. We may be able to maximize the sustainable yield with an appropriate level of fishing. This hypothesis depends on the assumptions that adult fish stocks are resource limited and that the individual growth rate would be reduced at high natural population densities. As population size is reduced, the resources are more abundant and reproductive output may be higher. Thus, an intermediate-sized fish population might produce the most fish in the long run. Text Box 18.1 provides an illustration of one model that may be used to demonstrate the expectation of a maximum sustainable yield.

Size limits also can be imposed to regulate the yield.

If the mesh size of a fishing net is too small, fishes will be caught before they can reproduce even once. On the other hand, the proportional somatic growth of a fish greatly slows with increasing age and size. If larger fishes are culled from the population, food will become available to the smaller fishes, which grow more rapidly per unit body weight. There is thus an intermediate optimal minimal-body-size limit for fishing.

There are several problems with mesh limits. First, it is difficult to define available resources and the optimal body size. Second, no net catches only one species, and there are different optimal sizes for different species. Finally, nets are not rigid, and larger animals may slip through a net designed for a smaller minimum body size.

Problems with the Maximum Sustainable Yield Concept and the Degradation of Fish Trophic Webs

✳ Current thinking suggests that the concept of maximum sustainable yield may not be useful in fisheries management.

The concept of maximum sustainable yield presupposes a rather precise relationship between population growth and population size. Can this theory be used in practice? First, let's appreciate the problem. Even if the population model is obeyed with precision, we must have accurate measurements of the parameters, and these include population size, reproductive rate, somatic growth rate, and mortality. In practice, obtaining accurate measurements is very difficult, a fact that increases the margin for interpretation of optimal fishing pressure. Those who want to fish more are liable to be biased toward estimates of high fish population size, whereas those biased toward control will look to population estimates at the lower end of the scale. A second problem is that the models usually exclude random fluctuations or causes extending from factors other than those considered by the fishery model. However, departures from fishery model expectations are common, suggesting that there is much uncertainty in applying the model.

A more important problem with the maximum-sustainable-yield concept involves the range of political pressures that lead to overexploitation (see the discussion on the blue whale fishery that follows). In

TEXT BOX 18.1 A Simple Model to Explain the Maximum Sustainable Yield

Presumably, a resource-limited growth rate would also imply the reduction of reproduction as resources become more and more limiting. The following logistic equation (see Note) can relate the growth rate of a population to a hypothetical maximum population that available resources can maintain (Box Figure 18.2a):

$$\frac{dN}{dt} = rN\frac{(K - N)}{K}$$

where dN/dt is the rate of change of population size or somatic growth, N is the population size, r is a rate of growth when resources are limitless, and K is the maximum population that available resources can support.

Box Figure 18.2b shows dN/dt as a function of N. When $N = 0$, the growth rate must be zero, but the growth rate is also zero when $N = K$. In between these two numbers, growth rate is positive. If we did nothing and resources were limiting, we might expect the population to converge on K. So what might be the effect of fishing if we reduced the population below K? Note that the growth rate is at a maximum at $K/2$. We would therefore gain more if the population were fished down to $K/2$, rather than letting the population grow to the level of K. It may seem counterintuitive at first, but if the population size is greater than $K/2$, eventual fish production would be increased by fishing more and reducing the stock to the size for which growth is maximal. According to the logistic model, the maximum sustainable yield, therefore, is set at $K/2$.

Note: The equation for dN/dt has been used widely in population models, but it is not clear that all populations behave according to it. It is used here only to illustrate how population growth may be maximized at intermediate population size.

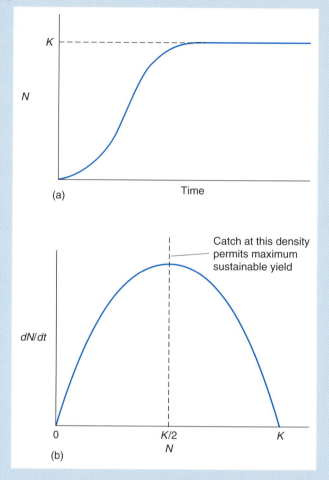

Box Fig. 18.2   (a) Population change over time according to the logistic model. (b) Population growth rate (dN/dt) as a function of population size N, according to the model.

a thoughtful article, Ludwig, Hilborn, and Walters[4] point out that there is a ratchet effect in fishery exploitation. When fish stocks increase, more boats are built and more fishing occurs. However, during periods of constant fish stock, few boats are decommissioned. Then a further period of good fishing years encourages more investment. When the fishery starts to collapse, the societal cost of reducing the work force or the number of active boats is enormous. Before effective management is possible, the overly large fishery fleet and work force will cause further collapse of the fishery. Ludwig and colleagues suggest that the pressures on decision makers relying upon already inadequate scientific models are so great that useful scientific management recommendations cannot be expected. Unfortunately, the collapse of many fisheries around the world proves the authors right, although they point out that controls on open-seas harvesting and some attention to spawning grounds have allowed Pacific salmon populations to increase over the last 30 years.

[4] See Ludwig and others, 1993, in Further Reading, Fisheries.

✳ Overfishing has caused major alterations in the trophic structure of the water column.

Free access to the open seas and government subsidy of fishing fleets around the world have resulted in worldwide collapse of a wide variety of fisheries. Japanese boats, for example, still fish intensively in South Australia for the southern bluefin tuna, despite its relative rarity. Because investment in large open-sea ships entails large loans, it is difficult politically for a nation to discourage use of such equipment, once acquired.

We can use our ecological knowledge to predict a fundamental change in water column communities throughout the world. We have learned that there must, of necessity, be less biomass at the top of a food web, owing to inefficiencies in trophic transfer from trophic levels below (see Chapter 10). Species at the top of the food web, moreover, are usually large in body size, since they are often very mobile predators that must seize and ingest large prey. With low biomass and large body size, top predators are usually few in number. If you think of the number of mountain lions versus the number of deer versus the number of mice, you will get the idea. Also, such top predators are desirable fish for humans, and include various species of tuna, salmon, and so on.

If fishing focuses on these relatively rare top predators, we should expect fishing collapses to occur much more rapidly (too bad we don't prefer to eat bacteria!). Just this collapse has been found by Daniel Pauly and colleagues[5] in a worldwide analysis of the trophic structure of fisheries. They used a worldwide data set on fisheries, compiled by the Food and Agriculture Organization of the United Nations. In the past 45 years, fish landings have shifted from large piscivorous fishes toward planktivorous fishes and invertebrates. On the one hand, fishing out the top predators may actually increase the prey species, which would increase the latter's stock size. But there is a flaw in this sort of thinking. First, it is clear that the fishing pressure on piscivorous fish (tuna and salmon) is unsustainable. Second, we like to eat these species. Third, there is no reason to believe that we won't overfish the lower trophic levels as well, given the increase in world population.

Causes and Cures of Stock Reduction

✳ Stock reduction can result from random variation as well as from environmental change; overfishing would be superposed on the effects of these factors.

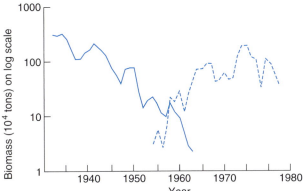

Fig. 18.10 Increase of the anchovy (dashed curve), following the decline of the Pacific sardine (solid curve), off the coast of California.

Fishery managers are concerned with the causes of the great fluctuations in fish populations. Such fluctuations are commonplace in estuarine, shelf, and open-ocean fisheries. One of the most spectacular changes ever recorded illustrates the difficulties in understanding the population changes. In the 1920s and 1930s the Pacific sardine was landed in the thousands of tons, yet the population declined sharply in the 1940s. After this decline, the anchovy increased greatly in abundance (Figure 18.10). A relaxation of fishing pressure did not result in a major recovery of sardines. Competition between the two species, changes in water temperature, and other factors have all been implicated in the decline, but no factor has been convincingly identified. A similar case can be made for various fish stocks of the North Sea, including herring, cod, and mackerel. Fishing, climate, increased zooplanktonic food for larvae, and pollution have all been suggested as factors, but the one clear fact is the presence of strong fluctuations in population size.

Fish populations consist of many age classes that grow and reproduce simultaneously. Most species, with exception of some like Pacific salmon, spawn more than once, usually on an annual basis. The individual factors that collectively affect each year class may have a profound effect on population size. Variation in recruitment, for example, may have a great effect on the subsequent age structure. If a given year class is extraordinarily successful, it will appear as a

---

[5] See Pauly and others, 1998, in Further Reading, Fisheries.

major peak in the size structure of the population and will contribute many more young than will other year classes. Strong fluctuation in recruitment will cause great perturbations in the age structure of subsequent years in the population. Year classes can often be traced through several years as a size peak (as was shown in Figure 18.5).

Random fluctuations in recruitment and mortality may be a major background variation in fish populations. It may be that such fluctuations have specific causes, but these causes may be so complex and varied that they cannot be identified individually, and their effects may be indistinguishable from random variation. Because of this potential, population biologists are often asked to estimate the degree of fluctuation of the total population and of the year-class composition, if the variation is random. Such estimation is especially important because random fluctuations alone may bring the population down to a very low level. Under such circumstances, the additional imposition of fishing mortality would be quite dangerous for the stock.

Environmental change in factors such as temperature and salinity may be important in the decline of major fisheries. A modest increase of salinity probably contributed greatly to the decline of the oyster fishery of Great South Bay, Long Island, New York, in the 1950s. In the Pacific, the periodic El Niño events (see Chapter 2) bring warm surface water to the normally cool inshore waters of the Pacific coast of subtropical and tropical North and South America. In Peruvian waters, the arrival of warm waters causes tremendous mortality in the fish and shellfish stocks.

* Fish stocks characterized by long generation times, small clutches of eggs, and fewer spawnings over time are the most vulnerable to overfishing.

It is rather easy to see that some fish stocks are potentially much more vulnerable than others, owing to their life history characteristics. Clearly, fish species with short generation time (which, for our purposes, is the time from birth to age of reproduction), multiple spawnings during adult life, and many offspring produced per female will have a greater chance of rebounding from low population levels. Species with long generation times, few spawnings, and small clutch size may be very vulnerable to combinations of environmental change and fishing pressure. Shark species are clearly vulnerable, as they typically produce very few pups and have typically long generation times, relative to bony fish species (Table 18.1). Most species feed at the tops of food chains, and therefore relatively small populations can be supported relative to members of lower food chain levels (Chapter 10). These general characteristics have resulted in surprisingly rapid declines in shark populations as various species have become new fisheries.

* Fishing has initiated or accelerated the decline of many stocks.

Human fishing is a major source of decline for fish stocks. In many instances, overfishing has led to drastic depletions of stocks, and fisheries have been closed down as a result. In Great South Bay, New York, a rich fishery of the hard clam *Mercenaria mercenaria* was overexploited by thousands of clammers who worked the bay with small boats and hand tongs. Owing to this, the landings have decreased markedly over the last few decades, and the largest company ceased to fish for clams in 1999. In North America, many fish stocks are now at very low levels, especially on the Atlantic coast. In Newfoundland, cod fishing has essentially ceased, and stocks of bottom-

Table 18.1 Life characteristics of some sharks, in comparison to Atlantic cod

	WHITE SHARK CARCHARODON CARCHARIAS	SANDBAR CARCHARHINUS PLUMBEUS	SCALLOPED HAMMERHEAD SPHYMA LEWINI	SPINY DOGFISH SQUALUS ACANTHIAS	ATLANTIC COD GADUS MORHUA
Age to maturity (y)	M, 9–10; F, 12–14	M, 13–16	M, 4–10; F, 4–15	M, 6–14; F, 10–12	M, 2–4
Litter size	2–10 pups	8–13 pups	12–40 pups	2–14 pups	2–11 million eggs
Reproductive frequency	Biennial (?)	Biennial	?	Biennial	Annual

Source: Data from Klimley, 1999.

associated fish off New England and Nova Scotia are dangerously low. Below, as an example of overfishing, the drastic decline of the whales feeding in the Antarctic Ocean will be discussed. This pattern has been common in fisheries from the estuaries to the blue ocean.

Overfishing can often be identified as a decrease in catch per unit effort. As the stock is overfished, more boats may be deployed, but the fish caught per boat per day decreases. This has been a common trend in the cod fisheries of the North Atlantic. One of the most compelling pieces of evidence for depletion of stocks from fishing is the trend of fish landings after World Wars I and II. During both world wars, fishing was understandably reduced. But after each war, the catch increased tremendously. This suggested that the war periods allowed enough time for the fish stocks to recover in numbers.

Overfishing has arisen from the use of long-ranging ships and technological advances that permit efficient catching and preservation. A song captures the old way: "Haul in the nets, same old fisherman, never catches more than he knows he can sell in a day." This pattern ended as large motor-driven trawlers began to ply the seas. After World War II, trawlers threw their nets over the stern. Later the fleets developed the capacity to freeze the fishes at sea, rather than bringing them back on crushed ice. In the years since 1950, a general trend from short-ranging to long-ranging fishing expeditions developed. Off the shores of North America, for example, Japanese and Soviet trawlers represented a dominant part of fishing and certainly took more fish than the small inshore fishing boats that once dominated the coast. A satellite photograph of Canadian and other trawlers on the Nova Scotia shelf would nicely define the map location of the shelf–slope break, where the offshore fishing was concentrated (Figure 18.11).

Ironically, protection of U.S. fisheries from foreign exploitation has not helped at all. In 1976 the passage of the federal Fisheries Conservation and Management Act (also known as the Magnuson Act) restricted foreign fishing to the outside of a perimeter greater than 200 miles from U.S. coastlines and established a series of eight regional fishing commissions to help in regulation of domestic fishing. While this momentarily protected continental shelf fishing grounds from foreign exploitation, no significant lim-

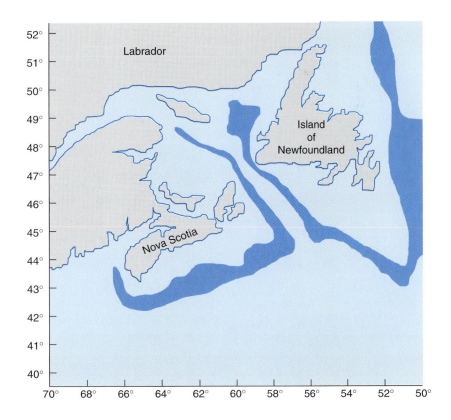

Fig. 18.11 Concentration of Canadian and other fishing vessels near the shelf–slope break off Nova Scotia and Newfoundland in the 1970s.

its were imposed on domestic fishers, even as the catches were clearly declining in the 1980s. Some attention was paid to protection of spawning grounds and some size limits were imposed, but the overall effect of management failed to curtail overfishing. The size and number of U.S. fishing boats increased dramatically, and the stocks declined precipitously, to the point that no region of the contiguous lower 48 United States is currently in good condition. In 1994, for example, salmon fishing was curtailed in the Pacific northwest, and fishing on the Georges Bank of New England (Figure 18.12) was essentially stopped by federal proclamation more recently.

✳ Quotas or transferred quotas might produce a sustainable fishery.

To sustain a fishery is quite difficult, given the strong natural fluctuations of natural populations against which additional fishing pressure must be regulated. As we have mentioned, merely excluding some groups from fishing does not usually solve the prob-

lem, as other groups fill in the void. Fishing furthermore is becoming more and more sophisticated, with satellite navigation, sonar detection, and other technologies aiding exploitation.

A good population model is needed whose objective is a sustainable population that can continue to be exploited. Unfortunately such a model must include natural variation, which may make the imposition of a quota difficult. For example, El Niño of 1982–1983 was followed by a crash of the "loco" fishery, a large gastropod, *Concholepas concholepas*, found on Chilean rocky coasts. This fishery was intensely worked for many years and it is likely, but unproven, that the collapse was assured by the unavailability of a stock to replenish the population when the environment improved.

Minimally, one expects that the quota will vary with time, which creates difficulties with deployment of ships. One interesting approach has been the **individual transferable quota** (ITQ) system, under which licenses are given to individuals, who can sell their personal quotas to other fishers. This approach

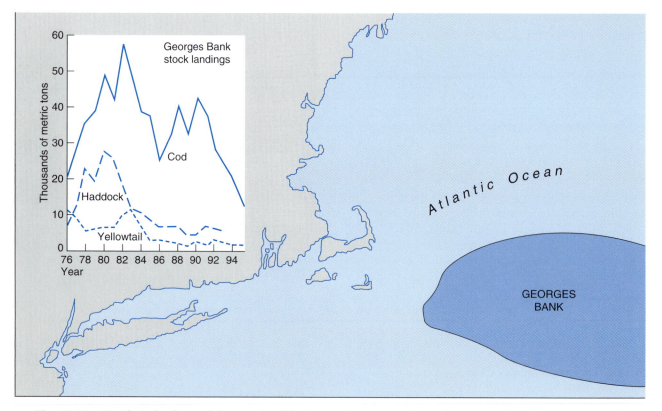

Fig. 18.12   Trends in landings of three major fisheries on Georges Bank, on the New England continental shelf.

has been applied in New Zealand, British Columbia (Canada), and Australia, particularly to invertebrate fisheries, and it might give the economic flexibility needed in deployment of fishing boats and gear.

We shall discuss another approach, marine protected areas and no-take zones, in a later section of this chapter.

Overexploitation of Whales: A Case History

✳ **Whaling began as a shore-based fishery and then developed into an open-ocean fishery.**

Whaling is one of the most romanticized of human endeavors. In our folklore, a sinister and powerful monster appears suddenly on the horizon, and the brave men of the longboats move out after it, perhaps to their death. In the nineteenth century, American whaling ships ranged far and wide, from the coast of Greenland to the antipodes, searching for and killing the profitable beasts. In those days, the sperm whale was sought for its oil, bones, and sperm for perfume. Other kinds of whale were boiled down for oil and their flesh might be taken for food. "Whalebone," a bonelike material taken from baleen whales, was prized for women's corsets.

Whales may be classified into two major groups (see Chapter 8 for more detail). The baleen whales are larger and feed on small fish and larger zooplankton such as krill. The baleen, a large bony structure suspended from the mouth, strains smaller animals from the water. The toothed whales (e.g., the sperm whale) differ in having a more standard set of mammalian teeth. They hunt larger fish and sometimes smaller marine mammals such as seals and otters. Gestation in the larger whales is nearly a year, and blue whale newborns are about 6 m long.

At first, whaling was a shore-based endeavor. Fishermen hunted the Biscayan right whale nearly to extinction as early as the seventeenth century, merely with harpoons thrown from skiffs that quickly returned to shore. On the south shore of Long Island, white settlers and natives harpooned whales from rowboats launched in the surf. With the advent of large and swift sailing vessels in the early nineteenth century, voyages often lasted for several years, and larger open-ocean whales were hunted and processed at sea. Whales were sighted in the distance as water spouted through the blowhole ("Thar she blows!" was the cry of the spotter, when he saw a whale spout-

ing), and men in longboats rowed out to harpoon the whales by hand. After the kill, the whales were tied alongships and were butchered and cooked onboard. If the killing was a dangerous outing, neither was it exactly safe to stand on the slippery, blubbery carcass and butcher the enormous animal.

We can only imagine the fantastic numbers of whales that the first Antarctic explorers must have encountered, for such numbers exist no more. At first, in the earliest part of the nineteenth century, whales migrating from the Antarctic were caught when they arrived in some of their breeding grounds, south of the Australian continent. The whales were hunted by "local" convicts who had been transported from England, and also by whalers who had traveled from North America. Shore stations were especially effective in reducing the numbers of the humpback whale, which bred in bays in Australia and New Zealand.

A number of technological advances set the stage for modern whaling in the Antarctic seas. In the 1860s a Norwegian sea captain invented the cannon-powered harpoon, and whaling crews began to pump air into carcasses to keep them afloat during butchering. Also a technique of hunting was developed that was devastating in its efficiency. A series of smaller **catcher ships** searched out whales. When an individual whale had been harpooned and killed, it was delivered to a larger **factory ship**, equipped with a **stern slipway** for hauling in the whale. The whale was then cut up and processed on board for oil. (That was the main produce obtained from Antarctic whales until after World War II, when fishers began to save the meat for use in pet food and for human consumption.) Even the canning was done onboard the factory ship. This method was first fully developed in about 1925, when the first factory ship equipped with a stern slipway operated in the Antarctic Ocean.

✳ **Open-ocean fishing technology resulted in the decline of blue-water whale populations.**

The effectiveness of the catcher–factory ship system led to hunting for blue whales in preference to the other species (fin and sei whales), owing to its large size (Figure 18.13). By the 1930s, blue whales were already declining. This decline continued until blue whales eventually reached very low numbers. A 1937 agreement set a minimum size limit on blue whales and other species, and prohibited the killing of whales

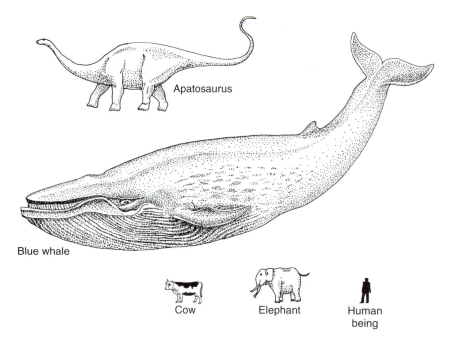

Apatosaurus

Blue whale

Cow   **Elephant**   **Human being**

Fig. 18.13 The size of a 100-foot-long blue whale in relation to some other creatures.

that had calves. A season was also set. No upper limit was placed on whaling in general, however, and the number of catcher ships increased, with no increase in the total blue whale catch. In other words, the total landings remained the same, but the catch per unit effort decreased. Subsequent international whaling conferences set limits on the number of catcher ships, whaling season, size, and number. A **blue whale unit** (**BWU**), established in 1944, set the following catch equalities: 1 blue whale = 2 fin whales = 2.5 humpback whales = 6 sei whales. Limits were set in terms of these blue whale units. This was not particularly good for the blue whale itself, because whalers included in their limit as many blue whales as possible. The BWU was thought necessary to protect whales in general on the grounds that hunters would not be so impractical as to pass up any whale simply because it was not a blue whale. To do so would have wasted too much ship time.

The International Whaling Commission (**IWC**) was established in 1946. It had the advantage of forcing the representatives of whaling nations to continue to meet to establish quotas and fishing seasons. The charter of the commission permitted nations to ignore the limits, however. This laxity was a political necessity because whaling was a high-seas fishery in the Antarctic and, therefore, no nation's law regulated fishing. This led to accommodations that were

political in nature, and the limits were set above a sustainable yield. The commission also failed at first to set limits for each individual nation, although this was later done by means of negotiations outside the formal proceedings of the IWC. The early lack of individual country limits encouraged very intensive fishing effort.

Over time, the catcher ships continued to increase in number and size, and the blue whale stocks declined precipitously through the 1950s. Individual national limits were set in 1962, and fishing for blue whales was stopped in 1965, but blue whales had already ceased to be abundant enough to be taken in significant numbers. By the early 1960s, it had become impossible for the whaling nations to catch as many whales as allowed by the quotas adopted by the IWC. The fin whale also began to decrease greatly, and fishing effort was shifted to sei whales, which were useful for their meat. Quotas were reduced, but not enough to permit the fin or blue whales to recover very well.

In recent years, there has been tremendous international pressure to ban Antarctic whaling altogether. Many conservationists have pointed out the dangerous position of the blue whale and fin whale stocks, and in 1986 an agreement was finally reached by the IWC that imposed a moratorium on whaling. Both Norway and Japan, however, have objections to the moratorium especially with regard to the minke whale. Japan still does some whaling, ostensibly for scientific purposes. In 1994, the IWC established a sanctuary throughout much of the Antarctic Ocean, but Japan objected and continues "scientific" whaling, even within the sanctuary. See Hot Topics in Marine Biology 18.1 for the latest on tracing processed whale meat to its source, even to the very whale that was killed.

It is very difficult to regulate an open-ocean fishery such as whaling. No jurisdiction controls the high seas effectively, and the International Whaling Commission has always had to recognize the possibility that any or all of the major whaling nations could (and sometimes did) walk out and ignore the proceedings and established quotas. Because the major whaling nations had invested heavily in ships and land facilities, it was difficult to dissuade them from whaling heavily in the Antarctic. Ships cost money, and many families and communities had become dependent on the income from the whaling. Although many conservationist groups were quite right in pointing out the overexploitation, no one was more keenly aware of the problem than the scientific advisers and whaling commissioners, who nevertheless found it difficult to take a stand that was strongly in favor of greatly reduced quotas. Indeed, the quotas recommended by a scientific committee were often ignored in favor of ones that could not even be met by the fishermen.

Other Types of Degradation

✳ Fish stocks can be affected by human degradation of water quality or fish habitats, or by killing of fish by means other than direct fishing.

Although overfishing began to take its toll early in this century, other factors have combined to reduce fish stocks. In many bays and estuaries, the introduction of industrial wastes and sewage has degraded water quality substantially (see Chapter 19 for more details). As a consequence, many fisheries have collapsed. In the most dramatic cases, pesticides have eliminated entire populations. For example, the spraying of the insecticide kepone in the Chesapeake Bay region had drastic effects on crab populations, collapsing the fishery for several years. Other toxic substances, such as heavy metals and various organic compounds, have contaminated fish and shellfish populations. Probably the major water quality change, however, is in the dissolved oxygen content. As discussed in Chapter 19, inputs of sewage result in eventual drastic declines in oxygen, which is essential for fish. These effects are most extreme in warm summers and in water bodies with sluggish circulation. In the summer of 1987, for example, an exceptionally hot and calm summer resulted in the development of anoxia in large areas of Long Island Sound, New York. Large kills of fish and lobsters soon followed as a result in the western part of the sound. Many of the inshore quiet bays of the world have such periods of anoxia, causing extensive fish mortality and emigration.

Human disturbance of either spawning grounds or migration routes is also a major cause of the decline of fish populations. In the Pacific Northwest region of North America, dams and deforestation have caused major reductions in salmon stocks. Dams interrupt the migration route, and juvenile salmon often swim through hydropower turbine intakes, causing extensive damage and mortality. Migrating juveniles often prefer the sluggish parts of lakes behind dams, and these areas are often filled with predators. The extensive clear-cutting of forests in areas such as British Columbia also has drastic effects, because trees normally prevent erosion. Following a period of clear-cutting of trees, soil erosion increases greatly, and gravel spawning beds are often ruined by influxes of soft sediment. In many coastal areas, salt marshes have been filled in to allow coastal construction. The salt marsh creeks are nursery grounds for a variety of fishes, and many of these areas have been eliminated in the eastern United States.

Many activities cause direct mortality in fish populations. The intakes of power plants are often the site of extensive larval and adult fish mortality. In many of the power plants, water is taken in to cool the turbines, and warm water is piped out into estuaries and coastal areas. The pump intake region **entrains** fishes and draws them through the intake pipes. When the larvae or adults approach the intake pipes, the sucking force may **impinge** them on the intakes, although a variety of diverting screens have been placed on intakes to prevent this.

Another human-imposed source of fishing mortality is the incidental killing of undesirable species of fishes, known as **bycatch**. We mentioned earlier the killing of dolphins, which are trapped in nets during

HOT TOPICS IN MARINE BIOLOGY 18.2 Dams and Water Withdrawal: A Case of the People Versus the Fish

No single resource is more important to humans than water. Water is not only crucial to agriculture and necessary for drinking, but it is also a major source of electrical power generation. To serve these needs, water is often diverted from rivers and estuaries and its rate of flow is altered. To meet the needs of agriculture, river water is diverted to irrigation pipes in order to water cropland. In the former Soviet Union, the diversion of water from rivers that once flowed into the inland Sea of Azov has created one of the greatest environmental catastrophes in history. Water flow into the sea has been reduced to a trickle, and thousands of square kilometers once covered by water productive with fishes are now a dusty wasteland.

Although nothing in the United States rivals the scale of destruction in the Sea of Azov, agricultural diversions have hurt our coastal fisheries. Over the last 50 years, the fisheries of San Francisco Bay have declined to very low levels, as water has been removed to water the farms of the central valleys of California. The Columbia River, which empties into the sea at the Oregon–Washington border, was once one of the greatest salmon rivers in the world. An extensive series of dams along its course now provides hydroelectricity, and water is withdrawn for irrigation. Because of extensive damming along the upper Columbia River and its tributary the Snake, salmon juveniles experience high mortality, despite trucking of juveniles around dams. As a result, one salmon species no longer runs to its traditional spawning grounds in the Snake River (a tributary that extends to Idaho). Others are highly endangered and have been classified in 1999 by the U.S. government for protection under the Endangered Species Act (Box Figure 18.3). While various interests will conflict, the dismantling of several earthen dams has been suggested as the only course of action that will allow the river to run free so that salmon can migrate normally.

In Texas, dams have diverted water for agricultural use, but they have also endangered coastal fisheries. The cutoff of fresh water brings salty water upstream into estuaries and alters necessary low-salinity spawning and feeding grounds. The reduction in flow is exacerbated by industrial pollution, which becomes more concentrated and toxic to fishes in the sluggish waters. One bright spot is the Apalachicola Bay in the Florida Panhandle, where a successful environmental movement led to the abandonment of plans to dam the drainage system feeding into the bay. This was crucial in saving oyster grounds and crab spawning beds, although other factors have hurt the bay more recently.

Rivers are also a major source of drinking water, and the Hudson River in New York State has been a recent battleground between municipal water authorities and fishermen. The freshwater needs of New York City have risen steadily and cannot be met during drought periods. Up to a few years ago, New York City relied mainly on a magnificent reservoir system originating in the Catskill Mountains, as well as on some of the flow of the Delaware River. However, the rising needs have led to use of water from the Hudson River during drought emergencies. So far, this has happened three times, and there are plans to tap this source again. Fisheries ecologists have argued that water withdrawal will kill fish larvae and may endanger the still-healthy fisheries of the Hudson. New York City is beginning to adopt conservation measures such as water metering, but the request for diversion still stands. Ironically, water flow in the summer of 1999 was extremely low, moving salty water so far north that communities that normally use freshwater Hudson for drinking water nearly lost their supply.

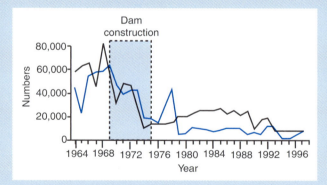

Box Fig. 18.3 Wild salmon counted on the upper Snake River, Washington: blue line, spring–summer Chinook; black line, summer steelhead. (Idaho Department of Fish and Game.)

fishing for tuna. Many gill nets catch large numbers of fishes of inappropriate species that are killed in the process of recovery of the nets. Fishery legislation has required that fishing nets have trap doors, which allow the escape of sea turtles that would otherwise drown in the nets.

One of the most potentially profound effects on living resources is exerted by bottom trawlers, which are dragged along the bottom to catch bottom fish, scallops, and other resources associated with the bottom. In heavily fished areas, just about every square meter of bottom is dragged a few times a year. Many bottoms with gravel and a rich epifauna of bryozoans, hydroids, and echinoderms will be turned over, leaving a smoother soft-bottom interface with smaller soft-bottom invertebrates. The effects of such strong changes are obvious for the benthic communities, but the feedbacks on the resource species are poorly understood. If a bottom-feeding fish depends upon smaller mobile prey, then it might suffer greatly for lack of a variegated cobbly bottom that shelters smaller fish and shrimp. On the other hand, some fish might find a bare mud bottom to be a more suitable habitat. At this stage we certainly know that bottom trawling exerts profound effects on the bottom,[6] but the total effect on fisheries remains to be understood. It will be very important to understand the recovery time, which will help set a trawling intensity quota, if necessary. On the Grand Banks, recovery may be as much as a year.[7]

Marine Protected Areas

✳ A marine protected area is a geographic conservation unit designed to protect crucial communities and to provide reproductive reserves for fisheries that hopefully will disperse over wider areas.

As human disturbance and fishery pressure increase throughout the ocean, especially along the coastline, more and more marine communities are being disturbed, with loss of structural habitats and species. It is probably impossible to stop the impacts everywhere, but given an appropriate justification, crucial areas might be protected sooner. This is the philosophy behind **marine protected areas**, an approach to conservation that focuses preservation efforts on a relatively small proportion of the total geographic range of a species or a community type. Rules in such areas range from mild restrictions on fishing or ecosystem damage to strict rules prohibiting fishing or taking of marine specimens.

In the United States, the Marine Sanctuaries Act enables designated areas to be subject to management plans that usually allow commercial exploitation of resource species but otherwise help to reduce ecosystem damage. For example, the Stellwagen Bank sanctuary, located off the Massachusetts coast, is designed to protect a crucial habitat for whales, which visit the bank to feed. Continual assessments are made of entanglements of marine mammals with fishing gear, the effects of pollution, and even the effects of whale watching, since noise and boat traffic may have negative effects. Fishing is still permitted, but rules set by the New England and Mid-Atlantic Fishery Management Councils protect spawning grounds and juveniles.

The establishment of marine protected areas can be justified from two points of view:

1. Protection of areas crucial to the maintenance and even population expansion of fishery species
2. Protection of very diverse structural habitats, such as coral reefs, or other communities that are deemed by society to be of importance for economic, educational, or aesthetic reasons

✳ Marine protected areas for marine resource species are based on the concept of metapopulations.

In Chapter 3 we discussed metapopulations, which are a set of interconnected local populations of a single species. Some local populations may be **source populations**; that is, individuals from these local populations move to other local populations by means of larval spread and adult dispersal and are aided by high reproduction. By contrast, **sink populations** receive individuals dispersing from others, but do not provide dispersing individuals to other local populations. A simple example would be a location to which fish would disperse, whereupon all would be consumed by predators. A sink population, incidentally could also be produced by human predators, so con-

---

[6] See Thrush and others, 1998, in Further Reading, Fisheries.
[7] See Schwinghammer and others, 1998, in Further Reading, Fisheries.

servation policy can be integrated into metapopulation dynamics.

Consider the following example: Lord Howe Island, Australia, is the southernmost coral reef in the world and quite separated from the nearest coral reef tract, the Great Barrier Reef, to the north. Lord Howe, moreover, is in relatively cold waters, reaching as low as 17°C in winter. It is therefore possible that local coral reproduction is limited. If so, it would have to be sustained by supply of coral larvae via currents from the main coral reef tract, which lies to the north. Lord Howe reef would therefore be a sink.

To sustain the Lord Howe populations of corals, it would be necessary to protect upcurrent supplies of larvae to the north. As it turns out, Peter Harrison discovered that mass spawning on Lord Howe occurs only a couple of months later than on more northern reefs. Lord Howe is probably sustained by local reproduction but perhaps also by supplies from outside. This would lead to a different protection strategy. It would be necessary to determine the proportion of larval supply that comes from the north.

The metapopulation approach therefore leads to useful hypotheses about coral reef populations. This example demonstrates that much information on reproduction, dispersal, and survival is needed to design marine protected areas. The main principle is to find source populations to protect. Otherwise, sinks cannot be sustained. Sinks, as mentioned, could be areas that are fished. To be sustainable, fishery areas must include protected source areas, from which larvae or adults could disperse into exploited zones. Crucial areas include spawning habitats and juvenile nursery grounds.

✳ Structural habitats are often endangered by human use and especially tourism. A multitiered strategy is essential for protection.

Structural habitats often depend upon the maintenance of a suitable substratum and the species that construct a biological landscape, such as corals in coral reefs, or kelps in kelp forests. Therefore, conservation efforts must include the metapopulation concept but also pay attention to the integrity of the structural habitat. Coral reefs are an excellent example. They are endangered from many directions today, but tourist visits are especially worrisome. With tourists come organic pollution and direct dis-

turbance of the reef through diving, boat anchor damage, taking of rare live marine specimens, and even hammering and blasting out coral colonies.[8]

Many coral reefs have been protected by means of marine protected areas, but the meaning of protected varies from place to place. The Looe Key National Marine Sanctuary in the Florida Keys prohibits fishing, anchoring, spearfishing, and taking of live marine specimens. Even more impressive is the Marine Park, which encompasses nearly the whole coastline of Bonaire, Netherlands Antilles, just north of the Venezuela coast. Boats longer than 12 feet cannot anchor on the reef but must attach to moorings that are regularly spaced along the coast. Collecting specimens, including by means of spearfishing, is prohibited. Conservation takes a three-tiered approach, called **zoning**. Some areas are completely closed to any visits, others are open to researchers, and the rest can be visited by nature tourists. This strategy has proven very successful, especially because of the consciousness with regard to conservation on the island.[9]

The Great Barrier Reef of Australia, probably the largest marine protected area in the world, is zoned according to multiple uses.[10] Zones are (1) for scientific research only, (2) for marine national park (regulated) use by tourists, and (3) for general use (includes commercial and recreational fishing, but with some regulation). Success of conservation efforts has been variable, and further regulations have subdivided tourist and general-use zones. Many local areas are more affected by crown-of-thorns starfish outbreaks (see Chapter 15) and bleaching than by local effects. Nevertheless the Great Barrier Reef is a large conservation management unit, and its protected status represents a great achievement of Australian public opinion.[11]

✳ No-take marine protected areas constitute the strongest protection.

Marine protected areas may also be **no-take areas**, where no organisms of one or more species may be exploited. The no-take concept can be applied to cru-

[8] See Luttinger, 1997, in Further Reading, Marine Protected Areas.
[9] See this Web site for more: http://www.bmp.org/park/
[10] See Laffoley, 1995, in Further Reading, Marine Protected Areas.
[11] See Sapp, 1999, in Further Reading, Marine Protected Areas.

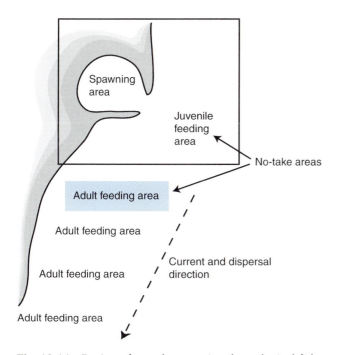

Spawning area

Juvenile feeding area

No-take areas

Adult feeding area

Adult feeding area

Current and dispersal direction

Adult feeding area

Adult feeding area

Fig. 18.14 Design of no-take areas in a hypothetical fishery. The spawning and juvenile feeding areas and one adult feeding ground are no-take zones, to allow a minimum population to complete a life cycle. Other areas might be fished, although quotas might be set on the amount of the take.

cial areas that give a minimum chance to sustain a fishery by allowing completion of a life cycle by a substantial part of the population. Figure 18.14 shows one model that could be used for management of a marine fish. The spawning and juvenile feeding grounds are protected completely, as is one of the adult feeding areas. Other areas may be exploited, but the minimum configuration might provide a sustainable fish population even if fishing overexploitation occurred suddenly in the other adult feeding areas.

No-take marine protected areas are difficult to implement because most fisheries policy is based upon expectations of damage to an entire stock by a certain amount of fishing. Indeed, spatial issues are rarely addressed, except in occasional regulations to protect spawning beds. Once a stock has been identified, managers are keen to set limits on fishing, but this does not involve designating areas from which no fish can be taken. To justify the establishment of no-take areas, it will be necessary to provide numerical models of the benefits of sustainability.

Mariculture

General Principles of Mariculture

✳ **In mariculture, the habitats of some natural populations can be simulated, changed for convenience of harvest, or enhanced to increase yields.**

Mariculture is a somewhat loose term, covering all techniques by which marine organisms are reared for most of their lives under controlled conditions in seawater directly in or connected to the sea. In the crudest form of mariculture, animals are transplanted to habitats that are optimal for growth (e.g., transplantation of hatchery-reared fishes to natural habitats). In intermediate cases, organisms are reared throughout their life cycle, but much of the rearing is in open-ocean pens, and conditions are not controlled exactly (e.g., sea ranching of salmon). In the most controlled case, the mariculturist controls the rearing environment completely and may attempt to provide completely defined foods (e.g., indoor aquarium hatcheries of striped bass in Connecticut).

Successful mariculture requires the proper choice of a species for rearing. The following characteristics are desirable.

1. **Desirability as food.** The species should be already known as a desired food item, or at least should be salable, with some publicity. Alternatively, the species should be usable as some sort of food product, such as food for another palatable species.
2. **Uncomplicated reproduction.** The organism should be relatively easy to propagate, or young organisms should be easy to obtain.
3. **Hardiness.** The species should be resistant to handling and changes in environmental conditions, and should be adaptable to different substrata.
3. **Disease resistance.** Diseases and parasites should be controllable, to minimize mortality.
4. **High growth rate per unit area.** The organisms should be able to grow rapidly in limited culture areas.
5. **Readily met food requirements.** Feeding the organisms should be easy and cheap. Animals that are high on the food chain are liable to require higher-cost protein foods.

6. **Readily met habitat requirements**. The physical habitat should be easy to duplicate in the mariculture system. Preferably the organisms should be able to grow in relatively high-density systems. There should be relatively low levels of aggressive behavior, and the organisms should be resistant to poisoning by waste products.

7. **Monoculture or polyculture**. It should be considered whether one species is to be grown alone (monoculture) or whether several species will grow most efficiently when placed in the same system (polyculture). For example, some mariculture habitats may be innately complex, and a polyculture, permitting several marketable species to grow in the assemblage of microhabitats, would be more efficient.

8. **Marketability**. The chosen species should be easy to market, accessible to markets, and of a presentable growth form to consumers.

9. **Minimal ecological effects**. The mariculture system should have few detrimental effects on the surrounding environment.

When these requirements are considered, many species turn out to be uneconomical to culture. The New England lobster *Homarus americanus* is a good example. This species is trapped extensively in the waters of the western Atlantic and is a highly prized food. In the late nineteenth century, the U.S. Fisheries Commission began to rear and release planktonic larvae. This program was discontinued when it was decided that this approach was not sufficiently enhancing the fishery. Later attempts at rearing adults to eating size have been relatively unsuccessful from the economic standpoint, although much has been learned about improving diet and growth conditions. Relatively slow growth and the need for large and separate pens probably make lobster mariculture economically inviable, although the development of better foods and rearing at higher temperature may help.

By contrast to this sad story, the shrimp *Macrobrachium rosenbergii*, which thrives in both brackish and fresh water, proved ideal for culture, as have marine shrimp. Food for the animals is easy to prepare (brine shrimp eggs, fish flesh, etc.), survival is satisfactory, and the animals grow rapidly. Rearing is accomplished in ponds that are relatively easy to maintain. Freshwater crayfish are also reared in ponds. Crayfish farms were established by the French before 1900, and the animals are now actively farmed in Louisiana. Unfortunately, side effects of shrimp farming can be severe. In southeast Asia thousands of hectares of mangrove forests and marshes have been destroyed in order to establish shrimp farms. Striped bass (*Brevoortia tyrannus*) are emerging as a rearable and highly desirable fish species, and active attempts are being made to scale up sturgeon mariculture.

Some Organisms Useful in Mariculture

* Mollusk mariculture systems enhance the availability of substratum and are located in areas of high phytoplankton supply.

Mollusks, principally bivalves, have been among the most successful mariculture organisms. This has much to do with the rapid growth of some species, combined with the ability to place animals in areas of high phytoplanktonic food supply. All the major cultured species are suspension feeders. A number of species of oysters and mussels have been the mainstay of the industry throughout the world.

The culture of bivalves involves two main steps: (1) rearing of larvae through settlement and metamorphosis into juvenile animals, called spat or seed, and (2) rearing of adults, usually in rafts or attached to suspended poles or ropes. Spat can be collected in natural habitats, but they are often reared from the egg (and sperm) in hatcheries. By contrast, commercially viable adult rearing is always done in bays and estuaries.

MUSSELS

Mussel culture is best developed in France and became established quite by accident. Patrick Walton, an Irish sailor, was shipwrecked in the year 1235 on the Atlantic coast of France. To snare birds for food, he placed a net attached to poles on a mudflat. Instead of attracting birds, the poles were colonized by settling mussel larvae, which grew far more rapidly than those on the mudflat itself. This is the basis of the modern method, employed mainly in Brittany, France. Ropes are placed near mussel beds and larvae settle on the ropes and grow to a size of 5–10 mm. Following this, the ropes are transported to the adult rearing site, where poles are driven into the mud. The ropes are wrapped in a spiral around the poles, and the mussels then commence to grow rapidly. The bottom of the pole is sheathed in smooth plastic, to prevent predators from climbing the poles. The mussels are thinned out to allow maximum growth, and the harvest is taken about one year later.

A variant of this technique is to use ropes suspended from rafts. In Japan, such rafts are often placed in bays with strong tidal currents, a tactic that exposed them to a rich phytoplankton food supply. Mussel culture is also a well-developed industry in Spain and in the Philippines. In the United States, mussel culture has become an especially important part of the coastal economy of Maine.

OYSTERS

The oyster is the most profitable bivalve employed in mariculture (Figure 18.15). Many different species have been used successfully in a wide variety of climates. Oyster culture is quite ancient: it is known to go back to the ancient Romans. In the past few decades, oyster culture has been expanded, owing to

Fig. 18.15 Oysters (*Crassostrea gigas*) are cultured in racks suspended from floats in a tidally flushed bay (top) in San Juan Island, Washington. Newly settled spat are sorted (lower left) to get the fastest-growing individuals (middle right), and these are raised in suspended plastic trays until ready for market (lower right). (Photographs by the author.)

the high price of oysters and to the decline of many natural oyster beds, which reside in environmentally vulnerable estuarine mudflats. Some of the great oyster grounds have disappeared completely in the last century. In the nineteenth century, Staten Island, New York, harbored one of the greatest oyster grounds in the world. Alas, pollution has taken its toll, and seagulls congregate around one of the largest landfills in the world, which now sits near the former oyster grounds. Although oysters are a delicacy in North America, they are a staple food in the Orient.

Like most bivalve mollusks, all species of oysters (members of the Family Ostreidae) have planktotrophic larvae, which swim in the water column for a few weeks. Culture therefore involves the settling and metamorphosis of larvae and the rearing of adults. Spat either are reared in hatcheries or are collected on hard substrata that are placed in natural oyster beds. Dead shells are a common substratum, but plastic and wood are also employed. Once spat have been collected, they are transferred to the adult growth area. In Japan, this may be a raft in quiet water or suspended rice ropes in more open waters. In France, oysters are placed on mudflats, to ensure that the shell develops a pleasing ovoid shape, so that the oysters may be served raw on the half-shell. There is not such a limitation in Japan, where the oysters are shucked before marketing.

Oysters have a variety of predators, but drilling snails and starfish are the two most important. Disease is a far more important problem. At present a poorly understood disease, MSX, is devastating the naturally occurring (noncultured) oysters of Chesapeake Bay and was responsible for a massive decline of noncultured oysters in Delaware Bay. In MSX, the oysters are parasitized by an amoeboid protozoan, but little is known of its life history. Many have lobbied for the introduction of an exotic oyster species, which might be more resistant to the parasite. Others argue that such tactics have often had disastrous unanticipated consequences, such as the introduction of other undesirable species and parasites.

Effects of Harmful Algal Blooms

* Red tides and other nuisance blooms can devastate fisheries, particularly bivalve–mollusk fisheries.

In Chapter 9, we discussed the dense red tides, dominated by toxic dinoflagellates, that sometimes develop in coastal waters. Such red tides, as well other harmful algal blooms, can severely affect shellfish un-

der mariculture (as well as noncultured shellfisheries). As mentioned in Chapter 9, the retention of toxins produced in phytoplankton blooms is not the same for all suspension-feeding mollusks. Whereas some toxins disappear from shellfish soon after the bloom has subsided, other toxins may persist for many months. The toxins are a health hazard, but they are also a source of severe economic loss. In 1972, a major red tide in New England caused the closure of hundreds of miles of coastline to shellfish exploitation. The presence of saxitoxin, the source of paralytic shellfish poisoning, has caused continuous closure of butter clam fisheries in the Pacific Northwest and in Alaskan waters. Sandra Shumway[12] points out that toxic algal blooms are a special problem for mariculturists and shellfisheries alike, because even a brief event erodes consumer confidence, which is very difficult to restore.

* Algal species causing harmful blooms have a range of effects—from shading, to irritating delicate organs of cultured organisms, to producing toxins that may have severe effects on cultivated organisms and human consumers.

Harmful algal blooms have a range of effects on cultured fish and shellfish and upon people. Some diatom species of the genus *Chaetoceros* have spines that irritate the gills of fish and shellfish. They may cause lesions and encourage infections in fish raised in pens. The chrysophyte alga causing northeastern U.S. so-called brown tides, *Aureococcus anophageffferens*, has direct toxic effects upon bivalves such as the bay scallop, *Argopecten irradians*. It also is the source of very high turbidity, which reduces light input into estuaries and causes harm to other phytoplankton and to attached benthic sea grasses (an important attachment site for juvenile scallops).

Other phytoplankton species are even more worrisome because they produce toxins that are greatly harmful to resource species, which may accumulate the toxins and transfer them to people, who also may be harmed. Most famous among these are the dinoflagellates responsible for **red tides**, which are a complex of blooms produced by a variety of dinoflagellate species. Most notably, members of the genera *Alexandrium*, *Gymnodinium*, and *Pyrodinium* cause **Paralytic Shellfish Poisoning** (PSP) by produc-

---

[12] See Shumway, 1990, in Further Reading, Mariculture.

ing any one of 18 neurotoxins that generally bind to nerve cell membrane-associated sodium channels. When humans consume PSP-contaminated shellfish, the toxins cause immediate numbness in the mouth area, and they also can cause strong gastrointestinal and respiratory symptoms, sometimes including death. Many species of bivalves mollusks concentrate the PSP toxin saxitoxin, although they do so to different degrees and concentrate it in different tissues. Accumulation and depuration rates vary with species, season, tissue, and location. A group of species of the diatom *Pseudonitzchia* has toxic strains that produce **domoic acid,** inducing **Amnesic Shellfish Poisoning** (ASP), which can cause amnesia, neurological damage, and even death when humans consume ASP-contaminated shellfish. Toxic strains were first identified in Prince Edward Island in eastern Canada[13] but have since been found worldwide. Another important group in this ignominious list is **Neurotoxic Shellfish Poisoning,** caused by brevetoxin, which is produced by the dinoflagellate *Gymnodinium breve.* Brevetoxin concentrates in shellfish, but it also can be breathed by humans in aerosols near the coastline, causing gastrointestinal and neurological problems, as well as asthma-like symptoms.

The dinoflagellate *Pfiesteria piscicida* is one of the most fascinating and troubling harmful algal bloom organisms, and it appears to have been increasing in extent and abundance in recent years. It was first observed in North Carolina estuaries, but has also been the source of fishery closures in Chesapeake Bay.

Dinoflagellates often have a number of life history stages, but *Pfiesteria piscicida* has 28 stages of encysted, amoeboid, and mobile biflagellate forms, which is surely an extreme case. As in other dinoflagellates, encysted stages are in the bottom, but they can apparently rapidly develop into biflagellate swimming forms that emerge from the bottom and attack fish. Using an as-yet unidentified toxin and a well-developed feeding organ known as a peduncle, they attack the skin of fish and leave bloody lesions, especially on the thin-skinned menhaden, *Brevoortia tyrannus. Pfiesteria* has been implicated in at least half of the large-scale fish kills observed in North Carolina inshore waters.[14] Worst of all, the toxin is very dangerous to humans, causing a number of symptoms including dementia and loss of vigor.

✳ **Harmful algal blooms have greatly increased in frequency and in geographic extent over the past 30 years.**

Toxic algal blooms are on the rise, and the frequency and number of localities have increased substantially over the past few decades.[15] The possible reasons for this change can be divided into three categories:

1. An increase in coastal nutrient input and general pollution.
2. An increase of arrival of new algal recruits, which has been exacerbated by the transport of organisms for mariculture, and an increase of general oceanic transport.
3. Increased mariculture, which results in local release of nutrients, disturbance of local habitats, and stimulation of harmful algal bloom growth.

Physical oceanographic processes may be an important external force in causing local harmful algal blooms. Red tide dinoflagellates have resting cysts that survive in sediments, probably for many years. Storms turn over the bottom and can stimulate the cysts to transform into active planktonic dinoflagellate stages. There are know geographically localized hot spots, where cysts concentrate and provide the source for later red tides. Even changes of current direction may cause harmful blooms. Mussel culture in the estuaries of northwestern Spain depends upon a nearly continuous upwelling system, which delivers nutrients that fuel rich phytoplankton blooms. Occasionally the upwelling collapses and offshore winds bring water to shore, including water dominated by red tide dinoflagellates.[16] Outbreaks of the dinoflagellate *Pfiesteria piscicida* occur when water flow is very slow, allowing the encysted dinoflagellates to detect fish in the water column above, emerge, and attack the fish.[17]

Seaweed Mariculture

✳ **Mariculture of seaweeds may be useful in the production of products such as agglutinants, food, and organic matter suitable for methane production.**

Seaweeds are consumed widely as food and are used for a variety of products. The Japanese use a seaweed

---

[13] See Subba Rao and others, 1988, in Further Reading, Mariculture.
[14] See Burkholder and others, 1999, in Further Reading, Mariculture.
[15] See Smayda, 1992, in Further Reading, Mariculture.
[16] See Tilston and others, 1994, in Further Reading, Mariculture.
[17] See Burkholder and others, 1999, in Further Reading, Mariculture.

food (called nori) derived from the red seaweed *Porphyra* in sushi and other foods, and it has become popular in the United States and Canada. In recent years, seaweed has been identified as a "health food," and many people have been harvesting seaweeds, especially on the west coast of the United States. This may lead to more pressure to culture seaweed in the United States, although seaweed culture is now very rare in this country, except for research purposes. The high nutritional value of seaweed is not just an illusion of some health-food enthusiasts. Nori is richer in protein than rice, and is more digestible. Seaweeds are also the source of a number of useful substances, including the agglutinant carrageenan, which is used in a number of food products. It has also been suggested that seaweed be grown as a source of methane for fuel. This has not passed much beyond an experimental idea.

Seaweed culturing is well developed in the Orient, particularly in Japan and China. Nori cultivation in Japan goes back to the seventeenth century and involves the collection of spores in nets or on branches of bamboo. The branches are then moved to an area suitable for rapid seaweed growth (e.g., a water body rich in dissolved nutrients such as an estuary). The thalli are harvested frequently and are cut and pressed into sheets for market. The best growth is in spring or early summer, when *Porphyra* grows in the form of a leafy thallus. In June the thalli begin to senesce and gametes are eventually formed, which unite to form a different type of spore. These are the source of a later life stage, the conchocoelis stage, which appears on oyster shells as a red film. This is the stage that produces the conchospores, a haploid stage that settles and germinates to form the leafy thallus that is eventually harvested. The discovery of the conchospore has been used to improve the efficiency of the hatchery rearing of nori. This has greatly increased the efficiency of culture because nori growers can immerse their nets in a tank filled with spores, thus eliminating the uncertainty of recruitment in the natural habitat.

In the eastern Pacific, kelp farming has become a major means of livelihood. The kelps *Macrocystis* and *Nereocystis* grow rapidly, sometimes as much as 15 cm a day. As discussed in Chapter 15, kelp forests develop near the coastline and are among the most productive of benthic marine habitats. The kelps are harvested mainly for alginates, used in a number of foods. Harvesting usually involves boats that have a conveyor belt with cutter blades, which cut the stipes a few feet below the water surface. Kelp populations are enhanced by the culturing of the small gametophyte stage, which produces the sporophytes, whose early stages are cast from boats or applied directly by divers onto the bottom.

Fish Ranching

✳ **Fish ranching, or farming, is a major means of rearing fin fishes such as salmon.**

Because it is mere hunting and gathering, fishing seems rather primitive. Of course, many sophisticated methods have been developed to estimate the size of stock, and various theories have had uneven success in management applications. To use an analogy to the changes that occurred from primitive to neolithic society, it may be more efficient to change from hunting to farming. The principal advantage of farming, including fish farming, is the management of a precise area to maximize the productivity and health of all life stages of the food organism.

Practical experience and research have combined to increase greatly the yield per acre of nearly all terrestrial crops and livestock. Although such control is somewhat less practical in the vast expanse of the open ocean, many marine animal species have been farmed profitably. Nevertheless, fish farming continues to account for a minor proportion of the total fish catch.

Successful fish farming requires species that are relatively easy to breed and rear, which excludes the vast majority of species. The best species must have low mortality, high growth efficiencies, and low susceptibility to disease and parasites. In addition, they should require relatively small areas that can be contained by means of inexpensive fencing. There is also a taste factor to be considered, because every society has become accustomed to the taste of a somewhat different suite of species. The spiny dogfish, for example, is consumed in Europe but is virtually untouched in the United States (although this species is being exported from New England to the United Kingdom). By comparison to the relatively narrow taste of the American public, the Japanese have a well-developed palate for a wide variety of invertebrate, vertebrate, and plant species.

Fish farming has long been well developed in the Orient. A variety of freshwater species are reared in ponds and streams. Multispecies culture has been developed where several distinct microhabitats occur, such as quiet- and running-water habitats. Salmon ranching is one of the most successful forms of fish farming, and it is widespread in North America and in the British Isles, Norway, and Ireland. Hatcheries are used to rear fry, which are introduced as young fishes into pens suspended in shallow-water coastal areas. A variety of artificial foods enhance growth, and fishes are raised to market size in the pens. While keeping the pens free of fouling organisms presents some problems, growth is excellent and the technique is spreading.

The ranching of salmon conflicts with the objective of restoring wild fishes to their former natural levels, before overfishing and water withdrawal take their tolls. Many ranched salmon escape, and it has been estimated that about one-third of Norwegian salmon originated from salmon pens around the North Atlantic and North Sea. If the ranched salmon displace the wild stocks, the eventual result will be the reduction of the genetic diversity of salmon.

Use of Genetic Manipulation

✳ **Genetic manipulation proves useful in improving performance of organisms in mariculture systems.**

In agricultural systems, genetic approaches have long been used to enhance production, disease resistance, and general vigor. Such approaches are in their infancy in mariculture systems. At present, there are a large number of hatcheries for fishes and shellfish, and these can be the source of stocks for crossbreeding. The crossbreeding of stocks from different areas, combined with selection for desirable traits, can be used to fashion a population, by selective breeding, into one with high growth and survival characteristics. This approach is useful because one or more desirable traits may initially be fixed in one population and another set of desirable traits may be present in others. It is the combination of such traits that has been a source of vigorous stocks of agricultural plants and stock animals. Modern molecular techniques of gene transfer have also begun to be used to enhance growth of farmed salmon.

Further Reading

FISHERIES

Baker, C. S., and S. R. Palumbi. 1994. Which whales are hunted? A molecular genetic approach to monitoring whaling. *Science*, v. 265, pp. 1538–1539.

Castilla, J. C. 1996. Chilean resources of benthic invertebrates: Fishery, collapses, stock rebuilding and the role of coastal management areas and national parks. In D. A. Hancock, D. C. Smith, A. Grant, and J. P. Beumer, eds., *Second World Fisheries Congress; Developing and Sustaining World Fisheries Resources.* Canberra, Australia: CSIRO, pp. 130–135.

Cipriano, F., and S. R. Palumbi. 1999. Genetic tracking of a protected whale. *Nature*, v. 397, pp. 307–308.

Cushing, D. 1988. *The Provident Sea.* Cambridge: Cambridge University Press.

Dayton, P. K., S. F. Thrush, M. T. Agardy, and R. J. Hofman. 1995. Environmental effects of marine fishing. *Aquatic Conservation: Marine and Freshwater Ecosystems*, v. 5, pp. 205–232.

Dethier, M. N., D. O. Duggins, and T. F. Mumford Jr. 1989. Harvesting of non-traditional marine resources in Washington State: Trends and concerns. *Northwest Environment*, v. 5, pp. 71–87.

Dewees, C. M. 1996. Summary of individual quota systems and their effects on New Zealand and British Columbia fisheries. *National Academy of Science International Conference on Ecosystem Management for Sustainable Marine Fisheries.* National Research Council, Ocean Studies Board, February 19–24, Monterey, CA.

Everhart, W. H., A. W. Eipper, and W. D. Youngs. 1975. *Fishery Science.* Ithaca, NY: Cornell University Press.

Gulland, J. A. 1974. *The Management of Marine Fisheries.* Seattle: University of Washington Press.

King, M. G. 1994. Fisheries. In L. S. Hammond and R. N. Synnot, eds., *Marine Biology.* Sydney, Australia: Longman Cheshire.

Klimley, P. 1999. Sharks beware. *American Scientist*, v. 87, pp. 488–491.

Kurlansky, M. 1997. *Cod: A Biography of the Fish That Changed the World.* New York: Walker and Co.

Lembi, C. A., and J. R. Waaland, eds., 1988. *Algae and Human Affairs.* Cambridge: Cambridge University Press.

Ludwig, D., R. Hilborn, and C. Walters. 1993. Uncertainty, resource exploitation and conservation: Lessons from history. *Science*, v. 260, pp. 17–18.

Mangel, M. 1993. Effects of high-seas driftnet fisheries on the northern right whale dolphin *Lissodelphis borealis. Ecological Applications*, v. 3, pp. 221–229.

Orensanz, J. M., A. M. Parma, and O. O. Iribarne. 1991. Population dynamics and management of natural stocks. In S. E. Shumway, ed., *Scallops: Biology, Ecology and Aquaculture.* Amsterdam: Elsevier, pp. 625–713.

Paerl, H. W. 1988. Nuisance phytoplankton blooms in coastal, estuarine, and island waters. *Limnology and Oceanography*, v. 33, pp. 823–847.

Pauly, D., V. Christensen, J. Dalsgaard, R. Froese, and F. Torres Jr. 1998. Fishing down marine food webs. *Science*, v. 279, pp. 860–863.

Rothschild, B. J. 1986. *Dynamics of Fish Populations*. Cambridge, MA: Harvard University Press.

Sainsburg, J. C. 1971. *Commercial Fishing Methods*. London: Fishing News Books.

Schwinghammer, P., D. C. Gordon, Jr., T. W. Rowell, J. Prena, D. L. McKeown, G. Sonnichsen, and J. Guigne. 1998. Effects of experimental otter trawling on surficial sediment properties of a sandy-bottom ecosystem on the Grand Banks of Newfoundland. *Conservation Biology*, v. 12, pp. 1215–1222.

Thrush, S. F., J. E. Hewitt, V. J. Cummings, P. K. Dayton, M. Cryer, S. J. Turner, G. A. Funnell, R. G. Budd, C. J. Milburn, and M. R. Wilkinson. 1998. Disturbance of the marine benthic habitat by commercial fishing: Impacts at the scale of the fishery. *Ecological Applications*, v. 8, pp. 866–879.

MARINE PROTECTED AREAS

Alder, J. 1996. Have tropical marine protected areas worked? An initial analysis of their success. *Coastal Management*, v. 24, pp. 97–114.

Ballantine, W. J. 1996. 'No-take' marine reserve networks support fisheries. In D. A. Hancock, D. C. Smith, A. Grant, and J. P. Beumer, eds., *Second World Fisheries Congress; Developing and Sustaining World Fisheries Resources*. Canberra, Australia: CSIRO, pp. 702–706.

Castilla, J. C., and R. Durán. 1985. Human exclusion from the rocky intertidal zone of central Chile: The effects on *Concholepas concholepas* (Gastropoda). *Oikos*, v. 45, pp. 391–399.

Dugan, J. E., and G. E. Davis. 1993. Applications of marine refugia to coastal fisheries management. *Canadian Journal of Fisheries and Aquatic Sciences*, v. 50, pp. 2049–2042.

Gubbay, S., ed. 1995. *Marine Protected Areas: Principles and Techniques for Management*. London: Chapman & Hall.

Laffoley, D. 1995. Techniques for managing marine protected areas: Zoning. In S. Gubbay, ed., *Marine Protected Areas: Principles and Techniques for Management*. London: Chapman & Hall, pp. 103–118.

Luttinger, N. 1997. Community-based coral reef conservation in the Bay Islands of Honduras. *Ocean & Coastal Management*. v. 36, pp. 11–22.

Puotinen, M. L. 1994. Designing effective baseline monitoring programs for the Great Barrier Reef Marine Park, Queensland, Australia. *Coastal Management*, 1994. v. 22, pp. 391–398.

Polunin, N. V. C., M. K. Halim, K. Kvalvagnaes, 1983. Bali Barat: An Indonesian marine protected area and its resources. *Biological Conservation*. 1983. v. 25, pp. 171–191.

Roberts, C. M., and N. V. C. Polunin. 1991. Are marine reserves effective in management of reef fisheries? *Reviews of Fish Biology and Fisheries*, v. 1, pp. 65–91.

Sapp, J. 1999. *What is Natural?: Coral Reef Crisis*. New York, Oxford University Press.

MARICULTURE

Anderson, D. M. 1997. Bloom dynamics of toxic *Alexandrium* species in the northeastern United States. *Limnology and Oceanography*, v. 42, pp. 1009–1022.

Anderson, D. M., D. M. Kulis, B. A. Kaefer, and E. Berdalet. 1999. Detection of the toxic dinoflagellate *Alexandrium fundyense* (Deinophyceae) with oligonucleotides and antibody probes: Variation in labeling intensity with physiological condition. *Journal of Phycology*, v. 35, pp. 870–883.

Bardach, J. E., J. H. Ryther, and W. O. McLarney. 1972. *Aquaculture, The Farming of Freshwater and Marine Organisms*. New York: Wiley-Interscience.

Bird, K. T., and P. H. Benson, eds. 1987. *Seaweed Cultivation for Renewable Resources*. Amsterdam: Elsevier.

Bricelj, V. M., and S. E. Shumway. 1998. Paralytic shellfish toxins in bivalve molluscs: Occurrence, transfer kinetics, and biotransformation. *Reviews in Fisheries Science*, v. 6, pp. 315–383.

Burkholder, J. M., M. A. Mallin, and H. B. Glasgow, Jr. 1999. Fish kills, bottom-water hypoxia, and the toxic *Pfiesteria* complex in the Neuse River and Estuary. *Marine Ecology—Progress Series*, v. 179, pp. 301–310.

Laubier, A., and L. Laubier. 1993. Marine crustacean farming: Present status and perspectives. *Aquatic Living Resources*, v. 6, pp. 319–329.

Naylor, R. L., et al. 2000. Effect of aquaculture on world fish supplies. *Nature*, v. 405, pp. 1017–1024.

Primavera, J. H. 1993. A critical review of shrimp pond culture in the Philippines. *Reviews in Fisheries Science*, v. 1, pp. 151–201.

Shumway, S. E. 1990. A review of the effects of algal blooms on shellfish and aquaculture. *Journal of the World Aquaculture Society*, v. 21, pp. 65–104.

Smayda, T. 1992. Global epidemic of noxious phytoplankton blooms and food chain consequences in large ecosystems. In K. Sherman, L. M. Alexander, and B. D. Gold, *Food Chains, Yields, Models, and Management of Large Marine Ecosystems*. Boulder CO: Westview, pp. 275–307.

Subba Rao, D. V., M. A. Quilliam, and R. Pocklington. 1988. Domoic acid—A neurotoxic amino acid produced by the marine diatom Nitszchia pungens in culture. *Canadian Journal of Fisheries and Aquatic Science*, v. 45, pp. 2076–2079.

Sullivan, W. 1988. Methods of analysis for DSP and PSP toxins in shellfish—A review. *Journal of Shellfish Research*, v. 7, pp. 587–596.

Tilston, G. H., F. G. Figueiras, and F. Fraga. 1994. Upwelling-downwelling sequences in the generation of red tides in a coastal upwelling system. *Marine Ecology—Progress Series*, v. 112, pp. 241–253.

Review Questions

1. Why is stock identification so crucial in fisheries management?

2. How is the size of a fishery stock usually assessed?

3. Draw a diagram illustrating an expected relationship between stock size and recruitment in the following reproductive season.

4. Why might the expected stock recruitment relationship you have diagrammed fail to develop?

5. What is the basis of the concept of maximum sustainable yield? Why has the effectiveness of this concept as a tool in fisheries management been criticized?

6. How did technology contribute to the decline of open-ocean whale stocks?

7. Why was the Blue/Whale Unit adopted? Why did it not a represent a good management tool?

8. What characteristics are desirable in choosing a species useful for mariculture?

9. How do harmful algal blooms affect the economy of shellfish mariculture?

10. Many fishers are resistant to switching from fishing to mariculture. Why do you think that this is so?

11. What are some benefits and possible problems connected with fish ranching?

12. How might genetic approaches improve fish and invertebrate mariculture?

19

Marine Pollution

It is hard to believe that many of the most polluted coastal waters once brimmed over with fish and shellfish. In the early nineteenth century, the waters of metropolitan New York City were a culinary delight. Raritan Bay, now bordering the largest landfill in the world, once harbored some of the richest oyster beds in America, and the East River was a paradise for sports fishing. Commercial shell fishing and fin fishing were major enterprises. Needless to say, times have changed. The arrival of the Industrial Revolution and the sudden flowering of New York City as a major commercial center brought along major habitat deterioration. The oyster beds are gone, and any remaining shellfish are dangerous to eat. Although fishes are far more abundant than many politicians usually acknowledge, they are often tainted with toxic substances, and those depending upon the bottom are reduced in abundance and diversity. This pattern of decline is not unique to the New York region; it is typical of urbanized coasts throughout the world.

Human Effects on the Marine Environment

✳ **Complex interactions of human impacts often make it difficult to understand the role of various pollutants in degrading the marine environment.**

As obvious as biotic degradation may seem to be, both its causes and cures are sometimes obscure. For one thing, polluted areas usually suffer from several types of human insult. These combine in complex ways to cause biotic degradation, making difficult both the quantification of the degradation and management decisions on which problem to address. There is also the question of cost and motivation. Although most people would like a clean environment, the will or the ability to pay for it may be lacking. Sometimes private developers who are willing to pay strict attention to possible biological problems find it prohibitively expensive to make up for the cumulative impact of past development and other human activity. Sometimes, an appropriate political climate for stewarding environmental cleanliness is lacking. In many instances, environmentalist organizations have used the courts to obtain action on particular problems that would not otherwise have been addressed.

Man's effects on the marine environment may be divided into the following general categories:

1. Alteration of bottom substrate through dredging, changing of shoreline structures, and filling
2. Introduction of toxic substances dangerous either to marine life or to human beings
3. Release of sewage rich in nutrients for marine microorganisms

4. Heating and release of heated water by power plants

* Pollution may be long term (chronic) or short term (acute).

It is convenient to divide the effects of pollution into **long-term** (**chronic**) and **short-term** (**acute**) effects. Chronic pollution involves the introduction of a toxic substance or other anthropogenic factor, often continuously and in fairly low levels, causing a degradation of the environment. Inputs of nutrients derived from sewage are a good example. Short-term inputs may have sharp effects, but these may dissipate with time. An oil spill is an example of a short-term toxic input. At first, oil often has catastrophic effects on a marine biota, but these effects may gradually be ameliorated as the oil breaks down.

* Pollution may come from point sources or from a variety of geographic points.

In some cases, pollution comes from a **point source**, such as a single sewer pipe or factory wastewater outfall. In such cases, the concentration of the substance or the intensity of the effect (e.g., temperature near a power plant outfall) should decline with increasing distance from the point source. The nature of the decline depends on the physicochemical properties of the substance or factor, the water currents and the sedimentary environment, and the rate of introduction of the substance or factor. Such cases are relatively simple in terms of identification and management because a regulatory agency can find the source and monitor the spatial extent of its effects. By contrast, **non–point source effects** cannot be attributed to any single spot. Runoff following rain is a good example of a nonpoint source; toxic substances and fertilizer-derived nutrients may then be swept into a basin over a broad extent of the coast. Such sources are far more difficult to manage because the source is difficult to define geographically.

* The effects of human activities can be difficult to assess objectively, but changes in overall diversity, the presence and physiological condition of bioassay species, and the degree of evolved resistance can be used for monitoring.

Although certain anthropogenic effects are obvious (e.g., those of an oil spill), it is often difficult to assess in an objective way the degree of degradation of a marine habitat. This fact stems partly from the tremendous temporal variability of marine populations. To detect a change, it is essential to have a backlog of data on the normal background populations as they existed before human beings affected the environment. These sorts of data are rather rare, and unless the change is sudden and severe, it is often difficult to demonstrate that marine populations have declined. In some cases, studies made over time use different sampling techniques, making comparisons problematic. For example, it is difficult to assess changes in soft-bottom benthic populations in New York Harbor, owing to the use of different mesh sizes in different surveys.

When appropriate baseline data are available, four criteria may be used to gauge human impact on a marine environment. These are changes in diversity, changes in relative abundance of species (community structure), the physiological condition of bioassay species, and the degree of evolved resistance in species residing in the area under consideration.

Diversity can be estimated in terms of the number of species or by a relative abundance component. Diversity is said to increase if the number of species increases, or if the abundances of the species are more evenly distributed. For example, a low-diversity habitat would consist of very few species, with strong dominance by one species. Although one can clearly find exceptions, diversity tends to decline in strongly polluted habitats (Figure 19.1). In polluted habitats, only species resistant to environmental change are likely to persist. These are often a small subset of the total species pool that includes species capable of rapid colonization in strongly disturbed habitats; such species are termed **opportunists**. Marine communities subject to pollution often resemble natural assemblages strongly affected by closely related physical disturbance factors. While diversity can be monitored to assess pollution, usually the effects of pollution can be detected with much more sensitivity by using statistical approaches to measure species abundance changes and patterns of association among species. In all cases it is essential to compare polluted habitats with unaffected habitats that are otherwise similar in substratum, depth, current strength, and so on.

While total species richness can be assessed, pollution may also have impacts on community structure. Top predators in food webs are of necessity

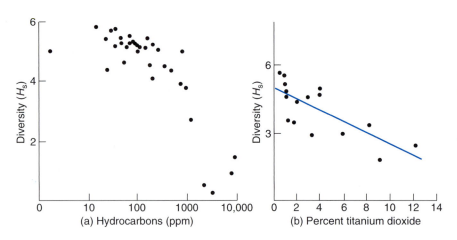

Fig. 19.1 Reduction of diversity of benthic macrofauna (a) along a gradient of increasing sediment hydrocarbon concentration near an oil platform and (b) with increasing concentration of titanium dioxide waste in a Norwegian fjord; H_S is a measure of diversity that increases with the number of species and the relative evenness of abundance of the species. (After Gray, 1989.)

lowest in abundance and might be most easily eliminated by a pollution event. Such a loss might cascade downward through the food web, causing changes in species abundances and competitive interactions. Rare species in general are likely to become extinct more than common species, which might increase the resemblance between species composition at different sites.

In some cases, certain vulnerable species are used as indices of pollution. The absence of the species, reduced reproduction, or impaired physiological performance may be used as evidence of environmental degradation. For this purpose, much applied biological research has been concentrated on a few common marine species. For example, the marine mussel *Mytilus edulis* has been investigated extensively, and the effects upon it of varying food, temperature, and toxic substances have been measured. Crabs are also studied extensively, especially owing to their vulnerability to pesticides. This is especially true of crab larvae, which are usually sensitive to concentrations of toxins far below those that affect the adults.

The use of indicator species has led to the concept of **bioassay**, which is the measurement of some parameter in a "standard" bioassay species. The bioassay species is usually selected for its abundance and ease of rearing. A population is exposed to a range of concentrations of a toxic substance, for example. Mortality rate, uptake rate of the toxic substance, or impairment of physiological function can then be measured. For example, the effect of metals can be estimated in terms of its effects on lysosomes, which are important cellular organelles in which protein degradation takes place. More commonly, the concentration at which half an experimental population is killed is measured.

Bioassays can be used in either of two ways: (1) to test the toxicity of a given chemical, water, or sediment or (2) to measure various physiological, biochemical, or morphological properties for use in estimating the degree of effect of a given substance in the environment on species collected in the field. The first approach was mentioned in the preceding paragraph, which described the exposure of laboratory populations to a toxic substance or suspected toxic water or sediment. However, by the time of testing, field populations may have already responded to toxic substances, perhaps allowing us to measure some indicative property, such as abnormal body weight or development of abnormal structures. For example, in many gastropods one can find females with extensive development of male anatomical parts including a penis and a vas deferens. Such individuals occur in greater frequency in the presence of toxic metals, and the percentage of females with such male development turns out to be a good indicator of the toxic effects of tributyltin (TBT), the antifouling paint once used extensively on boats, as mentioned in Chapter 18. The decline of the gastropod *Nucella lapillus* in southwestern England has been related to the use of TBT by measurement of the frequency of male pseudogenitalia in field populations.[1] As it turns out, the development of a vas deferens can block the genital pore and therefore curtail the reproductive output of a female snail.

The introduction of a toxic substance may kill off

[1] See Bryan and others, 1986, in Further Reading.

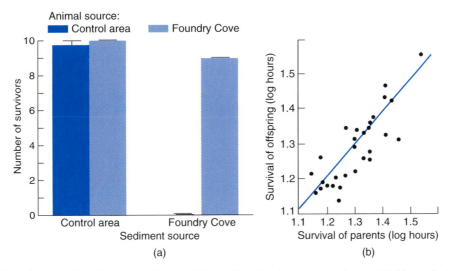

Fig. 19.2 (a) When the aquatic oligochaete *Limnodrilus hoffmeisteri* was taken from a highly cadmium-polluted cove, it was found to have much greater resistance to cadmium than worms from an unpolluted control area. Its survival was nearly the same in metal-polluted sediment as in clean sediment, whereas the worms from the unpolluted habitat soon died when exposed to the metal-polluted sediments. (b) The correlation of resistance to cadmium between parents and offspring indicates a strong genetic component to the resistance trait. (After Klerks and Levinton, 1989a, 1989b, 1992.)

only a fraction of a species population. If there is genetic variability for resistance to the toxic substance, then the population may gradually evolve resistance. After a period of time, the average individual will not be sensitive to the toxic substance. The degree of resistance can therefore be used as an index of the biological effect of a substance introduced by human beings (Figure 19.2b).

Toxic Substances

* When a marine organism is exposed to a toxic substance, the toxic substance may not increase in concentration, or it may indeed continue to increase. Some toxic substances are magnified in concentration in organisms, as the toxin is transferred through the food web.

Unfortunately, a wide variety of toxic substances have been released into the marine environment, some with catastrophic consequences for marine life and for man. In many cases, the catastrophe was unanticipated. For example, the use of the pesticide kepone in the Chesapeake Bay region devastated the crab fishery in the 1960s. Similarly, a release of polychlorinated biphenyls (PCBs) into the the Hudson

River, New York State caused the closure of the entire striped bass fishery, because the fish took up the PCBs, which are believed to be carcinogenic.

As mentioned earlier, toxic substances may be introduced in a single large pulse or chronically, perhaps at low levels. A single pulse of input may be followed by gradual biological availability, especially if the toxic substance fails to break down rapidly. This is the case for PCBs, extremely toxic compounds that are broken down very slowly by bacteria. In the Hudson River, sediment erosion, transport, and deposition tend to spread PCBs from localized hot spots throughout the estuarine system.

The biological uptake pattern may also differ. **Noncumulative toxic substances** do not increase in concentration in the body over time, even if the organism is exposed chronically to the substance. By contrast, **cumulative toxic substances** continue to increase in concentration and may be found most abundantly in a single tissue. For example, cadmium tends to increase over time in the digestive gland of crabs. The accumulation of toxic substances may result in **food chain magnification** of concentrations in animals living at higher trophic levels (e.g., carnivores as opposed to herbivores). This occurs only if there is a specific biological mechanism of tissue concentration and if an animal living at a given trophic level consumes large quantities of organisms at lower levels. Furthermore, the rate of loss of the toxic substance must be low. Such magnification occurred in the case of uptake of the pesticide DDT by birds of prey, as will be discussed later. In some cases, a substance under consideration that does not affect the organism in which it is concentrated will have strong effects on the organism's predator.

Toxic Metals

✶ Metals are often cumulative toxins and have strong effects when consumed by human beings.

For thousands of years, a variety of metals have been released by human beings into the marine environment. Mining is a major source of metals, and mines in estuaries have been a major source of pollution. A wide variety of industrial processes release metals—for example, before regulatory efforts, the manufacture of wood pulp released mercury. Metals also are found in great quantities in sewage; in urban areas, this is probably the single greatest source of metal pollution. In agricultural areas, metals are sometimes released as components of insecticides and fungicides. Mercury and copper have been components of antifouling paints, and the toxic effects of tributyltin are recognized widely.

Although the effects of metals on human beings have been studied extensively, relatively little is known about the specific physiological effects on marine organisms. Metals such as zinc and copper are known to denature many proteins and are therefore fundamental poisons. Copper also may bind to blood pigments and impair their function. Many metals are known to increase mortality of a wide variety of species.

Metals are often sequestered in intracellular organelles. Marine organisms also produce metal-binding proteins, such as metallothionein, which combine with metals allow their deposition in chemically less reactive forms and apparently aid in reducing the exposure of cell constituents to chemical reactions with metals.

Mercury is probably the most notoriously toxic metal. It is also toxic when attached to a short-carbon-chain alkyl group in the form of methylmercury. Apparently some bacterially mediated processes in seawater transform either ionic or elemental mercury into this toxic form. Its principal effect on human beings is through strong disruption of nervous function. The odd behavior of the Mad Hat-

ter in *Alice in Wonderland* constitutes an allusion, which would have been obvious to Lewis Carroll's contemporaries, of the effects of the mercury poisoning that was so common among hatmakers in mid-nineteenth-century England. In Minamata, Japan, the industrial release of mercury, which was acquired by fishes and shellfish, led to large outbreaks of nervous disorders and deaths from the 1950s to the 1970s. Unfortunately, mercury passes across the placenta, and many women who had eaten mercury-laden fish bore children who had the affliction that came to be known as Minamata disease. Some other coastal populations have been affected in much the same way.

Cadmium is another fairly common metal, although its sources are often more diffuse than is the case for mercury. Cadmium may enter in sewage, but it also comes from outfalls in electroplating factories and battery manufacturing plants. In Foundry Cove, a bay adjacent to the Hudson River estuary, a battery factory outfall caused cadmium concentrations in bottom sediments to be in the range of 1–25 percent! The most notable toxic effect on human beings is on kidney function, but larger accumulations can lead to bone deformities and severe pain. Cadmium is found to be concentrated in rice in certain areas in Japan, but crabs and shellfish also are contaminated. At present, cadmium is in very high concentrations in blue crabs in the Hudson River, and regular eating of the crabs will produce pathological renal effects.

The badly contaminated Foundry Cove of the Hudson River is remarkable for the fact that benthic animals were dense there, despite common sediment cadmium concentrations of 10,000 ppm. The oligochaete *Limnodrilus hoffmeisteri* apparently evolved a resistance to high concentrations of cadmium (Figure 19.2a). Individuals of this species from other areas died soon after being introduced into Foundry Cove sediments, whereas indigenous worms survived well. The resistance seems to have a genetic basis and may be related to the higher production of a metal-binding protein and the ability to precipitate cadmium sulfide in intracellular organelles. The cove was cleaned up partially in 1994.

Lead was used in Roman times in water pipes, pottery, and coins. In the industrial era, lead has been used extensively in lead storage batteries and paints, and as an additive to motor fuels. Lead is also found in fossil fuels and is therefore emitted during burning in power plants. Lead is most common as an air pollutant, and there has been a movement in recent years to reduce lead in internal combustion engine fuels. The action is the result of knowledge of the strongly toxic effects of lead on the nervous system, and especially on the mental development of young children.

Lead is now found in relatively high concentrations in estuarine and marine sediments adjacent to urban areas. The most probable sources are release from industrial pipes, atmospheric deposition, washout of gasoline-ridden wastewater into storm sewers, direct disposal of fuels into seawater, and sewage release.

While other toxic substances are also monitored, metals are a prime target for study in the Mussel Watch Program (an international monitoring study that analyzes mussels in coastal inhabitants on both sides of the Atlantic). Mussels are analyzed for a variety of metals to judge whether bioavailable toxic substances are entering coastal marine ecosystems. Owing to the differences in chemistry among metals, an increase of concentration must be interpreted carefully.

Pesticides

✳ **Pesticides are usually designed to kill terrestrial insects, but they are washed into coastal waters and are often toxic to marine life.**

Pesticides include a wide variety of compounds used typically to kill insects harmful to crops. Because of the large spatial scale of pest infestations, most pesticides are applied in large amounts (nearly one bil-lion kilograms per year in the United States) and in great variety (there are about a thousand different types). Although some pesticides are very effective indeed, most target species have sufficient genetic variability of resistance to evolve immunity rapidly. At first, the frequency of resistant variants may be very low, but unlike nonresistant individuals, the resistant individuals will survive and reproduce. The evolution of resistance has tended to magnify both the toxicity and the variety of pesticide deployment. Despite this arms race, the insects have been winning by and large, and crop damage has steadily increased over the last few decades. Herbicides are effective in killing undesirable plants, but they also are very toxic to human beings.

There is not enough space here to describe the variety of pesticides in any detail, but several have been extremely harmful to marine organisms. To assess the potential harm of a pesticide, the solubility and chemical characteristics related to mobility, toxicity, and rate of degradation must be known. Substances that are easily mobilized in runoff have generally toxic effects in many organisms, and those that are slow to degrade are liable to be the most dangerous.

Chlorinated hydrocarbons (DDT, dieldrin, chlordane) have the most dangerous combination of harmful properties. DDT's degradation rate is on the order of years, and it washes readily from salt marshes into adjacent shallow estuarine and marine bottoms. DDT (Figure 19.3) was used widely as a means of eliminating the *Anopheles* mosquito, the carrier of malaria. It was later used as a general insecticide on

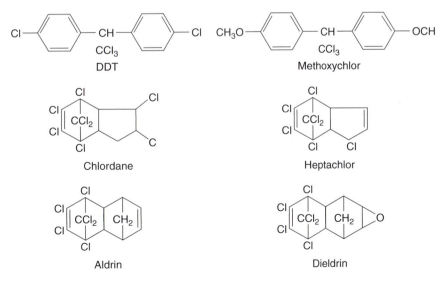

Fig. 19.3 DDT and related compounds.

many crops. In the 1960s the general decline of a large number of seabird species was noticed, especially those at the top of food chains, such as the peregrine falcon, the Bermuda petrel, the brown pelican, and the osprey. DDT and a few related compounds, which are very soluble in fat, were found to be magnified up the food chain to these birds, which ate large numbers of contaminated fishes. DDT and related residues disrupt reproduction and especially eggshell construction, to the degree that shells became too thin to permit normal egg development. Many species declined catastrophically until widespread bans on DDT use were imposed by Western industrialized nations. In recent years, as DDT has

Hot Topics in Marine Biology 19.1 The Plight of the Baymen

Fishing has never been an easy life, but recent events have spelled the end for many traditional coastal fishing communities. Many survived for hundreds of years, until mid-twentieth-century pollution, habitat disruption, and overfishing took their toll.

The baymen of the south fork of Long Island have lived as a distinct community since the eighteenth century. In the 1700s, they rowed their dories into the surf and hunted everything from whales to gigantic striped bass, which teemed in the waters offshore. The community has remained distinct from the rest of largely suburban Long Island to the extent that the baymen have their own distinct accent, dating back to colonial days.

Up to a few years ago, a good living could be had from launching surf boats and fishing for striped bass. In addition, baymen fished for scallops in Peconic Bay, one of the richest scallop beds in the world. The swift currents brought algal food, and dense eelgrass provided sites for larval settlement. Over the last couple of decades, it became more and more difficult to exploit the formerly rich clam fishery; overfishing had taken its toll. Even the largest clamming company on Long Island gave up on the clam beds in 1999.

Two other important events collapsed all remaining major fishing for these stalwart fishermen. Baymen caught striped bass by means of haul-seine nets, which were set in a curve by small boats offshore and then hauled into shore by trucks. Both commercial fishermen like the baymen and recreational anglers feasted on the bass stock, which had reached a peak of abundance in the 1970s. Then the effects of overfishing set in, and the stock declined drastically into the 1980s.

In the nearby Hudson River, PCBs were released and allowed to spill over a dam and contaminate the entire river system. This was a major problem for striped bass, which spent much of their lives migrating into the river to spawn and feed. Eventually, Long Island striped bass developed PCB concentrations over the 2 ppm limit, and the New York State Department of Health was compelled to prohibit commercial and recreational fishing. Ironically, this and stringent controls on the Chesapeake Bay stocks led to a major resurgence of Hudson River striped bass, which did not seem to be affected by PCBs significantly. Most recently, PCB concentrations in stripers have declined, and restrictions on fishing have been relaxed. However, recreational fisher groups have lobbied strongly against restoring rights to the baymen to haul-seine fish again. According to fisheries biologist John Waldman of the Hudson River Foundation, there appears to be no real danger that the modest effects of haul-seining near shore will damage the fishery much, but the recreational anglers have still been able curtail such fishing, although the more damaging technique of gill netting will still be permitted. The PCB episode was a major blow to the baymen, but more were to come.

In Peconic Bay, in the summer of 1987, the water began to turn chocolate brown. A tiny algal species, barely known previously to science, began to grow to extremely high densities. Scallops choked on this excess because particle densities were so high and because the diatom was apparently only poorly digestible. Two summers of this so-called brown tide killed off all the scallops in the bay. Although a reseeding program is beginning to bring back the population, it may be too late for the baymen, who depended upon the scallops as an important part of their income. Proud fishermen sold their houses to be able to continue to run their boats, but even this was not enough. Many have moved away, and others now pump gas and drive school buses instead of keeping up a tradition that connects us with the Long Island of 300 years ago. As the popular Long Island–born songwriter Billy Joel puts it:

I was a bayman like my father was before.
Can't make a living as a bayman any more.
There ain't much future for a man who works the sea.
There ain't no Island left for Islanders like me.

degraded in the U.S. marine environment, peregrine falcons, osprey, and bald eagles have all recovered strongly.

Many of the pesticides directed at insects are also toxic to their arthropodan relatives, the crustaceans. Spraying of insecticide on coastal agricultural areas and on marshes for mosquito control may therefore have unfortunate consequences. The spraying of the insecticide kepone caused the closure of the James River (Chesapeake Bay watershed) to fishing and devastated the blue crab population. Other insecticides, such as mirex, harm crabs, especially during larval development. Even when marine invertebrate populations are not affected, they may sequester the toxins, which may pose a danger to human beings. Dioxin, for example, is a contaminant of some herbicides and is believed to be carcinogenic to people. There have been reports of dioxin in fish and shellfish. Dioxin in the sediment interferes with the reproductive cycle of estuarine fish.

PCBs

* Polychlorinated biphenyls (PCBs) derive from industrial activities and have proven to pose a major toxicity problem in estuarine environments.

Polychlorinated biphenyls are a class of compounds that have been used extensively as lubricants in various types of industrial machinery. Throughout the world, PCBs have been released into coastal waters and have been found as a contaminant of invertebrates, fishes, and marine mammals. PCBs cause carcinomas in mice and are therefore thought to be a danger to human beings. These substances are particularly a problem because of their very high toxicity and chemical stability. Although they can be degraded by marine bacteria, the process is very slow.

PCBs have been discovered in a wide variety of commercially captured fishes, such as bluefish and striped bass in the New York and southern New England region. They have been implicated in reproductive failures and reduced populations of seals in the North Sea–Baltic Sea regions. A release of PCBs in the Hudson River from a General Electric Corporation facility resulted in high loads of PCBs in Hudson River sediments, the contamination of fishes, and the shutdown of the Hudson River striped bass fishery. The substances have also turned up in fish caught by native Americans in Alaska, perhaps owing to exposure of fishes that spawned in industrially polluted

rivers in the former Soviet Union. It is not clear whether the fishes take up PCBs primarily from contaminated prey or directly from solution. In either case, the sediments are so toxic that it is not clear that dredged materials can be safely dumped.

Oil Pollution

* Oil pollution can have both short-term and long-lasting effects on communities and individual species.

In the past 50 years, oil pollution has become a major problem in the coastal zone. Drilling, transport, and burning have all led to additions of oil to marine environments. The following are the major sources of oil pollution: (1) leaks from marine terminals and in harbors, (2) leaks from offshore drilling, (3) leaks from and breakup of oil tankers and barges, and (4) washout of oil from settled areas into storm drains and direct washout to the shoreline.

The wreck of the tanker *Torrey Canyon* off the English coast in 1967 was the first oil spill that awakened the international community to the dangers of oil transport. In a few days after the ship cracked up on the rocks, 80 tons of crude oil was released. The remaining 40 tons burned after the Royal Air Force made a bombing run over the site. To break up the oil slick, detergents were sprayed onto the sea surface. Both the oil and the detergents devastated seabirds and shore invertebrates. It was clear that tanker accidents could have a devastating effect on marine life.

More recently, in 1978, a large stretch of the Brittany coast was devastated by the wreck of the tanker *Amoco Cadiz*. The tanker spilled about 200,000 tons of oil over more than 300 km of coast. The result was devastation to seabirds, soft-bottom benthos, and oyster beds (Figure 19.4). After two decades recovery apears nearly complete.

In the United States, an accidental leak in 1969 from a well offshore near Santa Barbara, California, affected marine life in the coastal zone. The effects of the spill were not studied adequately. In the same year, however, a spill from the relatively small barge *Florida* off Cape Cod, Massachusetts, was studied intensively by a team of benthic ecologists and chemists led by Howard Sanders of the Woods Hole Oceanographic Institution. This barge carried 14,000 barrels of oil (1 barrel = 159 liters), a quarter of which was spilled in the relatively small Wild Harbor area. Since the load was number 2 diesel fuel oil, the concen-

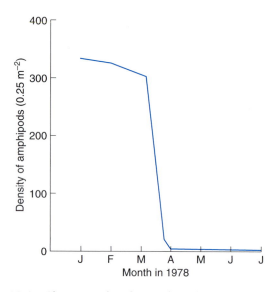

Fig. 19.4 Change in abundance of amphipods in the bottom sediments following the wreck of the *Amoco Cadiz* off the coast of Brittany.

tration of relatively toxic aromatic hydrocarbons was high, about 41 percent.

The spill had devastating effects on the benthic community, as was demonstrated by temporal sampling. A diverse community of clams and polychaetes crashed and was replaced by a very few species. In particular, the polychaete *Capitella capitata* came to dominate the intertidal and subtidal soft bottoms. It took several years for the oil to lose its effects. Much of the oil was buried, only to spread during winter storms that eroded the cover. Toxic substances, such as aromatic hydrocarbons, were found in shellfish more than one year after the spill. This suggested a lingering danger to human beings who might consume the shellfish. Reproduction of shellfish such as mussels was impaired strongly.

Major oil spills have not abated. In March 1989, the tanker *Exxon Valdez* hit a reef in Prince William Sound, Alaska, and spilled about 11 million gallons of oil, the worst spill in U.S. history. Thousands of marine mammals and seabirds were killed, and hundreds of miles of shoreline were covered with oil. The cleanup probably did more damage to rocky-shore communities than the oil spill itself, but toxic substances from the oil still remain in sediments and had a negative impact on anadromous fish reproduction for several years after the spill. Even Antarctica has

not been spared. Also in 1989, an Argentinean tanker capsized and spilled about 200,000 gallons of oil. Thousands of penguins, seals, and seabirds were killed. The extent of the damage from the 1989 accidents is still not entirely known. In the case of the *Amoco Cadiz* spill on the Brittany coast, benthic populations eventually recovered. It is not clear how fast recovery will be in the colder waters of Alaska and Antarctica, where oil breakdown may be slower.

Some of the complex effects of oil spills were well studied by scientists at the Smithsonian Tropical Research Institution following an oil spill on the Caribbean side of Panama in 1986 (Figure 19.5). Luckily, the bay in which it occurred, Bahia Las Minas, had been surveyed for years before the spill washed oil ashore. Initially the spill destroyed approximately 7 percent of the mangroves on the shoreline. The decay of the mangrove trees caused them to dislodge and roll about the shore, causing damage to the benthos. The oil also damaged sea grass meadows, which caused rhizome mats to disappear. The sediment was no longer held in place and was transported seaward, which increased the turbidity and sedimentation and reduced the survival of nearby corals. Oil seeped into the mangrove forest sediment, but the rainy season soon came, and storms and freshwater flow caused the oil to be eroded and transported offshore, which caused more damage to sea grasses and corals. Normally, the mangrove ecosystem traps sediment, which helps clarify the water and benefits the nearby coral reefs. The oil spill set off a cascade of ecological effects that still can be observed today.

Despite the notoriety of major tanker accidents, most oil is probably spilled during delivery of oil to harbor terminals. Spills occur when values malfunc-

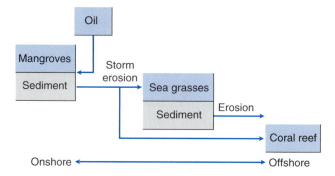

Fig. 19.5 The cascading effects of oil derived from a 1986 spill along the shores of Bahia Las Minas in Panama.

tion and when workers attempt to pump more oil into a tank than it can hold. U.S. law requires a set of containment booms to surround any marine loading area, but not all countries have legislation like this. Because of the lack of such a precaution, an August 1999 release from open valves of the tanker *Laura D'Amato* in Sydney, Australia, resulted in a spill of as much as 300,000 liters of Saudi Arabian crude oil along the shores of Sydney Harbor (Figure 19.6). The spill oiled thousands of shorebirds, but dispersed after a few days from the shoreline. Chronic releases are important in increasing the concentrations of toxic substances, such as polycyclic aromatic hydrocarbons, in marine sediments (see later).

Components and Effects of Oil

✳ **The effect of oil varies with chemical composition and the affected organisms.**

Oils may have the following components:

1. **Paraffins**. Straight or branched-chain alkanes that are stable, saturated compounds having the formula C_nH_{2n+2}.

2. **Naphthenes**. Cycloparaffins that are saturated but whose chain ends are joined to form a ring structure.

3. **Aromatics**. Unsaturated cyclic compounds that are based on the benzene ring, with resonating double bonds, and six fewer hydrogen atoms per ring than the corresponding naphthane. Often toxic, aromatics have been implicated in cancers.

4. **Olefins**. Alkenes, or unsaturated noncyclic compounds with two or fewer hydrogen atoms for each carbon atom. Olefins have straight or branched chains; they are not found in crude oil.

The effect of oil varies with oil chemistry and the organisms affected. Crude oil is widely regarded as the least toxic. Refined oil is rich in toxic substances such as aromatic compounds. Crude oil usually has less than 5 percent aromatics, but fuel oil may have 40–50 percent aromatics. The toxic compounds in oil are known to impair cell membrane function and may impair behavior in a wide variety of organisms. As mentioned earlier, reproduction can be impaired in

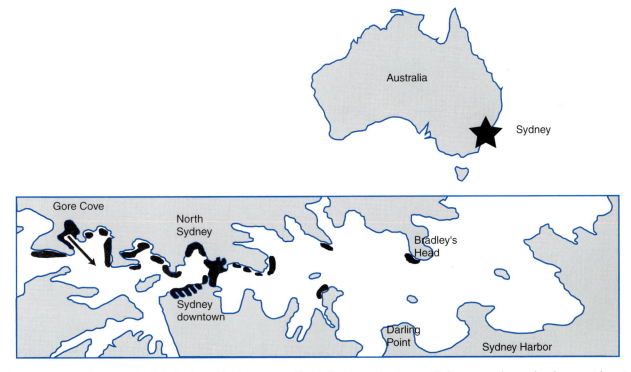

Fig. 19.6 Distribution of oil in Sydney Harbor, Australia, following a harbor spill from a tanker unloading accident in August 1999. (After report in *Sydney Morning Herald*.)

invertebrates exposed to these substances. Survival and development of fish eggs and larvae are also affected negatively. Phytoplankton production can also be reduced.

Oil has an especially devastating effect on seabirds. Birds maintain a high and constant body temperature. Feathers act partially as insulation. The fluffy down feathers provide an air space for insulation, and the air is sealed in by contour feathers. The barbules interlock efficiently, and the hydrophobic surface of the feathers helps to keep water from collapsing the downy layer beneath (Figure 19.7). Unfortunately, oil readily coats the surface of the contour feathers and collapses their interlock. Seabirds that come into contact with oil therefore soon lose their insulation and are likely to die of exposure. The oil also impedes flight, and the birds often ingest toxic oil while preening. (Some birds such as puffins are attracted to oil, as if they expect food to be found on the surface.) These are some of the reasons why oil spills are usually followed by frantic efforts by conservationists to clean oil from birds, but these efforts are usually in vain (Figure 19.8). Both the *Torrey Canyon* and *Amoco Cadiz* spills caused the majority of the affected Atlantic puffins and other diving birds to cease breeding.

Oil Containment and Dispersal

✳ Oil spills can be contained with floats and are sometimes dispersed with emulsifiers or naturally by storms.

Because of the devastating effects of oil, a variety of containment and dispersal methods have been developed. In active ports, oil spills may occur every day, and tankers now routinely have floating pens to contain any spilled oil. As mentioned earlier, chemical dispersants have been used extensively, but they often damage marine life further. Surfactants are usually lipophilic (oil-compatible) molecules with a hydrophilic (water-compatible) group at one end. This structure acts to emulsify oil—that is, to break it up. Although these materials are often toxic, they are used in moderation to break up oil slicks, to prevent seabird mortality. In some cases, wave action breaks up the oil, so even though it may harm the benthos, it does no damage to diving birds at the surface.

Oil tankers have become larger in the last few decades, and many so-called supertankers exceed 200,000 barrels in capacity. Although it is not possible to prevent all collisions, double-walled construction and individual oil compartments help

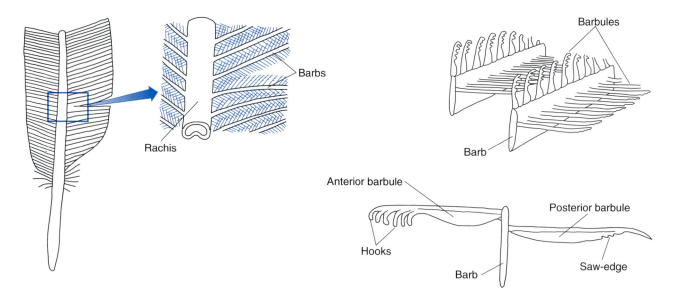

Fig. 19.7 The structure of a contour feather, showing the way in which the barbules are hooked together to seal the spaces between barbs. This interlocking system prevents water from penetrating to the downy layer of feathers beneath. (From Nelson-Smith, 1973.)

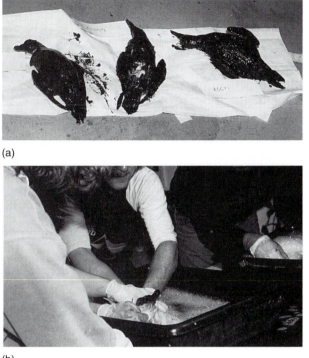

(a)

(b)

Fig. 19.8 (a) Birds washed, but covered with oil from an oil spill. (b) Rescue workers attempting to wash oil from a seabird. (Courtesy of Sam Sadove.)

reduce leakage. A series of international agreements has helped the process of compensation for nations victimized by tanker oil spills, but the strongly international nature of oil transport and the concept of freedom of the high seas make for difficulties in regulation and enforcement. The *Torrey Canyon*, for example, was registered in Liberia, owned by an American company, and under charter to a British concern. The wreck was technically in international waters, even though the oil affected both the British and French coasts. Such international complications prevent simple legal redress.

In many cases, the origin of oil spills cannot be isolated. Tankers often spill oil at sea, and minor spills from many small vessels combine to make large-scale oil pollution problems in ports. There has been considerable effort devoted to developing methods of analysis that might fingerprint oil, so that a community could identify the source of the spill. The most promising fingerprinting techniques involve chemical analysis of the hydrocarbons in oil. A technique known as gas chromatography—mass spectrometry is especially effective because it focuses on a number of compounds whose relative proportions can be used to pinpoint a particular oil with respect to its type and origin. Triterpenes and stearanes are the most useful diagnostic compounds.

PAHs

* Polycyclic aromatic hydrocarbons (PAHs) are derivatives from fossil fuels. They are known to be carcinogenic in mammals and are major contaminants in coastal marine environments.

PAHs derive from both point and nonpoint sources, including sewage systems, runoff, various oil spills, and burning of fossil fuels. They range greatly in molecular weight (there are mainly two-, three-, and four-ringed PAHs) and are adsorbed readily onto sedimentary particles, owing to their hydrophobic properties. PAHs therefore have become concentrated in coastal sediments and are widespread in near-shore bottoms in urban zones. PAHs can be toxic to benthic invertebrates, and various tissue abnormalities and cancers in fishes have been related to these compounds. Some PAHs can induce cancers in laboratory mammals, and their occurrence in resource species has troubled environmental managers. Bacteria in sediments degrade PAHs, but the rate of degradation is much slower for the higher molecular weight forms.

* PAHs and PCBs severely disrupt endocrine function in vertebrates and may be a major cause of reproductive failure.

PAHs and PCBs both appear to strongly affect reproductive cycles, particularly of fishes. They act as **endocrine disrupters**, exerting their influence by mimicking the effect of endogenous hormones such as estrogens and androgens, antagonizing the effects of endogenous hormones, altering the pattern of synthesis and metabolism of normal hormones, and modifying hormonal receptor levels. In vertebrates the secretion of gonadotropin-releasing hormone and follitropin from the hypothalamus and pituitary glands stimulates ovarian follicle growth and estradiol synthesis in the female. In oviparous vertebrates including fish, the release of estradiol from the ovary causes the liver to produce large amounts of vitellogenin, a lipoprotein precursor for egg yolk. Circulating vitellogenin and hormone levels in natural fish populations during the reproductive season have been shown to be a promising indicator of incipient reproductive dysfunction. High vitellogenin concentrations in male fish from contaminated environments have been interpreted as indicative of the estrogenic properties of contaminants.

Induction of gene expression of cytochrome P450-1A by PAHs and coplanar PCBs has been found to one of the most sensitive if not the most sensitive response and has been used extensively in field investigations. The induction of the gene CYP1A is known to play a role in both detoxification and activation of toxic compounds and is also significant for health of the fish.[2]

Nutrient Input and Eutrophication

✳ **Agricultural activities and sewage add nutrients, as well as disease organisms, to the water.**

Agricultural activities and sewage release cause a great deal of damage to marine life and contaminate fishes and shellfish. The major impact is the result of nutrient release, which indirectly reduces water quality and will be discussed in this section. With sewage and animal waste, undesirable microorganisms are also released into the marine environment. Pathogens such as hepatitis viruses and the bacterium *Salmonella,* often concentrated by suspension feeders such as clams and mussels, may be the cause of a variety of diseases. Outbreaks of dysentery and other diseases can be common in areas where people collect shellfish near sewage outfalls and in heavily populated areas with septic tanks. Local environmental agencies usually count the number of **coliform bacteria** in seawater. These bacteria are associated with the human gut and are therefore believed to be correlated with release of other pathogens. The recent problems in the northeastern United States with ocean dumping of medical waste raise the possibility of contamination with a variety of highly virulent diseases.

Nutrient Sources

✳ **Human activities result in large additions of dissolved nutrients to coastal waters.**

Eutrophication is the addition of dissolved nutrients to a water body, resulting in large increases in phytoplankton production and microbial activity. As freshwater ponds fill in with sediment, eutrophication occurs naturally, owing to the large influx of runoff and decaying plant matter into the relatively small bodies of water (i.e., the ponds). In coastal waters, eutrophication is due to several nutrient sources related to human activity. The following are the principal known sources:

1. Point sources such as sewage treatment outfall pipes
2. Point sources such as storm sewer overflows, which may be connected to sewage pipe systems
3. Commercial fertilizer, which is applied heavily to agricultural lands, and runoff of which adds large amounts of nitrate and ammonium ions to surface- and groundwaters
4. Animal waste, which is a significant contributor to nitrogen in runoff, especially in rural agricultural areas

Figure 19.9 shows an estimate of the relative contributions of nitrogen addition to the Chesapeake Bay watershed. Note that agricultural sources dominate, but there is a surprisingly large contribution from atmospheric deposition, which will be discussed shortly. Point sources, such as sewage outfalls, comprise a minor percentage. This is not true in areas such as New York Harbor and Boston Harbor, where point sources such as combined sewer–storm pipe outfalls and sewage treatment plant outfalls are the major contributors of nitrogen. In New York Harbor, nutrient supply is so plentiful that light and temperature, rather than nutrients, are the limiting factors for phytoplankton.

✳ **The atmosphere can be a major source of nutrient addition to coastal bays.**

Recently, a new source of nitrogen has been suggested that may be a significant contributor in coastal areas of industrialized countries: fossil fuel combustion, which is a major potential source of nitrogen oxide emissions. These gaseous emissions are eventually returned to the earth as soluble nitrates in wet or dry precipitation. The material becomes part of the now-famous acid rain, whose sulfur components may reduce the pH values of some lake and estuarine waters to the point of toxicity to fishes. The nitrates de-

---

[2] See Haasch and others, 1993, in Further Reading.

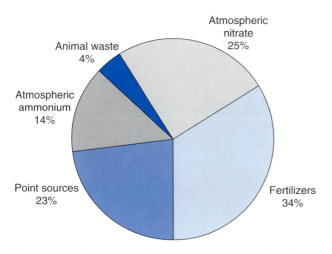

Fig. 19.9 Relative contributions of nitrogen to the Chesapeake Bay watershed. (Courtesy of Environmental Defense Fund.)

posited by the rain are worrisome partially because they may stimulate primary production, which leads to eutrophication. Hans Paerl and colleagues[3] studied this effect by observing the effects of adding of rain-water to containers with phytoplankton from coastal North Carolina. Rain-water addition strongly stimulated phytoplankton growth. This effect is likely to be especially strong in more offshore shelf waters, where phytoplankton primary production is strongly limited by nitrogen concentration.

Atmospheric precipitation of nitrogen may constitute as much as 39 percent of the total nitrogen addition to Chesapeake Bay. About 14 percent can be explained by agricultural sources, but most of the rest originates from fossil fuel power plants. Given the current trends of fossil fuel consumption, unless stringent conservation efforts result in reduction of fuel consumption, these additions are likely only to increase.

Effects of Added Nutrients

✳ Nutrient stimulation of primary production often results in hypoxia or anoxia.

Nutrient additions of nitrate and ammonia stimulate phytoplankton growth, as discussed in Chapter 9. At modest levels, one might expect higher water column and benthic production. These nutrient additions also stimulate bacterial production, however, especially in water columns where light is reduced. This is a problem in rivers, where high particle loads reduce light

penetration. As large populations of phytoplankton and bacteria build up, there is little likelihood that a zooplankton population will graze the material. In the shallow waters of southern San Francisco Bay, the benthic suspension feeders can often keep up with the strongly eutrophied waters, and phytoplankton populations do not build up. This is not the case, however, in deeper estuaries with high nutrient loads and sluggish circulation. During the late spring and summer, when the water column stabilizes, much of the phytoplankton may die. Subsequent degradation by aerobic bacteria strongly reduces the level of dissolved oxygen, which is essential for nearly all animals (Figure 19.10). **Hypoxia** is the condition of strongly depleted dissolved oxygen. Recall that anoxia is the complete absence of oxygen. Both hypoxia and anoxia may be accompanied by significant concentrations of hydrogen sulfide, which is toxic to many marine organisms. The lowered oxygen and the presence of hydrogen sulfide cause mass mortality of fishes and benthos. The rise of bacterial degradation is also exacerbated by the turbidity due to dense phytoplankton, whose shade prevents primary production in deeper water.

In the summer of 1987, the western part of Long Island Sound, New York, experienced a hot, windless summer, and the nutrient load caused extensive hypoxia and anoxia. Divers reported large numbers of dead animals, and the stench could be smelled even through diving masks. There were also large mass kills of many commercially important species. Hypoxia is a common problem in urban harbors throughout the world. Such areas contain benthic communities with very low diversity and strong dominance by opportunistic species.

Ocean dumping of solid organic waste may also cause extensive damage to marine life. In the United States, this practice was confined to the New York metropolitan area, though in 1988 legislation was passed that ended the practice there as well. Solid residues from sewage treatment plants were transported and dumped approximately 100 miles out to sea, and the material caused a large part of the seabed on the continental shelf to be anoxic and devoid of clams and polychaetes. In former years, dumping occurred within 12 miles of shore, and the "dead sea" bed thus produced occasionally migrated to within a mile of the south shore of Long Island. The devasta-

[3] See Paerl and others, 1990, in Further Reading.

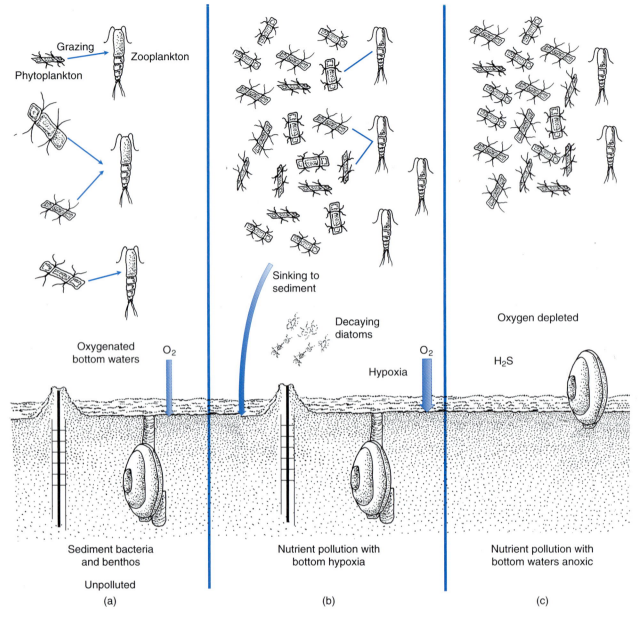

Fig. 19.10   Development of hypoxia in an estuary. (a) Normal situation: phytoplankton is grazed and bottoms waters are oxygenated. (b) Nutrient input from sewage stimulates phytoplankton growth, and some dead phytoplankton sinks to bottom waters; bacterial decomposition reduces oxygen, and other material sinks to bottom sediment, where more oxygen is consumed from bottom waters. (c) Oxygen is removed from bottom waters and benthos dies.

tion came from smothering by the dumping spoils, but also from the anoxia generated by bacterial degradation of the organic waste. In the far-offshore site, material did not affect the bottom to the same degree because it was more highly dispersed as it sank.

Abating Eutrophication

✳ Eutrophication can be abated by eliminating ocean dumping of solid sewage waste and better treatment of sewage before waste-waters are released into the coastal zone.

Hypoxia and anoxia events in bays and estuaries have been frequent enough to draw public attention to the problem of nutrient additions, and sewage treatment has generally improved throughout the United States. Sewage treatment plants (Figure 19.11) typically practice **primary treatment**, in which solids are intercepted by screens, or **secondary treatment**, in which more toxic nitrogenous organic compounds and colloids are stirred in aerobic tanks so that only phosphates, nitrates, and ammonia will be released into coastal waters. The solid residue must then be disposed of. Very few treatment plants carry out **tertiary treatment**, in which even dissolved phosphates, nitrates, and ammonia are removed. Various anaerobic tanks may be employed to enable microbial removal of the dissolved nitrogen as gas, and iron is used to combine with phosphate; however, such practices are very expensive. These costly tertiary treatment methods, moreover, produce large amounts of sludge, which must be disposed of. Many municipal outfalls in the world do not even have primary treatment. The recent increases in hypoxia and anoxia events have stimulated interest in the expansion of secondary and tertiary treatments. Thanks to the improvement of sewage treatment, New York Harbor has improved steadily over the last 50 years (Figure 19.12).

The relative importance of nitrogen and phosphorus becomes an important issue in tertiary treatment. Phosphorus is much cheaper to remove than nitrogen. In most marine waters, nitrogen is generally believed to be the limiting nutrient. In estuaries, there

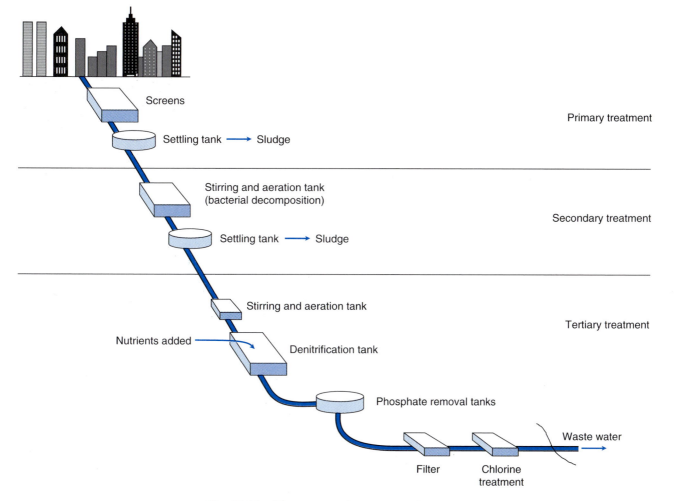

Fig. 19.11 Three types of treatment of sewage.

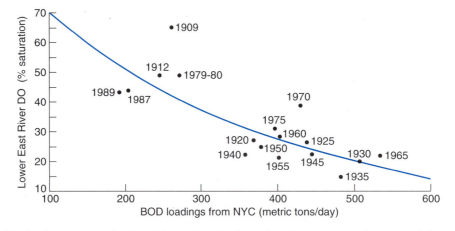

Fig. 19.12   The dissolved oxygen in the East River, New York Harbor–Estuary, as a function of the particulate organic matter entering the harbor from sewage (measured in terms of the potential biological oxygen demand, or BOD, that the material places on the harbor waters). As BOD loading has declined, oxygen in harbor waters has increased. The years 1909 and 1912 probably represent times when material was released into marshlands and not directly through pipes into the harbor—hence, the lower BOD loading. (Courtesy of Dennis Suszkowski, Hudson River Foundation.)

is more room for dispute because phosphorus is known to be limiting in freshwater rivers. An extensive study of lower Chesapeake Bay shows that nitrogen is limiting there. This is a major problem, because nitrogen does not come primarily from point sources, such as sewage treatment plants. Control, therefore, will be very difficult.

It is likely that dumping of sewage sludge will soon end, but there is no agreement on an alternative means of disposal. Some have suggested incineration, but without an elaborate system of filters and scrubbers, this would release large amounts of toxic material into the atmosphere. Management techniques have been developed to convert sewage sludge into

HOT TOPICS IN MARINE BIOLOGY 19.2 Getting the Oxygen Back into Our Coastal Water: Should the Price Be Paid?

Over the last hundred years, the human population in the coastal zones has exploded. About 15 million people inhabit the New York metropolitan area, and a similarly sized population lives in the southern California region. With people comes the inevitable sewage, which contains great concentrations of carbon, nitrogen, and toxic metals. In New York, the sewage is oxidized chemically in the water and as a result New York Harbor has had low oxygen concentrations. However, things have been improving steadily for the last 30 years, since the installation of primary sewage treatment (Figure 19.11). Secondary treatment is on the way, and this will yield further improvements. New York Harbor now has a "problem" that disappeared there a century ago: shipworms (boring bivalves of the genus *Teredo*) that could not survive the low-oxygen conditions

are infesting wooden pilings! A plan has been implemented to jacket pilings in plastic, which will cost many millions of dollars. It is said that further improvements await the investment of approximately a billion dollars. Should we pay the cost to reoxygenate our waters?

The nutrients, especially nitrogen, are stimulating phytoplankton growth, and phytoplankton decomposition causes hypoxia in the water. Coastal waters from Boston to Los Angeles are suffering, but the technology to recover nitrogen is very expensive and relies upon complex microbial digestion processes. Communities of the states of New York and Connecticut are now being faced with the alternative of spending hundreds of millions of dollars, curbing development, or both, to prevent further degradation of the water. Should we pay the cost?

compost for agricultural use; this may well be the only practicable alternative. Deep-ocean dumping is still being considered as a possible means of disposal.

Thermal Pollution

✳ Power-generating stations require water for heating and as a result kill aquatic life, by entraining and impingement.

Conventional and nuclear power stations require large amounts of water to transport heat from the power-generating system. Two popular means of heat dissipation are used. In the first, cold water is taken from a bay or river and passed through the plant; the heated water is then circulated through large cooling towers, such as the ones at the famous Three Mile Island nuclear facility in Pennsylvania. The towers then radiate heat into the overlying atmosphere. The relatively cool water is next returned to the river, or, alternatively, it can be recycled in a closed system between the power plant and the cooling tower. The second major approach involves removing water from a bay or estuary, passing it through the plant, and returning it quite hot to the environment. Typically, the water is returned to a man-made embayment, and then passes out to the adjacent estuary or coastline while still approximately 10–15°C over ambient temperature.

The uptake of large volumes of water at intakes creates the problems of **entrainment** and **impingement**. Eggs, larvae, and juveniles may be entrained or moved through the intake pipe, passed through the power plant system, heated suddenly, and returned to the open water through an outflow pipe. A significant amount of mortality is often associated with this passage. If the fishes are larger, they may be impinged, or trapped, on the intake screens, which often kills them. The released warm-water effluent may also form a plume that moves out into the open water. A warm-water plume may attract fish that should instead migrate away from the otherwise cooling waters in the autumn. If the plume should dissipate during a storm, the fishes probably would die owing to the cold shock. This sort of effect is probably less important than entrainment and impingement.

Impingement results from the strong pressure created by water intake pumps. Various screens have been designed, partially to reduce fish kills but mainly to keep pipes from clogging. A net placed in the surrounding area can be helpful, but such nets often become fouled with algae and cannot be deployed in summer. At intake sites, the angle screen is a popular design because it may divert fish along itself to an exit pipe, supposedly reducing fish mortality. Ian Fletcher[4] analyzed the effectiveness of this design and found that it did little to reduce fish mortality. It was believed that some component of the flow would move at an angle along the screen, but Fletcher showed that the large pumps brought water directly through the screen, thus imping the fish. Recall from Chapter 4 the paramount rule of hydrodynamics: It is impossible to cross flow lines if you are entrained in the flow. Because the flow lines are determined by the large pumps, the flow lines move directly across the screen, and thus the fish cannot cross the flow lines. Attempts are being made to design a better screen system that attracts fish to the exit pipes. Alternatively, a system has been developed that traps fish on the screen, then causes them to drop into buckets of water, which are transported to an outflow pipe.

✳ Thermal emissions may also affect plant production.

Thermal pollution may inhibit phytoplankton growth and change the character of plant communities in the vicinity of warm-water outfalls. In the Turkey Point nuclear generating station outfall in south Florida, turtle grass was replaced by blue-green cyanobacteria. The hot waters were often too extreme for all but the hardy blue-greens. Usually, thermal effluents are a certain number of degrees above ambient and therefore exert their strongest effects when summer water temperatures are maximal. In summer at the power plant at Eaton's Neck on Long Island, water is heated 15°C above ambient and phytoplankton production is reduced in the hot effluent waters. The heating of water in spring increases algal biomass, although it lowers diversity. In summer, the blue-green cyanobacteria take over. These effects, however, are strongly localized, especially when the outfall mixes with coastal waters of high current energy.

[4] See Fletcher, 1985, in Further Reading.

Global Environmental Change and the Ocean

Global Warming

✳ Atmospheric quantities of carbon dioxide are increased strongly because of the burning of fossil fuels and less so owing to deforestation; these factors may be warming the earth's surface.

Human activities have reached a point at which they may be altering the earth's global climate. Since the Industrial Revolution in the nineteenth century, industrial activity has greatly accelerated the burning of fuels, particularly fossil fuels, such as coal and petroleum products. Mainly as a result of this activity, approximately 25 percent has been added to the storehouse of carbon dioxide in the earth's atmosphere. Measurements from Mauna Loa, Hawaii, demonstrate an increase from 315 ppm in 1958 to over 350 ppm in 1988 (Figure 19.13). It is believed that the atmospheric concentration was 280 ppm in the nineteenth century, as the Industrial Revolution began. Most scientists believe that current trends could cause a doubling of the carbon dioxide in the atmosphere over the next 100 years.

Deforestation is a secondary source of carbon release. Trees are net absorbers of carbon owing to photosynthesis, which absorbs far more carbon dioxide than is released during respiration (yes, trees breathe). Many of the tropical forests in southeast Asia, Africa, and South America are being cut down to clear land for agriculture and to provide wood for sale to the developed world. Worse than that, much of the for-

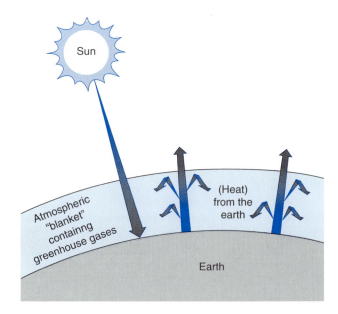

Fig. 19.14   The greenhouse effect.

est is being burned, which releases yet more carbon dioxide into the atmosphere. Nevertheless, this release is probably less than 10 percent of the total fossil fuel release, for most of which developed nations are responsible.

The problem with this release is the heat-absorbing capacity of carbon dioxide, which traps heat near the earth's surface. The short-wavelength radiation of sunlight is converted to long-wavelength radiation, which cannot leave the atmosphere (Figure 19.14). This sequence is known as the **greenhouse effect** because of the role of greenhouses as heat traps. As the atmosphere becomes more and more carbon dioxide rich, more and more heat is trapped in it. Other gases, including methane, ozone, and chlorofluorocarbons, also contribute significantly to the heat-trapping effect. Of these, only chlorofluorocarbons, constituents of cooling fluids and some solid materials, will surely be reduced in production over the next few decades owing to their negative effects on the UV-protecting ozone layer. Many have predicted that continued emissions of greenhouse gases will lead to global warming. Although most scientists agree that global warming is likely, there is some controversy about what has occurred in terms of warming over the past hundred years. There is convincing evidence from overall world temperature records that the earth's air temperature has increased over that period, perhaps by about one-half degree Celsius. Because there were

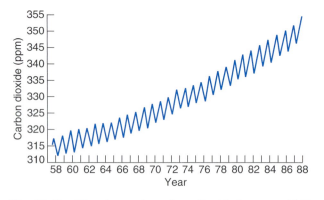

Fig. 19.13   The change in carbon dioxide between 1958 and 1988, as monitored at an observatory at Mauna Loa, Hawaii.

some decade-long periods of no increase in temperature, even though industrial burning remained steady, it is unclear how precisely to relate temperature trends to human activity. Nevertheless, there is cause for alarm because there is reason to believe that global climate may change significantly over the twenty-first century. Even a couple of degrees on average may translate into major local changes. Significant changes in rainfall and the frequency of storms (such as hurricanes) could cause major social disruptions.

Unfortunately, the magnitude and even the direction of climate change are not easy to predict. Cloud cover is a major factor in climate. Clouds reflect short-wave length radiation, reducing atmospheric heating, but prevent the back-radiation of long-wave length radiation, which traps heat. On balance, clouds tend to cool the earth's surface. Current computer models suggest that global warming will reduce cloud cover, further heating the earth. However, tiny aerosol particles containing sulfate are major sources of cloud nuclei, and industrial emissions of sulfur may thus enhance cloud cover. Although this effect is evidenced in the coastal zone, a study by Paul Falkowski and others[5] suggests that the effect on the ocean as a whole is minimal. Phytoplankton production may also affect cloud cover, since certain species produce the volatile substance dimethyl sulfide (DMS). When DMS enters the atmosphere, particles and clouds tend to form. Any stimulation of oceanic production by additions of CO_2 may therefore reduce the effects of global warming.

Hypothesized Effects on Sea Level and Circulation

✳ **Increased temperature may cause sea-level rise and major changes in oceanic circulation.**

Unfortunately, it is difficult to predict the specific effects of global warming. Certainly, sea-level rise is one strong possibility, for warming could partially melt the large volumes of glacial ice, adding a significant amount of water to the ocean, and also expanding the volume of the already liquid water. Although the great glaciers are in retreat at the present time in geological history, further melting could raise sea levels substantially, and some predictions suggest a rise of about 1 m in the next 50–100 years. One meter of sea-level rise, combined with storms,

would result in the piercing of many of the barrier bar systems in the world. Low-lying cities and estuaries would also suffer. Consider those who get their fresh water from the freshwater parts of estuaries: a sea-level rise might make the water too salty. There is widespread disagreement over the exact magnitude of sea-level rise, but the worst-case scenarios are very worrisome. Coral reefs seem particularly vulnerable because they typically grow upward at rates of 10 mm per year at best, which is disturbingly close to some estimates of sea-level rise. Grigg and Epp[6] suggested that sea-level rise owing to global warming might outstrip coral growth, which would trigger a worldwide catastrophe. Even if the reef were able to grow, increasing sea temperatures might cause extensive coral bleaching and contraction of reef systems. El Niño periods often exceed the thermal limits of hermatypic corals, and general warming could lead to catastrophe. It is not clear, however, that the thermal effects would overwhelm all reef species; some might survive better than others, perhaps coming to dominate a new reef community able to survive warmer temperatures.

It is also possible that global warming will cause major changes in oceanic circulation. Warming may cause significant changes in oceanic winds, and this may lead to changes in the intensity of ocean currents. Some nearshore regions are very productive because of upwelling, but these systems and their fisheries might collapse if currents are reduced or changed. A study of the North Atlantic has shown that changes in winds and currents sometimes come very fast, and large local changes in temperature have occurred in the past during climate shifts.

Effects on Biological Productivity

✳ **Increased temperature and carbon dioxide may increase biological productivity, especially in nutrient-enriched estuaries.**

Any prediction of how global warming might affect production in the ocean must be extremely speculative. Carbon dioxide increases can stimulate plant growth, which may increase primary production in waters in which nitrogen (for example) is not strongly

[5] See Falkowski and others, 1992, in Further Reading.
[6] See Grigg and Epp, 1989, in Further Reading.

HOT TOPICS IN MARINE BIOLOGY 19.3 The Greenhouse Effect: An Oceanographic Fix?

Phytoplankton grow if there are enough nutrients and light. In most situations, phytoplankton growth is nutrient limited, and nitrogen has generally been believed to be the limiting nutrient of overall marine primary production. This orthodox view, however, was challenged successfully by the late John Martin, who discovered an apparent relationship between primary production in the northeastern Pacific and the abundance of iron.[7] Martin added iron to bottles of phytoplankton, and primary production increased greatly. This finding may prove crucial in our understanding of global climate change. Data from cores taken in Antarctic glacial ice show a relationship between aluminum and atmospheric carbon dioxide. When aluminum is up, carbon dioxide is down. Because aluminum is often correlated with iron, we may have found the smoking gun of global climate control. An iron increase may stimulate primary production, which would increase the import of carbon dioxide from the atmosphere, in order to fuel photosynthesis. In this scenario, as organic carbon particles sank into the ocean, carbon dioxide would decrease and the greenhouse effect would decline as well.

In the Antarctic Ocean, phytoplankton growth seems not to be limited by nitrogen or phosphorus. Surface waters downwell and carry much unused dissolved nitrogen below the depths at which photosynthesis is possible. Some other nutrient must be limiting phytoplankton growth. That appears to be iron, which is an essential, if more minor, element, required in the synthesis of a number of proteins. The Antarctic Ocean may be able to be far more productive.

This potential could prove the source of inspiration for a massive experiment in global ecology. Over the last century, the burning of fossil fuels has increased the earth's atmospheric carbon dioxide by about a quarter, and there is good evidence that the earth warmed over that period. Was this due to burning of fossil fuels? We can't be sure, but it is dangerous to take the risk of inaction. Global warming will turn formerly rich and moist agricultural zones into deserts and will help melt glacial ice and drown many of the lowlands and coastal cities of the world. Deforestation is also a major problem, because trees absorb carbon dioxide, and they are being removed and even burned at a rapid rate throughout the tropics. What to do?

Aside from energy conservation and reforestation, oceanographers have hit upon an idea to use the Antarctic Ocean as part of a global cure. The solution is to add thousands of tons of ground iron to the surface water, assuming that this will stimulate phytoplankton growth. The phytoplankton will grow, use up atmospheric carbon dioxide, and then sink to deeper waters, taking the carbon out of circulation from the world's cycle. This idea sounds good, but it is possible that some other factor limits primary production in the Antarctic Ocean. If it develops, for example, that the low light levels in the Southern Ocean limit production, iron may not enhance production significantly. However, an experiment performed in 1993[8] in the equatorial Pacific showed that iron addition strongly stimulated photosynthesis, so at least the hypothetical effect of iron on a large scale has been confirmed. The biological details may be important, since some types of phytoplankton (e.g., nitrogen-fixing bacteria) are far more dependent upon iron than other groups.

Should such a program of iron addition be carried out on a massive scale? Is there any danger? Will stimulation of the phytoplankton cause nuisance algal blooms? Would the Antarctic phytoplankton be dominated by forms too small to be eaten by krill? These issues remain to be studied.

[7] See Martin, 1991, in Further Reading.
[8] See Behrenfield and others, 1996, in Further Reading.

limiting. One can imagine a scenario in which currents increase and bring deep-water nitrogen to the surface, while increased carbon dioxide increases photosynthesis. In some polluted estuaries (e.g., New York Harbor), nutrients are not limiting during the year, and primary production increases as a response to increasing temperature. Under such circumstances, global warming might increase primary production, which in turn would cause more waters to be hy-poxic, as explained earlier in this chapter. All in all, the potential problems created by global warming are very worrisome, if only because of what is still unknown. It is nearly impossible right now to map global change onto local areas.

✳ Increase of greenhouse gases and global warming could intensify coastal upwelling and increase primary production.

Any process affected by temperature change may be affected by global warming. One of the most likely factors to be affected is the balance of winds along the coastline. As discussed in Chapter 9, upwelling and enhanced primary productivity occur in regions where wind stress brings nutrient-rich deep water to the surface. Along the California coastline, northerly and northwesterly winds cause the offshore transport of surface water, a process that is in turn balanced by a rise of nutrient-rich deep waters. Global warming can accelerate such a process because daytime temperature on the adjacent land will increase and nighttime cooling will decrease. This overall temperature increase will accentuate the temperature difference between the land and the adjacent cool ocean, which will intensify winds and upwelling. Andrew Bakun[9] has demonstrated that wind speeds in upwelling regions have steadily increased since the 1950s, when data first began to be collected, and this finding is consistent with the steady increase in greenhouse gases. An increase in primary productivity has not been measured but likely has occurred, because upwelling of nutrient-rich deep water is the limiting factor in these systems. Whether the changes have, or will have, an enhancing effect on fisheries is not yet clear.

✱ Changes in primary production may occur in the open ocean over a few decades, but there is no evidence at present that primary production has increased to any degree over the last 70 years or so.

The effects of anthropogenic (human-derived) additions of carbon dioxide may be related to changes in oceanic primary production, but we have few long-term data sets that can be used to judge the relative interactions of CO_2 addition, local change of nutrient input, and changes of climate that can be attributed to factors other than global warming. Human activities add about 7 gigatons of carbon to the atmosphere each year. Phytoplankton fix 35–50 gigatons, so they could exert a considerable effect on the global carbon cycle, especially with regard to the human addition. If the phytoplankton increase in production and sink to the seabed, where the carbon becomes buried in the sediment, a considerable amount of anthropogenically derived carbon could be removed from the atmosphere and surface waters. So the question arises: Has phytoplankton produc-

tivity changed over the time during which industrial and other CO_2 release activity was on the rise?

There is good evidence for swings in production over decades. A study by E. L. Venrick[10] and colleagues showed that chlorophyll *a* concentration increased over two decades in the open-ocean North Pacific gyre. During this period, winter winds increased, and this probably caused upwelling of nutrient-rich water from below the thermocline. From this study we know that primary production levels are not fixed completely, and changes can occur on the scale of decades. A similar pattern of increase has been found in the past decade or so in the North Atlantic and in the North Sea.[11]

Longer-term data are difficult to obtain, because widespread studies of primary production were done only from the 1960s onward. There is a simple substitute, however. In the nineteenth century, Secchi, an astronomer-monk, devised a white disk, which he lowered into the water until it could no longer be seen (the Secchi depth). This depth is a surprisingly good index of primary production, because the deeper the disk could be seen, the lower is the phytoplankton density (especially in the open ocean, where other particles, such as sedimentary grains, are rare). One must correct for such factors as latitude and season, which is an indication of the local light intensity. The advantage of these data is that there is a geographically complete record going back at least 70 years, because Secchi depths have been recorded during that time by nearly all oceanographic vessels.

Falkowski and Wilson[12] compiled Secchi disk data that cover the past 70 years in the North Pacific Ocean. They found no significant changes in Secchi depth, which suggests that there have not been long-term changes in primary production there. This may suggest that global increases of carbon dioxide have failed to have an impact on open-ocean primary production. Other evidence suggests no increase in sea temperature over the past couple of decades. Thus, global warming may be a potential threat, especially for coral reefs, but there is no strong alarm signal to be discerned as yet.

[9] See Bakun, 1990, in Further Reading.
[10] See Venrick and others, 1987, in Further Reading.
[11] See Reid and others, 1998, in Further Reading.
[12] See Falkowski and Wilson, 1992, in Further Reading.

Further Reading

American Institute of Biological Sciences. 1976. *Symposium on Sources, Effects, and Sinks of Hydrocarbons in the Aquatic Environment*, American University, Washington, DC, 1976. Washington, DC: American Institute of Biological Sciences.

American Institute of Biological Sciences. 1978. *Conference on Assessment of Ecological Impacts of Oil Spills, Keystone, Colorado, 1978.* Washington, DC: American Institute of Biological Sciences.

Anderson, J. W., J. M. Neff, B. A. Cox, H. E. Tatem, and G. M. Hightower. 1974. Characteristics of dispersions and water-soluble extracts of crude and refined oils and their toxicity to estuarine crustaceans and fish. *Marine Biology*, v. 27, pp. 75–88.

Bakun, A. 1990. Global climate change and intensification of coastal ocean upwelling. *Science*, v. 247, pp. 198–201.

Bayne, B. L., D. A. Brown, K. Burns, D. R. Dixon, I. Ivanovici, D. R. Livingstone, D. M. Lowe, M. N. Moore, A. R. D. Stebbing, and J. Widdows. 1985. *The Effects of Stress and Pollution on Marine Animals.* New York: Praeger Scientific.

Behrenfield, M. J., A. J. Bale, Z. S. Kolber, J. Aiken, and P. G. Falkowski. 1996. Confirmation of iron limitation of phytoplankton photosynthesis in the equatorial Pacific Ocean. *Nature*, v. 383, pp. 508–511.

Bitman, J., H. C. Cecil, and G. F. Fries. 1970. DDT-induced inhibition of avian shell gland carbonic anhydrase: A mechanism for thin eggshells. *Science*, v. 168, pp. 594–596.

Blumer, M., and J. Sass. 1972. Oil pollution: Persistence and degradation of spilled fuel oil. *Science*, v. 176, pp. 1120–1122.

Bryan, G. W., P. E. Gibbs, L. G. Hummerstone, and G. R. Burt. 1986. The decline of the gastropod *Nucella lapillus* around south-western England: Evidence for the effect of tributyltin from antifouling paints. *Journal of the Marine Biological Association of the United Kingdom*, v. 66, pp. 611–640.

Bue, B. G., S. Sharr, and J. E. Seeb. 1998. Evidence of damage to pink salmon populations inhabiting Prince William Sound, Alaska, two generations after the *Exxon Valdez* oil spill. *Transactions of the American Fisheries Society*, v. 127, pp. 35–43.

Clark, R. B. 1986. *Marine Pollution.* Oxford: Clarendon Press.

Dauvin, J.-C. 1998. The fine sand *Abra alba* community of the Bay of Mordaix twenty years after the *Amoco Cadiz* oil spill. *Marine Pollution Bulletin*, v. 36, pp. 669–676.

Falkowski, P. G., and C. Wilson. 1992. Phytoplankton productivity in the North Pacific Ocean since 1900 and implications for absorption of anthropogenic CO_2. *Nature*, v. 358, pp. 741–743.

Falkowski, P. G., Y. Kim, Z. Kolber, C. Wilson, C. Wirick, and R. Cess. 1992. Natural versus anthropogenic factors affecting low-level cloud albedo over the North Atlantic. *Science*, v. 256, pp. 1311–1313.

Farrington, J. W. 1991. Biogeochemical processes governing exposure and uptake of organic pollutant compounds in aquatic organisms. *Environmental Health Perspectives*, v. 90, pp. 75–84.

Fletcher, R. I. 1985. Risk analysis for fish diversion experiments. Pumped intake systems. *Transactions of the American Fisheries Societies*, v. 114, pp. 652–694.

Forbes, V. E., and T. L. Forbes. 1994. *Ecotoxicology in Theory and Practice.* London: Chapman & Hall.

Gilfillan, E. S., D. S. Page, E. J. Harner, and P. D. Boehm. 1995. Shoreline ecology program for Prince William Sound, Alaska, following the *Exxon Valdez* oil spill. 3. Biology. In P. G. Wells, J. N. Butler, and J. S. Hughes, eds., *Exxon Valdez Oil Spill: Fate and Effects in Alaskan Waters.* Philadelphia: American Society for Testing and Materials, pp. 398–481.

Gray, J. S. 1989. Effects of environmental stress on species rich assemblages. *Biological Journal of the Linnaean Society*, v. 37, pp. 19–32.

Grigg, R. W., and D. Epp. 1989. Critical depth for the survival of coral islands: Effects on the Hawaiian archipelago. *Science*, v. 243, pp. 638–641.

Haasch, M. L., R. Prince, P. J. Wejksnora, K. R. Cooper and J. J. Lech. 1993. Caged and wild fish: Induction of hepatic cytochrome P450 (CYP1A1) as an environmental biomonitor. *Environmental Toxicology and Chemistry*, v. 12, pp. 885–895.

Houghton, R. A., and G. M. Woodwell. 1989. Global climatic change. *Scientific American*, v. 260, pp. 36–44.

Jones, P. D., T. M. L. Wigley, and P. B. Wright. 1986. Global temperature variations between 1861 and 1984. *Nature*, v. 322, pp. 430–434.

Kamrin, M. A., and R. K. Ringer. 1994. PCB residues in mammals: A review. *Toxicology and Environmental Chemistry*, v. 41, pp. 63–84.

Klerks, P. L., and J. S. Levinton. 1989a. Rapid evolution of resistance to extreme metal pollution in a benthic oligochaete. *Biological Bulletin*, v. 176, pp. 135–141.

Klerks, P. L., and J. S. Levinton. 1989b. Effects of heavy metals in a polluted aquatic ecosystem. In S. A. Levin, J. R. Kelley, and M. A. Harvell, eds., *Ecotoxicology: Problems and Approaches.* New York: Springer-Verlag, pp. 41–67.

Klerks, P., and J. S. Levinton. 1992. Evolution of resistance and changes in community composition in metal-polluted environments. In R. Dallinger and P. S. Rainbow, eds., *Ecotoxicology of Metals in Invertebrates*. Boca Raton, FL: Lewis Publishers, pp. 223–241.

Klerks, P. L., and J. S. Weis. 1987. Genetic adaptation to heavy metals in aquatic organisms: A review. *Environmental Pollution*, v. 45, pp. 173–205.

Levin, S. A., M. A. Harwell, J. R. Kelly, and K. D. Kimball, eds. 1988. *Ecotoxicology: Problems and Approaches*. Berlin: Springer-Verlag.

Mann, K. H., and R. B. Clark. 1978. Long-term effects of oil spills on marine intertidal communities. *Journal of the Fisheries Research Board of Canada*, v. 35, pp. 791–816.

Martin, J. H. 1991. The case for iron. *Limnology and Oceanography*, v. 36, pp. 1793–1802.

McElroy, A. E., J. W. Farrington, and J. M. Teal. 1989. Bioavailability of polycyclic aromatic hydrocarbons in the aquatic environment. In U. Varanasi, ed., *Metabolism of Polycyclic Aromatic Hydrocarbons in the Aquatic Environment*. Boca Raton, FL: CRC Press, pp. 1–39.

National Research Council. 1985. *Oil in the Sea: Inputs, Fates, and Effects*. Washington DC: National Academy Press.

Nelson-Smith, A. 1973. *Oil Pollution and Marine Ecology*. New York: Plenum.

Paerl, H. W., J. Rudek, and M. A. Mallin. 1990. Stimulation of phytoplankton production in coastal waters by natural rainfall inputs: Nutritional and trophic implications. *Marine Ecology Progress Series*, v. 107, pp. 247–254.

Peakall, D. B. 1979. Eggshell thinning and DDE residue levels among peregrine falcons *Falco peregrinus*: A global perspective. *Ibis*, v. 121, pp. 200–204.

Peakall, D. B. 1993. DDE-induced eggshell thinning: An environmental detective story. *Environmental Reviews*, v. 1, pp. 13–20.

Pearson, T. H., and R. Rosenberg. 1978. Macrobenthic succession in relation to organic enrichment and pollution of the marine environment. *Oceanography and Marine Biology Annual Review*, v. 16, pp. 229–311.

Reid, P. C., M. Edwards, H. G. Hunt, and A. J. Warner. 1998. Phytoplankton change in the North Atlantic. *Nature*, v. 391, p. 546.

Roberts, A. E., D. R. Hill, and E. C. Tifft Jr. 1982. Evaluation of New York Bight lobsters for PCBs, DDT, petroleum hydrocarbons, mercury, and cadmium. *Bulletin of Environmental Contamination and Toxicology*, v. 29, pp. 711–718.

Sami, S., M. Faisal, and R. J. Huggett. 1993. Effects of laboratory exposure to sediments contaminated with polycyclic aromatic hydrocarbons on the hemocytes of the American oyster *Crassostrea virginica*. *Marine Environmental Research*, v. 35, pp. 131–135.

Sanders, H. L., J. F. Grassle, G. R. Hampson, L. S. Morse, S. Garner-Price, and C. C. Jones. 1980. Anatomy of an oil spill: Long-term effects from the grounding of the barge *Florida* off West Falmouth, Massachusetts. *Journal of Marine Research*, v. 38, pp. 265–380.

Shaw, D. G. 1992. The *Exxon Valdez* oil-spill: Ecological and social consequences. *Environmental Conservation*, v. 19, pp. 253–258.

Smith, R. C., B. B. Prezelin, K. S. Baker, R. R. Bidigare, N. P. Boucher, T. Coley, D. Karentz, S. MacIntyre, H. A. Matlick, D. Menzies, M. Ondrusek, Z. Wan, and K. J. Waters. 1992. Ozone depletion: Ultraviolet radiation and phytoplankton biology in Antarctic waters. *Science*, v. 255, pp. 952–959.

Tatem, H. E., B. A. Cox, and J. W. Anderson. 1978. The toxicity of oils and petroleum hydrocarbons to estuarine crustaceans. *Estuarine and Coastal Shelf Science*, v. 6, pp. 365–373.

Venrick, E. L., J. A. McGowan, D. R. Cayan, and T. L. Hayward. 1987. Climate and chlorophyll *a*: Long-term trends in the Central North Pacific Ocean. *Science*, v. 238, pp. 70–72.

Vernberg, W. B., A. Calabrese, F. P. Thurberg, and F. J. Vernberg. 1979. *Marine Pollution: Functional Responses*. New York: Academic Press.

Vincent, W. F., and S. Roy. 1993. Solar ultraviolet-B radiation and aquatic primary production: Damage, protection, and recovery. *Environmental Reviews*, v. 1, pp. 1–12.

Watson, A., P. Liss, and R. Duce. 1991. Design of a small-scale in situ iron fertilization experiment. *Limnology and Oceanography*, v. 36, pp. 1960–1965.

Williams, T. M., R. A. Kastelein, R. W. Davis, and J. A. Thomas. 1988. The effects of oil contamination and cleaning on sea otters (*Enhydra lutris*). 1. Thermoregulatory implications based on pelt studies. *Canadian Journal of Zoology*, v. 66, pp. 2776–2781.

Wolfe, D. A., and T. P. O'Connor, eds., 1988. *Urban Wastes in Coastal Marine Environments*. Malabar, FL: Krieger.

Wurster, C. F. 1968. DDT reduces photosynthesis by marine phytoplankton. *Science*, v. 159, pp. 1474–1475.

Wurster, C. F., and D. B. Wingate. 1968. DDT residues and declining reproduction in the Bermuda petrel. *Science*, v. 159, pp. 179–181.

Review Questions

1. Distinguish between point sources and nonpoint sources of pollution.

2. Name three general approaches that can be used to assess the overall effects of pollution in a given habitat.

3. Why are pesticides, which are designed to kill terrestrial arthropods, often quite dangerous in marine ecosystems?

4. Why is mercury a particular danger?

5. What effects does spilled oil have on marine environments?

6. Why does oil composition and type matter with regard to toxic effects?

7. Why has the treatment of oil spills sometimes been as damaging as the oil spills themselves?

8. Nutrient input increases primary production, which should support more fish. So why are people upset about increasing sewage input throughout the coastal ocean?

9. How does nutrient enrichment lead to hypoxia?

10. In what type of marine environment is atmospheric deposition of nutrients likely to have the greatest effects?

11. Why are plants that generate electric power a potential danger to fish stocks?

12. Which marine environments do you think are more vulnerable to pollution: Tropical ones or those in high latitudes? Why?

13. Why has carbon dioxide increased greatly in the atmosphere in the last 100 years or so?

14. How might carbon dioxide increase affect global climate? Oceanic circulation?

15. Is there any evidence that the greenhouse effect has caused any increase in oceanic temperatures in the past few decades? Explain your answer.

Glossary

Abyssal plain. The deep-ocean floor, an expanse of low relief at depths of 4,000–6,000 m

Abyssopelagic zone. The 4,000–6000 m depth zone, seaward of the shelf–slope break

Acclimation. Given a change of a single parameter, a readjustment of the physiology of an organism, reaching a new steady state

Age structure. The relative abundance of different age classes in a population

Aggregated spatial distribution. A case of individuals in a space occurring in clusters too dense to be explained by chance

Ahermatypic. Non-reef-building (referring to scleractinian corals)

Allele. One of several variants that can occupy a locus on a chromosome

Allopatric speciation. The differentiation of geographically isolated populations into distinct species

Allozyme. A variant of an enzyme type. These may be variants of a specific enzyme (e.g., cytochrome *c*) that are the products of a single genetic locus

Alternative stable states. Two configurations of species abundances in a habitat, each of which tends to be stable under certain conditions. Usually it takes a large disturbance to cause a shift from one stable state to another.

Amensal. Negatively affecting one or several species

Amino acids. The basic structural units of proteins

Anadromous fish. Fish that spends most of its life feeding in the open ocean but that migrates to spawn in fresh water

Anoxic. Lacking oxygen

Arrow worms. Members of the phylum Chaetognatha; a group of planktonic carnivores

Asexual reproduction. Reproduction of the individual without the production of gametes and zygotes

Assimilation efficiency. The fraction of ingested food that is absorbed and used in metabolism

Assortative mating. The mating of a given genotype with another genotype at a frequency disproportionate to that expected from a series of random encounters

Atoll. A horseshoe or circular array of islands, capping a coral reef system perched around an oceanic volcanic seamount

Attenuation (of light). Diminution of light intensity; explained, in the ocean, in terms of absorption and scattering

Autotrophic algae. Algae capable of photosynthesis and growth using only dissolved inorganic nutrients

Auxotrophic algae. Algae requiring a few organically derived substances, such as vitamins, along with dissolved inorganic nutrients, for photosynthesis

Bathypelagic zone. The 2,000–4,000 m depth zone seaward of the shelf–slope break

Benthic–pelagic coupling. The cycling of nutrients between the bottom sediments and overlying water column

Berm. A broad area of low relief in the upper part of a beach

Between-habitat comparison. A contrast of diversity in two localities of differing habitat type (e.g., sand vs. mud bottoms)

Biodiversity. *See* Species richness

Biogenically reworked zone. The depth zone, within a sediment, that is actively burrowed by benthic organisms

Biogenic graded bedding. A regular change of sediment median grain size with depth below the sediment–water interface due to the activities of burrowing organisms

Bioluminescence. Light emission, often as flashes, by many marine organisms

Biomass. *See* Standing crop

Blood pigment. A molecule (e.g., hemoglobin) used by an organism to transport oxygen efficiently, usually in a circulatory system

Bloom (phytoplankton) A population burst of phytoplankton that remains within a defined part of the water column

Bohr effect. When blood pH decreases, the ability of hemoglobin to bind to oxygen decreases; an adaptation to release oxygen in the oxygen-starved tissues in capillaries where respiratory carbon dioxide lowers blood pH

Boreal. Pertaining to the Northern Hemisphere, north temperate zone

Boring. Capable of penetrating a solid substratum by scraping or chemical dissolution

Boundary layer. A layer of fluid near a surface, where flow is affected by viscous properties of the fluid. At the surface, fluid velocity must be zero, and the boundary layer is a thin film that depends upon surface texture, fluid velocity in the "mainstream of flow," and fluid mass properties such as salinity

Brackish sea. Semienclosed water body of large extent in which tidal stirring and seaward flow of fresh water do not exert a strong enough mixing effect to prevent the body of water from having its own internal circulation pattern

Browsers. Organisms that feed by scraping thin layers of living organisms from the surface of the substratum (e.g., periwinkles feeding on rock surface diatom films; urchins scraping a thin, filmy sponge colony from a rock)

Bycatch. Fish and other animals caught unintentionally during fishing for a preferred species

Calcareous. Made of calcium carbonate

Carrying capacity. The total number of individuals of a population that a given environment can sustain

Carnivore. An organism that captures and consumes animals

Catadromous fish. Fish that spawns in seawater but feed and spends most of its life in estuarine or fresh water

Chaetognaths. *See* Arrow worms.

Character displacement. A pattern in which two species with overlapping ecological requirements differ more when they co-occur than when they do not. The difference is usually in a morphological feature related to resource exploitation, as in the case of head size, which may be related to prey size

Chemosynthesis. Primary production of organic matter, using various substances instead of light as an energy source; confined to a few groups of microorganisms

Chlorinity. Grams of chloride ions per 1,000 grams of seawater

Chloroplast. In eukaryotic organisms, the cellular organelle in which photosynthesis takes place

Cladogram. A treelike diagram showing evolutionary relationships. Any two branch tips sharing the same immediate node are most closely related. All taxa that can be traced directly to one node (i.e., are "upstream of a node") are said to be members of a monophyletic group.

Coastal reef. A coral reef occurring near and parallel to a coastline

Comb jellies. Members of the phylum Ctenophora, a group of gelatinous forms feeding on smaller zooplankton

Commensal. Having benefit for one member of a two-species association but neither positive nor negative effect on the other

Compensation depth. The depth of the compensation light intensity

Compensation light intensity. The light intensity at which oxygen evolved from a photosynthesizing organism equals that consumed in its respiration

Competition. An interaction between or among two or more individuals or species in which exploitation of resources by one affects any others negatively

Complex life cycle. A life cycle that consists of several distinct stages (e.g., larva and adult)

Conformer. An organism whose physiological state (e.g., body temperature) is identical to, and varies identically with, that of the external environment

Continental drift. Horizontal movement of continents located in plates moving via seafloor spreading

Continental shelf. A broad expanse of ocean bottom sloping gently seaward from the shoreline to the shelf–slope break at a depth of 100–200 m

Continental slope. *See* Slope

Convergence. The contact at the sea surface between two water masses converging, one plunging below the other

Copepod. Order of crustaceans found often in the plankton

Coprophagy. Feeding on fecal material

Coral reef. A wave-resistant structure resulting from cementation processes and the skeletal construction of hermatypic corals, calcareous algae, and other calcium carbonate–secreting organisms

Corer. Tubular benthic sampling device that is plunged into the bottom to obtain a vertically oriented cylindrical sample

Coriolis effect. The deflection of air or water bodies, relative to the solid earth beneath, as a result of the earth's eastward rotation

Countercurrent exchange mechanism. Mechanism by which two vessels are set side by side, with fluid flowing in opposite directions, allowing efficient uptake and retention of heat, oxygen, or gas, depending upon the type of exchanger

Countershading. Condition of organisms in the water column that are dark colored on top but light colored on the bottom

Counterillumination. Having bioluminescent organs that are concentrated on the ventral surface to increase the effect of countershading (see also countershading)

Critical depth. That depth above which total integrated photosynthetic rate equals total integrated respiration of photosynthesizers

Critical salinity. A salinity of approximately 5–8o/oo that marks a minimum of species richness in an estuarine system

Ctenophora. *See* Comb jellies

Daily estuary. An estuary in which tidal movements cause substantial changes in salinity at any one location on a daily basis

Deep layer. The layer extending from the lowest part of the thermocline to the bottom

Deep-scattering layer. Well-defined horizon in the ocean that reflects sonar; indicates a layer usually consisting of fishes, squid, or other larger zooplankton

Demographic. Referring to numerical characteristics of a population (e.g., population size, age structure)

Density (seawater). Grams of seawater per milliliter of fluid (g mL^{-1}L)

Density-dependent factors. Factors, such as resource availability, that vary with population density

Deposit feeder. An organism that derives its nutrition by consuming some fraction of a soft sediment

Detritus. Particulate material that enters into a marine or aquatic system; if derived from decaying organic matter it is organic detritus

Diatom. Dominant planktonic algal form with siliceous test, occurring as a single cell or as a chain of cells

Diffusion. The net movement of units of a substance from areas of higher concentration to areas of lower concentration of that substance

Digestion efficiency. The fraction of living food that does not survive passage through a predator's gut

Dinoflagellate. Dominant planktonic algal form, occurring as a single cell, often biflagellate

Directional selection. Preferential change in a population, favoring the increase in frequency of one allele over another

Dissolved organic matter. Dissolved molecules derived from degradation of dead organisms or excretion of molecules synthesized by organisms

Disturbance. A rapid change in an environment that greatly alters a previously persistent biological community

Diversity. A parameter describing, in combination, the species richness and the evenness of a collection of species. Diversity is often used as a synonym for species richness

Diversity gradient. A regular change in diversity correlated with a geographic space or gradient of some environmental factor

Ekman circulation. Movement of surface water at an angle from the wind, as a result of the Coriolis effect

El Niño–Southern Oscillation (ENSO). Condition in which warm surface water moves into the eastern Pacific, collapsing upwelling and increasing surface water temperatures and precipitation along the west coast of North and South America

Emigration. The departure of individuals from a given area

Endosymbiotic. Being symbiotic and living within the body of an individual of the associated species

Environmental stress. Variously defined as (a) an environmental change to which an organism cannot acclimate and (b) an environmental change that increases the probability of death

Epibenthic (epifaunal or epifloral). Living on the surface of the bottom

Epidemic spawning. Simultaneous shedding of gametes by a large number of individuals

Epipelagic zone. The 0–150 m depth zone, seaward of the shelf–slope break

Epiphyte. Microalgal organism living on a surface (e.g., on a seaweed frond)

Estuarine flow. Seaward flow of low-salinity surface water over a deeper layer of higher salinity

Estuarine realms. Large coastal water regions that have geographic continuity, are bounded landward by a stretch of coastline with freshwater input, and are bounded seaward by a salinity front

Estuary. A semienclosed body of water that has a free connection with the open sea and within which seawater is diluted measurably with fresh water derived from land drainage

Euphausiid. Member of an order of holoplanktonic crustaceans

Eutrophic. Water bodies or habitats having high concentrations of nutrients

Evenness. The component of diversity accounting for the degree to which all species are equal in abundance, as opposed to strong dominance by one or a few species

Fecal pellets. *See* Pellets.

Fecundity. The number of eggs produced per female per unit time (often: per spawning season)

Foliose coral. A coral whose skeletal form approximates that of a broad, flattened plate

Food chain. An abstraction describing the network of feeding relationships in a community as a series of links of trophic levels, such as primary producers, herbivores, and primary carnivores

Food chain efficiency. Amount of energy of some other quantity extracted from a trophic level, divided by the amount of energy produced by the next-lower trophic level

Food web. A network describing the feeding interactions of the species in an area

Foraminifera. Protozoan group, individuals of which usually secrete a calcareous test; both planktonic and benthic representatives

Founder principle. A small colonizing population is genetically unrepresentative of the source of population.

Freshet. An increase of water flow into an estuary during the late winter or spring, owing to increased precipitation and snow melt in the watershed

Front. A major discontinuity separating ocean currents and water masses in any combination

Fugitive species. A species adapted to colonize newly disturbed habitats

Gametophyte. Haploid stage in the life cycle of a plant

Generation time. The time period from birth to average age of reproduction

Genetic drift. Changes in allele frequencies that can be ascribed to random effects

Genetic locus. A location on a chromosome (possibly of a diploid organism with variants that segregate according to the rules of Mendelian heredity)

Genetic polymorphism. Presence of several genetically controlled variants in a population

Genotype. The genetic makeup of an organism, with respect to a given genetic locus, the alleles it carries

Genus (plural: genera). The level of the taxonomic hierarchy above the species but below the family level

Geotactic. Moving in response to the earth's gravitational field

Geotrophic flow. Movement of water in the oceans as a combined response to the Coriolis effect and gravitational forces created by an uneven sea surface

GIS: Geographic Information System. A system that allows automatic location of information suitable for mapping. Usually involves a software system that assembles geographic position data and other data (e.g., type of bottom sediment) for use in creating in order to create a map. Data on processes (e.g., current speed) can be incorporated to make a geographic model of flow.

Global warming. Predicted increase in the earth's oceanic and atmospheric temperatures owing to additions of carbon dioxide to the atmosphere as a result of human activities

GPS: Global Positioning System. An electronic device that uses positioning signals from satellites to locate precisely latitude and longitude. Now used nearly exclusively for locating ship sampling stations at sea, but also useful for locations near- and onshore.

Grab. Benthic sampling device with two or more curved metal plates designed to converge when the sampler hits bottom and take a specified volume of bottom sediment

Grazer. A predator that consumes organisms far smaller than itself (e.g., copepods graze on diatoms)

Greenhouse effect. Carbon dioxide traps solar-derived heat in the atmosphere near the earth

Gregarious settling. Settlement of larvae that have been attracted to members of their own species

Gross primary productivity. The total primary production, not counting the loss in respiration

Guild. A group of species, possibly unrelated taxonomically, that exploit overlapping resources

Gyre. Major cyclonic surface current systems in the oceans

Halocline. Depth zone within which salinity changes maximally

Harmful algal bloom. The proliferation of a toxic or nuisance algal species that causes a negative impact to natural resources or humans

Herbivore. An organism that consumes plants

Heritable character. A morphological character whose given state can be explained partially in terms of the genotype of the individual

Hermaphrodite. An individual capable of producing both eggs and sperm during its lifetime

Hermatypic. Reef-building

Heterotrophic algae. Algae that take up organic molecules as a primary source of nutrition

Heterozygote. With respect to a given genetic locus, a diploid individual carrying two different alleles

Highly stratified estuary. An estuary having a distinct surface layer of water that is fresh or very low in salinity, capping a deeper layer of more oceanic water, higher in salinity

Histogram. A multiple-bar diagram representing the frequency distribution of a group as a function of some variable. The frequency of each class is proportional to the length of its associated bar.

Holoplankton. Organisms spending all their life in the water column, not on or in the seabed

Homeotherm. An organism that regulates its body temperature despite changes in the external environmental temperature

Homozygote. With respect to a given genetic locus, a diploid individual carrying two identical alleles

Hydrographic. Referring to the arrangement and movement of bodies of water, such as currents and water masses

Hydrothermal vents. Sites in the deep-ocean floor where hot, sulfur-rich water is released from geothermally heated rock

Hypothesis. A refutable statement about one or a series of phenomena

Infaunal. Living within a soft sediment and being large enough to displace sedimentary grains

Interspecific competition. Condition in which one species's exploitation of a limiting resource negatively affects another species

Interstitial. Living in the pore spaces among sedimentary grains in a soft sediment

Isotonic. Having the same overall concentration of dissolved substances as a given reference solution

Keystone species. A predator at the top of a food web, or discrete subweb, capable of consuming organisms of more than one trophic level beneath it

Laminar flow. The movement of an entire fluid that is regular and with parallel streamlines.

Larva. A discrete stage in many species, beginning with zygote formation and ending with metamorphosis

Larvacea. A group of planktonic tunicates that secrete a gelatinous house, used to strain unsuitable particles (large particles are rejected). An inner filter apparatus of the house, the so-called food trap or particle-collecting apparatus, is used to retain food particles.

LD$_{50}$. The value of a given experimental variable required to cause 50 percent mortality.

Leaching. The loss of soluble material from decaying organisms

Lecithotrophic larva. A planktonic-dispersing larva that lives off yolk supplied via the egg

Leeward. The side of an island opposite the one facing a prevailing wind

Life table. A table summarizing statistics of a population, such as survival and reproduction, all broken down according to age classes

Litter. Accumulations of dead leaves in various states of fragmentation and decomposition

Locus. *See* Genetic locus.

Logistic population growth. Population growth that is modulated by the population size relative to carrying capacity: population growth declines as population approaches carrying capacity and is negative when population size is greater than carrying capacity

Longshore current. A current moving parallel to a shoreline

Macrobenthos (macrofauna or macroflora). Benthic organisms (animals or plants) whose shortest dimension is greater than or equal to 0.5 mm

Macrofauna. Animals whose shortest dimension is greater than or equal to 0.5 mm

Macrophyte. An individual alga large enough to be seen easily with the unaided eye

Macroplankton. Planktonic organisms that are 200–2,000 μm in size

Mainstream flow. The flow in a part of the fluid (e.g., in a tidal creek) that is well above the bottom or well away from a surface and essentially not under the influence of the boundary layer. *See* Boundary layer.

Mangel. *See* Mangrove forest.

Mangrove forest. A shoreline ecosystem dominated by mangrove trees, with associated mudflats

Marine snow. Fragile organic aggregates, resulting from the collision of dissolved organic molecules or from the degradation of gelatinous substances such as larvacean houses; usually enriched with microorganisms

Marine protected area. A geographic conservation unit designed to protect crucial communities and to provide reproductive reserves for fisheries that hopefully will disperse over wider areas

Maximum sustainable yield. In fisheries biology, the maximum catch obtainable per unit time under the appropriate fishing rate

Meiobenthos (meiofauna or meioflora). Benthic organisms (animals or plants) whose shortest dimension is less than 0.5 mm but greater than or equal to 0.1 mm

Meiofauna. Animals whose shortest dimension is less than 0.5 mm but greater than or equal to 0.1 mm

Megaplankton. Planktonic organisms that are greater than or equal to 2,000 μm in size

Meroplankton. Organisms that spend part of their time in the plankton but also spend time in the benthos (e.g., planktonic larvae of benthic invertebrates)

Mesopelagic. The 150–2,000 m depth zone, seaward of the shelf–slope break

Metabolic rate. The overall rate of biochemical reactions in an organism. Often estimated by rate of oxygen consumption in aerobes

Metamorphosis. Major developmental change as a larva develops into an immature adult

Metapopulation. A group of interconnected subpopulations, usually of subequal size. The features of individuals now found in one subpopulation might have been determined by conditions affecting them when they were located in another subpopulation.

Microbenthos (microfauna or microflora). Benthic organisms (animals or plants) whose shortest dimension is less than 0.1 mm

Microfauna. Animals whose shortest dimension is less than 0.1 mm

Mixing depth. The water depth to which wind energy evenly mixes the water column

Mixoplankton. Planktonic organisms that can be classified at several trophic levels. For example, some ciliates can be photosynthetic but also can ingest other plankton and are heterotrophic.

Moderately stratified estuary. An estuary in which seaward flow of surface low-salinity water and moderate vertical mixing result in a modest vertical salinity gradient

Monophyletic. Refers to a group of species whose members all have a single common ancestral species

Mucus bag suspension feeder. Suspension feeder employing a sheet or bag of mucus to trap particles nonselectively

Mutualism. An interaction between two species in which both derive some benefit

Mutualistic. Conferring reciprocal benefit to individuals of two different associated species

Nannoplankton. Planktonic organisms that are 2–20 μm in size

Neap tides. Tides occurring when the vertical range is minimal

Nekton. Organisms with swimming abilities that permit them to move actively through the water column and to move against currents

Neritic. Seawater environments landward of the shelf–slope break

Net primary productivity. Total primary production, minus the amount consumed in respiration

Neuston. Planktonic organisms associated with the air–water interface.

Niche. A general term referring to the range of environmental space occupied by a species

Niche overlap. An overlap in resource requirements by two species

Niche partitioning. Differential utilization of resources by cooccurring species

Nitrogen fixation. The conversion of gaseous nitrogen to nitrate by specialized bacteria

No-take reserves. Geographic areas where by law no one is allowed to fish or collect biological specimens; rule may apply to one or all species

Nuisance bloom. A rapid increase of one or only a few species of phytoplankton, resulting in densities high enough to cause discoloration of the surface water, possible increase of toxins, and degradation of water quality aspects such as dissolved oxygen

Nutrient cycling. The pattern of transfer of nutrients between the components of a food web

Nutrients. The constituents required by organisms for maintenance and growth (we use this term in this book in application to plants)

Oceanic. Associated with seawater environments seaward of the shelf–slope break

Oceanic ridge. A sinuous ridge rising from the deep-sea floor

Oligotrophic. Refers to water bodies or habitats with low concentrations of nutrients

Omnivory. Being able to feed in more than one distinct way (e.g., an organism capable of carnivory and herbivory)

Optimal foraging theory. A theory designed to predict the foraging behavior that maximizes food intake per unit time

Organic. Deriving from living organisms

Organic nutrients. Nutrients in the form of molecules synthesized by or originating from other organisms

Osmoconformer. An organism whose body fluids change directly with a change in the concentrations of dissolved ions in the external medium

Osmoregulator. An organism that regulates the concentration of dissolved ions in its body fluids irrespective of changes in the external medium

Osmosis. The movement of pure water across a membrane from a compartment with relatively low dissolved ions to a compartment with higher concentrations of dissolved ions

Outwelling. The outflow of nutrients from an estuary or salt marsh system to shelf waters

Overdominance. Selection favoring heterozygotes

Oxygen dissociation curve. A curve showing the percentage of saturation of a blood pigment, such as hemoglobin, as a function of oxygen concentration of the fluid

Oxygen minimum layer. A depth zone, usually below the thermocline, in which dissolved oxygen is minimal

Oxygen technique (primary productivity). The estimation of primary productivity by the measurement of the rate of oxygen increase

Parapatric speciation. The differentiation into distinct species of populations experiencing some gene flow

Parasite. An organism living on or in, and negatively affecting, another organism

Particulate organic matter. Particulate material in the sea derived from the decomposition of the nonmineral constituents of living organisms

Patchiness. A condition in which organisms occur in aggregations

Pelagic. Living in the water column seaward of the shelf–slope break

Pellets. Compacted aggregations of particles resulting from either egestion (fecal pellets) or the burrow-constructing activities of marine organisms

Penetration anchor. In hydraulically burrowing organisms, any device used to penetrate and gain an initial purchase on the sediment so that the body can be thrust in farther

Peptides. Chains of amino acids; often portions of a protein molecule

pH. Measure of the acidity or basicity of water ($-\log10$ of the activity of hydrogen ions in water)

Phenotypic plasticity. The capacity of an individual to produce different phenotypes under different conditions; nongenetic potential variability within the range of a single individual

Photic zone. The depth zone in the ocean extending from the surface to that depth permitting photosynthesis

Photorespiration. Enhanced respiration of plants in the light relative to dark respiration

Photosynthate. A substance synthesized in the process of photosynthesis

Photosynthetic quotient. In photosynthesis, the moles of oxygen produced, divided by the moles of carbon dioxide assimilated

Photosynthetic rate. The rate of conversion of dissolved carbon dioxide and bicarbonate ion to photosynthetic product

Phototactic. Moving in response to light

Physiological race. A geographically defined population of a species that is physiologically distinct from other populations

Phytoplankton. The photosynthesizing organisms residing in the plankton

Planktivorous. Feeding on planktonic organisms

Plankton. Organisms living suspended in the water column and incapable of moving against water currents

Planktotrophic larva. Plankton-dispersing larva that derives its nourishment by feeding in the plankton

Planula. The planktonic larval form produced by scleractinian corals and coelenterates

Plate. Major section of the earth's crust, bounded by such features as midoceanic ridges

Pleistocene epoch. Period of time, going back to approximately 2 million years before the present, in which alternating periods of glaciation and deglaciation have dominated the earth's climate

Pleuston. Refers to plankton that have a float protruding above the sea surface (e.g., Portuguese man-of-war)

Poikilotherm. An organism whose body temperature is identical to that of the external environment

Polyp. An individual of a solitary coelenterate or one member of a coelenterate colony

Polyphyletic. Refers to a group of species that do not have one common ancestor species

POM. Particulate organic matter

Population density. Number of individuals per unit area or volume

Porifera. The phylum comprising the sponges

Predation. The consumption of one organism by another

Predator. An organism that consumes another living organism (carnivores and herbivores are both predators by this definition)

Primary producer. An organism capable of using the energy derived from light or a chemical substance to manufacture energy-rich organic compounds

Primary production. The production of living matter by photosynthesizing organisms or by chemosynthesizing organisms; usually expressed as grams of carbon per square meter per year $(g \ C \ m^{-2} \ y^{-1})$

Province. A geographically defined area with a characteristic set of species or characteristic percentage representation by given species

Protein polymorphism. Presence of several variants of a protein of a given type (e.g., a certain enzyme, such as carboxylase) in a population

Pseudofeces. Material rejected by suspension feeders or deposit feeders as potential food before entering the gut

Pteropods. Group of holoplanktonic gastropods

Pycnocline. Depth zone within which seawater density changes maximally

Q_{10} Increase of metabolic rate with an increase of 10°C

Quantitative genetics. The study of the genetic basis of traits, usually explained in terms of the interaction of a group of genes with the environment

r. The intrinsic rate of increase of a population

Radiocarbon technique (primary productivity). The estimation of primary productivity by the measurement of radiocarbon uptake

Radiolaria. Protistan phylum whose members are planktonic and secrete an often elaborate siliceous test

Random spatial distribution. Situation in which individuals are randomly distributed in a space; probability of an individual's being located at any given point is the same irrespective of location in the space

Recruitment. The residue of those larvae that have (1) dispersed; (2) settled at the adult site; (3) made some final movements toward the adult habitat; (4) metamorphosed successfully; and (5) survived to be detected by the observer

Redox potential discontinuity: RPD. That depth below the sediment–water interface marking the transition from chemically oxidative to reducing processes

Red tide. A dense outburst of phytoplankton (usually dinoflagellates) often coloring water red brown

Refuge. A device by which an individual can avoid predation

Regulator. An organism that can maintain constant some aspect of its physiology (e.g., body temperature) despite different and changing properties of the external environment

Renewable resource. A resource that can be regenerated (e.g., a growing diatom population that is being exploited by a copepod)

Reproductive effort. The fraction of assimilated nutrients that are devoted to reproductive behavior and gamete production

Resource. A commodity that is required by an organism and is potentially in short supply

Respiration. Consumption of oxygen in the process of aerobic metabolism

Respiratory pigment. A molecule, polymer, or other complex adapted to bind and transport oxygen efficiently, usually in a circulatory system (e.g., hemoglobin)

Respiratory quotient. The ratio of moles of carbon dioxide produced to oxygen consumed in respiration

Rete mirabile. A countercurrent exchange structure of capillaries that allows gas uptake in a fish swim bladder

Reverse Bohr effect. Effect that occurs when lactate builds up in the blood of certain invertebrates and pH decreases, increasing the affinity of hemocyanin for oxygen

Reynolds number: *Re*. A number that represents the relative importance of viscous forces and inertial forces in a fluid. As *Re* increases, inertial forces become more important. In seawater, *Re* increases with increasing water velocity and with the size of the object in the water

Rip current. A concentrated rapid current moving offshore from a beach fronting a longshore current

Rise. Bottom of low relief at the base of the continental slope

ROV. Abbreviation for remotely operated vehicle, usually a submersible tethered to a ship, with facilities for video, remote sampling by grabbing arms, and precise navigation

Salinity. Number of grams of dissolved salts in 1,000 g of seawater

Salps. A group of pelagic tunicates (Phylum Urochordata), either colonial or solitary, with buccal and atrial siphons on opposite sides of the body

Salt marsh. A coastal habitat consisting of salt-resistant plants residing in an organic-rich sediment accreting toward sea level

Scavenger. An organism that feeds on dead or decomposing animals or macrophytes

Scleractinia. Order of coelenterates, usually producing calcareous skeletons with hexameral symmetry

Scope for growth. The surplus of energy available for growth beyond that required for maintenance

Scyphozoa. The true jellyfish, members of the phylum Cnidaria

Seafloor spreading. The horizontal movement of oceanic crust

Seasonal estuary. An estuary in which salinity at any one geographic point changes seasonally (e.g., decreases during the spring melt)

Seaward. Side of an island that faces the direction of wave action generated either by winds or by currents generated by more indirect forces

Secondary production. The production of living material per unit area (or volume) per unit time by herbivores; usually expressed as grams carbon per meter square per year (g C m^{-2} y^{-1})

Selection. A change in allele frequency over time in a population

Sequential hermaphrodite. An individual that sequentially produces male and then female gametes or vice versa

Sessile. Immobile because of an attachment to a substratum

Seston. Particulate matter suspended in seawater

Setules. Chitinous projections from copepod maxillipeds that trap food particles

Shelf–slope break. Line marking a change from the gently inclined continental shelf to the much steeper depth gradient of the continental slope

Sibling species. Closely related species that are so similar that they are nearly indistinguishable morphologically

Siphonophores. A group of specialized hydrozoan cnidarians, consisting of large planktonic polymorphic colonies

Sled. A benthic sampling device designed to slide along the sediment surface, digging into the bottom to a depth of at most a few centimeters

Slope. A steep-sloping bottom extending seaward from the edge of the continental shelf and downward toward the rise

Snow. *See* Marine snow.

Somatic growth. Growth of the body, exclusive of gametes

Sorting (of a sediment). The range of scatter of particle sizes about the median grain size of a sediment

Space limited. Description of a situation in which space is a limiting resource

Spatial autocorrelation. A situation in which some parameter at any location (e.g., population density) can be predicted through a knowledge of the values of the parameter in other locations

Spatial distribution. The arrangement of individuals in a space

Speciation. The process of formation of new species

Species. A population or group of populations that are in reproductive contact but are reproductively isolated from all other populations

Species–area effect. A regular logarithmic relationship between the number of species in a confined geographic area (e.g., an island) and the area in which the species occur

Species richness. The number of species in an area or biological collection

Sporophyte. Diploid stage in the life cycle of a plant

Spring diatom increase. The major rapid population increase of diatoms, occurring in the spring in temperate–boreal latitudes

Spring tides. Fortnightly tides occurring when the vertical tidal range is maximal

Stability–time hypothesis. Hypothesis that states that higher diversity occurs in habitats that are ancient and stable environmentally

Standing crop. The amount of living material per unit area or volume; may be expressed as grams of carbon, total dry weight, and so on

Stock recruitment models. Fishery models that predict the amount of juvenile recruitment as a function of the parent stock

Stratification. In benthos, the presence of different infaunal species at distinct respective horizons below the sediment–water interface

Subtropical. Refers to the portion of the temperate zone closest to the equator

Succession. A predictable ordering of a dominance of a species or groups of species following the opening of an environment to biological colonization

Surface layer. The layer of the ocean extending from the surface to a depth above which the ocean is homogeneous due to wind mixing

Survivorship curve. The curve describing changes of mortality rate as a function of age

Suspension feeder. An organism that feeds by capturing particles suspended in the water column

Swash rider. An invertebrate that can migrate up- and downshore with the rising and falling tide to maintain station at a level that is moist but not overly washed by the waves

Teleplanic larva. Larva capable of dispersal over long distances, such as across oceans

Temperate. Pertaining to the latitudinal belt between 23° 27' and 66° 33' north or south latitude

Tentacle–tube foot suspension feeder. Suspension feeder that traps particles on distinct tentacles or tube feet (in echinoderms)

Terminal anchor. In hydraulically burrowing organisms: any device used to anchor the leading portion of the burrower, permitting muscular contraction to drag the rest of the body into the sediment

Territoriality. Defense of a specified location against intruders

Tertiary production. The production of living material per unit area (or volume) per unit time by organisms consuming the herbivores; usually expressed as grams of carbon per meter square per year (g C m^{-2} y^{-1})

Test. Refers to a hard skeleton or shell, usually of a microscopic planktonic organism such as a foraminiferan or a radiolarian.

Thermocline. Depth zone within which temperature changes maximally

Thermohaline circulation. Movement of seawater that is controlled by density differences that are largely explained in terms of temperature and salinity

Tidal current. A water current generated by regularly varying tidal forces

Tides. Periodic movement of water resulting from gravitational attraction between the earth, sun, and moon

Trade winds. Persistent winds at low latitudes in both Northern and Southern Hemispheres, blowing toward the west and the equator

Trench. Deep and sinuous depression in the ocean floor, usually seaward of a continental margin or an arcuate group of volcanic islands

Trophic level. In a food chain, a level containing organisms of identical feeding habits with respect to the chain (e.g., herbivores)

Tropical. Being within the latitudinal zone bounded by the two tropics (23° 27' north and south latitude)

Turbidity. The weight of particulate matter per unit volume of seawater

Ultraplankton. Planktonic organisms that are less than 2 μm in size

Uniform spatial distribution. Situation in which individuals are more evenly spread in space than would be expected on the basis of chance alone

Upwelling. The movement of nutrient-rich water from a specified depth to the surface

Vents. *See* Hydrothermal vents.

Vertically homogeneous estuary. An estuary in which, at any given location, wind or tidal mixing homogenizes salinity throughout the water column

Vitamin. Chemical substances required in trace concentrations acting as a cofactor with enzymes in catalyzing biochemical reactions

Viviparous (development). Refers to development of an organism through the juvenile stage within a parent

Wash zone. The depth zone in which sediments are disturbed by wave action near the shoreline

Water mass. A body of water that maintains its identity and can be characterized by such parameters as temperature and salinity

Watershed. The land area that is drained by a river or estuary and its tributaries

Westerlies (prevailing westerlies). Persistent eastward–equatorward winds in midlatitudes in both the Northern and Southern Hemispheres

Windward. The side of an island that faces a prevailing wind

Within-habitat comparison. A contrast of diversity between two localities of similar habitat type

Wrack zone. A bank of accumulated litter at the strandline

Year–class effect. The common domination of a species population by individuals recruited in one reproductive season

Zonation. Occurrence of single species or groups of species in recognizable bands that might delineate a range of water depth or a range of height in the intertidal zone

Zooplankton. Animal members of the plankton

Zooxanthellae. A group of dinoflagellates living endosymbiotically in association with one of a variety of invertebrate groups (e.g., corals)

Marine Biology Journals

Some day you may want to write a research paper in one of many subjects in marine biology. The specialty journals listed here are mainly concerned with marine biological subjects. You may also wish to consult the Marine Biology Web Page (*http://life.bio. sunysb.edu/marinebio/mbweb.html*). Here you will find a variety of resources, including reference lists for a number of marine biology subjects.

American Journal of Zoology
American Malacological Bulletin
American Naturalist
Biological Bulletin (Woods Hole)
Biological Reviews (Cambridge)
Botanica Marina
Canadian Journal of Fisheries and Aquatic Sciences
Coral Reefs
Ecological Applications
Ecological Monographs
Ecology
Estuaries
Functional Ecology
Gulf of Mexico Science
Helgoländer Wissenschaften Meeresuchusungen
ICES Journal of Marine Science
Invertebrate Biology
Invertebrate Reproduction and Development

Journal of Animal Ecology
Journal of Crustacean Biology
Journal of Experimental Marine Biology and Ecology
Journal of Marine Research
Journal of Molluscan Studies
Journal of Phycology
Journal of Plankton Research
Journal of the Marine Biological Association of the United Kingdom
Journal of Shellfish Research
Marine Biology
Marine Ecology—Progress Series
Marine and Freshwater Research (Australia)
Marine Pollution Bulletin
Nature
Oceanography and Marine Biology Annual Review
Oecologia
Oikos
Ophelia (Denmark)
Pacific Science
Paleobiology
Phycologia
Quarterly Review of Biology
Sarsia
Science
Trends in Ecology and Evolution

Index